P9-CBT-195

ADVANCES IN CHEMICAL PHYSICS

VOLUME 100

EDITORIAL BOARD

Advances in CHEMICAL PHYSICS

Edited by

I. PRIGOGINE

Center for Studies in Statistical Mechanics and Complex Systems
The University of Texas
Austin, Texas
and
International Solvay Institutes
Université Libre de Bruxelles
Brussels, Belgium

and

STUART A. RICE

Department of Chemistry
and
The James Franck Institute
The University of Chicago
Chicago, Illinois

VOLUME 100

AN INTERSCIENCE® PUBLICATION
JOHN WILEY & SONS, INC.
NEW YORK • CHICHESTER • WEINHEIM • BRISBANE • SINGAPORE • TORONTO

This text is printed on acid-free paper.

An Interscience® Publication

Library of Congress Catalog Number: 58-9935

ISBN 0-471-17458-0

Printed in the United States of America

10 9 8 7 6 5 4 3 2 1

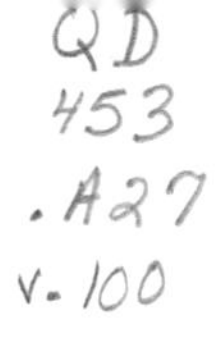

CONTRIBUTORS TO VOLUME 100

GUY ASHKENAZI, Department of Physical Chemistry and the Fritz Haber Center for Molecular Dynamics, The Hebrew University, Jerusalem, Israel

URI BANIN, Department of Physical Chemistry and the Fritz Haber Center for Molecular Dynamics, The Hebrew University, Jerusalem, Israel

F. BARAS, Centre for Nonlinear Phenomena and Complex Systems, Université Libre de Bruxelles, Brussels, Belgium

ALLON BARTANA, Department of Physical Chemistry and the Fritz Haber Center for Molecular Dynamics, The Hebrew University, Jerusalem, Israel

JEFFREY A. CINA, Department of Chemistry, University of Oregon, Eugene, Oregon

W. T. COFFEY, School of Engineering, Department of Electrical and Electronic Engineering, Trinity College, Dublin, Ireland

WOLFGANG DOMCKE, Institute of Physical and Theoretical Chemistry, Technical University of Munich, Garching, Germany

L. J. GEOGHEHAN, Department of Applied Mathematics and Theoretical Physics, The Queen's University of Belfast, Belfast, Northern Ireland

RONNIE KOSLOFF, Department of Physical Chemistry and the Fritz Haber Center for Molecular Dynamics, The Hebrew University, Jerusalem, Israel

M. MALEK MANSOUR, Centre for Nonlinear Phenomena and Complex Systems, Université Libre de Bruxelles, Brussels, Belgium

MICHEL MARESCHAL, Centre for Nonlinear Phenomena and Complex Systems, Université Libre de Bruxelles, Brussels, Belgium

B. MULLIGAN, School of Engineering, Department of Electrical and Electronic Engineering, Trinity College, Dublin, Ireland

SANFORD RUHMAN, Department of Physical Chemistry and the Fritz Haber Center for Molecular Dynamics, The Hebrew University, Jerusalem, Israel

GERHARD STOCK, Institute of Physical and Theoretical Chemistry, Technical University of Munich, Garching, Germany

LOWELL W. UNGAR, Department of Chemistry and The James Franck Institute, The University of Chicago, Chicago, Illinois

INTRODUCTION

Few of us can any longer keep up with the flood of scientific literature, even in specialized subfields. Any attempt to do more and be broadly educated with respect to a large domain of science has the appearance of tilting at windmills. Yet the synthesis of ideas drawn from different subjects into new, powerful, general concepts is as valuable as ever, and the desire to remain educated persists in all scientists. This series, *Advances in Chemical Physics*, is devoted to helping the reader obtain general information about a wide variety of topics in chemical physics, a field that we interpret very broadly. Our intent is to have experts present comprehensive analyses of subjects of interest and to encourage the expression of individual points of view. We hope that this approach to the presentation of an overview of a subject will both stimulate new research and serve as a personalized learning text for beginners in a field.

I. PRIGOGINE
STUART A. RICE

CONTENTS

Theory of Ultrafast Nonadiabatic Excited-State Processes and Their Spectroscopic Detection in Real Time 1

By Wolfgang Domcke and Gerhard Stock

Short-Time Fluorescence Stokes Shift Dynamics 171

By Lowell W. Ungar and Jeffrey A. Cina

Quantum Description of the Impulsive Photodissociation Dynamics of I_3^- in Solution 229

By Guy Ashkenazi, Uri Banin, Allon Bartana, Ronnie Kosloff, and Sanford Ruhman

Microscopic Simulations of Complex Flows 317

By Michel Mareschal

Microscopic Simulations of Chemical Instabilities 393

By F. Baras and M. Malek Mansour

Differential Recurrence Relations for Non-Axially Symmetric Rotational Fokker-Planck Equations 475

By L. J. Geoghegan, W. T. Coffey, and B. Mulligan

Author Index 643

Subject Index 663

ADVANCES IN CHEMICAL PHYSICS

VOLUME 100

THEORY OF ULTRAFAST NONADIABATIC EXCITED-STATE PROCESSES AND THEIR SPECTROSCOPIC DETECTION IN REAL TIME

WOLFGANG DOMCKE AND GERHARD STOCK

Institute of Physical and Theoretical Chemistry, Technical University of Munich, D-85748 Garching, Germany

CONTENTS

I. Introduction
II. Basic Concepts of Intramolecular Non-Born-Oppenheimer Dynamics
  A. Adiabatic and Diabatic Electronic Representations
  B. Explicit Construction of Diabatic Electronic States
  C. Model Hamiltonians
    1. General Aspects
    2. Normal-Mode Taylor Expansion
    3. A Model of Nonadiabatic Photoisomerization
III. Time-Dependent Methodology
  A. Representation of the State Vector
  B. Propagation of the State Vector
IV. Aspects of Ultrafast Intramolecular Non-Born-Oppenheimer Dynamics
  A. Wave Packets
  B. Electronic Population Dynamics
  C. Observables of the Nuclear Dynamics
V. Basic Concepts of Femtosecond Spectroscopy
  A. Preparation and Detection of Nonstationary States
  B. Spectroscopic Signals
    1. Transient Transmittance
    2. Time-Resolved Fluorescence
    3. Photon Echo
    4. Time-Resolved Ionization and Fragment Detection
VI. Evaluation of the Nonlinear Polarization
  A. Perturbative Approach

Advances in Chemical Physics, Volume 100, Edited by I. Prigogine and Stuart A. Rice.
ISBN 0-471-17458-0

1. General Aspects
2. Characterization of Spectroscopic Processes
3. Numerical Evaluation of Multitime Correlation Functions
4. Discussion

B. Nonperturbative Approach
1. Directional Dependence of the Nonlinear Response
2. Example: The One-Dimensional Harmonic Oscillator

VII. Femtosecond Spectroscopy of Ultrafast Nonadiabatic Excited-State Processes
A. Real-Time Detection of Internal-Conversion Processes
B. Real-Time Detection of Nonadiabatic Photoisomerization Processes
C. Discussion of Femtosecond Experiments

VIII. Quantum-Mechanical and Semiclassical Treatment of Systems with Many Degrees of Freedom
A. Real-Time Path-Integral Approach
1. Construction of the Discretized Path Integral for Vibronic-Coupling Models
2. Evaluation of Path-Integral Expressions

B. Reduced Density-Matrix Approach
C. Semiclassical Description of Nonadiabatic Transitions
D. Classical Modeling of Internal Conversion and Photoisomerization
1. The $\tilde{C} \rightarrow \tilde{B} \rightarrow \tilde{X}$ Internal-Conversion Process in the Benzene Cation
2. Photoisomerization Dynamics of a Chromophore in Solution
3. Classical Modeling of Femtosecond Experiments

IX. Note Added in Proof
Acknowledgments
References

I. INTRODUCTION

The electronic excitation of a polyatomic system by the absorption of a photon generally initializes a complex sequence of dynamical processes. The investigation of these processes has been a topic of extensive research for several decades. In principle, photoinduced processes are of interest in virtually all systems, e.g., atoms, molecules, clusters, surfaces, liquids or crystals. To reasonably limit the scope of the present discussion, we shall focus on photoinduced processes in either isolated polyatomic molecules (for example, aromatic systems) or polyatomic chromophores in solution. The theory of the photoinduced dynamics of small polyatomic molecules (in particular, triatomic molecules) in the gas phase has been comprehensively discussed in the recent monograph of Schinke [1] and can therefore largely be excluded from the present discussion. We shall rather be concerned with larger polyatomic systems, for which the photoinduced dynamics is typically dominated by dissipative processes even in the absence of intermolecular collisions.

The photoinduced processes in molecules are conventionally classified as photophysical, if the molecule retains its chemical identity; or photochemical, if the molecule undergoes a chemical change. The photophysical processes include radiative electronic transitions (fluorescence, phosphorescence), as well as the so-called radiationless electronic transitions (internal conversion and intersystem crossing). Photoinduced electron-transfer processes may also be considered as a photophysical phenomenon. Examples of photochemical processes are photodissociation and photoinduced isomerization reactions. Internal-conversion, intersystem-crossing and electron-transfer processes involve, by definition, a transition between different Born–Oppenheimer (BO) adiabatic electronic potential-energy (PE) surfaces. The same is presumably true for most photochemical reactions. Transitions between different adiabatic electronic PE surfaces induced by non-BO couplings are, in fact, the rule rather than the exception in the photoinduced dynamics of polyatomic molecules. For a review of these basic and well-established concepts, we refer to existing monographs and textbooks, e.g., [2–5].

Photophysical and photochemical processes cover a wide range of timescales. The kinetics of reactive and radiationless processes at microsecond and nanosecond timescales has been investigated for decades. With the advent of picosecond and femtosecond laser technology, the interest has shifted towards the study of the kinetics of the very fastest molecular processes. Reviews of this rapidly expanding field of research, nowadays referred to as "femtochemistry", can be found in recent conference reports and books [6–10]. A few arbitrarily selected examples may be quoted to illustrate the enormous potential of real-time spectroscopy at femtosecond timescales. The ultrafast bond-breaking dynamics of diatomic and triatomic systems has been monitored in real time, using femtosecond laser spectroscopy with fluorescence detection [11, 12]. The observation of femtosecond quantum beats in the pump-probe spectra of dye molecules in solution [13] and photosynthetic reaction centers [14] has revealed the relevance of vibrational coherence effects at sub-picosecond timescales even in complex systems. Nonexponential optical phase relaxation of chromophores in liquids has been resolved in real time, using photon-echo techniques [15, 16]. The *cis-trans* photoisomerization dynamics of rhodopsin has been monitored in real time [17]. Even the fastest intramolecular electron-transfer processes, e.g., in mixed-valence compounds or in photosynthetic reaction centers, can now explicitly be observed in real time [18, 19]. Finally, femtosecond laser spectroscopy with mass spectrometric ion detection has provided direct insight into the ultrafast excited-state dynamics of metal clusters [20, 21]. Even this incomplete list of recent results makes clear that femtosecond time-resolved spectroscopy is in the process of revolutionizing our knowledge of ultrafast processes in molecules, clusters, and condensed matter [6–10].

From a mechanistic point of view, photophysical and photochemical processes are very complex, irrespective of whether we are considering an isolated large molecule or cluster, or a chromophore in solution. To characterize the theoretical approach on which the present article is based, it is helpful to distinguish three categories of interactions involved in photoinduced dynamics.

1. *Strong Intramolecular Couplings.* In this category we subsume strong electron-vibrational interactions (e.g., gradients of excited-state PE surfaces with respect to Condon active normal modes), strong anharmonic mode–mode couplings as well as strong electronic interstate couplings (i.e., non-BO couplings). An exact (nonperturbative) treatment of these strong couplings is considered to be essential.
2. *System-Environment Coupling.* The separation of the problem into a "system" and an "environment" is not unique, but is inherent to essentially all qualitative models in the field. This terminology is obvious in the case of a chromophore in solution, for example. When considering isolated polyatomic molecules, the system Hamiltonian may include the active degrees of freedom, which are strongly coupled to the optical transition, while the inactive vibrational modes represent the environment. If, as is usually the case, only very few degrees of freedom are included in the system, the system–environment coupling is essential to account for the dissipative nature of photoinduced dynamics. If the system-environment separation is properly chosen, the mutual coupling will be weak and perturbation theory for the system–environment coupling may be applied.
3. *Radiation–Matter Coupling.* The radiation–matter coupling is generally weaker than the couplings within the material system. In the standard approach to nonlinear spectroscopy, the radiation–matter coupling is treated in perturbation theory.

The approach to femtosecond molecular dynamics and spectroscopy described in this article focuses on strong intramolecular couplings (type 1) and emphasizes, in particular, the exact (nonperturbative) treatment of these couplings. The extremely short lifetimes (in the 10 fs range) that have been reported, for example, for higher excited valence states of aromatic systems [22–24] and intramolecular charge-transfer states [18] imply the existence of very strong intramolecular non-BO couplings which cannot adequately be treated by perturbation theory, i.e., the Golden Rule approach. On the other hand, it can be expected that, even in large molecules, only relatively few degrees of freedom will give rise to the strongest interactions. This suggests the possibility of an accurate and systematic modeling of the strongest intramolecular couplings in terms of realistic (i.e., molecule-specific) model Hamiltonians, defined in terms of appropriate nuclear kinetic-energy operators and PE functions.

Of particular relevance in this approach are conical intersections of electronic PE surfaces [25]. Conical intersections are points (actually hypersurfaces in the multidimensional nuclear coordinate space) of exact degeneracy of adiabatic PE surfaces [26, 27]. It is now increasingly realized that conical intersections are a very common phenomenon in polyatomic systems [28–34]. The divergence of the non-BO coupling element at the seam of intersection and the pronounced anharmonicity of the adiabatic PE surfaces cause very strong electronic interstate couplings as well as mode–mode couplings; they therefore dominate the dynamics at the shortest timescales [35]. Exact (numerical) time-dependent wave-packet calculations for models involving conically intersecting PE surfaces have shown that the fundamental dissipative processes of population and phase relaxation are already clearly expressed in systems involving three strongly coupled modes [36–39]. This strongly supports the idea that conical intersections, also referred to as "photochemical funnels" [3, 34] provide the microscopic mechanism for ultrafast nonadiabatic relaxation processes in polyatomic molecules [40].

In contrast to the established theory of radiationless transitions in polyatomic molecules [4, 41–43], which is formulated in terms of tiers of zero-order BO levels, we favor an approach that is formulated in terms of electronic PE surfaces and wave functions and thus firmly founded on *ab initio* electronic structure theory. This opens the possibility of a first-principles treatment of the initial stages of ultrafast internal-conversion processes, employing accurate *ab initio* PE surfaces and non-BO couplings [44, 45]. The formulation of photoinduced dynamics in terms of wave-packet dynamics on coupled PE surfaces will naturally lead, moreover, to a unified description of photophysical and photochemical processes. A basic assumption of this modeling of femtosecond processes is the existence of a separation of timescales: only a few active modes, which are strongly coupled to the optical transition, are assumed to dominate the dynamics at the shortest timescales; the coupling of the active modes with the large number of inactive modes (interaction 2) is believed to be weaker and to become relevant only at longer timescales. It can therefore be neglected in a first approximation when considering photoinduced dynamics at the shortest timescales.

Such a reduced-dimensionality approach, motivated by a separation of timescales, may not be appropriate in all cases. There may be systems where it is not possible to identify a reasonably small number of active modes. Moreover, one eventually would like to understand subsequent stages of the relaxation process at a microscopic level also. The computational cost of exact (numerical) time-dependent wave-packet calculations grows exponentially with the number of degrees of freedom, and such calculations are, therefore, not possible for large systems. Thus, one has to look for alternative computational strategies that allow the treatment of the time-dependent dynamics of truly multidimensional quantum systems. A few of these methods, which appear particularly promising in the present context, will be discussed in Section VIII.

As already mentioned above, the interaction of the molecular system with the laser field (interaction 3) does not necessarily require a nonperturbative treatment. In spectroscopic applications, the field–matter interaction serves as a *probe* of the molecular dynamics, but is not supposed to alter the properties of the system. For spectroscopic pump-probe experiments, it is thus usually justified to evaluate the electric polarization in the lowest nonvanishing order of perturbation theory. The perturbative approach is most advantageous in cases where the multi-time nonlinear system response functions [46] can be evaluated analytically, e.g., for few-level systems [47], or for the harmonic oscillator [48, 49]. In a more realistic description of the molecular dynamics, on the other hand, where a numerical wave-packet propagation is required for the treatment of the material dynamics, the exact inclusion of the time-dependent field–matter interaction requires no major additional effort and is, in fact, more straightforward than the cumbersome numerical evaluation of multiple integrals involving multi-time response functions. The nonperturbative evaluation of the polarization [50] thus appears to be the method of choice for most of the applications we have in mind here.

Detailed simulations of femtosecond time-resolved experiments on the basis of molecule-specific PE functions and couplings have so far largely been restricted to simple systems, mostly diatomic molecules. Stimulated by the pioneering experiments of the Zewail group [12, 51], the predissociation of the $\tilde{A}$ state of NaI, for example, has been analyzed by several workers [52–58]. Coherent vibrational dynamics of the diatomic molecules Na_2 and I_2 has extensively been investigated in several laboratories [21, 59–61], and detailed theoretical simulations of the spectra have been performed [62–64]. Solvent effects on the coherent wave-packet dynamics of I_2 also have been theoretically investigated [65–67]. Another theoretically well-characterized example is the photodissociation of the quasi-diatomic system ICN [11, 68]. Very recently, femtosecond time-resolved ionization spectra of Na_3 [69, 70] have quantitatively been interpreted in terms of Jahn–Teller and pseudo-Jahn–Teller induced excited-state dynamics [71–73].

The majority of theoretical studies of femtosecond molecular spectroscopy are concerned with the real-time detection of the dynamics of complex systems, e.g., large molecules or chromophores in solution, and adopt a different type of approach. The description of the material dynamics is generally based on a very simple model for the "system" that couples to the laser field, e.g., a few-level system or a harmonic oscillator. Dissipation is introduced by coupling the system to a thermal bath, either at a phenomenological level (optical Bloch equations [74]) or in a microscopic manner [75–78].

Strong intramolecular non-BO and mode–mode couplings are not considered in this approach; in terms of the classification introduced above, the focus is on interactions of type 2 rather than type 1. Mitsunaga and Tang were among the first to formulate a description of molecular femtosecond pump-probe spectroscopy based on the optical Bloch equations [47]. This type of approach has been further elaborated, among others, by Lin and collaborators [55, 79]. A time-dependent wave-packet description of femtosecond spectroscopy based on harmonic oscillators and the phenomenological inclusion of optical dephasing effects has been formulated by Pollard, Lee, and Mathies [49, 80]. Mukamel and collaborators have elaborated a comprehensive density-matrix description of femtosecond spectroscopy, adopting either few-level systems, damped harmonic oscillators, or harmonic multimode models to represent the material dynamics [48, 81–84]. Recently, this work has been reviewed in a comprehensive monograph [84], which also provides a general introduction to modern nonlinear spectroscopy. Non-BO effects are usually not explicitly considered in this approach. An exception are recent works by Jean [85], May and collaborators [86], and Tanimura and Mukamel [87], where electronic interstate coupling has been included in addition to vibrational dissipation.

As discussed above, we shall focus in this article on the modeling of the non-BO dynamics associated with relatively few strongly coupled active degrees of freedom of a polyatomic molecule. Out primary goal is to elaborate a fully microscopic understanding of the initial stages of photoisomerization and internal-conversion processes. In particular, we focus on the non-BO dynamics associated with photochemical funnels and investigate the possibility of detection of these processes in real time. As specific examples, we have considered in previous work the ultrafast internal-conversion dynamics induced by conical intersections in the benzene cation [88], pyrazine [44, 89–92], and in ozone [93]. The interplay of photoisomerization and internal-conversion dynamics has more recently been investigated for simple models of photochemical funnels involving a large-amplitude torsional degree of freedom [39, 94].

The article is organized as follows. In Section II.A the basic definitions of the theory are introduced. In Section II.B the explicit *ab initio* computation of diabatic electronic PE surfaces is reviewed. A discussion of this somewhat technical problem is considered essential in order to make clear that the present approach is explicitly founded on *ab initio* electronic-structure theory. In Section II.C we discuss concepts for the construction of model Hamiltonians that are suitable to describe the dynamical processes associated with photochemical funnels at the quantum mechanical level. An essential ingredient of the present approach are efficient numerical methods for the solution of the time-dependent Schrödinger equation, possibly including the interaction of the system with an external time-dependent field. These computational techniques are discussed in Section III. Section IV gives a brief review of the most essential aspects of the time-dependent dynamics on conically intersecting PE surfaces. The microscopic modeling of the molecular non-BO dynamics provides the basis for the theoretical description of femtosecond spectroscopy, which is developed in Sections V and VI, adopting either a perturbative or a nonperturbative evaluation of the electric polarization. In Section VII the practical implementation of these concepts is illustrated for the example of the $S_1(n\pi^*)$–$S_2(\pi\pi^*)$ conical intersection in pyrazine as well as for a model describing a photochemical isomerization reaction. Finally, Section VIII addresses the computational treatment of nonadiabatic processes in truly multidimensional systems.

II. BASIC CONCEPTS OF INTRAMOLECULAR NON-BORN–OPPENHEIMER DYNAMICS

A. Adiabatic and Diabatic Electronic Representations

The coupled equations resulting from the expansion of the molecular wave function in Born-Oppenheimer (BO) adiabatic electronic states constitute the formal basis of most theoretical treatments of dynamic processes in molecules. These equations also provide the point of departure for the present developments.

Denoting electronic and nuclear coordinates collectively by **r** and **R**, respectively, the molecular Hamiltonian is written as

$$H_M = T_N + T_E + V(\mathbf{r}, \mathbf{R}) \tag{2.1}$$

Here T_E and T_N are the kinetic-energy operators of electrons and nuclei, respectively, and $V(\mathbf{r}, \mathbf{R})$ is the potential energy.

The BO adiabatic electronic states $\tilde{\psi}_n(\mathbf{r}, \mathbf{R})$ are introduced as the solutions of the fixed-nuclei electronic eigenvalue problem [95]

$$\{T_E + V(\mathbf{r}, \mathbf{R})\}\tilde{\psi}_n(\mathbf{r}, \mathbf{R}) = V_n(\mathbf{R})\tilde{\psi}_n(\mathbf{r}, \mathbf{R}) \tag{2.2}$$

They depend *parametrically* on the nuclear coordinates **R**. The full time-independent molecular wave function $\Psi(\mathbf{r}, \mathbf{R})$ is expressed as an expansion in terms of adiabatic electronic states

$$\Psi(\mathbf{r}, \mathbf{R}) = \sum_n \tilde{\chi}_n(\mathbf{R})\tilde{\psi}_n(\mathbf{r}, \mathbf{R}) \tag{2.3}$$

Insertion of Eq. (2.3) into the time-independent Schrödinger equation yields the coupled equations [95]

$$\{T_N + V_n(\mathbf{R}) - E\}\tilde{\chi}_n(\mathbf{R}) = \sum_n \Lambda_{nm}\tilde{\chi}_m(\mathbf{R}) \tag{2.4a}$$

where

$$\Lambda_{nm} = -\int d\mathbf{r}\,\tilde{\psi}_n(\mathbf{r},\mathbf{R})[T_n, \tilde{\psi}_m(\mathbf{r},\mathbf{R})] \tag{2.4b}$$

The Λ_{nm}, which are operators in nuclear coordinate space, represent the nonadiabatic coupling effects in the adiabatic electronic representation. In terms of first-order and second-order derivative couplings [96]

$$\tilde{f}_{nm}^{(j)}(\mathbf{R}) = \int d\mathbf{r}\,\tilde{\psi}_n^*(\mathbf{r},\mathbf{R})\,\frac{\partial}{\partial R_j}\,\tilde{\psi}_m(\mathbf{r},\mathbf{R}) \tag{2.5a}$$

$$\tilde{h}_{nm}^{(j)}(\mathbf{R}) = \int d\mathbf{r}\,\tilde{\psi}_n^*(\mathbf{r},\mathbf{R})\,\frac{\partial^2}{\partial R_j^2}\,\tilde{\psi}_m(\mathbf{r},\mathbf{R}) \tag{2.5b}$$

the Λ_{nm} read

$$\Lambda_{nm} = -\sum_j \frac{\hbar^2}{M_j}\tilde{f}_{nm}^{(j)}\frac{\partial}{\partial R_j} - \sum_j \frac{\hbar^2}{2M_j}\tilde{h}_{nm}^{(j)} \tag{2.6}$$

where the M_j are nuclear masses.

Neglect of the nonadiabatic coupling operators Λ_{nm} in Eq. (2.4) leads to uncoupled equations

$$\{T_N + V_n(\mathbf{R}) - E\}\tilde{\chi}_n(\mathbf{R}) = 0 \tag{2.7}$$

which define the well-known BO adiabatic approximation [95].

It is well established, and therefore need not be discussed in detail here, that the BO adiabatic approximation is appropriate when the separation of electronic energy levels is large compared with typical energy spacings of the nuclear motion. Corrections to the single-surface approximation of Eq. (2.7) can be obtained by evaluating explicitly the Λ_{nm} and solving the coupled equations (2.4). Recent advances in electronic structure theory [97–99] and systematic application of analytic-derivative techniques have greatly facilitated the computation of derivative couplings [100–102]. An authoritative discussion of the computational treatment of nonadiabatic interactions in the BO adiabatic representation can be found in the review article of Lengsfield and Yarkony [96]. The possibility of time-dependent wave-packet propagation in the adiabatic electronic representation has been discussed by Parlant and Yarkony [103], but so far no multidimensional applications of this method have been reported.

While nonadiabatic couplings are weak for well-separated electronic states, they can become very strong and even singular when two electronic energy surfaces closely approach or intersect each other. A well-known example is the dynamical Jahn–Teller effect, which involves a singularity of the nonadiabatic coupling element at the point of degeneracy of the electronic energy surfaces [35, 104]. Other examples will be discussed below. Since divergent couplings are a nuisance for computational work, one usually prefers in such situations an alternative expansion of the molecular wave function in terms of so-called *diabatic* electronic states [105–110]. Such an expansion avoids singular coupling elements.

The elimination of derivative couplings is trivial in the (hypothetical) case of a complete electronic basis set. Since the solutions of the electronic Schrödinger equation at any nuclear geometry $\mathbf{R}_0$ define a complete set, the total wave function can be expanded in this complete set of $\mathbf{R}$-independent electronic basis states, which has been termed "crude adiabatic" basis [104, 111]. The convergence of this expansion is extremely slow, however, and the crude adiabatic expansion is therefore useless in practice.

The electronic basis functions for a practical expansion should be free of those singular variations with respect to $\mathbf{R}$ which are caused by interactions between closely spaced electronic states. They should, on the other hand, behave adiabatically with respect to interactions between energetically well separated electronic states in order to allow for a reasonably compact expansion.

Let us, for the moment, assume that a set $\{\psi_n(\mathbf{r}, \mathbf{R})\}$ of such diabatic electronic basis states is given. The expansion of the molecular wave function then reads

$$\Psi(\mathbf{r}, \mathbf{R}) = \sum_n \chi_n(\mathbf{R})\psi_n(\mathbf{r}, \mathbf{R}) \tag{2.8}$$

Assuming that derivative couplings in the diabatic basis are negligible, the coupled equations in the diabatic representation read

$$(T_N + U_{nn}(\mathbf{R}) - E)\chi_n(\mathbf{R}) = \sum_m U_{nm}(\mathbf{R})\chi_m(\mathbf{R}) \tag{2.9}$$

The $U_{nm}(\mathbf{R})$ are matrix elements of the electronic Hamiltonian in the diabatic basis

$$U_{nm}(\mathbf{R}) = \int d\mathbf{r}\, \psi_n^*(\mathbf{r}, \mathbf{R})[T_E + V(\mathbf{r}, \mathbf{R})]\psi_m(\mathbf{r}, \mathbf{R}) \tag{2.10}$$

The derivative couplings Λ_{nm} of the adiabatic representation have thus been replaced by potential-energy (PE) couplings in the diabatic representation.

The fact that diabatic basis sets are neither unique nor strictly $\mathbf{R}$-independent has lead to some confusion in the literature. Recently, a comprehensive and clear discussion of the concept of diabatic states has been given in review articles by Sidis [112] and Pacher, Cederbaum, and Köppel [113]. The review of Sidis [112] includes a brief account of the history of the subject in the context of atomic and molecular collisions. Pacher et al. [113] discuss in detail the gauge theoretical formulation of the non-BO coupling in the adiabatic representation (see also ref. [114]). We refer to these articles for a detailed analysis of the concept of diabatic states and a discussion of the earlier literature. In what follows we briefly discuss possible methods for the explicit construction of diabatic states, emphasizing practical aspects of the *ab initio* calculation of diabatic basis sets for polyatomic systems. We shall only be concerned here with finite-rank electronic representations. Infinite diabatic basis sets and diabatic continua are relevant for the theoretical treatment of photoionization and electron-molecule collision dynamics. Methods for the construction of diabatic electronic continuum states have been discussed, for example, by O'Malley [107] and Domcke [115].

B. Explicit Construction of Diabatic Electronic States

From the practical point of view, the many proposals for the construction of diabatic states can be divided into two categories. Methods of category (i) assume that the derivative couplings in the adiabatic representation are given, and seek a suitable unitary transformation which at least partially eliminates these couplings. Methods of category (ii) aim at the direct construction of smoothly varying electronic states, bypassing the formidable task of computing the derivative couplings. As usually applied, these direct methods are approximate, as it is assumed that the derivative couplings in the diabatic basis are negligible.

The formal basis of approach (i) has been elaborated by Smith [108], Baer [109, 110] and Mead and Truhlar [116]. Starting from the adiabatic electronic representation defined by Eq. (2.2), one introduces diabatic states via a unitary transformation within a *finite subspace*

$$\psi_n(\mathbf{r}, \mathbf{R}) = \sum_{i=1}^{N} \tilde{\psi}_i(\mathbf{r}, \mathbf{R}) X_{in}(\mathbf{R}) \tag{2.11}$$

The requirement that the first-order derivative couplings should vanish in the diabatic representation for all nuclear coordinates R_j $(j = 1 \ldots M)$

$$\int d\mathbf{r}\, \psi_n^*(\mathbf{r}, \mathbf{R}) \frac{\partial}{\partial R_j} \psi_m(\mathbf{r}, \mathbf{R}) = 0 \tag{2.12}$$

leads to first-order differential equations for the transformation matrix $\mathbf{X}$

$$\partial \mathbf{X}/\partial R_j + \tilde{\mathbf{f}}^{(j)} \mathbf{X} = 0 \tag{2.13}$$

where $\tilde{\mathbf{f}}^{(j)}$ is the matrix of first-order derivative couplings defined in Eq. (2.5a). The necessary and sufficient conditions for a unique solution of Eq. (2.13) to exist are [109, 112]

$$\partial \tilde{\mathbf{f}}^{(i)}/\partial R_j - \partial \tilde{\mathbf{f}}^{(j)}/\partial R_i = [\tilde{\mathbf{f}}^{(i)}, \tilde{\mathbf{f}}^{(j)}] \tag{2.14}$$

Equation (2.14) is trivially fulfilled for a single nuclear degree of freedom, e.g., in the case of radial motion of a diatomic system. For polyatomic systems, it has been shown that Eq. (2.14) cannot be fulfilled for a finite subspace of electronic states [116]: the variation of the $\tilde{\mathbf{f}}^{(j)}$ with respect to $R_1, \ldots, R_M$ involves contributions from basis states lying outside the subspace, which violate Eq. (2.14). The strict fulfillment of Eq. (2.14) in the polyatomic case would require a complete electronic basis set, which leads back to the trivial and practically useless case of a crude adiabatic basis. In polyatomic systems it is thus impossible to transform a finite subset of adiabatic basis states into an equivalent subset of strictly diabatic basis states [116].

Equation (2.13) nevertheless provides the basis for the construction of useful diabatic representations. The adiabatic-to-diabatic transformation matrix $\mathbf{X}$ obtained by integration of Eq. (2.13) depends on the path of integration. According to the above discussion, this is a common property of diabatic bases in polyatomic systems [112, 113, 116, 117]. Alternatively, one may consider the minimization of a suitable norm of the matrices [118]

$$\mathbf{X}^{\dagger}\, \partial \mathbf{X}/\partial R_j + \mathbf{X}^{\dagger}\, \tilde{\mathbf{f}}^{(j)}\mathbf{X}$$

A representative application of the Smith–Baer approach at the *ab initio* level has recently been reported for the example of the conically intersecting $1^1A''$ and $2^1A''$ states of ozone [102].

Despite recent progress with the *ab initio* evaluation of derivative couplings [96, 102], the computation of the $\tilde{\mathbf{f}}^{(j)}$ and $\tilde{\mathbf{h}}^{(j)}$ as a function of all nuclear coordinates represents a serious computational bottleneck for larger systems. Methods of category (ii), which attempt to construct diabatic states directly, without prior determination of the derivative couplings, thus appear preferable for polyatomic applications. A variety of such direct methods has been proposed in the literature. These differ in the choice of the criterion adopted to monitor the smoothness of electronic wave functions. A commonly employed approach is to enforce the smoothness of a suitable electronic property, e.g., the dipole moment [119], quadrupole moment [120], or transition dipole moment [121], as a function of the nuclear coordinates. Alternatively, the smoothness of the configurational character of a configuration-interaction (CI) or multiconfiguration self-consistent-field (MCSCF) wave function may be required [122–124]. Quite generally, diabatic states may be defined by requiring maximum overlap of wave functions at nearby geometries [125, 126].

The flexibility of modern methods of electronic structure theory can be exploited to define a fairly general recipe for the *ab initio* construction of diabatic states. As mentioned above, diabatization is particularly relevant when we have to deal with conical intersections or weakly avoided crossings of electronic states. In such situations of near degeneracy of electronic states, a variational *ansatz* of the complete-active-space self-consistent-field (CASSCF) type is most appropriate [97, 98]. The state-averaged CASSCF functional, in particular, allows for a balanced description of the quasi-degenerate states under consideration [96, 127, 128]. Dynamical electron correlation effects can be taken into account by a subsequent multireference configuration interaction (MRCI) calculation [99] or by multireference perturbation theory, such as the CASPT2 method [129].

It has recently been demonstrated that the CASSCF method defines a fairly general and practically useful framework for the direct construction of diabatic states [130–132]. The redundancy of the variational parameters of the CASSCF wave function, i.e., the invariance of the CASSCF energy functional with respect to orbital rotations within the active and external spaces (see [97]), leaves sufficient flexibility to enforce smoothness of the CASSCF orbitals as a function of the nuclear coordinates. Diabatic orbitals may be introduced by a unitary transformation of the adiabatic orbitals, requiring that these orbitals resemble frozen orbitals (i.e., orbitals that are independent of the nuclear geometry except for the movement of the basis functions with the atoms [133, 134]) as closely as possible. This condition, together with the constraint that the CSSCF energy functional must be invariant, is equivalent to the block-diagonalization [135, 136] of the generalized Fock matrix (for canonical orbitals) or the density matrix (for natural orbitals) in the basis of frozen molecular orbitals [132]. Alternatively, a maximum overlap criterion may be employed to enforce smoothness of the CASSCF orbitals [131].

Given these diabatic orbitals, diabatic CI coefficients are obtained in a second step by a block-diagonalization of the Hamiltonian matrix in the configuration-state-function representation [125, 130, 137]. The latter prescription implies that the CASSCF wave function retains, as much as possible, its configurational character. Altogether this construction guarantees (for a given atomic basis set and a given choice of the active space) minimum variation of molecular-orbital and CI coefficients as a function of the nuclear geometry. Hence, the *ab initio* calculations can be performed on a much coarser grid than would be required for the computation of the derivative couplings in the adiabatic representation. Applications of this method have been reported for three-dimensional PE surfaces of triatomic systems [131, 132]. Applications to truly polyatomic systems (benzene, pyrazine) have also been reported [130].

In the special case of conical intersections, which is of particular interest in the present context, the adiabatic-to-diabatic transformation has the important property that it eliminates the singular part of the nonadiabatic coupling at the point of intersection. Smooth variations of the molecular-orbital and CI coefficients, arising from configuration interaction with energetically well separated configurations, are retained in the diabatic wave functions. The residual derivative couplings in the diabatic representation are thus not zero, but should be negligible for all practical purposes. If an essentially exact treatment of the dynamics is desired, the residual derivative couplings may explicitly be evaluated. Since these residual couplings are small and slowly varying, their computation is technically easier than the computation of the original adiabatic derivative couplings in the vicinity of conical intersections.

In the case of conical intersections there exists an alternative and even simpler method for the elimination of the singular part of the nonadiabatic couplings. This method exploits the fact that in an electronic two-state system the diabatic-to-adiabatic mixing angle is fully determined by the adiabatic PE surfaces in the vicinity of the crossing. In the subspace of two electronic states the matrix $\mathbf{X}$ of Eq. (2.11) is a two-dimensional rotation matrix (we assume, as usual, real-valued electronic wave functions) [35, 104, 116,

138]

$$\mathbf{X}(\mathbf{R}) = \begin{pmatrix} \cos\,\theta(\mathbf{R}) & \sin\,\theta(\mathbf{R}) \\ -\sin\,\theta(\mathbf{R}) & \cos\,\theta(\mathbf{R}) \end{pmatrix} \tag{2.15}$$

The diabatic representation is thus completely determined by a single real function $\theta(\mathbf{R})$, the mixing angle, which is given by

$$\theta(\mathbf{R}) = \tfrac{1}{2}\tan^{-1}[2H_{12}(\mathbf{R})/(H_{22}(\mathbf{R}) - H_{11}(\mathbf{R}))] \tag{2.16}$$

The H_{ij} are the Hamiltonian matrix elements in the diabatic representation. In the adiabatic representation, the fixed-nuclei electronic Hamiltonian is diagonal with elements

$$V_{1,2}(\mathbf{R}) = \tfrac{1}{2}(H_{11}(\mathbf{R}) + H_{22}(\mathbf{R})) \pm \tfrac{1}{2}[(H_{11}(\mathbf{R}) - H_{22}(\mathbf{R}))^2 + 4H_{12}(\mathbf{R})^2]^{1/2} \tag{2.17}$$

In general, Eqs. (2.15–2.17) are of little value in practice, since the mixing angle $\theta(\mathbf{R})$ to be determined depends on the unknown diabatic Hamiltonian matrix elements $H_{ij}(\mathbf{R})$. In the specific case of a conical intersection of two electronic surfaces, however, the $H_{ij}(\mathbf{R})$ may locally be expanded in a Taylor series around the point of intersection [104, 138]. It can be shown that the expansion coefficients can directly be related to the gradients of the *adiabatic* energy surfaces at the intersection. This procedure thus gives the mixing angle $\theta(\mathbf{R})$ in the vicinity of the intersection according to Eq. (2.16), which may be employed to construct diabatic energy surfaces from the given adiabatic surfaces. It can be shown that the singular part of the nonadiabatic coupling is completely eliminated in this diabatic representation. For details of this approach we refer the reader to ref. [45].

Let us finally illustrate these concepts by an example. The conical intersection of the $S_1(n\pi^*)$ and $S_2(\pi\pi^*)$ surfaces of pyrazine has recently been characterized by extensive calculations at the CASSCF/MRCI level [45]. This system, to which we shall also refer in subsequent sections of this article, provides a clear demonstration of the importance of a diabatic representation for multidimensional conical intersections.

It suffices to consider, for the purpose of illustration, the coordinate Q_{10a} of B_{1g} symmetry, which couples the $S_1(^1B_{3u})$ and $S_2(^1B_{2u})$ states of pyrazine in first order, as well as one of the totally symmetric coordinates (Q_{6a}), which induces a crossing of the S_1 and S_2 PE surfaces. The vibrational coordinates are defined as the normal coordinates of the electronic ground state of pyrazine [45]. Figure 1(*a*) gives a close-up view of the conical intersection of the adiabatic S_1 and S_2 surfaces of pyrazine, calculated at the CASSCF level (see ref. [45] for details). The conical shape of the PE surfaces in the vicinity of the intersection is clearly visible. The *ab initio* diabatic potentials (diagonal elements H_{11}, H_{22}) and the diabatic coupling function H_{12} are shown in Fig. 1(*b,c*). In contrast to the adiabatic surfaces, which exhibit a nondifferentiable cusp at the intersection, the diabatic surfaces are smooth functions of the nuclear coordinates. The diabatic potentials can thus adequately be represented by a low-order Taylor expansion.

The mixing angle $\theta(Q_{6a}, Q_{10a})$ defined in Eq. (2.16) is displayed in Fig. 2(*a*). It exhibits the well-known topological phase ("Berry's phase") of the adiabatic electronic wave functions, enforced by the intersection of the adiabatic potentials [114]. The derivatives $\partial\theta/\partial Q_{6a}$ and $\partial\theta/\partial Q_{10a}$, which determine the singular part of the first-order nonadiabatic coupling, are shown in Fig. 2(*b,c*), respectively. This singular part of the nonadiabatic coupling is eliminated by the transformation to the diabatic representation.

The nonadiabatic character of electronic wave functions is also reflected in the transition-dipole-moment functions, which are defined as

$$\boldsymbol{\mu}_{nm}(\mathbf{R}) = \int d\mathbf{r}\, \psi_n^*(\mathbf{r}, \mathbf{R}) \boldsymbol{\mu} \psi_m(\mathbf{r}, \mathbf{R}) \tag{2.18}$$

Here

$$\boldsymbol{\mu} = -\sum_{i=1}^{N} e\mathbf{r}_i \tag{2.19}$$

is the electronic dipole operator and the ψ are either adiabatic or diabatic electronic wave functions. As discussed in detail in ref. [45], the adiabatic S_0–S_1 and S_0–S_2 transition-dipole-moment functions of pyrazine vary rapidly in the vicinity of the intersection. The transformation to the diabatic basis eliminates these rapid variations.

Figure 3 displays, as an example, the μ_{10}^x element, i.e., the x component of the S_1–S_0 transition dipole moment (the molecule lies in the yz plane, the z axis going through the N atoms). The transition dipole moment in the adiabatic representation, shown in Fig. 4(*a*), varies rapidly as a function of Q_{6a} and Q_{10a}, reflecting the geometry-dependent mixing of the $S_1(n\pi^*)$ and $S_2(\pi\pi^*)$ wave functions in the vicinity of the intersection [cf. Fig. 2(*a*)]. The same element obtained in the diabatic representation, shown in Fig. 3(*b*), is virtually constant [note the different ordinate scales in Fig. 3(*a* and *b*)]. In this

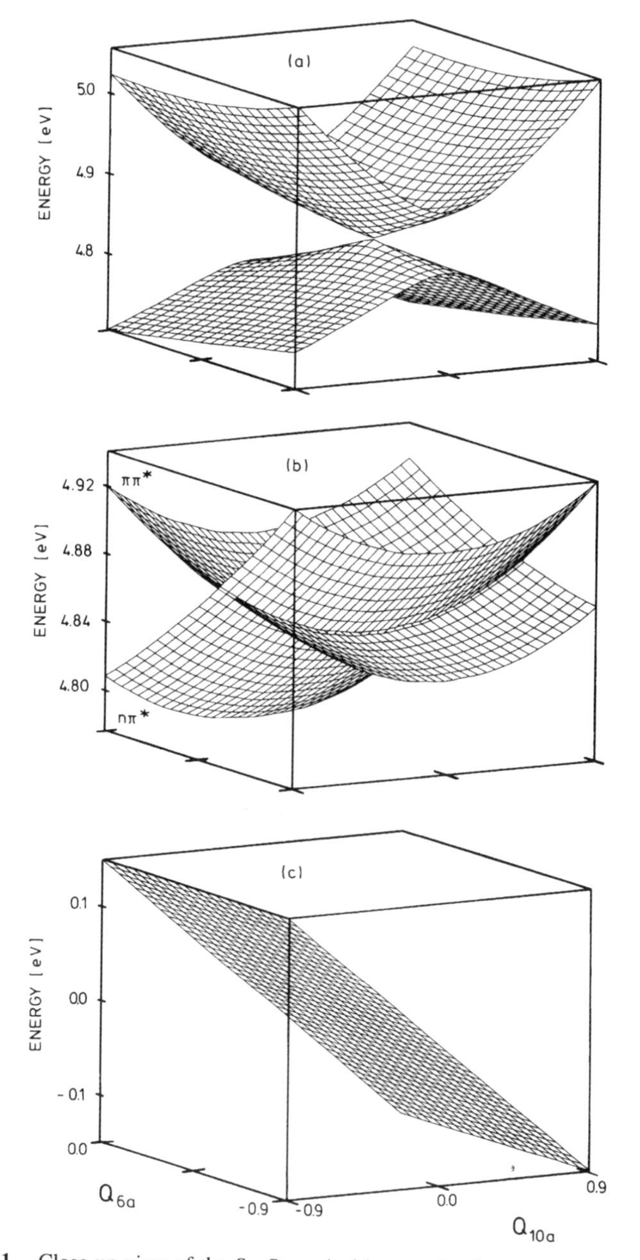

Figure 1. Close-up view of the S_1–S_2 conical intersection in pyrazine in the Q_{6a}–Q_{10a} space, calculated at the CASSCF level [45]. Adiabatic surfaces are shown in (*a*), diabatic surfaces in (*b*). The diabatic coupling element is shown in (*c*).

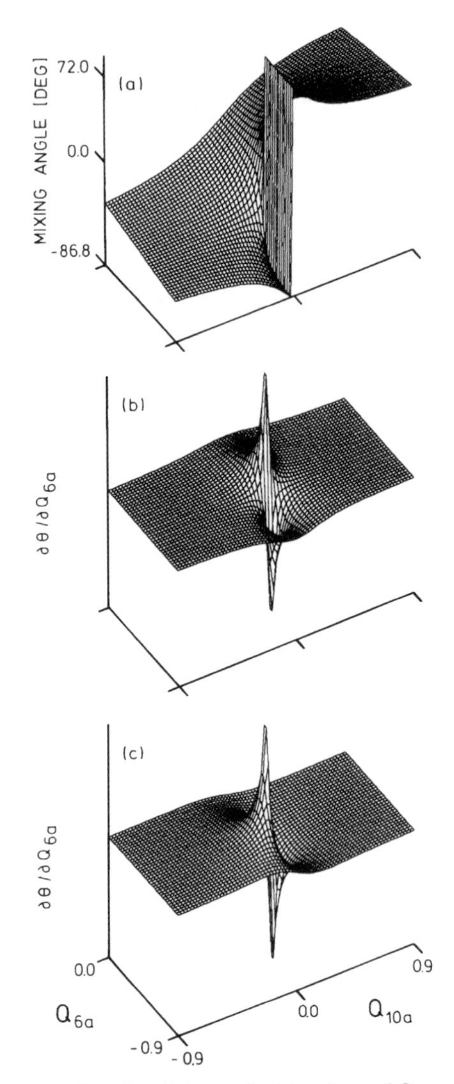

Figure 2. Diabatic-to-adiabatic mixing angle (*a*) and non-BO coupling elements (*b*), (*c*) near the S_1–S_2 conical intersection in pyrazine, obtained from the *ab initio* diabatic potentials shown in Fig. 1.

example, a low-order Taylor expansion of the transition dipole moment is obviously an excellent approximation in the diabatic representation. Clearly, a low-order Taylor expansion would be meaningless for the adiabatic transition dipole moment of Fig. 3(*a*).

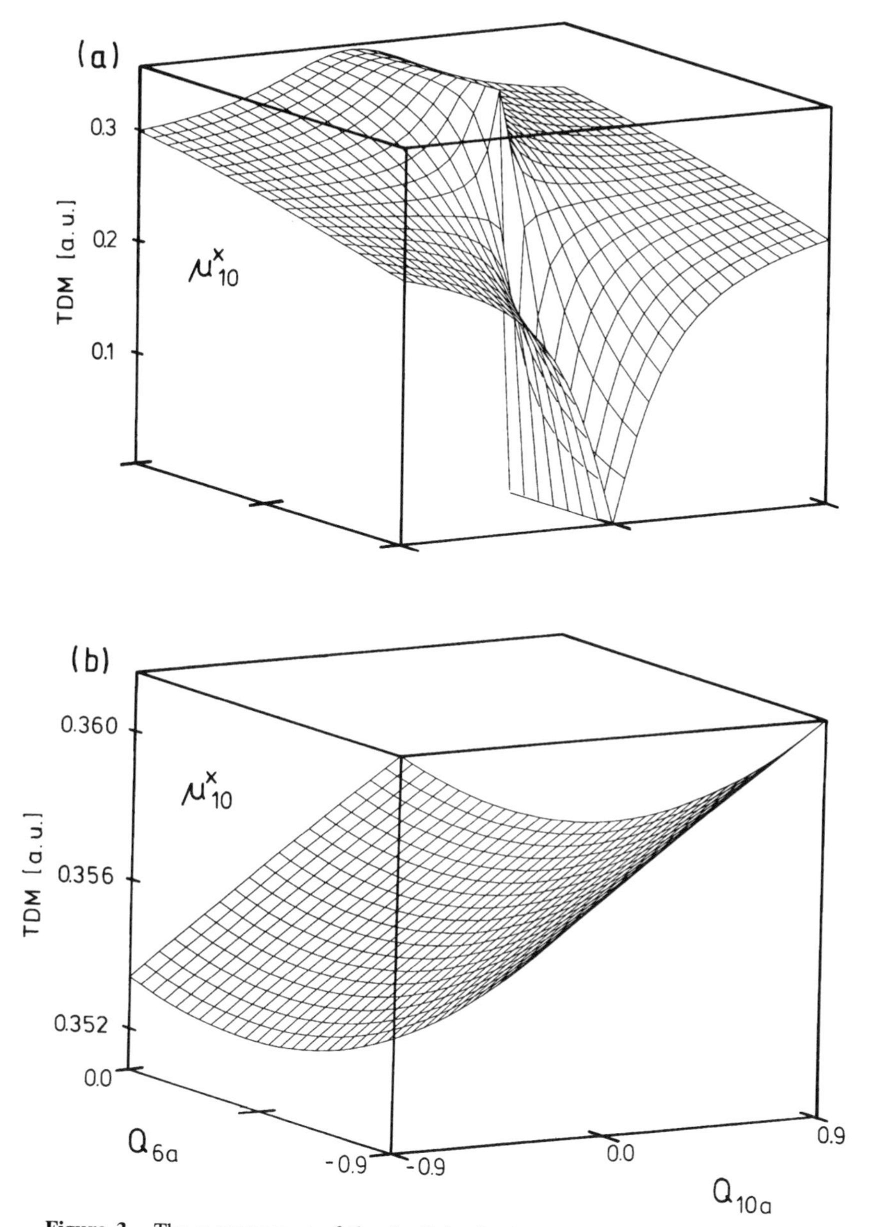

Figure 3. The x component of the S_0–$S_1(n\pi^*)$ *ab initio* transition dipole moment of pyrazine is shown as a function of Q_{6a} and Q_{10a} in the adiabatic (a) and diabatic (b) representations [45]. Note the different ordinate scales: on the scale of (a), the transition dipole moment function of (b) is essentially a constant.

In summary, the diabatic representation is an essential concept for the treatment of non-BO dynamics in polyatomic systems. In multidimensional curve-crossing problems, the diabatic representation reflects more clearly the essential physics of the problem. The diabatic representation thus facilitates the construction of suitable model Hamiltonians. As has been emphasized above, the smoothness (with respect to variations of the nuclear geometry) of diabatic PE functions as well as electronic wave functions is an essential aspect, allowing the approximation of relevant quantities by rapidly converging expansions.

In vibronic-coupling theory, the concept of the diabatic representation has been employed in a heuristic manner for a long time (see, e.g., [35, 104, 111, 139, 140]). As discussed above, more recent developments have shown how diabatic electronic wave functions can systematically be constructed on the basis of accurate *ab initio* calculations for polyatomic systems. Such calculations probably will never be truly straightforward, since the very concept of a diabatic representation is problem-specific. It can nevertheless be expected that the explicit *ab initio* computation of diabatic PE functions and coupling elements of polyatomic molecules will no longer represent a major obstacle for future applications.

C. Model Hamiltonians

1. General Aspects

a. PE Surfaces

The development of realistic global descriptions of BO PE surfaces of polyatomic molecules is a difficult and largely unsolved problem. This is particularly true if the dynamics is governed by more than one BO surface and thus a mathematically consistent modeling of several energy surfaces and coupling elements is required. A global description of the PE surfaces is generally necessary for the treatment of fragmentation or photochemical rearrangement processes. The problem is simpler if the amplitudes of all degrees of freedom are restricted, as may be the case in photophysical relaxation processes. In the latter situation, polynomial expansions represent a powerful and generally applicable tool to characterize the PE surfaces in the relevant local region of nuclear coordinate space.

Triatomic systems possess just three internal degrees of freedom. In this case it is possible, in principle, to map out the complete energy surface on a sufficiently dense three-dimensional grid, employing state-of-the-art electronic structure theory. Spline fitting may possibly be used to interpolate between *ab initio* grid points. High-quality *ab initio* surfaces of this type are nowadays available for the electronic ground state and for a few excited states of a number of triatomic systems. The recent monograph on photodissociation dynamics by Schinke [1] gives an excellent account of the systematic use of such complete *ab initio* surfaces in dynamics calculations of triatomic systems. Strong non-BO effects associated with surface intersections have recently been treated in this type of approach (see refs. [141–143] for representative examples).

For polyatomic systems with more than four atoms, such a systematic and complete mapping of the global PE surfaces is not possible. One has to resort to analytical representations of the PE surfaces, characterized by relatively few parameters, which can in turn be determined by fitting *ab initio* calculations. Finding analytic representations of polyatomic BO energy surfaces is a central problem of chemical reaction dynamics. The extensive work in this area has been reviewed, for example, by Murrell et al. [144], Truhlar et al. [145], Varandas [146], and Schatz [147]. Most of this work refers to triatomic systems and is concerned with the representation of single-valued BO surfaces. It seems that very little is known about the representation of global multisheeted surfaces in molecules with more than three atoms.

In principle, one could avoid the construction of global analytic representations of PE surfaces by evaluating the adiabatic potential and derivatives of the electronic wave function (the derivative couplings) at every point where they are required in a classical or quantum-mechanical dynamical calculation. In practice, however, the computation of the energy and the wave function by *ab initio* electronic-structure theory is much too time-consuming, especially for excited electronic states. Progress has recently been made with dynamical calculations on the electronic ground-state surface, which treat the nuclear motion classically and evaluate the electronic potential "on the fly," employing relatively inexpensive *ab initio* methods such as Hartree–Fock or density functional theory [148–150]. Applications of such techniques to excited-state problems have been very limited [151], and no applications to coupled electronic-state dynamics are known to the authors. At the semiempirical level, the "on the fly" approach appears more tractable and has been employed for the modeling of nonadiabatic processes within classical surface-hopping approximations [152].

The reaction-path Hamiltonian model of Miller, Handy, and Adams [153] represents another approach to chemical dynamics that avoids the construction of global PE surfaces. In this approach the classical or quantum-mechanical Hamiltonian is expressed in terms of the curvilinear steepest-descent reaction coordinates S, as well as S-dependent normal coordinates that describe vibrations orthogonal to the reaction path. As far as we are aware of, dynamical calculations based on the reaction-path Hamiltonian have not been performed for photochemical applications.

The concept that we want to elaborate in this article is based on the construction of model Hamiltonians for multidimensional curve-crossing problems in a diabatic electronic representation. As discussed in the preceding section, the diabatic representation generally exhibits the basic physics of curve-crossing problems more clearly than the standard BO adiabatic representation. The Jahn–Teller effect in degenerate electronic states [104, 140] represents a well-known example: the diabatic PE surfaces can be modeled in terms of a low-order Taylor expansion, even though the shape of the adiabatic surface in the vicinity of the intersection is complicated and defies a low-order polynomial representation.

In order to be able to deal with photochemical dynamics, we have to consider the modeling of *global* diabatic PE functions and coupling matrix elements. Given a basis of diabatic electronic states $\{|\psi_n\rangle\}$ as defined in Section II.A, the molecular Hamiltonian is written as

$$H_M = \sum_n |\psi_n\rangle(T_N + V_n(\mathbf{R}))\langle\psi_n| + \sum_{nn'} |\psi_n\rangle V_{nn'}(\mathbf{R})\langle\psi_{n'}| \tag{2.20}$$

It is understood that we shall neglect the residual derivative couplings in the diabatic representation, i.e., the nuclear kinetic-energy operator T_N in Eq. (2.20) acts only on the vibrational part of the wave function. In most applications that we have in mind, the $|\psi_n\rangle$ represent a few of the lowest singlet states of the molecule. In some applications (see Section VII) the set $\{|\psi_n\rangle\}$ may also include high Rydberg states as well as ionization continua.

An essential aspect in the construction of model Hamiltonians for chemical dynamics is the choice of the nuclear coordinates $\mathbf{R}$. The form of the electronic PE, as well as the kinetic-energy expression, depends on the choice of coordinates; coordinate systems may be selected to simplify either the construction of PE functions or the construction of the kinetic-energy operator. For classical trajectory calculations, where the computational cost scales only linearly with the dimension, Cartesian or mass-weighted Cartesian coordinates are an obvious choice, since the kinetic-energy expression is trivial. In quantum-mechanical treatments, the computational cost grows much more rapidly with the dimension of the problem, and the elimination of translation and overall rotation is essential. It thus appears natural to formulate the Hamiltonian in internal coordinates as defined, for example, by Wilson et al. [154]. Internal coordinates defined in terms of bond lengths and bond angles are best suited for the modeling of the PE function. Their drawback is the possibly very complicated form of the quantum-mechanical nuclear kinetic energy operator (see below).

A modeling of general polyatomic BO PE functions in terms of bond length and bond angle internal coordinates, which draws on well-established concepts of the molecular mechanics approach [155], has been extensively discussed by Thompson and collaborators in the context of classical trajectory simulations of reaction and intramolecular vibrational relaxation processes [156–162]. For example, the effect of isomerization dynamics on vibrational relaxation processes has been investigated for systems that undergo large-amplitude torsional motion [162]. In this approach PE functions are modeled by an expansion in terms of internal displacement coordinates around local minima, employing the harmonic approximation for bends, Morse functions for stretching coordinates, and truncated cosine series for torsional modes. Switching functions [147] are employed to smoothly join the expansions for different isomers [158, 160, 162]. This methodology may be employed to construct, in a straightforward manner, expressions for the diabatic PE functions $V_n(\mathbf{R})$ and coupling elements $V_{nn'}(\mathbf{R})$ in Eq. (2.20). If, as is usually the case, the local minima are associated with different diabatic sheets, each sheet may be represented by an expansion about its minimum. The diagonalization of the diabatic PE matrix then yields global adiabatic PE functions without the need of introducing switching functions (see, for example [163, 164]). As an example, a simple model of a photochemical funnel induced by a large-amplitude torsional motion will be discussed below (Section C.3).

b. Kinetic-Energy Operator

The derivation of the exact (in the nonrelativistic limit and within the BO approximation) quantum-mechanical kinetic-energy operator in internal coordinates is straightforward in principle but cumbersome in practice. The complete expressions for kinetic-energy operators in internal coordinates for triatomic and some types of tetra-atomic molecules have only recently been derived with the help of computer algebra [165, 166].

There exist two equivalent approaches for the derivation of the quantum mechanical kinetic-energy expression in internal coordinates. In the first approach, one starts with the Lagrangian form of the kinetic energy in terms of classical velocities. Using successive transformations, one obtains [154]

$$T_N = \tfrac{1}{2}\mathbf{P}^{\mathrm{T}}\mathbf{G}(\mathbf{R})\mathbf{P} \tag{2.21}$$

where the P_i are the momenta conjugate to the internal coordinates R_i, and $\mathbf{G}(\mathbf{R})$ is the well-known $\mathbf{G}$ matrix of vibrational spectroscopy [154] defined in Eq. (2.24) below. The Podolsky formalism [167] yields the quantum-mechanical operator

$$T_N = -\frac{\hbar^2}{2} g^{1/4} \sum_{ij=1}^{M} \frac{\partial}{\partial R_i} g^{-1/2} G_{ij} \frac{\partial}{\partial R_j} g^{1/4} \tag{2.22}$$

where $g = \det(G_{ij})$ and $M = 3N - 6$ is the number of internal coordinates for a nonlinear molecule.

The second approach, pursued by Sutcliffe [168] and Handy [165], starts from the Laplacian in Cartesian coordinates. Choosing a set of M internal coordinates and applying twice the chain rule, the vibrational part of T_N is expressed as

$$T_N = -\frac{\hbar^2}{2}\left[\sum_{jk}^{M} G_{jk} \frac{\partial^2}{\partial R_j \partial R_k} + \sum_{j}^{M} h_j \frac{\partial}{\partial R_j}\right] \tag{2.23}$$

where

$$G_{jk} = \sum_{i}^{3N} \frac{1}{m_i} \frac{\partial R_j}{\partial X_i} \frac{\partial R_k}{\partial X_i} \tag{2.24}$$

$$h_j = \sum_{i}^{3N} \frac{1}{m_i} \frac{\partial^2 R_j}{\partial X_i^2} \tag{2.25}$$

Employing symbolic computer algebra, Handy and collaborators have derived exact vibrational and rotational-vibrational kinetic-energy operators for various choices of internal coordinates in triatomic and tetra-atomic molecules [165, 166]. A simplification of this procedure has recently been discussed by Lukka [169]. The possibility of extending these techniques to larger molecules has been demonstrated by Császár and Handy, who have obtained the exact expression for the vibrational ($J = 0$) kinetic-energy operator for sequentially bonded N-atomic molecules [170].

The developments just discussed have been motivated by the interest in accurate variational calculations of vibration-rotation energy levels in the electronic ground state of triatomic and tetra-atomic molecules, see [171–173] for representative applications. When considering, on the other hand, the photochemical dynamics of large molecules, especially the short-time (sub-picosecond) dynamics, it will generally not be necessary to explicitly include all internal nuclear degrees of freedom. As discussed in detail in Section I, we can expect that often only a few nuclear degrees of freedom are strongly coupled to the electronic transition and are thus of relevance at the shortest timescales. In the modeling of the short-time dynamics, it may thus be appropriate to freeze the inactive degrees of freedom by the imposition of appropriate constraints. A systematic procedure for the construction of classical and quantum mechanical kinetic-energy expressions for such constrained model systems has been discussed by Hadder and Frederick [174] (see also [154]). In this case it is essential to introduce the constraints, in the form of constant bond lengths and bond angles, into the classical Lagrangian form of the kinetic energy before transforming to the Hamiltonian form. The Podolsky expression, Eq. (2.22), finally gives the correct quantum mechanical operator for constrained motion [174].

In their most general form, these concepts have not yet been implemented for the description of nonadiabatic excited-state dynamics of molecules with more than three atoms. The existing calculations which include nonadiabatic effects for larger polyatomic systems have all been performed with simplified forms of the kinetic energy and have usually also assumed very simple PE functions. In the following two subsections we discuss two representative examples of such simplified models of polyatomic coupled electronic-state dynamics. The first example is the well-established vibronic-coupling approach, based on a Taylor expansion of diabatic PE functions in terms of normal-mode displacements. This approach is expected to be appropriate for the description of photophysical dynamics associated with conical intersections, but does not allow the description of photochemically reactive dynamics involving a large-amplitude motion. As a first step towards a microscopic description of photochemical dynamics, we subsequently consider the modeling of a photochemical funnel that is induced by large-amplitude torsional motion of a carbon-carbon double bond.

2. *Normal-Mode Taylor Expansion*

The construction of appropriate kinetic-energy operators and PE functions simplifies considerably if we can assume that no large-amplitude motions are involved. In this case we can define a suitable reference geometry of the system (for example, the equilibrium geometry of the electronic ground state) and adopt internal displacement coordinates to describe small vibrations about the reference geometry. It is then a standard procedure to introduce normal coordinates by the simultaneous diagonalization of the kinematic matrix (taken at the reference geometry) and the matrix of force constants for the electronic ground state [154]. It is convenient to define dimensionless normal coordinates Q_j via

$$Q_j = (\hbar/\omega_j)^{1/2} q_j \tag{2.26}$$

where the q_j are the normal coordinates as defined in [154] and ω_j are harmonic vibrational frequencies. The kinetic-energy operator then takes the simple form

$$T_N = -\frac{1}{2} \sum_{j=1}^{M} \hbar\omega_j \partial^2/\partial Q_j^2 \tag{2.27}$$

and the PE function of the electronic ground state, expanded up to second order in the Q_j, reads

$$V_0 = \frac{1}{2} \sum_{j=1}^{M} \hbar\omega_j Q_j^2 \tag{2.28}$$

The excited state PE functions and diabatic interstate coupling functions are likewise approximated by Taylor expansions. The following expansions define the first-order (κ, λ) and second-order (γ) electronic-vibrational coupling constants [175]

$$V_n(\mathbf{Q}) = V_0(\mathbf{Q}) + \sum_{j=1}^{M} \kappa_j^{(n)} Q_j + \sum_{i,j=1}^{M} \gamma_{ij}^{(n)} Q_i Q_j + \ldots \tag{2.29}$$

$$V_{nn'}(\mathbf{Q}) = V_{nn'}(\mathbf{O}) + \sum_{j=1}^{M} \lambda_j^{(nn')} Q_j + \ldots \tag{2.30}$$

Note that in Eqs. (2.28–2.30) all PE functions have been expanded in a single set of normal coordinates (the normal coordinates of the electronic ground state). The $\kappa_j^{(n)}$ represent the gradients of the excited-state PE functions at the equilibrium geometry of the ground state and are nonzero for totally symmetric vibrational modes. Since these modes "tune" the energy gap of different electronic states and may lead to intersections of diabatic PE functions, they have been termed "tuning modes" [35]. The normal modes for which $\lambda_j^{nn'} \neq 0$ are responsible for electronic interstate coupling and have therefore been termed "coupling modes" [35].

The explicit expressions [Eqs. (2.27–2.30)], in conjunction with Eq. (2.20), define a class of model Hamiltonians that may be called vibronic-coupling Hamiltonians. The underlying concept is rooted in early treatments of the dynamical Jahn–Teller effect [104]. It has systematically been extended and extensively applied by many workers. Since extensive reviews of the vibronic-coupling concept and its applications can be found in the literature [35, 140, 176–178], we need not go into details here.

Model Hamiltonians constructed according to Eqs. (2.27–2.30) are particularly well suited for the calculation of low-resolution absorption spectra, photoelectron spectra and resonance-Raman spectra of polyatomic molecules. As is well known [1, 175, 179], these spectra are largely determined by the short-time dynamics in the excited state, which, in turn, is governed by the shape of the PE functions within the so-called Franck–Condon zone of the optical transition. In this limited range of nuclear geometries, the multidimensional PE functions are generally well approximated by the Taylor expansions in Eqs. (2.29) and (2.30).

The modeling of the conical intersection of the $S_1(^1B_{3u})$ and $S_2(^1B_{2u})$ states of pyrazine [45, 180, 181] provides a good illustration of the application of the vibronic-coupling concept. Elementary group theory shows that $V_{12}(0) = 0$ in this case, and that $\lambda_j^{(12)}$ can be nonzero only for the mode Q_{10a} of B_{1g} symmetry. The gradients $\kappa_j^{(n)}, n = 1, 2$, can be nonzero for the five totally symmetric normal modes $\nu_1, \nu_2, \nu_{6a}, \nu_{8a}, \nu_{9a}$ of pyrazine. The *ab initio* calculation of cuts of the S_0, S_1, S_2 PE surfaces along the five totally symmetry coordinates has shown that these functions are well approximated by the harmonic expansion of Eq. (2.29) over a wide range of nuclear geometries [181]. The harmonic vibrational frequencies ω_j follow directly from a normal-mode analysis of the electronic ground-state surface. The interstate vibronic-coupling constant $\lambda_{10a}^{(12)}$ is given by the gradient of $V_{12}(\mathbf{Q})$ with respect to Q_{10}, see Fig. 1(*c*). The $\kappa_j^{(n)}$ are given by the gradients of $V_n(\mathbf{Q})$ with respect to totally symmetric normal coordinates Q_j at the reference geometry $\mathbf{Q} = 0$, see Fig. 1(*b*). In ref. [45] accurate values for these constants have been obtained at the CASSCF/MRCI level. Electronic absorption and resonance Raman spectra obtained with this vibronic-coupling model Hamiltonian are in excellent agreement with experiment [45, 182]. We shall return to this *ab initio* based model of vibronic coupling in pyrazine and the associated dynamics in the discussion of real-time pump-probe spectra (Section VII).

3. *Model of Nonadiabatic Photoisomerization*

As an example of the modeling of photochemically reactive dynamics, let us consider the twisting of a carbon-carbon double bond in an unsaturated hydrocarbon molecule. It is generally believed that the bond twisting can lead to a degeneracy or near degeneracy of S_0 and S_1 electronic states in the perpendicularly twisted configuration and that this mechanism is responsible for ultrafast internal conversion to the ground state [3, 5]. We shall briefly consider here the modeling of such a photochemical funnel [3] following ref. [39].

Let us consider a single large-amplitude coordinate described by the torsional angle φ. The remaining nonreactive degrees of freedom are represented by a set of small-amplitude displacement coordinates. For convenience, we adopt normal coordinates Q_j, $j = 1, \ldots, M - 1$, defined at the equilibrium geometry of the electronic ground state for the description of the nonreactive coordinates. Assuming the reduced moment of inertia for the torsional motion to be constant (see ref. [183] for a generalization of this *ansatz*) and neglecting kinematic torsion-vibration couplings, the kinetic-energy operator is in the simplest approximation given by

$$T_N = -\frac{\hbar^2}{2I}\frac{\partial^2}{\partial\varphi^2} - \frac{1}{2}\sum_{j=1}^{M-1}\hbar\omega_j\frac{\partial^2}{\partial Q_j^2} \tag{2.31}$$

More rigorous kinetic-energy operators can be derived for specific systems, for example, the torsion of ethylenic molecules [184].

The PEs as a function of the torsional angle are periodic and are thus represented by a Fourier series [183]

$$V_n(\varphi) = \sum_{k=0}^{\infty} V_k^{(n)}\cos(k\varphi) \tag{2.32}$$

while the Taylor expansions in Eqs. (2.28–2.30) are employed for the remaining degrees of freedom. Assuming, for simplicity, a single mode Q_c that can couple the S_0 and S_1 states in first order and a set of $M - 2$ totally symmetric and thus Condon active modes Q_i, the diabatic PE functions read [39]

$$V_n(\varphi, Q_c, Q_1, Q_2 \ldots) = V_n(\varphi) + \frac{\hbar\omega_c}{2}Q_c^2 + \sum_{i=1}^{M-2}\frac{\hbar\omega_i}{2}Q_i^2 \tag{2.33}$$

$$+ \sum_{i=1}^{M-2}\kappa_i^{(n)}Q_i + \sum_{ij=1}^{M-2}\gamma_{ij}^{(n)}Q_iQ_j \qquad n = 0, 1$$

$$V_{01} = \lambda Q_c \tag{2.34}$$

To recover the basic aspects of the problem, it is sufficient to consider just the leading terms of the Fourier expansion. The torsional parts of the potentials are approximated as

$$V_0(\varphi) = E_0 + \tfrac{1}{2}V_0[1 - \cos(k\varphi)] \tag{2.35}$$

$$V_1(\varphi) = E_1 - \tfrac{1}{2}V_1[1 - \cos(k\varphi)] \tag{2.36}$$

where $k = 1$ for 2π-periodic potentials or $k = 2$ for π-periodic potentials.

The potential model of Eqs. (2.33–2.36) is highly simplified, because it neglects, apart from the bilinear terms leading to Duschinsky mixing of the normal modes, any nonseparability of the modes on the diabatic surfaces. Mode coupling effects arise solely from the intersection of diabatic terms and their interaction through the coupling mode. The intersection and interaction of the diabatic potentials leads to a pronounced anharmonic distortion of the adiabatic PE surfaces. This is illustrated in Fig. 4 for a two-dimensional model including the torsional coordinate φ and the coupling coordinate Q_c [39]. The adiabatic surfaces avoid crossing for $Q_c \neq 0$, which results in a conical intersection of the adiabatic PE surfaces. Despite the simple separable form [Eq. (2.33)] of the diabatic PE functions [Fig. 4(*a*)], the adiabatic PE surfaces [Fig. 4(*b*)] are seen to be manifestly nonseparable. This "vibronically induced" anharmonicity, as well as the singular non-BO coupling, are essential features of photochemical funnels.

Diabatic model potentials of the type in Eqs. (2.32–2.36) thus appear appropriate to model essential aspects of photochemical funnels. On the other hand, generic mode–mode coupling effects such as cubic or quartic anharmonicities or effects of large amplitude motion on vibrational relaxation processes as discussed by Thompson and collaborators [158, 162] are neglected. Moreover, kinematic mode–mode coupling effects are neglected as a consequence of the simplified form of the kinetic-energy operator in Eq. (2.31). As will be seen below (Section IV), such simplified models of photochemical funnels are nevertheless able to describe electronic population decay, irreversible isomerization dynamics, mode-to-mode energy transfer, and coherence decay of vibrational motion on femtosecond timescales. It is believed that these phenomena are characteristic of the initial stage of a typical ultrafast internal-conversion process. At longer timescales it may be necessary to consider more exact kinetic-energy operators and more sophisticated potential models, as discussed in Section II.C.1.

III. TIME-DEPENDENT METHODOLOGY

In this section we are concerned with technical aspects of the solution of the time-dependent Schrödinger equation. The nonstationary dynamics of the molecular system as such is described by the time-dependent Schrödinger equation with the molecular Hamiltonian H_M

$$i\hbar \frac{\partial}{\partial t} |\Psi(t)\rangle = H_M |\Psi(t)\rangle \tag{3.1}$$

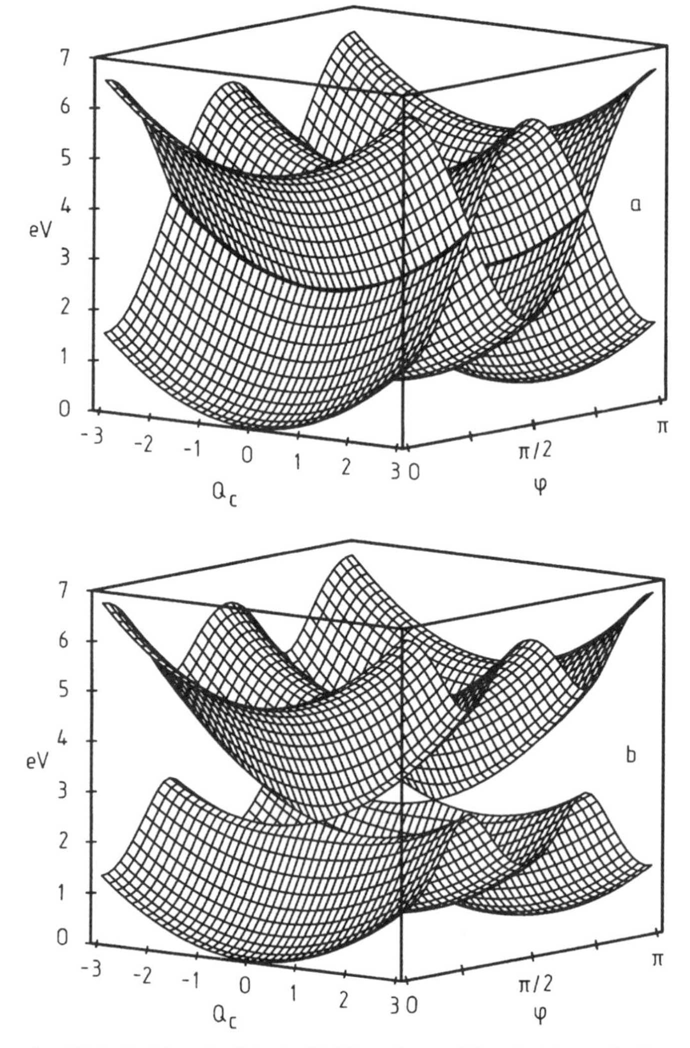

Figure 4. Diabatic (*a*) and adiabatic (*b*) PE surfaces of the photoisomerization model in the two-dimensional space spanned by the torsional angle φ and the coupling coordinate Q_c.

The Hamiltonian H_M is assumed to be given as a few-state few-mode model Hamiltonian as specified by Eq. (2.20). Equation (3.1) has to be solved as an initial-value problem in time for some given initial condition $|\Psi(0)\rangle$. Explicit solutions of Eq. (3.1) for sufficiently complex systems can be of considerable interest as they can reveal elementary features of ultrafast radiationless processes in molecules (see Section IV). In addition, the well-established evaluation of real-time pump-probe signals via perturbation theory with respect to the radiation-matter interaction [74] ultimately requires the solution of Eq. (3.1) to characterize the material dynamics (see Section VI.A).

Alternatively, we may consider the equation of motion of the molecular system, including its interaction with an external time-dependent electric field

$$i\hbar \frac{\partial}{\partial t} |\Psi(t)\rangle = H(t)|\Psi(t)\rangle \tag{3.2a}$$

$$H(t) = H_M + H_{\text{int}}(t), \tag{3.2b}$$

$$H_{\text{int}}(t) = -\sum_{n>m} |\psi_n\rangle \mathbf{E}(t) \cdot \boldsymbol{\mu}_{nm} \langle \psi_m| + \text{h.c.} \tag{3.2c}$$

The $\boldsymbol{\mu}_{nm}$ are electronic transition dipole moments and have been defined in Eq. (2.18). The solution of the explicitly time-dependent problem of Eq. (3.2a) is required if we wish to consider the effect of strong external fields on the molecular dynamics, in particular in the context of a control of the molecular dynamics [185]. Even in the weak-field case, the nonperturbative evaluation of the electric polarization via the solution of Eq. (3.2a) can in some cases be technically preferable, see Section VI.B.

If H_M is electronically diagonal (no vibronic coupling) and the coupling of the electronic states occurs only through the radiation field, one may simplify the numerical solution of Eq. (3.2a) by transformation to an interaction representation. Decomposing $H(t)$ into the electronically diagonal part H_M and the nondiagonal part $H_{\text{int}}(t)$, the equation of motion in the interaction representation becomes

$$i\hbar \frac{\partial}{\partial t} |\Psi^{(I)}(t)\rangle = H^{(I)}_{\text{int}}(t)|\Psi^{(I)}(t)\rangle \tag{3.3a}$$

$$|\Psi^{(I)}(t)\rangle = e^{iH_M t/\hbar}|\Psi(t)\rangle \tag{3.3b}$$

$$H^{(I)}_{\text{int}}(t) = e^{iH_M t/\hbar} H_{\text{int}}(t) e^{-iH_M t/\hbar} \tag{3.3c}$$

By this transformation rapid oscillations of $|\Psi(t)\rangle$ that reflect the electronic energy gap are eliminated. If the laser field is resonant with the electronic transition, the problem may further be simplified by introducing the rotating-wave approximation, which implies the neglect of the counter-rotating term in $H^{(I)}_{\text{int}}(t)$ [186]. It has generally been found that the rotating-wave approximation is very good for resonant excitation [50, 91, 187]. If the states coupled by the laser field are simultaneously coupled by a strong intramolecular interaction, the transformation to the interaction representation is of no advantage.

The partial differential equation in Eq. (3.1) or Eq. (3.3a) is converted into a numerically tractable problem by a discretization of space and time. (If the electronic basis includes a continuum, an additional discretization of the continuum may be required; see, for example, ref. [91].) The discretization of the problem in Eq. (3.1) may be decomposed into two steps. In the first step, the spatial dependence of the wave packet $\Psi(\mathbf{R}, t)$ is discretized by an expansion of $\Psi(\mathbf{R}, t)$ in a set of basis functions or by a representation on a spatial grid. As a result, the partial differential equation, Eq. (3.1), is converted into a set of coupled first-order ordinary differential equations. In the second step, one of the many available numerical algorithms is used to propagate the solution in time, employing a suitable numerical propagator.

We shall not consider here the obvious option of expanding $|\Psi(t)\rangle$ in terms of eigenstates of H_M. The computation of $|\Psi(t)\rangle$ via this strategy requires the solution of the eigenvalue problem of H_M (all eigenvalues and eigenvectors), which represents a serious computational bottleneck for multidimensional systems.

A. Representation of the State Vector

Let us expand $|\Psi(t)\rangle$ in a complete set of time-independent basis states. For the class of Hamiltonians given by Eq. (2.20) a complete basis can be constructed as the direct product of (diabatic) electronic basis states $\{|\psi_n\rangle\}$ and suitable orthonormal basis states $\{\chi_{v_j}\}, v_j = 0, 1, 2, \ldots$ for each nuclear degree of freedom, $j = 1, 2, 3, \ldots$

$$|\Psi(t)\rangle = \sum_n \sum_{v_1 v_2 v_3 \ldots} C_{n, v_1 v_2 v_3 \ldots}(t) |\psi_n\rangle |\chi_{v_1}\rangle |\chi_{v_2}\rangle |\chi_{v_3}\rangle \ldots \tag{3.4}$$

The time-dependent state vector $|\Psi(t)\rangle$ is thus represented by the countably infinite set of coefficients $C_{n, v_1 v_2 v_3 \ldots}(t)$. In practice, the basis for each vibrational degree of freedom has to be truncated, yielding a vector

$$\mathbf{C}(t) = \{C_{n, v_1 v_2 v_3 \ldots}(t)\} \tag{3.5}$$

of finite dimension. We further introduce the representation of the Hamiltonian in the direct-product basis

$$\mathbf{H} = \{H_{n, v_1 v_2 v_3 \ldots ; n', v'_1 v'_2 v'_3 \ldots}\} \tag{3.6}$$

$$H_{n,v_1v_2v_3\ldots;n',v_1'v_2'v_3'\ldots} = \langle\chi_{v_1}|\langle\chi_{v_2}|\langle\chi_{v_3}| \\ \ldots\ \langle\psi_n|H|\psi_{n'}\rangle|\chi_{v_1'}\rangle|\chi_{v_2'}\rangle|\chi_{v_3'}\rangle\ \ldots \tag{3.7}$$

Here H stands for either H_M or $H(t)$.

The expansion of Eq. (3.4) converts the time-dependent Schrödinger equations of Eqs. (3.1) or (3.3a) into a set of coupled first-order ordinary differential equations

$$i\hbar\dot{\mathbf{C}}(t) = \mathbf{H}\mathbf{C}(t) \tag{3.8}$$

If H includes the interaction with an external field as in Eq. (3.3a), the matrix $\mathbf{H}$ is time-dependent. All methods that are nowadays employed to solve the time-dependent Schrödinger equation for multidimensional systems involve the solution of such a set of first-order differential equations. The various methods differ in the specific choice of the basis and in the choice of the approximation that is employed to propagate the solution in time.

The choice of the basis functions to be employed in the expansion of Eq. (3.4) depends, of course, on the problem under consideration. If the nuclear kinetic-energy operator and the diabatic potential-energy (PE) functions are of simple form, i.e., factorizable and given by simple analytic functions for each degree of freedom, the multidimensional matrix elements of H can be evaluated analytically. Well-known examples are harmonic PE functions for small-amplitude vibrations, Morse potentials for bond stretching PE functions, or low-order Fourier series for torsional PE functions. In these cases one may choose harmonic-oscillator functions, Morse eigenstates, and sine and cosine functions (so-called free-rotor functions), respectively, as problem-adapted basis functions, which allow an analytic evaluation of the relevant matrix elements. Since no error (apart from finite-precision arithmetic) is involved in the computation of matrix elements and since the evaluation of analytic expressions is usually fast, the numerical wave-packet propagation procedure is particularly efficient for these problems. In some cases though, e.g., for matrix elements with Morse wave functions, the evaluation of integrals by numerical quadrature can be easier and numerically more stable than the evaluation of cumbersome analytic expressions.

This type of approach has been extensively pursued in vibronic-coupling theory for model Hamiltonians based on normal-mode Taylor expansions, see ref. [35] for a review and refs. [36, 38, 188, 189] for representative applications. Owing to the numerical efficiency of this approach, vibronic-coupling problems with up to seven nonseparable vibrational degrees of freedom could be treated over an extended time range [182, 189]. Apart from the calculation of low-resolution photoelectron spectra [189], absorption spectra and resonance Raman spectra [182], a fully microscopic picture of ultrafast electronic decay dynamics has emerged from these studies [36, 37, 190] (see below). Recently, intramolecular electron-transfer processes have been treated in a similar manner [191].

A related area is the time-dependent treatment of vibrational relaxation processes on a single Born–Oppenheimer (BO) surface caused by cubic and quartic anharmonic couplings. The corresponding coupling matrix elements can be evaluated analytically in a direct-product basis of harmonic-oscillator or Morse functions, allowing accurate and fast numerical calculations. Systems with many degrees of freedom have been treated in this context (see, for example, refs. [192–195]).

To illustrate the direct-product-basis expansion approach in the context of non-BO and reactive dynamics, let us consider the example of a two-surface three-mode model of photoisomerization dynamics as discussed in Section II.C.3. The model involves a torsional coordinate (angle φ) as the reactive mode, an interstate coupling coordinate Q_c and an additional tuning coordinate Q_1, see Eqs. (2.33) and (2.34). We choose harmonic oscillator basis functions

$$\chi_v(Q) = (\pi^{1/2} v! 2^v)^{-1/2} H_v(Q) e^{-(1/2)Q^2} \tag{3.9}$$

for the quasi-harmonic degrees of freedom (Q_c and Q_1) and free-rotor basis functions [183] for the torsional degree of freedom. The free-rotor basis functions are

$$\chi_n^{(+)}(\varphi) = \pi^{-1/2}[1 + (2^{-1/2} - 1)\delta_{no}] \cos(n\varphi), \qquad n = 0, 1, 2 \ldots \tag{3.10a}$$

$$\chi_n^{(-)}(\varphi) = \pi^{-1/2} \sin(n\varphi), \qquad n = 1, 2, 3 \ldots \tag{3.10b}$$

The expansion of the time-dependent state vector of the two-state three-mode system thus reads

$$|\Psi(t)\rangle = |\Psi^{(+)}(t)\rangle + |\Psi^{(-)}(t)\rangle \tag{3.11a}$$

$$|\Psi^{(\pm)}(t)\rangle = \sum_{k=0,1} \sum_{n} \sum_{v_c} \sum_{v_1} C_{k,nv_cv_1}^{(\pm)}(t) |\psi_k\rangle |\chi_n^{(\pm)}\rangle |\chi_{v_c}\rangle |\chi_{v_1}\rangle \tag{3.11b}$$

The Hamiltonian defined by Eqs. (2.33) and (2.34) is an even function of φ and thus cannot couple even ($\chi_n^{(+)}$) with odd ($\chi_n^{(-)}$) free-rotor functions. The components $|\Psi^{(+)}(t)\rangle$ and $|\Psi^{(-)}(t)\rangle$ of the state vector can thus be propagated independently of each other.

As a consequence of the simple form of the kinetic-energy operator and the diabatic PE functions in Eqs. (2.33, 2.34) and the choice of problem-adapted basis functions, the matrix elements of the Hamiltonian can be evaluated analytically in this case. The simplifications introduced in the construction of the Hamiltonian operator are reflected in a transparent structure of the Hamiltonian matrix **H**. In this particular example, the matrix elements of the torsional Hamiltonian $\langle \chi_n^{(\pm)}|H_t|\chi_{n'}^{(\pm)}\rangle, H_t = T_\varphi + V(\varphi)$, in the free-rotor basis vanish, except when $|n - n'| = 2$ or $n + n' = 2$, i.e., the matrix representation of H_t is a banded matrix [183]. Due to the absence of odd-even couplings, the basis can be reorganized to yield a tridiagonal representation of H_t for each diabatic surface. The restriction to linear coupling in Q_c and Q_1 also leads to tridiagonal representations in these modes. As a result, the representation of H_M in the direct-product basis is a sparse matrix with a transparent pattern of nonzero elements. The resulting structure of the Hamiltonian matrix $\mathbf{H}^{(\pm)}$ is schematically shown in Fig. 5. Since the coupling mode Q_c is assumed to be nontotally symmetric in this example, $\mathbf{H}^{(\pm)}$ once more decouples into two blocks, corresponding to vibronic levels that transform according to the irreducible representations Γ_0 and Γ_1 of the two electronic states (see ref. [39] for more details). The simple structure of **H** arises both from the assumed *factorizability* of H_M in the diabatic representation as well

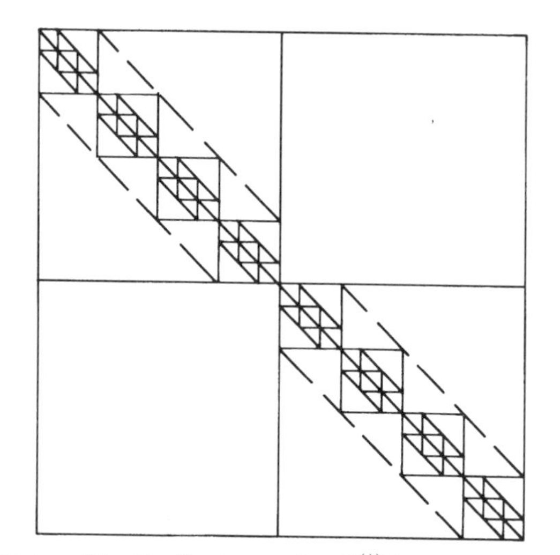

Figure 5. Structure of the Hamiltonian matrices $\mathbf{H}^{(\pm)}$ for the example of the three-dimensional photoisomerization model.

as the *sparsity* of the representation in each mode (tridiagonal matrices) and allows for an extremely efficient computation of the matrix-vector product. As emphasized by Bramley and Carrington, the former is more critical than the latter for computational efficiency [172].

The number of basis states for each mode has to be chosen such that converged results are obtained for the quantities of interest. Propagation on coupled surfaces generally implies large excess energy ($\simeq 5$ eV in the present example) and thus requires large basis sets and care in controlling the basis truncation error. In the example considered here, the appropriate maximum occupation numbers are $N_t = 98$ for the torsional mode, $N_c = 24$ for the coupling mode, and $N_1 = 22$ for the additional tuning mode, resulting in a dimension of 51774 for each sub-block of $\mathbf{H}^{(+)}$ [39].

The dimension of the Hamiltonian matrix $\mathbf{H}$ and the coefficient vector $\mathbf{C}(t)$ will generally be very large for multidimensional problems. Applications involving up to seven nonseparable vibrational modes have been reported, and correspond to Hamiltonian matrices of dimension 10^6 or 10^7 [182, 189]. The transparent structure of $\mathbf{H}$ allows the matrix-vector operation on the right-hand side (rhs) of Eq. (3.8) to be coded directly without explicit generation and storage of $\mathbf{H}$. In addition, the sparsity of $\mathbf{H}$ minimizes the number of floating-point operations. The factorizability and sparsity of $\mathbf{H}$ may also be exploited for vibronic energy-level calculations by employing the iterative Lanczos algorithm [189, 196]. The latter calculations are, in fact, somewhat less demanding since only real vectors are involved, whereas the time-dependent coefficient vector $\mathbf{C}(t)$ is complex.

If the multidimensional PE functions or the kinetic-energy operator are of complicated form, the evaluation of the multidimensional matrix elements in Eq. (3.7) becomes a nontrivial problem. To deal with this problem, the pseudo-spectral method, also known as discrete variational representation (DVR) [197–199] is nowadays commonly employed. It represents a technically convenient and widely applicable approach to evaluate the necessary matrix elements of the Hamiltonian for multidimensional problems. While anticipated by early one-dimensional [200, 201] and two-dimensional [202] specific applications, the generality of the DVR scheme has been revealed by Light and collaborators [198, 203–205]. The application of pseudo-spectral methods in the context of time-dependent wave-packet propagation has been pioneered by Feit and Fleck [206] and Kosloff [207–209].

Briefly stated, the pseudo-spectral approach involves a set of orthonormal basis functions $\{\chi_v(R)\}$ for each nuclear degree of freedom. The grid points are determined by the basis set: they can be obtained, for example, by diagonalizing the matrix representation of the position operator in the global function basis. To begin with, the exact spatial wave function $\Psi(R)$ is approximated by the expansion

$$\overline{\Psi}(R) = \sum_{v} a_v \chi_v(R) \tag{3.12}$$

In the pseudo-spectral approach, it is required that $\overline{\Psi}$ matches the exact solution $\Psi(R)$ at the set of grid points $\{R_\alpha\}$

$$\overline{\Psi}(R_\alpha) = \Psi(R_\alpha) = \sum_{v} a_v \chi_v(R_\alpha) \tag{3.13}$$

The orthogonality of the basis functions allows direct inversion and thus yields the coefficients (assuming unit weight factors)

$$a_v = \sum_{\alpha} \Psi(R_\alpha)\chi_v^*(R_\alpha) \tag{3.14}$$

In the pseudo-spectral approach, the action of any local operator such as the potential V on the wave function is described in the grid representation. This means, if

$$\Phi(R) = V\Psi(R) \tag{3.15}$$

then

$$\Phi(R_\alpha) = V(R_\alpha)\Psi(R_\alpha) \tag{3.16}$$

The action of nonlocal operators, such as the derivatives contained in the nuclear kinetic-energy operator, is evaluated in the global function space. In all practically relevant cases, these matrix elements are given by analytic expressions. The matrix **T** with elements

$$T_{v\alpha} = \langle \chi_v | R_\alpha \rangle \tag{3.17}$$

serves as the unitary transformation matrix between the grid representation and the global basis representation. In the case of a multidimensional product basis, **T** is simply given by the tensorial product of the individual **T** matrices [37, 172, 202, 203]. It can be shown that the DVR scheme is equivalent to an evaluation of matrix elements by Gaussian quadrature [197, 199]. For a more detailed discussion of the DVR method we refer to recent reviews [205, 208, 210] and representative applications, mostly rovibronic energy-level calculations [172, 173, 211–213]. Representative applications of the

DVR method to time-dependent wave-packet propagation on coupled surfaces can be found in [141, 214–217].

A special case of the pseudo-spectral representation is the popular Fourier method [218, 219]. In this case the sampling points are equally spaced and the basis functions are plane waves with periodic boundary conditions. The transformation between grid and basis representation can be effected by the fast-Fourier-transform (FFT) algorithm, which renders this method numerically efficient. The Fourier method is tailored for Cartesian coordinates. If curvilinear coordinates are employed, the technique needs to be modified. Generalizations of the Fourier method for curvilinear coordinates have been discussed in the literature [220–222].

In the area of rovibronic energy-level calculations on the electronic ground-state surface, the concept of basis-set contraction has found widespread application [193, 203, 223–226]. While these schemes allow for a considerable reduction in the size of the basis required for converged calculations, they generally spoil the simple structure and sparsity of the Hamiltonian matrix. So far, contraction methods have found little application in time-dependent wave-packet calculations. Basis-contraction techniques are presumably less efficient for reactive problems, which are usually treated via the time-dependent approach, than for rovibronic bound-state problems. Moreover, there is the difficulty of devising convergence criteria, in particular for driven systems. More research in this direction appears necessary.

A common feature of all numerical quantum-mechanical wave-packet methods is the exponential growth of the computational cost with the number of degrees of freedom. With the techniques discussed in this section, coupled-surface wave-packet calculations are routinely possible for systems with three or four nuclear degrees of freedom. In favorable cases, time-dependent vibronic-coupling calculations for systems with seven nuclear degrees of freedom have been performed. If one wants to tackle truly multidimensional problems, it is necessary to consider alternative strategies that exhibit a more favorable scaling of the computational cost with the size of the system. A few of these alternative strategies will be discussed in Section VIII.

B. Propagation of the State Vector

Equation (3.8) represents a system of coupled first-order differential equations with either constant or variable coefficients, and has to be solved as an initial-value problem. This is a standard problem of numerical analysis that has been extensively covered in numerous textbooks, e.g., refs. [227–230]. From the point of view of numerical computation, the task of solving Eq. (3.8) is a comparatively simple one because of the unitarity of the Schrödinger propagator, i.e., the absence of exponentially increasing or decreasing solutions. In fact, any of the standard methods of numerical analysis, for example, Runge–Kutta, Bulirsch–Stoer, or predictor-corrector methods, is appropriate. The nature of the problem and the choice of representation of $|\Psi\rangle$ enter only indirectly through the dimension and sparsity of the Hamiltonian matrix, which in turn determine the computational cost of the matrix–vector operation on the right-hand side of Eq. (3.8).

Over the years many different algorithms have been introduced for the purpose of solving the time-dependent Schrödinger equation for molecular problems. McCullough and Wyatt [231] were among the first to adopt a time-dependent methodology for a molecular scattering problem. They employed a Crank–Nicholson scheme to propagate the wave function in a grid representation. Askar and Cakmak [232], Kosloff and Kosloff [219], and Leforestier [233] implemented the second-order differencing method. The popular split-operator technique has been introduced by Feit et al. [218]. Predictor-corrector methods have been employed by Kulander [234] and Köppel et al. [35, 188]. Tal-Ezer and Kosloff [235] have introduced the Chebysheff propagation scheme, which also is widely used today. The Lanczos tridiagonalization procedure has been adapted to provide a class of short-time propagators that have been found useful for time-dependent wave-packet propagation [236, 237]. Standard Runge–Kutta methods of fourth or fifth order with fixed or variable time step also have been shown to be suitable for time-dependent wave-packet propagation [39, 93]. Symplectic propagators, derived from the analogy of wave-packet dynamics with classical mechanics, have been discussed by Gray and Verosky [238].

A detailed coverage of these various algorithms is not intended here. A comparative discussion of time-dependent propagation methods in the context of nuclear dynamics within the BO approximation can be found in articles by Kosloff [207, 209], Leforestier et al. [237], and Truong et al. [239]. Leforestier et al., in particular, have discussed in detail the relative merits and drawbacks of the second-order differencing, fourth-order differencing, split-operator, Chebysheff, and Lanczos propagator methods for the example of a one-dimensional model problem. We shall present a brief discussion of the most commonly employed methods (split-operator, Chebysheff, Lanczos, predictor-corrector, Runge-Kutta, Bulirsch-Stoer). For a detailed description of these algorithms, we refer to the quoted original papers and textbooks [227–230].

The split-operator technique of Feit et al. [218] is based on the symmetric second-order disentangling of the time-evolution operator ($\varepsilon = t/\hbar$)

$$e^{-iH\varepsilon} = e^{-iT_N\varepsilon/2}e^{-iV\varepsilon}e^{-iT_N\varepsilon/2} + O(\varepsilon^3) \tag{3.18}$$

It is usually employed in conjunction with a plane-wave DVR. This way, the action of T_N and V are evaluated in their diagonal representations [218,

219]. Many authors have used this method to propagate wave packets on a single BO PE surface. The method has been applied to coupled-surface dynamics by Mele and Socolar [240], Alvarellos and Metiu [241], Engel and Metiu [52], Choi and Light [53], Broeckhove et al. [242], Fernández and Micha [243], and Garraway and Suominen [244]. The algorithm is not suitable for problems with complicated kinetic energy operators that exhibit a mixed dependence on positions and momenta. Higher-order generalizations of Eq. (3.18) have been proposed [245, 246].

Another widely used method for wave-function propagation with a time-independent Hamiltonian is the Chebysheff propagation scheme [207, 235]. It is a large-time-step propagation method, based on an expansion of the time-evolution operator in terms of complex Chebysheff polynomials. The latter are evaluated recursively. The Chebysheff propagator has been shown to be efficient and suitable for accurate long-time propagation [237]. The method is not suitable if one needs the wave function on a dense mesh in time, e.g., if a Fourier transformation or a convolution integral with laser pulses is subsequently to be performed. Extensions of the Chebysheff method for explicitly time-dependent problems have recently been proposed [247, 248].

In recent years, short-time propagators based on Lanczos tridiagonalization have become popular [215, 236, 237, 249]. The Lanczos scheme has the interesting property that it tailors the approximate propagator to both the operator and the initial wave function, (see ref. [237] for a detailed discussion). The Lanczos propagation method has been found to be efficient for multidimensional non-BO problems [190, 215] and can also be applied for explicitly time-dependent Hamiltonians [250].

The predictor-corrector scheme is one of the standard methods for the solution of differential equations [251]. The predictor-corrector integrator is not necessarily the most efficient choice, but it is reliable and can yield accurate results when the time step is sufficiently small. A fourth-order Milne-Hammings predictor-corrector [227] has been extensively applied to propagate the wave function for multimode vibronic-coupling problems (see, e.g., refs. [36, 188, 189]). As a fixed-time-step method, the predictor-corrector method is not suitable for explicitly time-dependent Hamiltonians.

In more recent applications, we have propagated the state vector with a fourth-order Runge–Kutta–Merson scheme with variable time step, as implemented in the NAG Fortran library. This propagation method is typically somewhat faster than the predictor-corrector (if one requires the same accuracy of the norm of the wave function) and needs less memory. The step-size control eliminates the need to determine the time step by a trial-and-error procedure. Variable time-step Runge–Kutta schemes are also suitable for propagation with time-dependent Hamiltonians.

Finally, we mention the Bulirsch–Stoer scheme, which is a controlled-step-size method. It appears particularly suitable for explicitly time-dependent problems. The Bulirsch–Stoer method with judicious step-size control has been implemented in [91] for a multidimensional driven non-BO problem and shown to be computationally efficient and to yield accurate results.

To summarize, the following two key factors should be considered when choosing a propagation method: (1) The complexity of the problem, i.e., the cost of the evaluation of the right-hand side of Eq. (3.8). In the case of a high-dimensional problem, sophisticated high-order integrators that require many rhs evaluations per time step might be slower than simpler integration schemes. (2) The time-dependence of the wave-function coefficients and the Hamiltonian matrix. For problems involving several timescales and, in particular, for explicitly time-dependent problems, adaptive step-size control has been found to be essential.

IV. ASPECTS OF ULTRAFAST INTRAMOLECULAR NON-BORN–OPPENHEIMER DYNAMICS

In this section we survey characteristic features of time-dependent wave-packet dynamics on coupled potential-energy (PE) surfaces at femtosecond timescales, as revealed by calculations for representative model systems. We shall focus on the dynamics associated with conical intersections of PE surfaces. We shall primarily be concerned with a microscopic description of relaxational dynamics at ultrafast timescales, in particular internal-conversion dynamics and its interplay with isomerization dynamics. A closely related topic is ultrafast electron-transfer dynamics, which is traditionally modeled in terms of a coupled electronic two-level system that is linearly coupled to an infinite set of harmonic oscillators [252]. The latter type of problem, which has extensively been investigated in the chemical physics literature, will be considered in some detail in Section VIII.

We shall exclude from the present discussion more or less direct photodissociation processes on conically intersecting PE surfaces. Direct photodissociation dynamics on coupled surfaces has been investigated in detail for a number of triatomic systems, e.g. H_3^+ [214, 253], HCN^+ [215], ICN [216, 254] and H_2S [141, 255]. A well-studied and representative example is the photodissociation of H_2S through the first absorption band involving the ${}^1B_1$ and ${}^1A_2$ electronic states, which are coupled via a conical intersection [141]. It has been shown that the population of the optically excited diabatic ${}^1B_1$ state decays within less than 40 fs as a consequence of nonadiabatic coupling to the dissociative ${}^1A_2$ surface [141, 255]. In contrast to such directly dissociative dynamics, we shall primarily be concerned here with relaxational processes in bound multidimensional systems, which may extend up to a picosecond time range.

A. Wave Packets

The numerical techniques described in the preceding section yield the time-dependent coefficient vector $\mathbf{C}(t)$. Given $\mathbf{C}(t)$, the component of the wave packet on the nth *diabatic* surface

$$\Psi_n(R_1, R_2, R_3, \ldots, t) = \langle R_1 | \langle R_2 | \langle R_3 | \ldots \langle \psi_n | \Psi(t) \rangle \tag{4.1}$$

can immediately be constructed

$$\Psi_n(R_1, R_2, R_3, \ldots, t) = \sum_{v_1 v_2 v_3 \ldots} C_{n, v_1 v_2 v_3 \ldots}(t) \chi_{v_1}(R_1) \chi_{v_2}(R_2) \chi_{v_3}(R_3) \ldots \tag{4.2}$$

For interpretative purposes, we shall also be interested in the projections of the wave packet on *adiabatic* electronic states, that is

$$\tilde{\Psi}_n(R_1, R_2, R_3, \ldots, t) = \langle R_1 | \langle R_2 | \langle R_3 | \ldots \langle \tilde{\psi}_n | \Psi(t) \rangle \tag{4.3}$$

Since [cf. Eq. (2.11)]

$$\left| \tilde{\psi}_n \right\rangle = \sum_i X_{ni}(R_1, R_2, R_3, \ldots) \left| \psi_i \right\rangle \tag{4.4}$$

it is possible to calculate the adiabatically projected wave packets from the diabatic projections. To effect the transformation, one needs the representation of $X_{ni}(R_1, R_2, R_3, \ldots)$ in the direct-product basis. Exploiting the fact that the necessary transformations can be performed sequentially for each mode, the construction of adiabatically projected wave packets is feasible for systems with up to four modes, see refs. [37, 39, 91, 214] for details.

For systems with just one or two degrees of freedom, the explicit consideration of time-dependent wave packets as defined in Eqs. (4.1) and (4.3) can be illuminating. The book of Schinke on photodissociation dynamics contains nice examples [1]. If the problem involves three or more nuclear degrees of freedom, on the other hand, a reduced description that condenses the information carried by the wave packet is desirable. Such reduced descriptions are obtained by integrating the probability density over some of the degrees of freedom. The relevance of reduced descriptions for complex systems is based on the fact that an experimental measurement will not yield the complete quantum mechanical wave functions, but rather partially integrated information, e.g., the population of an electronic state or the population of a particular isomer. The very concept of an isomerization reaction of a complex molecule, for example, implies a reduced description.

Let, for example, R_1 denote the reaction coordinate of interest in a multidimensional system. The corresponding reduced probability density is defined as

$$P_n(R_1, t) = \int dR_2 \int dR_3 \ldots |\Psi_n(R_1, R_2, R_3, \ldots, t)|^2 \tag{4.5}$$

$P_n(R_1, t)$ carries the information on the spatial distribution of the wave packet with respect to R_1 on the nth diabatic surface.

Let us consider, as a first example, a nonadiabatic photoisomerization model of the type discussed in Section II.C.3. The model [39] involves three nuclear degrees of freedom. In addition to the torsional degree of freedom, which plays the role of the reaction coordinate, the model includes a harmonic vibration ν_c, which couples the diabatic S_0 and S_1 states in first order and an additional harmonic tuning mode ν_t. A two-dimensional representation of the PE surfaces as a function of the torsional angle and the coupling coordinate has been shown in Fig. 4 for a closely related model. To define the initial condition, we assume that the system interacts at $t = 0$ with an ultrashort laser pulse which prepares a localized wave packet at the saddle of the S_1 surface (see Section V.A for a more detailed discussion of the preparation process).

Figure 6 displays the diabatically projected reduced torsional probability densities $P_1(\varphi, t)$ (a) and $P_0(\varphi, t)$ (b), respectively [94]. Figure 6(a) shows the initial broadening of the torsional wave packet on the inverted S_1 surface, which is followed by damped quasiperiodic torsional motion. On the diabatic S_0 surface, on the other hand, the wave packet is largely localized near the potential minima at $\varphi = 0$ and $\varphi = \pi$. It is seen that the S_0 wave packet is quasi-periodically created and fed back to the S_1 surface during the transient torsional motion. Beyond about 200 fs the distributions on the S_1 and S_0 surfaces become quasi-stationary. In the long time limit, this model predicts equal population of both diabatic surfaces (see below).

Figure 6 provides a detailed microscopic visualization of internal-conversion dynamics triggered by large-amplitude torsional motion. It can be seen that the internal-conversion process results in rather broad, delocalized probability distributions within less than 100 fs. The quantum dynamical calculations thus indicate that the description of the internal-conversion dynamics by a single classical trajectory, as is often proposed in the chemical literature, may not be appropriate.

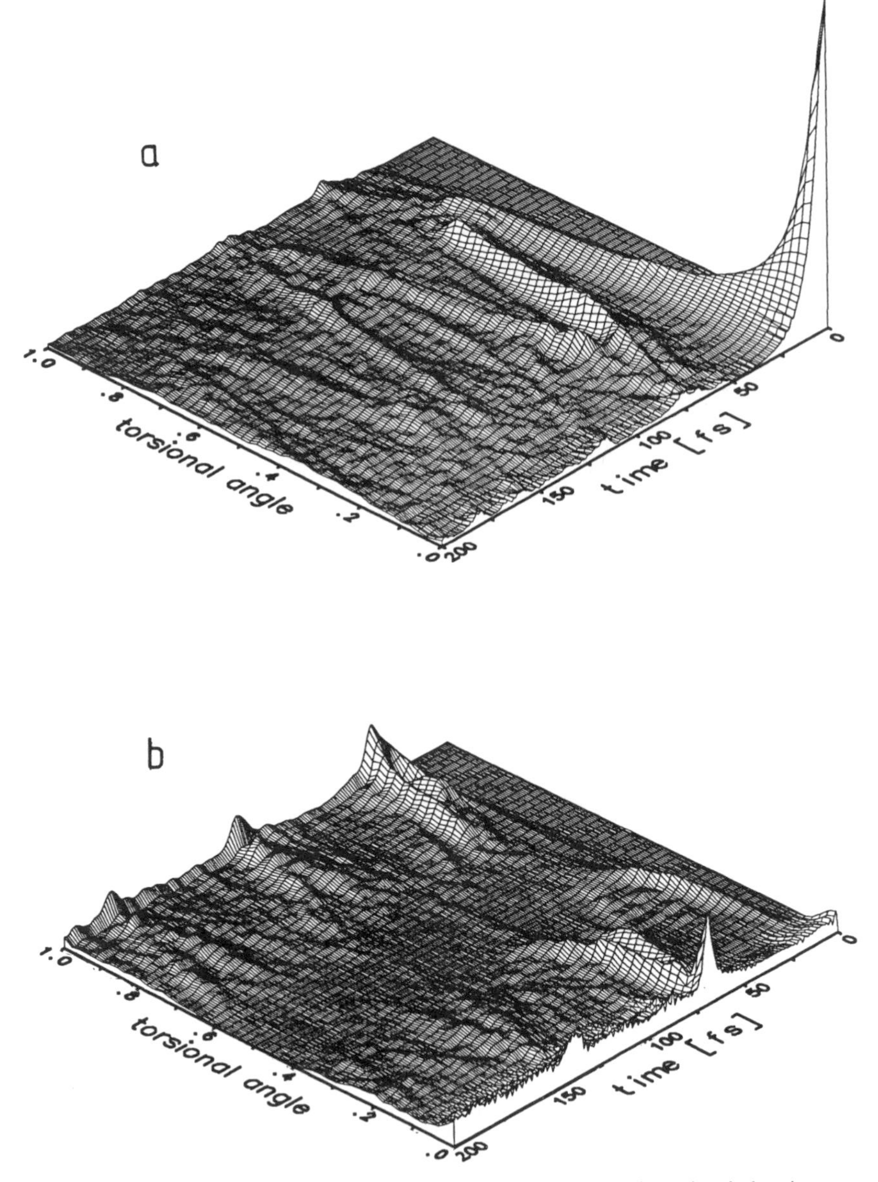

Figure 6. Reduced torsional probability densities for the three-dimensional photoisomerization model, projected on (*a*) the diabatic S_1 surface and (*b*) the diabatic S_0 surface.

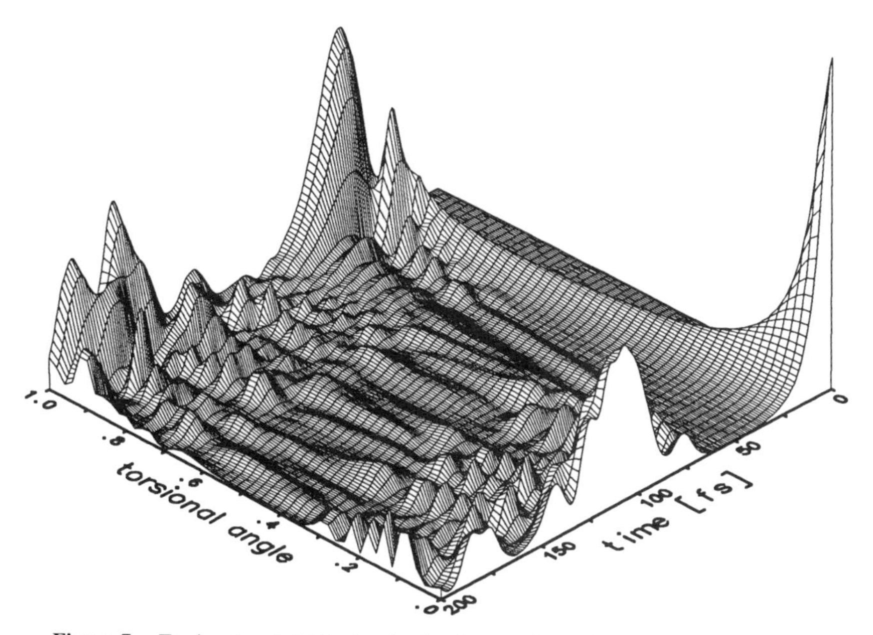

Figure 7. Torsional probability density for the one-dimensional model system (torsional mode only).

It is instructive to contrast the dynamics of the three-dimensional system with the dynamics of the one-dimensional model (torsional mode only). In the absence of vibronic coupling, the wave packet prepared in the S_1 state remains there forever (we neglect spontaneous radiative processes). The torsional probability density of the one-dimensional model is shown in Fig. 7 [94]. It is seen that the Gaussian-like wave packet created at $t = 0$ at the maximum of the S_1 surface spreads rapidly and relocalizes quasi-periodically at the potential maxima ($\varphi = 0, \pi$). At later times, destructive interference between outgoing and reflected components of the wave packet in the periodic torsional potential leads to a very complex torsional probability distribution.

The comparison of Figs. 6 and 7 provides the first clue to the dramatic effect of a conical intersection on the nature of the dynamics. The conical intersection not only causes a rapid transition of population between electronic states, but also has a pronounced effect on the dynamics of individual modes. The latter effect arises from the substantial anharmonicity of the *adiabatic* PE surfaces in the vicinity of the intersection [see, for example, Fig. 4(*b*)], which leads to particularly strong mode–mode coupling effects. In Fig. 6(*a*) the effect of "intramolecular friction" by energy transfer from the torsional mode to other degrees of freedom is clearly visible. The strong mode–mode couplings lead to a rapid loss of phase coherence in the wave function. The pronounced interference effects, which dominate the probability density of the one-dimensional system (Fig. 7) are thus largely absent in the three-dimensional system. This finding, as well as other observations to be discussed shortly, show that genuinely relaxational dynamics can be observed for quantum systems with just three strongly coupled degrees of freedom.

In Fig. 6, we have focused on the photochemically reactive coordinate. Alternatively, we may consider reduced probability distributions for the coupling mode to reveal the special role of this mode [36, 37]. As an example, we show in Fig. 8 reduced probability distributions of the coupling mode for a three-dimensional model of the conically intersecting $\tilde{X}$ and $\tilde{A}$ states in the ethylene cation [35, 256, 257]. These are taken from the work of Manthe and Köppel [38] (full curves). The figure shows that the probability distribution of the coupling mode, which initially is peaked at $Q_c = 0$, broadens and splits for this strongly coupled ($\lambda/\omega_c > 1$) system (the dashed and dash-dotted curves show wave packets for weaker vibronic coupling, where this effect is absent). The splitting of the wave packet in the coupling coordinate reflects the pronounced distortion of the lower adiabatic energy surface, on which the wave packet mainly resides after the first passage through the intersection. The conical spike expels the wave packet from the region of strong nonadiabatic coupling, which effectively prevents the backtransfer of population. This picture explains in simple intuitive terms the largely unidirectional character of population transfer through a conical intersection [38].

A specific aspect of wave-packet dynamics near conical intersections is the importance of geometric-phase effects in the adiabatic representation [104, 114]. In the context of reaction dynamics, the relevance of geometric-phase effects has extensively been discussed for the fundamental $H + H_2$ reaction (e.g., ref. [258]). Cina et al. have considered the possibility of a spectroscopic detection of geometric-phase effects via interferometric two-pulse experiments for models of the $E \times E$ Jahn–Teller effect [259]. For the example of the $\tilde{B}$ state of the sodium trimer, Schön and Köppel have explicitly demonstrated the manifestation of geometric-phase effects in the time-dependent wave-packet dynamics [72]. Figure 9 shows their comparison of the wave-packet dynamics in the degenerate bending mode of Na_3 for the case of a Jahn–Teller and a pseudo-Jahn–Teller model with identical adiabatic PE surfaces. It is seen that geometric-phase effects are significant at longer times ($t > 500$ fs in this example).

The visualization of wave packets on conically intersecting PE surfaces is certainly instructive and yields the most detailed insight into the dynamics of the ultrafast internal-conversion process. In most applications to complex

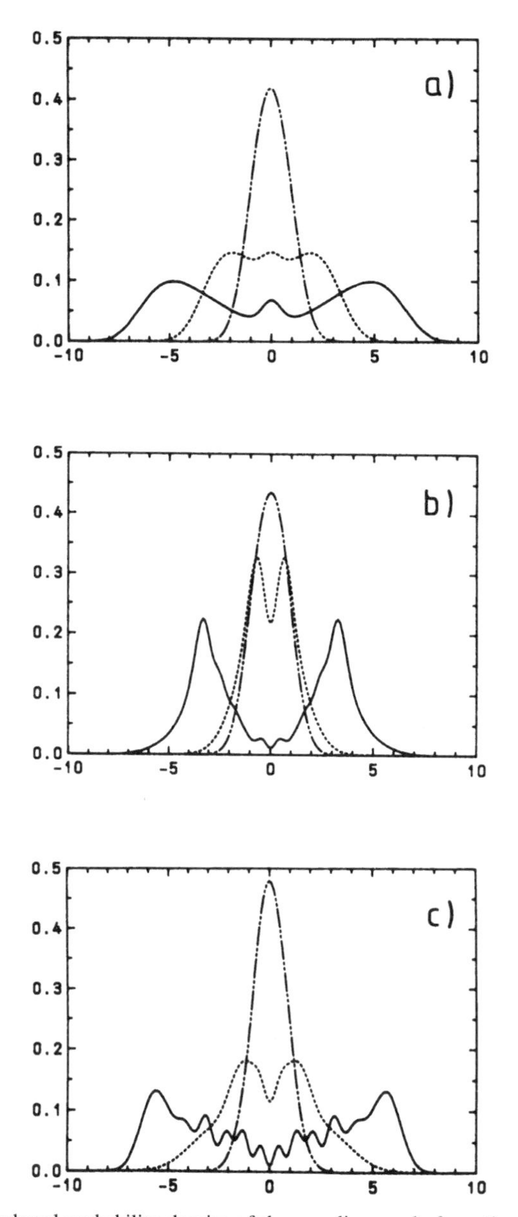

Figure 8. Reduced probability density of the coupling mode for a three-mode model of $C_2H_4^+$ at (*a*) 16, (*b*) 32, and (*c*) 64 fs. The dotted and dash-dotted curves represent results for weaker vibronic coupling. Reproduced with permission from [38].

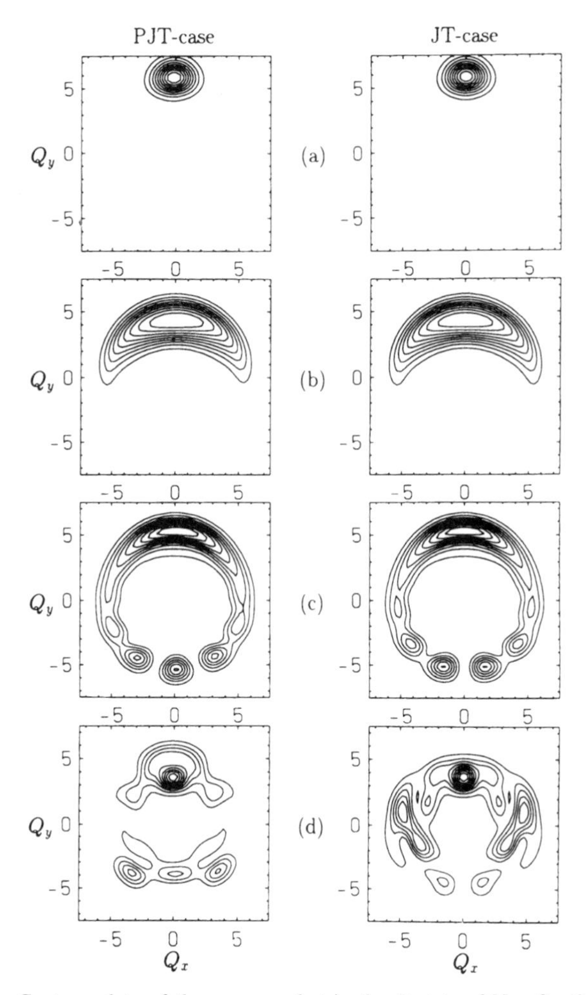

Figure 9. Contour plots of the wave packet in the B state of Na_3 for the Jahn–Teller (right) and pseudo-Jahn-Teller (left) models at (*a*) $t = 0$, (*b*) $t = 209$ fs, (*c*) $t = 500$ fs, (*d*) $t = 1755$ fs. The difference between the Jahn–Teller and pseudo-Jahn–Teller dynamics becomes significant at longer times. Reproduced with permission from [72].

systems, however, one is not interested in the most detailed description; one would prefer a compact description of the process in terms of a few expectation values that can directly be correlated with experimentally detectable transient signals. With this in mind, we shall consider various "observables" of the intramolecular dynamics. An observable is understood here as the expectation value of a Hermitian operator with the time-dependent wave function. As such, it is not necessarily a measurable quantity. We shall see in Section VII, however, that there exists a close relationship between such idealized observables and measurable time-resolved pump-probe signals.

B. Electronic Population Dynamics

The quantity of primary interest for the description of radiationless electronic transitions is the time-dependent population probability of excited electronic states. The population $P_n(t)$ of the nth diabatic electronic state is defined as the expectation value of the projection operator

$$\hat{P}_n^{\text{di}} = |\psi_n\rangle\langle\psi_n| \tag{4.6}$$

i.e.,

$$P_n^{\text{di}}(t) = \langle\Psi(t)|\hat{P}_n^{\text{di}}|\Psi(t)\rangle \tag{4.7}$$

The superscript "di" refers to the fact that the projector in Eq. (4.6) is defined in the diabatic electronic representation. Alternatively, we may consider the time-dependent population of *adiabatic* electronic states, i.e.,

$$P_n^{\text{ad}}(t) = \langle\Psi(t)|\hat{P}_n^{\text{ad}}|\Psi(t)\rangle \tag{4.8}$$

$$\hat{P}_n^{\text{ad}} = |\tilde{\psi}_n\rangle\langle\tilde{\psi}_n| \tag{4.9}$$

Equivalently, these electronic population probabilties may be expressed as the integral of the probability density over all nuclear degrees of freedom, i.e.,

$$P_n^{\text{di}}(t) = \int dR_1 \int dR_2 \int dR_3 \ldots |\Psi_n(R_1, R_2, R_3, \ldots, t)|^2 \tag{4.10}$$

$$P_n^{\text{ad}}(t) = \int dR_1 \int dR_2 \int dR_3 \ldots |\tilde{\Psi}_n(R_1, R_2, R_3, \ldots, t)|^2 \tag{4.11}$$

From Eqs. (4.10) and (4.11) it is obvious that the electronic populations are just the diagonal elements of the reduced electronic density matrix, defined as the trace of the complete density matrix over the nuclear degrees of freedom.

Since $\hat{P}_n^{\text{di}}$ commutes with H_M in the absence of interstate couplings $V_{nn'}$, [cf. Eq. (2.20)], $P_n^{\text{di}}(t)$ becomes a constant of motion in this limit. Correspondingly, $P_n^{\text{ad}}(t)$ becomes a constant of motion if the nonadiabatic coupling operators defined in Eq. (2.6) are neglected. The populations $P_n(t)$ are thus a direct measure of nonadiabatic transitions in either the diabatic or the adiabatic representation [35].

The transition rate out of the electronic state n (in either the diabatic or adiabatic representation) is given by

$$k_n(t) = -\dot{P}_n(t)/P_n(t) \tag{4.12}$$

In a nonperturbative treatment of radiationless decay, $P_n(t)$ is generally not simply an exponential function; the nonperturbative rate $k_n(t)$ is thus time-dependent. It should be stressed that $P_n(t)$, and thus $k_n(t)$, can be measured explicitly, at least in principle, in a femtosecond time-resolved experiment. There is thus no need to invoke a long time limit or to perform an average over time to obtain time-independent rates.

The population probabilities $P_n(t)$ defined in Eqs. (4.6–4.11) should not be confused with the population probabilties that have been considered in the extensive earlier literature on radiationless transitions in polyatomic molecules, see [4, 43] for reviews. There, the population of a single "bright" (i.e., optically accessible from the electronic ground state) zero-order Born–Oppenheimer (BO) level is considered. Here, in contrast, we define the electronic population as the sum of the populations of all levels within a given (diabatic or adiabatic) electronic state [35]. These different definitions are adapted to different regimes of timescales of the material dynamics. If nonadiabatic interactions are relatively weak, and radiationless transitions relatively slow, the concept of zero-order BO levels is useful; the populations of these levels can be prepared and probed using suitable laser pulses (typically of nanosecond duration). If the nonadiabatic transitions occur on femtosecond timescales, the preparation of individual zero-order BO levels is no longer possible. The total population of an electronic state then becomes the appropriate concept for the interpretation of time-resolved experiments [88, 89].

As a representative example, let us consider the ultrafast population dynamics associated with the S_1–S_2 conical intersection in pyrazine. We already have referred to this system in Sections II.B and II.C.2. A two-dimensional view of the conical intersection has been displayed in Fig. 1. Initially, an empirical model involving three vibrational modes $(\nu_1, \nu_{6a}, \nu_{10a})$ has been proposed [180]. Subsequently, a refined model involving four degrees of freedom $(\nu_1, \nu_{6a}, \nu_{9a}, \nu_{10a})$ has been derived from elaborate *ab initio* electronic-structure calculations [45, 181]. We shall consider here the dynamics of the original three-dimensional model, which has been extensively investigated by several authors [36, 38, 89, 91, 180, 249, 260]. The dynamics of the more recent four-dimensional model is very similar [44].

We assume preparation of the S_2 state at $t = 0$ by an ideally short laser pulse. The solid line in Fig. 10 represents the time-dependent population probability of the *diabatic* S_2 state. It exhibits an initial decay on a timescale of $\approx$ 20 fs, followed by quasi-periodic recurrences of the population, which are damped on a timescale of a few hundred femtoseconds. Beyond $\approx$ 500 fs the S_2 population probability becomes quasi-stationary, fluctuating statistically around its asymptotic limit of $\approx$ 0.3.

The dotted curve in Fig. 10 shows the time-dependent population of the *adiabatic* S_2 state. The population $P_2^{\text{ad}}(t)$ is seen to decay even faster than $P_2^{\text{di}}(t)$ and to attain an asymptotic value of $\approx$ 0.1. Ninety percent of the population thus relax to the S_1 state, and this happens essentially within a single vibrational period of the system. The higher asymptotic value of P_2^{di} can be qualitatively interpreted as representing the admixture of the upper diabatic state in the wave packet that has relaxed to the minimum of the lower adiabatic surface [261]. The finite asymptotic value of $P_2^{\text{ad}}(t)$ is a consequence of the restricted phase space of this model. The population $P_2^{\text{ad}}(t)$ is expected to relax to zero for systems with many degrees of freedom (see Section VIII.D).

It is again instructive to compare the dynamics of the three-mode model ($\nu_{10a}, \nu_{6a}, \nu_1$) with the dynamics obtained in one-dimensional (coupling mode ν_{10a} only) and two-dimensional (ν_{10a}, ν_{6a}) models. Figure 11 shows the population of the adiabatic S_2 state for the one-dimensional and two-dimensional models. It is seen that virtually no population dynamics takes place when only the coupling mode is considered. When, in addition, the tuning mode ν_{6a} is included, and a conical intersection is thus formed, a pro-

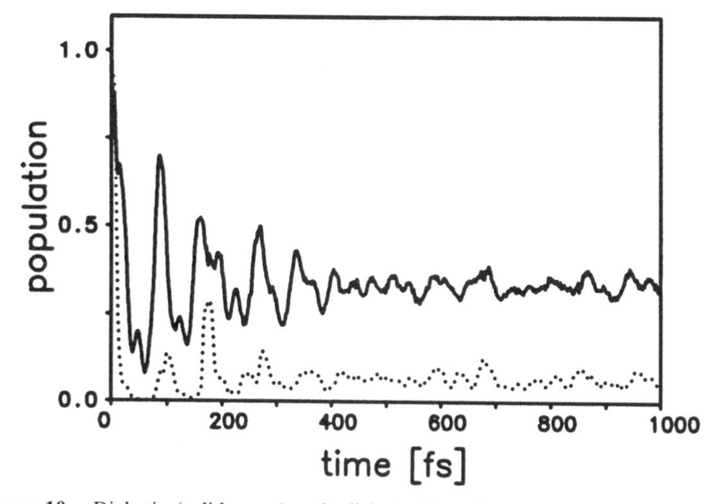

Figure 10. Diabatic (solid curve) and adiabatic (dotted curve) population probability of the S_2 state for the three-mode S_1–S_2 conical intersection model of pyrazine.

nounced and irreversible electronic population dynamics results. The fluctuations of $P_2^{\rm ad}$ around the asymptotic value of ≈ 0.2 are still rather large in the two-dimensional model. Upon inclusion of a third mode, the fluctuations are strongly suppressed and the asymptotic limit of $P_2^{\rm ad}$ decreases to ≈ 0.1 (Fig. 10). Figure 11 provides a convincing demonstration that the conical intersection is the key for the development of electronic population relaxation in isolated molecules at femtosecond time scales.

Figure 12 shows the time-dependent population probabilities $P_1^{\rm di}(t)$ (full curve) and $P_1^{\rm ad}(t)$ (dashed curve) of the three-dimensional nondiabatic photoisomerization model referred to in the preceding subsection. The population $P_1^{\rm di}(t)$ simply represents the norm of the reduced wave packet shown in Fig. 6(*a*). It is seen that $P_1^{\rm di}(t)$ exhibits a rapid initial decrease, which corresponds to the first passage through the intersection, and then approaches an asymptotic limit of 0.52. The adiabatic population probability $P_1^{\rm ad}(t)$ exhibits an additional relaxation process on a timescale of a few hundred femtoseconds and approaches an asymptotic value of 0.23. In this model, the internal-conversion process is slower and less complete than for the S_1–S_2 model of pyrazine (Fig. 10). The higher asymptotic value of $P_1^{\rm ad}(t)$ in the photoisomerization model reflects the much larger vertical electronic energy gap (5.0 eV, compared to 0.9 eV in the pyrazine model): it is more difficult to dissipate this large excess energy in a three-mode system. As discussed above for pyrazine, calculations for one-dimensional and two-dimensional models have confirmed the essential role of the conical intersection of the adiabatic

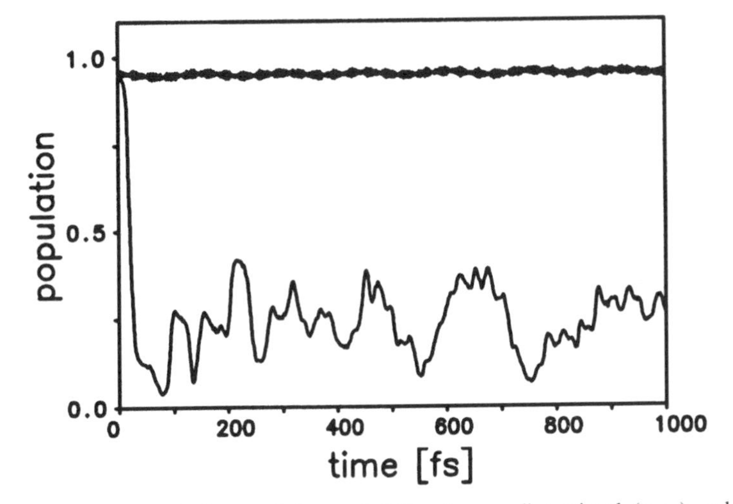

Figure 11. Adiabatic S_2 population probability for one-dimensional (ν_{10a}) and two-dimensional (ν_{10a}, ν_{6a}) models of S_1–S_2 vibronic coupling in pyrazine.

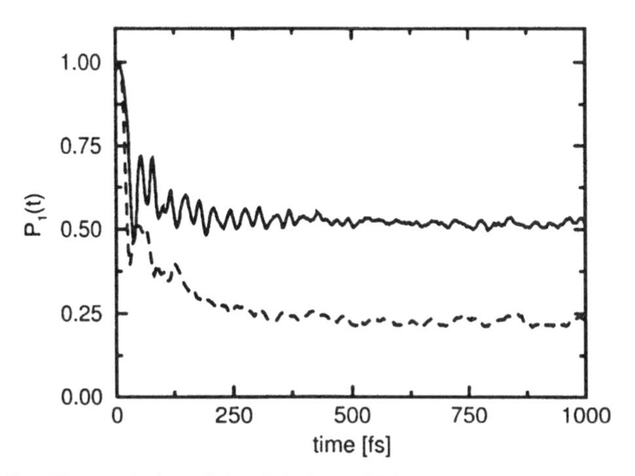

Figure 12. Time evolution of the diabatic (solid line) and adiabatic (dahsed line) electronic populations for the three-dimensional photoisomerization model.

PE surfaces [39]. Calculations for a four-mode system [39] as well as multimode systems [262] indicate that $P_2^{\mathrm{ad}}(t)$ relaxes towards zero for sufficiently large systems.

Figures 10 and 12 clearly reveal the irreversible relaxation character of the electronic population dynamics triggered by conical intersections. The time evolution of the wave function is of course reversible, being determined by the time-dependent Schrödinger equation with a Hermitean Hamiltonian. The irreversibility (on the timescale of interest) enters through the reduction process, i.e., the integration over part of the system. The observations which we have discussed here are general: very similar results have been found for other systems exhibiting conical intersections, e.g., $C_2H_4^+$ [37, 38, 188, 256], NO_2 [38] and $C_6H_6^+$ [88, 189, 190]. In all cases that have been studied in detail, it has been found that a minimum of three strongly coupled nuclear degrees of freedom is required to obtain clearly developed relaxational behavior of electronic populations [36, 39].

The radiationless decay of the benzene cation prepared in the $\tilde{C}^2A_{2u}$ state [263] appears to represent a particularly interesting case. In addition to the $\tilde{C}^2A_{2u}$–$\tilde{B}^2E_{2g}$ pseudo-Jahn–Teller coupling [40, 189] and the $\tilde{B}^2E_{2g}$–$\tilde{X}^2E_{1g}$ vibronic coupling [190], there are strong multimode Jahn–Teller couplings within the $\tilde{B}$ and $\tilde{X}$ states, which lead to very effective mode mixing [264, 265]. The sequential $\tilde{C} \rightarrow \tilde{B} \rightarrow \tilde{X}$ internal-conversion process has been investigated at the quantum-mechanical level for a simplified model by Köppel [190]. In a classical description, this process has been simulated with inclusion of all active vibrational modes [266, 267]. Time scales of $\approx$ 10 fs for the $\tilde{C} \rightarrow \tilde{X}$ transition and $\approx$ 100 fs for the $\tilde{B} \rightarrow \tilde{X}$ transition have been found (see Section VIII.D). According to these predictions, the internal-conversion process in the benzene cation is one of the fastest intramolecular relaxation processes known to date.

C. Observables of the Nuclear Dynamics

Ultrafast internal-conversion processes involve a complex interplay of electronic and nuclear motions. The visualization of time-dependent wave packets, as discussed in Section IV.A, is a possible way of revealing this interplay. Alternatively, we may consider time-dependent expectation values of appropriate operators that reflect the most relevant properties of the nuclear dynamics.

As is well known in nonequilibrium statistical mechanics, it is necessary to distinguish between energy-transfer processes and phase-relaxation processes in the description of the dynamics of complex systems. To monitor the coherence of vibrational motion, we may consider the expectation values of the position and momentum operators

$$\langle R_i \rangle_t = \langle \Psi(t) | R_i | \Psi(t) \rangle \tag{4.13}$$

$$\langle P_i \rangle_t = \langle \Psi(t) | P_i | \Psi(t) \rangle \tag{4.14}$$

Here R_i is the vibrational coordinate and P_i the conjugate momentum of the ith mode. Note that the definition (4.13, 4.14) implies the trace over the electronic subsystem. We thus do not differentiate the nuclear dynamics with respect to individual electronic surfaces [36].

For uncoupled harmonic systems in the absence of environmental dissipation, the expectation values $\langle Q_i \rangle_t, \langle P_i \rangle_t$ of normal coordinates and momenta evolve periodically in time, corresponding to undamped coherent vibrational motion. In vibronically coupled systems, on the other hand, the oscillation of $\langle Q_i \rangle_t$ and $\langle P_i \rangle_t$ is typically damped on femtosecond timescales. Figure 13 shows this phenomenon for the tuning modes ν_1 and ν_{6a} of the three-mode pyrazine model. The same phenomenon has been found for the harmonic tuning modes of three- and four-dimensional photoisomerization models [39]. It should be pointed out that the decay of the amplitude of $\langle Q_i \rangle_t$ and $\langle P_i \rangle_t$ is not related to the dissipation of vibrational energy. In fact the energy of the tuning modes increases during the internal-conversion process, owing to the interconversion of electronic energy into vibrational energy (see below). The damping of the amplitude seen in Fig. 13 reflects, rather, a vibrational pure dephasing process, that is, the loss of phase coherence of the initially prepared vibrational wave packet. This femtosecond vibrational dephasing process, which is faster than typical vibrational relaxation processes in isolated electronic states of polyatomic molecules [192, 268–270], is another man-

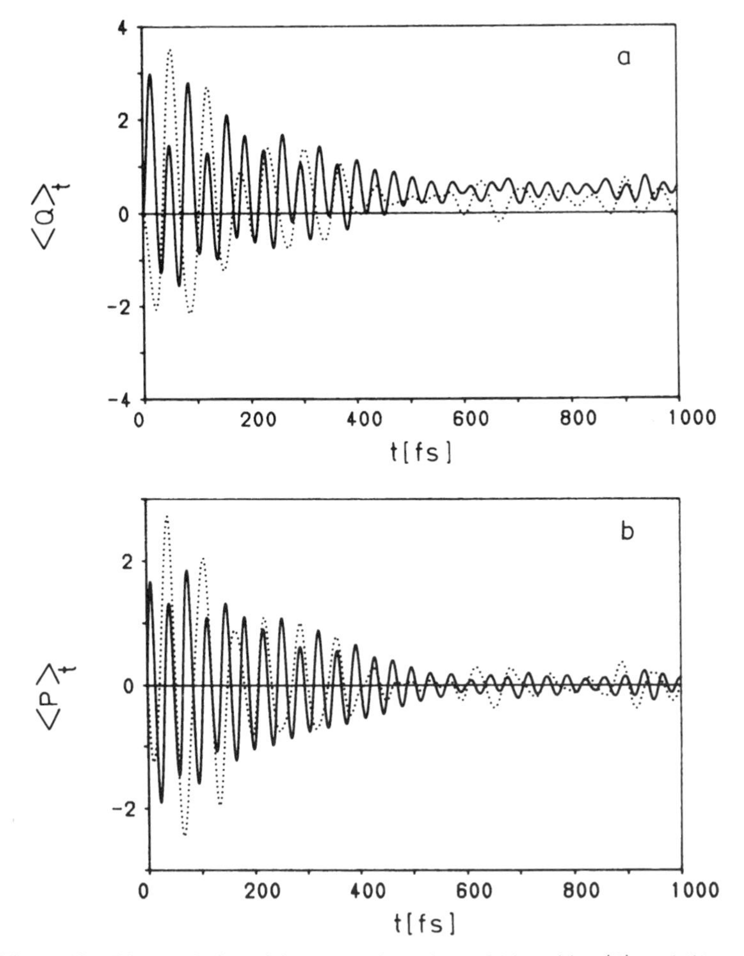

Figure 13. Time evolution of the expectation values of (*a*) position $\langle Q \rangle_t$ and (*b*) momentum $\langle P \rangle_t$ for the tuning modes ν_1 (solid line) and ν_{6a} (dotted line) of the three-mode pyrazine model.

ifestation of the pronounced mode–mode coupling effects in systems with conically intersecting PE surfaces.

Figure 14 shows an alternative visualization of the vibrational dephasing effect for the same system: the trajectory of the mean values $\langle Q_1 \rangle_t$ and $\langle Q_{6a} \rangle_t$ in the Q_1, Q_{6a} plane is shown for a time interval of 1 ps. The dashed line in Fig. 14 indicates the seam of intersection of the S_1 and S_2 surfaces. It is seen that the vibrational trajectory, which starts at $Q_1 = Q_{6a} = 0$ at $t = 0$, crosses the conical seam a few times before the system is irreversibly trapped by the vibrational dephasing process.

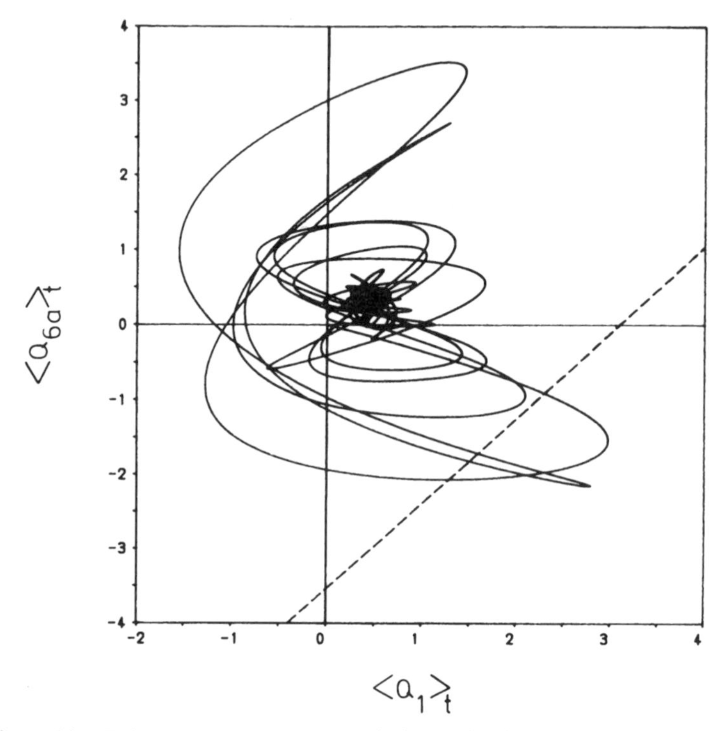

Figure 14. Trajectory of the mean values $\langle Q_1 \rangle_t$ and $\langle Q_{6a} \rangle_t$ in the Q_1, Q_{6a} plane for the three-mode pyrazine model. The time range is 1 ps. The dashed line represents the seam of intersection of the S_1 and S_2 surfaces.

As discussed in detail in [36], the vibrational dephasing process represents the origin of the irreversible time evolution of the electronic population (Fig. 10). The initial quasi-periodic recurrences of $P_2^{\mathrm{di}}(t)$ reflect the driving of electronic population by initially coherent vibrational motion in the tuning modes ν_1 and ν_{6a}. The vibrational dephasing process destroys the coherence of vibrational motion and thus irreversibly traps the electronic populations [36].

Let us next consider vibrational energy relaxation. To characterize these processes, we shall consider the energy content of individual nuclear degrees of freedom. The definition of individual mode energies is to some extent arbitrary, since there exists, in general, no unique decomposition of the total vibrational energy into single-mode contributions. The concept of time-dependent vibrational energy expectation values for individual modes can nevertheless be very useful for a qualitative analysis of ultrafast internal-conversion processes [36, 188, 262, 266].

As an example, we consider the vibronic-coupling model Hamiltonian defined by Eqs. (2.20, 2.27–2.30), assuming two coupled electronic states ($n = 1, 2$) and $\gamma_{ij}^{(n)} = 0$. For a totally symmetric tuning mode Q_i, the energy content $E_i(t)$ can be defined as [36, 188]

$$E_i(t) = \sum_{n=1,2} \langle \Psi(t)|\psi_n\rangle \left[-\frac{\hbar\omega_i}{2} \frac{\partial^2}{\partial Q_i^2} + \frac{\hbar\omega_i}{2} \left(Q_i + \frac{\kappa_i^{(n)}}{\hbar\omega_i} \right)^2 \right] \langle \psi_n|\Psi(t)\rangle \tag{4.15}$$

To define the energy content of a coupling mode Q_c, we rewrite the Hamiltonian in the unitarily transformed diabatic basis

$$|\psi_\pm\rangle = 2^{-1/2}(|\psi_1\rangle \pm |\psi_2\rangle) \tag{4.16}$$

This transformation shifts the coupling term λQ_c [cf. Eq. (2.30)] into the diagonal. The expectation value of the energy of the coupling mode is thus given by

$$E_c(t) = \sum_{k=\pm} \langle \Psi(t)|\psi_k\rangle \left[-\frac{\hbar\omega_c}{2} \frac{\partial^2}{\partial Q_c^2} + \frac{\hbar\omega_c}{2} \left(Q_c + \frac{\lambda^{(k)}}{\hbar\omega_c} \right)^2 \right] \langle \psi_k|\Psi(t)\rangle \tag{4.17}$$

with $\lambda^{(\pm)} = \pm\lambda$. For a torsional mode, which has been introduced in Section II.C.3 as an example of a photochemically reactive mode [cf. Eq. (2.32)], the expectation value of the energy can be defined as

$$E_t(t) = \sum_{k=0,1} \langle \Psi(t)|\psi_k\rangle \left(-\frac{\hbar^2}{2I} \frac{\partial^2}{\partial \varphi^2} + V^{(k)}(\varphi) + (V_1 - E_1)\delta_{k1} \right) \langle \psi_k|\Psi(t)\rangle \tag{4.18}$$

Figure 15 shows the expectation values of the vibrational energy for the three modes ν_1, ν_{6a} and ν_{10a} of the pyrazine model. The rapid initial increase of $\langle E_i\rangle_t$, in particular for the modes ν_{6a} and ν_{10a}, shows that the interconversion of electronic to vibrational energy occurs essentially on the timescale of a single vibrational period. After about 500 fs, the mode energies become stationary. In this model, as well as in other few-mode vibronic-coupling models, we do not find equipartition of vibrational energy in the long time

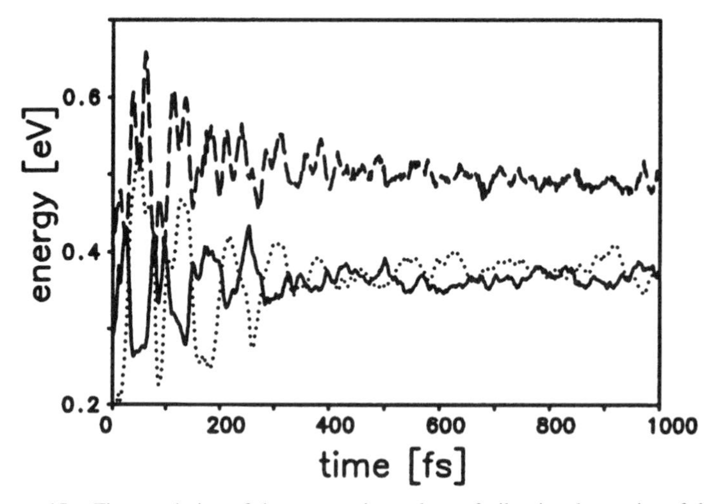

Figure 15. Time evolution of the expectation values of vibrational energies of the three modes ν_1 (solid line), ν_{6a} (dotted line), and ν_{10a} (dash-dotted line) of the pyrazine model.

limit. It can be expected that additional coupling mechanisms not included in the model will ultimately lead to complete equilibration of the vibrational energy in real systems. The model calculations nevertheless provide a hint that vibrational energy equilibration may be incomplete in internal-conversion systems at sub-picosecond timescales. Recent experimental observations for several systems support this idea (e.g., refs. [271, 272]).

The vibrational energy expectation values for the three-mode nonadiabatic photoisomerization model are shown in Fig. 16 [39]. In this case the large electronic excess energy (5.0 eV) is initially stored in the torsional mode. It is seen that energy is transferred from the highly excited torsional mode to the other modes on a 100 fs timescale. Again, the energy equilibration remains incomplete in this model. It would be interesting to investigate these energy redistribution processes for more realistic three- or four-dimensional photoisomerization models to see the effect of additional anharmonic as well as kinematic couplings. Mode-to-mode energy transfer processes will be further analyzed in Section VIII.D, in the framework of a semiclassical modeling of nonadiabatic photoisomerization dynamics.

Finally, we consider the characterization of chemically reactive dynamics. To define reactant/product population probabilities and corresponding transition rates, we have to introduce imaginary hypersurfaces separating the configuration spaces of reactants and products. As an example, consider the photoisomerization model repeatedly referred to above, which involves a single "reactive" torsional degree of freedom with minima of the S_0 PE surface at φ

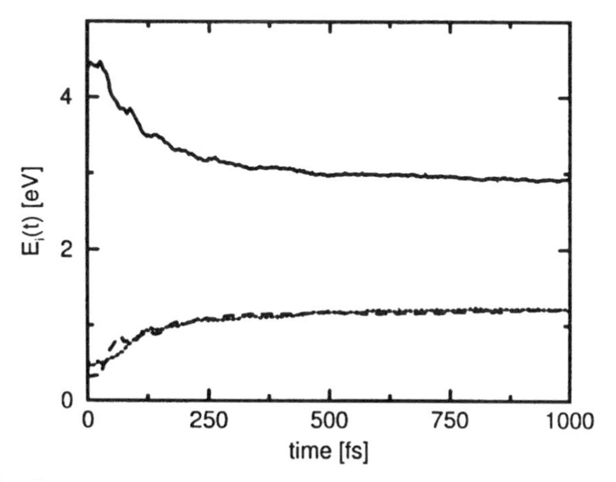

Figure 16. Time-dependent energy content of the torsional mode (solid curve), the coupling mode (dashed curve), and the tuning mode (dotted curve) for the three-dimensional photoisomerization model.

$= 0$ and $\varphi = \pi$, which may be defined as the "*trans*" and "*cis*" conformations of the molecule. Projection operators defining the "*trans*" $(\pi/2 < \varphi < \pi/2)$ and "*cis*" $(\pi/2 < \varphi < 3\pi/2)$ conformations are thus

$$\hat{P}_{trans} = \theta(\pi/2 - \varphi)\,\theta(\varphi + \pi/2) \tag{4.19a}$$

$$\hat{P}_{cis} = \theta(3\pi/2 - \varphi)\,\theta(\varphi - \pi/2) \tag{4.19b}$$

The time-dependent "*trans*" and "*cis*" population probabilities are then given by

$$P_{trans}(t) = \langle \Psi(t)|\hat{P}_{trans}|\Psi(t)\rangle \tag{4.20a}$$

$$P_{cis}(t) = \langle \Psi(t)|\hat{P}_{cis}|\Psi(t)\rangle \tag{4.20b}$$

Nonperturbative time-dependent transition rates can be defined analogously to Eq. (4.12). To characterize a photochemical process in more detail, we may furthermore consider the joint probability of populating a given electronic state as well as a given conformer, for example,

$$P^{\text{di}}_{n,trans} = \langle \Psi(t)|\hat{P}^{\text{di}}_n \hat{P}_{trans}|\Psi(t)\rangle \tag{4.21}$$

Figure 17(a) shows, as an illustration, the time-dependent *trans* population probability for the three-dimensional photoisomerization model, assuming excitation of the *trans* conformer at $t = 0$ to the inverted S_1 surface (cf. Fig. 4). Figure 17(*b*) shows, for comparison, $P_{trans}(t)$ for the one-dimensional model (torsion only). In the one-dimensional case, the wave packet relocalizes periodically at the *trans* conformation (cf. Fig. 7) and $P_{trans}(t)$ oscillates quasi-periodically. In the three-dimensional case, the torsional quasi-periodicity is destroyed already during the first passage through the conical intersection and $P_{trans}(t)$ quickly becomes stationary (the transient oscillations in Fig. 17(*a*) arise from the harmonic tuning mode). As expected for this completely symmetric system, asymptotically we have $P_{trans} = P_{cis} = 0.5$. Although this model is somewhat academic, it clearly reveals the fundamental difference between reversible torsional motion in the one-dimensional case and irreversible photoisomerization in the three-dimensional case. A model with more interesting *cis*/*trans* population dynamics will be con-

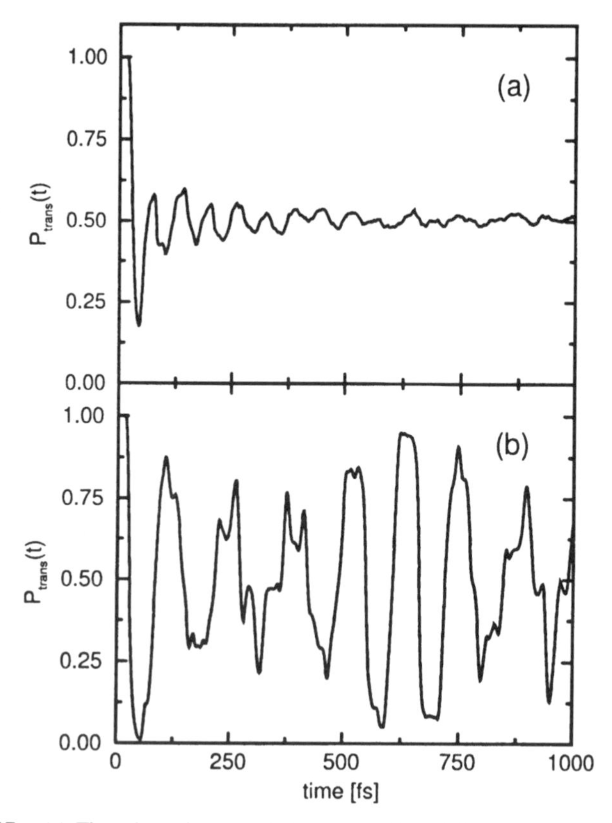

Figure 17. (*a*) Time-dependent population probability of the *trans* conformer obtained for the three-dimensional photoisomerization model. (*b*) The *trans* population probability of the one-dimensional model (torsional mode only).

sidered in Section VII.B in the context of real-time detection of photoisomerization dynamics.

To summarize, conical intersections have a profound impact on the intramolecular dynamics at femtosecond timescales. The few examples discussed here illustrate the interesting interplay of coherent vibrational motion, vibrational energy relaxation, and electronic transitions, within a fully microscopic quantum mechanical description. It is remarkable that irreversible population and phase relaxation processes are so clearly developed in system with just three or four nuclear degrees of freedom. This makes first-principles quantum dynamical studies of ultrafast electronic decay processes possible, and thus opens the perspective of unravelling in detail the most elementary steps of photochemical dynamics in polyatomic molecules.

V. BASIC CONCEPTS OF FEMTOSECOND SPECTROSCOPY

It is widely realized that femtosecond spectroscopy of complex systems is strongly dependent on theoretical support. Given the experimental data alone, it is generally not possible to decide how many electronic states and vibrational modes are involved and what type of dynamics (e.g., intra- or intermolecular dynamics, electronic decay, or vibrational relaxation) is observed. Moreover, the signals usually depend in a complex manner both on the material dynamics as well as on laser pulse properties. In order to establish a connection between the experimental signals that are measured and the molecular dynamics that is to be observed, we discuss in the following two sections the theoretical description of femtosecond experiments. As the focus of this article is on femtosecond spectroscopy of nonadiabatic photoinduced dynamics, we will particularly consider time-resolved spectroscopic techniques that are suitable to monitor these processes. To set the stage for a more general discussion of the potential of modern time-resolved spectroscopy, it is nevertheless important to give a brief overview of various spectroscopic techniques.

A. Preparation and Detection of Nonstationary States

There exists a large variety of spectroscopic techniques that employ ultrashort laser pulses. These methods may differ, for example, in the detection mechanism and in the number and properties of laser fields, and will, in general, monitor different aspects of the dynamics of the molecular system. Most of these experiments are of the pump-probe (PP) type, that is, the molecular system is prepared by a first laser pulse (the "pump") into a nonstationary state, the time evolution of which is interrogated by a time-delayed second laser pulse (the "probe"). It is important to distinguish between resonant and nonresonant excitation of the system. In the latter case it is not possible to establish a *population* in the excited electronic state that survives the duration of the pump field. As a consequence, nonresonant excitation gives only rise to "Raman-like" emission, which is known to essentially reflect the dynamics in the electronic ground state [49]. Being mainly interested in the observation of excited-state dynamics, we therefore focus on resonant excitation of the molecule, which results in both resonance Raman as well as stimulated-emission contributions to the overall signal [74]. Aspects of electronic excitation by a pulsed laser field have been discussed, for example, in refs. [68, 273, 274].

While the *preparation* of the system by an ultrashort laser pulse is common to most femtosecond experiments, there exists a wealth of different techniques to *probe* the dynamics initiated by the pump. Let us consider a standard two-pulse PP experiment and let us assume that at $t = 0$ the system has been excited by the first pulse into the nonstationary state $|\Psi_P\rangle$. In order to probe the dynamics of the excited-state wave function $|\Psi_P(t)\rangle = e^{-iH_M t}|\Psi_P\rangle$, the second pulse is used to project the evolving state $|\Psi_P(t)\rangle$ onto a known final state $|\Psi_f\rangle$. A major criterion to distinguish the probe mechanisms is thus given by the kind of final states $|\Psi_f\rangle$ involved in the experiment. Possible choices for the electronic part of $|\Psi_f\rangle$ are the electronic ground state (through time-resolved fluorescence spectroscopy [275–279] and time-resolved stimulated-emission pumping[1]); higher-lying electronic states (through excited-state absorption [8, 11, 12, 51, 283]); and the cationic ground state, as well as high-Rydberg states (through pulsed ionization with ion or photoelectron detection [21, 59, 69, 70, 284–295]). Alternatively, one may measure the transient infrared absorption of the sample [296–299]. In the case of reactive systems, it is often advantageous to probe the dynamics of the products (instead of the parent molecule), which may be accomplished, for example, through excited-state absorption with subsequent detection of the laser-induced fluorescence [8, 11, 12, 51].

While experimental considerations are beyond the scope of this article (various PP techniques have been extensively discussed in the works of Zewail (e.g., refs. [8, 51]) and in ref. [300]), we wish to address some theoretical issues concerning the choice of the detection mechanism. First note that, to facilitate a microscopic interpretation of the PP signals, *the final state must be well known*. This is often the case for the electronic and ionic ground states, but generally not so for the excited states of polyatomic molecules. Second, it is clear that each final electronic state is associated with different Franck–Condon factors and selection rules, i.e., depending on the spe-

[1]Femtosecond stimulated-emission pumping represents a standard technique in the condensed phase (see, for example, refs. [13, 280, 281]), but has only recently been demonstrated in the gas phase [282].

cific system under consideration, each probe mechanism will monitor different aspects of the dynamics. Finally, it should be stressed that, in general, different detection techniques also monitor different dynamical processes. For example, transient transmittance spectroscopy reflects both ground-state dynamics (via impulsive resonance Raman scattering), as well as excited-state dynamics (via stimulated-emission and excited-state absorption). Time-resolved fluorescence spectroscopy, on the other hand, exclusively monitors the dynamics of the initially prepared excited state, thus facilitating the interpretation of the experimental data.

Besides various detection *mechanisms* (e.g., stimulated emission or ionization), there exist numerous possible detection *schemes*. For example, we may either directly detect the emitted polarization ($\propto \mathbf{PP}^*$, so-called homodyne detection[2]), thus measuring the decay of the *electronic coherence* via the photon-echo effect, or we may employ a heterodyne detection scheme ($\propto \mathbf{EP}^*$), thus monitoring the time evolution of the *electronic populations* in the ground and excited electronic states via resonance Raman and stimulated-emission processes. Furthermore one may use polarization-sensitive detection techniques (transient birefringence and dichroism spectroscopy [280, 301–303]); employ frequency-integrated (see, e.g., ref. [13]) or dispersed (see, e.g., ref. [281]) detection of the emission; and use laser fields with definite phase relation [61, 64]. On top of that, there are modern coherent multi-pulse techniques, which combine several of the above mentioned options. For example, phase-locked heterodyne-detected four-pulse photon-echo experiments [304] make it possible to monitor all three time evolutions inherent to the third-order polarization, namely, the electronic coherence decay induced by the pump field, the dynamics of the system occuring after the preparation by the pump, and the electronic coherence decay induced by the probe field. For a recent theoretical survey of various spectroscopic detection schemes, see the book of Mukamel [84].

In this review we will focus mainly on the discussion of experiments that monitor the time-evolution of the electronic excited-state dynamics, which is accomplished, for example, by transient transmittance, time-resolved fluorescence, and time-resolved ionization spectroscopy. This is because these techniques have the potential to directly monitor the photochemical excited-state processes that we wish to investigate. It should be stressed, however, that within the nonperturbative approach outlined in Section VI.B any of the above-mentioned spectroscopic signals is readily obtained. In other words, the hard part of the calculation is the computation of the time-dependent molecular wave function; the subsequent time integrations or Fourier trans-

[2] In the time-dependent formulation adopted here, "proportional to $\boldsymbol{E}^n$" means $\propto \int dt_n \ldots \int dt_2 \int dt_1\, \boldsymbol{E}(t_n) \ldots \boldsymbol{E}(t_2)\boldsymbol{E}(t_1)$.

formations to obtain the spectroscopic signals are a comparatively easy matter.

B. Spectroscopic Signals

In this section, we introduce the definitions of time-resolved spectroscopic signals of interest. To clarify the approximations involved and to discuss the inherent physical limitations when asking for temporal and frequency resolution at the same time, we derive expressions for various time- and frequency-resolved PP signals such as transient transmittance [80, 92] and time-resolved fluorescence spectra [305]. Furthermore, the signals for two- and three-pulse photon echos, time-resolved ionization, and fragment detection are defined.

1. *Transient Transmittance*

We are concerned with the interaction of a molecular system with the electric field $\mathbf{E}(\mathbf{x}, t)$. Within the electric dipole approximation the response of the molecular system to the field $\mathbf{E}(\mathbf{x}, t)$ is completely described by the electric polarization $\mathbf{P}(\mathbf{x}, t)$, which is defined as the expectation value of the electronic dipole moment of the system [74]. $\mathbf{E}(\mathbf{x}, t)$ is considered as a classical field

$$\mathbf{E}(\mathbf{x}, t) = \sum_{i=1,2} \mathbf{E}_i(\mathbf{x}, t) \tag{5.1a}$$

$$\mathbf{E}_i(\mathbf{x}, t) = \boldsymbol{E}_i(\mathbf{x}, t) \exp(i\mathbf{k}_i\mathbf{x}) + \text{c.c.} \tag{5.1b}$$

$$\boldsymbol{E}_i(\mathbf{x}, t) = \boldsymbol{\varepsilon}_i E_i(\mathbf{x}, t) \tag{5.1c}$$

$$E_i(\mathbf{x}, t) = \mathcal{E}_i(\mathbf{x}, t) e^{-i\omega_i t} \tag{5.1d}$$

consisting of the pump field $\boldsymbol{E}_1$ and the probe field $\boldsymbol{E}_2$, which are characterized by wave vector $\mathbf{k}_i$, laser frequency ω_i, polarization vector $\boldsymbol{\varepsilon}_i$, and pulse envelope function $\mathcal{E}_i$.[3]

The experimental PP signal is usually defined as the difference between the transmittance of the probe pulse with and without the preparation of the sample by the pump pulse. To calculate the electric field of the probe pulse after passing the sample, we need to consider the propagation of the probe in the nonlinearly polarized medium. Assuming that the spatial extent of the pulse is large compared to its wavelength, we may invoke the slowly varying envelope approximation. This reduces the second-order wave equation for

[3] Throughout this article we use Roman type-setting for real fields (e.g., $\mathbf{E}, \mathbf{P}$), italics for complex fields ($\boldsymbol{E}, \boldsymbol{P}$), and calligraphics for the corresponding envelope functions ($\mathcal{E}, \mathcal{P}$).

the probe field $\mathbf{E}_2$ to a first-order differential equation for the corresponding envelope function $\mathcal{E}_2$ [74]

$$\left(\frac{\partial}{\partial z} + \frac{1}{v}\frac{\partial}{\partial t}\right)\mathcal{E}_2(z,t) = \frac{2\pi i\omega_2}{nc}\,\mathcal{P}(z,t) \tag{5.2}$$

where $n = v/c$ is the refractive index of the medium, and it has been assumed that the pulses propagate along the z-axis. It should be noted that the nonlinear polarization $\mathcal{P}(z,t)$ is itself a function of the local probe field $\mathcal{E}_2(z,t)$ in the sample. To properly describe the propagation of the classical field $\mathcal{E}_2(z,t)$ through the medium characterized by $\mathcal{P}(z,t)$, we thus have to solve the coupled field–matter equations [i.e., Eq. (5.2) for $\mathcal{E}_2(z,t)$ and the Schrödinger equation to obtain $\mathcal{P}(z,t)$] in a self-consistent way [306].

In many practical cases, however, it is justified to neglect pulse-propagation effects and ignore the time-derivative at the left-hand side of Eq. (5.2). The situation furthermore simplifies considerably if one assumes an *optically thin sample*, that is, the incident electric field that induces the polarization passes, by definition, through the medium unchanged. Hence the total electric field at the end of the sample ($z = l$) is simply given as a sum of the incident field and the polarization [92, 280, 307].

$$\mathcal{E}_2(l,t) = \mathcal{E}_2(0,t) + \frac{2\pi i\omega_2 l}{nc}\,\mathcal{P}(0,t) \tag{5.3}$$

Alternatively, the depletion of the incident pulse may be taken into account by assuming a simple (e.g., linear) dependence of the polarization $\mathcal{P}$ on the field $\mathcal{E}_2$, thus obtaining a Lambert–Beer-like description [cf. Eq. (5.9)].

Note that the first two approximations employed (classical external field and dipole approximation) are usually well justified in molecular physics, whereas the slowly varying envelope approximation may approach the limits of validity in the case of femtosecond pulses with optical frequencies. The validity of the assumption of an optically thin sample, i.e., the complete neglect of pulse propagation effects, depends on the specific experiment under consideration. To simplify the notation, we henceforth suppress the $\mathbf{x}$-dependence of polarizations and fields, and also drop the prefactor of the polarization in Eq. (5.3).

From classical electrodynamics it is well known that the rate of dissipated energy of an electric field $\mathbf{E}(t)$ in a medium that is characterized by its polarization $\mathbf{P}(t)$ is given by $\mathbf{E}(t)\cdot\dot{\mathbf{P}}(t)$ [74]. Assuming a photodetector at the end of the sample that counts all photons being emitted in $\mathbf{k}_2$ direction, the PP signal is proportional to the total energy dissipated (or gained) by the probe pulse in the medium

$$I = \int_{-\infty}^{\infty} dt\, \dot{\mathbf{E}}_2(t) \cdot \tilde{\mathbf{P}}(t) \tag{5.4}$$

Here $\tilde{\mathbf{P}}(t) = \mathbf{P}_{\text{pump on}} - \mathbf{P}_{\text{pump off}}$ denotes the difference of the polarizations radiating along $\mathbf{k}_2$ with and without the preparation of the sample by the pump pulse, and we have performed a partial integration. The sign convention is such that the signal is negative for absorption (disappearance of photons) and positive for emission (creation of photons). From Eq. (5.4) the signal depends on the pulse delay time Δt as well as on the properties of the laser pulses (e.g., pulse durations τ_i and laser frequencies ω_i), and may be considered as a time- and frequency-dependent signal $I_{In}(\omega_2, \Delta t)$ when recorded as a function of Δt and the probe carrier frequency ω_2. As Eq. (5.4) is (inherently) integrated over all emission frequencies, it will be referred to as *integral* PP signal. It is instructive to rewrite the integral PP spectrum in Eq. (5.4) as

$$I_{In}(\omega_2, \Delta t) = 2\omega_2 \,\text{Im} \int_{-\infty}^{\infty} dt\, E_2(t) \tilde{P}^*(t) \tag{5.5}$$

where we have employed the slowly-varying envelope approximation and the rotating-wave approximation [186], thus neglecting $\dot{E}_2$ contributions and terms with rapidly oscillating ($\propto e^{2i\omega_2 t}$) integrands, respectively. Although both approximations may approach the limits of validity in the case of intense femtosecond laser pulses with optical frequencies, it has been found in numerical studies that deviations between the complete signal of Eq. (5.4) and the approximated signal of Eq. (5.5) are usually minor [50].

In order to obtain the spectrum of the emitted field, the probe pulse is frequency resolved by a spectrometer after it has passed the sample (see, e.g., ref. [281]). Because the PP signal is measured as the time-integrated energy rate [cf. Eq. (5.4)], the corresponding spectrum may be considered as stationary, although it inherently depends on the delay time Δt. Thus, in contrast to the case of time-resolved fluorescence (see next section), the effects of the spectrometer need not be considered in the theoretical description [308], and we may define the *dispersed* PP signal as the intensity of the Fourier transform of the total emitted field, Eq. (5.3), yielding [80, 92]

$$I_D(\omega, \Delta t) = 2\omega \,\text{Im}\, \mathbf{E}_2(\omega) \tilde{\mathbf{P}}^*(\omega) \tag{5.6}$$

where

$$\mathbf{E}_2(\omega) = \int_{-\infty}^{\infty} dt\, e^{i\omega t}\mathbf{E}_2(t) \tag{5.7}$$

$$\tilde{\mathbf{P}}(\omega) = \int_{-\infty}^{\infty} dt\, e^{i\omega t}\tilde{\mathbf{P}}(t) \tag{5.8}$$

denote the Fourier transform of the incident probe field and the polarization, respectively. Combining the definitions in Eqs. (5.4–5.8), it is easily verified that integration of the dispersed PP signal in Eq. (5.6) over all emission frequencies ω again yields the integral PP signal [Eq. (5.4)]. It should be noted that the experimental transmittance spectrum is often alternatively defined by normalizing the dispersed PP spectrum in Eq. (5.6) to the intensity of the incident field [80, 281], which makes it possible to express the PP spectrum in terms of an exponential law

$$I_D(\omega, \Delta t) = |\mathbf{E}_2(\omega)|^2 e^{-\Gamma(\omega, \Delta t)l}$$

$$\Gamma(\omega, \Delta t) = \frac{4\pi\omega}{nc}\ \mathrm{Im}\ \tilde{\mathbf{P}}(\omega)/\mathbf{E}_2(\omega) \tag{5.9}$$

It is interesting to note that the simultaneous time and frequency resolution of the *dispersed* transmittance spectrum $I_D(\omega, \Delta t)$ is not limited by the well-known time-frequency uncertainty principle. This is because the frequency resolution of this signal is determined by the polarization decay time [cf. Eq. (5.8)], which is independent of the pulse duration representing the time resolution. In the case of the *integral* transmittance spectrum, on the other hand, time and frequency resolution are Fourier limited. This is because the duration of the pulses defines the duration of the measurement, which in turn determines the frequency resolution of the experiment [92].

The definitions of the dispersed transmittance spectrum $I_D(\omega, \Delta t)$ [Eq. (5.6)] and the integral transmittance spectrum $I_{In}(\omega_2, \Delta t)$ [Eq. (5.4)] connect these experimental signals to the macroscopic polarization difference $\tilde{\mathbf{P}}(t)$. Because we have neglected all pulse propagation effects, both the incident electric fields and the polarization are approximated by plane waves in **x**-space, i.e., the macroscopic polarization is simply given by an orientational average over the microscopic polarizations of the individual molecules in a unit volume. Describing the molecular system as an electronic two-state system, the orientational averaging of $\tilde{\mathbf{P}}(t)$ simply amounts to the multplication with a constant, i.e., the macroscopic and microscopic polarizations are proportional to each other. Assuming, on the other hand, a model system with several (≥ 2) optically bright electronic states, the averaging process in general depends on the relative orientations of the individual transition dipole moments. In this case it is necessary to explicitly perform the orientational average [309].

We note in passing that, within the assumption of an optically thin sample, one obtains equivalent expressions for the spectroscopic signals if one starts the derivation from a "system point of view" instead of the "electric field point of view" adopted above. In the former approach, adopted by several workers [48, 55, 250, 310], one considers the time derivative of the expectation value of the total Hamiltonian $H(t)$ [consisting of the molecular system H_M and the field-matter interaction Hamiltonian $H_{\text{int}}(t)$, cf. Eq. (3.2)], yielding

$$d/dt\langle H(t)\rangle = -\dot{\mathbf{E}}(t)\cdot\tilde{\mathbf{P}}(t) \tag{5.10}$$

A comparison of Eqs. (5.4) and (5.10) reassures us that the energy loss of the molecular system, given by time integration over the quantity in Eq. (5.10), is identical to the energy gain of the electric field in Eq. (5.4).

2. *Time-Resolved Fluorescence*

With the development of fluorescence upconversion techniques [275], which nowadays provide femtosecond time resolution [276–279, 304], it is also possible to directly measure the time evolution of the spontaneous emission following the excitation of the sample by the pump pulse. In this method the fluorescence is collected and focused onto a nonlinear crystal, where it is superposed with the probe beam in order to perform upconversion. Time resolution is achieved because the probe pulse creates a "time gate" for the spontaneous emission, i.e., the fluorescence is only measured within the duration of the probe. Frequency resolution is achieved by subsequently dispersing the upconverted signal in a monochromator. Although fluorescence detection provides less photon yield than stimulated techniques, it has the desirable feature of exclusively monitoring the time evolution in the initially excited electronic states (cf. discussion above).

The theoretical description of time-resolved fluorescence spectroscopy requires (1) calculating the spontaneous emission of the molecular system and (2) taking into account the detection process (i.e., upconversion and dispersion). The first task is usually performed within time-dependent perturbation theory (see Section VI.A), treating the radiation field quantum-mechanically [85, 311, 312] or classically [305]. The nonlinear upconversion process, on the other hand, is usually not explicitly described, but is taken into account in a phenomenological manner. The following derivation of the time- and frequency-resolved fluorescence spectrum is based on the work of Kowalczyk *et al.* [305], who have extended the theory of Eberly and Wódkiewicz [308] to describe femtosecond fluorescence upconversion experiments.

Let $E_S(t)$ denote the electric field that is spontaneously emitted by the sample. It is clear that the photodetector behind crystal and monochromator does not see the entire field $E_S(t)$ but a filtered version of it, denoted by $E_F(t)$. Assuming that the photodetector counts all photons of the field $E_F(t)$, the time-resolved fluorescence signal can be defined as

$$I_F(\omega_f, \Delta t) = \omega_f^3 \int_{-\infty}^{\infty} dt_2 \int_{-\infty}^{\infty} dt_1 \langle \hat{E}_F(t_2) \hat{E}_F^{\dagger}(t_1) \rangle \tag{5.11}$$

where ω_f denotes the setting frequency of the filter to be described below, and $\hat{E}_F(t)$ denotes the quantum-mechanical field operator. Since we are not interested in quantum effects of the light field, it is sufficient to treat the radiation field classically, thus yielding

$$I_F(\omega_f, \Delta t) = \omega_f^3 \int_{-\infty}^{\infty} dt \, |E_F(t)|^2 \tag{5.12}$$

The filtered field $E_F(t)$ differs from the incident fluorescence emission $E_S(t)$ in two ways. First, due to the upconversion process, $E_F(t)$ is only generated within the duration of the probe pulse. The nonlinear crystal thus represents a "time gate" for the incident field $E_S(t)$. After the crystal, this results in the field

$$E(t) = G(t, \Delta t) E_S(t) \tag{5.13}$$

where the gating function $G(t, \Delta t)$ is usually chosen to be of Gaussian form $\exp[-\alpha(t - \Delta t)^2]$. Second, the spectrometer following the crystal represents a frequency filter $F(\omega, \omega_f)$, which in the time domain corresponds to convolution of the field with a filter function $F(t, \omega_f) = \exp[-(i\omega_f + \gamma)t]$, where γ reflects the limited frequency resolution due to the filter [308]. The filtered field measured by the photodetector is thus given by

$$E_F(t) = \int_{-\infty}^{t} dt' \; F(t - t', \omega_f) G(t', \Delta t) E_S(t') \tag{5.14}$$

It is clear from Eq. (5.14) that the actually achievable frequency resolution is determined by the duration of the probe pulse (through $G(t, \Delta t)$) as well as by the instrumental width γ. In the case of femtosecond pulses, however, γ will be typically much smaller than the width $\approx \sqrt{\alpha}$ introduced by the pulse duration. In contrast to the dispersed transmittance spectrum $I_D(\omega, \Delta t)$ discussed above, the time and frequency resolution of the dispersed time-resolved fluorescence spectrum $I_F(\omega_f, \Delta t)$ is thus Fourier limited.

3. *Photon Echo*

As mentioned above, transient transmittance and fluorescence spectroscopy are heterodyne-detection techniques ($\propto \mathbf{EP}^*$) and monitor the time evolution of the electronic excited-state populations. To give a representative example of a homodyne-detection technique ($\propto \mathbf{PP}^*$) that measures the time evolution of the electronic coherence, let us consider the photon-echo experiment [313]. The two-pulse photon-echo signal is given by the time-integrated coherent emission in the background-free direction $2\mathbf{k}_2 - \mathbf{k}_1$

$$I_{PE}(\Delta t) = \int_{-\infty}^{\infty} d\tau \; |\mathbf{P}_{2\mathbf{k}_2 - \mathbf{k}_1}(\tau, \Delta t)|^2 \tag{5.15}$$

Applying fluorescence upconversion techniques with sub-picosecond time resolution, it is possible to monitor in addition to the signal in Eq. (5.15) the time evolution of the emission [314]. Assuming an ultrashort detection pulse, the three-pulse photon-echo signal is given by

$$I_{PE}(\tau, \Delta t) = |\mathbf{P}_{2\mathbf{k}_2 - \mathbf{k}_1}(\tau, \Delta t)|^2 \tag{5.16}$$

Photon-echo experiments allow us to discriminate homogeneous and inhomogeneous phase relaxation processes, and have therefore proven to be valuable for the investigation of electronic coherence dynamics in the condensed phase (see, for example, refs. [15, 315]). As is demonstrated in Section VII.B for the example of a simple model of photoisomerization, however, the measurement of the electronic coherence decay provides little information on the photochemical reaction dynamics of the system. For an extensive discussion of various coherent spectroscopic techniques see, for example, refs. [48, 84, 315, 316]).

4. *Time-Resolved Ionization and Fragment Detection*

The majority of spectroscopic techniques considered in this article are based on the detection of photons. Thus the calculation of the electric polarization represents the central task of the theoretical description. In femtosecond ionization spectroscopy, on the other hand, charged particles (i.e., ions or photoelectrons) are detected, i.e., the measured signals are obtained as the count rate of charged particles [21, 59, 69, 70, 284–295]. In the case of reactive systems, it is often preferable to probe the dynamics of the products instead of the parent molecule. This may once again be achieved through ionizing the fragments [21, 287], or through excited-state absorption with subsequent detection of the laser-induced fluorescence (LIF) [8, 11, 12, 51]. The main advantage of ionization experiments clearly is the extreme sensitivity of particle detectors, which make gas-phase experiments on systems with small quantum yield possible. Measurements of the ion yield with femtosecond laser pulses have been reported, for example, in refs. [21, 284, 287–293]. It has been shown by these measurements that femtosecond time-resolved ionization is a powerful tool for studying the dynamics of ultrafast internal-conversion processes in gas-phase molecules. In combination with mass-selective ion detection, moreover, femtosecond ionization experiments can yield detailed information on fragmentation reactions following the primary excitation process [21, 287, 289, 290, 292, 293].

To model these experiments, one usually identifies the count rate of the detector with the field-induced populations of the electronic or ionic states under consideration. To give an example for product detection through the measurement of the LIF, let us consider the NaI experiment of Zewail and coworkers [11, 51]. In this experiment the photoinduced predissociation of NaI is observed through two-photon excitation and subsequent monitoring of the LIF of the sodium fragment. Assuming that the LIF emission is directly proportional to the field-induced population in the excited electronic state $|\psi_{\mathrm{Na^*I}}\rangle$, the LIF signal is given by [52, 62]

$$I_{LIF}(\Delta t) = \langle \Psi_{PP}(t) | \psi_{\mathrm{Na^*I}} \rangle \langle \psi_{\mathrm{Na^*I}} | \Psi_{PP}(t) \rangle \tag{5.17}$$

where $|\Psi_{PP}(t)\rangle$ denotes the wave function of the molecular system after the interaction with the pump and probe fields, and Δt represents the delay time between the two pulses. The calculation of the LIF signal via Eq. (5.17) depends on the knowledge of the potential-energy surface of the excited state and, moreover, relies on the assumption that competing excitation pathways involving different electronic states can be excluded. Although straightforward in principle, the interpretation of LIF experiments therefore requires detailed information on the excited states of the molecular system.

To define time-resolved ionization signals, it is necessary to extend the diabatic model Hamiltonian in Eq. (2.20) to include at least a single ionization continuum. As a specific example, which will also be referred to in Section VII.A, consider a molecular system comprising the electronic ground state $|\psi_0\rangle$, two excited electronic states $|\psi_1\rangle, |\psi_2\rangle$ and two ionization continua $|\psi_1^{(k)}\rangle, |\psi_2^{(k)}\rangle$

$$H = \sum_{l=0}^{2} |\psi_l\rangle h_l \langle\psi_l| + (|\psi_1\rangle V_{12}\langle\psi_2| + \text{h.c.})$$

$$+ \sum_{l=1}^{2} \int_0^\infty dE_k |\psi_l^{(k)}\rangle(\tilde{h}_l + E_k)\langle\psi_l^{(k)}| \tag{5.18}$$

where $\tilde{h}_l$ represents the vibrational Hamiltonian of the ionic state $|\psi_l^{(k)}\rangle$ and the index k labels the energy E_k of the continuum electrons. Let us assume, for example, that the excited electronic state $|\psi_1\rangle$ is radiatively coupled to $|\psi_1^{(k)}\rangle$ and $|\psi_2\rangle$ to $|\psi_2^{(k)}\rangle$, respectively. In this model we neglect the Rydberg series converging towards the ionization thresholds and their possible autoionization. This simplifying assumption should be justified in particular for polyatomic systems, where autoionization is typically strongly quenched by fast intramolecular decay processes [317–319].

The simplest signal measured in time-resolved ionization experiments is the total ion yield following the action of the two pulses, which is given by the integrated population probability of the ionization continua in the limit $t \rightarrow \infty$ [91, 320, 321]

$$I_{\text{ion}}(\Delta t) = \sum_{l=1,2} \int_0^\infty dE_k \langle\Psi(t\rightarrow\infty)|\psi_l^{(k)}\rangle\langle\psi_l^{(k)}|\Psi(t\rightarrow\infty)\rangle \tag{5.19}$$

Alternatively, the energy spectrum of the photoelectrons may be measured. This signal is represented by the population probability density of the ionization continua [91, 320, 322]

$$I_{\text{ion}}(E_k, \Delta t) = \sum_{l=1,2} \langle\Psi(t\rightarrow\infty)|\psi_l^{(k)}\rangle\langle\psi_l^{(k)}|\Psi(t\rightarrow\infty)\rangle. \tag{5.20}$$

In the limiting case of $E_k \rightarrow 0$, Eq. (5.20) describes the zero-kinetic-energy (ZEKE) photoelectron signal. ZEKE photoelectron spectroscopy [323] is a background-free technique with high frequency-resolution, which recently has been combined with PP techniques with picosecond [285] and femtosecond [294] time resolution. Representative calculations of various ionization signals are presented in Section VII.A, where we shall discuss, in particular, the significance of this technique for the detection of ultrafast internal-conversion dynamics.

VI. EVALUATION OF THE NONLINEAR POLARIZATION

A. Perturbative Approach

Perturbation theory represents the standard approach to describe the interaction of matter and a radiation field [74, 84]. Within the scope of this article, there are two main reasons for employing a perturbative treatment. First, in virtually all experiments discussed here, the radiation–matter coupling is considerably weaker than the intramolecular couplings, thus justifying a perturbative treatment of the field–matter interaction to lowest order. In other words, in this article we exclusively consider the electric field as a measuring device to monitor the molecular dynamics. Second, and conceptually more important, the perturbative treatment allows us to classify and to distinguish various spectroscopic processes contributing to a given experiment (e.g., stimulated Raman, stimulated emission, photon echo). It should be stressed that this aspect of the perturbative description is crucial to facilitate the interpretation of the spectroscopic signals in terms of the dynamics of the system. The nonperturbative approach, introduced in the subsequent section, can be computationally advantageous, even if one does not focus on strong-field effects. As discussed below, however, in order to distinguish the various spectroscopic processes under consideration, the nonperturbative calculation still requires a perturbative analysis of the calculated signals.

1. *General Aspects*

Let us recall the definition of the total model Hamiltonian $H(t)$ [Eq. (3.2b)], which consists of the molecular system H_M and the field–matter interaction Hamiltonian $H_{\text{int}}(t)$

$$H(t) = H_M + H_{\text{int}}(t) \tag{6.1a}$$

$$H_M = \sum_k |\psi_k\rangle h_k \langle\psi_k| + \sum_{k \neq k'} |\psi_k\rangle V_{kk'} \langle\psi_{k'}| \tag{6.1b}$$

$$\begin{aligned} H_{\text{int}}(t) &= -\boldsymbol{\mu}\cdot\mathbf{E}(t) \\ &= -\sum_{k \neq k'} |\psi_k\rangle \boldsymbol{\mu}_{kk'}\cdot\mathbf{E}(t)\langle\psi_{k'}| + \text{h.c.} \end{aligned} \tag{6.1c}$$

By solving the time-dependent Schrödinger equation, Eq. (3.2a), with $H(t)$, we obtain the overall polarization $\mathbf{P}(t)$ as quantum-mechanical expectation value of the dipole operator

$$\mathbf{P}(t) = \langle \Psi(t) | \boldsymbol{\mu} | \Psi(t) \rangle$$
$$= 2 \operatorname{Re} \sum_{k \neq k'} \langle \Psi(t) | \psi_k \rangle \boldsymbol{\mu}_{kk'} \langle \psi_{k'} | \Psi(t) \rangle \tag{6.2}$$

Within the electric dipole approximation the response of the molecular system to an electric field is completely described by the electric polarization $\mathbf{P}(t)$, which therefore represents the central quantity of interest for the calculation of spectroscopic signals. For simplicity, we want to restrict ourselves to model systems with a single dipole-allowed electronic transition (say, the $|\psi_0\rangle$–$|\psi_2\rangle$ transition).

Time-dependent perturbation theory may be formulated within the density-matrix or the wave-function formalism, respectively. The density-matrix formalism is more general, in that it allows for the inclusion of finite temperature effects. Moreover, it is possible in density-matrix perturbation theory to include individual phenomenological relaxation constants for the populations (the diagonal elements of the density matrix) and the coherences (the off-diagonal elements of the density matrix), thus accounting for pure electronic dephasing effects [74, 84]. The concept of this work, however, is to account microscopically (not phenomenologically) for the relaxation dynamics of a molecular system. Dealing with intramolecular vibrational modes of typically relatively high frequencies, moreover, the thermal occupation of the vibrational modes can be neglected to a good approximation. In the following we thus employ standard wave-function perturbation theory to describe the nonlinear response of the system to the radiation field which, in particular, allows for a physically intuitive description of the various spectroscopic processes involved [49]. For completeness, the connection between wave-function and density-matrix formulations is illustrated by an example [Eq. (6.15)].

Initially, i.e., before the interaction with the laser field, the molecular system is assumed to be in its electronic and vibrational ground state $|\Psi_0\rangle$ (throughout this section $\hbar = 1$)

$$|\Psi^{(0)}(t)\rangle = e^{-iH_M t} |\Psi_0\rangle \tag{6.3}$$

Writing the time-dependent wave function as

$$|\Psi(t)\rangle = \sum_{N=0}^{\infty} |\Psi^{(N)}(t)\rangle \tag{6.4a}$$

$$|\Psi^{(N)}(t)\rangle = i \int_{-\infty}^{t} dt' \, e^{-iH_M(t-t')} \boldsymbol{\mu} \cdot \boldsymbol{E}(t') |\Psi^{(N-1)}(t')\rangle \tag{6.4b}$$

we obtain for the polarization

$$\mathbf{P}(t) = \sum_{N=0}^{\infty} \mathbf{P}^{(2N+1)}(t) \tag{6.5a}$$

$$\mathbf{P}^{(2N+1)}(t) = 2 \,\mathrm{Re} \sum_{i=0}^{N} \langle \Psi^{2(N-i)}(t) | \psi_0 \rangle \boldsymbol{\mu}_{02} \langle \psi_2 | \Psi^{(2i+1)}(t) \rangle \tag{6.5b}$$

Owing to the initial condition of Eq. (6.3), the linear ($\mathbf{P}^{(1)}$) and third-order ($\mathbf{P}^{(3)}$) polarizations are the first nonvanishing terms of the expansion. The total number of nonzero terms occurring in Eq. (6.5) depends crucially on whether the following conditions are fulfilled:

1. *Rotating-Wave Approximation (RWA).* Within the RWA [186] only resonant optical transitions (e.g., $\propto e^{i(\omega - h_2)t}$) are considered, and non-resonant transitions (e.g., $\propto e^{i(\omega + h_2)t}$) are disregarded. This reduces the number of terms by a factor of two *in each order* of the expansion in Eq. (6.4).
2. *Weak-Field Limit.* In the case of small laser intensities it is sufficient to consider the nonlinear response only to lowest (i.e., third) order.
3. *Non-Overlapping Laser Pulses.* If the laser fields do not overlap in time, we need to consider only one electric field $\mathbf{E}_i(t)$ in each time integration in Eq. (6.4b) instead of two. When speaking of non-overlapping pulses, we will always assume that the pump field $\mathbf{E}_1$ arrives before the probe field $\mathbf{E}_2$. In the perturbative treatment we will always assume non-overlapping laser fields, while the more general case where the probe pulse is coincident with or ahead of the pump pulse (so-called coherent artifact [80, 333]) will briefly be considered in the nonperturbative approach.
4. *Phase-Averaged Detection.* If the spectral signals [e.g., Eqs. (5.4) and (5.6)] are averaged over many laser shots [as it is the case in common pump-probe (PP) experiments], only signals proportional to $\mathbf{E}_1^{2n}\mathbf{E}_2^{2m}$ survive the inherent averaging over the individual phases of the fields. Using phase-locked pulses [61, 64, 316], however, it is also possible to monitor signals proportional to $\mathbf{E}_1^{2n+1}\mathbf{E}_2^{2m+1}$.

2. *Characterization of Spectroscopic Processes*

An important experimental criterion to discriminate the various spectroscopic processes arising in Eq. (6.5) is the emission direction $\mathbf{k}$ of the polarization. Because the incident electric fields $\mathbf{E}_1, \mathbf{E}_2$ radiate along their wave vectors $\mathbf{k}_1, \mathbf{k}_2$, the emission directions $\mathbf{k}$ of the thus induced polarization are given as linear combinations of $\mathbf{k}_1$ and $\mathbf{k}_2$. In the most general case (i.e., when none of the assumptions above hold) it is clear that we get an enormous number of contributions and corresponding emission directions $\mathbf{k}$ in the expansion in Eq. (6.5). Employing the RWA, however, it has been shown in ref. [50] that the polarization of $(2N + 1)$th order only radiates into the directions

$$\mathbf{k} = j(\mathbf{k}_2 - \mathbf{k}_1) + \mathbf{k}_2 \tag{6.6}$$

where j runs from $-(N + 1)$ to N. As a direct consequence of Eq. (6.6), we obtain in linear response only emission in the $\mathbf{k}_1$ and $\mathbf{k}_2$ directions, while the third-order polarizations may radiate along the directions $\mathbf{k}_1, \mathbf{k}_2, 2\mathbf{k}_2 - \mathbf{k}_1$, and $2\mathbf{k}_1 - \mathbf{k}_2$. Assuming, furthermore, non-overlapping pulses, it is easy to show that, even *in arbitrary order of the radiation field*, there is only emission in the directions $\mathbf{k}_1, \mathbf{k}_2$, and $2\mathbf{k}_2 - \mathbf{k}_1$ (i.e., regardless of $N, j = -1, 0, 1$) [50].

To illustrate these considerations and to elucidate the origin of the individual contributions, Fig. 18 shows a diagrammatic representation of various spectroscopic processes contributing to the electric polarization up to third order.[4] In the case of non-overlapping pulses and within the RWA, the evaluation of Eq. (6.5) yields two terms of the linear polarization propagating in $\mathbf{k}_1$ and $\mathbf{k}_2$ directions; and six, eight, and two terms of the third-order polarization propagating in $\mathbf{k}_1, \mathbf{k}_2$, and $2\mathbf{k}_2 - \mathbf{k}_1$ direction, respectively. In addition to the diagrams shown in Fig. 18, we get four terms $\propto \mathbf{E}_i^3$ that are independent of Δt, and also two terms in $\mathbf{k}_2$ that are obtained from the ones shown by simple permutations of the electric fields. The horizontal bars in each diagram represent the ground and excited electronic state of the two-state system [Eq. (6.1b)], respectively, the arrows correspond to electronic transitions due to the electric fields $\boldsymbol{E}_i$ and their complex conjugates, and the time moves forward from left to right. As discussed in Section V.A, polarizations are experimentally measured by using either a homodyne ($\propto |\boldsymbol{P}|^2$) or a heterodyne ($\propto \boldsymbol{P}^*\boldsymbol{E}_2$) detection scheme, i.e., the (heterodyne) PP signals defined above [Eqs. (5.4) and (5.6)] are obtained by adding an interaction with the probe field $\boldsymbol{E}_2$ at the right side of each diagram.

Let us first consider the diagrams in the $\mathbf{k}_2$ direction, describing the stim-

[4]Several workers (e.g., [48–50, 74, 80, 316]) have used double-Feynman diagrams [324] or simplified versions of it to visualize nonlinear spectroscopic processes.

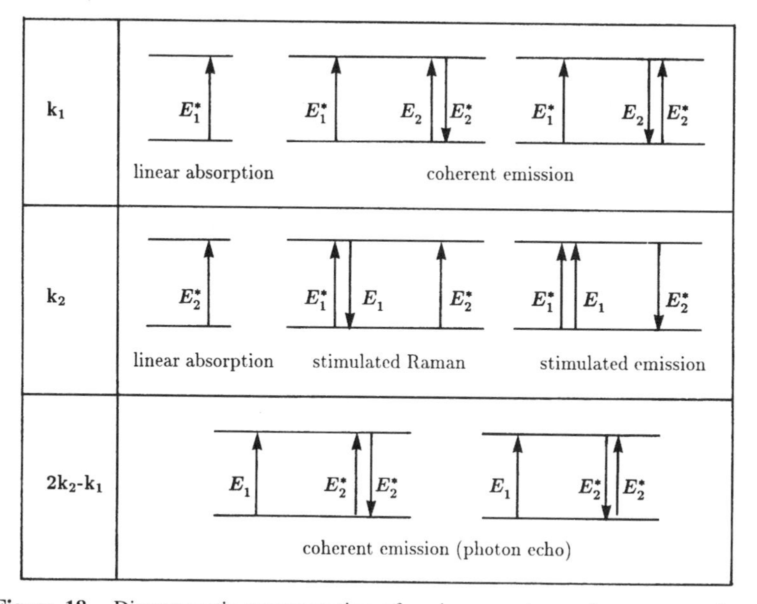

Figure 18. Diagrammatic representation of various spectroscopic processes that contribute to the electric polarization up to third order. Assuming non-overlapping pump ($\boldsymbol{E}_1$) and probe ($\boldsymbol{E}_2$) laser fields within the rotating-wave approximation, we obtain, regardless of the laser intensities, only emission into the directions $\mathbf{k}_1, \mathbf{k}_2$, and $2\mathbf{k}_2 - \mathbf{k}_1$.

ulated Raman and the stimulated emission contribution, respectively. Note that within the delay time Δt (i.e., after the interaction with the pump) the system propagates in either the electronic ground state (Raman case) or in the excited electronic state (stimulated emission case) [74]. As a consequence, it is possible to monitor the electronic population dynamics in a PP experiment [88–93]. Measuring the transient transmittance of the probe pulse as defined above, the heterodyne-detected signal in the $\mathbf{k}_2$ direction is proportional to $|\boldsymbol{E}_1|^2|\boldsymbol{E}_2|^2$, and will thus show up in a standard PP experiment with phase-averaged detection. The contributions in the $\mathbf{k}_1$ and $2\mathbf{k}_2 - \mathbf{k}_1$ directions, on the other hand, describe the time evolution of the system when it is in a electronic coherence. Coherent spectroscopic techniques such as photon-echo measurements usually employ a homodyne detection scheme and monitor the emission in the background-free direction $2\mathbf{k}_2 - \mathbf{k}_1$ [74, 84]. Being proportional to $\boldsymbol{E}_1^*|\boldsymbol{E}_2|^2\boldsymbol{E}_2$ and $\boldsymbol{E}_1\boldsymbol{E}_2^*|\boldsymbol{E}_2|^2$, respectively, the corresponding heterodyne-detected signals are phase-sensitive and therefore only measurable in a PP experiment employing phase-locked laser fields.

In what follows we want to derive and discuss expressions for the polarizations of various important spectroscopic processes. To simplify the notation we define

$$\hat{\mu}_{kk'} = |\psi_k\rangle \boldsymbol{\varepsilon}_i \cdot \boldsymbol{\mu}_{kk'} \langle \psi_{k'}| \tag{6.7}$$

and consider henceforth the projection P of the electric polarization **P** onto the polarization vector $\boldsymbol{\varepsilon}_i$ of the field E_i [cf. Eq. (5.1)]. Let us first consider the linear response $\mathrm{P}^{(1)}_{\mathbf{k}_i}(t)$ of the medium to the electric field $E_i(t)$ which is given by

$$\begin{aligned} \mathrm{P}^{(1)}_{\mathbf{k}_i}(t) &= 2\,\mathrm{Re}\langle \Psi^{(0)}(t)|\,\hat{\mu}_{02}\,|\Psi^{(1)}_{\mathbf{k}_i}(t)\rangle \\ &= 2\,\mathrm{Re}\; i e^{i\mathbf{k}_i\cdot\mathbf{x}} \int_{-\infty}^{t} dt'\; E_i(t') \\ &\quad \langle \Psi_0|\hat{\mu}_{02} e^{-iH_M(t-t')} \hat{\mu}_{20} e^{iH_M(t-t')}|\Psi_0\rangle \end{aligned} \tag{6.8}$$

Assuming a stationary laser field (i.e., $E_i(t) = e^{-i\omega_i t}$) and considering the *rate* of absorbed photons [rather than integrating over all absorbed photons as in Eq. (5.4)], one readily obtains the standard expression [179, 325, 326] for the continuous-wave (CW) absorption spectrum in Eq. (7.1) discussed below. Assuming pulsed laser fields, on the other hand, the *linear* polarization in $\mathbf{k}_1$ leads to a heterodyne signal that is proportional to $\boldsymbol{E}_1^*\boldsymbol{E}_2$ and thus an explicit function of the pulse delay time Δt. Being a first-order effect, this term is expected to dominate the phase-locked PP signal, but is absent in phase-averaged detection [90].

Due to the initial condition of Eq. (6.3), the third-order polarization $\mathrm{P}^{(3)}(t)$ represents the lowest-order nonlinear response of the medium to the electric field. As explained above, the third-order polarization in the $\mathbf{k}_2$ direction consists of resonance Raman and stimulated-emission (SE) contributions

$$\mathrm{P}^{(3)}_{\mathbf{k}_2}(t) = 2\,\mathrm{Re}(P_{RA}(t) + P_{SE}(t)) \tag{6.9}$$

Inserting the first-order and second-order wave functions [Eq. (6.4)] into Eq. (6.5) and arranging the electric fields according to the SE diagram in Fig. 18, we obtain for the SE polarization

$$\begin{aligned} P_{SE}(t) &= \langle \Psi^{(2)}_{\mathbf{k}_1-\mathbf{k}_2}(t)|\hat{\mu}_{02}|\Psi^{(1)}_{\mathbf{k}_1}(t)\rangle \\ &= i e^{i\mathbf{k}_2\cdot\mathbf{x}} \int_{-\infty}^{t} dt'\; E_2(t')\langle \Psi^{(1)}_{\mathbf{k}_1}(t')|\hat{\mu}_{20} e^{iH_M(t-t')} \hat{\mu}_{02} e^{-iH_M(t-t')}|\Psi^{(1)}_{\mathbf{k}_1}(t')\rangle \end{aligned} \tag{6.10}$$

where we have labeled the wave functions by the subscript $n\mathbf{k}_1 - m\mathbf{k}_2$ to indicate the interactions with the electric fields $\boldsymbol{E}_1^n, \boldsymbol{E}_2^{*m}$. Note that Eq. (6.10) resembles the expression for the linear polarization in Eq. (6.8), except that the initial ground-state wave function $|\Psi_0\rangle$ in Eq. (6.8) has been replaced by the nonstationary excited-state wave function $|\Psi_{\mathbf{k}_1}^{(1)}(t)\rangle$ in Eq. (6.10). In analogy to Eq. (6.8), which describes the linear response of a stationary system, one may thus interpret the SE polarization (6.10) as the *linear* response of a *nonstationary* system [49]. The SE polarization in Eq. (6.10) therefore reflects the time evolution of the excited-state wave function $|\Psi_{\mathbf{k}_1}^{(1)}(t)\rangle$. Similarly we obtain for the Raman contribution

$$
\begin{aligned}
P_{RA}(t) &= \langle\Psi^{(0)}(t)|\hat{\mu}_{02}|\Psi_{\mathbf{k}_2}^{(3)}(t)\rangle + \langle\Psi^{(2)}(t)|\hat{\mu}_{02}|\Psi_{\mathbf{k}_2}^{(1)}(t)\rangle \\
&= ie^{i\mathbf{k}_2\cdot\mathbf{x}} \int_{-\infty}^{t} dt'\, E_2(t')\big(\langle\Psi^{(0)}(t')|e^{iH_M(t-t')}\hat{\mu}_{02}e^{-iH_M(t-t')}\hat{\mu}_{20}|\Psi^{(2)}(t')\rangle \\
&\quad + \langle\Psi^{(2)}(t')|e^{iH_M(t-t')}\hat{\mu}_{02}e^{-iH_M(t-t')}\hat{\mu}_{20}|\Psi^{(0)}(t')\rangle\big) \qquad (6.11)
\end{aligned}
$$

Note that $|\Psi^{(2)}(t')\rangle$ represents the nonstationary ground-state wave function prepared by the pump pulse. The Raman polarization thus reflects the time evolution of the ground-state wave function $|\Psi^{(2)}(t')\rangle$ [49].

As an example for coherent emission, consider the photon echo in the $2\mathbf{k}_2 - \mathbf{k}_1$ direction, the polarization of which is given by

$$
\begin{aligned}
\mathrm{P}_{2\mathbf{k}_2-\mathbf{k}_1}^{(3)}(t) &= 2\,\mathrm{Re}\langle\Psi_{\mathbf{k}_1-\mathbf{k}_2}^{(2)}(t)|\hat{\mu}_{02}|\Psi_{\mathbf{k}_2}^{(1)}(t)\rangle \\
&= 2\,\mathrm{Re}\, ie^{i\mathbf{k}_2\cdot\mathbf{x}} \int_{-\infty}^{t} dt'\, E_2(t')\big(\langle\Psi_{\mathbf{k}_1}^{(1)}(t')|\hat{\mu}_{20} \\
&\quad \cdot e^{iH_M(t-t')}\hat{\mu}_{02}e^{-iH_M(t-t')}|\Psi_{k_2}^{(1)}(t')\rangle \\
&\quad + \langle\Psi_{\mathbf{k}_1-\mathbf{k}_2}^{(2)}(t')|e^{iH_M(t-t')}\hat{\mu}_{02}e^{-iH_M(t-t')}\hat{\mu}_{20}|\Psi^{(0)}(t')\rangle\big) \qquad (6.12)
\end{aligned}
$$

Contrary to the SE and Raman polarizations, which reflect the dynamics of the system in the ground and excited electronic state, respectively, the photon-echo polarization reflects the decay of the electronic coherence [84].

To make contact with a widely used alternative formulation of nonlinear spectroscopy, we rewrite, for example, the SE polarization in Eq. (6.10) as

$$P_{SE}(t) = i^3 e^{i\mathbf{k}_2 \cdot \mathbf{x}} \int_{-\infty}^{t} dt_3 \int_{-\infty}^{t_3} dt_2 \int_{-\infty}^{t_2} dt_1 \, E_2(t_3)$$

$$\big(E_1^*(t_2) E_1(t_1) R_1(t - t_3, t_3 - t_2, t_2 - t_1)$$

$$+ E_1(t_2) E_1^*(t_1) R_2(t - t_3, t_3 - t_2, t_2 - t_1)\big) \tag{6.13}$$

Here

$$R_1(t_3, t_2, t_1) = \langle \Psi_0 | e^{iH_M t_1} \hat{\mu}_{02} e^{iH_M t_2} \hat{\mu}_{20} e^{iH_M t_3} \hat{\mu}_{02} e^{-iH_M (t_1 + t_2 + t_3)} \hat{\mu}_{20} | \Psi_0 \rangle \tag{6.14a}$$

$$R_2(t_3, t_2, t_1) = \langle \Psi_0 | \hat{\mu}_{02} e^{iH_M (t_1 + t_2)} \hat{\mu}_{20} e^{iH_M t_3} \hat{\mu}_{02} e^{-iH_M (t_2 + t_3)} \hat{\mu}_{20} e^{-iH_M t_1} | \Psi_0 \rangle \tag{6.14b}$$

are nonlinear dipole response functions, which account for the molecular dynamics associated with excited-state emission processes. In a similar way, it is possible to express any contribution to the polarization $P^{(N)}(t)$ in terms of the corresponding nonlinear response function $R_i(t_N, t_{N-1}, \ldots, t_2, t_1)$. Because the nonlinear response functions carry the complete dynamical information of a given spectroscopic process, the response-function formalism allows us to decompose the computation of the polarization into the calculation of purely molecular quantities (i.e., $R_i(t_3, t_2, t_1)$) and subsequent time integrations. The characterization of nonlinear optical processes in terms of nonlinear response functions has been extensively discussed by Mukamel [46, 84, 316].

It is interesting to compare Eq. (6.13), which has been obtained by wave-function perturbation theory for an isolated system, to the corresponding expression obtained by density-matrix perturbation theory within the optical Bloch approach [77]. The density-matrix formulation differs from the wave-function formulation in that (1) one averages over an initial finite-temperature distribution ρ_0 instead of averaging over a single initial state $|\Psi_0\rangle$ (corresponding to the case of zero temperature), and (2) one obtains a damping factor accounting for additional dissipation effects

$$e^{-(t - t_3 + t_2 - t_1)/T_2} e^{-(t_3 - t_2)/T_1} = e^{-(t + t_3 + t_2 + t_1)/2T_1} e^{-(t - t_3 + t_2 - t_1)/T^*} \tag{6.15}$$

Here, as usual, T_1 denotes the excited-state population decay time, T^* is the so-called pure electronic dephasing time, and $1/T_2 = 1/2T_1 + 1/T^*$ represents the total dephasing rate of the optical transition [74]. While phenomenological electronic population decay is readily introduced into the wave-function formulation through an effective Hamiltonian *ansatz* $H_{eff} = H_M - i/2T_1$, the description of phenomenological pure electronic dephasing effects requires the density-matrix formulation.

3. Numerical Evaluation of Multitime Correlation Functions

The formulation of spectroscopic processes in terms of nonlinear response functions is useful if the response functions $R_i(t_3, t_2, t_1)$ can be evaluated analytically. This is possible, in particular, in the case of N-level systems (where N has to be small) and harmonic-oscillator (HO) problems. Separable HO models are a popular approach to describe the multidimensional vibrational dynamics of molecular systems. Furthermore, the HO propagator can be extended to describe damped motion (the "Brownian oscillator" model [328, 329]), thus including the effect of the solvent friction in a phenomenological manner. A representative example for the HO approach is the work of Pollard et al., who have successfully modeled impulsive resonance-Raman spectra of bacteriorhodopsin using a 29-mode harmonic potential-energy (PE) surface derived from a prior resonance-Raman intensity analysis [327].

Because transient Raman spectra essentially reflect the electronic ground-state dynamics of the system, they are often well described by a model representing uncoupled harmonic oscillators. In this work, however, we are concerned with the description of excited-state dynamics of polyatomic molecules. These processes are, in many cases, characterized by ultrafast electronic and vibrational relaxation processes, which are beyond a simple modeling by uncoupled oscillators. In such cases, the corresponding nonlinear response functions cannot be evaluated analytically, and are therefore not useful for the practical evaluation of the nonlinear polarization.

To illustrate the computational aspects of the problem, let us again consider the SE polarization of Eq. (6.10). Assuming non-overlapping laser fields, we may rewrite the two-time correlation function [i.e., the term in brackets in Eq. (6.10)] as

$$C_{SE}(t,t') = \langle \Psi_P | e^{iH_M t'} \hat{\mu}_{20} e^{iH_M (t-t')} \hat{\mu}_{02} e^{-iH_M t} | \Psi_P \rangle \tag{6.16}$$

where

$$|\Psi_P\rangle = i \int_{-\infty}^{\infty} dt\, E_1(t) e^{iH_M t} \hat{\mu}_{20} e^{-iH_M t} |\Psi_0\rangle \tag{6.17}$$

represents the nonstationary excited-state wave function prepared by the pump pulse. Note that the assumption of non-overlapping laser pulses reduces the number of time dependencies from three in Eq. (6.14) to two in Eq. (6.16). This represents an essential simplification of the problem if the correlation functions have to be evaluated numerically.

To calculate the time evolution of $C_{SE}(t, t')$, one needs to (1) adopt a discretization of the wave function that is suitable for the problem under consideration (see Section III.A), and (2) evaluate Eq. (6.16) by using a standard time-dependent propagation method (see Section III.B). In general, it is not possible to evaluate $C_{SE}(t, t')$ by inserting the eigenstates of H_M, as in most relevant cases the direct diagonalization of H_M is not feasible. Hence we need to stay in the *eigenstate-free* representation adopted above and compute $C_{SE}(t, t')$ by successive time propagations. This means that for each time pair $t, t'(t > t')$, one first propagates the initial state vector $|\Psi_P\rangle$ on the excited PE surfaces up to the time t to obtain the vectors $\langle\Psi_1| = \langle\Psi_P|e^{iH_M t'}$ and $|\Psi_2\rangle = e^{-iH_M t}|\Psi_P\rangle$, then propagates $|\Psi_2\rangle$ on the electronic ground-state PE surface to obtain the vector $|\Psi_3\rangle = e^{iH_M(t-t')}\hat{\mu}_{02}|\Psi_2\rangle$, and finally calculates the overlap integral $\langle\Psi_1|\hat{\mu}_{20}|\Psi_3\rangle$ to obtain the correlation function $C_{SE}(t, t')$. Although straightforward in principle, this procedure is obviously cumbersome and rather time-consuming in practice. In the general case where a numerical wave-function propagation is required in the ground as well as in the excited electronic states, it therefore appears advantageous to employ the nonperturbative evaluation scheme introduced in Section VI.B, thus avoiding the tedious computation of multi-time correlation functions.

The evaluation of multi-time correlation functions simplifies considerably, however, if we assume that only the excited electronic states $|\psi_1\rangle$ and $|\psi_2\rangle$ are coupled through the nonadiabatic coupling V_{12} in Eq. (6.1b), whereas the coupling to the energetically well-separated electronic ground state $|\psi_0\rangle$ may be neglected to a first approximation. This means that only the dynamics in the *excited* electronic states needs to be treated in terms of an eigenstate-free wave-packet propagation. The nuclear dynamics in the electronic *ground* state, on the other hand, may be described in terms of unperturbed basis functions (e.g., free-rotor or HO functions), the eigenstates $|\mathbf{v}\rangle$ and eigenvalues $\varepsilon_{\mathbf{v}}$ of which are known analytically. Inserting the sum over vibrational eigenstates of the electronic ground state $\sum_{\mathbf{v}} |\mathbf{v}\rangle\langle\mathbf{v}|$ into Eq. (6.16), we thus obtain

$$C_{SE}(t, t') = \sum_{\mathbf{v}} D_{\mathbf{v}}(t) D_{\mathbf{v}}^*(t') \tag{6.18}$$

$$D_{\mathbf{v}}(t) = e^{i\varepsilon_{\mathbf{v}} t}\langle\mathbf{v}|\langle\psi_0|\hat{\mu}_{02} e^{-iH_M t}|\Psi_P\rangle \tag{6.19}$$

Owing to the fact that the model Hamiltonian in Eq. (6.1b) is diagonal in the electronic ground state, the evaluation of the two-time correlation function $C_{SE}(t, t')$ has been reduced to a *single* propagation of the excited-state wave function. Note that the functions $D_{\mathrm{v}}(t)$ represent the matrix elements of the Heisenberg dipole operator $\hat{\mu}_{02}(t) = e^{iH_M t}\hat{\mu}_{02}e^{-iH_M t}$ taken with respect to the initially prepared wave function $|\Psi_P\rangle$ and the final state $|\psi_0\rangle|\mathbf{v}\rangle$. In the case of an ultrashort pump pulse we have $|\Psi_P\rangle = \hat{\mu}_{20}|\Psi_0\rangle$, and we may define the correlation functions

$$\Phi_{\mathrm{v}}(t) = \langle \mathbf{v}|\langle\psi_0|\hat{\mu}_{02}e^{-iH_M t}\hat{\mu}_{20}|\Psi_0\rangle \tag{6.20}$$

which exclusively reflect the intramolecular dynamics of the system and do not depend on properties of the laser fields. As is shown below, all spectroscopic signals of interest can be expressed in terms of the correlation functions $D_{\mathrm{v}}(t)$ or $\Phi_{\mathrm{v}}(t)$, thus requiring only a single numerical wave-packet propagation. As the nonperturbative computation of a typical time- and frequency-resolved PP spectrum in general requires many numerical propagations, the perturbative approach appears to be favorable in these cases.

4. *Discussion*

The interpretation of time-resolved experiments is complicated by the fact that the observed signals generally depend on both the system dynamics and the properties of the laser fields. It is therefore helpful to consider idealized limiting cases, where the measured signals, at least in principle, become independent of the experimental details and reflect solely the dynamics of the molecular system. In time-resolved spectroscopy a natural choice is the assumption of infinitely short pulses (δ-function pulses), which in practice means that the duration of the measurement has to be short compared to the system dynamics under consideration. To define the limiting case of ultrashort pulses (henceforth referred to as the *impulsive limit*)[5] more precisely, let us characterize the molecular system by two timescales, namely, the period corresponding to the electronic excitation energy Δ ($\hbar/\Delta \approx 1$ fs) and the period of a typical vibrational motion with frequency ω ($\hbar/\omega \approx 100$ fs). The impulsive limit is realized if (1) the laser frequencies are approximately resonant with the electronic excitation energy and (2) the pulse duration is long compared to $\hbar/\Delta$ but short compared to $\hbar/\omega$. Note that a laser pulse meeting these conditions coherently excites all vibrational levels of a given electronic state according to the Condon principle, which is precisely the initial condition that has been chosen in Section IV for the calculation of the time-dependent observables.

[5]The impulsive limit has been discussed, for example, in refs. [47–49, 88, 92].

In the following discussion, we consider explicit expressions for various PP signals defined in Section V.B, assuming the case of an electronic three-state system with uncoupled electronic ground state and a single transition-dipole moment (μ_{02}). We shall discuss, in particular, the interpretation of these signals in the impulsive limit. Let us first consider the integral transmittance spectrum in Eq. (5.5), which is given as sum of the Raman (I_{IRA}) and stimulated emission (I_{ISE}) contributions

$$I_{In}(\omega_2, \Delta t) = I_{IRA}(\omega_2, \Delta t) + I_{ISE}(\omega_2, \Delta t)$$
$$= 2\omega_2 \, \mathrm{Im} \int_{-\infty}^{\infty} dt \, E_2(t)(P_{RA}^*(t) + P_{SE}^*(t)) \tag{6.21}$$

Insertion of Eq. (6.11) into Eq. (6.21) yields the integral Raman signal

$$I_{IRA}(\omega_2, \Delta t) = 2\omega_2 \, \mathrm{Re} \int_{-\infty}^{\infty} dt \int_{-\infty}^{t} dt_3 \int_{-\infty}^{t_3} dt_2 \int_{-\infty}^{t_2} dt_1 \, E_2(t) E_2^*(t_3)$$
$$\times \sum_{\mathbf{v}} [E_1^*(t_2) E_1(t_1) e^{i\varepsilon_0(t_3 - t_1)} e^{i\varepsilon_{\mathbf{v}}(t - t_2)} \Phi_{\mathbf{v}}^*(t - t_3)$$
$$\cdot \Phi_{\mathbf{v}}(t_2 - t_1) + E_1(t_2) E_1^*(t_1) e^{-i\varepsilon_0(t - t_1)} e^{i\varepsilon_{\mathbf{v}}(t_3 - t_2)}$$
$$\cdot \Phi_{\mathbf{v}}^*(t - t_3) \Phi_{\mathbf{v}}^*(t_2 - t_1)] \tag{6.22}$$

Assuming non-overlapping laser pulses with Gaussian envelope functions

$$\mathcal{E}_i(t) = N_i \exp[-\alpha_i (t - t_i)^2] \tag{6.23}$$

with $N_i = (4 \ln 2/\pi)^{1/2} E_i/\tau_i, \alpha_i = 4 \ln 2/\tau_i^2$, and $t_1 = 0$, $t_2 = \Delta t$, Eq. (6.22) can be rewritten as

$$I_{IRA}(\omega_2, \Delta t) = 2\omega_2 \, \mathrm{Re} \sum_{\mathbf{v}} \left(e^{i(\varepsilon_{\mathbf{v}} + \varepsilon_0)\Delta t} S_{\mathbf{v}}^*(\omega_2) S_{\mathbf{v}}(\omega_1) \right.$$
$$\left. + \, e^{i(\varepsilon_{\mathbf{v}} - \varepsilon_0)\Delta t} Q_{\mathbf{v}}(\omega_2) T_{\mathbf{v}}(\omega_1) \right) \tag{6.24}$$

where

$$S_{\mathbf{v}}(\omega_i) = \int_{-\infty}^{\infty} dt \int_{-\infty}^{t} dt' \, \mathcal{E}^*(t) \mathcal{E}(t') e^{i\omega_i(t - t')} e^{-i\varepsilon_{\mathbf{v}} t} e^{-i\varepsilon_0 t'} \Phi_{\mathbf{v}}(t - t') \tag{6.25a}$$

$$Q_{\rm v}(\omega_i) = \int_{-\infty}^{\infty} dt \int_{-\infty}^{t} dt' \, \mathcal{E}(t)\mathcal{E}^*(t')e^{-i\omega_i(t-t')}e^{-i\varepsilon_0 t}e^{i\varepsilon_{\rm v} t'}\Phi_{\rm v}^*(t-t') \tag{6.25b}$$

$$T_{\rm v}(\omega_i) = \int_{-\infty}^{\infty} dt \int_{-\infty}^{t} dt' \, \mathcal{E}(t)\mathcal{E}^*(t')e^{-i\omega_i(t-t')}e^{-i\varepsilon_{\rm v} t}e^{i\varepsilon_0 t'}\Phi_{\rm v}^*(t-t') \tag{6.25c}$$

and $\mathcal{E}(t)$ denotes a Gaussian envelope function centered at $t = 0$. Note that, despite its lengthy appearance, the numerical evaluation of Eq. (6.24) is fairly easy. This is because only a single and rather short (of the order of the pulse duration τ) numerical excited-state propagation is required to obtain the functions $S_{\rm v}, Q_{\rm v}$, and $T_{\rm v}$. On the other hand, it is clear that the time-resolved Raman signal cannot yield much information on the excited-state processes occurring in the molecule. The time dependence of $I_{IRA}(\omega_2, \Delta t)$ mostly reflects ground-state dynamics, which by construction of the model in Eq. (6.1b) is rather simple. In the limiting case of impulsive excitation we obtain $S_{\rm v}, Q_{\rm v}, T_{\rm v} \propto \delta_{{\rm v},0}$, i.e., the Raman signal becomes stationary for ideally short laser pulses.

Let us next consider the SE contribution to the transient transmittance spectrum in Eq. (6.21). Inserting Eq. (6.10) into Eq. (6.21), the integral SE PP signal can be written in the suggestive form

$$I_{ISE}(\omega_2, \Delta t) = \omega_2 \sum_{\rm v} \left| \int_{-\infty}^{\infty} dt \, \mathcal{E}^*(t)e^{i\omega_2 t} D_{\rm v}(t + \Delta t) \right|^2 \tag{6.26}$$

Equation (6.26) illustrates the fact that the SE PP spectrum essentially monitors the time evolution and the frequency distribution of the Heisenberg dipole operator $D_{\rm v}(t) = \langle {\rm v}|\langle\psi_0|e^{iH_M t}\hat{\mu}_{02}e^{-iH_M t}|\Psi_P\rangle$. The probe pulse duration defines a time window (centered at Δt) and a uncertainty-related frequency window (centered at ω_2), which, by virtue of the time integration, determine the time and frequency resolution of the integral PP spectrum [92]. In the impulsive limit we obtain

$$I_{ISE}^{\delta}(\omega_2, \Delta t) = \omega_2 \langle\Psi_P|e^{iH_M t}\hat{\mu}_{20}\hat{\mu}_{02}e^{-iH_M t}|\Psi_P\rangle \propto P_2^{\rm di}(\Delta t) \tag{6.27}$$

i.e., the impulsive integral PP signal is proportional to the diabatic electronic population probability $P_2^{\rm di}(t)$ defined in Eq. (4.7). Equations (6.26) and (6.27) are the most important results of this section. Note that both expressions resemble the schematic description of a PP experiment given in Section V.A. The nonstationary state to be monitored by the experiment is given by the excited-state wave function $e^{-iH_M t}|\Psi_P\rangle$, and the expectation value determined by the measurement is obtained by projecting this state vector via the dipole-transition operator $\hat{\mu}_{02}$ onto the electronic ground state. As anticipated above, the impulsive integral PP signal measures the electronic population remaining in the initially excited electronic state $|\psi_2\rangle$, thus directly monitoring the non-Born–Oppenheimer dynamics of the system [88–93].

It is interesting to compare the integral SE spectrum Eq. (6.26) to the time-resolved fluorescence spectrum defined in Eq. (5.12). As pointed out above, both signals are Fourier-limited with respect to their time and frequency resolution, and both reflect the excited-state dynamics of the system. In the far field limit, the correlation function of the spontaneous field is given by the dipole correlation function [305], i.e.,

$$\langle E_S(t) E_S^{\dagger}(t')\rangle = \sum_{\mathrm{v}} D_{\mathrm{v}}(t) D_{\mathrm{v}}^{*}(t') \tag{6.28}$$

We thus obtain for the time-resolved fluorescence signal

$$I_F(\omega_f, \Delta t) = \omega_f^3 \int_{-\infty}^{\infty} dt \sum_{\mathrm{v}} \left| \int_{-\infty}^{t} dt' \mathcal{E}^{*}(t') e^{i\omega_f t'} D_{\mathrm{v}}(t' + \Delta t) \right|^2 \tag{6.29}$$

where the gating function of Eq. (5.13) of the upconversion process is represented by the Gaussian function $\mathcal{E}(t)$, and the instrumental width γ of the spectrometer has been neglected. Comparison of Eqs. (6.26) and (6.29) reveals that the expressions for the integral SE spectrum and the time-resolved fluorescence spectrum are indeed quite similar, and are thus expected to result in qualitatively similar signals.

Although the above formulae for the PP signals represent compact expressions that are well suited for numerical evaluation, they still have to be evaluated for each individual frequency ω_2 or ω_f. Being interested in the transient emission *spectrum*, one may circumvent this cumbersome procedure by first calculating the polarization $P^{\delta}(t, \Delta t)$ for an impulsive probe pulse. The polarization $P(t, \Delta t)$, induced by a finite probe pulse $E_2(t)$ centered at the delay time Δt, can then be simply obtained by [92]

$$P(t, \Delta t) = \int_{-\infty}^{t} dt'\, E_2(t') P^{\delta}(t', \Delta t) \tag{6.30}$$

It is physically more suggestive, however, if one calculates the *spectrum* $P^{\delta}(\omega, \Delta t)$ of the impulsive polarization, that is, the Fourier transform of $P^{\delta}(t, \Delta t)$ with respect to t. Inserting Eq. (6.30) into Eqs. (5.8) and (5.6), the dispersed transmittance spectrum for finite probe pulses can be written as [49, 327]

$$I_D(\omega, \Delta t) = 2\omega \text{ Im } E_2(\omega) \int_{-\infty}^{\infty} dt\, E_2(t) P^{\delta}(\omega, t) \tag{6.31}$$

Considering the SE contribution, the impulsive dispersed transmittance spectrum reads

$$\begin{aligned} I^{\delta}_{DSE}(\omega, \Delta t) &= 2\omega \text{ Im } e^{i\omega t} P^{\delta}(\omega, \Delta t) \\ &= 2\omega \text{ Re} \sum_{\mathbf{v}} D^{*}_{\mathbf{v}}(\Delta t) \int_{-\infty}^{\infty} dt\, e^{i\omega t} D_{\mathbf{v}}(t + \Delta t) \end{aligned} \tag{6.32}$$

Given at the time $t = 0$ a nonstationary state $|\Psi_P\rangle$, the impulsive spectrum of Eq. (6.32) represents the time- and frequency resolved SE response of the system, which is independent of properties of the laser fields. Integrating Eq. (6.32) over all emission frequencies ω, one again obtains the electronic population probability, Eq. (6.27), of the initially excited electronic state. In complete analogy to Eq. (6.32), one may also evaluate the Raman contribution to the transmittance signal.

The combination of Eqs. (6.32) and (6.31) thus suggests a convenient perturbative evaluation scheme for transient spectra: One first calculates the impulsive transmittance spectrum $\propto P^{\delta}(\omega, \Delta t)$, and subsequently obtains, via Eq. (6.31), the desired PP signals for probe pulses of arbitrary frequency and duration. While there are several ways to perform the necessary time integrations in a intelligent way, it should be emphasized that for multidimensional systems the numerical excited-state propagation represents the most time-consuming part of the calculation. The particular choice of a time- and frequency-resolved spectrum (e.g., $I_{In}(\omega, \Delta t), I_F(\omega, \Delta t)$, or $I_D(\omega, \Delta t)$) thus mostly depends on interpretational preferences (e.g., for nonadiabatic processes $I^{\delta}_{In}(\omega, \Delta t)$ represents a key quantity) and, of course, the specific experiment that one wishes to simulate.

B. Nonperturbative Approach

Besides the commonly applied perturbative evaluation of spectroscopic signals, nonperturbative descriptions have also been considered recently [50, 91, 94, 250, 310, 330–332]. Assuming, as usual, an optically thin sample, the only formal difference between the two formulations is whether the electric polarization $\mathbf{P}(t)$ is evaluated in a perturbative or nonperturbative manner. As a consequence, the definitions of time- and frequency-resolved PP signals are completely equivalent in both formulations. Besides the obvious point that one needs to propagate the wave function with a time-dependent Hamiltonian instead of a time-independent one, the calculation of PP signals appears to be rather similar in both formulations.

The nonperturbative calculation of PP spectra, however, involves an additional, albeit only technical, problem. As explained above, the overall polarization consists of several contributions

$$\mathbf{P}(t) = 2 \operatorname{Re} \sum_{n,m} P_{nm}(t) \exp[i(n\mathbf{k}_1 + m\mathbf{k}_2)\mathbf{x}] \tag{6.33}$$

which, experimentally as well as in a perturbative calculation, are easily distinguished via the direction of the wave vector, $\mathbf{k} = n\mathbf{k}_1 + m\mathbf{k}_2$, of the emitted radiation. For example, the coherent emission due to the photon-echo effect is observed in the directions $2\mathbf{k}_2 - \mathbf{k}_1$ and $2\mathbf{k}_1 - \mathbf{k}_2$, while the transmission of the pump and probe pulse due to stimulated emission and resonance Raman processes is observed in the $\mathbf{k}_1$ and $\mathbf{k}_2$ directions, respectively (cf. discussion of Fig. 18 on p. 77).

In a nonperturbative calculation, however, we obtain, by evaluating Eq. (6.2), the *overall* polarization of the system, which is given as sum of all these processes. Furthermore, for the sake of simplicity one usually assumes a *coherent* electric field in the theoretical description. The pump and probe pulses are thus phase-locked, giving rise to a number of contributions to the polarization [61, 64, 316], which are not observed in standard PP experiments, where the signal is incoherently averaged over many laser shots. Thus, although we know that in the weak-field limit the overall polarization reduces to the results from lowest-order perturbation theory, we still have to distinguish the different contributions to $\mathbf{P}(t)$ in order to calculate a specific experimental PP signal. In most nonperturbative PP calculations this problem has been circumvented by (1) considering fragmentation cross sections (ions or dissociation fragments) instead of photon detection [21, 91, 250], (2) considering the time evolution of the system (following pulsed excitation) instead of PP signals [310, 331], or (3) assuming a strongly repulsive excited PE surface, and restricting oneself to the time-dependent dynamics in the electronic ground state [332].

Following ref. [50], we present a general approach to extract the individual contributions and spectroscopic signals from the overall polarization, and discuss in detail the simplifications that arise when the usual assumptions (i.e., non-overlapping and weak laser fields, slowly varying envelope assumption, and RWA) are invoked.

1. Directional Dependence of the Nonlinear Response

The basic idea of how to determine the directional dependence of the nonlinear polarization in a nonperturbative calculation is rather simple and is most easily demonstrated by an example. Let us first consider non-overlapping laser fields within the RWA, and let us suppose that we want only to discriminate the (phase-insensitive) $\mathbf{k}_2$ emission from the (phase-sensitive) $\mathbf{k}_1$ and $2\mathbf{k}_2 - \mathbf{k}_1$ contributions. Rewriting the overall polarization as

$$P(\phi) = \sum_n P_n \exp(in\phi) \tag{6.34}$$

where $n\phi = n\mathbf{k}_1\mathbf{x}$ denotes the phase the polarization acquires through the interaction with the pump field, it is clear that the terms P_{-1}, P_0, and P_1 represent the polarization in the $2\mathbf{k}_2 - \mathbf{k}_1, \mathbf{k}_2$, and $\mathbf{k}_1$ directions, respectively. Performing two separate calculations of the overall polarization $\mathbf{P}(\phi)$ employing the phases $\phi_1 = 0$, $\phi_2 = \pi$, it is easy to see that the polarizations in $\mathbf{k}_2$ and $\mathbf{k}_1, 2\mathbf{k}_2 - \mathbf{k}_1$ are given by the linear combinations

$$\mathbf{P}_{\mathbf{k}_2} = \mathrm{Re}(P(0) + P(\pi)) \tag{6.35a}$$

$$\mathbf{P}_{\mathbf{k}_1, 2\mathbf{k}_2 - \mathbf{k}_1} = \mathrm{Re}[P(0) - P(\pi)] \tag{6.35b}$$

As a second example, allow for overlapping laser fields within the RWA and ask for the polarizations up to third order into all possible directions, i.e., into $\mathbf{k}_2, \mathbf{k}_1, 2\mathbf{k}_2 - \mathbf{k}_1$, and $2\mathbf{k}_1 - \mathbf{k}_2$. The calculation of the overall polarization $P(\phi)$ for the phases $\phi_k = k\pi/2$ ($k = 0, 1, 2, 3$) yields a linear system of four equations for the P_n in Eq. (6.34), the solution of which yields the desired polarizations

$$\mathbf{P}_{\mathbf{k}_2} = \tfrac{1}{2}\mathrm{Re}[P(0) + P(\pi/2) + P(\pi) + P(3\pi/2)] \tag{6.36a}$$

$$\mathbf{P}_{\mathbf{k}_1} = \tfrac{1}{2}\mathrm{Re}[P(0) - P(\pi)] + \tfrac{1}{2}\mathrm{Im}[P(\pi/2) - P(3\pi/2)] \tag{6.36b}$$

$$\mathbf{P}_{2\mathbf{k}_2 - \mathbf{k}_1} = \tfrac{1}{2}\mathrm{Re}[P(0) - P(\pi)] + \tfrac{1}{2}\mathrm{Im}[P(3\pi/2) - P(\pi/2)] \tag{6.36c}$$

$$\mathbf{P}_{2\mathbf{k}_1 - \mathbf{k}_2} = \tfrac{1}{2}\mathrm{Re}[P(0) - P(\pi/2) + P(\pi) - P(3\pi/2)] \tag{6.36d}$$

From the above examples the generalization of the procedure to cases including arbitrary nonlinear processes of any order is straightforward. To calculate the contributions of the polarization in N directions, one has to perform N calculations of the overall polarization $\boldsymbol{P}(\phi)$ with different phases ϕ, and solve the resulting linear system of equations for the $\boldsymbol{P}_n$ in Eq. (6.34). Beyond the RWA, however, one needs to generalize the *ansatz* in Eq. (6.34) and "tag" both the interactions with the pump and the probe fields with phases $\phi^{(1)} = \mathbf{k}_1\mathbf{x}$ and $\phi^{(2)} = \mathbf{k}_2\mathbf{x}$, respectively, thus performing a two-dimensional Fourier analysis of the overall polarization.

As in the experimental case, we thus obtain the nonlinear response of the molecular system resolved in the directions of emission, but summed up over all contributions in each direction. Assuming non-overlapping laser fields within the RWA, it is moreover possible to separate the PP signals arising from the stimulated Raman and the stimulated emission contribution, respectively [50]. This important feature, which is rather helpful in the interpretation of complex PP spectra (see below), stems from the fact that after the interaction with the pump pulse we may separately consider the electronic ground-state component $\langle\psi_0|\Psi(t)\rangle$ or excited-state component $\langle\psi_1|\Psi(t)\rangle$ of the total wave function.

2. *Example: The One-Dimensional Harmonic Oscillator*

It is instructive to demonstrate the computational procedure outlined above for the case of a simple and well-understood model problem. Let us consider a one-dimensional harmonic oscillator that is characterized by its vibrational period $T = 2\pi/\omega = 65$ fs and the dimensionless coordinate shift $\kappa/\omega = 2$, describing the displacement of the excited-state geometry with respect to the electronic ground-state geometry. We furthermore assume Gaussian laser pulses of duration $\tau_1 = \tau_2 = 10$ fs that are resonant with the optical transition (i.e., $\omega_1 = \omega_2 = E_1 - E_0 = 3$ eV), and choose the amplitudes of the field–matter interactions as $|\boldsymbol{\mu}\cdot\boldsymbol{E}_1| = 0.04$ eV and $|\boldsymbol{\mu}\cdot\boldsymbol{E}_2| = 0.01$ eV, respectively. This results in an excited-state population of 7% after the preparation of the system by the pump pulse.

For this model, Fig. 19 shows the modulus of various contributions to the electronic polarization, plotted as a function of the delay time Δt and of the emission time after the probe pulse, $\tau = t - \Delta t$. The overall polarization shown in Fig. 19(a) is clearly dominated by the linear response of the system. The linear response to the probe field $\mathbf{E}_2$ manifests itself as wave trains traveling parallel to the Δt axis. The corresponding emission is peaked at times $\tau = nT = 0$, 65 fs, 130 fs, ... (clearly seen for $\Delta t < 0$, where the $\mathbf{E}_2$ emission does not yet interfere with the $\mathbf{E}_1$ emission). Due to the $\mathbf{P}(\Delta t, t - \Delta t)$ representation chosen, the linear response to the pump field $\mathbf{E}_1$ results in wave trains traveling diagonally from right to left. In order to distinguish linear and nonlinear response, we calculate the polarizations for the cases in which only the pump pulse or the probe pulse interacts with the system,

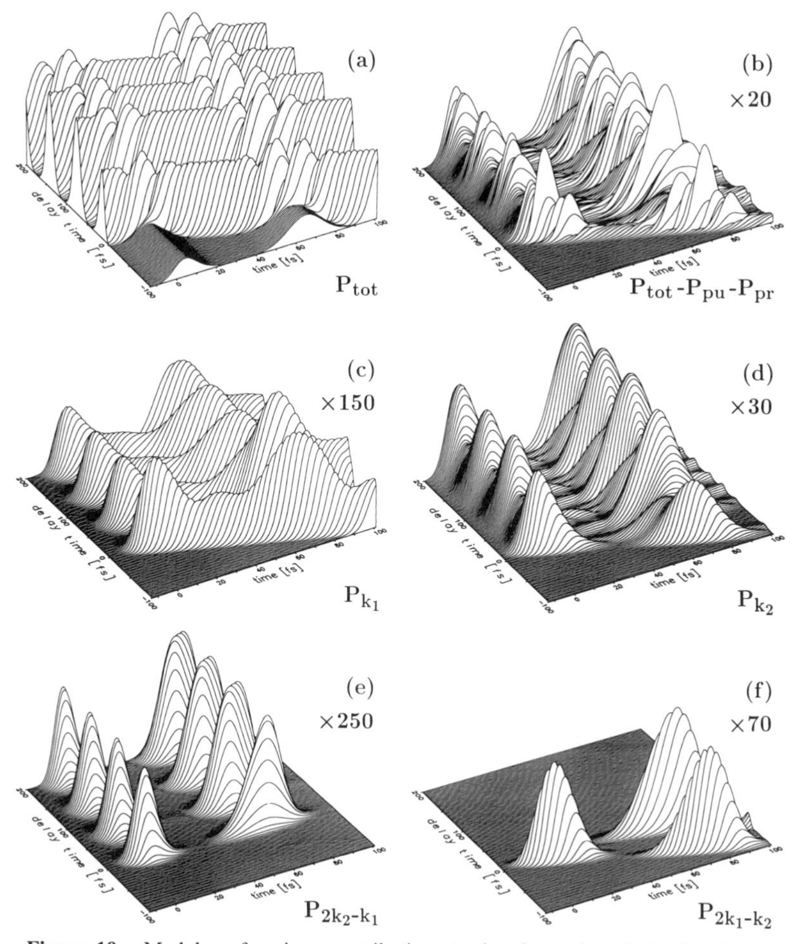

Figure 19. Modulus of various contributions to the electronic polarization for a one-dimensional shifted harmonic oscillator, plotted as a function of the delay time Δt and the emission time $\tau = t - \Delta t$. (*a*) The *overall* polarization $\mathbf{P}_{\text{tot}}$ as obtained in a nonperturbative calculation. (*b*) The *total nonlinear* polarization, which is defined as $\mathbf{P}_{NL} = \mathbf{P}_{\text{tot}} - \mathbf{P}_{\text{pump only}} - \mathbf{P}_{\text{probe only}}$. Assuming weak laser fields within the rotating-wave approximation, the polarization $\mathbf{P}_{NL}$ radiates exclusively in four directions. (*c*)–(*f*) The corresponding direction-resolved polarizations radiating in directions $\mathbf{k}_1$ (*c*), $\mathbf{k}_2$ (*d*) $2\mathbf{k}_2 - \mathbf{k}_1$ (*e*) and $2\mathbf{k}_1 - \mathbf{k}_2$ (*f*).

respectively, and subtract these polarizations from the overall polarization. This yields the total nonlinear polarization $\mathbf{P}_{\text{NL}} = \mathbf{P}_{\text{tot}} - \mathbf{P}_{\text{pump only}} - \mathbf{P}_{\text{probe only}}$ as shown in Fig. 19(*b*). The total nonlinear polarization $\mathbf{P}_{\text{NL}}$ is about a factor of 20 weaker than the overall polarization $\mathbf{P}_{\text{tot}}$ shown in Fig. 19(*a*), where it is only noticeable as little humps on top of the wave trains. The emission is mainly peaked at delay times and emission times $\Delta t, \tau = nT$, but also exhibits complex additional structures, arising from the superposition of the individual contributions to the polarization $\mathbf{P}_{NL}$.

Employing the computational procedure defined by Eq. (6.36), we finally obtain the direction-resolved polarizations, radiating in directions $\mathbf{k}_1$ (*c*), $\mathbf{k}_2$ (*d*), $2\mathbf{k}_2 - \mathbf{k}_1$ (*e*), and $2\mathbf{k}_1 - \mathbf{k}_2$ (*f*), respectively. It should be emphasized that for the field intensities considered here, the RWA, as well as the assumption of only four directions of the emission [cf. Eq. (6.6)], represent excellent approximations. As the probe field is taken to be a factor of four weaker than the pump field, the contributions proportional to $\mathbf{E}_1\mathbf{E}_2^2$ (i.e., $\mathbf{P}_{\mathbf{k}_1}, \mathbf{P}_{2\mathbf{k}_2-\mathbf{k}_1}$) are consequently smaller (also roughly by a factor of four) than the contributions proportional to $\mathbf{E}_2\mathbf{E}_1^2$ (i.e., $\mathbf{P}_{\mathbf{k}_2}, \mathbf{P}_{2\mathbf{k}_1-\mathbf{k}_2}$). Note, furthermore, that the first peaks (i.e., for $\tau \approx 0$) are somewhat smaller than the following ones (for $\tau \approx nT$), owing to the fact that the polarization still needs to be built up during the interaction with the pump field.

The polarizations $\mathbf{P}_{2\mathbf{k}_2-\mathbf{k}_1}$ and $\mathbf{P}_{2\mathbf{k}_1-\mathbf{k}_2}$ represent the well-known two-pulse photon-echo emission, emerging for $\Delta t \gtrsim 0$ in direction $2\mathbf{k}_2 - \mathbf{k}_1$ and for $\Delta t \lesssim 0$ in direction $2\mathbf{k}_1 - \mathbf{k}_2$ [48, 84]. The name photon echo refers to the recurrences (echos) exhibited at times $\tau = \Delta t = nT$, reflecting the coherent rephasing of the emission. In the case of an undamped harmonic oscillator (i.e., in the absence of any dephasing processes), the system dynamics is harmonic, and one therefore obtains rephasings whenever τ *or* Δt equals multiples of the vibrational period T. The polarizations radiating in $\mathbf{k}_1$ and $\mathbf{k}_2$ are seen to exhibit more complicated structures, which arise from the fact that several spectroscopic processes contribute in these directions. Recalling the discussion of Fig. 18 on p. 77, we obtain coherent emission in $\mathbf{k}_1$ direction, while the (phase-insensitive) emission in $\mathbf{k}_2$ direction consists of the stimulated Raman and stimulated emission contributions. The complex structures arising when the probe pulse is coincident with or ahead of the pump pulse correspond to the so-called coherent artifact, which has been discussed by several authors [48, 80, 333].

As pointed out above, the $\mathbf{k}_2$ emission represents the only contribution to the overall polarization that is measured in a typical PP experiment with phase-averaged detection. To give an example of the connection between polarization and measured PP signal, Fig. 20(*a*) shows the integral PP spectrum $I_{In}(\omega_2, \Delta t)$ obtained by nonperturbative evaluation of $\mathbf{P}_{\mathbf{k}_2}$ (for many probe carrier frequenices ω_2) and subsequent time integration over the probe field $\mathbf{E}_2$ [cf. Eq. (5.4)]. As has been discussed by several authors [49, 48, 55, 80, 90], the transient transmission spectrum monitors the time evolution of the nonstationary wave function prepared by the pump pulse, which manifests itself through nuclear wave-packet motion in the electronic ground state (Raman contribution) and in the excited electronic state (stimulated-emission contribution). As noted above, in the case of non-overlapping laser fields

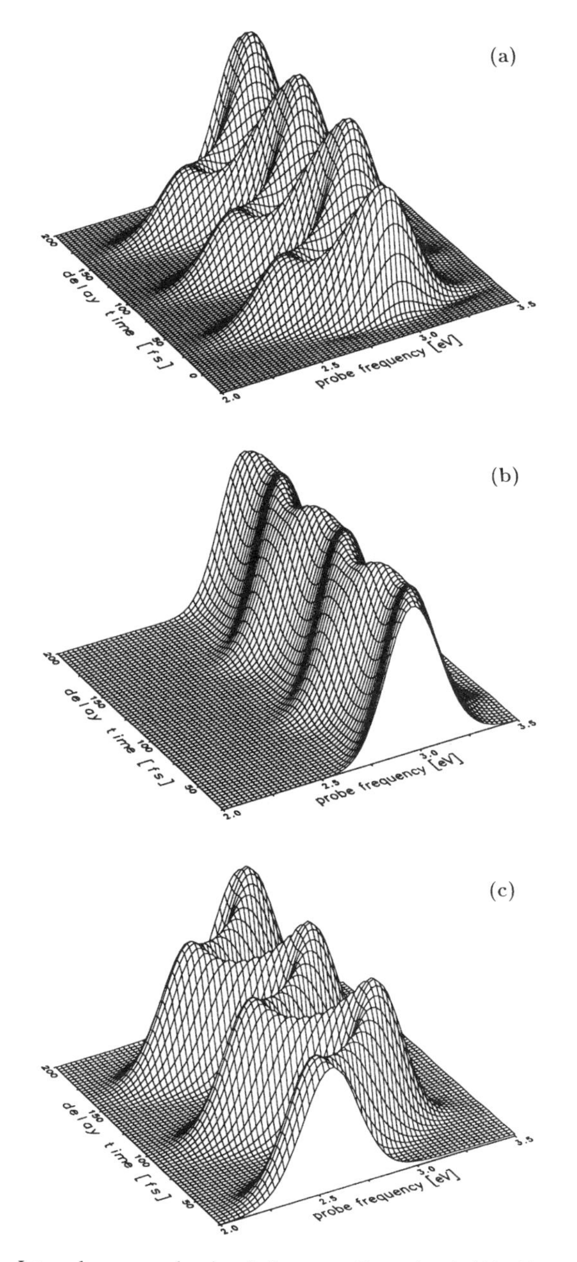

Figure 20. Integral pump-probe signals for a one-dimensional shifted harmonic oscillator as a function of the delay time Δt and the probe carrier frequency ω_2, obtained for Gaussian laser pulses of 10-fs duration. The total signal (*a*) consists of the stimulated Raman contribution (*b*) and the stimulated-emission contribution (*c*).

it is possible to distinguish between these two contributions, thus yielding separately the Raman signal, shown in Fig. 20(*b*), and the stimulated-emission signal, shown in Fig. 20(*c*). It is seen that the wave-packet motion in the exited electronic state is considerably more pronounced than the wave-packet motion in the ground state. This is due to the fact that the maximal displacement of the ground-state wave-packet is determined by the interaction time with the pump pulse (i.e., for a ideally short pump pulse the Raman contribution becomes stationary), while the displacement of the excited-state wave-packet is mainly determined by the coordinate shift of the exited-state geometry, resulting in a maximal red shift of the emission at the classical turning points $\Delta t = (n + \frac{1}{2})T$.

Figures 19 and 20 illustrate the nonperturbative approach to calculating PP signals that are simultaneously resolved in frequency, time, and direction of emission. The procedure is straightforward and clearly analogous to experiment, but involves several steps. If one is interested in the spectroscopic response of a harmonic oscillator, one would certainly prefer the standard perturbative approach, because the corresponding nonlinear response functions can be evaluated analytically. Considering cases where the description of the molecular system requires a numerical wave-packet propagation, on the other hand, the exact treatment of the field-matter interaction is often more straightforward than the cumbersome numerical evaluation of multi-time system response functions and allows, furthermore, the treatment of strong-field effects.

VII. FEMTOSECOND SPECTROSCOPY OF ULTRAFAST NONADIABATIC EXCITED-STATE PROCESSES

In this section we present detailed numerical investigations of femtosecond time-resolved spectra, which serve as an illustration of the formal theory developed in Sections V and VI. The applications are based on the model Hamiltonians as defined in Sections II.C.2 and II.C.3, as well as on the time-dependent wave-packet methods as described in Section III. We first address the real-time detection of photophysical dynamics, namely the ultrafast $S_2 \rightarrow S_1$ internal-conversion process associated with the S_1–S_2 conical intersection in pyrazine. This system represents the most extensively and carefully investigated example of dynamical effects associated with a conical intersection in a polyatomic molecule. Subsequently, the real-time detection of ultrafast photochemical dynamics will be analyzed for an idealized model of a nonadiabatic isomerization reaction.

A. Real-Time Detection of Internal-Conversion Processes

Experimental femtosecond time-resolved optical spectra of pyrazine and similar systems are not available at present. The existing information on the spectroscopy of the valence excited states of pyrazine is based on condensed-phase or gas-phase CW absorption spectra, liquid-phase or gas-phase resonance Raman (RR) spectra, or laser-induced fluorescence and resonant multiphoton ionization spectra in supersonic jets, see ref. [334] for a review. Before proceeding with simulations of the femtosecond spectroscopy of the ultrafast $S_2 \rightarrow S_1$ internal-conversion process in pyrazine, it is appropriate to briefly compare the CW spectroscopy of the *ab initio* model [45] with the available experimental spectra.

Let us first consider the S_2 absorption spectrum of pyrazine. In the time-dependent formalism [179, 325, 326], the absorption cross section is given by

$$\sigma_A(\omega) = \frac{2}{3}\,\pi\,\frac{\alpha}{e^2}\,2\omega\,\mathrm{Re}\int_0^\infty dt\; e^{i(\omega+\varepsilon_0)t}e^{-t/T_2}\Phi_0(t) \tag{7.1}$$

where α is the fine structure constant and ε_0 denotes the energy of the initially populated vibrational ground state. The autocorrelation function $\Phi_0(t)$ defined in Eq. (6.20) represents the time-dependent overlap of the excited-state wave function with its initial value. T_2 is a phenomenological optical dephasing constant that accounts for the effects of the weakly coupled modes of pyrazine, which are not explicitly included in the model. The calculations reported below have been performed for an extended *ab initio* based model, which includes, in addition to $\nu_1, \nu_{6a}, \nu_{9a}, \nu_{10a}$, the modes ν_4, ν_{8a}, and ν_{14}. For a description of the model and computational details, see ref. [182].

In Fig. 21 the room temperature gas-phase absorption profile of the S_2 band of pyrazine [335], Fig. 21(*a*), is compared with the calculated CW absorption spectrum [182], Fig. 21(*b*). The stick spectrum in Fig. 21(*b*) represents the vibronic energy levels and absorption intensities of the seven-mode *ab initio* model [182], and has been obtained by diagonalizing the vibronic Hamiltonian matrix using the Lanczos algorithm [196]. The very high density of vibronic levels reflects the dissolution of the S_2 Born–Oppenheimer levels into the dense manifold of levels of the lower-lying S_1 surface. The envelope in Fig. 21(*b*), which agrees nearly quantitatively with the experimental spectrum, has been obtained by including a phenomenological optical dephasing ($T_2 = 35$ fs). It should be noted that the irregularly spaced and diffuse structures of the absorption profile cannot simply be assigned in terms of harmonic vibrations in the S_2 state. These structures reflect the short-time dynamics (on a timescale of tens of femtoseconds) on the upper adiabatic potential-energy (PE) sheet of the S_1–S_2 conical intersection in pyrazine (see [36] for more details).

The RR cross section for orientationally averaged scatterers, integrated over the scattered bandwidth and all scattered directions, can be written as

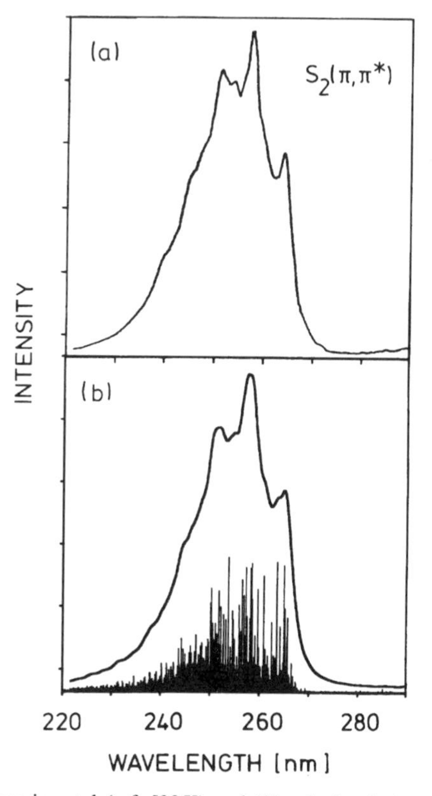

Figure 21. (*a*) Experimental (ref. [335]) and (*b*) calculated absorption spectrum of the $S_2(\pi\pi^*)$ state of pyrazine. The stick spectrum in (*b*) represents the vibronic energy levels and absorption intensities of the seven-mode model; the envelope is obtained by including a phenomenological electronic dephasing (T_2 = 35 fs).

[336, 337]

$$\sigma_{v0}(\omega_l) = \frac{8}{9}\,\pi\left(\frac{\alpha}{e^2c}\right)^2 \omega_s^3\omega_l \left|\int_0^\infty dt\; e^{i(\omega_l+\varepsilon_0)t}e^{-t/T_2}\Phi_v(t)\right|^2 \tag{7.2}$$

where ω_l and ω_s denote the frequencies of the laser and the spontaneous emission, repectively. The time-dependent correlation functions $\Phi_v(t)$ defined in Eq. (6.20) represent the time-dependent overlap of the excited-state wave function with the final vibrational state, and thus reflect the intramolecular excited-state dynamics of the system in these particular vibrational modes. Note that in Eq. (7.2) we have assumed that only the S_2 state of pyrazine is optically accessible, i.e., we have neglected the weak oscillator strength of the S_1 state. In the case that both excited electronic states carry oscillator strength, interference effects of the radiation emitted from the two electronic states have to be taken into account (see ref. [309] for a detailed discussion).

Recently, gas-phase RR spectra of pyrazine have been measured by Swinney and Hudson for a number of excitation wavelengths within the S_2 absorption band [182]. These spectra provide a crucial test of the accuracy of the theoretical model, at least at short time scales (because the information inherent in RR spectra is limited by the optical dephasing time). Figure 22 is a representative example of the comparison of the *ab initio* predicted spectra [182] with the experiment, showing the experimental [Fig. 22(*a*)] and theoretical [Fig. 22(*b*)] RR spectra of pyrazine for an excitation wavelength of 266 nm, near the origin of the S_2 absorption band. For better comparison

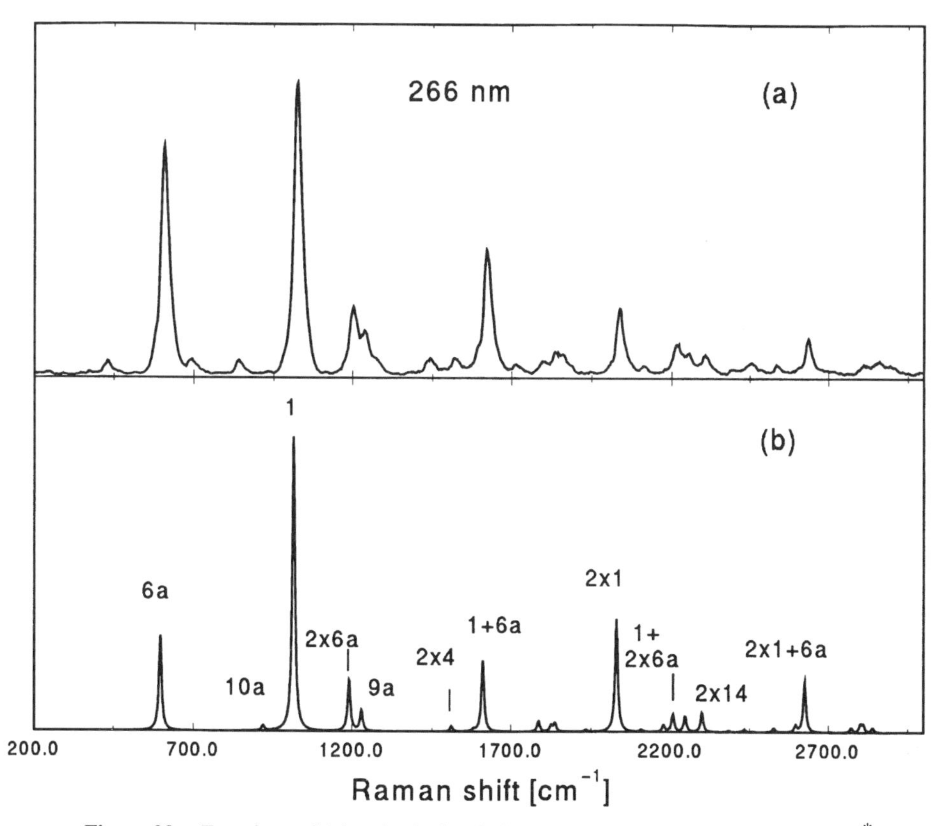

Figure 22. Experimental (*a*) and calculated (*b*) resonance Raman spectra of the $S_2(\pi\pi^*)$ state of pyrazine obtained for the excitation wavelength $\lambda = 266$ nm.

with experimental data, the computed intensities have been plotted at the energetic positions determined from measured ground-state fundamentals. A linewidth of 10 cm^{-1} full-width half-maximum (FWHM) has been assumed in the theoretical spectra. It is seen that the theoretical spectrum reproduces quantitatively all major and most minor peaks of the experiment. The overall agreement between theory and experiment is excellent. The agreement obtained for other excitation wavelengths is similarly good, although it deteriorates for excitations at the blue end of the absorption band [182]. In this case it is to be expected that higher excited electronic states have to be considered in the calculation, i.e., the electronic three-state model may break down. It is interesting to note that the nontotally symmetric coupling mode ν_{10a} is hardly observed under resonant excitation of the S_2 state of pyrazine, although it appears prominently in the S_1 RR spectrum [337].

It may be concluded from this brief comparison of CW absorption and RR spectra with experiment that the S_1–S_2 vibronic–coupling model derived from CASSCF/MRCI electronic-structure calculations [45] provides an accurate description of the short-time dynamics on the conically intersecting S_1 and S_2 PE surfaces of pyrazine. The ultrafast nonadiabatic dynamics occurring after optical excitation of the S_2 state is implicit in the complete dissolution of the S_2 level structure in the dense manifold of the vibrational levels of the S_1 state. The comparison of the *ab initio* model with experiment confirms the adequacy of describing the intramolecular short-time dynamics in terms of relatively few active degrees of freedom. The effect of the many inactive vibrations appears to be well approximated by a single (that is, energy-independent) optical dephasing rate $1/T_2$ within the energy range of the S_2 absorption band. It should be noted that such a phenomenological modeling of the optical dephasing processes is primarily necessary for the computation of CW spectra. When real-time spectroscopy with ultrashort pulses is considered, frequency-integrated pump-probe (PP) spectra are not affected by optical dephasing processes (cf. the discussion in Section V). Such spectra thus can be evaluated in a completely microscopic manner from an *ab initio* model Hamiltonian without the need of invoking phenomenological relaxation terms.

As a first example of *real-time* spectroscopy of ultrafast S_1–S_2 dynamics in pyrazine, Fig. 23 shows the integral transient transmittance spectrum [Eq. (5.4)] of the three-mode model of pyrazine (see Section IV), plotted as a function of the delay time Δt and the central frequency ω_2 of the probe pulse. We have assumed Gaussian laser pulses of low intensity and 20 fs duration.[6]

[6]Throughout this article, we mean by the duration τ of a laser pulse the FWHM width of the electric *field* as defined in Eq. (6.23). The FWHM width τ' of the corresponding *intensity* distribution is for a Gaussian pulse envelope given by $\tau' = \tau/\sqrt{2}$.

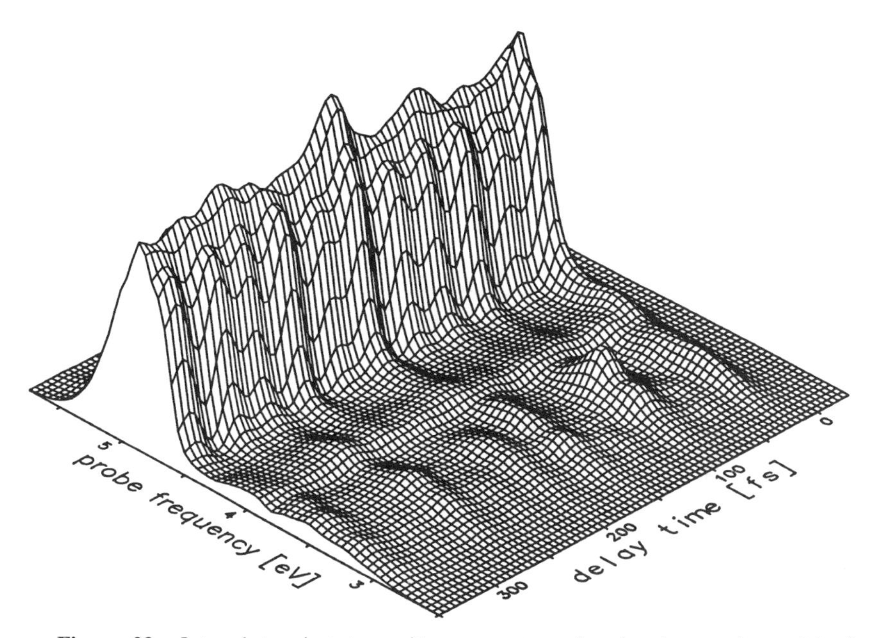

Figure 23. Integral transient transmittance spectrum for the three-mode model of pyrazine, obtained for Gaussian laser pulses of 20-fs duration. The spectrum consists of two components, the dominant resonance Raman contribution centered at $\omega_2 \approx 4.8$ eV and the weaker red-shifted stimulated-emission contribution centered at $\omega_2 \approx 3.4$ eV.

The PP signal has been calculated numerically exactly through nonperturbative evaluation of the electronic polarization, thus taking explicitly into account overlapping laser fields, non-RWA contributions, and higher-order corrections. Within the limits of our model, the PP signal shown in Fig. 23 may therefore be regarded as a direct simulation of a possible experiment. After an initial peak at delay times where the pump and the probe fields overlap, the PP signal is seen to split up into two components: a dominant, slightly oscillating feature centered at the electronic gap Δ of the S_0–S_2 transition ($\omega_2 \approx 4.8$ eV), and a weaker red-shifted contribution, centered at $\omega_2 \approx 3.4$ eV. As is shown below, the first component at $\omega_2 \approx \Delta$ is mostly due to stimulated RR scattering, thus reflecting coherent harmonic motion in the electronic ground state. The red-shifted component, on the other hand, represents the stimulated-emission contribution to the signal. It exhibits complex oscillatory structures and reflects nonadiabatic wave-packet dynamics in the coupled excited electronic states.

To distinguish ground-state and excited-state dynamics, one would like to separately consider the Raman and the stimulated-emission contribution. Experimentally, this can be done either by considering the time-resolved fluo-

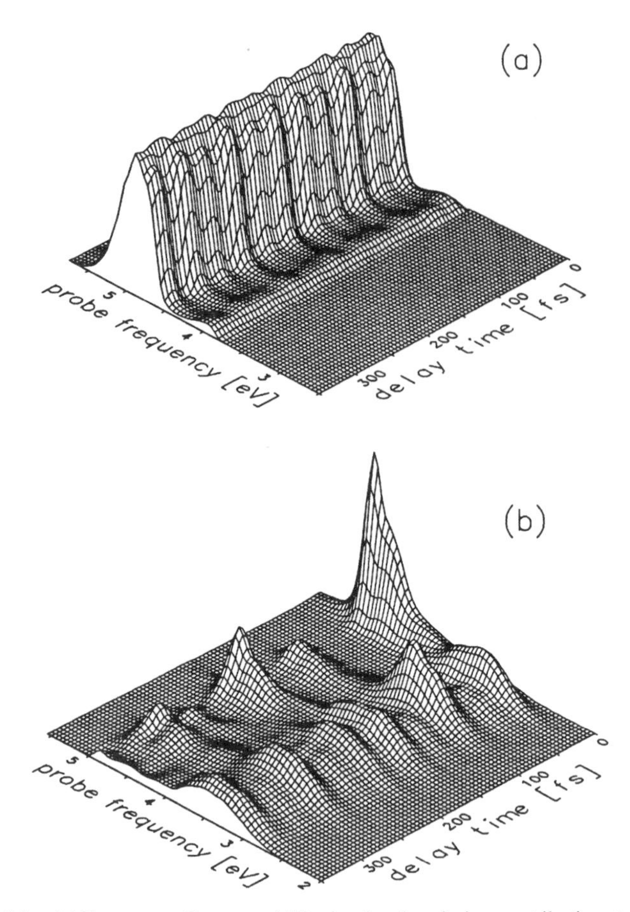

Figure 24. (*a*) Resonance Raman and (*b*) stimulated-emission contributions to the integral transient transmittance spectrum shown in Fig. 23.

rescence spectrum, or by using a polarization-sensitive detection scheme that suppresses the ground-state contribution to the transmittance spectrum [303]. To calculate individual contributions to the total PP signal, the perturbative evaluation of the electronic polarization is often advantageous.[7] Figure 24 (*a*) shows the Raman and Fig. 24(*b*) the stimulated-emission contribution to the transient transmittance spectrum. The stimulated Raman contribution is seen to manifest itself as broad, only slightly oscillating background emission located at the CW absorption bands of pyrazine. It is interesting to note

[7]For the three-state model of pyrazine with uncoupled electronic ground state the perturbative calculation requires less CPU time by a factor of about hundred than the nonperturbative one (cf. the discussion in Section VI.A.3).

that, although we have exclusively considered the S_0–S_2 transition dipole moment in the calculation, there is a weak Raman emission centered at the S_1 absorption band ($\omega_2 \approx 4$ eV) due to intensity borrowing induced by vibronic coupling. In contrast to the quasi-stationary appearance of the Raman contribution, the stimulated-emission spectrum [Fig. 24(*b*)] exhibits an irreversible initial decay on a timescale of about 20 fs, accompanied by a significant red-shift of the emission. As discussed below, the dynamical red-shift and the coherent oscillations of the stimulated emission reflect the ultrafast $S_2 \rightarrow S_1$ internal-conversion process in pyrazine.

Being mainly interested in the dynamics of the excited electronic states, we focus in the following on the excited-state contribution to the PP spectrum. Figures 24 and 25 compare three different excited-state PP signals, namely the integral stimulated-emission spectrum [Fig. 24(*b*)], the time-resolved fluorescence spectrum, [Fig. 25(*a*)], and the dispersed stimulated-emission spectrum [Fig. 25(*b*)]. The signals have been obtained by evaluating Eqs. (6.26), (6.29), and (6.31), respectively. As may be expected from the comparison of Eqs. (6.26) and (6.29), the integral SE spectrum and the time-resolved fluorescence spectrum are rather similar. Because of the ω^3 prefactor, though, the spontaneous emission for low frequencies is somewhat weaker than the stimulated emission. It should be noted that the time-resolved fluorescence spectrum requires only probe pulses of a single color (to clock the upconversion process), while the frequency resolution of the integral transmittance spectrum is given by using probe pulses with different carrier frequencies ω_2. The dispersed transmittance spectrum in Fig. 25(*b*) has been evaluated for two probe carrier frequencies, $\omega_2 = 4.8$ eV and $\omega_2 = 3.4$ eV. For emission frequencies $\omega \approx \omega_2$, the dispersed spectrum closely resembles the integral and the fluorescence spectra. Owing to the limited spectral width of 20-fs laser pulses, however, the dispersed transmittance spectrum for a single probe frequency ω_2 reveals only a small fraction of the total emission spectrum. In other words, even when dispersing the transient transmittance spectrum, one still needs probe pulses of several colors to probe the complete emission spectrum. From Figs. 24 and 25 one can conclude that, for the model under consideration, the three different excited-state PP spectra yield essentially equivalent information on the molecular dynamics.

In the following exposition, we wish to investigate in more detail the monitoring of the ultrafast excited-state relaxation dynamics, which is exhibited, for example, in the stimulated-emission spectrum [Fig. 24(*a*)]. After a rapid initial decay within ≈ 20 fs, the PP spectrum splits up into a dominant red-shifted component and a weaker blue part of the emission. The red-shift of the emission signal reflects the transition from the initially prepared S_2 levels to high vibrational levels of the S_1 surface, from which, according to the Condon principle, emission can be stimulated only to correspondingly

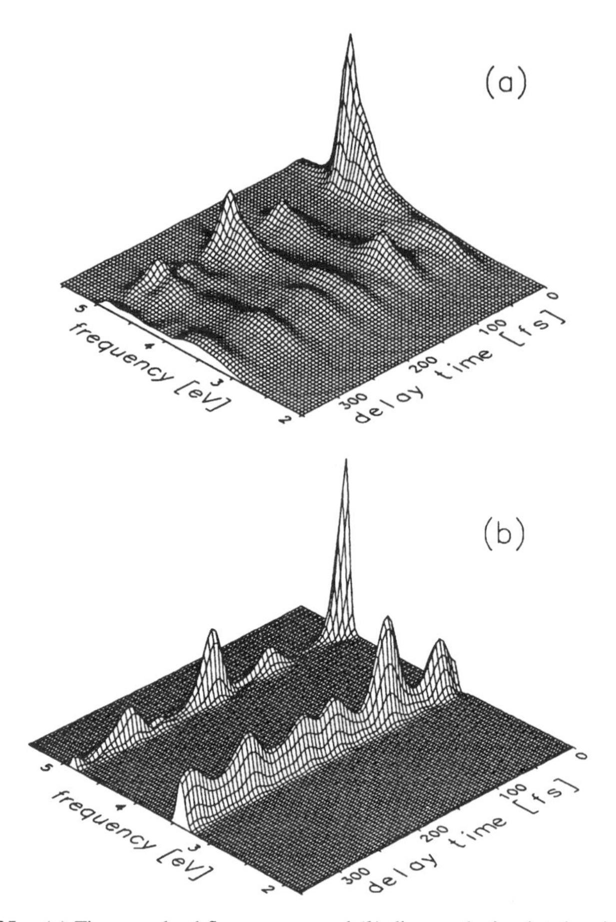

Figure 25. (*a*) Time-resolved fluorescence and (*b*) dispersed stimulated-emission spectra for the three-mode model of pyrazine, obtained for Gaussian laser pulses of 20-fs duration. The dispersed spectrum (*b*) has been evaluated for two probe carrier frequenices, $\omega_2 = 4.8$ eV and $\omega_2 = 3.4$ eV.

high levels of S_0 [89, 90]. The blue part of the spectrum, on the other hand, reflects the recurrences of the excited-state wave packet to its initial position [89, 338]. This is most easily seen from Eq. (6.26), which shows that blue wavelengths of the probe favor stimulated emission into the initial vibrational state $|\mathbf{0}\rangle$. For a blue-shifted probe laser, the autocorrelation function $\Phi_0(t)$ [Eq. (6.20)] thus gives the main contribution to the stimulated-emission signal.

To illustrate this effect, Fig. 26 compares the integral PP signal for $\omega_2 \approx 4.8$ eV to the squared modulus of the autocorrelation function $\Phi_0(t)$.

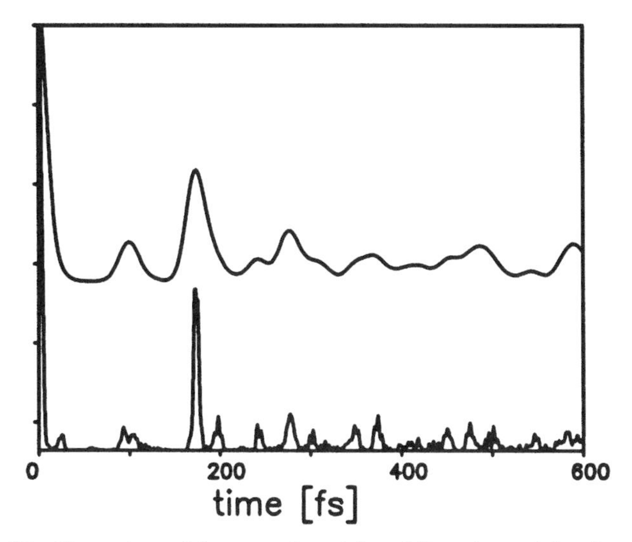

Figure 26. Comparison of the squared modulus of the autocorrelation function (lower curve) with the integral stimulated-emission signal [Fig. 24(*b*)] for a blue-shifted probe laser frequency of $\omega_2 \approx 4.8$ eV. To facilitate the comparison, the upper curve is shifted by 0.2.

Although the limited time resolution achievable with 20-fs pulses tends to smooth out the finer details of the autocorrelation function, the PP signal clearly matches the main recurrences of $\Phi_0(t)$. It is interesting to note that the autocorrelation function $\Phi_0(t)$ measured by the blue-shifted stimulated-emission spectrum is not necessarily equivalent to the Fourier transform of the CW absorption spectrum [Eq. (7.1)], which is given by $e^{-t/T_2}\Phi_0(t)$. Like photon-echo experiments, which discriminate inhomogeneous dephasing processes, detuned time-resolved stimulated-emission spectroscopy allows us to eliminate dephasing processes, which are slow on the experimental timescale of interest, thus yielding "sub-linewidth" information on the optical transition under consideration [338].

It has been shown in Section VI.A that in the limit of ultrashort laser pulses the stimulated-emission PP signal is proportional to the population probability of the initially excited diabatic state ($|\psi_2\rangle$). As has been emphasized in Section IV, the electronic population probability $P_2^{\text{di}}(t)$ represents a key quantity in the discussion of internal-conversion processes, as it directly reflects the non-Born–Oppenheimer dynamics (in the absence of vibronic coupling, $P_2^{\text{di}}(t) = const.$) It is therefore interesting to investigate to what extent this intramolecular quantity can be measured in a realistic PP experiment with finite laser pulses [89]. It is clear from Eqs. (6.26) and (6.27) that the detection of $P_2^{\text{di}}(t)$ is facilitated if a probe pulse is employed that stimulates a major part of the excited-state vibrational levels into the electronic ground state, that is, the probe laser should be tuned to the maximum of the emission band. Figure 27(*a*) compares the diabatic population probability $P_2^{\mathrm{di}}(t)$ with a cut of the stimulated-emission spectrum for $\omega_2 \approx 3.4$ eV, i.e., at the center of the red-shifted emission band. Apart from the first 20 fs where the probe laser is not resonant with the emission [cf. Fig. 24(*b*)], the PP signal is seen to capture the overall time evolution of electronic population probability. As anticipated in the discussion above, PP experiments have the potential to directly monitor electronic populations and thus non-Born–Oppenheimer dynamics in real time.

Figure 27(*a*) again demonstrates that laser pulses of finite duration tend

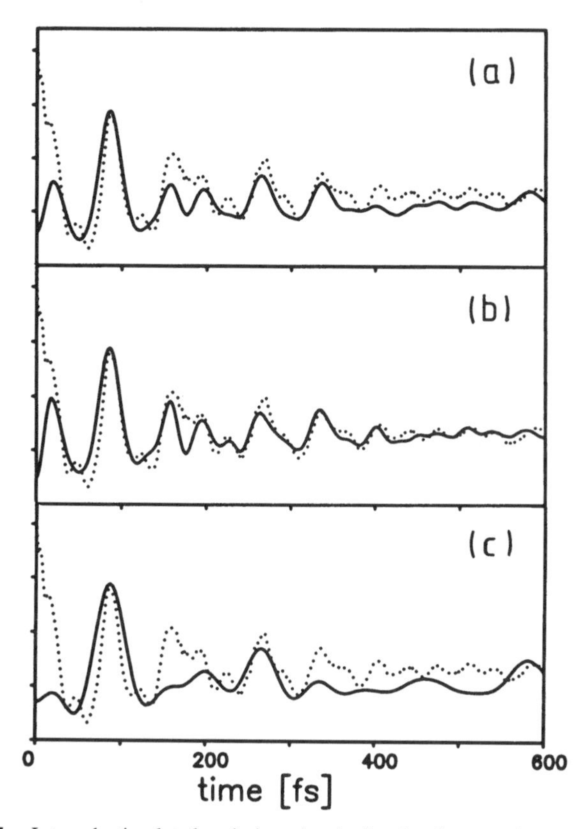

Figure 27. Integral stimulated-emission signals for the three-mode model of pyrazine (solid lines), obtained for pulse durations (*a*) $\tau_1 = \tau_2 = 20$ fs, (*b*) $\tau_1 = 0$, $\tau_2 = 20$ fs, and (*c*) $\tau_1 = \tau_2 = 40$ fs. It is seen that laser pulses of 20-fs duration are sufficiently short to monitor the time evolution of the diabatic electronic population probability (dotted lines).

to smooth out the details of molecular time-dependent observables. To give a representative example of the dependence of the PP signals on the pulse duration, Fig. 27 compares PP signals obtained for different pulse durations: Fig. 27(a) $\tau_1 = \tau_2 = 20$ fs, Fig. 27(b) $\tau_1 = 0$, $\tau_2 = 20$ fs, and Fig. 27(c) $\tau_1 = \tau_2 = 40$ fs. It is interesting to note that impulsive preparation of the molecular system with a δ-function pulse [Fig. 27(b)] results only in minor changes of the PP signal. This indicates that in the present case the impulsive limit is virtually achieved by resonant 20-fs pulses, as the pulse duration is shorter than the characteristic (e.g., vibrational) time scales of the molecular system [44, 92]. The time resolution achieved with pulses of 40-fs duration [Fig. 27(c)] starts to become inadequate to monitor the quantum beats of the electronic population in real time. Nevertheless, the very existence of partial recurrences of the electronic population can still be observed.

The preceding discussion has shown that the decay of the emission spectrum as a function of time qualitatively reflects the electronic dynamics of the system. The frequency distribution of the signal, on the other hand, reflects the evolution of the vibrational character of the system. As has been demonstrated in ref. [44], the time-dependent frequency distribution can be interpreted in terms of the excited-state wave-packet motion, which has been considered in Section IV (cf. Figs. 13 and 14). Qualitatively, the redshifted peaks at $\Delta t \approx 21$ fs, 87 fs, 156 fs can be attributed to the times when the Q_1–Q_{6a} trajectory in Fig. 14 is close to the diabatic S_2 potential-energy minimum and far from the Franck–Condon region. The blue-shifted peaks at $\Delta t \approx 100$ fs and 174 fs on the other hand, reflect the recurrences of the autocorrelation function, i.e., when the wave packet is close to the Franck–Condon region. For larger times the vibronic coupling causes a considerable distortion of the initially well-defined wave-packet. For $\Delta t \approx 270$ fs, for example, the emission spectrum simultaneously exhibits a peak at both sides of the spectrum, indicating a splitting of the wave packet. This makes an analysis of the spectrum in terms of mean positions meaningless.

Let us finally consider the possibility of time-resolved ionization spectroscopy of internal-conversion dynamics. In the case of pyrazine, the calculation of ionization signals requires taking into account the first two ionization continua I_0 and I_1, which are radiatively coupled to the S_1 and S_2, states respectively [91]. For details regarding the model system, representation of the ionization continua, and nonperturbative evaluation method, the reader is referred to ref. [91]. As a first example, let us consider the total ionization signal $I_{\text{ion}}(\Delta t)$, which has been defined in Eq. (5.19). In general, the total ionization probability is approximately zero for probe frequencies well below the ionization thresholds and becomes almost independent of delay time well above the thresholds. An interesting case emerges, however, if we employ a probe frequency that is ≈ 1.5 eV larger than the energy difference between the vertical S_2 excitation energy and the vertical ionization threshold of the first ion-core state I_0. Figure 28 shows that in this case the total ionization signal monitors the population probability $P_1^{\rm di}(t) = 1 - P_2^{\rm di}(t)$ of the diabatic S_1 state. A Gaussian probe pulse of 16 fs duration has been assumed. Recall that in the case of stimulated-emission spectroscopy into the electronic ground state a red-shift of the probe frequency is necessary to detect the diabatic population probability $P_2^{\rm di}(t)$. As discussed above, this red-shift reflects the conversion from initially prepared S_2 levels into high vibrational levels of the S_1 surface. Analogously, in ionization spectroscopy a *blue-shift* of the probe frequency is required to achieve resonant excitation of the $I_0 \leftarrow S_1$ transition, thus monitoring the population of the diabatic S_1 state. The $I_1 \leftarrow S_2$ ionizing transition, on the other hand, is strongly suppressed after $\approx$ 20 fs because (1) the larger relative displacement of the I_1 and S_2 potential-energy minima along the totally symmetric mode ν_{6a} results in poor Franck–Condon overlaps compared to the $I_1 \leftarrow S_1$ transition, and (2) the I_1 state is energetically higher-lying than the I_0 state. During the first 20 fs, however, the $I_1 \leftarrow S_2$ transition dominates, i.e., $I_{\rm ion}(\Delta t)$ mainly reflects the ultrafast $S_2 \rightarrow S_1$ initial decay of $P_2(t)$.

As a further example of a spectroscopic technique that is resolved in time and energy, Fig. 29 shows the time-resolved photoelectron spectrum $I_{\rm ion}(E_k, \Delta t)$, which is given as a function of the electron energy E_k and the delay time Δt [Eq. (5.20)]. Impulsive excitation and a probe pulse of 16 fs duration and a carrier frequency of $\omega_2 = 7.49$ eV has been assumed [91]. In contrast to the situation discussed in Fig. 28, this probe frequency ensures

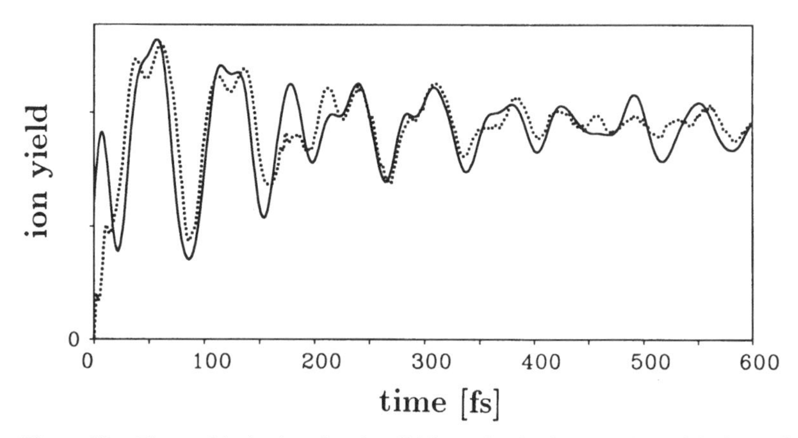

Figure 28. The total ionization signal (solid lines) for the three-mode model of pyrazine, obtained for Gaussian laser pulses of 16-fs duration. Assuming a probe frequency that is $\approx$ 1.5 eV larger than the energy difference between the S_2 excitation energy and the ionization threshold, the ionization signal monitors the population of the diabatic S_1 state (dotted lines).

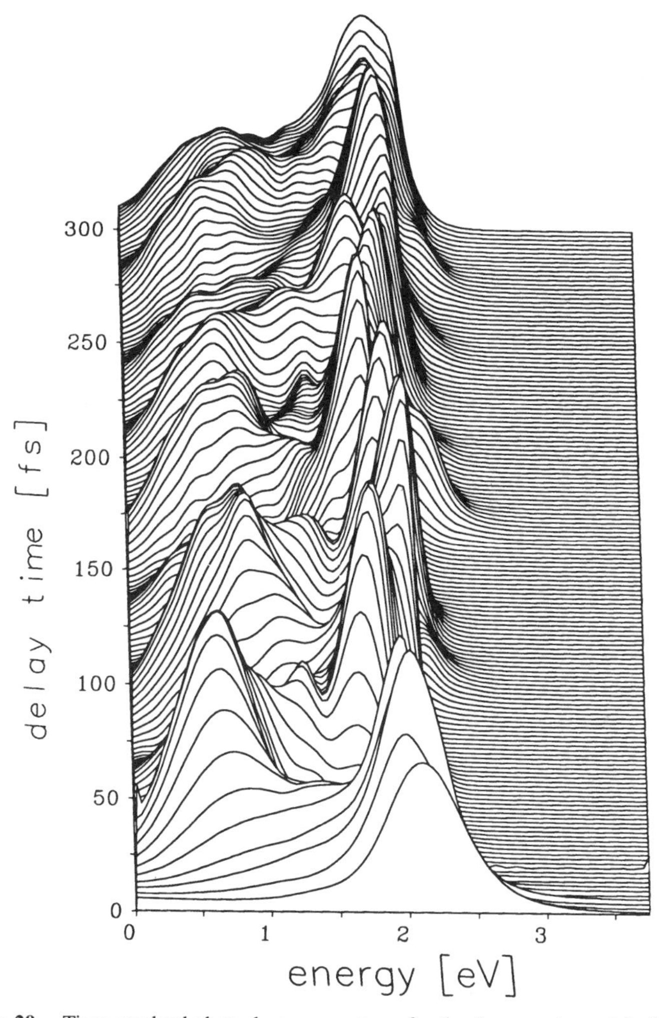

Figure 29. Time-resolved photoelectron spectrum for the three-mode model of pyrazine, plotted as a function of the electron energy E_k and the delay time Δt.

that both the $I_0 \leftarrow S_1$ and the $I_1 \leftarrow S_2$ transitions are energetically possible. The spectrum reveals complex structure, reflecting the vibrational and electronic relaxation dynamics of the vibronically coupled S_1–S_2 system. In a fashion similar to that described above for the stimulated-emission signal, it is again possible to interpret the decay of the spectrum as a function of time in terms of the electronic dynamics, while the time-dependent frequency distribution reflects the vibrational dynamics of the system [91]. As has been shown by Meier and Engel [322, 339] for one-dimensional model systems, the time-resolved photoelectron spectrum in particular allows us to monitor the nuclear wave-packet motion on the excited electronic PE surfaces, including the direct observation of the shape, recurrences, and splittings of the wave packet. These features and the excellent sensitivity that may be achieved in photoelectron detection render femtosecond photoelectron spectroscopy a promising technique for the investigation of ultrafast excited-state dynamics.

Unfortunately, no femtosecond experiments on pyrazine have yet been reported. Nevertheless, the investigation of this system, which has been exceptionally well characterized theoretically, certainly provides a demonstration of the potential of femtosecond techniques. The qualitative insight gained in these studies will guide the more general discussion of femtosecond experiments in Section VII.C.

B. Real-Time Detection of Nonadiabatic Photoisomerization Processes

As a first step towards the modeling of photoinduced *cis-trans* isomerization in unsaturated hydrocarbons, we have introduced in Section II.C a class of model Hamiltonians for photoisomerization dynamics, which involve a large-amplitude torsional motion, a vibronically active mode, and totally symmetric modes that modulate the energy gap of the interacting states. By employing exact time-dependent wave-packet calculations, it has been shown in Section IV that a three-mode model is suitable for revealing basic aspects of photoisomerization and internal-conversion dynamics. In the following presentation, we wish to investigate the possibility of observing these processes in real time. To facilitate the distinction between reactant and product dynamics, a model with unsymmetric torsional potentials [i.e., $V_0 > V_1$ in Eq. (2.36)] and a periodicity of 2π is employed here. For details, see ref. [50].

To visualize the basic photochemical aspects of the model, Fig. 30 shows a schematic view of the adiabatic PE surfaces plotted as a function of the torsional angle φ and the coupling mode Q_c. Note that the excited-state potential is inverted, i.e., it exhibits a saddle at the ground-state equilibrium geometries ($\varphi = 0, \pm\pi$). Let us assume that at the time $t = 0$ the molecular system is excited by an ultrashort laser pulse to the upper PE surface. In the course of the subsequent photoreaction, the excited-state wave packet propagates on the multidimensional upper PE surface, until it decays through the conical intersection to the lower-lying S_0 state. As a consequence of vibrational relaxation on the adiabatic energy surface of the electronic ground state, the molecular system will eventually localize in the *cis* and the *trans* conformations, i.e., around $\varphi = 0$ and $\varphi = \pi$, respectively. Note that the parameters of

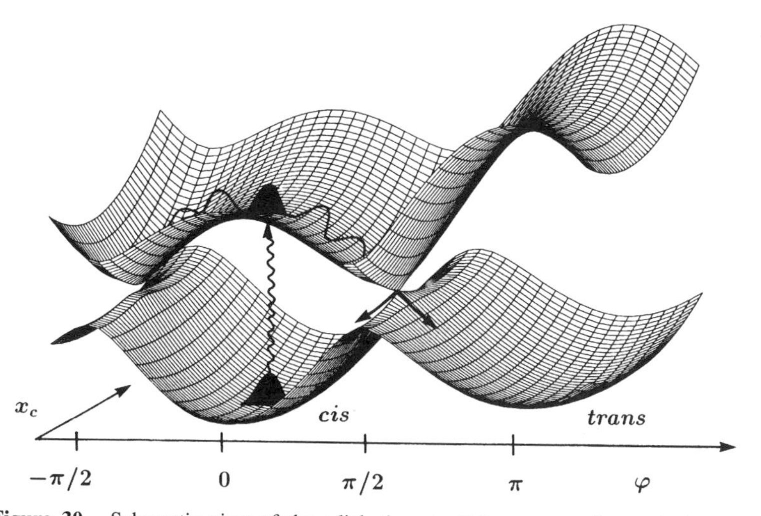

Figure 30. Schematic view of the adiabatic potential-energy surfaces of photoisomerization model plotted as a function of the torsional angle φ and the coupling mode x_c. We assume that the molecular system is initially in its electronic and vibrational ground state, approximately represented by a Gaussian wave packet localized at $\varphi, x_c = 0$. At the time t = 0 the molecule is excited by an ultrashort laser pulse to the upper potential-energy surface, and subsequently undergoes ultrafast isomerization and internal conversion into the adiabatic electronic ground state.

the model are chosen such that the vertical electronic energy gap at $\varphi = 0$ ($V^{(1)} - V^{(0)} = 2.75$ eV) is smaller than the energy gap at $\varphi = \pm\pi$ ($V^{(1)} - V^{(0)}$ $= 4.75$ eV). Applying a second, time-delayed laser pulse to probe the photodynamics initiated by the first pulse, it is thus clear from Fig. 30 that electronic transitions occurring at $\varphi \approx 0$ (the absorption of the reactants) should be resonant to a laser frequency of ≈ 2.75 eV, while electronic transitions occurring at $\varphi \approx \pm\pi$ (the absorption of the products) would be expected for laser frequencies of ≈ 4.75 eV.

To confirm these qualitative considerations, we have performed a series of nonperturbative calculations, considering various wavelengths and pulse durations of the laser fields and studying integral and dispersed PP spectra as well as coherent photon-echo signals. As has been discussed in Section VI, the nonperturbative evaluation of the signals is advantageous in cases where the electronic states are coupled by both intramolecular and radiative interactions. In all calculations, it has been assumed that the molecular system is resonantly excited by the pump pulse, i.e., $\omega_1 = E_1 - E_0$.

To get a first impression of the excited-state dynamics of the system, Fig. 31 shows the integral PP spectrum [Eq. (5.4)] obtained with Gaussian laser pulses of 6-fs duration. As the Raman contribution mainly results in a struc-

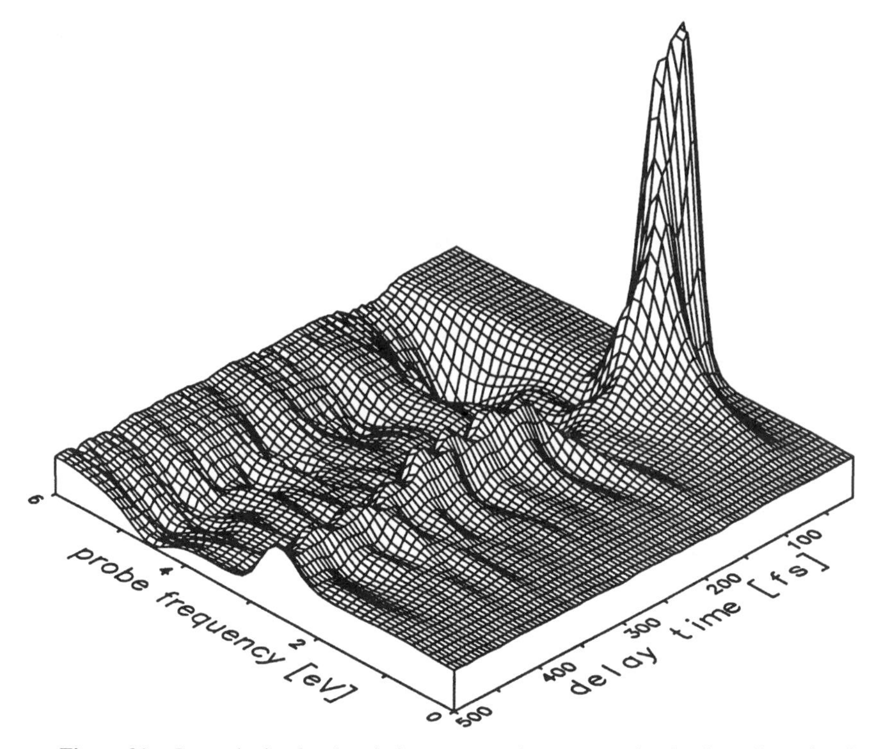

Figure 31. Integral stimulated-emission pump-probe spectrum for the three-dimensional model of *cis-trans* isomerization, obtained for Gaussian laser pulses of 6-fs duration.

tureless background emission around $\omega_2 \approx E_1$ [cf. Fig. 32(a)], it has been subtracted in Fig. 31. For zero delay time, $\Delta t \approx 0$, the PP spectrum resembles the stationary absorption spectrum of the S_1 state. Beginning with $\Delta t \gtrsim 100$ fs, the PP signal is seen to split up into two components centered at $\omega_2 \approx 2.5$ eV and $\omega_2 \approx 4.5$ eV, respectively. As has been anticipated in the discussion of Fig. 30, the first component describes the stimulated emission ($S_1 \rightarrow S_0$) and absorption ($S_0 \rightarrow S_1$) of the reactants, while the latter component describes the absorption ($S_0 \rightarrow S_1$) of the products. Recall that by definition in Eq. (5.4) the signal is positive for emission processes and negative for absorption processes. In the following we will refer to the two emission/absorption bands as "reactant channel" and "product channel," respectively.

To investigate the nonadiabatic photoisomerization dynamics exhibited by the overview spectrum in Fig. 31 in some more detail, let us consider the dispersed PP spectra employing probe frequencies that are resonant to the vertical electronic transition energy of the reactant and product channel, respectively. To study the system response for longer laser pulses, we choose

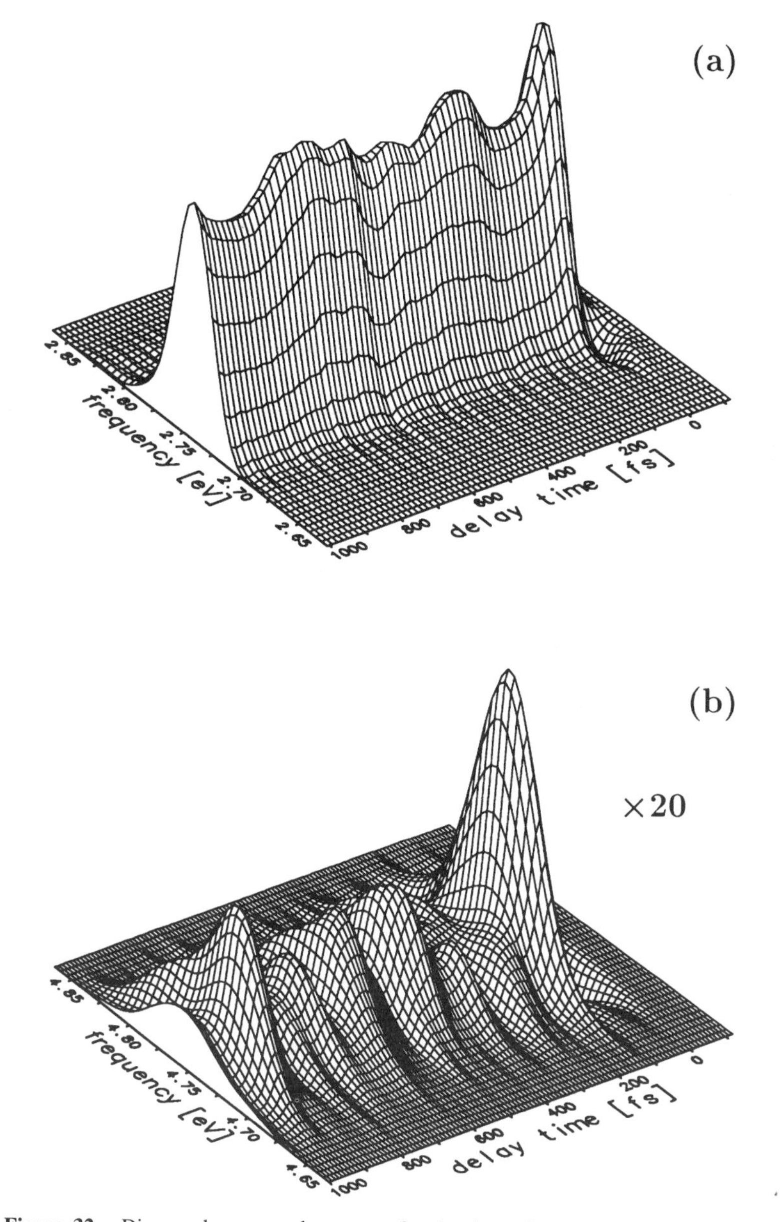

Figure 32. Dispersed pump-probe spectra for the three-dimensional model of *cis-trans* isomerization, assuming Gaussian pulses of 40-fs duration. In (*a*) the probe pulse is resonant to the electronic gap at the torsional angle $\varphi = 0$ ($\omega_2 = V^{(1)}(0) - V^{(0)}(0) = 2.75$ eV), thus detecting the dynamics of the reactant channel. In (*b*) the carrier frequency of the probe pulse is $\omega_2 = V^{(1)}(\pi) - V^{(0)}(\pi) = 4.75$ eV, thus detecting the dynamics of the product channel.

pulses of 40-fs duration in this example. As result, Fig. 32 shows the first picosecond of the time evolution of the corresponding dispersed transmittance spectra, obtained for $\omega_2 = 2.75$ eV [Fig. 32(*a*)] and $\omega_2 = 4.75$ eV [Fig. 32(*b*)], respectively. The reactant channel [Fig. 32(*a*)] is seen to be dominated by the Raman contribution, which manifests itself as broad background emission. The slow beating in the signal results from the superposition of the Raman and stimulated emission contributions. To better see the coherent response of the product channel, the spectrum in Fig. 32(*b*) has been inverted and enlarged by a factor of 20. After a somewhat delayed onset, the product channel is seen to exhibit complex structures in time and frequency, reflecting coherent wave-packet motion in the ground state of the product conformation.

As an aid for the interpretation of PP signals, let us again invoke the time-dependent intramolecular observables introduced in Section IV. To describe nonadiabatic isomerization dynamics, a useful quantity is, for example, the joint probability $P^{\text{ad}}_{1,cis}(t)$ of populating the adiabatic S_1 state as well as the *cis* conformer [cf. Eq. (4.21)]. Recalling that we have attributed the $S_1 \rightarrow S_0$ stimulated emission close to the equilibrium geometry of the S_0 state (i.e., for $\varphi = 0$) to the reactant channel, it is intuitively clear that this PP signal should resemble $P^{\text{ad}}_{1,cis}(t)$. In the same way, it may be expected that the $S_0 \rightarrow S_1$ absorption of the product channel monitors the population probability $P^{\text{ad}}_{0,trans}(t)$. Figure 33 confirms these conjectures by showing cuts of constant probe frequency of the integral PP spectrum $I(\omega_2, \Delta t)$, assuming pulses of a duration of 6 fs [Fig. 33(*a*)–(*d*)] and 40 fs [Fig. (*e*) and (*f*)], respectively. The PP signals in Fig. 33(*a*) and (*b*) show cuts that have been obtained at the center of the reactant emission spectrum ($\omega_2 = 2.5$ eV) and the product absorption spectrum in Fig. 31 ($\omega_2 = 4.5$ eV), respectively. It is seen that the integral PP signals with 6-fs pulses monitor most of the features of the field-free intramolecular observables $P^{\text{ad}}_{1,cis}(t)$ and $P^{\text{ad}}_{0,trans}(t)$. When the probe laser frequency is tuned in resonance with the electronic transition energy of the two channels, i.e., $\omega_2 = 2.75$ eV [Fig. 31(*c*)] or $\omega_2 = 4.75$ eV [Fig. 31(*d*)], the agreement between the PP signals and the population probabilities deteriorates. In particular, the signal pertaining to the reactant channel exhibits a coherent beating with the frequency of the tuning mode, which is not observed in $P^{\text{ad}}_{1,cis}(t)$. Finally, the comparison of PP signals and population probabilities shown in Fig. 31(*e* and *f*) demonstrate that the qualitative features of $P^{\text{ad}}_{1,cis}(t)$ and $P^{\text{ad}}_{0,trans}(t)$ are still observable with pulses of 40-fs duration. Due to the dominant Raman contribution to the reactant signal in the case of longer pulse durations, however, the comparison is worse for $P^{\text{ad}}_{1,trans}(t)$.

Let us finally investigate to what extent the nonadiabatic isomerization dynamics, clearly exhibited in the PP signal, is monitored by standard coher-

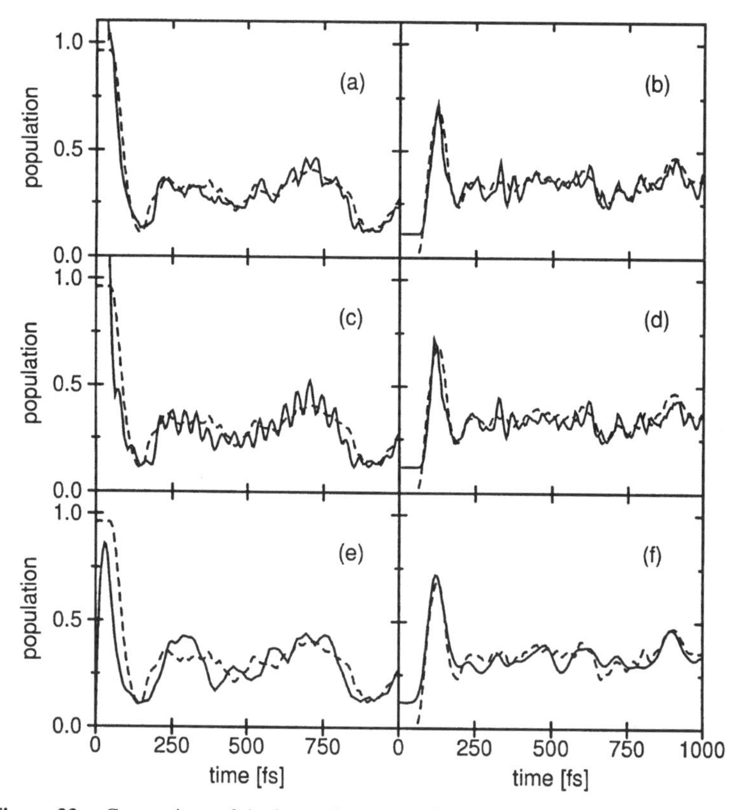

Figure 33. Comparison of the integral pump-probe signal (solid lines) with the adiabatic population probabilities $P^{\mathrm{ad}}_{1\,cis}(t)$ and $P^{\mathrm{ad}}_{0\,trans}(t)$ (dashed lines). The pump-probe signals on the left side are obtained for probe frequencies resonant to the reactant channel [$\omega_2 = 2.5$ eV (*a*), 2.75 eV (*c*), (*e*)] and are in good agreement with $P^{\mathrm{ad}}_{1\,cis}(t)$. The pump-probe signals on the right side are obtained for probe frequencies resonant to the product channel [$\omega_2 = 4.5$ eV (*b*), 4.75 eV (*d*), (*f*)] and match the population probability $P^{\mathrm{ad}}_{0\,trans}(t)$.

ent techniques. As an example of the coherent response of the model system, Fig. 34 shows the two-time photon-echo signal [Eq. (5.16)] for 6-fs pulses with frequencies $\omega_1 = \omega_2 = E_1 - E_0$. The photon-echo signal exhibits oscillations with the period of the tuning mode ($\approx$ 41 fs), which are damped on a timescale corresponding to a total electronic dephasing time of $T_2 \approx 100$ fs. It is interesting to note that Fig. 34 resembles the photon-echo signal for a simple one-dimensional harmonic oscillator that is homogeneously broadened [84, 315, 316]. This is to say, for the present model the nonadiabatic isomerization is solely reflected in a generic homogeneous electronic dephasing process. The same trend may be seen by comparing the squared modulus

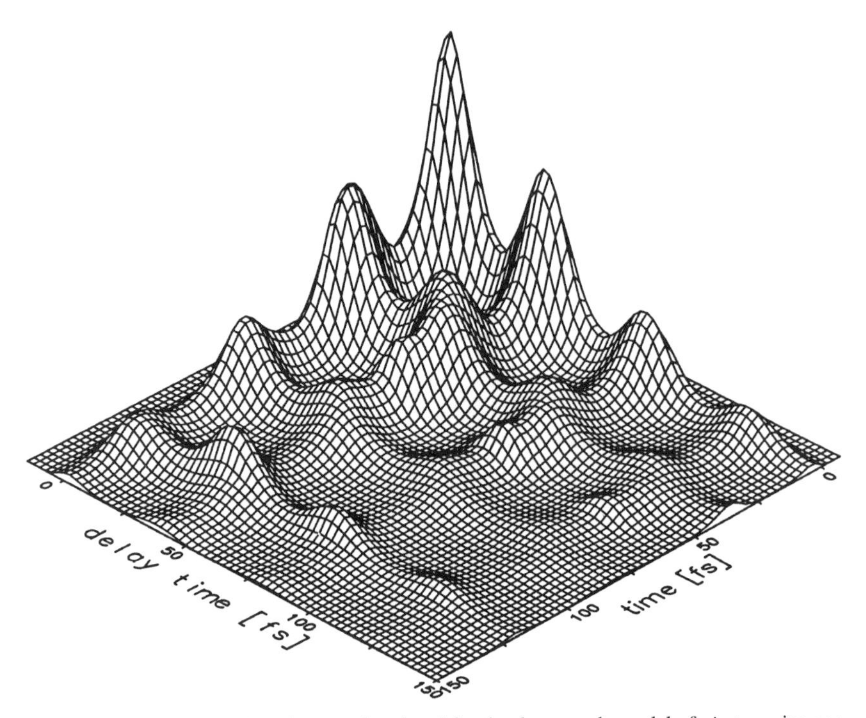

Figure 34. Three-pulse photon-echo signal for the three-mode model of *cis-trans* isomerization, plotted as a function of the delay time Δt and the emission time $\tau = t - \Delta t$. We have assumed resonant ($\omega_1 = \omega_2 = 2.75$ eV) laser pulses of 6-fs duration.

of the autocorrelation function $\Phi_0(t)$ [Eq. (6.20)] with the time-integrated photon-echo signal [Eq. (5.15)] as shown in Fig. 35. Apart from the rapid oscillations of $\Phi_0(t)$ with the frequency $\omega = E_1 - E_0$ that are not resolved with 6-fs pulses, the time evolution of the autocorrelation function qualitatively matches the decay of the photon echo, thus reconfirming that the electronic transition is essentially homogeneously broadened. It should be noted, however, that it has been shown in photon-echo simulations employing a simple model that the low-frequency large-amplitude motion along the reaction coordinate on a single PE surface may also result in characteristic inhomogeneous dephasing of the electronic transition [338].

C. Discussion of Femtosecond Experiments

As an illustration of the general concepts outlined above, we briefly discuss recent femtosecond experiments on molecules exhibiting ultrafast internal-conversion and isomerization dynamics. As a representative study of ultrafast internal-conversion processes we consider recent experiments on tri-

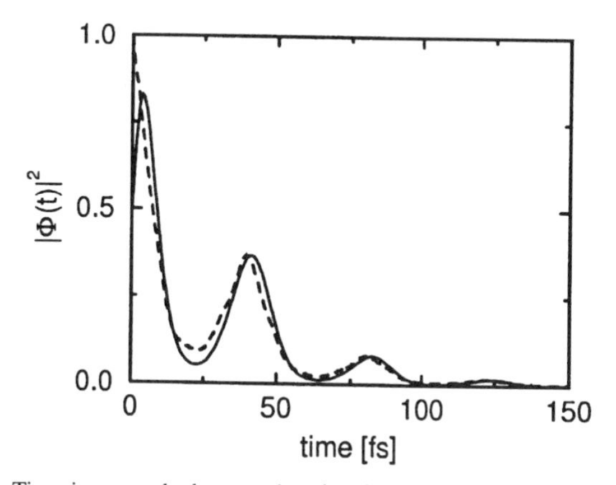

Figure 35. Time-integrated photon-echo signal (solid line) obtained for resonant 6-fs pulses compared to the squared modulus of the autocorrelation function (dashed line).

phenylmethane dyes such as malachite green [13, 15, 276, 280, 281]. To discuss the time-resolved detection of nonadiabatic isomerization processes, we consider femtosecond experiments on substituted ethylenes such as stilbene [277, 342–344] and tetraphenylethylene [345], and on the retinal chromophore in rhodopsin [17, 346, 347]. Finally, the potential of femtosecond time-resolved photoionization experiments is illustrated by referring to recent experiments on hexatriene [293, 295]. We briefly summarize the main experimental results and make an attempt to analyze the observed femtosecond dynamics in the light of our numerical studies outlined above.

Inspired by the experiments of Tang and coworkers [13], a number of femtosecond experiments have been performed on triphenylmethane dyes, including the measurements of transient transmittance [281], transient birefringence and dichroism [280], time-resolved fluorescence [276], and femtosecond photon echoes [15]. The PP signals typically reveal a biexponential decay with time constants of ≈20–150 fs and ≈2–10 ps, respectively. Furthermore, it has been found that the initial decay is superposed by oscillations with a period of ≈ 150 fs, which are damped on a timescale of ≈ 200 fs. There have been a number of speculations about the origin of these oscillations, including the interpretation as quantum beats pertaining to molecular eigenstates [13, 47], as vibrational motion in the ground or excited electronic states [340], and as electronic population dynamics triggered by excited-state vibrational motion [88, 89].

An important clue to the interpretation of these data is provided by resonance Raman experiments, which suggest that the period of ≈ 150 fs corresponds to the breathing mode of triphenylmethane dyes [341]. This mode is strongly Raman active, and is isolated in frequency between weakly active higher modes and overdamped torsion modes at lower frequencies [341]. As has been shown above, resonance Raman spectroscopy allows us to identify vibrational modes involving a large excited-state coordinate shift with respect to the equilibrium geometry in the electronic ground state. For this reason, we may expect that modes which are strongly Raman active contribute considerably to the initial step of photoinduced dynamics. The time-resolved observation of a Raman-active mode does not indicate, however, whether the vibrational motion takes place in the ground or the excited electronic state.

Mokhtari, Chebira, and Chesnoy have reported time-resolved fluorescence experiments with a 80 fs time resolution, which showed that the oscillations in malachite green and nile blue are absent in the fluoresence signals, while they are retained in the case of oxazine 1 [276]. This indicates that the oscillations in malachite green and nile blue are due to wave-packet motion in the electronic ground state, whereas the latter case corresponds to coherent wave-packet motion in the excited electronic state. The example again underlines the importance of spectroscopic techniques, such as time-resolved fluorescence [276], polarization-sensitive detection [303], and time-resolved ionization [293], which allow the discrimination of ground-state and excited-state contributions to the PP signal. In the case of malachite green, though, Shank and coworkers have presented transient transmittance experiments with a 10-fs time resolution, which assigned a red-shifted component of the spectrum to stimulated emission, i.e., excited-state dynamics [281]. The PP signal at this wavelength exhibits damped oscillations with a period of $\approx$ 20 fs, which could not be resolved in the fluorescence experiment [276]. To summarize the above experimental results on dye molecules, the time-resolved measurements suggest that the overall decay of the excited-state PP signals can be roughly characterized by three timescales, that is, (1) a rapid initial decay occurring within $\approx$ 10–10^2 fs, (2) the timescale of vibrational dephasing occurring within $\approx$ 10^2–10^3 fs, and (3) an incoherent relaxation on a picosecond timescale.

Let us try to discuss the above-defined three phases of excited-state relaxation in the light of the above reported numerical studies for the S_1–S_2 vibronic-coupling model system of pyrazine.[8] It has been shown in Fig. 24 that the calculated excited-state PP signals exhibit a rapid initial decay on a timescale of $\approx$ 20 fs. By analysis of the signals in terms of time-dependent molecular observables, it was found that this initial relaxation reflects the ultrafast decay of the electronic population into the lower-lying electronic state [cf. Fig. 10]. Considering internal conversion as an chemical reaction that is completed when, say, 90% of the electronic population is in the lower

[8]Similar numerical investigations have also be performed for $C_2H_4^+$ [36, 38], NO_2 [38], $C_6H_6^+$ [88, 190], and ozone [93].

adiabatic state, one may conclude that the internal conversion process takes place within less than 100 fs. After this time in most of the model systems studied, there are only minor recurrences of the adiabatic population probability [38]. In practice, however, the observation of the initial relaxation component is often hampered by the limited time resolution achieved in the experiment. In the case of malachite green, for example, initial relaxation times of 60 fs [13], 20 fs [15] and 150 fs [276] have been reported, depending on the detection scheme and the time resolution achieved.

It has been shown in Fig. 10 that the *adiabatic* electronic population decays within less than 100 fs, whereas the *diabatic* population probability exhibits prominent quasi-periodic recurrences that are damped out within $\approx$ 500 fs. This finding is quite important for the interpretation of time-resolved spectroscopic data, because in the vicinity of a conical intersection the electronic dipole transition operator is only smooth in the diabatic representation. This means, the excited-state PP signal monitors *diabatic* electronic populations, i.e., transitions between diabatic electronic states that are triggered by nuclear wave-packet motion on coupled PE surfaces [88, 89].

It can thus be expected that, in large molecules, ultrafast internal conversion and vibrational relaxation of modes that are strongly coupled to the electronic transition are generally closely interrelated. This is because, on the one hand, the intersections of PE surfaces are the crucial condition for ultrafast internal conversion to occur, and on the other hand, on a femtosecond timescale, only relatively few degrees of freedom will give rise to the strongest interactions. For these reasons we believe that, as was demonstrated for the simple model of pyrazine, coherent transients of ultrafast decaying excited-state PP signals are due to wave-packet motion on intersecting PE surfaces, thus reflecting electronic population dynamics as well as vibrational dynamics. Simulations on a variety of model systems have shown that the vibrational dephasing process accompanying the electronic relaxation is typically completed after 5–20 periods, and that a single mode or several vibrational modes may determine the beating frequency of the diabatic population [36, 38].

It is interesting to note that the first two phases of excited-state relaxation (i.e., initial decay and vibrational dephasing) are virtually independent of the properties of the solvent, whereas the subsequent picosecond relaxation process has been found to clearly depend on the viscosity of the solvent [13, 280]. On this timescale, it is to be expected that the remaining inter- and intramolecular degrees of freedom will contribute to the time evolution of the system. The picosecond relaxation has been identified to reflect the redistribution of the electronic excess energy into the nuclear degrees of freedom [348]. It should be noted that the first step of this energy redistribution already takes place during the vibrational dephasing of the strongly coupled modes. As demonstrated in Figs. 15 and 16, this process is reflected in an increase of the mean energy content of these modes. The inclusion of solvent effects will be discussed in Section VIII.

Let us finally turn to a discussion of molecular systems that, besides photophysical relaxation, undergo a photochemical reaction upon electronic excitation. While many of the features discussed above for nonreactive systems will be quite similar for reactive systems, the latter also offer the possibility of monitoring the disappearance of the reactants as well as the appearance of the photoproducts in real time.

A beautiful example of ultrafast *cis-trans* isomerization is provided by the experiment of Shank and collaborators on the retinal chromophore in rhodopsin [17, 346, 347]. Using resonant pump pulses of 35-fs duration and probe pulses of variable wavelength and 10-fs duration, they have reported measurements of the transient transmittance of rhodopsin. The time-resolved spectra exhibit two separated absorption/emission bands, which were interpreted as the bleaching of the ground-state reactant and the transient absorption of the photoproducts, which are formed within 200 fs [17, 346]. Furthermore, since they found that both the reactants channel and the products channel exhibit coherent oscillations of the PP signals, they concluded that the photoiosomerization of rhodopsin is a vibrationally coherent photochemical reaction [347].

The comparison of the experimental results on rhodopsin with the above-reported model simulations shown in Figs. 31 and 32 reveals, for the first few hundred femtoseconds, a remarkable agreement of the photoinduced relaxation dynamics of a biophysical system and a three-mode model. In both cases we obtain a rise time of the product absorption of 200 fs, which is mostly determined by the inertial moment of the reaction mode and largely independent of the solvent. Similar time constants for the formation of the photoproduct have been reported for *cis*-stilbene [277, 342–344], tetraphenylethylene [345], and 9-anthryl-trifluoromethylketone [279]. Coherent transients of the photoproducts have been also observed in the *cis-trans* isomerizations of stilbene [344] and tetraphenylethylene [345]. The subsequent picosecond relaxation process in these systems reflects the cooling of the vibrationally hot photoproducts in the solvent, a process which has been discussed in detail in ref. [342]. It is interesting to note that, contrary to the picture adopted in Rice-Ramsberger-Kassel-Marcus (RRKM) theory of thermal reactions, the intramolecular vibrational energy redistribution actually is slow compared to the timescale of the photoreaction. Classical simulations of this effect and its manifestation in time- and frequency-resolved PP spectra are discussed in Section VIII.D.

We finally mention a recent experiment that illustrates the significant potential of femtosecond time-resolved photoionization and photoelectron spectroscopy for unraveling the microscopic mechanisms of ultrafast internal-conversion dynamics in polyatomic molecules. By measuring the ion yield in a PP experiment with femtosecond laser pulses, Hayden and collaborators have directly observed the individual steps of the internal-conversion dynamics following excitation of the lowest optically allowed $S_2(^1B_{1u})$ state of *cis* and *trans* hexatriene [293, 295]. By detecting the population of the $S_1(^1A_g)$ state as a transient intermediate, the $S_2 \rightarrow S_1$ internal-conversion process could clearly be separated from the subsequent $S_1 \rightarrow S_0$ internal-conversion process. The observation of different S_1 lifetimes in *cis* and *trans* hexatriene, respectively, excludes rapid interconversion between both isomers on the S_1 surface. Recently, Cyr and Hayden have reported the first photoelectron spectra generated with femtosecond laser pulses [295]. As previously emphasized in theoretical studies [320, 322], such spectra provide a direct real-time mapping of the time-dependent wave-packet dynamics in the intermediate electronic state. In *cis* hexatriene, the red-shift and narrowing of the photoelectron energy distribution directly reflect the $S_2 \rightarrow S_1$ internal-conversion process on a timescale less than 100 fs, in close analogy to the simulations discussed in Section VII for the $S_2 \rightarrow S_1$ internal-conversion process in pyrazine (Fig. 29). It represents a challenge for the theory to reproduce theoretically the femtosecond time-resolved photoelectron data for *cis* and *trans* hexatriene [295] in terms of wave-packet dynamics on coupled *ab initio* S_1 and S_2 PE surfaces.

In conclusion, the comparison of experimental and calculated data on systems undergoing ultrafast internal conversion and isomerization has been found to support the concept adopted above, i.e., that within the first few hundred femtoseconds only a few strongly coupled intramolecular degrees of freedom determine the time evolution of the system. The PP simulations of nonadiabatically coupled model systems involving only three nuclear degrees of freedom are able to reproduce the initial decay and the coherent relaxation that has been observed in these systems. To account for the subsequent picosecond relaxation, the models have to be extended to include remaining weakly-coupled inter- and intramolecular degrees of freedom.

VIII. QUANTUM MECHANICAL AND SEMICLASSICAL TREATMENT OF SYSTEMS WITH MANY DEGREES OF FREEDOM

A common feature of all numerical quantum mechanical wave-packet methods is the exponential growth of the computational cost with the number of degrees of freedom, the so-called basis-set dilemma. This property limits the application of conventional time-dependent wave-packet methods as discussed in Section III to relatively small systems. Alternative methods that exhibit a different scaling of the computational cost are obviously required if one wants to tackle truly multidimensional problems.

The development of computational methods for the quantum dynamics of large systems is an active area of research. Most of the contemporary developments can be associated with one of the following methodical categories: time-dependent self-consistent field approximations, path-integral methods, reduced density-matrix descriptions, or semiclassical approximations.

The most straightforward and most widely applicable approximation that eliminates the exponential growth of the computational cost in time-dependent quantum dynamics is the time-dependent self-consistent field (TDSCF) approximation. The TDSCF approximation follows from a variational *ansatz* for the time-dependent wave function, which factorizes the wave function, either completely or partly, with respect to the degrees of freedom [349, 350]. It provides the basis for widely employed mixed classical-quantum descriptions of condensed-phase spectroscopy and cluster dynamics [351–353], (see also Section VIII.C below). Recently multiconfiguration extensions of the TDSCF approximation such as the multiconfiguration time-dependent Hartree (MCTDH) method of Meyer and collaborators have been developed [354–358]. Applications of TDSCF and MC-TDSCF methods to multidimensional surface-crossing problems have been reported in refs. [260, 354, 358, 359]. We shall not describe the MCTDH method here, since detailed expositions can be found elsewhere [357, 358].

In the following three subsections we briefly review recent developments in the application of path-integral techniques, reduced-density-matrix methods, and semiclassical methods to the femtosecond dynamics and spectroscopy of coupled electronic states.

A. Real-Time Path-Integral Approach

The path-integral (PI) formalism introduced by Feynman [360, 361] represents an elegant alternative formulation of quantum mechanics. As is well established in several areas of chemical and condensed-matter physics, the PI approach is well suited to dealing with the thermodynamics (calculation of the partition function) as well as the real-time dynamics (calculation of time-dependent correlation functions) of quantum systems with many degrees of freedom. In many applications, the exponential growth of the computational cost with the number of degrees of freedom can be eliminated in the PI formulation, at the expense of introducing a sum over histories in discretized time [361, 362].

In the context of electronically nonadiabatic dynamics involving few electronic states and many nuclear degrees of freedom, PI techniques have so far nearly exclusively been applied to the so-called spin-boson problem. The spin-boson Hamiltonian in its widely known form [252, 363] describes a system of two states with constant interstate coupling, which are in turn linearly coupled to an infinite set of harmonic oscillators. This model has been extensively used to develop a rigorous microscopic theory of electron-transfer dynamics in condensed media [363–368]. The spin-boson model also has been adapted to describe nonadiabatic excited-state dynamics in the context of molecular electronic spectroscopy [369–371].

In the well-known PI formulation of the spin-boson problem (see [252, 363] for reviews), time-dependent electronic site correlation functions or population probabilities are expressed as a sum over histories in electronic-state space in discretized time. The contribution of each single path is completely factorized with respect to the vibrational modes and can be interpreted as a product of propagators of driven harmonic oscillators. The summation over harmonic-oscillator degrees of freedom can be performed explicitly in this representation, yielding the well-known nonlocal influence functional first introduced by Feynman and Vernon [372]. The dimensionality dilemma of wave function quantum mechanics has thus been eliminated in the PI formulation, at the expense of introducing a summation over histories in electronic-state space.

Spin-boson PI theory provides the basis for rigorous analytical investigations of quantum dynamics in a dissipative environment, see, e.g., refs. [252, 363]. In numerical applications, the spin-boson PI has been evaluated by brute force summation of paths [369, 373]. As a consequence of the exponential growth of the number of paths with propagation time, such computations are limited to extremely short timescales. To overcome the latter limitation, the application of Monte Carlo sampling techniques has been investigated by several authors [374, 375]. The evaluation of real-time PIs by Monte Carlo methods is a tedious problem owing to the cancellation of oscillatory contributions and the resulting large statistical sampling error (the so-called sign problem [376]). It seems, however, that considerable progress has recently been achieved in overcoming the sign problem by partial analytic summation and filtering techniques [367, 368, 375, 377]. Very recently, alternative approximation methods have been developed that are free of statistical errors and potentially allow the treatment of long-time dynamics with numerical PI methods [378–383].

It appears natural to generalize the real-time PI methodology, which has proven extremely powerful in application to the spin-boson problem, to more general few-state multi-mode vibronic coupling problems, in order to treat electronically nonadiabatic dynamics in a more general context in terms of a sum over histories in electronic-state space. In the following discussion we briefly review our recent attempts to develop a computationally practical PI treatment for general two-state multi-mode vibronic-coupling models [384–386]. The problem is two-fold: The first part consists of expressing the quantities of interest (e.g., time-dependent correlation functions or population probabilities) in terms of a sum over paths in electronic-state space, such that the contribution of each path is completely factorized with respect to the vibrational modes. The second part of the problem consists of devising approximation methods that eliminate the exponential growth of the computational cost with propagation time. This has been achieved by a partial resummation of paths, combined with a recursive evaluation of partial sums [378, 379, 384]. With these techniques, essentially exact calculations become possible for vibronic-coupling problems that have many more degrees of freedom than can be treated by conventional wave-packet methods.

1. Construction of the Discretized Path Integral for Vibronic-Coupling Models

We shall be concerned with multi-mode vibronic-coupling models as defined in Section II.C.2. It is helpful to start with the special case of two electronic states that are linearly coupled by a single vibronically active mode, the coupling mode. In addition, we allow for many tuning modes, which modulate the energy gap of the interacting states and are therefore not separable from the vibronic problem. As discussed in Section II.C.2, this scenario represents the simplest generic model of a conical intersection of adiabatic potential-energy (PE) surfaces [35]. The S_1–S_2 vibronic-coupling problem in pyrazine (cf. Sections II.B,C and VII.A) is of this type. Since the interstate coupling is represented by a nuclear degree of freedom with an associated infinite-dimensional Hilbert space, the model is more general than the spin-boson model, where the interstate coupling is a *c*-number. The generalization of the theory required in the case of several coupling modes will be discussed subsequently.

As the simplest example of a response function, the correlation function $\Phi_v^{(22)}(t)$ defined in Eq. (6.20) will be considered. The construction of a PI expression for $\Phi_v^{(22)}(t)$ via the well-known Trotter break-up procedure [387, 388] is analogous to the spin-boson case [369]. Rewriting the time-development operator

$$U(t) = \exp(-iH't) \qquad H' = H/\hbar \tag{8.1}$$

as

$$U(t) = (U(\varepsilon))^N \qquad \varepsilon = t/N \tag{8.2}$$

and inserting a complete set of electronic states between all factors, one obtains

$$\Phi_{\mathbf{v}}^{22}(N\varepsilon) = \langle \mathbf{v} | \sum_{n_1 \ldots n_{N-1} = 1,2} U_{2n_{N-1}}(\varepsilon) \ldots U_{n_2 n_1}(\varepsilon) U_{n_1 2}(\varepsilon) | \mathbf{0} \rangle \tag{8.3}$$

where $U_{rs}(\varepsilon)$ denotes an electronic matrix element of $U(\varepsilon)$. Each sequence $\mathbf{n} = (n_1, n_2, \ldots, n_{N-1})$ in Eq. (8.3) can be interpreted as a path in electronic-state space. The elementary propagators $U_{rs}(\varepsilon)$ in Eq. (8.3) are now replaced by some suitable low-order approximation. Writing the Hamiltonian as

$$H = H_c + H_t \tag{8.4}$$

where H_c and H_t are the parts of the vibronic-coupling Hamiltonian representing the coupling and tuning modes, respectively, and employing the symmetrized second-order approximation [389, 390]

$$U(\varepsilon) = \exp(-iH_t'\varepsilon/2)\exp(-iH_c'\varepsilon)\exp(-iH_t'\varepsilon/2) + O(\varepsilon^3) \tag{8.5}$$

one achieves the disentanglement of the propagators of the coupling and tuning modes for each path. Denoting the propagator matrix element of each mode for a path $\mathbf{n}$ by $R_{vo}(\mathbf{n})$, the PI can be written compactly as

$$\Phi_{\mathbf{v}}^{(22)}(N\varepsilon) = \sum_{\{\mathbf{n}\}} R_{v_0 0}(\mathbf{n}) \prod_{j=1}^{M-1} R_{v_j 0}(\mathbf{n}) \tag{8.6}$$

Eq. (8.6) reveals explicitly the factorization of the contribution of each path with respect to the vibrational modes.

The factorization of the right-hand side of Eq. (8.6) allows the sequential computation of the contribution of each mode, i.e., the computational cost increases only linearly with the number M of vibrational degrees of freedom, and the computer storage demands are independent of M. The price to be paid is the summation over 2^{N-1} electronic paths in Eq. (8.6). The computational cost thus increases exponentially with the number N of time slices.

The matrix elements of the operators $\exp(-iH_c'\varepsilon)$ and $\exp(-iH_t'\varepsilon)$ in Eq. (8.5) can easily be evaluated to any desired numerical accuracy (recall that H_t is separable in the absence of the coupling mode). The elementary time step is thus solely determined by the $(O\varepsilon^3)$ error arising from the disentangling of the modes in Eq. (8.5). Each single mode is propagated exactly, independent of the time step ε. This feature of the formalism is essential for numerical applications and allows the use of relatively large time steps, typically of the order of 1 fs [384]. This nevertheless requires the evaluation of path sums with many hundreds of time slices, if we wish to evaluate correlation functions and population probabilities over a time range of a few hundred femtoseconds. The summation over paths in Eq. (8.6) thus represents a challenging computational problem.

We now briefly address the more general case of a vibronic-coupling problem with several coupling as well as tuning modes. In this case the PI construction leading to Eq. (8.6) does not result in a complete factorization of individual path contributions with respect to the vibrational modes. A solution to this problem has recently been found by invoking a homomorphism of two-state evolution operators with a subset of four-state evolution operators [386]. One thus obtains (see ref. [386] for details) a PI representation of the time-dependent correlation function of a general two-state multi-mode vibronic-coupling problem

$$\Phi_{\mathbf{v}}^{(22)}(N\varepsilon) = 2^{-(N+1)} \sum_{\{\boldsymbol{\nu}\}} \mathrm{sign}(\boldsymbol{\nu}) R^{el}(\boldsymbol{\nu}) \prod_j R_{v_j 0}(\boldsymbol{\nu}) \tag{8.7}$$

Each path is now labelled by a vector $\boldsymbol{\nu} = (\nu_1, \nu_2, \ldots, \nu_{N+1})$, where $\nu_i = 1, 2, 3, 4$. At the expense of extending the path sum, for N time slices, from 2^{N-1} terms in Eq. (8.6) to 4^N terms in Eq. (8.7), a complete factorization of the vibrational propagator has again been achieved.

In contrast to the PI construction leading to Eq. (8.6), the coupling and tuning modes are treated equivalently in the PI expression, Eq. (8.7). In Eq. (8.7), the contribution of each mode is given as the propagator of a driven harmonic oscillator

$$R_{v_j 0}(\boldsymbol{\nu}) = \langle v_j | T \exp\left(-\frac{i}{\hbar} \int_0^t dt' \, \tilde{h}_j(\boldsymbol{\nu}, t') \right) |0\rangle \tag{8.8}$$

$$\tilde{h}_j(\boldsymbol{\nu}, t) = h_j + f(\boldsymbol{\nu}, t) c_j(t) Q_j \tag{8.9}$$

The function $c_j(t)$ represents the modulus of the force acting on the jth oscillator. The definition of sign($\boldsymbol{\nu}$) and the time-dependent force $f(\boldsymbol{\nu}, t)$, both of which depend on the global path, is somewhat involved, see ref. [386] for details.

The matrix elements [Eq. (8.8)] of linearly driven harmonic oscillators can be evaluated using the Magnus formula [391, 392]. It is then possible, furthermore, to perform explicitly the summation over vibrational modes for each term of the PI, yielding an influence-functional expression for multimode vibronic-coupling problems. For the autocorrelation function, for example, we finally obtain [386]

$$\Phi_{\mathbf{0}}^{(22)}(N\varepsilon) = 2^{-(N+1)} \sum_{\{\boldsymbol{\nu}\}} \operatorname{sign}(\boldsymbol{\nu}) \cdot \exp\left(I(\boldsymbol{\nu}) - i\varepsilon\Delta/2 \sum_{l=0}^{N-1} f_{2l+1}(\boldsymbol{\nu}) \right) \tag{8.10}$$

where $\hbar\Delta$ is the electronic energy gap and $I(\boldsymbol{\nu})$ is the influence functional

$$I(\boldsymbol{\nu}) = -\int_0^t dt_1 \int_0^{t_1} dt_2\, f(\boldsymbol{\nu}, t_1) f(\boldsymbol{\nu}, t_2) K(t_1, t_2) \tag{8.11}$$

The kernel $K(t_1, t_2)$ is given by

$$K(t_1, t_2) = \sum_j c_j(t_1) c_j(t_2) \exp[-i\omega_j(t_1 - t_2)] \tag{8.12}$$

An analogous influence-functional expression can be derived for the time-dependent population probability $P_2(t)$, see ref. [386].

In the influence-functional representation of the PI, the vibrational modes are integrated out, and one can therefore handle systems with an infinite number of degrees of freedom. Equations (8.10–8.12) generalize the influence-functional idea of Feynman and Vernon [372] to a wide class of vibronic problems. The results can provide the basis, for example, for a rigorous quantum dynamical treatment of the multimode dynamical Jahn–Teller effect [140, 177, 393] in large molecules and crystals.

2. *Evaluation of Path-Integral Expressions*

As has been discussed above, the PI formulation eliminates, in principle, the need of direct-product basis or grid representations, which restricts wave-function-based treatments to systems with few degrees of freedom. On the other hand, the summation over 2^{N-1} paths for a two-state system with N time slices (or 4^N paths in the more general case, see above) represents a serious computational problem. The exponential growth of the computational effort with N limits brute-force computations of the path sum to extremely short timescales. While this may be satisfactory in some cases [369, 373, 394, 395], it is in general necessary to devise approximation methods that extend the tractable time range beyond that accessible to direct numerical path summation.

As an alternative to the widely used Monte Carlo sampling methods, which are plagued with the sign problem in real-time applications, a new method for the evaluation of discretized PIs for spin-boson and vibronic-coupling problems has recently been introduced [379, 380, 384, 385], which we briefly review below. The basic idea is to replace the sum over paths, which becomes computationally intractable for more than about 30 time steps, by a manageable sum over subsets (so-called classes) of paths. The contribution of each class is given by the product of the combinatorial weight of the class and the average of the multimode propagator, taken over all paths within the class. The class averages are evaluated by recursion with respect to the length of the path. In practical applications, approximations have to be introduced in evaluating the averages of propagator products, but the method is free of statistical errors. The resulting computational scheme has the interesting property of being intermediate between conventional time-dependent wave-packet propagation and direct numerical summation over paths.

To illustrate the basic ideas, let us return to Eq. (8.6) for the correlation function in the special case of a single coupling mode. We denote a class of paths by the symbol $\gamma_g, g = 1, \ldots, G$, where G is the number of classes. Introducing the average over paths within a class, $\langle \ldots \rangle_{\gamma_g}$, and the number of paths within a class, $Z(\gamma_g)$, the path sum is rewritten as

$$\Phi^{(rs)}_{\mathbf{vv}'}(N\varepsilon) = \sum_{g=1}^{G} Z(\gamma_g) \left\langle R_{v_0 v_0'} \prod_{j=1}^{M-1} R_{v_j v_j'}(\mathbf{n}) \right\rangle_{\gamma_g} \tag{8.13}$$

Each single-mode functional $R_{v_j v_j'}(\mathbf{n})$ can now be written as its average value plus the deviation from the average

$$R_{v_j v_j'}(\mathbf{n}) = \langle R_{v_j v_j'} \rangle + \delta R_{v_j v_j'}(\mathbf{n}) \tag{8.14}$$

Insertion of Eq. (8.14) into Eq. (8.13) yields the following expansion of the PI in terms of single-mode averages and mode correlations [384]

$$\begin{aligned} \Phi^{(rs)}_{\mathbf{vv}'}(N\varepsilon) = \sum_{g=1}^{G} Z(\gamma_g) \Bigg[& \langle R_{v_0 v_0'} \rangle_{\gamma_g} \prod_{j=1}^{M-1} \langle R_{v_j v_j'} \rangle_{\gamma_g} \\ & + \sum_{j} \langle \delta R_{v_0 v_0'} \delta R_{v_j v_j'} \rangle_{\gamma_g} \prod_{i \neq j} \langle R_{v_i v_i'} \rangle_{\gamma_g} \\ & + \sum_{lj} \langle \delta R_{v_l v_l'} \delta R_{v_j v_j'} \rangle_{\gamma_g} \langle R_{v_0 v_0'} \rangle_{\gamma_g} \prod_{i \neq l,j} \langle R_{v_i v_i'} \rangle_{\gamma_g} \\ & + \cdots \Bigg] \end{aligned} \tag{8.15}$$

While the propagator factorizes for each individual path, this is no longer true for class averages. The averaging over classes introduces correlations between vibrational modes. The *dynamical* mode–mode couplings, which have formally been eliminated in the construction of the PI [Eq. (8.6)], now appear in the form of *statistical* correlations.

The convergence of the correlation expansion [Eq. (8.15)] depends, of course, on the definition of the classes. In principle, the classes should be chosen such that the fluctuations of path contributions are kept as small as possible. A more refined grouping of paths into classes with similar contributions will result in a more rapid convergence of the expansion, Eq. (8.15). It has been shown that the most important criterion for distinguishing paths is the cumulative time a path stays in either electronic state [385]. By this criterion paths are grouped into classes $\gamma_{l_1 l_2}$, where l_i denotes the number of intervals in electronic state i ($i = 1, 2$). This classification accounts for the fact that the vibrational dynamics evolves on different PE surfaces in either diabatic state.

A finer classification may be obtained by discriminating the paths, in addition to l_1 and l_2, by the number k of hoppings between electronic states [384]. These classes are denoted by $\gamma_{kl_1 l_2}$. The additional discrimination with respect to k is suggested by analogy with perturbation theory, i.e., the contributions of paths with the same k exhibit the same scaling with the interstate coupling strength. Therefore fluctuations within the $\gamma_{kl_1 l_2}$ classes can be expected to be smaller than fluctuations within the $\gamma_{l_1 l_2}$ classes. The combinatiorial multiplicities of $\gamma_{l_1 l_2}$ and $\gamma_{kl_1 l_2}$ classes can be found in [385].

The single-mode averages and the mode correlations appearing in Eq. (8.15) can be evaluated by a simple recursion scheme [384]. This scheme follows from the fact that each path of length N can be constructed from a path of length $N - 1$ by adding either a hopping interval or an interval of straight propagation in either electronic state. For the example of $\gamma_{kl_1 l_2}$ classes, the recursion relation for single-mode averages reads ($s = 1, 2$)

$$\begin{aligned} \langle \boldsymbol{R}_j^{(1s)} \rangle_{kl_1 l_2} = Z(\gamma_{kl_1 l_2})^{-1} [& Z(\gamma_{k, l_1 - 1, l_2}) \boldsymbol{U}_j^{(11)} \langle \boldsymbol{R}_j^{(1s)} \rangle_{k, l_1 - 1, l_2} \\ & + Z(\gamma_{k-1, l_1 l_2}) \boldsymbol{U}_j^{(12)} \langle \boldsymbol{R}_j^{(2s)} \rangle_{k-1, l_1 l_2}] \end{aligned} \tag{8.16}$$

The $\boldsymbol{R}_j^{(rs)}$ and $\boldsymbol{U}_j^{(rs)}$ are matrices in a single-mode vibrational basis representation. The relation for $r = 2$ is obtained by interchanging the indices 1 and 2. Analogous recursion relations hold for correlations of two modes, three modes, etc.

The correlation expansion, Eq. (8.15), together with the recursive computation of averages and correlations represents a novel approach to evaluating

PIs. In contrast to most of the established methods for the numerical evaluation of PIs, which rely on global integration methods, the PI is evaluated here iteratively, much as in familiar time-dependent wave-packet propagation schemes discussed in Section III. However, the PI approach exhibits a different scaling of the computational cost with respect to the number of modes and the propagation time. In the zeroth order of the correlation expansion, the computational cost scales linearly with M, as in classical mechanics. In second order of the expansion, a propagation has to be performed for each pair of modes, and the effort thus rises as M^2. In general, the cost of an nth-order computation for a system with M modes scales as $\binom{M}{n}$. The scaling of the computational cost with respect to the propagation time depends on the class structure. For $\gamma_{l_1 l_2}$ classes, the cost increases with N^2, while $\gamma_{k l_1 l_2}$ classes lead to a N^3 scaling of the numerical effort. In some sense, the PI scheme described above can be viewed as an interpolation between time-dependent wave-function propagation on the one hand and direct numerical path summation on the other hand. In particular, the method offers considerable flexibility in tailoring the computational scheme according to the demands of the problem.

The efficiency and accuracy of the recursive PI method has been tested for the two-state four-mode model of S_1–S_2 vibronic coupling in pyrazine, referred to in Sections II.B and II.C.2 above [384, 385]. This model, for which exact numerical wave-packet calculations are available [44, 45], provides a challenging test for the PI approach. As an example, Fig. 36 shows the real part of the autocorrelation function $\Phi_{\mathbf{0}}^{(22)}(t)$ of the upper (S_2) electronic state. Path-integral results obtained with the $\gamma_{k l_1 l_2}$ classification of paths are compared with the exact reference (full line). Even in the zeroth order of the correlation expansion, which has been termed the class-average factorization approximation [384], a qualitatively correct result is obtained (dashed line). The result obtained in the second order of the correlation expansion with $\gamma_{k l_1 l_2}$ classes is shown by diamonds in Fig. 36. This result agrees to within drawing accuracy with the exact reference up to 80 fs. In this calculation, a time step of 0.7 fs has been employed, corresponding to a maximum of 147 intervals (see ref. [385] for more details). Figure 36 demonstrates that the recursive PI method represents an accurate computational method if an appropriate class structure and a sufficiently high order of the correlation expansion is employed.

For a few-mode system, the numerical effort required for an accurate PI integral calculation is rather higher than the effort for time-dependent wave-packet propagation using the methods of Section III. The computational effort of the PI approach, however, scales much more favorably with the dimension of the system and therefore makes it possible to carry out multimode calculations that are definitely impossible with wave-function-

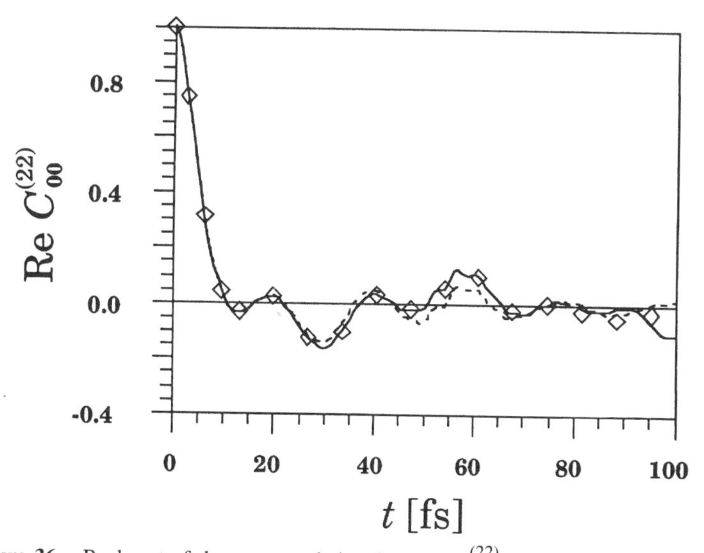

Figure 36. Real part of the autocorrelation function $\Phi_{00}^{(22)}(t)$ for the S_2 electronic state of the four-mode S_1–S_2 vibronic-coupling model of pyrazine. Path integral results obtained in zeroth order (dashed line) and second order (diamonds) of the correlation expansion are compared with the exact reference (solid line).

based methods. To demonstrate this fact, the PI method has been employed to compute the absorption spectrum of a 24-mode vibronic-coupling model [384, 385]. This model has been obtained by adding 20 weakly coupling tuning modes to the 4-mode vibronic-coupling model. The resulting absorption spectrum, obtained by Fourier transformation of the autocorrelation functions $\Phi_0^{(11)}(t)$ and $\Phi_0^{(22)}(t)$, is displayed in Fig. 37. The lower-energy part (representing the S_1 state, including intensity borrowing lines from the S_2 state) has been obtained via a second-order $\gamma_{l_1 l_2}$ calculation; the higher-energy part (representing the S_2 state), via a second-order $\gamma_{k l_1 l_2}$ calculation (see ref. [385] for details). In contrast to the few-mode model calculation for pyrazine shown in Fig. 21, no phenomenological optical dephasing has been assumed. In the 24-mode model, rapid optical dephasing in the energy range of the S_2 band arises in a microscopic manner due to multimode correlation effects. This leads to a nearly structureless absorption profile. In the energy range of the S_1 state, these ultrafast dephasing effects are absent, resulting in a sharply structured spectrum. It can be concluded that the PI method allows the study of energy-dependent optical dephasing processes in isolated large molecules within a purely microscopic framework.

The recursive path-class averaging approach, described above for the example of a vibronic-coupling system with a single coupling mode, can

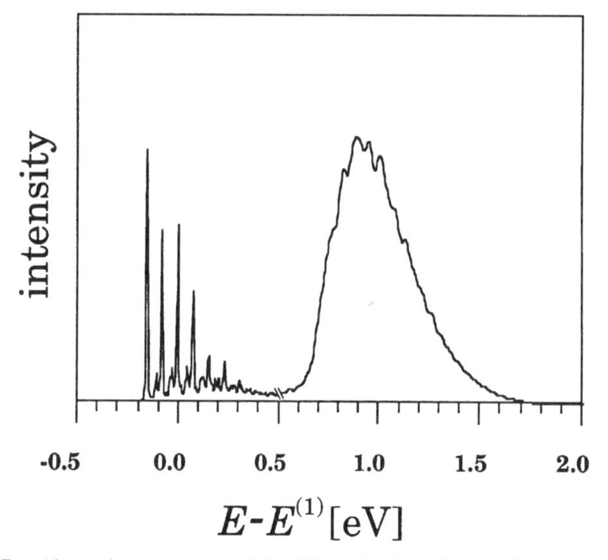

Figure 37. Absorption spectrum of the 24-mode vibronic-coupling model. The lower-energy part has been obtained via a second-order $\gamma_{l_1 l_2}$ calculation, the higher-energy part by a second-order $\gamma_{k l_1 l_2}$ calculation.

straightforwardly be extended to evaluate the PI for more general vibronic-coupling problems. It suggests itself to classify the paths in this case according to the cumulative propagation times l_1, l_2, l_3, l_4 in either of the four levels of the fictitious four-state system.

The correlation expansion [Eq. (8.15)] is, of course, restricted to systems with a finite number of modes. However, the basic idea of evaluating PIs by averaging over classes of paths and employing recurrence relations for the evaluation of averages can be combined with the influence-functional formalism for PI expressions involving linearly driven harmonic oscillators. Thus infinite-dimensional systems, such as the spin-boson model with Ohmic bath [252], become tractable by employing cumulant techniques for the evaluation of influence-functional averages [379]. An alternative iterative method for evaluating PIs for a low-dimensional system coupled to a dissipative bath has recently been developed by Makarov and Makri [381, 382]. This so-called tensor propagation method exploits the limited range of the memory kernel in cases where the system–bath coupling is described by a broad and structureless spectral function [382, 383]. When these techniques are applied to influence-functional PI expressions for general two-state multimode vibronic-coupling problems such as Eq. (8.10), a wide range of complex non-Born–Oppenheimer problems becomes accessible to a rigorous computational treatment.

To summarize, considerable progress has recently been made in developing computationally tractable real-time PI techniques for electronically nonadiabatic dynamics in systems with many degrees of freedom. The PI concept appears well suited to overcome the dimensionality dilemma of quantum mechanics for vibronic-coupling problems. As such it opens the way, for example, for a fully microscopic treatment of radiationless-transition dynamics. The available computationally tractable PI formulations are tailored according to the structure of the problem. For vibronic-coupling problems, for example, they rely on the fact that the dynamics on individual diabatic PE surfaces is assumed to be separable. Extension of these methods to a more general class of problems, e.g., multidimensional chemically reactive dynamics, is therefore not straightforward. Path-integral techniques are valuable, however, in providing rigorous results for the short-time dynamics of complex systems, which cannot be obtained by any other method. These results may serve as benchmarks to judge the performance of more approximate methods.

The interfacing of the PI formulation of vibronic-coupling dynamics with femtosecond time-resolved spectroscopy has not yet been elaborated. In the perturbative approach, multi-time response functions (see Section VI.A) have to be expressed as PIs, which is straightforward in principle. The extension of the path resummation techniques to these generalized response functions may be a tedious task, however. In the nonperturbative approach, explicitly time-dependent interstate couplings have to be incorporated into the PI formulation. Further developments of the theory are required in this area.

B. Reduced Density-Matrix Approach

The concept of partitioning a physical problem into a "system" and a "bath" plays an essential role in the description of photoinduced dynamics of large molecules or chromophores in solution. As explained in Section I, the vibrational modes that are strongly coupled to the electronic transition have to be treated explicitly and thus define the "system" degrees of freedom. The large number of vibrational modes, which are weakly or only indirectly coupled to the electronic transition, may be considered as the bath. The Hamiltonian of the material system is thus partitioned as

$$H = H^S + H^B + H^{SB}, \tag{8.17}$$

where H^S is the Hamiltonian of the isolated system, H^B the Hamiltonian of the bath, and H^{SB} describes their mutual interaction. Being primarily interested in observables referring to the system, we define the reduced density operator $\sigma(t)$ by taking the trace over the bath variables, i.e.,

$$\sigma(t) = \mathrm{Tr}_B(\rho(t)) \tag{8.18}$$

where

$$\rho(t) = e^{-iHt/\hbar}\rho(0)e^{iHt/\hbar} \tag{8.19}$$

is the density operator of the complete material system.

Using projection-operator techniques, a formally exact equation of motion can be derived for $\sigma(t)$ [396, 397]. The Nakajima–Zwanzig equation is too complicated, however, to be of practical utility. An approximate but tractable equation of motion for $\sigma(t)$ can be obtained by invoking a series of approximations that are clearly described, for example, in refs. [76, 78, 398, 399]. The derivation relies on time-dependent perturbation theory with respect to H^{SB}, the assumption that the bath remains in thermal equilibrium, as well as the Markovian approximation, which implies that the correlation time of the bath is short compared to relevant timescales of the system dynamics [78]. In the eigenstate representation of the system, one obtains the equation of motion [76, 78, 399]

$$\frac{\partial}{\partial t}\sigma_{ij}(t) = -i\omega_{ij}\sigma_{ij}(t) + \sum_{kl} R_{ijkl}\sigma_{kl}(t) \tag{8.20}$$

where $\omega_{ij} = (E_i - E_j)/\hbar$ and E_i, E_j are eigenvalues of H^S. The first term in Eq. (8.20) describes coherent dynamics of the isolated system. The Redfield relaxation tensor $\mathbf{R}$ accounts for the interaction of the quantum system with the thermal environment. The matrix elements of $\mathbf{R}$ are given by [78]

$$R_{ijkl} = \Gamma^{(+)}_{ljik} + \Gamma^{(-)}_{ljik} - \delta_{lj}\sum_r \Gamma^{(+)}_{irrk} - \delta_{ik}\sum_r \Gamma^{(-)}_{lrrj} \tag{8.21}$$

where

$$\Gamma^{(+)}_{ljik} = \hbar^{-2}\int_0^\infty dt\; e^{-i\omega_{ik}t}\langle H^{SB}_{lj}(t)H^{SB}_{ik}(0)\rangle_B \tag{8.22a}$$

$$\Gamma^{(-)}_{ljik} = \hbar^{-2}\int_0^\infty dt\; e^{-i\omega_{lj}t}\langle H^{SB}_{lj}(0)H^{SB}_{ik}(t)\rangle_B \tag{8.22b}$$

The brackets in Eqs. (8.22a, b) represent the trace over the bath in thermal equilibrium, and $H^{SB}(t)$ is defined as

$$H^{SB}(t) = \exp(iH^Bt/\hbar)H^{SB}\exp(-iH^Bt/\hbar) \tag{8.23}$$

It is seen that the matrix elements of the relaxation tensor are given by the Laplace transform of correlation functions of the system–bath coupling operator. For the evaluation of the matrix elements of H^{SB}, the complete set of eigenvectors of the system Hamiltonian is required.

Reduced-density-matrix (RDM) theory and simplified variants thereof are a well-established tool in many areas of physics and chemistry, e.g., in nuclear magnetic resonance (NMR) and optical spectroscopy. In NMR spectroscopy one is concerned with the driven dynamics of a few-level system (the nuclear spin) that is coupled to a thermal environment (the crystal lattice or the solvent) [76, 398, 399]. In optical spectroscopy, the celebrated optical Bloch equations describe the evolution of a few-level system in interaction with an electromagnetic field as well as a dissipative environment [74, 400]. The RDM equation of motion of the damped harmonic oscillator is widely employed to model dissipative effects in quantum optics [401, 402] and in molecular dynamics [403–405]. Mukamel and collaborators, for example, have developed a comprehensive description of femtosecond time-resolved spectroscopy that is based on the modeling of the material dynamics by a collection of damped harmonic oscillators [48, 84].

The familiar few-level optical Bloch equations [400] are not an appropriate model for the description of the real-time spectroscopy of ultrafast molecular processes. Dynamics on femtosecond timescales implies strong coupling and thus a complex level structure of the material system, and many of those levels are coherently coupled by the spectrally broad fields. A multilevel description of the material system is thus generally required [406, 407]. When RDM theory is applied to more complex systems such as those involving nonadiabatically coupled electronic states and several vibrational degrees of freedom, the scaling of the computational cost with the size of the system becomes a crucial issue. If the dimension of the relevant Hilbert space of H^S is N, there are N^2 elements of the reduced density matrix σ and N^4 elements of the relaxation tensor $\mathbf{R}$. The computational effort required for the evaluation of $\mathbf{R}$ and the propagation of σ thus scales as N^4. Moreover, the complete set of eigenvectors of H^S is required for the construction of $\mathbf{R}$, which represents another limitation for large N. The description of ultrafast nonadiabatic excited-state dynamics in condensed phases in terms of RDM theory thus represents a nontrivial computational problem.

The application of multilevel Redfield theory to coupled electronic-state dynamics, in particular electron-transfer processes, in condensed phases has recently been addressed by several authors [85, 86, 408–415]. In these models two diabatic electronic states, e.g., a neutral excited state and a charge-transfer state, are usually considered. These are diagonally coupled to a single vibrational mode, the reaction mode, and off-diagonally coupled by a constant electronic matrix element V. The reaction mode, which is usually taken as a harmonic mode, is in turn coupled to a thermal bath. This type of model is well suited for the investigation of the interplay of coherent vibrational dynamics, electronic surface-hopping processes, as well as energy and phase relaxation processes [411, 412]. In contrast to the widely employed Golden Rule description of electron-transfer, this approach treats the interstate coupling in a nonperturbative manner. It is thus suitable to deal with strong couplings that are typical for intramolecular electron-transfer processes or quasi-resonant charge transfer in bridge systems [412, 414].

Representative applications of proper Redfield theory in the eigenstate representation of the system, as defined in Eq. (8.20), are given in refs. [85, 412–415]. The bath is described classically by a fluctuating force [413, 416]. Such calculations are possible for systems with $\approx 10^2$ energy levels. They allow, for example, the treatment of multisite dissipative tunneling processes with a single reaction mode [414], but can hardly be extended to multimode vibronic-coupling systems, where N is larger by several orders of magnitude.

May, Schreiber, and collaborators have pursued the strategy of applying RDM theory in the diabatic state representation rather than the eigenstate representation of the system [86, 410, 411]. Denoting by $\mu, \nu, \ldots$ electronic-vibrational direct-product states in the diabatic representation, the RDM equation of motion takes the form

$$\frac{\partial}{\partial t} \sigma_{\mu\nu} = -i\omega_{\mu\nu}\sigma_{\mu\nu} - \frac{i}{\hbar} \sum_{k} (V_{\mu\kappa}\sigma_{\kappa\nu} - V_{\kappa\nu}\sigma_{\mu\kappa}) - \sum_{\kappa\lambda} R_{\mu\nu\kappa\lambda}\sigma_{\kappa\lambda} \quad (8.24)$$

Here

$$\omega_{\mu\nu} = (E_\mu - E_\nu)/\hbar \quad (8.25)$$

are differences of energy levels of the unperturbed diabatic surfaces and the $V_{\mu\nu}$ are matrix elements of the diabatic interstate coupling. The relaxation tensor **R** in Eq. (8.24) describes relaxation of the diabatic electronic-vibrational states through the coupling to the environment.

Equation (8.20) represents the more correct physical model, while Eq. (8.24) is more convenient from the computational point of view. The construction of the relaxation tensor in Eq. (8.24) is much easier, especially if the diabatic surfaces can be assumed to be harmonic, which is often a good approximation for electron-transfer processes. The model defined by Eq. (8.24) is thus very useful for the description of electron-transfer processes in a dissipative environment, if the interstate coupling V is not too strong. The deficiencies of the modeling of dissipation in the uncoupled diabatic representation have been studied in detail for the exactly solvable problem of the dissipative Jaynes–Cummings model [417].

If dissipation is introduced in the diabatic representation, as defined in Eq. (8.24), it is straightforward to extend the model to include several system modes, that is, to associate H^S with the multimode vibronic-coupling Hamiltonian discussed in Section II.C.2. Assuming, for simplicity, harmonic diabatic surfaces with identical vibrational frequencies and invoking the secular (or rotating-wave) approximation for the system-bath coupling [77, 401], the RDM equation of motion in operator notation takes the form [418]

$$\frac{\partial}{\partial t}\,\sigma(t) = -\frac{i}{\hbar}\,[H^S, \sigma(t)] + R(\sigma) \tag{8.26}$$

$$R(\sigma) = \sum_{ij}\sum_{k}\frac{\Gamma_k}{2}\,|\psi_i\rangle\lambda_k^{(ij)}\langle\psi_j| \tag{8.27}$$

$$\lambda_k^{(ij)} = b_k^{(i)\dagger}\,b_k^{(i)}\sigma^{(ij)} + \sigma^{(ij)}b_k^{(j)\dagger}\,b_k^{(j)} - 2b_k^{(i)}\sigma^{(ij)}b_k^{(j)\dagger} \tag{8.28}$$

Here $\sigma^{(ij)} = \langle\psi_i|\sigma|\psi_j\rangle$ are electronic matrix elements of the reduced density operator and the $b_k^{(i)}$ ($b_k^{(i)\dagger}$) are lowering (raising) operators for the kth oscillator in the diabatic state $|\psi_i\rangle$. The operator H^S is of the form (cf. Section II.C.2)

$$H^S = H_o + V \tag{8.29}$$

$$H_o = \sum_{i=1,2}|\psi_i\rangle\left\{E_i + \sum_k \hbar\omega_k\left(b_k^{(i)\dagger}\,b_k^{(i)} + \frac{1}{2}\right)\right\}\langle\psi_i| \tag{8.30}$$

$$V = |\psi_1\rangle V_{12}\langle\psi_2| + \text{h.c.} \tag{8.31}$$

Equations (8.26–8.28) generalize the vibronic-coupling models of Section II.C.2 to include damping of the harmonic-oscillator motion on the diabatic surfaces through coupling to a zero-temperature environment. The generalization to a finite-temperature environment is straightforward [77, 401]. In this model, the relaxation tensor is characterized by just a few parameters, namely a relaxation rate Γ_k for each vibrational mode. Equations (8.27, 8.28) are appropriate in the limit of weak damping. In the case of stronger system–bath coupling, nonsecular relaxation terms have to be included. The corresponding relaxation operator, which is equivalent to the classical Brownian oscillator model [404, 405] is also well known [419, 420] and allows the modeling of overdamped vibrational motion. As discussed above, the RDM model in Eq. (8.26) is limited to the case of weak to moderate interstate coupling V. The range of validity of Eqs. (8.26–8.28) extends, however, beyond the Golden Rule regime. It defines a model that is very useful for the analysis of the interplay of multimode intramolecular electron-transfer dynamics and environmental dissipation.

In contrast to Redfield theory in the system eigenstate representation as defined by Eq. (8.20), the construction of the Redfield tensor in Eq. (8.27) represents no problem even for multimode systems. The problem thus reduces to solving the RDM equation of motion, Eq. (8.26). At first sight, this appears to be an impossible task for system Hamiltonians with three or four nonseparable modes, where N is of the order of 10^4 to 10^6 (cf. Section III), and σ thus possesses 10^8 to 10^{12} elements. However, it has recently been shown for the example of master equations of the dissipative two-level system and the damped harmonic oscillator arising in quantum optics that the solution of the master equation for the RDM can be replaced by a stochastic wave-function propagation scheme [421–424]. This so-called Monte Carlo wave-function propagation or quantum-jump method (which can be viewed as an integration of the perturbation series for the density matrix using Monte Carlo importance-sampling techniques [425]) exhibits a considerably more favorable scaling of the computational cost with the size of the system than the conventional integration of the density-matrix equations of motion. An alternative wave-function description of the time evolution of open quantum systems is provided by the so-called quantum state diffusion formalism [426–428].

In the quantum-jump method, one starts with the definition of a non-Hermitian effective Hamiltonian [422]. For the master equation of Eqs. (8.26–8.28), the effective Hamiltonian reads [418]

$$H_{\text{eff}} = H^S - i\hbar \sum_k \frac{\Gamma_k}{2} |\psi_i\rangle b_k^{(i)\dagger} b_k^{(i)} \langle\psi_i| \tag{8.32}$$

The effective Hamiltonian is then employed to propagate the wave function according to the time-dependent Schrödinger equation

$$i\hbar \frac{\partial}{\partial t} |\Psi(t)\rangle = H_{\text{eff}} |\Psi(t)\rangle \tag{8.33}$$

By expansion in a basis set, cf. Eq. (3.4), Eq. (8.33) is converted into a set of coupled first-order differential equations, which are solved using one of the methods discussed in Section III.B.

Quantum jumps are required in order to restore the conservation of the trace of the density operator, see refs. [422, 423, 425] for details. In the

Monte Carlo algorithm proposed by Mølmer et al. [422] the deviation δp of the norm of the wave function from unity is evaluated after an appropriate time step and compared with a number ε, which is randomly chosen from the interval [0, 1]. If $\delta p > \varepsilon$, a quantum jump occurs in one of the vibrational modes, i.e., the lowering operator b_k acts on $\Psi(t)$. The relative probability of a quantum jump in mode k is given by $\Gamma_k / \sum_l \Gamma_l$. The wave function is then renormalized. If $\delta p < \varepsilon$, no quantum jump occurs and the wave function is just renormalized. Matrix elements of the density operator are finally obtained as an ensemble average of products of expansion coefficients over a sufficiently large number N_R of realizations

$$\sigma^{(ij)}_{v_1 v_2 \ldots v_1' v_2' \ldots}(t) = \overline{C^{(i)}_{v_1 v_2 \ldots}(t) C^{(j)^*}_{v_1' v_2' \ldots}(t)} \tag{8.34}$$

where the bar denotes the ensemble average. It can be shown that this procedure is equivalent, for $N_R \rightarrow \infty$, to the solution of the master equation, Eq. (8.26) [422].

In practice, a few hundred realizations have been found to be sufficient for obtaining reasonably converged results [418]. It is obvious that the Monte Carlo wave-function method is preferable to the propagation of the RDM equation of motion if the dimension N of the system Hilbert space exceeds the number of realizations. In particular, models of the type in Eq. (8.26), with several active vibrational modes in the system Hamiltonian, become computationally tractable with the stochastic wave-function method. At least for such models, the Monte Carlo wave-function propagation scheme opens the possibility of a numerical study of truly complex multi-level systems which are weakly coupled to a thermal environment.

To illustrate the application of these techniques, let us consider a model of a typical photoinduced electron-transfer process, where a charge-transfer state is populated by optical excitation from the electronic ground state. Internal conversion of the excited state to the ground state leads to ultrafast back-transfer of the electron. This type of process has extensively been studied, for example, in binuclear transition-metal complexes, see [18] and references therein. We shall consider a model for H^S which invovles three strongly coupled vibrational modes. The electronic inter-state coupling is taken as $V = 0.06$ eV and is small compared to the vibrational frequencies $\omega_1, \omega_2, \omega_3$. The system is weakly coupled to a bath at zero temperature, the damping rate being $\hbar\Gamma = 0.005$ eV for all three modes. For this choice of parameters, the assumptions made in the derivation of the model, Eq. (8.26–8.28), should be well justified (see ref. [429] for more details).

In this example, the dimension of the vector representing the wave function is $N = 6000$. The RDM thus possesses 3.6×10^7 elements. The direct numerical solution of the RDM equation of motion would be prohibitively expensive. With the Monte Carlo wave-function scheme, a few hundred propagations of a vector of dimension 6000 are required, which can be done within a few hours of CPU time on a modern workstation.

Fig. 38 shows, as a representative result, the population probability $P_1(t)$ of the diabatic charge-transfer state, assuming preparation of this state at $t = 0$ by vertical excitation from the ground state. The population dynamics of the three-mode system including dissipation (full line) is compared with the population dynamics of the isolated three-mode system (dashed line). It is seen that the model exhibits an ultrafast initial electron-transfer process which is unaffected by environmental effects. For long times, the population of the diabatic charge-transfer state in the isolated system fluctuates around 0.3, which is typical for few-mode vibronic-coupling systems (cf. Section IV). Upon inclusion of environmental damping, $P_1(t)$ relaxes to a very small asymptotic value. The additional decay of $P_1(t)$ due to environmental dissipation occurs on a timescale of $\approx$ 100 fs, which is considerably faster than the inverse of the damping rate Γ of the oscillators. This presumably reflects the fact that high vibrational levels of the ground state are populated in the ultrafast internal-conversion process, and these levels possess a shorter lifetime than low vibrational levels.

Similar studies of population relaxation dynamics of electronically coupled systems including dissipation have been reported by May, Schreiber, and collaborators [86, 409, 410] and Jean, Pollard, and Friesner [85, 412,

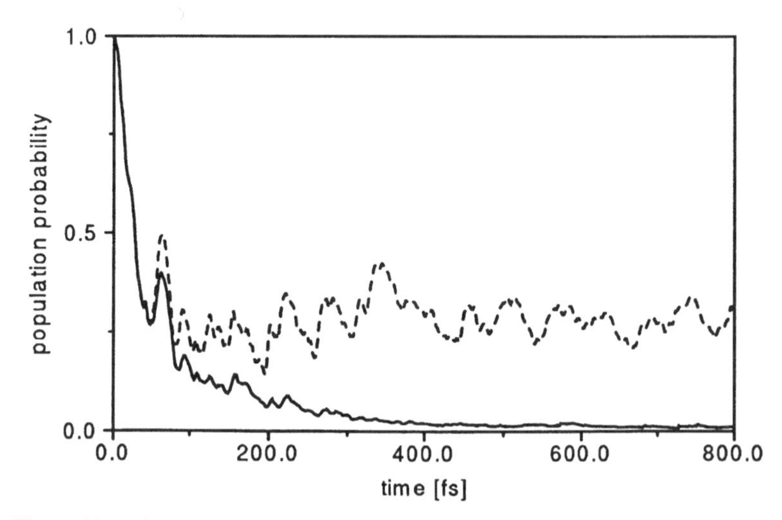

Figure 38. Time-dependent population probability of the diabatic charge transfer state, obtained for the three-mode model with (solid curve) and without (dashed curve) dissipation.

413, 415] for models with a single vibrational mode. A multimode description is probably necessary if one aims at a realistic description of ultrafast intramolecular charge transfer processes, in particular, if the model is to be based on *ab initio* PE surfaces. Although some of the multimode effects can be subsumed in the environmental relaxation operator in a phenomenological modeling approach, this may be in conflict with the requirement of weak system–bath coupling. It appears preferable to include all strongly coupled degrees of freedom explicitly in the system Hamiltonian, in order to ensure weak system–bath coupling.

It is straightforward to extend the RDM model of Eq. (8.24) to include the interaction of the material system with the radiation field. Clearly, the system–bath separation must be chosen such that the radiation field interacts only with the system and not with the bath. The equation of motion including the external driving field then reads

$$\frac{\partial}{\partial t}\sigma_{\mu\nu} = -i\omega_{\mu\nu}\sigma_{\mu\nu} - \frac{i}{\hbar}\sum_{\kappa}[(V_{\mu\kappa} - \boldsymbol{\mu}_{\mu\kappa}\cdot\boldsymbol{E}(t))\sigma_{\kappa\nu} - (V_{\kappa\nu} - \boldsymbol{\mu}_{\kappa\nu}\cdot\boldsymbol{E}(t))\sigma_{\mu\kappa}] - \sum_{\kappa\lambda} R_{\mu\nu\kappa\lambda}\sigma_{\kappa\lambda} \tag{8.35}$$

where $\boldsymbol{\mu}_{\kappa\lambda}$ are vibrational matrix elements of the transition-dipole-moment vector of Eq. (2.18) and $\boldsymbol{E}(t)$ is the external electric field.

As in the derivation of Eq. (8.24), it is required that the field-induced energy level shifts of the system be comparatively small. Equation (8.35) is thus appropriate to model the interaction of a dissipative system with laser fields of moderate intensity. If one aims at the description of strong field effects, the field has to be taken into account in the derivation of the relaxation tensor, employing, for example, the Floquet representation [430, 431].

Despite certain limitations, Eq. (8.35) provides a model that appears very suitable for the theoretical modeling of femtosecond PP spectra of molecules in a condensed-phase environment. As has been discussed in Section VI.B, a nonperturbative treatment of the laser field can be preferable, even for weak fields, if the system dynamics is complex and thus defies an analytic description. By combining the theory of Section VI.B with the RDM equation of motion [Eq. (8.35)], one obtains a model of femtosecond spectroscopy that describes the effects of intramolecular couplings, environmental dissipation, and field-induced transitions in a balanced manner. A related problem is the time-resolved spontaneous fluorescence of a vibronically coupled dissipative system, which has recently been treated by Jean [85].

C. Semiclassical Description of Nonadiabatic Transitions

In many areas of chemical physics, the application of classical mechanics has been found to be a powerful tool for the description of the dynamics of polyatomic systems [432, 433]. In cases where both the system under consideration and the observable to be calculated have an obvious classical analog (say, the translational-energy distribution after a scattering event), quasi-classical trajectory calculations are a well-established approach to simulate quantum dynamical processes. It is less clear, however, how one can incorporate into a classical theory quantum mechanical degrees of freedom that do not possess an obvious classical analog. To account for nonadiabatic processes such as internal conversion and intersystem crossing, one obviously has to go beyond the simple picture of running trajectories on a single Born–Oppenheimer PE surface.

The *formal* solution of this problem was given by Pechukas, who showed that the path–integral formalism provides a consistent formulation of the coupling of quantum and classical degrees of freedom [434]. Assuming that $|\psi_1\rangle, |\psi_2\rangle$ are the initial and final electronic states and $\mathbf{R}_1, \mathbf{R}_2$ are the corresponding initial and final nuclear positions of the system at times t_1, t_2, the propagator can be written as [434]

$$\langle \mathbf{R}_2 | \langle \psi_2 | e^{-i/\hbar H (t_2 - t_1)} | \psi_1 \rangle | \mathbf{R}_1 \rangle = \int_{\mathbf{R}_1}^{\mathbf{R}_2} \mathcal{D}\mathbf{R} \, K_{12}[\mathbf{R}(t)] \exp\left(i/\hbar \int_{t_1}^{t_2} dt \; T_{\mathbf{R}} \right) \tag{8.36}$$

where the path integral is over all possible nuclear paths from $\mathbf{R}_1$ to $\mathbf{R}_2$, and $T_{\mathbf{R}}$ denotes the kinetic energy of the nuclei. The function $K_{12}[\mathbf{R}(t)]$ represents the transition amplitude for the $|\psi_1\rangle \rightarrow |\psi_2\rangle$ electronic transition with the nuclei constrained to follow the path $\mathbf{R}(t)$. Equation (8.36) is exact and can be readily interpreted in terms of a two-step procedure for solving the problem [434–436]: One first chooses a nuclear trajectory $\mathbf{R}(t)$, for which the time-dependent electronic problem has to be solved to obtain $K_{12}[\mathbf{R}(t)]$. One then multiplies this by the phase factor associated with the nuclear kinetic energy and integrates over all possible nuclear paths. All approximate models discussed below perform, in one way or another, the first step of this procedure. In other words, these methods would actually be exact if one could perform the path integral over all possible nuclear paths.

Based on Pechukas' formulation, Miller and George developed a semiclassical theory of nonadiabatic transitions [435], which involves two main assumptions: a semiclassical approximation to the electronic transition amplitude $K_{12}[\mathbf{R}(t)]$ and a stationary phase-type approximation to evaluate the path integral. In this theory, nonadiabatic transitions are taken into account through the propagation of trajectories on analytically continued PE surfaces. This formulation requires the propagation of complex-valued trajectories in complex time. To make a transition, a trajectory follows a continuous path from the initial surface, passing through complex branch points where the PE surfaces intersect, and arrives at the final surface. A transition between adiabatic PE surfaces thus appears in this semiclassical theory as a classically forbidden process in the sense that the nuclear dynamics must follow a complex-valued trajectory [436, 437].

The formulations of Pechukas [434] and Miller and George [435] are beautiful theories that have been used, for example, to derive or justify more approximate descriptions such as the surface-hopping method [438–440]. Their computational value, however, is rather limited, because the practical implementation turns out to be more cumbersome than the corresponding exact quantum calculation. More approximate, but also more practical, methods that are suitable to treat large molecular systems have mostly been developed on empirical grounds. There are many mixed quantum-classical schemes: for example, various versions of the surface-hopping model of Tully and Preston [438–445] (for a recent review see ref. [446]) and a variety of classical-path models [447–461]. Both formulations share the idea that the electronic degrees of freedom are described quantum mechanically, while the nuclear degrees of freedom are described classically; however, they differ in the way the electronic system is coupled to the nuclear dynamics.

In the surface-hopping approach, classical trajectories are propagated on a single adiabatic PE surface until, according to some "hopping criterion," a transition probability $P_{0 \leftarrow 1}$ to another PE surface is calculated and, depending on the comparison of $P_{0 \leftarrow 1}$ with a random number, the trajectory "hops" to the other adiabatic surface. The many existing variants of the method differ mainly in choice and degree of sophistication of the hopping criteria, but share the problem that the hopping processes tend to destroy the electronic coherence after a relatively short time [440]. Classical-path methods, on the other hand, propagate the classical degrees of freedom self-consistently in an *averaged* potential, the value of which is determined by the instantaneous populations of the different quantum states. In this way, the electronic coherence is conserved, but one has to deal with problems arising from the approximation that the trajectory propagates in a mean potential instead of in a state-specific one.

Historically, the surface-hopping as well as the classical-path approach have been developed for scattering applications, where the molecular system is asymptotically in a single electronic state. These boundary conditions are naturally fulfilled by surface-hopping methods, whereas the mean-potential approximation inherent in classical-path methods may lead to unphysical results (see, e.g., the discussion in ref. [438]). The situation may be rather different, however, for the description of nonadiabatic bound-state dynamics such as the internal-conversion process of a molecule following photoexcitation. These processes may involve several transitions between electronic PE surfaces, thus rendering a correct treatment of the phase relation between individual electronic transitions quite important for a theoretical description. Moreover, time-dependent wave-packet calculations for multidimensional vibronic-coupling models have shown that the photoexcited system typically localizes in the lower *adiabatic* state, but is *diabatically* still in a mixture of states [37, 39]. This asymptotic behavior has been found to be well described by classical-path models, which are usually defined by representing the Hamiltonian in a diabatic electronic basis [266, 462–464]. Surface-hopping methods, on the other hand, have been extensively employed in molecular-dynamics simulations to describe charge-transfer processes in solution [439, 440, 465–467]. Only very recently have the first surface-hopping simulations of internal-conversion processes in polyatomic molecules been reported [152, 468]. Due to the lack of comparison with exact reference calculations, however, the validity of the surface-hopping approximation for the description of nonadiabatic bound-state relaxation dynamics is not yet clear.

To illustrate the similarities and differences of surface-hopping and classical-path methods, it is instructive to briefly consider the equations of motion of the two approaches. Classical-path models can be defined in the adiabatic [448, 450–452, 457, 459, 460] as well as in the diabatic [449–456, 458, 461] representation. Most numerical studies, however, have been performed using the diabatic representation, which avoids numerical problems due to the singularities of the nonadiabatic couplings [459]. Surface-hopping models, on the other hand, usually employ the adiabatic representation. There are two main reasons for this choice: (1) In the surface-hopping approach the nuclear trajectories are necessarily propagated on a *single* PE surface. A *single adiabatic* PE surface is expected to better represent the coupled-surfaces problem than a *single diabatic* one. (2) The picture of instantaneous hops only appears plausible if the interactions between the PE surfaces are localized. Nonadiabatic kinetic-energy couplings are usually better localized than the diabatic potential couplings.

Assuming that the nuclear motion can be described by the trajectory $\mathbf{R} = \mathbf{R}(t)$, the electronic Hamiltonian in Eq. (2.1) and the nonadiabatic couplings in Eq. (2.6) depend parametrically on $\mathbf{R}(t)$. In the adiabatic representation the wave function can be defined as

$$\Psi(\mathbf{r}, \mathbf{R}, t) = \sum_{k} a_k(t)\tilde{\psi}_k(\mathbf{r}, \mathbf{R}) \tag{8.37}$$

where the $a_k(t)$ are complex-valued expansion coefficients and the $\tilde{\psi}_k(\mathbf{r}, \mathbf{R})$ are adiabatic basis functions (see Section II.A). Insertion of Eq. (8.37) into the time-dependent Schrödinger equation in Eq. (3.1) yields the electronic equations of motion in the adiabatic representation

$$i\hbar\dot{a}_k = \tilde{V}_k(\mathbf{R})a_k + \sum_{k'} \Lambda_{kk'}(\mathbf{R})a_{k'} \tag{8.38}$$

where the $\tilde{V}_k(\mathbf{R})$ denote the adiabatic PE surfaces and $\Lambda_{kk'}(\mathbf{R})$ are the nonadiabatic coupling matrix elements defined in Eq. (2.6). To give the electronic equations of motion in the diabatic representation, we adopt diabatic basis states $\psi_k(\mathbf{r})$ and define the wave function as

$$\Psi(\mathbf{r}, \mathbf{R}, t) = \sum_{k} d_k(t)\psi_k(\mathbf{r}) \tag{8.39}$$

thus obtaining

$$i\hbar\dot{d}_k = V_k(\mathbf{R})d_k + \sum_{k'} V_{kk'}(\mathbf{R})d_{k'} \tag{8.40}$$

where $V_k(\mathbf{R})$ and $V_{kk'}(\mathbf{R})$ denote the diabatic potential matrix elements of the Hamiltonian in Eq. (2.20).

It should be pointed out that the electronic equations of motion are the same for both the surface-hopping model and the classical-path model. The main difference between the two methods arises from the way the electronic system is coupled to the nuclear dynamics, i.e., through the effective potential the nuclear trajectory experiences. In the surface-hopping scheme, the nuclear dynamics is governed by the equations of motion on a *single* adiabatic PE surface $\tilde{V}_k(\mathbf{R})$

$$M_j\ddot{R}_j = -\frac{\partial\tilde{V}_k(\mathbf{R})}{\partial R_j} \tag{8.41}$$

where the subscript j labels the vibrational degrees of freedom. In the classical-path scheme, on the other hand, the nuclear dynamics is governed by the quantum mechanically *averaged* Ehrenfest force [469]. Adopting, for simplicity, a diabatic representation, the nuclear equations of motion in the classical-path approximation read

$$M_j \ddot{R}_j = -\left\langle \Psi(t) \right| \frac{\partial H}{\partial R_j} \left| \Psi(t) \right\rangle$$

$$= -\sum_k |d_k|^2 \frac{\partial V_k(\mathbf{R})}{\partial R_j} - \sum_{kk'} d_k d_{k'}^* \frac{\partial V_{kk'}(\mathbf{R})}{\partial R_j} \tag{8.42}$$

A comparison of Eqs. (8.41) and (8.42) shows that in the surface-hopping approach the nonadiabatic couplings enter the nuclear dynamics only indirectly, via instantaneous hopping processes, while in the classical-path approach, the trajectory "feels" both the PE surfaces and the nonadiabatic couplings all the time. In both methods, the quantum nature of the initial state of the classically treated subsystem is simulated through quasi-classical sampling of the corresponding probability distribution, i.e., one performs a Monte Carlo integration over the nuclear initial conditions $\mathbf{R}, \dot{\mathbf{R}}$, which may be sampled, e.g., from the Wigner distribution [432].

To gain a sound theoretical understanding of the essential physics underlying mixed quantum-classical schemes, Stock [461, 464] has recently proposed a semiclassical TDSCF formulation that represents a generalization of existing classical-path theories. In this approach, the total density operator is approximated by a semiclassical *ansatz*, which couples the electronic degrees of freedom to the nuclear degrees of freedom in a self-consistent manner, whereby the vibrational density operator is described in terms of Gaussian wave packets. Employing the semiclassical TDSCF *ansatz*, the approximations leading to the classical-path equations of motion have been derived, thus revealing the criteria of validity and applicability of the method. It has been shown that, besides the inherent mean-potential approximation, the assumption of rapid randomization of nuclear phases represents the central approximation of the classical-path approach [461]. Efficient electronic relaxation [464] or coupling of the system to a quasi-continuous bath of vibrational modes [461], for example, have been found to be sufficient to ensure that the nuclear phases get randomized rapidly enough to be unimportant for the overall time evolution of the system. The validity of the mean-potential approximation has been shown to depend crucially on the form of the nonadiabatic coupling [464]. It has been found that the possible inadequacy of this approximation can be corrected to a certain extent by employing "Langer-type" corrections to the off-diagonal elements of the electronic density operator, as has been suggested earlier by Meyer and Miller in the theoretical framework of the classical electron analog model [451].

Apart from theoretical investigations, there also have been various numerical studies, which have demonstrated that the classical-path concept describes quite accurately ultrafast electronic and vibrational relaxation dynamics. Examples include the electron-transfer dynamics of the dissipative two-level system [461] and internal-conversion processes triggered by conically intersecting PE surfaces [266, 462, 463], as well as ultrafast nonadiabatic photoisomerization processes [262]. In comparison with exact quantum-mechanical calculations, classical-path models have been shown to reproduce, almost quantitatively, complex structures of time-dependent observables (e.g., electronic population probabilities and the mean positions and energy content of vibrational modes). They have also been used to calculate time- and frequency-resolved electronic spectra of these processes [262, 463].

The classical and semiclassical descriptions of nonadiabatic photodynamics represents a fascinating approach, the importance of which is also reflected in the rapid increase of theoretical work on this field. In particular, semiclassical TDSCF methods [456, 461, 464, 470, 471] and semiclassical propagation schemes employing the so-called initial-value representation [472–478] appear to be quite promising for future applications. As a demonstration of the potential of the classical-path scheme discussed above, some computational results of representative examples are presented in the following section.

D. Classical Modeling of Internal Conversion and Photoisomerization

In this section we will consider two examples, for which a classical-path description appears appropriate: (1) the $\tilde{C} \rightarrow \tilde{B} \rightarrow \tilde{X}$ internal conversion process in the benezene cation [190, 266, 464] and (2) several models describing nonadiabatic photoisomerization as introduced in Section II.C [39, 262]. We compare the classical simulations to exact quantum calculations and study how the dynamics changes when further vibrational degrees of freedom (e.g., remaining intramolecular modes or solvent degrees of freedom) are included in the model. Furthermore we present classical simulations of time- and frequency-resolved pump-probe spectra for these systems, and discuss briefly the prospects and limits of the classical approach.

1. The $\tilde{C} \rightarrow \tilde{B} \rightarrow \tilde{X}$ Internal-Conversion Process in the Benzene Cation

Recently Köppel, Domcke, and Cederbaum have presented quantum mechanical studies on the ultrafast $\tilde{C} \rightarrow \tilde{B} \rightarrow \tilde{X}$ electronic relaxation process in the benzene cation (Bz^+) [88, 189, 190]. As the $\tilde{X}^2E_{1g}$ and $\tilde{B}^2E_{2g}$ electronic states are degenerate, five electronic states have to be considered in the general model Hamiltonian in Eq. (2.20). Furthermore, in the case of Bz^+, one has to take into account three different kinds of vibronic interactions: the Jahn–Teller coupling within the $\tilde{X}$ and $\tilde{B}$ states through the modes ν_6–ν_9 (Wilson numbering), the coupling of the $\tilde{C}$ and $\tilde{B}$ states by the pseudo-Jahn–Teller active modes ν_{16} and ν_{17}, and the coupling of the $\tilde{B}$ and $\tilde{X}$ states by the modes ν_4 and ν_5 [190].

To be able to perform quantum-mechanical wave-packet propagations, Köppel has proposed a simplified model Hamiltonian, which neglects the presumably weaker coupled vibrational modes $\nu_2, \nu_5, \nu_7, \nu_9$, and ν_{16} and also ignores the degeneracy of the remaining modes ν_6, ν_8, ν_{17} and the $\tilde{X}$ and $\tilde{B}$ electronic states [190]. Large-scale dynamical quantum calculations have been reported employing the resulting three-state five-mode model [190]. These solve numerically exactly the time-dependent Schrödinger equation [190]. It has been assumed that the system is initially in the vibrational ground state of neutral benzene shifted up vertically to the $\tilde{C}$ PE surface, which corresponds to ionization from the electronic and vibrational ground state of neutral benzene.

In all classical simulations reported here we use a diabatic classical-path propagation scheme with dynamically corrected diabatic couplings, as described in ref. [464]. All parameters of the model Hamiltonians are taken from ref. [190]. For a further discussion of the classical methodology see refs. [266, 461, 464], and for details of the model Hamiltonians see refs. [189, 190].

The quantum and classical calculations of the diabatic population probabilities $P_k(t)$ for the 5-mode 3-state model of Bz^+ are compared in Fig. 39(*a* and *b*). The quantum calculation exhibits an oscillatory population decay of the initially excited $\tilde{C}$ electronic state, which is reflected in a rapid rise and similar beatings of the $\tilde{B}$ state population. After a somewhat delayed onset, the population of the $\tilde{X}$ electronic state exhibits a non-oscillatory rise and dominates over both the $\tilde{C}$ and $\tilde{B}$ electronic states after $\approx$ 120 fs. The classical calculation is seen to reproduce the features of the quantum reference, e.g., transient beatings and long-time limits, fairly well. Except for the amplitude of the quantum beats and the classically somewhat slow $\tilde{B} \rightarrow \tilde{X}$ decay, the agreement of quantum and classical calculations is almost quantitative. Considering the simplicity of the classical model (on a standard work-station the quantum calculations took 2 CPU days compared to 20 CPU minutes for the classical calculation), the accuracy of the classical-path propagation is remarkable.

Let us now consider the electronic relaxation dynamics of a more realistic model system for Bz^+, where the remaining vibrational modes $\nu_2, \nu_5, \nu_7, \nu_9$, and ν_{16} are included and the degeneracy of the (pseudo) Jahn–Teller modes and the $\tilde{X}$ and $\tilde{B}$ electronic states is explicitly taken into account. Figure 39(*c*) shows a classical calculation of the diabatic population probabilities obtained

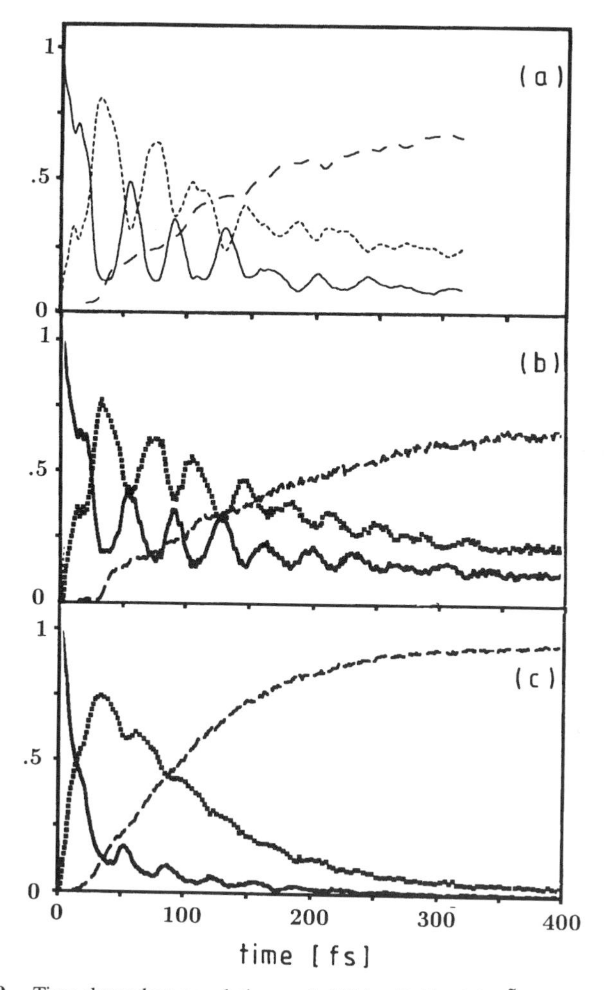

Figure 39. Time-dependent population probabilities $P_k(t)$ of the $\tilde{C}$ (solid line), $\tilde{B}$ (dotted line), and $\tilde{X}$ (dashed line) diabatic electronic states of the benzene cation. (*a*) Quantum and (*b*) classical calculation for the 5-mode 3-state model. (*c*) Classical calculation of the diabatic populations for the 16-mode 5-state model of the benzene cation.

for the resulting 16-mode 5-state model of Bz^+. A comparison of the 5-mode 3-state model Fig. 39(*b*) and the 16-mode 5-state model Fig. 39(*c*) reveals that the inclusion of additional electronic and vibrational degrees of freedom results in (1) a large suppression of the coherent quantum beating and (2) a faster and more complete internal-conversion process, i.e., within $\approx$ 300 fs the initially excited $\tilde{C}$ electronic state has completely decayed into the electronic ground state $\tilde{X}$ of the system.

A detailed computational study of the intramolecular vibrational relaxation dynamics of the two models of Bz^+ shows that the faster decay is mainly a consequence of the rather effective energy redistribution in the additional Jahn–Teller active modes ν_7 and ν_9 [464]. The $\tilde{C} \rightarrow \tilde{B} \rightarrow \tilde{X}$ internal conversion process in Bz^+ thus represents an interesting example of the highly efficient intramolecular energy redistribution in multimode Jahn–Teller systems. As has been discussed in detail in ref. [35] for the example of a two-mode Jahn–Teller model of BF_3^+, the mode–mode interaction results in a particularly strong distortion of the adiabatic PE surfaces, which in turn gives rise to the ultrafast electronic relaxation processes observed in these systems.

Figure 39 illustrates the transition from partially coherent, Fig. 39(*b*), to largely incoherent, Fig. 39(*c*), electronic relaxation, depending on the number of strongly coupled degrees of freedom incorporated in the system. The transition from coherent to incoherent relaxation behavior has been discussed in detail in the context of the dissipative two-level system [252], and was found to be well described by classical-path simulations [464]. Finally it should be pointed out that the internal conversion process in Bz^+ represents an example for a photoinduced relaxation process that, even on an ultrafast timescale, requires the inclusion of many intramolecular degrees of freedom. Such a situation renders an appropriate quantum-mechanical description very difficult.

2. *Photoisomerization Dynamics of a Chromophore in Solution*

One of the most appealing features of classical simulations is that an arbitrary form of the potential energy function can be used. For this reason, it is interesting to study to what extent the classical-path methodology introduced above works for model Hamiltonians with a more general form of the potential than in the previous example. In the following discussion, we focus on model systems describing large-amplitude motion associated with a unimolecular reaction. To check the validity of the classical-path model for this type of dynamics, we use the three-mode model for nonadiabatic photoisomerization introduced in Section II.C, and compare the classical simulations to the results of the exact quantum-mechanical propagation presented in Section IV.

Figure 40 shows (*a*) the diabatic and adiabatic population probabilities $P_k^{\text{di}}(t)$ and $P_k^{\text{ad}}(t)$, (*b*) the probability $P_{cis}(t)$, (*c*) and the mean vibrational energies $E_j(t)$ for a three-dimensional model of photoisomerization [39, 262]. As has been discussed in detail in Section IV, $P_{cis}(t)$ and the diabatic excited-state population $P_1^{\text{di}}(t)$ undergo a rapid initial decay, which is followed by transient oscillations. Asymptotically, both populations approach the value $\frac{1}{2}$. The adiabatic population probability $P_1^{\text{ad}}(t)$, on the other hand, is seen to decay within ≈ 250 fs to a value of ≈ 0.25. This finite value of $P_1^{\text{ad}}(\infty)$ reflects the rather low level density of the three-mode model. Figure 40(c) shows the vibrational energy transfer from the initially excited torsional coordinate (solid line) to the coupling mode (dashed line) and the tuning mode (dotted line), which occurs on the same timescale as isomerization and internal conversion. A comparison of the classical results on the right-hand side of Fig. 40 to the exact quantum calculations on the left-hand side reveals that the classical model is quite capable of reproducing the main features of the reference calculation. The asymptotic limits as well as transient structures of all time-dependent observables are captured at least semi-quantitatively. In particular, the classical simulation is seen to accurately describe the vibrational relaxation dynamics exhibited by the three-mode system.

Having checked the reliability of the classical-path description for a low-dimensional model of photoisomerization, we now want to ask the following question: Assuming that the model introduced above accounts for the three most important vibrational modes of the system, how will the inclusion of the remaining (presumably weakly coupling) nuclear degrees of freedom and interactions affect the overall time evolution of the molecular system? On the simplest level, the remaining intramolecular and intermolecular nuclear degrees of freedom of the problem may be described in terms of a harmonic "bath" that is bilinearly coupled to the "system" coordinate (i.e., the torsional mode). We thus wish to study the essential effects that arise when the three-mode model of photoisomerization (the "chromophore") is embedded in a simple environment (e.g., the solvent) [262].

As a representative example, Fig. 41 shows, for the first two picoseconds, the diabatic and adiabatic electronic population probabilities $P_1^{\text{di}}(t)$ and $P_1^{\text{ad}}(t)$ [Fig. 41(a)], the probability $P_{cis}(t)$ [Fig. 41(b)], the mean vibrational energies $E_j(t)$ [Fig. 41(c)], and the time-dependent vibrational energy distribution of the bath modes [Fig. 41(d)]. As in the three-mode calculation,[9] the adiabatic population probability undergoes a rapid initial decay followed by a small recurrence of the population at $t \approx 200$ fs. However, in contrast to the situation with the three-mode model, which for larger times fluctuates around 0.3, the adiabatic population of the full system is subsequently seen to decay to zero on a picosecond timescale. This electronic relaxation process is accompanied by an vibrational energy redistribution process shown in Fig. 41(c), which transfers energy from the torsional mode ($E_{\text{rot}}(t)$, upper solid lines) to the coupling mode ($E_c(t)$, dashed lines), to the tuning mode ($E_1(t)$, lower solid lines), and to the bath modes ($E_{\text{bath}}(t)$, dotted lines). It is seen that the rapid initial decay of $P_1^{\text{ad}}(t)$ coincides with the initial energy

[9]The three-mode model used in Fig. 41 is slightly different from the model shown in Fig. 40, see ref. [262].

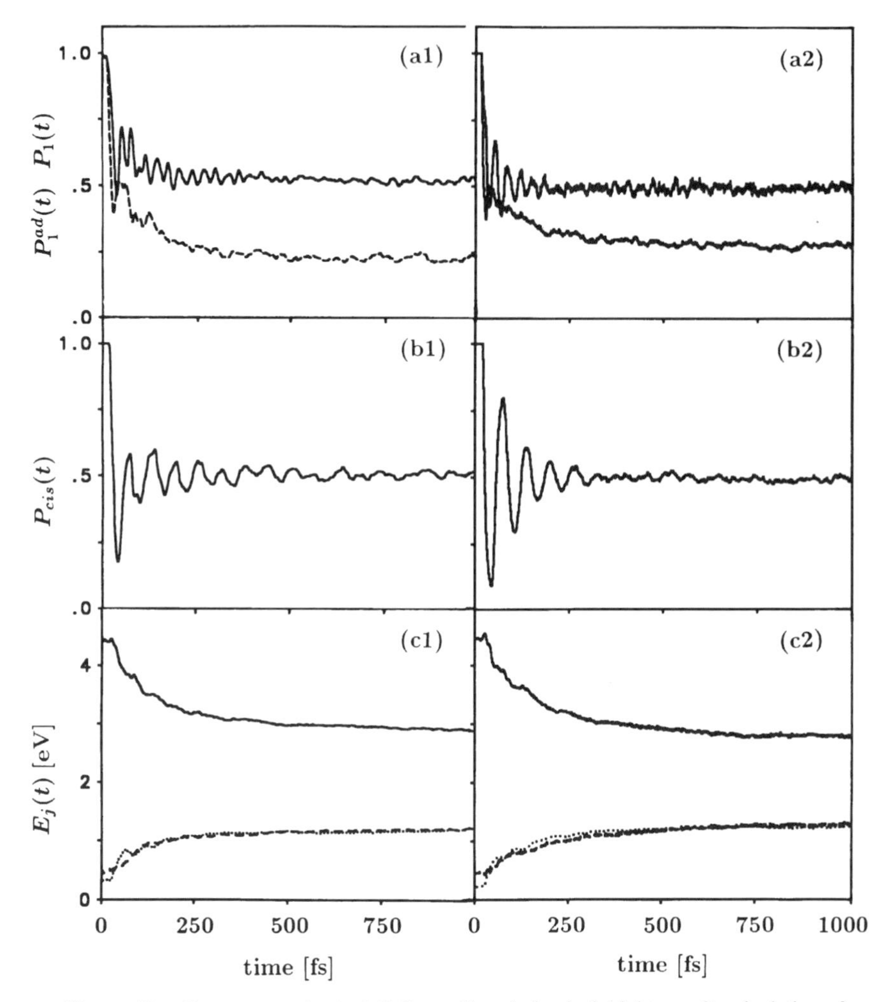

Figure 40. Quantum-mechanical (left panel) and classical (right panel) calculations for the three-dimensional model. Shown is the time evolution of the mean values of the diabatic population $P_k^{\text{di}}(t)$ [upper lines in (*a*)], the adiabatic populations $P_k^{\text{ad}}(t)$ [lower lines in (*a*)], and the probability $P_{cis}(t)$ (*b*). The vibrational mean energies $E_j(t)$ in panels (*c*) illustrate the vibrational energy redistribution from the torsional mode ($E_{\text{rot}}(t)$, solid lines) into the coupling mode ($E_c(t)$, dashed lines) and the tuning mode ($E_1(t)$, dotted line).

transfer into the coupling and tuning modes, while the picosecond decay of $P_1^{\text{ad}}(t)$ coincides with the slower energy transfer into the bath modes. In other words, for first few hundred femtoseconds the time evolution of the full system (chromophore and solvent) is quite similar to the case of the bare

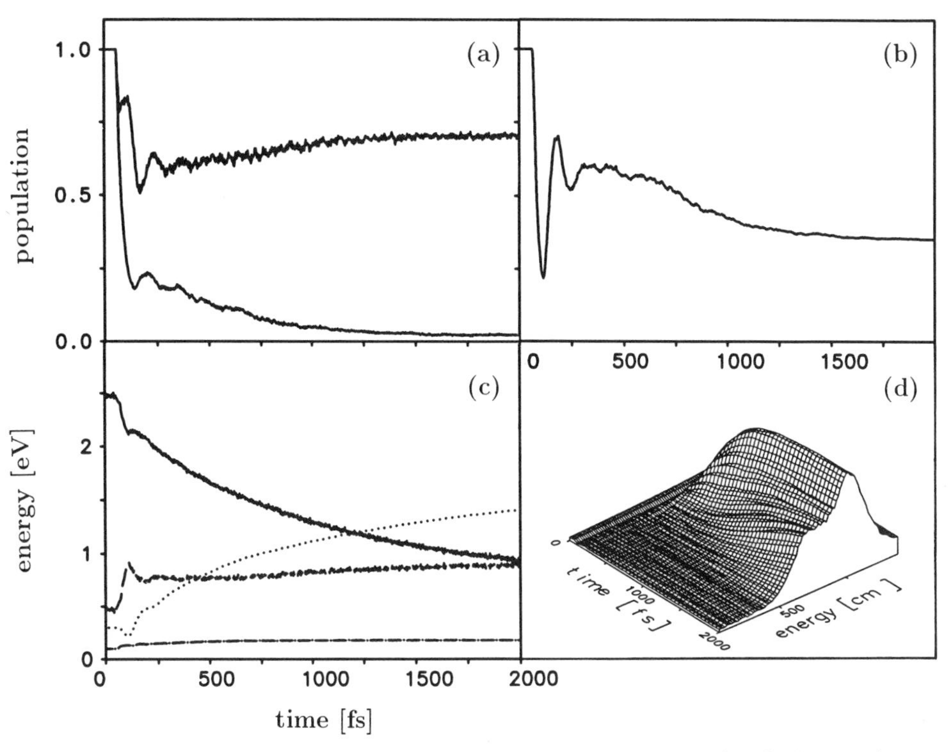

Figure 41. Classical simulation of the photoinduced relaxation dynamics of a three-mode chromophore embedded in a harmonic bath. Shown is the time evolution of the mean values of the diabatic population $P_k^{\mathrm{di}}(t)$ [upper lines in (*a*)], the adiabatic population $P_k^{\mathrm{ad}}(t)$ [lower lines in (*a*)], and the probability $P_{cis}(t)$ (*b*). The vibrational mean energies $E_j(t)$ in (*c*) illustrate the vibrational energy redistribution from the torsional mode ($E_{\mathrm{rot}}(t)$, upper solid line) into the coupling mode ($E_c(t)$, dashed line), the tuning mode ($E_1(t)$, lower solid line), and the sum of the bath modes ($E_{\mathrm{bath}}(t)$, dotted line). (*d*) Time-dependent distribution of vibrational energy in the bath modes.

three-mode model (chromophore only), thus supporting the general idea that only a few strongly coupled degrees of freedom determine the dynamics on the shortest time scales.

To study the vibrational energy transfer from the photoexcited chromophore to the environment in some more detail, Fig. 41(*d*) shows the time-dependent distribution of vibrational energy in the bath degrees of freedom. The vibrational energy distribution is seen to exhibit two maxima located at $\omega_1 \approx 240\ \mathrm{cm}^{-1}$ and $\omega_2 \approx 440\ \mathrm{cm}^{-1}$. The former peak develops within the first few hundred femtoseconds, while the latter exhibits a delayed onset. It is interesting to note that these resonance structures hardly depend on the properties of the bath modes, but rather are due to the torsional dynamics of the system. A closer examination shows that the two peaks of the vibrational energy distribution reflect (1) the initial motion of the wave packet on the S_1 PE surface, and (2) the subsequent quasi-harmonic motion with the averaged harmonic frequency of the torsional S_1 and S_2 potentials. While the general features are similar, numerical studies for a variety of model systems have shown that the details of the vibrational relaxation (e.g., the question of mode specificy or the decay of vibrational coherence) may be quite different. In particular, it has been found that the quantum yield of the *cis-trans* photoreaction (i.e., $P_{cis}(\infty)$) depends to a large extent on the specific chromophore-solvent coupling employed, as it governs directly the competition of the various relaxation pathways [464].

3. *Classical Modeling of Femtosecond Experiments*

Most of the semiclassical descriptions mentioned in Section VIII.C have been concerned with the calculation of nonadiabatic transition rates. Comparatively few studies have addressed the (semi)classical evaluation of electronic spectra [54, 307, 463, 479–486]. Recently, a classical formulation of the spectroscopy of nonadiabatic excited-state dynamics has been developed, which allows us to calculate the nonlinear polarization of vibronically coupled electronic states in a rather general way [463]. As has been discussed in detail [463], there are several ways to establish a quantum-classical correspondence. The simplest approach is to replace the dipole operator in the Heisenberg representation by its classical analog function [451], and to evaluate the quantum mechanical trace over the multi-time dipole correlation functions in terms of a classical phase-space integral over nuclear initial conditions. In the present context, we apply the theory of ref. [463] to the description of femtosecond spectroscopy of ultrafast nonadiabatic photoisomerization processes.

Let us first check the validity of the classical description for the calculation of time-resolved spectra. To this end we again adopt the above three-mode model of photoisomerization and compare the classical simulations to the exact quantum calculation shown in Fig. 31. Figure 42 shows the classical simulation of the time-resolved stimulated-emission spectrum of the three-mode model, plotted as a function of the pulse delay time Δt and the probe carrier frequency ω_2. We have assumed Gaussian laser pulses of 6-fs duration. As has been discussed in detail for the quantum mechanical calculation, the pump-probe spectrum for delay times $\Delta t \gtrsim 100$ fs is seen to split up into two components centered at $\omega_2 \approx 2.5$ eV and $\omega_2 \approx 4.5$ eV, respectively, reflecting the stimulated emission ($S_1 \rightarrow S_0$) of the reactants and the absorption ($S_0 \rightarrow S_1$) of the products. The classical simulation (Fig. 42) is seen to catch the overall features of the quantum mechanical calculation (Fig. 31).

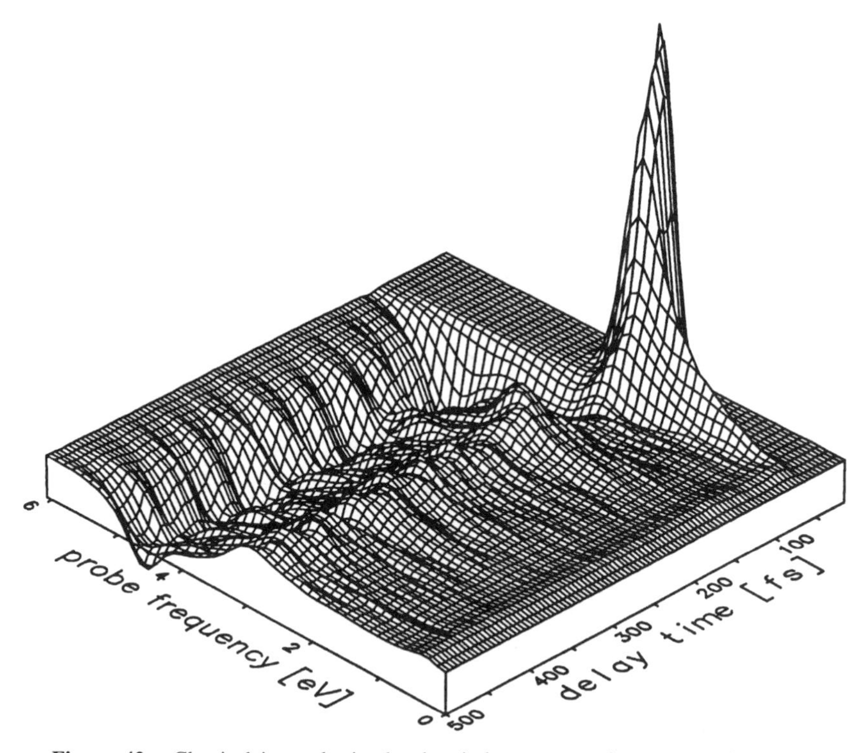

Figure 42. Classical integral stimulated-emission pump-probe spectrum for the three-mode model of photoisomerization, obtained for Gaussian laser pulses of 6-fs duration.

In particular, there is a reasonable agreement between the two approaches for the initial decay, the global time evolution of the reactant and product channel, and the width of the spectrum at a given time. The limits of the classical approximation are clearly seen, however, when one compares the detailed vibrational structures occurring in the absorption and the emission signals. Keeping in mind that the simulation of time- and frequency-resolved spectra is a much greater challenge for an approximative method than the calculation of a highly-averaged total reaction rate, the overall agreement of classical and quantum calculations is quite encouraging.

It is interesting to investigate the effect of weakly coupled bath degrees of freedom on the time-resolved signals. A simulation of the transient transmittance spectra reveals that initially ($\lesssim$ 300 fs) the time-resolved spectra pertaining to the three-mode model (chromophore only) and to the full system (chromophore and solvent) are rather similar [262]. For later times, however, the dynamics of the three-mode model becomes quasi-stationary, whereas the calculations for the full system reflect the redistribution of the

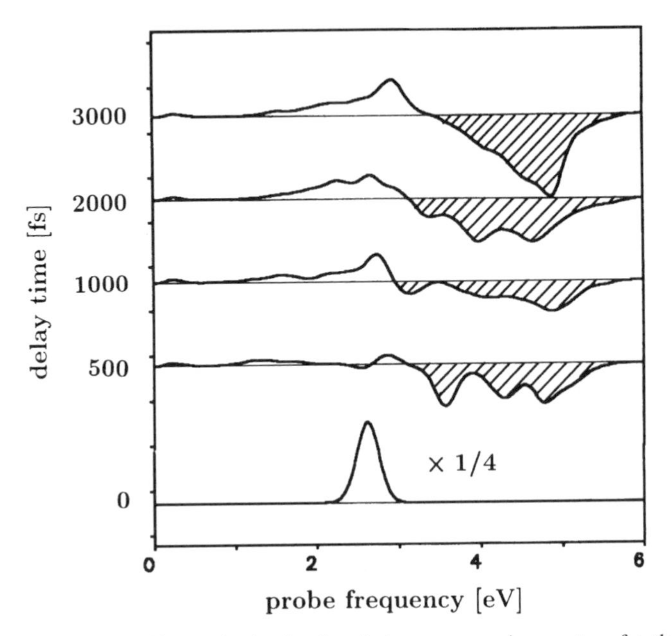

Figure 43. Classical integral stimulated-emission pump-probe spectra of a three-mode chromophore embedded in a harmonic bath, assuming Gaussian laser pulses of 10-fs duration.

excess energy of the reaction mode into the bath nuclear degrees of freedom. Figure 43 shows a classical calculation of the corresponding transient stimulated-emission spectrum, obtained for Gaussian laser pulses of 10-fs duration. As expected from elementary Frank–Condon and resonance-condition arguments, the simulation reveals that the cooling of the vibrationally hot photoproducts by the solvent is mainly reflected in a blue-shift and a narrowing of the width of the excited-state absorption spectrum [348].

It may be noted that the overall features of the pump-probe simulations shown in Figs. 42 and 43 are qualitatively quite similar to existing experimental data on ultrafast isomerization (cf. the discussion in Section VII.C). The microscopic description of representative photochemical problems, such as the photoisomerization of stilbene or rhodopsin in the condensed phase, represents a major challenge for future theoretical work. It can be expected that such simulations may soon become feasible by the use of mixed quantum-classical propagation schemes.

IX. NOTE ADDED IN PROOF

Since the completion of the manuscript, many experimental and theoretical papers addressing femtosecond excited-state dynamics and spectroscopy have been published. Particularly closely related to subjects covered in the present review is recent theoretical work on nonadiabatic torsional dynamics in 9-(N-carbazolyl)-anthracene [487] and on nonadiabatic photodissociation dynamics of ammonia [489] and organometallic complexes [488, 490]. Ferretti et al. [491] have recently obtained interesting quantum mechanical and classical results that illustrate the dynamics at a conical intersection.

ACKNOWLEDGMENTS

In this article, we have extensively drawn on results which have been obtained by former and present graduate students, in particular Stefan Krempl, Heiko Plöhn, Rudolf Schneider, Matthias Seel, Luis Seidner, Manfred Winterstetter, Brigitte Wolfseder, and Clemens Woywod, to whom we wish to express our gratitude for an intensive and pleasant collaboration. We further thank Andrzej Sobolewski and Tucker Carrington, Jr. for stimulating discussions. Continuous support by the Deutsche Forschungsgemeinschaft is gratefully acknowledged.

REFERENCES

1. R. Schinke, *Photodissociation Dynamics*, Cambridge University Press, Cambridge, 1993.
2. J. B. Birks, *Photophysics of Aromatic Molecules*, Wiley, New York, 1970.
3. J. Michl and V. Bonačić-Koutecký, *Electronic Aspects of Organic Photochemistry*, Wiley, New York, 1990.
4. E. S. Medvedev and V. I. Osherov, *Radiationless Transitions in Polyatomic Molecules*, Springer, Berlin, 1995.
5. M. Klessinger and J. Michl, *Excited States and Photochemistry of Organic Molecules*, VCH, New York, 1995.
6. Femtochemistry, Special Issue, Eds. J. Manz and A. W. Castleman, Jr., *J. Phys. Chem.*, **97** (1993).
7. *Ultrafast Phenomena IX*, P. F. Barbara, W. H. Knox, G. A. Mourou and A. H. Zewail, Eds., Springer, Berlin, 1994.
8. A. H. Zewail, *Femtochemistry—Ultrafast Dynamics of the Chemical Bond*, World Scientific, Singapore, 1994.
9. *Femtosecond Reaction Dynamics*, D. A. Wiersma, Ed., North Holland, Amsterdam, 1994.
10. *Femtosecond Chemistry*, J. Manz and L. Wöste, Eds., VCH, New York, 1994.
11. M. Dantus, M. J. Rosker, and A. H. Zewail, *J. Chem. Phys.*, **89,** 6128 (1988).
12. T. Rose, M. J. Rosker, and A. H. Zewail, *J. Chem. Phys.*, **88,** 6672 (1988); **91,** 7415 (1989); *Chem. Phys. Lett.*, **146,** 175 (1988).
13. M. J. Rosker, F. W. Wise, and C. L. Tang, *Phys. Rev. Lett.*, **57,** 321 (1986); F. W. Wise, M. J. Rosker, and C. L. Tang, *J. Chem. Phys.*, **86,** 2827 (1987).
14. M. H. Vos, J.-C. Lambry, S. J. Robles, D. C. Youvan, J. Breton, and J.-L. Martin, *Proc. Natl. Acad. Sci. USA*, **88,** 8885 (1991).
15. J.-Y. Bigot, M. T. Portella, R. W. Schoenlein, C. J. Bardeen, A. Migus, and C. V. Shank, *Phys. Rev. Lett*, **66,** 1138 (1991).
16. M. S. Pshenichnikov, K. Duppen, and D. A. Wiersma, *Phys. Rev. Lett.*, **74,** 674 (1995).

17. L. A. Peteanu, R. W. Schoenlein, R. A. Mathies, and C. V. Shank, *Proc. Natl. Acad. Sci. USA*, **90,** 11762 (1993).
18. P. J. Reid, C. Silva, P. F. Barbara, L. Karki, and J. T. Hupp, *J. Phys. Chem.*, **99,** 2609 (1995).
19. K. Dressler, E. Umlauf, S. Schmidt, P. Hamm, W. Zinth, S. Buchanan, and H. Michel, *Chem. Phys. Lett.*, **183,** 270 (1991).
20. B. Bühler, R. Thalweiser, and G. Gerber, *Chem. Phys. Lett.*, **188,** 247 (1992).
21. T. Baumert, C. Röttgermann, C. Rothenfusser, R. Thalweiser, V. Weiss, and G. Gerber, *Phys. Rev. Lett.*, **69,** 1512 (1992).
22. T. A. Gregory and S. Lipsky, *J. Chem. Phys.*, **65,** 296 (1976).
23. M. R. Topp, H.-B. Lin, and K.-J. Choi, *Chem. Phys.*, **60,** 47 (1981).
24. V. A. Lyubimtsev and V. L. Ermolaev, *Opt. Spectrosc.*, **61,** 320 (1986).
25. G. Herzberg and H. C. Longuet-Higgins, *Discuss. Faraday Soc.*, **35,** 77 (1963).
26. T. Carrington, *Accts. Chem. Res.*, **7,** 20 (1974).
27. L. Salem, *J. Am. Chem. Soc.*, **96,** 3486 (1974).
28. V. Bonačić-Koutecký and J. Michl, *Theor. Chim. Acta*, **68,** 45 (1985).
29. F. Bernardi, S. De, M. Olivucci, and M. A. Robb, *J. Am. Chem. Soc.*, **122,** 1737 (1990).
30. S. Xantheas, S. T. Elbert, and K. Ruedenberg, *Theor. Chim. Acta*, **78,** 365 (1991).
31. A. L. Sobolewski, C. Woywod, and W. Domcke, *J. Chem. Phys.*, **98,** 5627 (1993).
32. D. R. Yarkony, *J. Chem. Phys.*, **100,** 3639 (1994).
33. P. Celani, F. Bernardi, M. Olivucci, and M. A. Robb, *J. Chem. Phys.*, **102,** 5733 (1995).
34. M. Klessinger, *Angew. Chem.*, **107,** 597 (1995).
35. H. Köppel, W. Domcke, and L. S. Cederbaum, *Adv. Chem. Phys.*, **57,** 59 (1984).
36. R. Schneider, W. Domcke, and H. Köppel, *J. Chem. Phys.*, **92,** 1045 (1990).
37. U. Manthe and H. Köppel, *J. Chem. Phys.*, **93,** 345 (1990).
38. U. Manthe and H. Köppel, *J. Chem. Phys.*, **93,** 1658 (1990).
39. L. Seidner and W. Domcke, *Chem. Phys.*, **186,** 27 (1994).
40. W. Domcke, H. Köppel, and L. S. Cederbaum in *Stochasticity and Intramolecular Redistribution of Energy*, Eds. R. Lefebvre and S. Mukamel, Reidel, Dordrecht, 1987, p. 217.
41. G. W. Robinson, in *Excited States*, E. C. Lim., Ed., Academic Press, New York, 1974, Vol. 1, p. 1.
42. J. Jortner and S. Mukamel, in *Molecular Energy Transfer*, R. Levine, J. Jortner, Eds., Wiley, New York, 1976, p. 178.
43. K. F. Freed, in *Radiationless Processes*, F. K. Fong, Ed., Springer, Berlin, 1976, p.23.
44. G. Stock and W. Domcke, *J. Phys. Chem.*, **97,** 12466 (1993).
45. C. Woywod, W. Domcke, A. L. Sobolewski, and H.-J. Werner, *J. Chem. Phys.*, **100,** 1400 (1994).
46. S. Mukamel, *Phys. Rep.*, **93,** 1 (1982).
47. M. Mitsunaga and C. L. Tang, *Phys. Rev. A*, **35,** 1720 (1987).
48. S. Mukamel, *Ann. Rev. Phys. Chem.*, **41,** 647 (1990).
49. W. T. Pollard, S.-Y. Lee, and R. A. Mathies, *J. Chem. Phys.*, **92,** 4012 (1990).
50. L. Seidner, G. Stock, and W. Domcke, *J. Chem. Phys.*, **103,** 3998 (1995).
51. L. R. Khundkar and A. H. Zewail, *Ann. Rev. Phys. Chem.*, **41,** 15 (1990).

52. V. Engel and H. Metiu, *J. Chem. Phys.*, **90,** 6116 (1989); **91,** 1596 (1989).
53. S. E. Choi and J. C. Light, *J. Chem. Phys.*, **90,** 2602 (1989).
54. S.-Y. Lee, W. T. Pollard, and R. A. Mathies, *J. Chem. Phys.*, **90,** 6146 (1989).
55. S. H. Lin, B. Fain, and N. Hamer, *Adv. Chem. Phys.*, **79,** 133 (1990).
56. S. Chapman and M. S. Child, *J. Phys. Chem.*, **95,** 578 (1991).
57. Ch. Meier, V. Engel, and J. S. Briggs, *J. Chem. Phys.*, **95,** 7337 (1991).
58. H. Kono and Y. Fujimura, *Chem. Phys. Lett.*, **184,** 497 (1991).
59. T. Baumert, B. Bühler, R. Thalweiser, and G. Gerber, *Phys. Rev. Lett.*, **64,** 733 (1990).
60. R. M. Bowman, M. Dantus, and A. H. Zewail, *Chem. Phys. Lett.*, **161,** 297 (1989).
61. N. F. Scherer, R. J. Carlson, A. Mantro, M. Du, A. J. Ruggiero, V. Romero-Rochin, J. A. Cina, G. R. Fleming, and S. A. Rice, *J. Chem. Phys.*, **95,** 1487 (1991).
62. H. Metiu and V. Engel, *J. Chem. Phys.*, **93,** 5693 (1990).
63. M. Gruebele and A. H. Zewail, *J. Chem. Phys.*, **98,** 883 (1993).
64. N. F. Scherer, A. Mantro, L. D. Ziegler, M. Du, R. J. Carlson, J. A. Cina, and G. R. Fleming, *J. Chem. Phys.*, **96,** 4180 (1992).
65. Y. J. Yan, R. M. Whitnell, and K. R. Wilson, *Chem. Phys. Lett.*, **193,** 402 (1992).
66. R. Zadoyan, Z. Li, C. C. Martens, and V. A. Apkarian, *J. Chem. Phys.*, **101,** 6648 (1994); Z. Li, V. A. Apkarian, and C. C. Martens, *J. Phys. Chem.*, **99,** 7453 (1995).
67. Q. Liu, J.-K. Wang, and A. H. Zewail, *J. Phys. Chem.*, **99,** 11321 (1995).
68. S. O. Williams and D. G. Imre, *J. Phys. Chem.*, **92,** 6636, 6648 (1988).
69. T. Baumert, R. Thalweiser, and G. Gerber, *Chem. Phys. Lett.*, **209,** 29 (1993).
70. S. Rutz, K. Kobe, H. Kühling, E. Schreiber, and L. Wöste, *Z. Physik D*, **26,** 276 (1993).
71. A. J. Dobbyn and J. M. Hutson, *J. Phys. Chem.*, **98,** 11428 (1994).
72. G. Schön and H. Köppel, *Chem. Phys. Lett.*, **231,** 55 (1994); *J. Chem. Phys.*, **103,** 9292 (1995).
73. B. Reischl, *Chem. Phys. Lett.*, **239,** 173 (1995).
74. Y. R. Shen, *The Principles of Nonlinear Optics*, Wiley, New York, 1984.
75. U. Fano, *Rev. Mod. Phys.*, **29,** 174 (1957).
76. A. G. Redfield, *Adv. Magn. Reson.*, **1,** 1 (1965).
77. W. H. Louisell, *Quantum Statistical Properties of Radiation*, Wiley, New York, 1973.
78. K. Blum, *Density Matrix Theory and Applications*, Plenum Press, New York, 1981.
79. B. Fain, S. H. Lin, and V. Khidekel, *Phys. Rev. A*, **47,** 3222 (1993).
80. W. T. Pollard and R. A. Mathies, *Ann. Rev. Phys. Chem*, **43,** 497 (1992).
81. Y. J. Yan, L. E. Fried, and S. Mukamel, *J. Phys. Chem*, **93,** 8149 (1989).
82. Y. J. Yan and S. Mukamel, *Phys. Rev. A*, **41,** 6485 (1990).
83. W. B. Bosma, Y. J. Yan, and S. Mukamel, *J. Chem. Phys.*, **93,** 3863 (1990).
84. S. Mukamel, *Principles of Nonlinear Optical Spectroscopy*, Oxford University Press, New York, 1995.
85. J. M. Jean, *J. Chem. Phys.*, **101,** 10464 (1994).
86. O. Kühn, V. May, and M. Schreiber, *J. Chem. Phys.*, **101,** 10404 (1994).
87. Y. Tanimura and S. Mukamel, *J. Chem. Phys.*, **101,** 3049 (1994).
88. W. Domcke and H. Köppel, *Chem. Phys. Lett.*, **140,** 133 (1987).
89. G. Stock, R. Scheider, and W. Domcke, *J. Chem. Phys.*, **90,** 7184 (1989).

90. G. Stock and W. Domcke, *Opt. Soc. Am. B*, **7,** 1970 (1990).
91. M. Seel and W. Domcke, *J. Chem. Phys.*, **95,** 7806 (1991).
92. G. Stock and W. Domcke, *Phys. Rev. A*, **45,** 3032 (1992).
93. G. Stock, C. Woywod, and W. Domcke, *Chem. Phys. Lett.*, **200,** 163 (1992).
94. L. Seidner, G. Stock, and W. Domcke, *Chem. Phys. Lett.*, **228,** 665 (1994).
95. M. Born and K. Huang, *Dynamical Theory of Crystal Lattices*, Oxford University Press, London, 1954.
96. B. H. Lengsfield and D. R. Yarkony, *Adv. Chem. Phys.*, **82,** 1 (1992).
97. R. Shepard, *Adv. Chem. Phys.*, **69,** 63 (1987).
98. B. O. Roos, *Adv. Chem. Phys.*, **69,** 399 (1987).
99. H.-J. Werner, *Adv. Chem. Phys.*, **69,** 1 (1987).
100. B. H. Lengsfield, P. Saxe, and D. R. Yarkony, *J. Chem. Phys.*, **81,** 4549 (1984).
101. B. H. Lengsfield and D. R. Yarkony, *J. Chem. Phys.*, **84,** 348 (1986).
102. T. Neuheuser, N. Sukumar, and S. D. Peyerimhoff, *Chem. Phys.*, **194,** 45 (1995).
103. G. Parlant and D. R. Yarkony, *Int. J. Quant. Chem. Symp.*, **26,** 737 (1992).
104. H. C. Longuet-Higgins, in *Advances in Spectroscopy*, H. W. Thompson, Ed., Interscience, New York, 1961, Vol. 2, p. 429.
105. W. Lichten, *Phys. Rev.*, **131,** 229 (1963).
106. T. F. O'Malley, *Phys. Rev.*, **162,** 98 (1967).
107. T. F. O'Malley, *Adv. At. Mol. Phys.*, **7,** 223 (1971).
108. F. T. Smith, *Phys. Rev.*, **179,** 111 (1969).
109. M. Baer, *Chem. Phys. Lett.*, **35,** 112 (1975).
110. M. Baer, *Chem. Phys.*, **15,** 49 (1976).
111. C. J. Ballhausen and A. E. Hansen, *Ann. Rev. Phys. Chem.*, **23,** 15 (1972).
112. V. Sidis, *Adv. Chem. Phys.*, **82,** 73 (1992).
113. T. Pacher, L. S. Cederbaum, and H. Köppel, *Adv. Chem. Phys.*, **84,** 293 (1993).
114. C. A. Mead, *Rev. Mod. Phys.*, **64,** 51 (1992).
115. W. Domcke, *Phys. Rep.*, **208,** 97 (1991).
116. C. A. Mead and D. G. Truhlar, *J. Chem. Phys.*, **77,** 6090 (1982).
117. M. Baer, *Adv. Chem. Phys.*, **82,** 187 (1992).
118. T. Pacher, C. A. Mead, L. S. Cederbaum, and H. Köppel, *J. Chem. Phys.*, **91,** 7057 (1989).
119. H.-J. Werner and W. Meyer, *J. Chem. Phys.*, **74,** 5802 (1981).
120. C. Petrongolo, G. Hirsch, and R. J. Buenker, *Mol. Phys.*, **70,** 825 (1990).
121. M. Perić, S. D. Peyerimhoff, and R. J. Buenker, *Z. Physik D*, **24,** 177 (1992).
122. J. Hendeković, *Chem. Phys. Lett.*, **90,** 193 (1982).
123. M. Persico, in *Spectral Line Shapes*, Vol. 3, F. Rostas, Ed., De Gruyter, Berlin, 1985.
124. R. Cimiraglia, J.-P. Malrieu, M. Persico, and F. Spiegelmann, *J. Phys. B*, **18,** 3073 (1985).
125. F. X. Gadea and M. Pelissier, *J. Chem. Phys.*, **93,** 545 (1990).
126. I. D. Petsalakis, G. Theodorakopoulos, and C. A. Nicolaides, *Chem. Phys. Lett.*, **185,** 359 (1991).
127. K. Ruedenberg, L. M. Cheung and S. T. Elbert, *Int. J. Quant. Chem.*, **16,** 1069 (1979).
128. H.-J. Werner and W. Meyer, *J. Chem. Phys.*, **74,** 5794 (1981).

129. K. Anderson, P.-A. Malmquist, B. O. Roos, A. J. Sadlej, and K. J. Wolinski, *J. Phys. Chem.*, **94,** 5483 (1990).
130. W. Domcke, A. L. Sobolewski, and C. Woywod, *Chem. Phys. Lett.*, **203,** 220 (1993).
131. W. Domcke and C. Woywod, *Chem. Phys. Lett.*, **216,** 362 (1993).
132. W. Domcke, C. Woywod, and M. Stengle, *Chem. Phys. Lett.*, **226,** 257 (1994).
133. J. P. Gauyacq, *J. Phys. B*, **11,** 85 (1978).
134. J. B. Delos and W. R. Thorson, *J. Chem. Phys.*, **70,** 1774 (1979).
135. T. Pacher, L. S. Cederbaum, and H. Köppel, *J. Chem. Phys.*, **89,** 7367 (1988).
136. L. S. Cederbaum, J. Schirmer, and H.-D. Meyer, *J. Phys. A*, **22,** 2427 (1989).
137. T. Pacher, H. Köppel, and L. S. Cederbaum, *J. Chem. Phys.*, **95,** 6668 (1991).
138. M. Desouter-Lecomte, C. Galloy, J. C. Lorquet, and M. Vaz Pires, *J. Chem. Phys.*, **71,** 3661 (1979).
139. A. R. Gregory, W. H. Henneker, W. Siebrand, and M. Z. Zgierski, *J. Chem. Phys.*, **65,** 2071 (1976).
140. I. B. Bersuker and V. Z. Polinger, *Vibronic Interactions in Molecules and Crystals*, Springer, Berlin, 1989.
141. B. Heumann, K. Weide, R. Düren, and R. Schinke, *J. Chem. Phys.*, **98,** 5508 (1993).
142. H. Guo, *J. Chem. Phys.*, **96,** 6629 (1992).
143. A. Loettgers, A. Untch, M. Stumpf, R. Schinke, H.-J. Werner, C. Bauer, and P. Rosmus, *Chem. Phys. Lett.*, **230,** 290 (1994).
144. J. N. Murrell, S. Carter, S. C. Farantos, P. Huxley, and A. J. C. Varandas, *Molecular Potential Energy Functions*, Wiley, New York, 1984.
145. D. G. Truhlar, R. Steckler, and M. S. Gordon, *Chem. Rev.*, **87,** 217 (1987).
146. A. J. C. Varandas, *Adv. Chem. Phys.*, **74,** 255 (1988).
147. G. C. Schatz, *Rev. Mod. Phys.*, **61,** 669 (1989).
148. R. Car and M. Parinello, *Phys. Rev. Lett*, **55,** 2471 (1985).
149. B. Hartke and E. A. Carter, *Chem. Phys. Lett.*, **216,** 324 (1993).
150. J. K. Gregory and D. C. Clary, *Chem. Phys. Lett.*, **237,** 39 (1995).
151. J. C. Greer, R. Ahlrichs, and I. V. Hertel, *Z. Phys. D*, **18,** 413 (1991).
152. B. R. Smith, M. J. Bearpark, M. A. Robb, F. Bernardi, and M. Olivucci, *Chem. Phys. Lett.*, **242,** 27 (1995).
153. W. H. Miller, N. C. Handy, and J. E. Adams, *J. Chem. Phys.*, **72,** 99 (1980).
154. E. B. Wilson, Jr., J. C. Decius, and P. C. Cross, *Molecular Vibrations*, McGraw-Hill, New York, 1955.
155. U. Burkert and N. L. Allinger, *Molecular Mechanics*, American Chemical Society, Washington, DC, 1982, Vol. 177.
156. Y. Guan, G. C. Lynch, and D. L. Thompson, *J. Chem. Phys.*, **87,** 6957 (1987).
157. P. M. Agrawal, D. L. Thompson, and L. M. Raff, *J. Chem. Phys.*, **88,** 5948 (1988).
158. Y. Guan and D. L. Thompson, *Chem. Phys.*, **139,** 147 (1989).
159. A. Preiskorn and D. L. Thompson, *J. Chem. Phys.*, **91,** 2299 (1989).
160. T. D. Sewell and D. L. Thompson, *J. Chem. Phys.*, **93,** 4077 (1990).
161. Y. Qin and D. L. Thompson, *J. Chem. Phys.*, **96,** 1992 (1992).
162. C. C. Chambers and D. L. Thompson, *Chem. Phys. Lett.*, **218,** 166 (1994).
163. A. Warshel and R. M. Weiss, *J. Am. Chem. Soc.*, **102,** 6218 (1980).

164. W. H. Miller, in *Molecular Aspects of Biotechnology: Computational Models and Theories*, J. Bertrán, Ed., Kluwer, Dordrecht, 1992, p. 193.
165. N. C. Handy, *Mol. Phys.*, **61,** 207 (1987).
166. M. J. Bramley, W. H. Green, Jr., and N. C. Handy, *Mol. Phys.*, **73,** 1183 (1991).
167. B. Podolsky, *Phys. Rev.*, **32,** 812 (1928).
168. B. T. Sutcliffe, in *Current Aspects of Quantum Chemistry*, R. Carbo, Ed., Elsevier, Amsterdam, 1982.
169. T. J. Lukka, *J. Chem. Phys.*, **102,** 3945 (1995).
170. A. G. Császár and N. C. Handy, *J. Chem. Phys.*, **102,** 3962 (1995).
171. M. J. Bramley and N. C. Handy, *J. Chem. Phys.*, **98,** 1378 (1993).
172. M. J. Bramley and T. Carrington, Jr., *J. Chem. Phys.*, **99,** 8519 (1993).
173. M. J. Bramley, J. W. Tromp, T. Carrington, Jr., and G. C. Corey, *J. Chem. Phys.*, **100,** 6175 (1994).
174. J. E. Hadder and J. H. Frederick, *J. Chem. Phys.*, **97,** 3500 (1992).
175. L. S. Cederbaum and W. Domcke, *Adv. Chem. Phys.*, **36,** 205 (1977).
176. G. Fischer, *Vibronic Coupling*, Academic Press, New York, 1984.
177. I. B. Bersuker, *The Jahn-Teller Effect and Vibronic Interactions in Modern Chemistry*, Plenum Press, New York, 1984.
178. R. L. Whetten, G. S. Ezra, and E. R. Grant, *Ann. Rev. Phys. Chem.*, **36,** 277 (1985).
179. E. J. Heller, *Accts. Chem. Res.*, **14,** 368 (1981).
180. R. Schneider and W. Domcke, *Chem. Phys. Lett.*, **150,** 235 (1988).
181. L. Seidner, G. Stock, A. L. Sobolewski, and W. Domcke, *J. Chem. Phys.*, **96,** 5298 (1992).
182. G. Stock, C. Woywod, W. Domcke, T. Swinney, and B. Hudson, *J. Chem. Phys.*, **103,** 6851 (1995).
183. J. D. Lewis, T. B. Malloy, Jr., T. H. Chao, and J. Laane, *J. Mol. Struct.*, **12,** 427 (1972).
184. G. C. Groenenboom, PhD Thesis, Technical University of Eindhoven, 1991 (unpublished).
185. D. J. Tannor, R. Kosloff, and S. A. Rice, *J. Chem. Phys.*, **85,** 5805 (1986).
186. F. Bloch and A. Siegert, *Phys. Rev.*, **57,** 522 (1940).
187. R. Bavli and H. Metiu, *J. Chem. Phys.*, **98,** 6632 (1993).
188. H. Köppel, *Chem. Phys.*, **77,** 359 (1983).
189. H. Köppel, L. S. Cederbaum, and W. Domcke, *J. Chem. Phys.*, **89,** 2023 (1988).
190. H. Köppel, *Chem. Phys. Lett.*, **205,** 361 (1993).
191. A. D. Hammerich, A. Nitzan, and M. A. Ratner, *Theor. Chim. Acta*, **89,** 383 (1994).
192. M. Quack, *Ann. Rev. Phys. Chem.*, **41,** 839 (1990).
193. R. E. Wyatt, C. Iung, and C. Leforestier, *J. Chem. Phys.*, **97,** 3477 (1992).
194. C. Iung and R. E. Wyatt, *J. Chem. Phys.*, **99,** 2261 (1993).
195. S. A. Schofield, P. G. Wolynes, and R. E. Wyatt, *Phys. Rev. Lett.*, **74,** 3720 (1995).
196. J. K. Cullum and R. A. Willoughby, *Lanczos Algorithms for Large Sparse Eigenvalue Problems*, Birkhäuser, Basel, 1985, Vol. 1.
197. D. Gottlieb and S. A. Orszag, *Numerical Analysis of Spectral Methods: Theory and Applications*, SIAM, Philadelphia, 1977.
198. J. C. Light, I. P. Hamilton, and J. V. Lill, *J. Chem. Phys.*, **82,** 1400 (1985).

199. J. P. Boyd, *Chebyshev and Fourier Spectral Methods*, Springer, Berlin, 1989.
200. D. O. Harris, G. G. Engerholm, and W. D. Gwinn, *J. Chem. Phys.*, **43,** 1515 (1969).
201. A. S. Dickinson and P. R. Certain, *J. Chem. Phys.*, **49,** 4209 (1968).
202. W. Domcke, H. Köppel, and L. S. Cederbaum, *Mol. Phys.*, **43,** 851 (1981).
203. Z. Bačić, R. M. Whitnell, D. Brown, and J. C. Light, *Comput. Phys. Comm.*, **51,** 35 (1988).
204. R. M. Whitnell and J. C. Light, *J. Chem. Phys.*, **90,** 1774 (1989).
205. Z. Bačić and J. C. Light, *Ann. Rev. Phys. Chem.*, **40,** 469 (1989).
206. M. D. Feit and J. A. Fleck, Jr., *J. Chem. Phys.*, **78,** 301 (1983).
207. R. Kosloff, *J. Phys. Chem.*, **92,** 2087 (1988).
208. R. Kosloff, in *Numerical Grid Methods and Their Application to Schrödinger's Equation*, C. Cerjan, Ed., NATO ASI Series C 412, Kluwer, Dordrecht, 1993.
209. R. Kosloff, *Ann. Rev. Phys. Chem.*, **45,** 145 (1994).
210. T. Carrington, Jr., unpublished.
211. S. E. Choi and J. C. Light, *J. Chem. Phys.*, **97,** 7031 (1992).
212. J. A. Bentley, R. E. Wyatt, M. Menou, and C. Leforestier, *J. Chem. Phys.*, **97,** 4255 (1992).
213. H. Wei and T. Carrington, Jr., *J. Chem. Phys.*, **101,** 1343 (1994).
214. X.-P. Jiang, R. Heather, and H. Metiu, *J. Chem. Phys.*, **90,** 2555 (1989).
215. U. Manthe, H. Köppel, and S. Cederbaum, *J. Chem. Phys.*, **95,** 1708 (1991).
216. C. J. Williams, J. Quian, and D. J. Tannor, *J. Chem. Phys.*, **95,** 1721 (1991).
217. H. Guo, *Chem. Phys. Lett.*, **187,** 360 (1991); *J. Chem. Phys.*, **96,** 2731 (1992).
218. M. D. Feit, J. A. Fleck, Jr., and A. Steiger, *J. Comput. Phys.*, **47,** 412 (1982).
219. D. Kosloff and R. Kosloff, *J. Comput. Phys.*, **52,** 35 (1983).
220. F. Le Quéré and C. Leforestier, *J. Chem. Phys.*, **92,** 247 (1990).
221. C. E. Dateo, V. Engel, R. Almeida, and H. Metiu, *Comput. Phys. Comm.*, **63,** 435 (1991).
222. G. C. Corey and D. Lemoine, *J. Chem. Phys.*, **97,** 4115 (1992).
223. S. Carter and N. C. Handy, *Mol. Phys.*, **57,** 175 (1986).
224. J. Tennyson and J. R. Henderson, *J. Chem. Phys.*, **91,** 3815 (1989).
225. R. A. Friesner, J. A. Bentley, M. Menou, and C. Leforestier, *J. Chem. Phys.*, **99,** 324 (1993).
226. M. J. Bramley and T. Carrington, Jr., *J. Chem. Phys.*, **101,** 8494 (1994).
227. F. B. Hildebrand, *Introduction to Numerical Analysis*, McGraw-Hill, New York, 1974.
228. L. F. Shampine and L. K. Gordon, *Computer Solution of Ordinary Differential Equations: The Initial Value Problem*, W. H. Freeman, San Francisco, 1975.
229. J. Stoer and R. Bulirsch, *Einführung in die Numerische Mathematik II*, Springer, Berlin, 1978.
230. W. H. Press, B. P. Flannery, S. A. Teukolsky, and W. T. Vetterling, *Numerical Recipes*, Cambridge University Press, 1987.
231. E. A. McCullough, Jr. and R. E. Wyatt, *J. Chem. Phys.*, **51,** 1253 (1969).
232. A. Askar and A. S. Cakmak, *J. Chem. Phys.*, **68,** 2794 (1978).
233. C. Leforestier, *Chem. Phys.*, **68,** 2794 (1978).
234. K. C. Kulander, *J. Chem. Phys.*, **69,** 5064 (1978).

235. H. Tal-Ezer and R. Kosloff, *J. Chem. Phys.*, **81,** 3967 (1984).
236. T. J. Park and J. C. Light, *J. Chem. Phys.*, **85,** 5870 (1986).
237. C. Leforestier, R. H. Bisseling, C. Cerjan, M. D. Feit, R. Friesner, A. Guldberg, A. Hammerich, G. Jolicard, W. Karrlein, H.-D. Meyer, N. Lipkin, O. Roncero, and R. Kosloff, *J. Comput. Phys.*, **94,** 59 (1991).
238, S. K. Gray and J. M. Verosky, *J. Chem. Phys.*, **100,** 5011 (1994).
239. T. N. Truong, J. J. Tanner, P. Bala, J. A. McCammon, D. J. Kouri, B. Lesyng, and D. K. Hoffman, *J. Chem. Phys.*, **96,** 2077 (1992).
240. E. J. Mele and J. Socolar, *Int. J. Quant. Chem. Symp.*, **18,** 347 (1984).
241. J. Alvarellos and H. Metiu, *J. Chem. Phys.* **88,** 4957 (1988).
242. J. Broeckhove, B. Feyen, L. Lathowers, F. Arickx, and P. Van Leuven, *Chem. Phys. Lett.*, **174,** 504 (1990).
243. F. M. Fernández and D. A. Micha, *J. Chem. Phys.*, **97,** 8173 (1992).
244. B. M. Garraway and K.-A. Suominen, *Rep. Prog. Phys.*, **58,** 365 (1995).
245. H. De Raedt, *Comput. Phys. Rep.*, **7,** 1 (1987).
246. A. D. Bandrauk and H. Shen, *Chem. Phys. Lett.*, **176,** 428 (1991).
247. U. Peskin, R. Kosloff, and N. Moiseyev, *J. Chem. Phys.*, **100,** 8849 (1994).
248. G. Yao and R. E. Wyatt, *J. Chem. Phys.*, **101,** 1904 (1994).
249. A. Ferretti, A. Lami, and G. Villani, *Chem. Phys.*, **196,** 447 (1995).
250. A. D. Hammerich, R. Kosloff, and M. A. Ratner, *J. Chem. Phys.*, **97,** 6410 (1992).
251. W. H. Press and S. A. Teukolsky, *Computers in Physics*, **6,** 188 (1992).
252. A. J. Leggett, S. Chakravarty, A. T. Dorsey, M. P. A. Fisher, A. Garg, and W. Zwerger, *Rev. Mod. Phys.*, **59,** 1 (1987).
253. A. E. Orel and K. C. Kulander, *Chem. Phys. Lett.*, **146,** 428 (1988).
254. H. Guo and G. C. Schatz, *J. Chem. Phys.*, **92,** 1634 (1990).
255. B. Heumann and R. Schinke, *J. Chem. Phys.*, **101,** 7488 (1994).
256. H. Köppel, L. S. Cederbaum, and W. Domcke, *J. Chem. Phys.*, **77,** 2014 (1982).
257. H. Köppel, L. S. Cederbaum, W. Domcke, and S. S. Shaik, *Angew. Chem. Int. Ed. Engl.*, **22,** 210 (1983).
258. A. Kuppermann and Y. M. Wu, *Chem. Phys. Lett.*, **205,** 577 (1993).
259. J. A. Cina, T. J. Smith, Jr., and V. Romero-Rochin, *Adv. Chem. Phys.*, **83,** 1 (1993).
260. M. Durga Prasad, *Chem. Phys. Lett.*, **194,** 27 (1992).
261. H.-D. Meyer, *Chem. Phys.*, **82,** 199 (1983).
262. G. Stock, *J. Chem. Phys.*, **103,** 10015 (1995).
263. O. Braitbart, E. Castellucci, G. Dujardin, and S. Leach, *J. Chem. Phys.*, **87,** 4799 (1983).
264. J. Eiding, R. Schneider, W. Domcke, H. Köppel, and W. von Niessen, *Chem. Phys. Lett*, **177,** 345 (1991); **191,** 203 (1992).
265. J. Eiding, W. Domcke, W. Huber, and H.-P. Steinrück, *Chem. Phys. Lett.*, **180,** 133 (1991).
266. G. Stock, *Chem. Phys. Lett.*, **224,** 131 (1994).
267. G. Stock, *J. Chem. Phys.*, **103,** 2888 (1995).
268. V. E. Bondybey, *Ann. Rev. Phys. Chem.*, **35,** 591 (1984).
269. H. A. Stuchebrukhov, *Sov. Phys. JETP*, **64,** 1195 (1986).
270. T. Uzer, *Phys. Rep.*, **199,** 73 (1991).

271. R. J. Sension, S. T. Repinec, A. Z. Szarka, and R. M. Hochstrasser, *J. Chem. Phys.*, **98,** 6291 (1993).
272. S. K. Doorn, R. B. Dyer, P. O. Stoutland, and W. H. Woodruff, *J. Am. Chem. Soc.*, **115,** 6398 (1993).
273. M. V. Rama Krishna and R. D. Coalson, *Chem. Phys. Lett.*, **120,** 327 (1988).
274. H. Metiu and V. Engel, *J. Opt. Soc. Am. B*, **7,** 1709 (1990).
275. H. Mahr and M. D. Hirsch, *Opt. Commun.*, **13,** 96 (1975).
276. A. Mokhtari, A. Chebira, and J. Chesnoy, *J. Opt. Soc. Am. B*, **7,** 1551 (1990).
277. D. C. Todd, J. M. Jean, S. J. Rosenthal, A. J. Ruggiero, D. Yang, and G. R. Fleming, *J. Chem. Phys.*, **93,** 8658 (1990).
278. M. Du, S. J. Rosenthal, X. Xie, T. J. DiMagno, M. Schmidt, D. K. Hanson, M. Schiffer, J. R. Norris, and G. R. Fleming, *Proc. Natl. Acad. Sci. USA*, **89,** 8517 (1992).
279. A. E. Johnson, W. Jarzeba, G. C. Walker, and P. F. Barbara, *Israel J. Chem.*, **33,** 199 (1993).
280. J. Chesnoy and A. Mokhtari, *Phys. Rev. A*, **38,** 3566 (1988).
281. H. L. Fragnito, J. Y. Bigot, P. C. Becker, and C. V. Shank, *Chem. Phys. Lett.*, **160,** 101 (1989).
282. Y. Chen, L. Hunziker, P. Ludowise, and M. Morgen, *J. Chem. Phys.*, **97,** 2149 (1992).
283. P. Foggi, L. Pettini, I. Sànta, R. Righini, and S. Califano, *J. Phys. Chem.*, **99,** 7439 (1995).
284. J. M. Wiesenfeld and B. I. Greene, *Phys. Rev. Lett.*, **51,** 1745 (1983).
285. J. M. Smith, C. Lakshminarayan, and J. L. Knee, *J. Chem. Phys.*, **93,** 4475 (1990).
286. T. Baumert, M. Grosser, R. Thalweiser, and G. Gerber, *Phys. Rev. Lett.*, **67,** 3453 (1991).
287. T. Baumert, J. L. Herek, and A. H. Zewail, *J. Chem. Phys.*, **99,** 4430 (1993).
288. M. H. M. Janssen, M. Dantus, H. Guo, and A. H. Zewail, *Chem. Phys. Lett.*, **214,** 281 (1993).
289. H. Kühling, S. Rutz, K. Kobe, E. Schreiber, and L. Wöste, *J. Phys. Chem.*, **97,** 12500 (1993).
290. J. Purnell, S. Wei, S. A. Buzza, and A. W. Castleman, Jr., *J. Phys. Chem.*, **97,** 12530 (1993).
291. A. P. Baronavski and J. C. Owrutsky, *Chem. Phys. Lett.*, **221,** 419 (1994); *J. Phys. Chem.*, **99,** 10077 (1995).
292. W. Radloff, T. Freudenberg, H.-H. Ritze, V. Stert, K. Weyers, and F. Noak, *Chem. Phys. Lett.*, **245,** 400 (1995).
293. C. C. Hayden and D. W. Chandler, *J. Phys. Chem.*, **99,** 7897 (1995).
294. I. Fischer, D. M. Villeneuve, M. J. J. Vrakking, and A. Stolow, *J. Chem. Phys.*, **102,** 5566 (1995); M. J. J. Vrakking, I. Fischer, D. M. Villeneuve, and A. Stolow, *J. Chem. Phys.*, **103,** 4538 (1995).
295. D. R. Cyr and C. C. Hayden, *J. Phys. Chem.*, **104,** 771 (1996).
296. P. A. Afinrud, C. Han, T. Lian, and R. M. Hochstrasser, *J. Phys. Chem.*, **94,** 1180 (1990).
297. P. O. Stoutland, R. B. Dyer, and W. H. Woodruff, *Science*, **257,** 1913 (1992).
298. T. P. Dougherty and E. J. Heilweil, *J. Chem. Phys.*, **100,** 4006 (1994).
299. P. Hamm, M. Zurek, W. Mäntele, M. Meyer, H. Scheer, and W. Zinth, *Proc. Natl. Acad. Sci. USA*, **92,** 1826 (1995).
300. G. Beddard, *Rep. Prog. Phys.*, **56,** 63 (1993).

301. C. V. Shank and E. P. Ippen, *Appl. Phys. Lett.*, **26,** 62 (1975).
302. M. Cho, M. Du, N. F. Scherer, G. R. Fleming, and S. Mukamel, *J. Chem. Phys.*, **99,** 2410 (1993).
303. M. Morgen, W. Price, P. Ludowise, and Y. Chen, *J. Chem. Phys.*, **102,** 8780 (1995).
304. W. P. de Boeij, M. S. Pshenichnikov, and D. A. Wiersma, *Chem. Phys. Lett.*, **238,** 1 (1995).
305. P. Kowalczyk, C. Radzewicz, J. Mostowski, and I. A. Walmsley, *Phys. Rev. A*, **42,** 5622 (1990).
306. A. Icsevgi and W. E. Lamb, *Phys. Rev.*, **185,** 517 (1969).
307. R.E. Walkup, J. A. Misewich, J. H. Glownia, and P. P. Sorokin, *J. Chem. Phys.*, **94,** 3389 (1991).
308. J. H. Eberly and K. Wódkiewicz, *J. Opt. Soc. Am.*, **67,** 1252 (1977).
309. R. Heather and H. Metiu, *J. Chem. Phys.*, **90,** 6903 (1989).
310. G. Ebel and R. Schinke, *J. Phys. Chem.*, **101,** 1865 (1994).
311. H. Kono and Y. Fujimura, *J. Chem. Phys.*, **91,** 5960 (1989).
312. S. H. Lin, B. Fain, and C. Y. Yeh, *Phys. Rev. A*, **41,** 2718 (1990).
313. N. A. Kurnit, I. D. Abella, and S. R. Hartmann, *Phys. Rev. Lett.*, **13,** 567 (1964); I. D. Abella, N. A. Kurnit, and S. R. Hartmann, *Phys. Rev.*, **141,** 391 (1966).
314. W. H. Hesselink and D. A. Wiersma, *Chem. Phys. Lett.*, **56,** 227 (1978); *J. Chem. Phys.*, **73,** 648 (1980).
315. K. Duppen, E. T. J. Nibbering, and D. A. Wiersma, in *Femtosecond Reaction Dynamics*, D. A. Wiersma, Ed., North Holland, Amsterdam, 1994, p. 197.
316. M. Cho, N. F. Scherer, G. R. Fleming and S. Mukamel, *J. Chem. Phys.*, **96,** 5618 (1992).
317. J. Hager, M. A. Smith, and S. C. Wallace, *J. Chem. Phys.*, **84,** 6771 (1986).
318. A. Goto, M. Fujii, and M. Ito, *J. Chem. Phys.*, **91,** 2268 (1987).
319. A. Staib, W. Domcke and A. L. Sobolewski, *Chem. Phys. Lett.*, **162,** 336 (1989).
320. M. Seel and W. Domcke, *Chem. Phys.*, **151,** 59 (1991).
321. V. Engel, *Chem. Phys. Lett.*, **178,** 130 (1991).
322. C. Meier and V. Engel, *Chem. Phys. Lett.*, **212,** 691 (1993).
323. K. Müller-Dethlefs and E. W. Schlag, *Annu. Rev. Phys. Chem.*, **42,** 109 (1991).
324. T. K. Yee and T. K. Gustafson, *Phys. Rev. A*, **18,** 1597 (1978).
325. R. J. Kubo, *J. Phys. Soc. Japan*, **12,** 570 (1957).
326. R. G. Gordon, *Adv. Magn. Res.*, **3,** 1 (1968).
327. W. T. Pollard, S. L. Dexheimer, Q. Wang, L. A. Peteanu, C. V. Shank, and R. A. Mathies, *J. Phys. Chem.*, **96,** 6147 (1992).
328. Y. P. Gudkov, Y. T. Mazurenko, P. N. Pigurnov, and V. A. Smirnov, *Opt. Spectrosc.(USSR)*, **61,** 484 (1986).
329. W. B. Bosma, Y. J. Yan, and S. Mukamel, *Phys. Rev. A*, **42,** 6920 (1990).
330. T. Baumert, V. Engel, C. Meier, and G. Gerber, *Chem. Phys. Lett.*, **200,** 488 (1992).
331. M. Sugawara and Y. Fujimura, *Chem. Phys.*, **175,** 323 (1993).
332. U. Banin, A. Bartana, S. Ruhman, and R. Kosloff, *J. Chem. Phys.*, **101,** 8461 (1994).
333. H. A. Ferwerda, J. Terpstra, and D. A. Wiersma, *J. Chem. Phys.*, **91,** 3296 (1989).
334. K. K. Innes, I. G. Ross, and W. R. Moomaw, *J. Mol. Spectrosc.*, **132,** 492 (1988).

335. I. Yamazaki, T. Murao, T. Yamanaka, and K. Yoshihara, *Faraday Discuss. Chem. Soc.*, **75,** 395 (1983).
336. S.-Y. Lee and E. J. Heller, *J. Chem. Phys.*, **71,** 4777 (1979).
337. G. Stock and W. Domcke, *J. Chem. Phys.*, **93,** 5496 (1990).
338. G. Stock, *J. Chem. Phys.*, **101,** 246 (1994).
339. C. Meier and V. Engel, *J. Chem. Phys.*, **101,** 2673 (1994).
340. W. T. Pollard, H. L. Fragnito, J. Y. Bigot, C. V. Shank, and R. A. Mathies, *Chem. Phys. Lett.*, **168,** 239 (1990).
341. L. Angeloni, G. Smulevitch, and P. M. Marzocchi, *J. Raman Spectrosc.*, **8,** 305 (1979).
342. R. J. Sension, S. T. Repinec, A. Z. Szarka, and R. M. Hochstrasser, *J. Chem. Phys.*, **98,** 6291 (1993).
343. S. Pedersen, L. Bañares, and A. H. Zewail, *J. Chem. Phys.*, **97,** 8801 (1992).
344. A. Z. Szarka, N. Pugliano, D. K. Palit, and R. M. Hochstrasser, *Chem. Phys. Lett.*, **240,** 25 (1995).
345. E. Lenderink, K. Duppen, and D. A. Wiersma, *J. Phys. Chem.*, **99,** 8972 (1995).
346. R. W. Schoenlein, L. A. Peteanu, R. A. Mathies, and C. V. Shank, *Science*, **254,** 412 (1991).
347. Q. Wang, R. W. Schoenlein, L. A. Peteanu, R. A. Mathies, and C. V. Shank, *Science*, **266,** 422 (1994).
348. See, for example, T. Elsässer and W. Kaiser, *Annu. Rev. Phys. Chem.*, **42,** 83 (1991).
349. E. J. Heller, *J. Chem. Phys.*, **64,** 63 (1976).
350. R. B. Gerber, V. Buch, and M. A. Ratner, *J. Chem. Phys.*, **77,** 3022 (1982).
351. S. Sawada and H. Metiu, *J. Chem. Phys.*, **84,** 227 (1986); **84,** 6293 (1986); K. Haug and H. Metiu, *J. Chem. Phys.*, **97,** 4781 (1992); **99,** 6253 (1993).
352. E. Neria, A. Nitzan, R. N. Barnett, and U. Landman, *Phys. Rev. Lett.*, **67,** 1011 (1991); E. Neria and A. Nitzan, *J. Chem. Phys.*, **99,** 1109 (1993).
353. R. Alimi, A. Garcia-Vela, and R. B. Gerber, *J. Chem. Phys.*, **96,** 2034 (1992); Z. Li. and R. B. Gerber, *J. Chem. Phys.*, **99,** 8637 (1993).
354. Z. Kotler, A. Nitzan, and R. Kosloff, *Chem. Phys. Lett.*, **153,** 483 (1988).
355. H.-D. Meyer, U. Manthe, and L. S. Cederbaum, *Chem. Phys. Lett.*, **165,** 73 (1990).
356. J. Campos-Martinez and R. D. Coalson, *J. Chem. Phys.*, **93,** 4740 (1990).
357. U. Manthe, H.-D. Meyer, and L. S. Cederbaum, *J. Chem. Phys.*, **97,** 3199 (1992); **97,** 9062 (1992).
358. A. D. Hammerich, U. Manthe, R. Kosloff, H.-D. Meyer, and L. S. Cederbaum, *J. Chem. Phys.*, **101,** 5623 (1994).
359. J. Campos-Martinez, J. R. Waldeck, and R. D. Coalson, *J. Chem. Phys.*, **96,** 3613 (1992).
360. R. P. Feynman, *Rev. Mod. Phys.*, **20,** 367 (1948).
361. R. P. Feynman and A. R. Hibbs, *Quantum Mechanics and Path Integrals*, McGraw-Hill, New York, 1965.
362. L. S. Schulman, *Techniques and Applications of Path Integration*, Wiley, New York, 1981.
363. U. Weiss, *Quantum Dissipative Systems*, World Scientific, Singapore, 1993.
364. A. Garg, J. N. Onuchic, and V. Ambeokar, *J. Chem. Phys.*, **83,** 4491 (1985).
365. I. Rips and J. Jortner, *J. Chem. Phys.*, **87,** 2090 (1987).

366. J. S. Bader, R. A. Kuharski, and D. Chandler, *J. Chem. Phys.*, **93,** 230 (1990).
367. R. Egger and C. H. Mak, *J. Phys. Chem.*, **98,** 9903 (1994).
368. R. Egger, C. H. Mak, and U. Weiss, *J. Chem. Phys.*, **100,** 2651 (1994).
369. R. D. Coalson, *J. Chem. Phys.*, **86,** 995 (1987).
370. R. D. Coalson, *Phys. Rev. B*, **39,** 12052 (1989).
371. R. D. Coalson, *J. Chem. Phys.*, **94,** 1108 (1991).
372. R. P. Feynman and F. L. Vernon, *Ann. Phys. (N.Y.)*, **24,** 118 (1963).
373. D. E. Makarov and N. Makri, *Phys. Rev. A*, **48,** 3626 (1993).
374. E. C. Behrman and P. G. Wolynes, *J. Chem. Phys.*, **83,** 5863 (1985).
375. C. H. Mak and D. Chandler, *Phys. Rev. A*, **44,** 2352 (1991).
376. C. H. Mak and D. Chandler, *Phys. Rev. A*, **41,** 5709 (1990).
377. R. Egger and C. H. Mak, *Phys. Rev. B*, **50,** 15210 (1994).
378. M. Winterstetter and W. Domcke, *Phys. Rev. A*, **47,** 2838 (1993); **A, 48,** 4272 (1993).
379. M. Winterstetter and W. Domcke, *Chem. Phys. Lett.*, **236,** 445 (1995).
380. H. Plöhn, S. Krempl, M. Winterstetter, and W. Domcke, *Chem. Phys.*, **200,** 11 (1995).
381. D. E. Makarov and N. Makri, *Chem. Phys. Lett.*, **221,** 482 (1994).
382. N. Makri and D. E. Makarov, *J. Chem. Phys.*, **102,** 4600 (1995).
383. N. Makri, *J. Math. Phys.*, **36,** 2430 (1995).
384. S. Krempl, M. Winterstetter, H. Plöhn, and W. Domcke, *J. Chem. Phys.*, **100,** 926 (1994).
385. S. Krempl, M. Winterstetter, and W. Domcke, *J. Chem. Phys.*, **102,** 6499 (1995).
386. S. Krempl, W. Domcke, and M. Winterstetter, *Chem. Phys.*, **206,** 63 (1996).
387. M. F. Trotter, *Proc. Am. Math. Soc.*, **10,** 545 (1959).
388. E. Nelson, *J. Math. Phys.*, **5,** 332 (1964).
389. M. Suzuki, *Commun. Math. Phys.*, **51,** 183 (1978).
390. B. DeRaedt and H. DeRaedt, *Phys. Rev. A*, **28,** 3575 (1983).
391. W. Magnus, *Commun. Pure Appl. Math.*, **7,** 649 (1954).
392. P. Pechukas and J. C. Light, *J. Chem. Phys.,* **44,** 3897 (1966).
393. R. Englman, *The Jahn-Teller Effect*, John Wiley & Sons, New York, 1972.
394. M. Topaler and N. Makri, *J. Chem. Phys.*, **97,** 9001 (1992).
395. M. Topaler and N. Makri, *J. Chem. Phys.*, **101,** 7500 (1994).
396. S. Nakajima, *Progr. Theoret. Phys.*, **20,** 948 (1958).
397. R. Zwanzig, *J. Chem. Phys.*, **33,** 1338 (1960).
398. R. K. Wangsness and F. Bloch, *Phys. Rev.*, **89,** 728 (1953).
399. C. P. Slichter, *Principles of Magnetic Resonance*, Harper & Row, New York, 1963.
400. L. Allen and Z. H. Eberly, *Optical Resonance and Two-Level Atoms*, John Wiley & Sons, New York, 1975.
401. C. Cohen-Tannoudji, in *Frontiers in Laser Spectroscopy*, R. Balian, S. Haroche, and S. Liberman, Eds., North Holland, Amsterdam, 1977, Vol. 1, p. 7.
402. C. W. Gardiner, *Quantum Noise*, Springer, Berlin, 1991.
403. T. Takagahara, E. Hanamura, and R. Kubo, *J. Phys. Soc. Japan*, **44,** 728, 742 (1978).
404. Y. J. Yan and S. Mukamel, *J. Chem. Phys.*, **89,** 5160 (1988).
405. Y. Gu, A. Widom, and P. M. Champion, *J. Chem. Phys.*, **100,** 2547 (1994).
406. B. Fain, S. H. Lin, and W. X. Wu, *Phys. Rev. A*, **40,** 824 (1989).

407. B. Fain and S. H. Lin, *J. Chem. Phys.*, **93,** 6387 (1990).
408. V. May, *Chem. Phys. Lett.*, **170,** 543 (1990).
409. V. May and M. Schreiber, *Chem. Phys. Lett.*, **181,** 267 (1991).
410. V. May and M. Schreiber, *Phys. Rev. A*, **45,** 2868 (1992).
411. V. May, O. Kühn, and M. Schreiber, *J. Phys. Chem.*, **97,** 12591 (1993).
412. J. M. Jean, R. A. Friesner, and G. R. Fleming, *J. Chem. Phys.*, **96,** 5827 (1992).
413. W. T. Pollard and R. A. Friesner, *J. Chem. Phys.*, **100,** 5054 (1994).
414. A. K. Felts, W. T. Pollard, and R. A. Friesner, *J. Phys. Chem.*, **99,** 2929 (1995).
415. J. M. Jean and G. R. Fleming, *J. Chem. Phys.*, **103,** 2092 (1995).
416. F. E. Figuerido and R. M. Levy, *J. Chem. Phys.*, **97,** 703 (1992).
417. M. Murao and F. Shibata , *Physica A*, **217,** 348 (1995).
418. B. Wolfseder and W. Domcke, *Chem. Phys. Lett.*, **235,** 370 (1995).
419. G. S. Agarwal, *Phys. Rev. A*, **4,** 739 (1971).
420. S. Jang, *Nucl. Phys. A*, **499,** 250 (1989).
421. J. Dalibard, Y. Castin, and K. Mølmer, *Phys. Rev. Lett.*, **68,** 580 (1992).
422. K. Mølmer, Y. Castin, and J. Dalibard, *J. Opt. Soc. Am. B*, **10,** 524 (1993).
423. R. Dum, P. Zoller, and H. Ritsch, *Phys. Rev. A*, **45,** 4879 (1992).
424. C. W. Gardiner, A. S. Parkins, and P. Zoller, *Phys. Rev. A*, **46,** 4363 (1992).
425. M. Naraschewski and A. Schenzle, *Z. Physik D*, **33,** 79 (1995).
426. N. Gisin and I. C. Percival, *J. Phys. A*, **25,** 5677 (1992).
427. H. J. Carmichael, *An Open Systems Approach to Quantum Optics*, Springer, Berlin, 1993.
428. R. Schack, T. A. Brun, and I. C. Percival, *J. Phys. A*, **28,** 5401 (1995).
429. B. Wolfseder and W. Domcke, *Chem. Phys. Lett.*, **259,** 113 (1996).
430. R. Blümel, A. Buchleitner, R. Graham, L. Sirko, U. Smilansky, and H. Walter, *Phys. Rev. A*, **44,** 4521 (1991).
431. T. Dittrich, B. Oelschlägel, and P. Hänggi, *Europhys. Lett.*, **22,** 5 (1993).
432. L. M. Raff and D. L. Thompson, in *Theory of Chemical Reaction Dynamics*, M. Baer, Ed., Chemical Rubber, Boca Raton, FL, 1985, Vol. 3.
433. W. L. Hase, Ed., *Advances in Classical Trajectory Methods*, Jai Press, London, 1992, Vol. 1.
434. P. Pechukas, *Phys. Rev.*, **181,** 166 (1969); **181,** 174 (1969).
435. W. H. Miller and T. F. George, *J. Chem. Phys.*, **56,** 5637 (1972).
436. W. H. Miller, *Adv. Chem. Phys.*, **25,** 69 (1974).
437. E. E. Nikitin, *Theory of Elementary Atomic and Molecular Processes in Gases*, Clarendon, Oxford, 1974.
438. P. J. Kuntz, *J. Chem. Phys.*, **95,** 141 (1991).
439. F. J. Webster, P. J. Rossky, and R. A. Friesner, *Comput. Phys. Commun.*, **63,** 494 (1991).
440. D. F. Coker and L. Xiao, *J. Chem. Phys.*, **102,** 496 (1995).
441. J. C. Tully and R. K. Preston, *J. Chem. Phys.*, **55,** 562 (1971).
442. M. F. Herman, *J. Chem. Phys.*, **76,** 2949 (1982); **81,** 754 (1984); **81,** 764 (1984); **82,** 3666 (1985); M. F. Herman and J. C. Arce, *Chem. Phys.*, **183,** 335 (1994).
443. N. C. Blais and D. G. Truhlar, *J. Chem. Phys.*, **79,** 1334 (1983); N. C. Blais, D. G. Truhlar, and C. A. Mead, *J. Chem. Phys.*, **89,** 6204 (1988).
444. E. J. Heller and R. C. Brown, *J. Chem. Phys.*, **79,** 3336 (1983).

445. J. C. Tully, *J. Chem. Phys.*, **93,** 1061 (1990); S. Hammes-Schiffer and J. C. Tully, *J. Chem. Phys.*, **101,** 4657 (1994).
446. S. Chapman, *Adv. Chem. Phys.*, **82,** 423 (1992); D. F. Coker, in *Computer Simulations in Chemical Physics*, M. P. Allen and D. J. Tildesley, Eds., Kluwer, Dordrecht, 1993, p. 315.
447. N. F. Mott, *Proc. Cambridge Phil. Soc.*, **27,** 553 (1931).
448. W. R. Thorson, J. B. Delos, and S. A. Boorstein, *Phys. Rev. A*, **4,** 1052 (1971); J. B. Delos, W. R. Thorson, and S. K. Knudson, *Phys. Rev. A*, **6,** 709 (1972); J. B. Delos and W. R. Thorson, *Phys. Rev. A*, **6,** 720 (1972); **6,** 728 (1972).
449. G. D. Billing, *Chem. Phys. Lett.*, **30,** 391 (1975); *Comput. Phys. Rep.*, **1,** 237 (1984); *J. Chem. Phys.*, **99,** 5849 (1993).
450. W. H. Miller and C. W. McCurdy, *J. Chem. Phys.*, **69,** 5163 (1978).
451. H.-D. Meyer and W. H. Miller, *J. Chem. Phys.*, **70,** 3214 (1979).
452. H.-D. Meyer and W. H. Miller, *J. Chem. Phys.*, **71,** 2156 (1979); **72,** 2272 (1980).
453. D. A. Micha, *J. Chem. Phys.*, **78,** 7138 (1983).
454. D. J. Diestler, *J. Chem. Phys.*, **78,** 2240 (1983).
455. R. Graham and M. Höhnerbach, *Z. Phys. B*, **57,** 233 (1984).
456. S. Sawada and H. Metiu, *J. Chem. Phys.*, **84,** 227, 6293 (1986).
457. I. Cacelli, *Chem. Phys.*, **124,** 347 (1988); *J. Chem. Phys.*, **96,** 8439 (1992).
458. M. Amarouche, F. X. Gadea, and J. Durup, *Chem. Phys.*, **130,** 145 (1989).
459. J. M. Cohen and D. A. Micha, *J. Chem. Phys.*, **97,** 1038 (1992).
460. P. Bala, B, Lesyng, and J. A. McCammon, *Chem. Phys. Lett.*, **219,** 259 (1994); *Chem. Phys.*, **180,** 271 (1994).
461. G. Stock, *J. Chem. Phys.*, **103,** 1561 (1995).
462. H.-D. Meyer, *Chem. Phys.*, **82,** 199 (1983).
463. G. Stock and W. H. Miller, *Chem. Phys. Lett.*, **197,** 396 (1992); *J. Chem. Phys.*, **99,** 1545 (1993).
464. G.Stock, *J. Chem. Phys.*, **103,** 2888 (1995).
465. B. Space and D. F. Coker, *J. Chem. Phys.*, **96,** 652 (1992).
466. B. J. Schwartz and P. J. Rossky, *J. Chem. Phys.*, **101,** 6902 (1994).
467. A. Staib and D. Borgis, *J. Chem. Phys.*, **103,** 2642 (1995).
468. V. D. Vachev, J. H. Frederick, B. A. Grishanin, V. N. Zadkov, and N. I. Koroteev, *J. Phys. Chem.*, **99**, 5247 (1995).
469. P. Ehrenfest, *Z. Phys.*, **45,** 455 (1927).
470. E. Neria, A. Nitzan, R. N. Barnett and U. Landman, *Phys. Rev. Lett.*, **67,** 1011 (1991).
471. R. Alimi, A. Garcia-Vela, and R. B. Gerber, *J. Chem. Phys.*, **96,** 2034 (1992); Z. Li and R. B. Gerber, *J. Chem. Phys.*, **99,** 8637 (1993); P. Jungwirth and R. B. Gerber, *J. Chem. Phys.*, **103,** 6046 (1995); **104,** 5803 (1996).
472. W. H. Miller, *J. Chem. Phys.*, **53,** 1949 (1970); **53**, 3578 (1970); **95,** 9428 (1991).
473. M. F. Herman and E. Kluk, *Chem. Phys.*, **91,** 27 (1984); M. F. Herman, *Annu. Rev. Phys. Chem.*, **45,** 83 (1994).
474. E. J. Heller, *J. Chem. Phys.*, **94,** 2723 (1991); **95,** 9431 (1991); M. A. Sepúlveda, S. Tomsovic, and E. J. Heller, *Phys. Rev. Lett.*, **69,** 402 (1992).
475. K. G. Kay, *J. Chem. Phys.*, **100,** 4377 (1994); **100,** 4432 (1994); **101,** 2250 (1994).

476. G. Campolieti and P. Brumer, *J. Chem. Phys.*, **96,** 5969 (1992); *Phys. Rev. A*, **50,** 997 (1994).
477. T. Kinugawa, *Chem. Phys. Lett.*, **235,** 395 (1995).
478. A. R. Walton and D. E. Manolopoulos, *Chem. Phys. Lett.*, **244,** 448 (1995); *Mol. Phys.*, **87,** 961 (1996).
479. E. J. Heller, *J. Chem. Phys.*, **68,** 2066 (1978).
480. S. Mukamel, *J. Chem. Phys.*, **77,** 173 (1982); L. E. Fried and S. Mukamel, *J. Chem. Phys.*, **93,** 3063 (1990); *Adv. Chem. Phys.*, **84,** 435 (1993).
481. R. Bersohn and A. H. Zewail, *Ber. Bunsenges. Phys. Chem.*, **92,** 373 (1988).
482. P. J. Rossky, *J. Opt. Soc. Am. B*, **7,** 1728 (1990); B. J. Schwartz and P. J. Rossky, *J. Chem. Phys.*, **101,** 6917 (1994).
483. O. Zobey and G. Alber, *J. Phys. B*, **26,** 539 (1993).
484. M. A. Sepúlveda and S. Mukamel, *J. Chem. Phys.*, **102,** 9327 (1995).
485. T.J. Martinez, M. Ben-Nun, and G. Ashkenazi, *J. Chem. Phys.*, **104,** 2847 (1996).
486. Z. Li, J.-Y. Fang, and G. C. Martens, *J. Chem. Phys.*, **104,** 6919 (1996).
487. J. Manz, B. Proppe, and B. Schmidt, *Z. Phys. D*, **34,** 111 (1995).
488. K. Finger, C. Daniel, P. Saalfrank, and B. Schmidt, *J. Phys. Chem.*, **100,** 3368 (1996).
489. R. N. Dixon, *Mol. Phys.*, **88,** 949 (1996).
490. C. Daniel, M.C. Heitz, J. Manz, and C. Ribbing, *J. Chem. Phys.*, **102,** 905 (1995).
491. A. Ferretti, G. Granucci, A. Lami, M. Persico, and G. Villani, *J. Chem. Phys.*, **104,** 5517 (1996).

SHORT-TIME FLUORESCENCE STOKES SHIFT DYNAMICS

LOWELL W. UNGAR*

Department of Chemistry and The James Franck Institute, The University of Chicago, Chicago, IL 60637

JEFFREY A. CINA

Department of Chemistry, University of Oregon, Eugene, OR 97403

CONTENTS

I. Introduction
II. Theory
  A. System Excitation and Fluorescence Detection
  B. Time-Resolved Fluorescence Spectrum
  C. Deconvolution from Pulse Intensities
  D. Temporally Separated Pulses
III. Approximations
  A. Classical Franck Approximation
  B. Semiclassical Wave Packet Treatment of $\omega_{\text{peak}}(T')$
  C. Ground-State Linear Response
  D. Excited-State Linear Response
  E. Classical Approximation to $\Delta E(t)$
  F. Classical Linear Response
IV. Short-Time Dynamics of $\omega_{\text{peak}}(T')$
  A. Overlapping Pulses
  B. Short-Time Stokes Shift with Separated Pulses
    1. Wave Packet Picture
    2. Short-Time Expansion
    3. Rigid-Cage Approximation
    4. Difference-Potential-Only Approximation
V. Simulations
  A. Model System
  B. Effects of Pulse Duration, Overlap, and Center Frequency
  C. Tests of the Deconvolution Procedure

Advances in Chemical Physics, Volume 100, Edited by I. Prigogine and Stuart A. Rice.
ISBN 0-471-17458-0

*Present address: Department of Chemistry, University of Utah, Salt Lake City, UT 84112.

D. Classical, Wave Packet, and Linear Response Approximations to $\Delta E(t)$
1. Nonequilibrium Classical and Semiclassical Approximations
2. Ground and Excited State Linear Response Approximations
E. Instantaneous Normal Mode Treatment of Excited-State Dynamics
VI. Concluding Discussion
Appendix A. Stimulated Emission Amplitude for a Pump-Probe Signal
Appendix B. Doorway-Window Picture of the Observed Fluorescence Spectrum
Appendix C. Very Short Time and Very Long Time Behavior of the Peak Emission Frequency
Acknowledgments
References

I. INTRODUCTION

Ultrafast measurement of time- and frequency-resolved fluorescence has emerged as an effective way to probe the dynamics of both isolated molecules and chromophore-solvent systems on excited-state electronic potential energy surfaces [1–4]. In nonlinear optical experiments of this type an ultrashort light pulse pumps a chromophore to an excited electronic state. System fluorescence is then observed by a time- and frequency-gated detection process. In fluorescence upconversion, the technique that has achieved the best time resolution, the spontaneous fluorescence and a short upconversion pulse are combined in a nonlinear crystal, and the integrated intensity of a sum frequency is monitored. A time-resolved fluorescence spectrum can be constructed from the data for a range of time delays between the excitation pulse and the gate pulse, and a range of frequency windows determined by the gate pulse center frequency.

Wave packet motion in small molecules [2] and solvation dynamics of a polar solute in a polar solvent[1] are two among several seemingly very different molecular processes that have been investigated by fluorescence upconversion [9, 10]. Information about excited-state nuclear dynamics has been extracted from the time-resolved fluorescence spectrum in a variety of ways in the different experimental contexts. One quantity frequently derived from the full time-dependent fluorescence spectrum of systems that tend toward a steady state is the time-dependent Stokes shift,

$$S(t) = \frac{\omega_{\text{peak}}(t) - \omega_{\text{peak}}(\infty)}{\omega_{\text{peak}}(0) - \omega_{\text{peak}}(\infty)} \tag{1.1}$$

[1]We will not attempt to review experimental and theoretical work on ultrafast solvation dynamics. References [5–8] survey this rapidly progressing area, and discuss its implications for other aspects of chemical dynamics in solution.

The Stokes shift in Eq. (1.1) is the observed time-dependent peak frequency of emission, normalized to decay from one to zero. In probing wave-packet motion in a diatomic molecule [2] and spontaneous fluorescence from reaction centers at low temperature [4], it was deemed more useful to focus on fixed-frequency slices through the time-resolved fluorescence spectrum. If the light pulses are polarized, the anisotropy decay also can be followed [11, 12].

As the various experiments are closely related, and their usefulness is well established, it is worthwhile to give a unified treatment to and examine the connections among the different applications of fluorescence upconversion. Our purpose is in large part didactic—to clarify the origin of some general features in time-resolved fluorescence spectra and elaborate upon complicating details of the upconversion process. We focus on $\omega_{\text{peak}}(t)$ and the full time-dependent fluorescence spectrum, rather than on $S(t)$, because the systems of interest may not all tend toward a steady state on the timescale of an ultrafast experiment.

The basic theory of time- and frequency-resolved fluorescence spectra has already been expounded from several vantage points [13–19]. However, few detailed simulations have been conducted of the actual optical processes involved in time-dependent fluorescence Stokes shift measurements on molecular systems. Conversely, simulations of solvation dynamics usually focus on molecular quantities, especially the time-dependent transition energy $\Delta E(t)$, which is the average value of the potential difference between electronic states, $V_e(\underset{\sim}{r}) - V_g(\underset{\sim}{r})$, at time t after an abrupt transition from the ground to the excited state. $\Delta E(t)$ and $\omega_{\text{peak}}(t)$ generally are similar, and classical molecular dynamics simulations [20–24] and analytical studies [25–27] of the time-dependent transition energy (or solvation energy) have contributed a great deal to the current molecular-level understanding of the short-time Stokes shift. However, when the finite bandwidths of real pulses, and their temporal overlap at short delays, complicate the relationship between $\Delta E(t)$ and the measured peak emission frequency, it become necessary to simultaneously treat both molecular and optical processes at play in fluorescence upconversion.

The effects of pulse overlap also have been ignored in modeling fluorescence upconversion measurements of vibrational wave packet motion in a diatomic molecule [2]. Since the nuclear motion is quasi-periodic in this system, the "short-time dynamics" of nuclear motion can be examined at the first recurrence, when the optically prepared wave packet returns to the Franck–Condon region. By this time, the excitation and gate pulses are well separated. However, proper incorporation of pulse overlap at short interpulse delays will be necessary for polyatomic molecules or highly anharmonic diatomics, in which the nuclear motion is not quasi-periodic.

The effects of pulse overlap at short times also would seem difficult to side-step in fluorescence upconversion measurements on polar solvation. The recently observed rapid initial drop in peak emission frequency [1, 28] takes place on the ~ 10–100 fs timescale of libration in the first solvation shell—most of the solvation is completed before the pulses are well-separated.

Numerous other approximations have been used to streamline the simulation of time-resolved fluorescence spectra. Harmonic or few-level potentials are sometimes substituted for the true electronic potential energy surfaces [16]. Classical dynamics has been widely employed in studies of liquid-state solvation (an approximation that is almost certainly justified when the temperature is higher than the characteristic frequencies of solvent libration). Others have adopted linear response approximations that treat the response to a putatively instantaneous change in electronic potential in terms of equilibrium correlation functions in either the ground or excited electronic state [20, 21, 29]. Linear response has also been augmented with a further instantaneous normal mode approximation [30–32].

These simplifications are difficult to test directly on the systems to which they were applied, some of which have hundreds or thousands of degrees of freedom. Instead, we compare several approximate treatments of Stokes shift dynamics analytically and in a few-mode model system that mimics bulk polar solvation. This approach allows the fullest possible comparison between rigorously calculated time-dependent fluorescence signals and numerous simpler quantities characterizing nuclear motion in the excited electronic state. Although dielectric theories of the solvent [25–27], will not be directly considered in this paper, other approximations made in those treatments (such as linear response) will be evaluated.

The main contents of this paper are as follows: Section II sets forth a straightforward time-dependent perturbation treatment of time- and frequency-resolved fluorescence that is applicable to isolated or solvated chromophores in thermal equilibrium. Explicit treatment of system excitation and fluorescence detection leads to a formula for the observed time-dependent fluorescence spectrum that accounts for finite pulse bandwidths and includes the effects of pulse overlap at short delays. Our treatment of fluorescence upconversion requires excitation of a quantum mechanical three-level detector by the combined action of the fluorescence and a short gate pulse, thus capturing the inverse relationship between time and frequency resolution without overtly introducing the quantum radiation field. We further obtain a formula for a time-dependent fluorescence spectrum that has been deconvolved from the cross-correlation of excitation-pulse and gate-pulse intensities. For well-separated pulses, the deconvolved signal is expressed in terms of a prepared doorway operator (the excited-state density matrix increment) and a measured quantity (the window operator), analogous to operators previously used to describe pump-probe spectroscopy [33–35].

Section III entertains a number of approximations to the deconvolved time-dependent fluorescence spectrum and its peak frequency. We use the classical Franck approximation [36], which simplifies the action of suitably brief pulses, to clarify the fundamental distinction between the doorway operators appropriate to the observed and deconvolved fluorescence spectra. It is seen (and will later be illustrated) that as a composite operator derived from many distinct measurements, the doorway operator for the deconvolved spectrum is not a physically realizable nuclear state and is, therefore, released from the constraints of the uncertainty principle.

The window operator in the classical Franck approximation is used to identify detection conditions under which the time-dependent peak emission frequency gives a measure of the average electronic difference potential under the propagated density matrix increment. If the excitation pulse is short enough that nuclear motion during excitation can be neglected entirely (and deconvolution is unnecessary), then the average electronic difference potential in turn reduces to the widely studied time-dependent electronic transition energy.

We derive a semiclassical formula for the peak emission frequency in terms of the time-dependent moments of position and momentum that characterize the evolving nuclear distribution in the excited electronic state. This wave-packet treatment can incorporate some quantal aspects in simulations of fluorescence Stokes shift dynamics in moderate-sized systems. Simple expressions for the initial values of the moments show how the excited-state doorway operator is distorted from the equilibrium distribution.

Section III concludes by briefly describing several linear response and classical mechanical treatments of Stokes shift dynamics. Connections among them and conditions under which they can be expected to agree with the more general expressions are both investigated.

Section IV examines the short-time dynamics of the peak emission frequency in some detail. We derive a short-time expansion for the peak emission frequency during the excitation pulse, and show how it differs from the peak frequency for non-overlapping pulses. These results, together with a short-time expansion of our wave-packet expressions for the peak frequency (using the well-known equations of motion for a Gaussian nuclear distribution), clarify the physical origin of the ubiquitous temporal-Gaussian decay component in the Stokes shifts of both isolated and solvated chromophores. We examine two approximate treatments of the time-dependent transition energy: Maroncelli's rigid cage approximation [20] and the difference-potential-only approximation of Perera and Berkowitz [37]. It is shown that the former treatment gives the transition energy correctly at short times, while the latter does so only with an additional approximation.

Section V puts the approximate treatments of previous sections to the test with numerical simulations on a three-mode model system. The system studied consists of six acetonitrile molecules, which librate in a plane in response to a pump-induced change in the electric dipole moment of a chromophore. Effects of pulse duration, temporal overlap, and center frequency are graphically illustrated. Comparison of the time-dependent peak emission frequencies obtained from the observed and the deconvolved spectra gives an indication of the successes and limitations of the deconvolution procedure in extracting the ideal time-dependent transition energy from experimental observables. We compare the various approximate treatments of the time-dependent transition energy described in Section III, including semiclassical, classical nonequilibrium, and classical linear response approximations. We also introduce and test an instantaneous-normal-mode treatment of the excited-state dynamics.

A concluding discussion in Section VI assesses the implications of our findings for state-of-the-art simulations of fluorescence upconversion and other nonlinear optical signals, and points out several avenues for further investigation.

II. THEORY

A. System Excitation and Fluorescence Detection

We wish to calculate the time-resolved fluorescence spectra of simple systems, including the effects of a noninstantaneous excitation pulse and of detection via upconversion with a noninstantaneous gate pulse. The system is taken to include any relevant chromophore and solvent coordinates and two electronic states. Starting in thermal equilibrium in the ground electronic state, the system is electronically excited by an optical pulse. It then evolves adiabatically in the excited electronic state. Fluorescence from the excited state nuclear distribution during an upconversion time-window, which may partially coincide with the excitation pulse, is detected with transform-limited frequency resolution. The system-excitation, fluorescence, and detection processes are illustrated schematically in Fig. 1.

The Hamiltonian includes the system, a detector, their coupling via fluorescence, and time-dependent excitation– and gate–pulse interactions:

$$H(t) = H_s + V_{\text{exc}}(t) + V_{\text{fl}} + H_d + V_{\text{up}}(t) \tag{2.1}$$

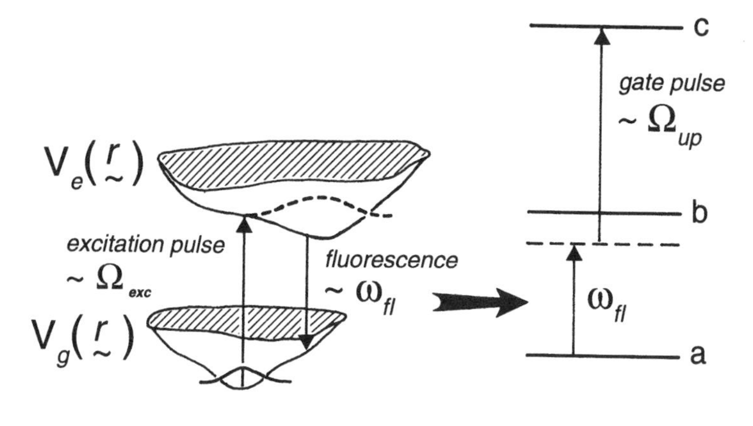

Figure 1. Schematic illustration of fluorescence upconversion following short-pulse excitation and excited-state propagation. An excitation pulse centered around frequency $\Omega_{\rm exc}$ promotes a wave packet to the excited potential surface V_e. After propagating, the system fluoresces at frequency $\omega_{\rm fl}$. The fluorescence is combined with a gate pulse of center frequency $\Omega_{\rm up}$ to doubly excite the detector. The detected fluorescence intensity at frequency $\omega_{\rm fl}$ is proportional to the population in the detector state $|c\rangle$.

The system Hamiltonian H_s comprises ground and excited adiabatic electronic states with associated nuclear Hamiltonians:

$$H_s = |g\rangle H_g \langle g| + |e\rangle H_e \langle e| \tag{2.2}$$

$$H_g = \tfrac{1}{2}\underset{\sim}{p}^2 + V_g(\underset{\sim}{r}) \tag{2.3}$$

$$H_e = \tfrac{1}{2}\underset{\sim}{p}^2 + V_e(\underset{\sim}{r}) \tag{2.4}$$

where $\underset{\sim}{r}$ and $\underset{\sim}{p}$ are nuclear coordinate and momentum operators in a mass-weighted coördinate system. We also will set $\hbar = 1$ for notational convenience.

The detector is a three-level system, so that it can undergo two-photon excitation by the combined action of system fluorescence and a gate pulse:

$$H_d = |a\rangle\varepsilon_a\langle a| + |b\rangle\varepsilon_b\langle b| + |c\rangle\varepsilon_c\langle c| \qquad \varepsilon_c - \varepsilon_b > \varepsilon_b - \varepsilon_a > 0 \tag{2.5}$$

Interaction between a Gaussian pulse and the system electronic dipole moment excites the system. Fluorescence is modeled as an interaction between the dipole moment operators of the system and the detector, and a Gaussian upconversion pulse time-gates the detector:

$$V_{\text{exc}}(t) = -\hat{\mu}_s E_{\text{exc}}(t) = -\hat{\mu}_s E_{\text{exc}} f_{\text{exc}}(t) \cos(\Omega_{\text{exc}} t) \tag{2.6}$$

$$V_{\text{fl}} = \eta_{\text{fl}} \hat{\mu}_s \hat{\mu}_d \tag{2.7}$$

$$V_{\text{up}}(t) = -\hat{\mu}_d E_{\text{up}}(t) = -\hat{\mu}_d E_{\text{up}} f_{\text{up}}(t - t_d) \cos(\Omega_{\text{up}} t) \tag{2.8}$$

The dipole moment operators of the system and the detector are given by

$$\hat{\mu}_s = \mu_s(|g\rangle\langle e| + |e\rangle\langle g|) \tag{2.9}$$

and

$$\hat{\mu}_d = \mu_d(|a\rangle\langle b| + |b\rangle\langle a| + |b\rangle\langle c| + |c\rangle\langle b|) \tag{2.10}$$

respectively. Each of the pulses in Eqs. (2.6) and (2.8) has an independent center frequency Ω_σ, length τ_σ (with intensity-FWHM $= \tau_\sigma 2(\ln 2)^{1/2}$, and Gaussian envelope $f_\sigma(t) = \exp(-t^2/2\tau_\sigma^2)$; t_d is the time delay between the excitation and gate pulses.

The system starts in thermal equilibrium in the electronic ground state, and the detector is initially in its lowest level, $|a\rangle$. The observed signal is proportional to the final population in the doubly excited detector state $|c\rangle$. Starting from a single nuclear eigenstate $|\psi_n\rangle$, where $\mathrm{H}_g|\psi_n\rangle = \mathrm{E}_n|\psi_n\rangle$, we will find a simple perturbative expression, valid for weak fields and slow fluorescence, for the amplitude in the detector c-state following electronic excitation, fluorescence, and interaction with the gate-pulse. This amplitude will then be squared to give the c-state population, and summed over thermally weighted nuclear eigenstates of the ground electronic manifold.

The calculation is easiest in the interaction representation, where the state $|\tilde{\Psi}(t)\rangle \equiv \exp[i(H_s + H_d)t]|\Psi(t)\rangle$ evolves from an initial state $|\Psi_0\rangle = |\psi_n\rangle|g\rangle|a\rangle$ according to $i(\partial/\partial t)|\tilde{\Psi}(t)\rangle = (\tilde{V}_{\text{exc}}(t) + \tilde{V}_{\text{fl}}(t) + \tilde{V}_{\text{up}}(t)|\tilde{\Psi}(t)\rangle$, with

$$\begin{aligned}\tilde{V}_{\text{exc}}(t) &\equiv \exp[i(H_s + H_d)t] V_{\text{exc}}(t) \exp[-i(H_s + H_d)t] \\ &\cong -\frac{\mu_s E_{\text{exc}}}{2} f_{\text{exc}}(t) \exp[i(H_e - \Omega_{\text{exc}})t]|e\rangle\langle g| \exp[-iH_g t] \\ &\quad + \text{h.c.}\end{aligned} \tag{2.11}$$

$$\tilde{V}_{\mathrm{fl}}(t) \equiv \exp[i(H_s + H_d)t]V_{\mathrm{fl}} \exp[-i(H_s + H_d)t]$$
$$\cong \eta_{\mathrm{fl}}\mu_s\mu_d \exp[i(H_g + \varepsilon_b)t]|g\rangle|b\rangle\langle a|\langle e| \exp[-i(H_e + \varepsilon_a)t] + \mathrm{h.c.} \tag{2.12}$$

$$\tilde{V}_{\mathrm{up}}(t) \equiv \exp[i(H_s + H_d)t]V_{\mathrm{up}}(t) \exp[-i(H_s + H_d)t]$$
$$\cong -\frac{\mu_d E_{\mathrm{up}}}{2} f_{\mathrm{up}}(t - t_d) \exp[i(\varepsilon_c - \varepsilon_b - \Omega_{\mathrm{up}})t]|c\rangle\langle b| + \mathrm{h.c.} \tag{2.13}$$

In each of the approximate equalities in Eqs. (2.11–2.13), we have made the conventional rotating wave approximation by neglecting energy-nonconserving terms that oscillate at electronic frequencies. The detector is chosen such that both the fluorescence and the upconversion pulse are nonresonant with both detector transitions, but the fluorescence is more nearly resonant with the $b \leftarrow a$ detector transition than with $c \leftarrow b$, and vice versa for the gate pulse. Hence the highly oscillatory terms in $\tilde{V}_{\mathrm{fl}}(t)$ that would drive the $c \leftarrow b$ transition may be neglected, as may contributions to $\tilde{V}_{\mathrm{up}}(t)$ that would couple to $b \leftarrow a$. (Relaxing the assumption that the gate pulse and the fluorescence couple preferentially to different detector transitions would affect only a constant prefactor.) Starting from the initial state $|\Psi_0\rangle = |\psi_n\rangle|g\rangle|a\rangle$, the detector level $|c\rangle$ can only be reached by the successive action of the excitation pulse, fluorescence, and the gate pulse. The amplitude in the c-state of the detector, with the system back in the electronic ground state, is then given by

$$\langle c|\langle g|\tilde{\Psi}\rangle = (-i)^3 \int_{-\infty}^{\infty} d\tau_3 \int_{-\infty}^{\tau_3} d\tau_2 \int_{-\infty}^{\tau_2} d\tau_1 \cdot \langle c|\langle g|\tilde{V}_{\mathrm{up}}(\tau_3)\tilde{V}_{\mathrm{fl}}(\tau_2)\tilde{V}_{\mathrm{exc}}(\tau_1)|\Psi_0\rangle$$
$$= i \int_{-\infty}^{\infty} d\tau \int_0^{\infty} d\tau' \int_0^{\tau'} d\tau'' \cdot \langle c|\langle g|\tilde{V}_{\mathrm{up}}(\tau)\tilde{V}_{\mathrm{fl}}(\tau - \tau'')\tilde{V}_{\mathrm{exc}}(\tau - \tau')|\Psi_0\rangle \tag{2.14}$$

In the second equality of Eq. (2.14) we have redefined the variables of integration in terms of the time of action τ of the gate pulse and the amounts τ' and τ'', respectively, by which excitation and fluorescence precede this.

This interaction operators in Eqs. (2.11–2.13) can be inserted in Eq. (2.14) to give

$$
\begin{aligned}
\langle c|\langle g|\tilde{\Psi}\rangle = i\mu_s^2\eta_{\text{fl}}\mu_d^2 \frac{E_{\text{exc}}E_{\text{up}}}{4} \int_{-\infty}^{\infty} d\tau \int_0^{\infty} d\tau' f_{\text{up}}(\tau - t_d) \\
\cdot f_{\text{exc}}(\tau - \tau')e^{i(\varepsilon_c - \varepsilon_a - \Omega_{\text{exc}} - \Omega_{\text{up}})\tau} e^{i\Omega_{\text{exc}}\tau'} \\
\cdot e^{iH_g\tau}\left[\int_0^{\tau'} d\tau'' e^{-i(\varepsilon_b - \varepsilon_a)\tau''} e^{-iH_g\tau''} e^{iH_e\tau''}\right] \\
\cdot e^{-iH_e\tau'} e^{iH_g(\tau' - \tau)}|\psi_n\rangle
\end{aligned}
\tag{2.15}
$$

As the difference between the system potentials is nonresonant with $\varepsilon_b - \varepsilon_a$ by assumption, we can make the replacement $e^{-iH_g\tau''}e^{iH_e\tau''} \approx e^{i\Delta E(t_d)\tau''}$ in the τ''-integrand, where $\Delta E(t_d)$ is the average transition energy at the center of the upconversion pulse. The integral in square brackets then becomes $i[\varepsilon_b - \varepsilon_a - \Delta E(t_d)]^{-1} \times \{\exp[-i(\varepsilon_b - \varepsilon_a - \Delta E(t_d))\tau'] - 1\}$. As the exponential term in this last expression oscillates at an electronic frequency, it does not survive the subsequent τ' integration. To a good approximation, we can therefore write

$$
\begin{aligned}
\langle c|\langle g|\tilde{\Psi}\rangle \cong \mu_s^2\eta_{\text{fl}}\mu_d^2 \frac{E_{\text{exc}}E_{\text{up}}}{4[\varepsilon_b - \varepsilon_a - \Delta E(t_d)]} \int_{-\infty}^{\infty} d\tau \int_0^{\infty} d\tau' f_{\text{up}}(\tau - t_d) \\
\cdot f_{\text{exc}}(\tau - \tau')e^{i(\omega - \Omega_{\text{exc}})\tau} e^{i\Omega_{\text{exc}}\tau'} e^{iH_g\tau} e^{-iH_e\tau'} e^{iH_g(\tau' - \tau)}|\psi_n\rangle
\end{aligned}
\tag{2.16}
$$

In Eq. (2.16) we have introduced the symbol $\omega \equiv \varepsilon_c - \varepsilon_a - \Omega_{\text{up}}$ for the nominal fluorescence frequency. The close similarity between the time-gated fluorescence amplitude of Eq. (2.16) and the probability amplitude for stimulated emission in a pump-probe experiment is detailed in Appendix A.

B. Time-Resolved Fluorescence Spectrum

The observed signal is proportional to the population left in detector state $|c\rangle$, from a system that starts in thermal equilibrium. Summing the c-state populations originating from all the nuclear eigenstates yields

$$
F(\omega, t_d) = \sum_n p_n \langle\tilde{\Psi}|g\rangle|c\rangle\langle c|\langle g|\tilde{\Psi}\rangle \tag{2.17}
$$

where the thermal occupation probabilities are $p_n = Z_g^{-1}\exp(-\beta E_n)$. Substitution of Eq. (2.16) in (2.17) leads (within a constant prefactor) to

$$F(\omega, t_d) = \int_{-\infty}^{\infty} d\tau \int_{-\infty}^{\infty} d\bar{\tau} \int_{0}^{\infty} d\tau' \int_{0}^{\infty} d\bar{\tau}'$$

$$\cdot f_{\text{up}}(\tau - t_d) f_{\text{up}}(\bar{\tau} - t_d) f_{\text{exc}}(\tau - \tau') f_{\text{exc}}(\bar{\tau} - \bar{\tau}')$$

$$\cdot\ e^{i(\omega - \Omega_{\text{exc}})(\tau - \bar{\tau})} e^{i\Omega_{\text{exc}}(\tau' - \bar{\tau}')}$$

$$\cdot\ \text{Tr}_{\text{nuc}}[e^{iH_g\tau} e^{-iH_e\tau'} e^{iH_g(\tau' - \tau)} \rho_g e^{-iH_g(\bar{\tau}' - \bar{\tau})} e^{iH_e\bar{\tau}'} e^{-iH_g\bar{\tau}}] \quad (2.18)$$

The equilibrium density operator appearing in Eq. (2.18) is $\rho_g = \sum |\psi_n\rangle p_n \langle\psi_n| \equiv Z_g^{-1}\exp(-\beta H_g)$, and $\text{Tr}_{\text{nuc}}[\ldots]$ denotes a trace over nuclear degrees of freedom. As a function of the interpulse delay and the nominal fluorescence frequency, $F(\omega, t_d)$ in Eq. (2.18) specifies the time-dependent fluorescence spectrum as it would actually be measured. It requires no approximations other than the use of low-order perturbation theory and the rotating wave approximation. Eq. (2.18) could also have been obtained from the general expressions given in refs. [13, 14] for the time-dependent physical spectrum of light.

C. Deconvolution from Pulse Intensities

In order to remove some of the "blurring" of the observed time-resolved spectrum due to finite pulse durations, it proves useful to rewrite Eq. (2.18) with new variables of integration. Using the averages, $T = \frac{1}{2}(\tau + \bar{\tau})$ and $T' = \frac{1}{2}(\tau' + \bar{\tau}')$, and the differences, $t = \tau - \bar{\tau}$ and $t' = \tau' - \bar{\tau}'$, we can put Eq. (2.18) in the form

$$F(\omega, t_d) = \int_{-\infty}^{\infty} dT' G(t_d - T') \tilde{F}(\omega, T') \quad (2.19)$$

Equation (2.19) expresses the observed spectrum as a convolution between the intensity cross-correlation function,

$$G(t_d - T') = \int_{-\infty}^{\infty} dT I_{\text{up}}(T - t_d) I_{\text{exc}}(T - T')$$

$$= \tau_{\text{up}}\tau_{\text{exc}} \sqrt{\frac{\pi}{\tau_{\text{up}}^2 + \tau_{\text{exc}}^2}} \exp\left(-\frac{(t_d - T')^2}{\tau_{\text{up}}^2 + \tau_{\text{exc}}^2}\right) \quad (2.20)$$

$$I_\sigma(t) = f_\sigma^2(t) \tag{2.21}$$

and a deconvolved spectrum, which vanishes for negative T' and, for positive values of its time argument, takes the form

$$\begin{aligned}\tilde{F}(\omega, T') = \mathrm{Tr}_{\mathrm{nuc}}\Bigg[\int_{-\infty}^{\infty} dt I_{\mathrm{up}}\left(\frac{t}{2}\right) U\left(\frac{t}{2}\right) U^\dagger\left(-\frac{t}{2}\right) e^{i\omega t} \\ \cdot\, e^{-iH_eT'}\left(\int_{-2T'}^{2T'} dt' I_{\mathrm{exc}}\left(\frac{t'-t}{2}\right) e^{i\Omega_{\mathrm{exc}}(t'-t)}\right. \\ \left.\cdot\, U\left(\frac{t'-t}{2}\right)\rho_g U^\dagger\left(-\frac{t'-t}{2}\right)\right) e^{iH_eT'}\Bigg]\end{aligned} \tag{2.22}$$

with

$$U(t) \equiv \exp(-iH_e t)\exp(iH_g t) \tag{2.23}$$

In Eq. (2.22), T' is the average time interval over which excited state dynamics occurs. While $\tilde{F}(\omega, T')$ is locally time-averaged over the width of the intensity cross-correlation to give the observed signal in Eq. (2.19), the deconvolved spectrum $\tilde{F}(\omega, T')$ itself depends on excitation and gate-pulse lengths and center frequencies. Indeed, the upconversion pulse must have finite bandwidth, and hence non-negligible duration, to allow frequency resolution [ω-dependence in Eq. (2.22)]. A time-dependent peak fluorescence frequency can be found from the maximum of either $F(\omega, t_d)$ or $\tilde{F}(\omega, T')$, and in general these two peak frequencies are different. The deconvolution procedure set forth here is an analytical realization, valid only for Gaussian pulses, of the standard deconvolution applied numerically to experimental time-resolved spectra [38, 39]. A careful numerical study of deconvolution and spectral reconstruction of simulated time-dependent emission data (on the hydrated electron) has been made by B. J. Schwartz and P. J. Rossky [40] with comparison to recent experimental studies on ultrafast electron solvation by Barbara and coworkers [41].

D. Temporally Separated Pulses

The general expression Eq. (2.22) for the deconvolved time-dependent fluorescence spectrum simplifies somewhat when the excitation pulse and gate pulse do not overlap in time. If t_d significantly exceeds τ_{exc} and τ_{up}, the limits of integration over t' can be extended to plus and minus infinity. In this case the integral over t' becomes t-independent, and Eq. (2.22) reduces to

$$\tilde{F}(\omega, T' \geq 0) = \mathrm{Tr}_{\mathrm{nuc}}[\tilde{W}(\omega) e^{-iH_e T'} \tilde{D}(\Omega_{\mathrm{exc}}) e^{iH_e T'}] \tag{2.24}$$

where

$$\tilde{D}(\Omega_{\mathrm{exc}}) = \int_{-\infty}^{\infty} dt' I_{\mathrm{exc}}\left(\frac{t'}{2}\right) e^{i\Omega_{\mathrm{exc}} t'} U\left(\frac{t'}{2}\right) \rho_g U^\dagger \left(-\frac{t'}{2}\right) \tag{2.25}$$

and

$$\tilde{W}(\omega) = \int_{-\infty}^{\infty} dt I_{\mathrm{up}}\left(\frac{t}{2}\right) e^{i\omega t} U\left(\frac{t}{2}\right) U^\dagger \left(-\frac{t}{2}\right) \tag{2.26}$$

Equation (2.24) represents the deconvolved time-dependent emission spectrum as the expectation value of a window operator $\tilde{W}(\omega)$ for the system in a "state" given by the nuclear density matrix increment $\tilde{D}(\Omega_{\mathrm{exc}})$ propagated for T' on the upper potential surface. Equations (2.24–2.26) will be used as starting points for several more approximate treatments of time-resolved fluorescence upconversion in subsequent sections.

A doorway-window picture can also be developed for the observed fluorescence spectrum. For the purpose of comparison, the key features are outlined in Appendix B. One basic difference between the observed and deconvolved fluorescence spectra has to do with the nature of the excited-state density matrix increment, or doorway operator, in each case. This distinction is most easily illustrated with the help of the approximate descriptions presented in the next section.

III. APPROXIMATIONS

A. Classical Franck Approximation

Very simple expressions can be obtained for the density matrix increment $\tilde{D}(\Omega_{\mathrm{exc}})$ and the window operator $\tilde{W}(\omega)$ in the case where the gate pulse and excitation pulse durations are somewhat shorter than the timescales of nuclear motion. With such short pulses, we can make the classical Franck approximation [36, 42] by assuming that H_e and H_g commute in Eq. (2.23). Then $U(t) \cong \exp(-i\Delta V t)$, where $\Delta V \equiv V_e(\underset{\sim}{r}) - V_g(\underset{\sim}{r})$ is the electronic difference potential. Substitution in Eqs. (2.25) and (2.26) and expansion of the integrands in powers of $(\Omega_{\text{exc}} - \Delta V)$ then give

$$\tilde{D}(\Omega_{\text{exc}}) \cong 2\tau_{\text{exc}}\sqrt{\pi}\left\{\rho_g - \frac{\tau_{\text{exc}}^2}{4}\left[(\Omega_{\text{exc}} - \Delta V)^2\rho_g + 2(\Omega_{\text{exc}} - \Delta V)\rho_g(\Omega_{\text{exc}} - \Delta V) + \rho_g(\Omega_{\text{exc}} - \Delta V)^2\right]\right\} \quad (3.1)$$

and

$$\tilde{W}(\omega) \cong 2\tau_{\text{up}}\sqrt{\pi}\exp[-(\omega - \Delta V)^2/\tau_{\text{up}}^2] \cong 2\tau_{\text{up}}\sqrt{\pi}[1 - \tau_{\text{up}}^2(\omega - \Delta V)^2] \quad (3.2)$$

for the density matrix increment and window operator, respectively, in the classical Franck approximation.[2]

From Eq. (3.2), $\partial\tilde{W}/\partial\omega \cong -4\tau_{\text{up}}^3\sqrt{\pi}(\omega - \Delta V)$, and it follows that the peak emission frequency is given by

$$\omega_{\text{peak}}(T') \cong \frac{\text{Tr}_{\text{nuc}}[\Delta V e^{-iH_eT'}\tilde{D}(\Omega_{\text{exc}})e^{iH_eT'}]}{\text{Tr}_{\text{nuc}}[\tilde{D}(\Omega_{\text{exc}})]} \quad (3.3)$$

under the classical Franck approximation. In the limiting situation where τ_{exc} is so short that nuclear motion is *entirely* negligible during the excitation pulse, the right-hand side of Eq. (3.1) is proportional to ρ_g, and Eq. (3.3) becomes simply

$$\omega_{\text{peak}}(t) \cong \text{Tr}_{\text{nuc}}[\Delta V e^{-iH_et}\rho_g e^{iH_et}] \equiv \Delta E(t) \quad (3.4)$$

The approximate equality in Eq. (3.4), between the peak-emission frequency and the time-dependent average transition energy has been the starting point for many simulations of the transient Stokes shift. It applies strictly only when the gate pulse is shorter, and the excitation pulse much shorter, than the characteristic timescales for nuclear motion. Under those very demanding conditions, deconvolution of Eq. (2.19) is almost unnecessary, and the time-argument T' is essentially equal to the interpulse delay t_d.

Returning to short but non-negligible pulse durations, the classical Franck

[2]One can obtain a classical Franck expression for the density matrix increment without expanding the exponentials by using the position representation

$$\langle \underset{\sim}{r}_1|\tilde{D}(\Omega_{\text{exc}})|\underset{\sim}{r}_2\rangle \cong 2\tau_{\text{exc}}\sqrt{\pi}\exp\{[2\Omega_{\text{exc}} - \Delta V(\underset{\sim}{r}_1) - \Delta V(\underset{\sim}{r}_2)]\tau_{\text{exc}}^2/2\}\langle \underset{\sim}{r}_1|\rho_g|\underset{\sim}{r}_2\rangle.$$

approximation can also be applied to the doorway and window operators of the observed spectrum, as discussed in Appendix B. The form of the window operator in Eq. (B.4) is essentially unchanged from Eq. (3.2). But the density matrix increment for the observed spectrum in the classical Franck approximation, Eq. (B.5), differs from Eq. (3.1), and this difference illustrates the fundamental distinction between the two quantities.

The doorway operator for the observed spectrum is the nuclear distribution prepared by an excitation pulse; it is a genuine nuclear state represented by a positive-definite density operator. On the other hand, $\tilde{D}(\Omega_{\text{exc}})$ is not a true nuclear state but a composite operator whose motion under H_e describes a combination of distinct measurements. While $D(\Omega_{\text{exc}})$ is positive-definite, $\tilde{D}(\Omega_{\text{exc}})$ may have some negative eigenvalues, corresponding to "negative populations" in some vibronic states.

A one-dimensional example under the classical Franck approximation can illustrate the non-positive-definite nature of $\tilde{D}(\Omega_{\text{exc}})$. If H_g and H_e are displaced harmonic oscillators, if only the $\nu = 0$ level of the ground electronic state is populated in ρ_g, and if Ω_{exc} is vertically resonant, then $\tilde{D}(\Omega_{\text{exc}})$ in a basis of vibrational levels in the electronic ground state has a positive population in $\nu = 0$ but has a negative population in $\nu = 1$ proportional to $\tau_{\text{exc}}^2(\partial\Delta V/\partial r)^2$. The density matrix increment $D(\Omega_{\text{exc}})$ of Eq. (B.2) has a positive population in $\nu = 0$ and none in $\nu = 1$ under the same conditions.

Because the density matrix increment for the deconvolved spectrum can be non-positive-definite, it can give rise to nonphysical "average values" of physical quantities besides the vibronic state populations. For instance, the deconvolved fluorescence spectrum, Eq. (2.24), which is the average value of the window operator, can exhibit negative intensities at some times and frequencies. But the most telling consequence of nonpositivity is that $\tilde{D}(\Omega_{\text{exc}})$ is relieved from the constraints of the uncertainty principle, which applies to pure states or incoherent sums of pure states with positive populations [43]. The following subsection illustrates that feature of $\tilde{D}(\Omega_{\text{exc}})$ and discusses its implications for deconvolved fluorescence spectra.

B. Semiclassical Wave Packet Treatment of $\omega_{\text{peak}}(T')$

A semiclassical wave-packet picture of the Stokes shift dynamics helps to clarify some characteristics of the initial response. Wave-packet methods [44] were pioneered by Heller for continuous-wave linear absorption [45] and Raman scattering [46] by pure states of isolated molecules, but they can be generalized to time-resolved nonlinear optical spectra from mixed states [47, 48]. These methods are applicable to short-time optical processes when the initial equilibrium state of the multidimensional system (including solvent coordinates and intramolecular chromophore modes) can be approximated as harmonic [47–49]. These wave-packet methods can also be modified to cope with anharmonicity [50] and with spatial coherence among multiple wells [51].

The semiclassical picture of the time-resolved Stokes shift used here is based on a Gaussian nuclear density matrix increment in the coordinate representation (see Appendix A of ref. [48]) or its Wigner-function equivalent. It applies when a harmonic equilibrium state is acted on by pulses treated under the classical Franck approximation (see later in this section), and propagated for short times in a locally quadratically approximated excited electronic state potential (see Section IV.B.1). The Gaussian density matrix increment is characterized by its first and second moments with respect to position and momentum [48]:

$$\bar{\underset{\sim}{r}}(T') = \langle \underset{\sim}{r} \rangle \tag{3.5}$$

$$\bar{\underset{\sim}{p}}(T') = \langle \underset{\sim}{p} \rangle \tag{3.6}$$

$$\underset{\approx}{\sigma}_r(T') = \langle (\underset{\sim}{r} - \bar{\underset{\sim}{r}})(\underset{\sim}{r} - \bar{\underset{\sim}{r}})^{\mathrm{T}} \rangle \tag{3.7}$$

$$\underset{\approx}{\sigma}_p(T') = \langle (\underset{\sim}{p} - \bar{\underset{\sim}{p}})(\underset{\sim}{p} - \bar{\underset{\sim}{p}})^{\mathrm{T}} \rangle \tag{3.8}$$

$$\underset{\approx}{\sigma}_{rp}(T') = \tfrac{1}{2} \langle (\underset{\sim}{r} - \bar{\underset{\sim}{r}})(\underset{\sim}{p} - \bar{\underset{\sim}{p}})^{\mathrm{T}} + ((\underset{\sim}{p} - \bar{\underset{\sim}{p}})(\underset{\sim}{r} - \bar{\underset{\sim}{r}})^{\mathrm{T}})^{\mathrm{T}} \rangle \tag{3.9}$$

In Eqs. (3.5–3.9), the superscript T denotes the transpose, and the unlabeled angular bracket represents a time-dependent average,

$$\langle \ldots \rangle \equiv \frac{\mathrm{Tr}_{\mathrm{nuc}}[\ldots \exp(-iH_e T')\tilde{D}(\Omega_{\mathrm{exc}}) \exp(iH_e T')]}{\mathrm{Tr}_{\mathrm{nuc}}[\tilde{D}(\Omega_{\mathrm{exc}})]} \tag{3.10}$$

While the propagated Gaussian density matrix increment can be expressed explicitly in terms of the time-dependent moments, Eqs. (3.5–3.9) [48], it is not necessary to do so here, as the time-development of experimental observables is expressed directly in terms of the moments themselves.

Using the angular brackets of Eq. (3.10), the peak emission frequency in the classical Franck approximation, Eq. (3.3), takes the compact form $\omega_{\mathrm{peak}}(T') = \langle \Delta V \rangle$. With a quadratic expansion of the difference potential about $\bar{\underset{\sim}{r}}(T')$,

$$\begin{aligned} \Delta V(\underset{\sim}{r}) \cong{}& \Delta V(\bar{\underset{\sim}{r}}(T')) + (\underset{\sim}{r} - \bar{\underset{\sim}{r}}(T'))^{\mathrm{T}} \cdot \underset{\sim}{\nabla} \Delta V(\bar{\underset{\sim}{r}}(T')) \\ &+ \tfrac{1}{2} (\underset{\sim}{r} - \bar{\underset{\sim}{r}}(T'))^{\mathrm{T}} \cdot \underset{\sim}{\nabla}\underset{\sim}{\nabla}^{\mathrm{T}} \Delta V(\bar{\underset{\sim}{r}}(T')) \cdot (\underset{\sim}{r} - \bar{\underset{\sim}{r}}(T')) \end{aligned} \tag{3.11}$$

The entity $\omega_{\text{peak}}(T')$ can be expressed in terms of the first and second moments of the time-dependent nuclear distribution in the excited electronic state:

$$\omega_{\text{peak}}(T') \cong \Delta V(\underset{\sim}{\bar{r}}(T')) + \tfrac{1}{2}\underset{\approx}{\sigma}_r(T') : \underset{\sim}{\nabla}\underset{\sim}{\nabla}^{\mathrm{T}}\Delta V(\underset{\sim}{\bar{r}}(T')) \tag{3.12}$$

Within its range of validity, Eq. (3.12) provides a simple physical picture of the Stokes shift dynamics. The first term on the right-hand side tracks the value of the electronic difference potential at the center of the moving wave packet, and the second term corrects for the finite spatial width of the nuclear distribution, using the curvature of the difference potential at the center of the wave packet.

The initial values of the parameters (3.5)–(3.9) in the excited electronic state can be found under the classical Franck expression for the density matrix increment $\tilde{D}(\Omega_{\text{exc}})$ [Eq. (3.1) and footnote 2] further approximated by expanding the difference potential through quadratic order in $\underset{\sim}{r}$. By comparison with the Gaussian density matrix in position representation or by direct calculation of the moments one obtains

$$\begin{aligned}\underset{\sim}{\bar{r}}(0) &= \langle \underset{\sim}{r}\rangle_g + 2\tau_{\text{exc}}^2\Omega_{\text{off}}\underset{\approx}{\sigma}_r(0)\cdot\underset{\sim}{\nabla}\Delta V(\langle\underset{\sim}{r}\rangle_g)\\ &\cong \langle \underset{\sim}{r}\rangle_g + 2\tau_{\text{exc}}^2\Omega_{\text{off}}(\underset{\approx}{\sigma}_r)_g\cdot\underset{\sim}{\nabla}\Delta V(\langle\underset{\sim}{r}\rangle_g)\end{aligned} \tag{3.13}$$

$$\underset{\sim}{\bar{p}}(0) = \langle \underset{\sim}{p}\rangle_g = 0 \tag{3.14}$$

$$\begin{aligned}\underset{\approx}{\sigma}_r(0) &= \{(\underset{\approx}{\sigma}_r)_g^{-1} + 2\tau_{\text{exc}}^2[\underset{\sim}{\nabla}\Delta V(\langle\underset{\sim}{r}\rangle_g)\underset{\sim}{\nabla}^{\mathrm{T}}\Delta V(\langle\underset{\sim}{r}\rangle_g)\\ &\quad - \Omega_{\text{off}}\underset{\sim}{\nabla}\underset{\sim}{\nabla}^{\mathrm{T}}\Delta V(\langle\underset{\sim}{r}\rangle_g)]\}^{-1}\\ &\cong (\underset{\approx}{\sigma}_r)_g - 2\tau_{\text{exc}}^2(\underset{\approx}{\sigma}_r)_g\cdot[\underset{\sim}{\nabla}\Delta V(\langle\underset{\sim}{r}\rangle_g)\underset{\sim}{\nabla}^{\mathrm{T}}\Delta V(\langle\underset{\sim}{r}\rangle_g)\\ &\quad - \Omega_{\text{off}}\underset{\sim}{\nabla}\underset{\sim}{\nabla}^{\mathrm{T}}\Delta V(\langle\underset{\sim}{r}\rangle_g)]\cdot(\underset{\approx}{\sigma}_r)_g\end{aligned} \tag{3.15}$$

$$\underset{\approx}{\sigma}_p(0) = (\underset{\approx}{\sigma}_p)_g - \frac{\tau_{\text{exc}}^2}{2}\,\Omega_{\text{off}}\underset{\sim}{\nabla}\underset{\sim}{\nabla}^{\mathrm{T}}\Delta V(\langle\underset{\sim}{r}\rangle_g) \tag{3.16}$$

$$\underset{\approx}{\sigma}_{rp}(0) = (\underset{\approx}{\sigma}_{rp})_g = 0 \tag{3.17}$$

In Eqs. (3.13), (3.15), and (3.16), the offset of the excitation-pulse center frequency from vertical resonance is denoted by

$$\Omega_{\text{off}} \equiv \Omega_{\text{exc}} - \Delta V(\langle\underset{\sim}{r}\rangle_g) \tag{3.18}$$

At the level of the classical Franck approximation, the moments in Eqs. (3.13–3.17) describe how the Gaussian density matrix corresponding to thermal equilibrium in the electronic ground state is distorted by electronic excitation with a pulse of non-negligible duration. The spatial center of the wave packet in Eq. (3.13) is displaced toward a region of resonance with the excitation pulse, and both the spatial and momentum widths are changed, Eqs. (3.15) and (3.16), respectively. Equation (3.14) states that the electronic excitation process does not impart momentum to the density matrix increment in the electronic excited state, which therefore has none at $T' = 0$.

Straightforward derivations also show that the moments of the doorway function for the *observed* spectrum, Eq. (B.5), are the same as in Eqs. (3.13–3.17), except that the momentum-width parameter takes the form,

$$\underset{\approx}{\sigma}_p(t_d = 0) = (\underset{\approx}{\sigma}_p)_g + \frac{\tau_{\text{exc}}^2}{2} [\underset{\sim}{\nabla}\Delta V(\langle \underset{\sim}{r} \rangle_g) \underset{\sim}{\nabla}^{\text{T}} \Delta V(\langle \underset{\sim}{r} \rangle_g) - \Omega_{\text{off}} \underset{\sim}{\nabla}\underset{\sim}{\nabla}^{\text{T}} \Delta V(\langle \underset{\sim}{r} \rangle_g)] \tag{3.19}$$

Equations (3.19) and (3.15) indicate that the momentum-width parameter of $D(\Omega_{\text{exc}})$ changes from its ground-state value in a way that tends to complement the change in position-width.

We now verify that the initial values of the coordinate and momentum widths of the density matrix increment $\tilde{D}(\Omega_{\text{exc}})$ can defy the uncertainty principle. Applying Eqs. (3.15) and (3.16) to a minimum uncertainty initial distribution ($\Delta r \Delta p = 0.5$) such as the displaced harmonic system mentioned in the previous subsection, resonantly excited from its zero-point level, yields $\Delta r \Delta p < 0.5$ at time zero for the doorway function of the deconvolved fluorescence spectrum. For the same system, Eqs. (3.19) and (3.15) show that the change in $\underset{\approx}{\sigma}_p$ compensates for the change in $\underset{\approx}{\sigma}_r$, and $\Delta r \Delta p = 0.5$ is unaltered (to low order) for the doorway function $D(\Omega_{\text{exc}})$ of the observed spectrum. This example makes clear how the density matrix increment for the deconvolved spectrum serves as a high-resolution probe of the excited-state potential surface: its initial momentum width is not enlarged to compensate for a spatial width narrowed by the finite power spectrum of the excitation pulse.

Numerical solution of the coupled equations of motion for the time-development of the parameters defined in Eqs. (3.5–3.9) with the initial conditions Eqs. (3.13–3.17) provides a semiclassical treatment of the time-dependent Stokes shift, which will be examined analytically in Section IV.B.1. A similar semiclassical treatment [51] will be tested numerically in Section V.D.1.

C. Ground-State Linear Response

If the change in equilibrium nuclear configuration between ground and excited states is sufficiently small, it is natural to consider the excited state propagation in terms of linear response to ΔV. To first order in the difference potential,

$$\exp(-iH_e t) \cong \exp[-i(H_g + \langle \Delta V \rangle_g)t]\left[1 - i\int_0^t d\tau(\Delta V(\tau) - \langle \Delta V \rangle_g)\right] \quad (3.20)$$

in which $\langle \cdots \rangle_g \equiv \mathrm{Tr}_{\mathrm{nuc}}[\cdots \rho_g]$, and $\Delta V(t) = \exp(iH_g t)\Delta V \exp(-iH_g t)$. Substituting Eq. (3.20) and its Hermitian conjugate in Eq. (3.4) gives the rudimentary quantum linear response expression

$$\Delta E(t) \cong \langle \Delta V \rangle_g - i\int_0^t d\tau \langle [\Delta V(\tau), \Delta V] \rangle_g \quad (3.21)$$

(where $[A, B] \equiv AB - BA$ is the commutator) for the time-dependent average transition energy following an abrupt change in potential, in terms of the equilibrium fluctuations of ΔV. Equation (3.21) is formally valid for propagation times much shorter than the inverse of $\langle (\Delta V - \langle \Delta V \rangle_g)^2 \rangle_g^{1/2}$.

A perturbative expression for the propagator, such as Eq. (3.20), could have been substituted directly in the general expression for $\tilde{F}(\omega, T')$. But the time-dependent average transition energy in the linear response approximation has been the subject of several (classical) simulations, and we will concentrate on that quantity in our numerical tests of linear response. The condition of small nuclear displacement, which favors linear response, also tends to slow the departure of the photo-excited nuclear distribution from the "Franck–Condon region" of $V_e(\underset{\sim}{r})$. Slower nuclear motion in the excited electronic state in turn makes less demanding the short-pulse criteria that produce the approximate equality in Eq. (3.4) between $\omega_{\mathrm{peak}}(t)$ and $\Delta E(t)$.

D. Excited-State Linear Response

As an alternative to ground-state linear response, which applies time-dependent perturbation theory to the excited-state propagator, one can apply thermodynamic perturbation theory to the equilibrium density matrix to obtain $\Delta E(t)$ in terms of excited state correlation functions. If the potentials are similar or the temperature ($T = 1/k_B\beta$) is high, so that $\beta(\Delta V - \langle \Delta V \rangle_e)$ is small [with $\langle \ldots \rangle_e \equiv \mathrm{Tr}_{\mathrm{nuc}}[\ldots \rho_e]$ being an equilibrium average with respect to $\rho_e \equiv Z_e^{-1} \exp(-\beta H_e)$] then

$$\rho_g = Z_g^{-1}\exp(-\beta H_g) \cong \rho_e\left(1 + \int_0^\beta d\beta'(\Delta V(-i\beta') - \langle\Delta V\rangle_e)\right) \tag{3.22}$$

where now $\Delta V(t) = \exp(iH_e t)\Delta V \exp(-iH_e t)$ is in the excited-state interaction representation. Substituting Eq. (3.22) in Eq. (3.4) gives

$$\Delta E(t) \cong \langle\Delta V\rangle_e + \int_0^\beta d\beta'(\langle\Delta V \Delta V(t + i\beta')\rangle_e - \langle\Delta V\rangle_e^2) \tag{3.23}$$

which expresses the time-dependent transition energy in terms of an excited-state equilibrium correlation function, albeit one with a complex time argument.

Strictly speaking, the excited-state linear response approximation is justified only if the range of transition energies accessed by pump-induced dynamics is small compared to $k_B T$. As the range of transition energies during propagation is at least as large as the steady-state Stokes shift, this condition implies that $\langle\Delta V\rangle_g - \langle\Delta V\rangle_e \ll k_B T$, an inequality that may fail to hold for chromophore-solvent systems chosen to exhibit a large Stokes shift. The fact that the integral in Eq. (3.21) vanishes for short times, while that in Eq. (3.23) vanishes at high temperatures or long times (for a system that approaches quasi-equilibrium in the excited state), illustrates the essential difference between the two treatments.

E. Classical Approximation to $\Delta E(t)$

As quantum calculations are difficult in large systems, it is often a practical necessity to assume classical dynamics instead. The classical approximation, like the excited state linear response approximation, tends to be justified in the limit of high temperature compared to the system frequencies. We first take the classical limit of Eq. (3.4). As the quantum operator $\Delta V(\underset{\sim}{r})$ is a function of coordinates only, the propagated operators are replaced by functions of classically propagated coordinates; classical averages are over a Boltzmann distribution of initial coordinates. Thus the classical expression for $\Delta E(t)$ is:

$$\Delta E(t) = \langle\Delta V(\underset{\sim}{r}(t))\rangle_g \equiv \int d^N p \int d^N r \rho_g(\underset{\sim}{p},\underset{\sim}{r})\Delta V(\underset{\sim}{r}(t)) \tag{3.24}$$

where $\rho_g(\underset{\sim}{p},\underset{\sim}{r}) = Z_g^{-1}\exp[-\beta(\underset{\sim}{p}^2/2 + V_g(\underset{\sim}{r}))]$ is the classical equilibrium distribution function in the ground electronic state, and $\underset{\sim}{r}(t)$ is propagated in the excited-state potential according to $d^2\underset{\sim}{r}/dt^2 = -\underset{\sim}{\nabla}V_e(\underset{\sim}{r}(t))$ starting from the initial conditions $\underset{\sim}{r}(0) = \underset{\sim}{r}$ and $d\underset{\sim}{r}(0)/dt = \underset{\sim}{p}$. Using Eq. (3.24), the time-dependent average transition energy can be calculated directly by nonequilibrium molecular dynamics (MD) simulation.

F. Classical Linear Response

To take the classical limit of the ground-state linear response expression, the equilibrium correlation function in Eq. (3.21) must be manipulated slightly. Using the property of equilibrium correlation functions $\langle AB(\tau)\rangle = \langle B(\tau - i\hbar\beta)A\rangle$ (we reintroduce $\hbar$ explicitly in this subsection only), the second term of Eq. (3.21) can be rewritten:[3]

$$-\frac{i}{\hbar}\int_0^t d\tau\langle[\Delta V(\tau), \Delta V]\rangle_g = \beta\int_0^t d\tau\left\langle \frac{\Delta V(\tau - i\hbar\beta) - \Delta V(\tau)}{-i\hbar\beta}\,\Delta V\right\rangle_g \tag{3.25}$$

Since in the classical limit $\hbar\beta$ is shorter than the timescale of nuclear motion, the ratio in the integrand of Eq. (3.25) becomes $d(\Delta V(\tau))/d\tau$, and the integral can be done directly. If the quantum mechanical autocorrelation function is then replaced by its classical counterpart, we obtain

$$\Delta E(t) \cong \langle \Delta V(\underset{\sim}{r})\rangle_g + \beta(\langle \Delta V(\underset{\sim}{r}(t))\Delta V(\underset{\sim}{r})\rangle_g - \langle \Delta V^2(\underset{\sim}{r})\rangle_g) \tag{3.26}$$

an elementary linear-response expression for the time-dependent transition energy under classical nuclear dynamics.

The classical limit of the excited-state linear response treatment of $\Delta E(t)$ also can be evaluated. As $\hbar\beta \to 0$ the second term on the right-hand side of Eq. (3.23) takes the approximate form

$$\int_0^\beta d\beta'(\langle \Delta V \Delta V(t + i\hbar\beta')\rangle_e - \langle \Delta V\rangle_e^2) \cong \beta(\langle \Delta V \Delta V(t)\rangle_e - \langle \Delta V\rangle_e^2) \tag{3.27}$$

yielding the classical limit:

$$\Delta E(t) \cong \langle \Delta V(\underset{\sim}{r})\rangle_e + \beta(\langle \Delta V(\underset{\sim}{r}(t))\Delta V(\underset{\sim}{r})\rangle_e - \langle \Delta V(\underset{\sim}{r})\rangle_e^2) \tag{3.28}$$

Equations (3.26) and (3.28), when normalized, are the usual linear response expressions for the Stokes shift, and can be calculated using equilibrium MD

[3]Both this derivation and Eq. (3.26) are implicit examples of the Kubo transform. See ref. [52], especially Eqs. (21.145) and (21.160).

simulations, MD simulations in conjunction with liquid-state theory [53], or (for hard bodies) kinetic theory [54, 55].

While the derivations of Eqs. (3.26) and (3.28) and their quantum mechanical antecedents in Eqs. (3.21) and (3.23) elucidate the assumptions underlying classical simulations of the Stokes shift, the conditions invoked in deriving those formulas are sufficient but not necessary. In a system of displaced harmonic oscillators, for example, both of the classical linear response expressions are exact [56] (in addition, the normalized Stokes shift is independent of temperature). The linear response and classical approximations will be tested below with numerical calculations on a model system.

IV. SHORT-TIME DYNAMICS OF $\omega_{\text{peak}}(T')$

The importance of "inertial" solvent motions in the behavior of the simulated and observed Stokes shifts at very short times has been the subject of considerable recent discussion [20, 21, 30]. Here we investigate some simple aspects of the initial Stokes shift dynamics using short-time expansions of quantum mechanical and semiclassical expressions for the peak fluorescence frequency and transition energy.

A. Overlapping Pulses: $T' \lesssim \langle(\Delta V - \langle\Delta V\rangle_g)^2\rangle_g^{-1/2} \lesssim \tau_{\text{exc}}$

Because of the interest in very short time Stokes shift dynamics, it is worth investigating the behavior of $\tilde{F}(\omega, T')$ for small T'. If the propagation time T' is shorter than both the absorption bandwidth and the excitation pulse duration, the right-hand side of Eq. (2.22) can be expanded in powers of T' (and then re-exponentiated) to yield a form that is correct through third order in T':

$$\tilde{F}(\omega, T') \cong 4T' \exp\left[-T'^2\left(\frac{1}{3\tau_{\text{exc}}^2} + \frac{4}{3}\langle(\Delta V - \Omega_{\text{exc}})^2\rangle_g\right)\right]$$
$$\cdot \int_{-\infty}^{\infty} dt I_{\text{up}}\left(\frac{t}{2}\right) I_{\text{exc}}\left(\frac{t}{2}\right) \exp[i(\omega - \Omega_{\text{exc}})t]$$
$$\cdot \exp\left[T'^2\left(-\frac{2it}{3\tau_{\text{exc}}^2}\langle\Delta V - \Omega_{\text{exc}}\rangle_g + \frac{t^2}{6\tau_{\text{exc}}^4}\right.\right.$$
$$\left.\left. + \frac{2}{3}\langle(\Delta V - \Omega_{\text{exc}})(\Delta V(t) - \Omega_{\text{exc}})\rangle_g\right)\right] \tag{4.1}$$

If at least one of the pulses is fairly short, $\Delta V(t)$ in the exponent can be expanded through second order, and then the integral over t can be done analytically. That treatment leads to a simple expression,

$$\omega_{\text{peak}}(T') \cong \Omega_{\text{exc}} + \frac{2T'^2}{3}\left(\frac{1}{\tau_{\text{exc}}^2}\langle \Delta V - \Omega_{\text{exc}}\rangle_g - \langle \Delta V[H_g, \Delta V]\rangle_g\right)$$

$$= \Omega_{\text{exc}} + \frac{2T'^2}{3}\left(\frac{1}{\tau_{\text{exc}}^2}\langle \Delta V - \Omega_{\text{exc}}\rangle_g - \frac{1}{2}\langle(\underset{\sim}{\nabla}^T \Delta V)\cdot(\underset{\sim}{\nabla}\Delta V)\rangle_g\right) \tag{4.2}$$

for the evolution of the peak emission frequency *during* the excitation pulse. Appendix C exhibits some numerical simulations of this very short time behavior. Later we will see that the quadratic form Eq. (4.2) differs in several respects from an analogous expression for the peak emission frequency following pulse overlap.

B. Short-Time Stokes Shift with Separated Pulses

1. *Wave Packet Picture*

A closed-form wave packet picture of the short-time Stokes shift dynamics is obtained by solving the parameter equations of motion [Eqs. (32a–f) of ref. [48] for a nuclear wave packet in the excited electronic state] for $\underset{\sim}{\bar{r}}(T')$ and $\underset{\approx}{\sigma}_r(T')$ through second order in T'. Using the fact that both $\underset{\sim}{\bar{p}}(0)$ and $\underset{\approx}{\sigma}_{rp}(0)$ vanish [see Eqs. (3.14) and (3.17)], we find that

$$\underset{\sim}{\bar{r}}(T') \cong \underset{\sim}{\bar{r}}(0) - \frac{T'^2}{2}\underset{\sim}{\nabla} V_e(\underset{\sim}{\bar{r}}(0)) \tag{4.3}$$

and

$$\underset{\approx}{\sigma}_r(T') \cong \underset{\approx}{\sigma}_r(0) + \frac{T'^2}{2}[2\underset{\approx}{\sigma}_p(0) - \underset{\approx}{\sigma}_r(0)\cdot \underset{\sim}{\nabla}\underset{\sim}{\nabla}^T V_e(\underset{\sim}{\bar{r}}(0))$$
$$- \underset{\sim}{\nabla}\underset{\sim}{\nabla}^T V_e(\underset{\sim}{\bar{r}}(0))\cdot\underset{\approx}{\sigma}_r(0))] \tag{4.4}$$

Substituting Eqs. (4.3) and (4.4) in Eq. (3.12) yields a concise formula for $\omega_{\text{peak}}(T')$, which is accurate within the Gaussian approximation through second order in time. It takes some account of the nonzero excitation pulse duration (finite spectral bandwidth), but does not accurately describe the peak emission frequency at the very shortest times when the excitation and gate pulses overlap.

In the wave-packet picture, the absence of linear dependence of $\omega_{\text{peak}}(T')$ on time can be ascribed to the initially vanishing time derivatives of both $\bar{\underset{\sim}{r}}$ and $\underset{\approx}{\sigma}_r$—the wave packet prepared in the excited electronic state may be distorted from ρ_g, but it is born static, and begins to move and spread only under the influence of the excited state potential. The quadratic change in the peak frequency usually is a decrease for two reasons. The center of the wave packet starts near the point of vertical resonance. As the gradient of the ground state potential is small here, the motion toward lower excited state potential [Eq. (4.3)] generally leads to a decreasing difference potential and hence a decreasing peak emission frequency [first term in Eq. (3.12); see Section V.B for an exception.] The short-time evolution of the coordinate width [Eq. (4.4)] has a similar effect but operates in a subtler way. To simplify this part of the description, we neglect distortion of the density matrix increment due to the excitation pulse. If the excited-state well is narrower than the ground-state well, with higher vibrational frequencies in certain modes, then the wave packet contracts in those directions. As the difference potential is larger in outlying areas of these modes, the contraction selectively eliminates higher transition frequencies [second term in Eq. (3.12)]. Conversely, an excited-state well that is wider in certain directions causes expansion of the wave packet along those coordinates, and leads to the inclusion of some lower transition energies from the wave packet's edges. Thus either spreading or contraction of the wave packet generally contributes to a decrease in the peak emission frequency of order T'^2.

2. *Short-Time Expansion for* $\tau_{\text{exc}}, \tau_{\text{up}} < t < \langle(\Delta V - \langle\Delta V\rangle_g)^2\rangle_g^{-1/2}$

It is interesting to consider the short-time Stokes shift of the peak emission frequency in the limiting case where the pulses are very short, and hence nonoverlapping, even for propagation times on the order of the inverse absorption bandwidth. As the pulses are very short, the peak emission frequency and the average electronic transition energy are approximately equal, according to Eq. (3.4). As we are considering time delays of, at most, on the order of the inverse absorption bandwidth, linear response is sufficient, and we may expand Eq. (3.21) to obtain

$$\Delta E(t) \cong \langle\Delta V\rangle_g - t^2\langle\Delta V[H_g, \Delta V]\rangle_g = \langle\Delta V\rangle_g - \frac{t^2}{2}\langle(\underset{\sim}{\nabla}^T\Delta V)\cdot(\underset{\sim}{\nabla}\Delta V)\rangle_g \quad (4.5)$$

Direct expansion of the nonequilibrium quantum expression of Eq. (3.4) in powers of time leads to the same result. The quadratic decrease in $\Delta E(t)$ is in accord with classical simulations and recent experiments, as well as with

Stratt and Cho's classical linear response result [30]. The time average of any other potential or function of the coordinates $f(\underset{\sim}{r})$ also can be expanded, as in Eq. (4.5), yielding $\langle f(t)\rangle_g \cong \langle f\rangle_g - t^2\langle(\underset{\sim}{\nabla}^T \Delta V) \cdot (\underset{\sim}{\nabla} f)\rangle_g/2$. Short-time-expansion and wave-packet treatments of the spectral *width* are given in ref. [57].

As a simple expression for the peak emission frequency of the time-dependent fluorescence spectrum, Eq. (4.5) is strictly valid only with extremely short excitation and gate pulses. It differs markedly from the peak emission frequency predicted by Eq. (4.2) for longer, overlapping pulses: while $\Delta E(t)$ is by definition independent of pulse characteristics, the initial value of $\omega_{\text{peak}}(0)$ in Eq. (4.2) depends upon the excitation pulse center frequency, and the quadratic term depends upon the excitation pulse center frequency and duration. In addition, the $(\nabla \Delta V)^2$ contributions to the quadratic terms in Eqs. (4.2) and (4.5) differ by a numerical prefactor. Similarities and differences between $\omega_{\text{peak}}(T')$ and the "underlying" $\Delta E(t)$ will be considered further in Sections V and VI.

3. *Rigid-Cage Approximation*

Maroncelli [20] has recently shed light on the physical nature of the short-time Stokes shift by performing classical simulations of polar solvation in which all solvent molecules but one are frozen, and hence unable to respond to a change in the dipole moment of a solute. His results demonstrated that the early time dynamics is single-particle in nature, arising from small amplitude motions within a preformed environment.

The validity of the rigid-cage approximation is not restricted to classical solvation, as can be seen by considering a fairly general situation in which the electronic difference potential is a sum of terms, each depending on only one coordinate:

$$\Delta V = \nu + \sum_j \nu_j(r_j) \tag{4.6}$$

Under a quantum mechanical rigid-cage approximation to the transition energy, the formula of Eq. (3.4) is replaced by

$$\Delta E_{\text{RC}}(t) = \nu + \sum_j \text{Tr}_{\text{nuc}}\{\nu_j \exp[-it(p_j^2/2 + V_e)]\rho_g \exp[it(p_j^2/2 + V_e)]\} \tag{4.7}$$

in which Eq. (4.6) is used for ΔV, and only one coordinate at a time is allowed to move in response to the instantaneous change in potential.

Expanding Eq. (4.7) through second order in the time variable gives

$$\Delta E_{\mathrm{RC}}(t) \cong \langle \Delta V \rangle_g + it \sum_j \mathrm{Tr}_{\mathrm{nuc}} \left\{ \left[\frac{p_j^2}{2} + V_e, \nu_j(r_j) \right] \rho_g \right\}$$
$$- \frac{t^2}{2} \sum_j \mathrm{Tr}_{\mathrm{nuc}} \left\{ \left[\frac{p_j^2}{2} + V_e, \left[\frac{p_j^2}{2} + V_e, \nu_j(r_j) \right] \right] \rho_g \right\} \quad (4.8)$$

By writing $V_e = V_g + \Delta V$ in (4.8) and carrying out some operator manipulations, one finds that the term linear in t vanishes, and

$$\Delta E_{\mathrm{RC}}(t) \cong \langle \Delta V \rangle_g - \frac{t^2}{2} \sum_j \mathrm{Tr}_{\mathrm{nuc}} \{ [\Delta V, [H_g, \nu_j]] \rho_g \} = \Delta E(t) \quad (4.9)$$

Thus, the rigid-cage approximation agrees with the time-dependent transition energy, Eq. (4.5), through second order in time if Eq. (4.6) is valid.

4. *Difference-Potential-Only Approximation*

An alternative treatment of the short-time excited-state dynamics, in which the solvent moves under the difference potential alone, rather than the full excited state potential, has been investigated by non-equilibrium classical molecular dynamics in a Stockmayer fluid [37]. Perera and Berkowitz showed that motion under a difference potential alone (that due to a change in charge on an atomic solute) correctly accounted for the inertial component of solvation.

It turns out, however, that the difference-potential-only approximation should in general depart from the exact short-time form of Eq. (4.5). In that approximation, the time-dependent average transition energy is given by

$$\Delta E_{\mathrm{DPO}}(t) = \mathrm{Tr}_{\mathrm{nuc}} \{ \Delta V \exp[-it(H_g - V_g + \Delta V)] \rho_g \exp[it(H_g - V_g + \Delta V)] \} \quad (4.10)$$

It is not necessary to assume that the difference potential is additive as in Eq. (4.6). Some operator algebra shows that Eq. (4.10) has the second-order expansion

$$\Delta E_{\mathrm{DPO}}(t) \cong \langle \Delta V \rangle_g - \frac{t^2}{2} \mathrm{Tr}_{\mathrm{nuc}} \{ [H_g - V_g + \Delta V, [H_g, \Delta V]] \rho_g \} \quad (4.11)$$

which can be written as

$$\Delta E_{\mathrm{DPO}}(t) = \Delta E(t) + \frac{t^2}{2} \langle (\underset{\sim}{\nabla}^T \Delta V) \cdot (\underset{\sim}{\nabla} V_g) \rangle_g \tag{4.12}$$

As $\Delta E_{\mathrm{DPO}}(t)$ and $\Delta E(t)$ can differ at second order in time if the difference potential $\Delta V(\underset{\sim}{r})$ has quadratic or higher-order contributions, the difference-potential-only approximation will not correctly capture the initial solvation dynamics in all cases.

V. SIMULATIONS

What are the effects of pulse characteristics on the observed fluorescence? How different are the observed and deconvolved time-dependent fluorescence spectra, and the peak emission frequencies derived from them? What are the consequences of neglecting pulse overlap in calculating the peak emission frequency at short times? How well do the various approximations outlined or reviewed in Section III apply in practice? Naturally, the answers to these questions will depend somewhat on the specific system and pulses. But it is difficult to test some of those questions directly on systems of the kind usually considered in molecular dynamics simulations of polar solvation, which consist of hundreds of nuclear coordinates. Instead, this section examines the time-dependent fluorescence spectrum, its peak frequency, and various approximations to the transition energy for a fairly simple model system that mimics several key features of a polar solute in a polar solvent. As the simulated system has only three modes, it cannot incorporate all the processes thought to be important in bulk solvation. But the ability to do rigorous nonequilibrium quantum dynamics simulations will help elucidate some of the factors that can affect the fluorescence Stokes shift.

Section V.A describes our model system, its continuous-wave absorption spectrum, and its observed time-dependent fluorescence spectrum prior to deconvolution. Section V.B examines the dependence on excitation-pulse and gate-pulse duration of the peak emission frequency obtained from the observed time-dependent spectrum. The deconvolution procedure outlined in Section II.C is tested for pulses of different length in Section V.C. Having demonstrated the (stringent) conditions for agreement between the time-dependent peak emission frequency and the time-dependent average transition energy, we compare the latter quantity with classical, semiclassical, and linear response approximations to it in Section V.D. Section V.E outlines an instantaneous normal mode approximation to the classical excited-state dynamics, and compares it to the exact classical dynamics of the trans-

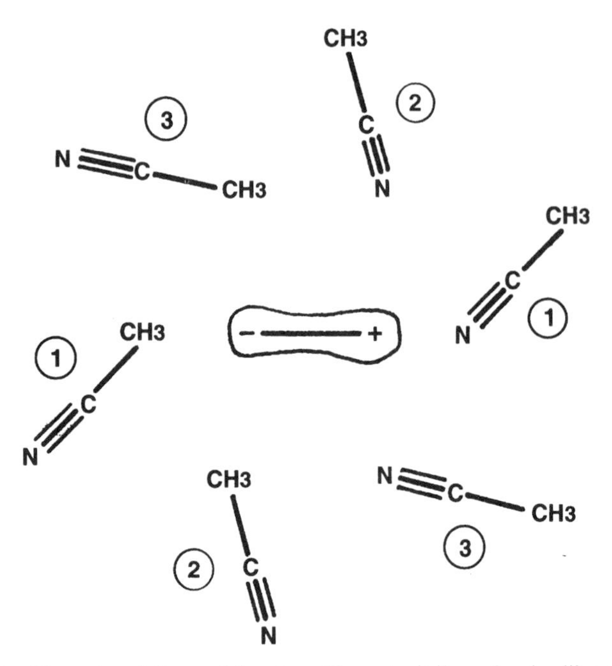

Figure 2. Geometry of the model system. Six acetonitrile molecules librate about fixed centers in a plane, and respond to a change, upon electronic excitation, in the dipole moment of a diatomic "solute." The "solvent" orientations shown correspond to the minimum of the electronic ground state potential. For solvent molecules 1, 2, 3, these angles are 47.0°, 103.9°, −12.6°, respectively. Positive angles represent counter-clockwise rotations from a horizontal reference position with the methyl group on the right. Displacements of these angles at the principal minimum of the excited-state potential are −2.14°, −0.48°, −1.77°.

ition energy. Simulations of the peak emission frequency at very short times (showing its detailed evolution during the excitation pulse) and at long times are reserved for Appendix C.

A. Model System

The model system, shown in Fig. 2, has three degrees of freedom, with motions reminiscent of librations in the first solvation shell of acetonitrile, which are thought to dominate the initial Stokes shift evolution.[4] Six CH_3CN molecules are arrayed in a plane around a fixed diatomic "solute," with fixed centers of mass 4.0 Å from the center of the solute and from each other

[4]However, the role of translational degrees of freedom has been investigated as well; see ref. [58].

(such that two of the CH_3CN molecules are 15 degrees off the solute dipole axis). The coordinates are the in-plane rotations of these "solvent" molecules around their centers of mass. The number of dynamical degrees of freedom is reduced from six to three by artificially constraining pairs of CH_3CN molecules at inverse positions relative to the solute to rotate in parallel. In our simulations the single angle variable describing the motion of two solvent molecules librating in tandem therefore carries twice the moment of inertia of a single acetonitrile molecule.

The atoms interact via Lennard–Jones and coulombic potentials. The potential parameters for the different sites are shown in Table I. Parameters for the three-site solvent molecules are taken from Edwards et al. [20, 59] and the Lennard–Jones potentials are constructed with the usual combination rules. The solute is intended to roughly resemble a dipolar dye molecule; its two sites are split by 2.0 Å and have identical Lennard–Jones parameters and opposite charge. The only difference between ground and excited electronic states is a change in charge of the two solute sites from 0.6 e to 1.0 e, changing the dipole moment from 5.76 D to 9.61 D. The solvent molecules each have a dipole moment of 4.12 D.

Several characteristics of the system can be extracted from the potential: the difference in the excited-state potential energy between the configurations for the lowest ground-state minimum and lowest excited-state minimum is $-174\ \text{cm}^{-1}$; this is approximately the solvation energy. The change in difference potential between these two configurations (roughly the steady-

TABLE I
Solvent and Solute Site Parameters for Lennard-Jones and Electrostatic Potential Interactions[a]

	σ_i	ϵ_i	q_i	m_i	radius
Solvent					
CH_3	3.6 Å	191 K	0.269e	15.0 amu	1.32 Å
C	3.4	50	0.129	12.0	−0.14
N	3.3	50	−0.398	14.0	−1.31
Solute					
+		0	0.6/1.0	∞	1.00
−		0	−0.6/−1.0	∞	−1.00

[a]The full potential is a sum over interactions between each site, $V(\underline{r}) = \sum_{i,j} V_{LJ}(r_{ij}) + V_{el}(r_{ij})$, with $V_{LJ}(r_{ij}) = 4\epsilon_{ij}[(\sigma_{ij}/r_{ij})^6 - (\sigma_{ij}/r_{ij})^{12}]$ and $V_{el}(r_{ij}) = -q_i q_j/r_{ij}$, where $r_{ij} = |r_i - r_j|$, $\epsilon_{ij} = \sqrt{\epsilon_i \epsilon_j}$, and $\sigma_{ij} = (\sigma_i + \sigma_j)/2$. "radius" in the table is the distance of the site from the center of mass of the molecule. The two charges q of the solute sites are the values in the ground and excited states. The relative positions of the molecules are shown in Fig. 2.

state Stokes shift) is $-343\ \mathrm{cm}^{-1}$. The frequencies of the normal modes at the ground state potential minimum are 264, 252, and 194 cm^{-1}, and those of the excited state minimum are 296, 265, and 188 cm^{-1}. The rotations are highly mixed in the normal modes. The orientational displacement upon excitation is small, especially in the second coordinate (see Fig. 2). While there are several wells in each state, only one is very significant at room temperature.

Figure 3 shows the electronic absorption spectrum of our model system. The zero of frequency corresponds to vertical electronic resonance from the configuration of the ground-state potential minimum. Small but significant excited-state displacements in all three normal modes account for the absence of a single dominant Franck–Condon progression. The absorption spectrum shown in Fig. 3 and all time-dependent signals presented below were calculated at room temperature (208 cm^{-1}). As 208 cm^{-1} is less than the frequency difference (steady-state Stokes shift) between the centers of gravity

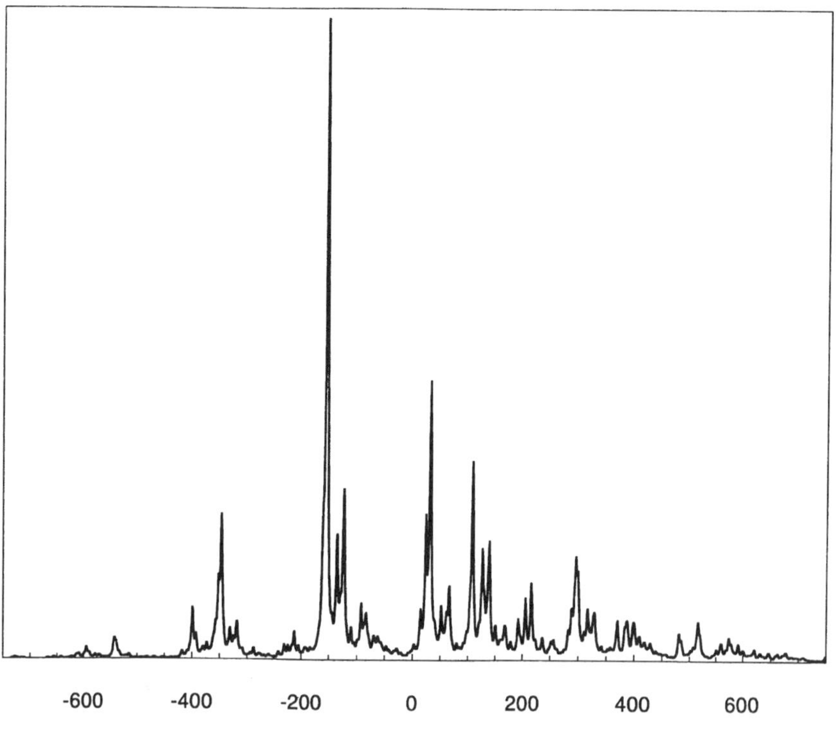

Figure 3. Absorption spectrum of the model system at room temperature. The vertical electronic transition has been shifted to 0 cm^{-1}. The intense peak at $-154.7\ \mathrm{cm}^{-1}$ is the zero–zero transition.

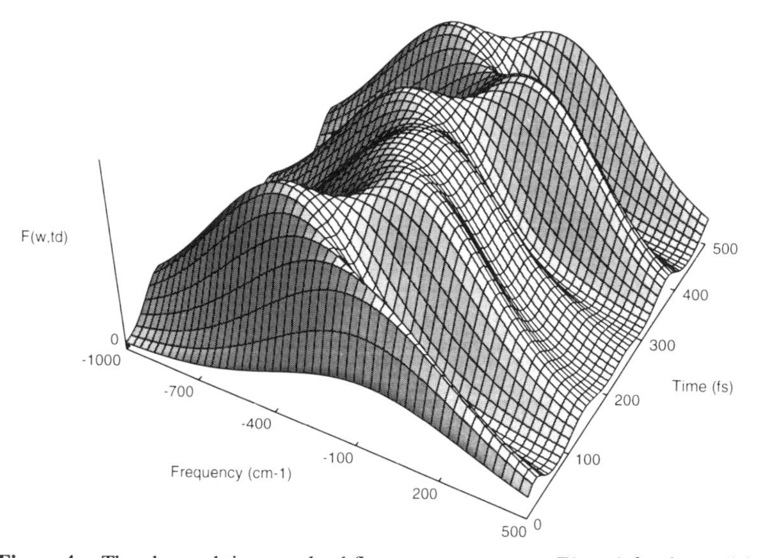

Figure 4. The observed time-resolved fluorescence spectrum $F(\omega, t_d)$ for the model system detected by upconversion with a short gate pulse ($\tau_{\text{up}} = 20$ fs) following nearly librationally instantaneous excitation ($\tau_{\text{exc}} = 10$ fs). This simulated spectrum corresponds to the experimentally observed signal; it includes the effect of overlap between the pulses, but has not been deconvolved from their intensity cross-correlation function.

of the fluorescence (not shown) and absorption spectra, we should anticipate readily discernible dynamics in the peak emission frequency following excitation with pulses short compared to the excited state periods (~130 fs).

As an example of numerically generated "raw data," Fig. 4 shows the observed time-dependent fluorescence spectrum of the model system generated with short excitation and gate pulses ($\tau_{\text{exc}} = 10$ fs and $\tau_{\text{up}} = 20$ fs). Calculation of $F(\omega, t_d)$ was accomplished by expanding Eq. (2.18) in eigenstates. The time-resolved spectrum corresponds to the experimental observable; it includes the effects of pulse overlap, but does not incorporate any deconvolution from the pulse intensities. The peak emission frequency as a function of interpulse delay (see, for example, Fig. 5) can be visualized as the projection on the time-frequency plane of the "ridge" in Fig. 4. Fluorescence can be observed for short negative values of the delay between the excitation pulse and the gate pulse (not shown), so it is no cause for concern that Fig. 4 exhibits nonzero values of $F(\omega, 0)$.

B. Effects of Pulse Duration, Overlap, and Center Frequency

The observed time-resolved fluorescence spectrum $F(\omega, t_d)$ and the peak emission frequency derived from it without prior deconvolution are highly dependent on the pulse lengths. Figure 5 shows the observed peak frequency as a function of t_d for the three-coordinate system described above with

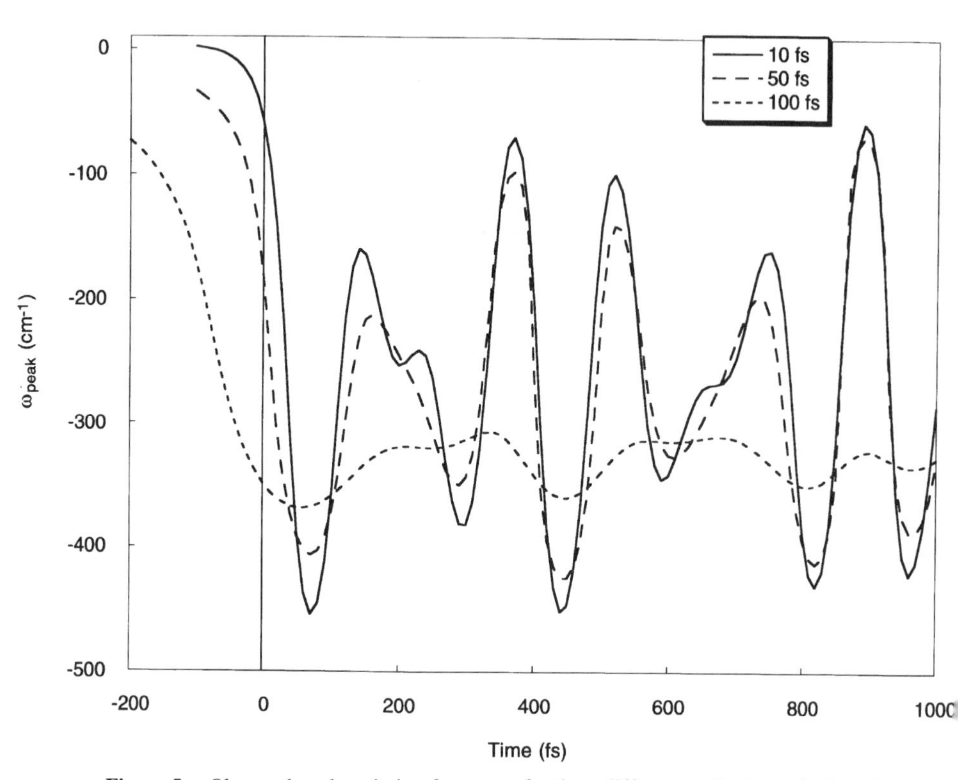

Figure 5. Observed peak emission frequency for three different excitation pulse lengths: $\tau_{\text{exc}} = 10$, 50, and 100 fs, as a function of inter-pulse delay t_d. Excitation pulses all had $\Omega_{\text{exc}} = 0$, corresponding to vertical resonance in the model system, and gate-pulse length was $\tau_{\text{up}} = 20$ fs. Notice the strong dependence of the initial peak frequency on the excitation pulse duration. The first of these traces, with $\tau_{\text{exc}} = 10$, is the time-dependent peak frequency of the fluorescence spectrum shown in Fig. 4.

$\tau_{\text{up}} = 20$ fs, $\Omega_{\text{exc}} = 0$ (corresponding to vertical resonance in the model system), and three different excitation pulse lengths: $\tau_{\text{exc}} = 10$, 50, and 100 fs.

The peak emission frequency at $t_d = 0$ and its decrease at short times show strong dependence on τ_{exc}. The emission frequencies are different at short times because the Stokes shift begins at negative time delays t_d, when the excitation and gate pulses first overlap (the evolution times in the excited state are always positive, even if the time delay between the pulse centers is zero or negative). For longer pulses more of the Stokes shift occurs before $t_d = 0$. In addition, there is a steady decrease in signal structure with increasing excitation-pulse duration. This loss of temporal resolution is a consequence of the changing form of the initial excited-state density matrix increment of Eq. (B.2) for the observed spectrum. With a 10-fs excitation pulse, much shorter than the librational periods, the initially prepared nuclear distribution is a slightly distorted copy of the equilibrium distribution in the electronic ground state. Each of the thermally weighted nuclear eigenstates of H_g is copied with reasonable fidelity into the excited electronic state, where it is made up of a coherent superposition of many nuclear eigenstates of H_e. These coherent superposition states (wave packets) move over the excited-state potential, and each one gives a (fairly similar) contribution to a strongly time-dependent peak emission frequency. On the other hand, a 100-fs excitation pulse has a spectral bandwidth on the order of the frequency spacing between nuclear eigenstates of H_e. Each of the nuclear eigenstates of H_g when excited by such a pulse is highly distorted, consisting primarily of a single, static eigenstate of H_e—their thermally-weighted sum gives a more nearly stationary $D(\Omega_{\rm exc})$. Notice that the weakly time dependent peak emission frequency following excitation with a 100-fs pulse is in the range expected for the Stokes-shifted fluorescence from an "equilibrated" distribution, despite the fact that the nuclear states prepared by that pulse are superpositions of a small number of eigenstates of H_e with energies very near the Franck–Condon energy (~170 $\rm cm^{-1}$) plus the eigenenergy of the original state in H_g.

Variations in the length of the gate pulse have effects similar, but not identical, to variations in the excitation pulse duration. In Fig. 6, the peak emission frequency is plotted versus interpulse delay for $\tau_{\rm exc}$ = 20 fs, $\Omega_{\rm exc}$ = 0, and three gate-pulse lengths, τ_{up} = 10, 50, and 100 fs. For each case, the resonant excitation pulse prepares an identical nonstationary density matrix increment in the excited electronic state. In the time domain, a longer gate pulse, like a longer excitation pulse, allows a wider range of excited state propagation times. A longer gate pulse also has less spectral width, and is therefore sensitive to emission from a smaller number of excited-state levels (to each ground-state librational level); thus the interference that would give rise to a time-dependent intensity is not observed. For a long enough upconversion pulse, the window function in Eq. (B.3) becomes selective for emission frequency in such a narrow range about ω that only spectrally isolated pairs of excited- and ground-state librational levels can contribute, causing a static signal.

In Fig. 6 the case where $\tau_{\rm up}$ = 50 fs illustrates an interesting feature that can arise when neither the time resolution nor the frequency resolution is high (compared respectively to the periods of libration or the frequency spacing between electronic-librational levels). Under those intermediate conditions, instantaneous jumps can be observed in the time-dependent peak emission frequency. The intensity of emission observed at a given frequency, which is due to population density in the resonant region of the

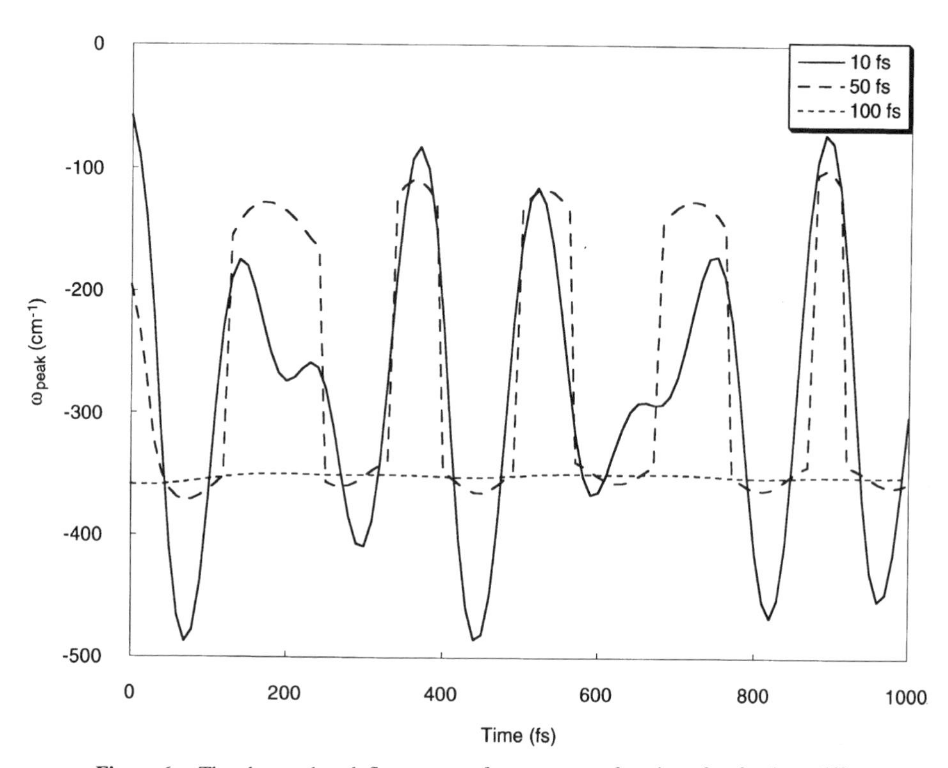

Figure 6. The observed peak fluorescence frequency as a function of t_d for three different gate-pulse widths: $\tau_{\text{up}} = 10$, 50, and 100 fs. The vertically resonant excitation pulse has $\tau_{\text{exc}} =$ 20 fs in each case.

potential, must change smoothly with time. The jumps in the peak emission frequency occur when there are multiple emission peaks due to a narrow bandwidth wave packet, and two peaks, each one with a smoothly evolving intensity, exchange prominence. While the observed peak emission frequencies all varied continuously in Fig. 5, intermediate-length excitation puluses can cause similar sudden jumps in the peak emission frequency (see also Fig. 10 below).

Possible effects of temporally overlapping pulses have often been neglected in simulations of time-dependent fluorescence signals. Figure 7 shows that pulse overlap can significantly influence the short-time Stokes shift. It compares the peak emission frequency for $\tau_{\text{up}} = 20$ fs, $\Omega_{\text{exc}} = 0$, and $\tau_{\text{exc}} = 50$, shown previously in Fig. 5, with that derived from the equivalent spectrum calculated neglecting the effects of pulse overlap [the formula for which is Eq. (B.1), applied as written even for t_d shorter than the pulse lengths]. Neglecting the effects of pulse overlap incorrectly allows the

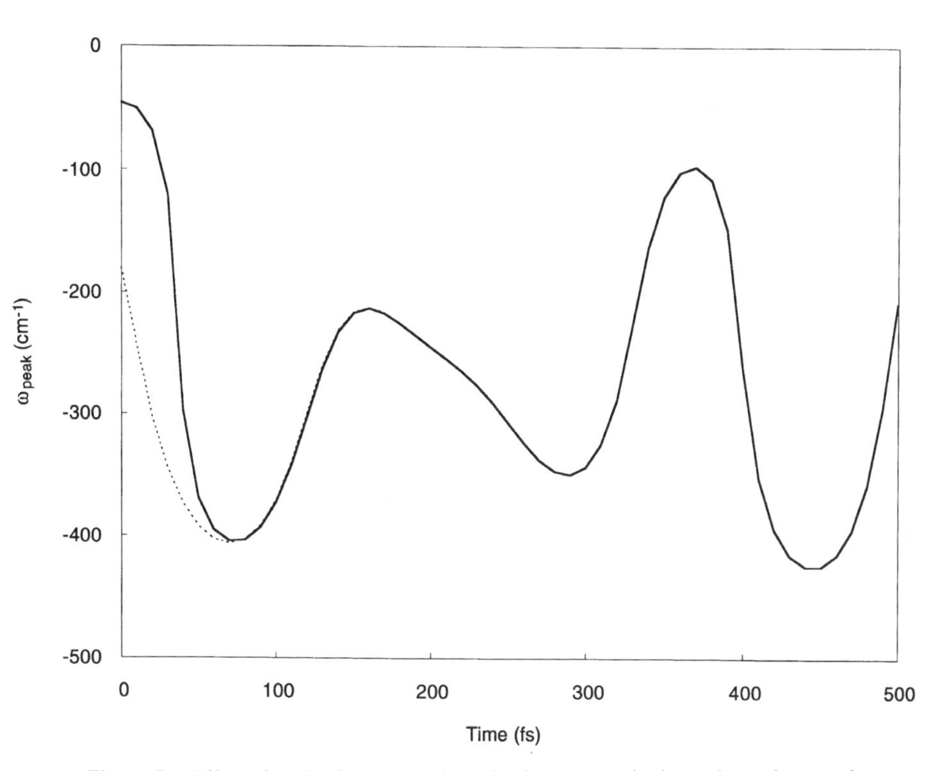

Figure 7. Effect of neglecting temporal overlap between excitation pulse and gate pulse. Observed peak fluorescence frequency for τ_{up} = 20 fs, Ω_{exc} = 0, and τ_{exc} = 50 fs with proper inclusion of pulse overlap (dotted line) is compared with that obtained from an "observed" time-dependent spectrum neglecting pulse overlap (solid line).

system to be excited after it fluoresces (with negative excited state propagation time in between). This is tantamount to allowing portions of the excitation pulse that follow the action of the gate pulse on the detector to contribute to the excited-state density matrix increment. Neglect of pulse overlap effects in Eq. (2.18) amounts to replacing the lower limits on both τ'– and $\bar{\tau}'$–integrations by $-\infty$. As the negative excitation-gate delays are shorter on average than the legitimate ones (for positive pulse delays), this artificial introduction of post-gate excitation postpones the Stokes shift, as seen in Fig. 7.

Some dependence of the observed peak emission frequency upon excitation-pulse center frequency should be expected as long as the pulse has non-negligible duration (finite bandwidth). Figure 8 exhibits an extreme case of this pulse frequency dependence—it compares the delay-dependent peak emission frequency for a vertically resonant excitation pulse with that from

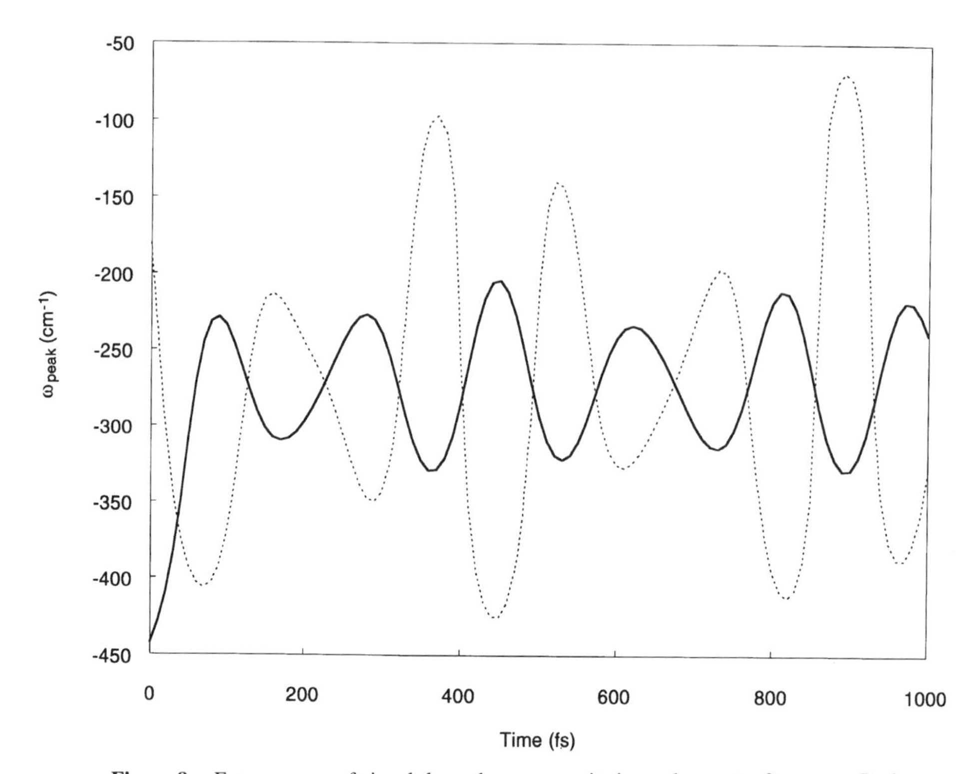

Figure 8. Extreme case of signal dependence on excitation-pulse center frequency. Peak emission frequency for a vertically resonant excitation pulse with $\tau_{\text{exc}} = 50$ fs and $\tau_{\text{up}} = 20$ fs (dotted line) is compared with that for an excitation pulse with center frequency 500 cm^{-1} below vertical resonance (solid line). Anti-Stokes shift at short times in the latter case implies initial wave packet motion in the direction of *increasing* electronic difference potential.

an excitation pulse of center frequency 500 cm^{-1} below vertical resonance, with $\tau_{\text{up}} = 20$ fs and $\tau_{\text{exc}} = 50$ fs in both cases. The anti-Stokes shift at short times in the latter case implies initial wave packet motion in the direction of increasing electronic difference potential. The subresonant pulse transfers to the excited electronic state only outlying portions of the equilibrium nuclear distribution, specifically portions that are located on the far side of the excited-state minimum from the distribution center. Subsequent motion brings this wave packet toward, rather than away from, the Franck–Condon region.

C. Tests of the Deconvolution Procedure

Some of the effects of noninstantaneous pulses can be countered by deconvolution. Figure 9 displays the peak-frequency of the observed fluorescence

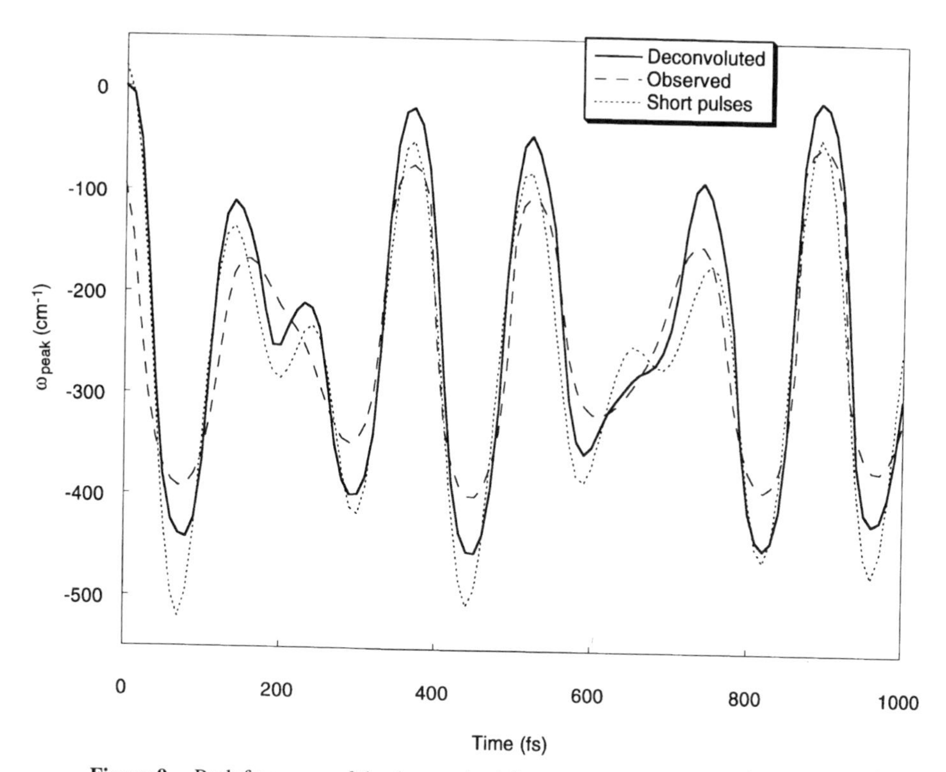

Figure 9. Peak frequency of the deconvolved fluorescence spectrum with $\tau_{\text{up}} = \tau_{\text{exc}} = 30$ fs and a vertically resonant excitation pulse is compared with the peak emission frequency from the corresponding observed spectrum prior to deconvolution, and the peak frequency from the observed time-dependent spectrum with $\tau_{\text{exc}} = 0$ fs and $\tau_{\text{up}} = 10$ fs.

spectrum, together with that from the deconvolved fluorescence spectrum in Eq. (2.22) for $\tau_{\text{up}} = \tau_{\text{exc}} = 30$ fs and a vertically resonant excitation pulse. Also shown is the peak emission frequency from a nearly "ideal" time-dependent spectrum with $\tau_{\text{exc}} = 0$ fs and $\tau_{\text{up}} = 10$ fs. In accord with Eq. (3.4), the peak frequency of the latter spectrum closely tracks the time-dependence of the average electronic transition energy; deconvolution would have little effect. But for longer pulses the deconvolution procedure clearly brings ω_{peak} more closely in line with the underlying time-dependent transition energy.

However, even with pulses shorter than a quarter librational period, $\omega_{\text{peak}}(T')$ from the deconvolved spectrum gives an imperfect rendering of $\Delta E(t)$. In Fig. 9 the initial peak emission frequency is $\omega_{\text{peak}}(0) = \Omega_{\text{exc}} = 0$, as predicted by Eq. (4.2), whereas $\Delta E(0)$ (and the initial value of the ideal observed spectrum) are greater than zero. The initial drop in $\omega_{\text{peak}}(T')$ to its first minimum is only 437 cm^{-1}, compared with the 535 cm^{-1} initial drop

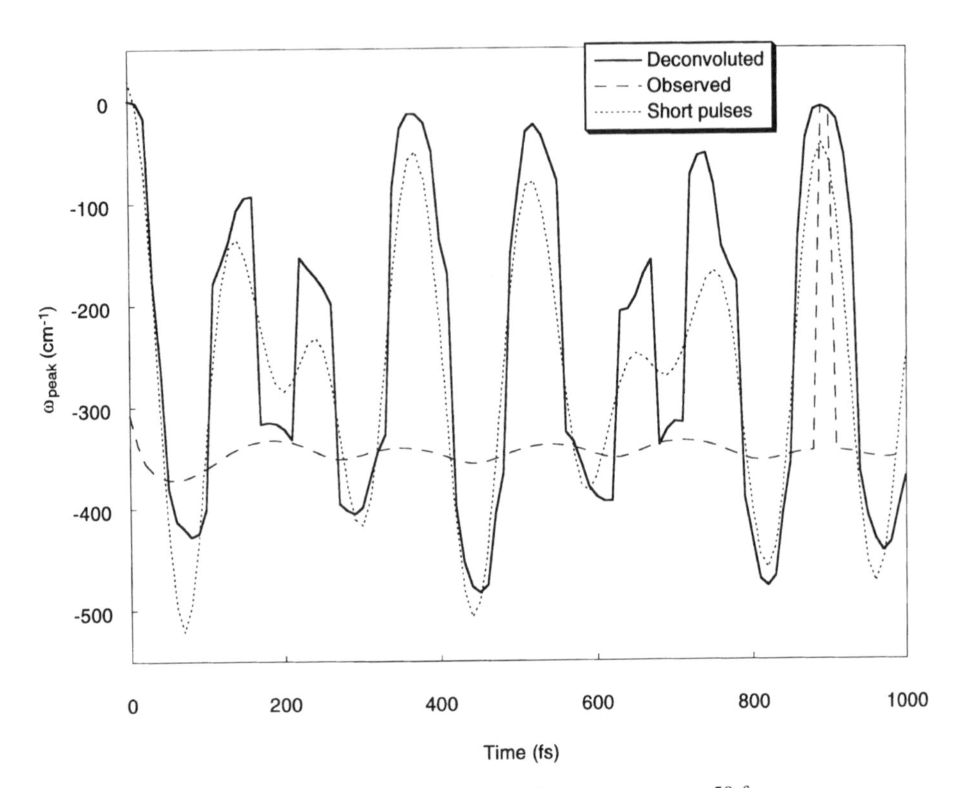

Figure 10. Same as Fig. 9, but for $\tau_{\text{up}} = \tau_{\text{exc}} = 50$ fs.

in the observed peak emission frequency calculated using $\tau_{\text{exc}} = 0$ fs and $\tau_{\text{up}} = 10$ fs (and the 587 cm^{-1} initial drop in $\Delta E(t)$ seen below in Fig. 11).

Figure 10 presents $\omega_{\text{peak}}(T')$ obtained from the more challenging deconvolution of $F(\omega, t_d)$ obtained with longer pulses, $\tau_{\text{up}} = \tau_{\text{exc}} = 50$ fs, $\Omega_{\text{exc}} = 0$. Given the complete lack of similarity between the time-dependent average transition energy and the peak emission frequency of $F(\omega, t_d)$ prior to deconvolution in this instance, we might not expect $\omega_{\text{peak}}(T')$ from the deconvolved spectrum to bear much resemblence to $\Delta E(t)$ either. Due to spectral bandwidth considerations alone, the excited state density matrix increment in Eq. (2.25) cannot be a very faithful copy of ρ_g in this case, nor can the frequency-derivative of the fluorescence window operator in Eq. (2.26) be simply related to the difference potential (as it would be under the classical Franck approximation—see Eq. (3.2) and following). Yet Fig. 10 shows that, apart from the appearance of some sudden jumps similar to those in Fig. 6, the peak emission frequency derived from $\tilde{F}(\omega, T')$ does recover the gross features of the underlying time-dependent transition energy.

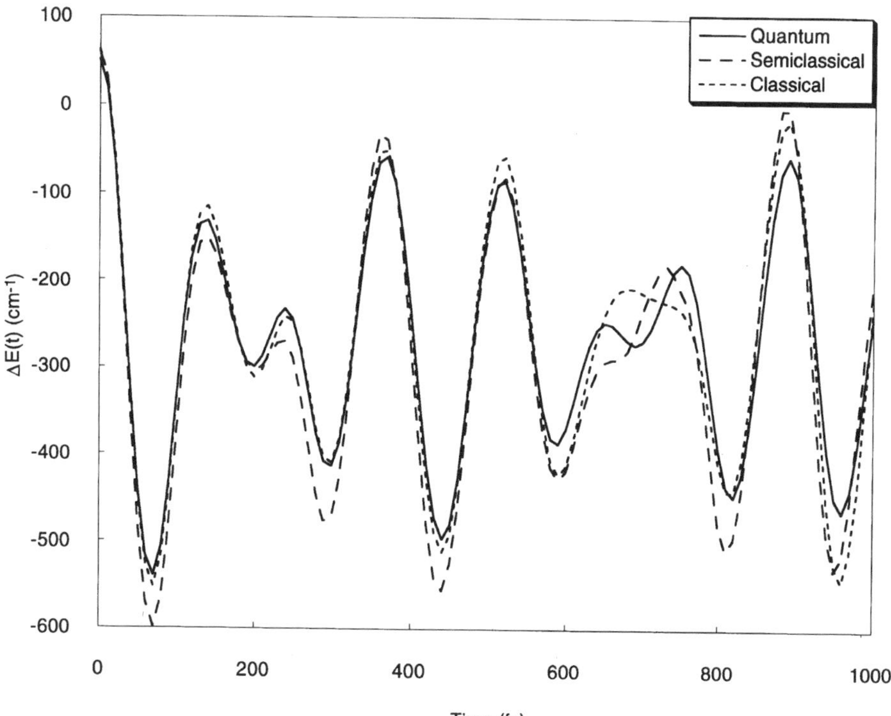

Figure 11. Time-dependent average electronic transition frequency following instantaneous excitation. Quantum mechanical calculation, semiclassical wave packet treatment, and classical dynamics calculations are compared.

Part of the success of the deconvolutions illustrated in Figs. 9 and 10 results from the fact that they are analytical deconvolutions from perfect Gaussian pulses of known length. Numerical deconvolution of experimental results should approximate these results but cannot be expected to achieve ideal results.

D. Classical, Wave Packet, and Linear Response Approximations to $\Delta E(t)$

The transition energy $\Delta E(t)$ is nearly equal to the peak fluorescence frequency $\omega_{\mathrm{peak}}(T')$ for very short pulses in our small model system (see Fig. 11). We now briefly carry through the calculation of several approximations to $\Delta E(t)$ itself based on nonequilibrium classical MD simulations, semiclassical wave-packet propagation, and equilibrium classical correlation functions (linear response treatment) in the excited electronic state and in the ground state. It is worth noting that some of the differences in simulated transition energy among these various approximations may not lead to noticeable changes in the normalized Stokes shifts, which are constrained to start at unity and decay to zero for a bulk system.

1. *Nonequilibrium Classical and Semiclassical Approximations*

Molecular dynamics simulations of solvation dynamics in bulk solutions assume that the nuclear motion of the solvent molecules is classical. In our three-dimensional model system at room temperature the time-dependent average transition energy is well approximated by the nonequilibrium classical expression of Eq. (3.24) (see Fig. 11). In nonequilibrium classical simulations at low temperature (not shown), the Stokes shift oscillations decay at a rate slightly slower than in quantum simulations because the classical initial population distribution is narrower. However, solvation experiments are most often carried out at room temperature, and modes with frequencies that are large compared to 300 K = 208 cm^{-1} are typically too fast to affect the dynamics observed with currently available time resolution.

A semiclassical wave-packet simulation of $\Delta E(t)$ is also shown in Fig. 11. For this calculation, the ground eigenstates are calculated in a basis of multidimensional harmonic oscillator eigenstates, as described in ref. [49]. The basis states are propagated in a locally quadratic expansion of the excited potential using variational equations of motion [49, 51]. The time-dependent average of the potential difference between ground and excited states, approximated as locally quadratic around the center of the propagated states, is found by numerical integration. This wave-packet method is closely related to the semiclassical approach described in Sections III.B and IV.B.1, but allows consideration of anharmonic ground-state potentials. The semiclassical transition energy is accurate at short times, but at longer times actually differs more from the fully quantum mechanical result than does the classical simulation. In fact, the semiclassical result at 300 K is similar to a low-temperature classical simulation (not shown) because the restricted wave-packet form constrains its spreading. In highly anharmonic excited potentials, semiclassical wave-packet calculations can fail miserably at longer time because the restricted wave-packet form causes averaging over inaccessible regions of the potential [60]. When these regions have very large difference potentials, the small misplaced probability density can have large consequences in the transition energy.

2. *Ground and Excited-State Linear Response Approximations*

The ground and excited-state linear response approximations to $\Delta E(t)$ using classical equilibrium MD simulations agree with the nonequilibrium classical simulation at different times (Fig. 12). Differences similar to those illustrated

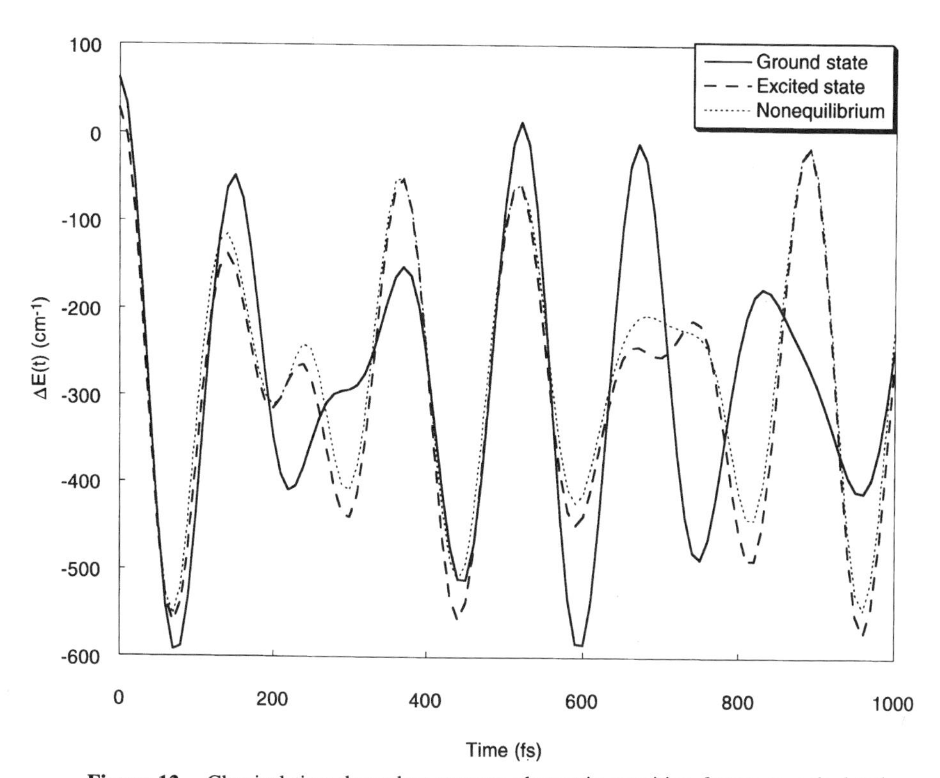

Figure 12. Classical time-dependent average electronic transition frequency calculated by ground-state linear response, excited-state linear response, and nonequilibrium classical molecular dynamics.

here, and their physical explanations, have been described in simulations of bulk solvents [20, 21, 23, 29], but it is informative to see the differences for a small system and to compare the results to the analytical expressions derived earlier.

The excited-state linear response approximation [Eq. (3.28)] is expected to be valid when the temperature is much larger than the range of the difference potential in the populated region during the propagation, so that the equilibrium distributions in the ground and excited states are similar. This condition is not well obeyed by the systems considered here or by bulk polar solvents that have been simulated. The most obvious inaccuracy of the excited-state linear response approximation is that it ignores "local heating" effects of the initial excitation—if there is a significant displacement between the ground and excited state minima, then the dynamics following a vertically resonant pulse are high in the potential, while the correlation function used to approximate the dynamics samples only fluctuations in a thermal equilibrium. In the three-coordinate system, the excited-state linear response approximation to $\Delta E(t)$ is fairly accurate but not exact, even at short times. At long times, if dissipation to a bulk solution were included in the simulations [61], the transition energy in both equilibrium and nonequilibrium simulations would approach the value it takes in thermal equilibrium in the excited state. Under excited-state linear response, the initial probability distribution is wrong but the dynamics are determined by a restricted portion of the correct potential, yielding an approximation that does not worsen with time.

Equilibrium simulations in the ground state using Eq. (3.26) (also shown in Fig. 12) are expected to be accurate at short times. The calculated value of the initial classical transition energy is exact. The initial decrease also is well modeled as long as the evolving distribution remains similar to the ground-state equilibrium distribution. However, over time the correlation functions reflect ground-state fluctuations, while the actual solvation dynamics occurs entirely in the excited state. In Fig. 12 the slightly different frequencies in the two electronic states become apparent as the undamped oscillations of the nonequilibrium and ground-state linear response signals fall out-of-phase by 800 fs.

E. Instantaneous Normal Mode Treatment of Excited-State Dynamics

Classical nonequilibrium dynamics can be approximated analytically from the equilibrium distribution of the system using instantaneous normal modes (INM) [30–32, 62].[5] In this treatment the system starts in a classical equilibrium distribution, but for each initial configuration $\underset{\sim}{r}$ is propagated in a harmonic approximation to the excited electronic state obtained from a quadratic expansion of the actual potential around the initial configuration. Note that here the INM approach is applied to the nonequilibrium classical expression, rather than the more typical equilibrium (linear response) INM calculation. The potential energy after propagation for time t in the excited state is then given by

$$V_e(\underset{\sim}{r}(t)) = V_e(\underset{\sim}{r}) + (\underset{\sim}{r}(t) - \underset{\sim}{r})^{\mathrm{T}} \cdot \underset{\sim}{\nabla} V_e(\underset{\sim}{r}) + \tfrac{1}{2}\,(\underset{\sim}{r}(t) - \underset{\sim}{r})^{\mathrm{T}} \cdot \underset{\sim}{\nabla}\underset{\sim}{\nabla}^{\mathrm{T}} V_e(\underset{\sim}{r}) \cdot (\underset{\sim}{r}(t) - \underset{\sim}{r}) \qquad (5.1)$$

[5]INM analysis has also been useful in characterizing the rotational, translational, and vibrational nature of the nuclear motions responsible for the short-time Stokes shift. See e.g. ref. [63].

An orthogonal matrix $\underset{\approx}{U}(\underset{\sim}{r})$ diagonalizes the force-constant matrix in Eq. (5.1):

$$\underset{\approx}{U}(\underset{\sim}{r}) \cdot \underset{\sim}{\nabla}\underset{\sim}{\nabla}^{\mathrm{T}} V_e(\underset{\sim}{r}) = \underset{\approx}{\omega}^2(\underset{\sim}{r}) \cdot \underset{\approx}{U}(\underset{\sim}{r}) \tag{5.2}$$

and the elements ω_α^2 of the diagonal matrix $\underset{\approx}{\omega}^2(\underset{\sim}{r})$ are the squares of the instantaneous normal mode frequencies for the given initial configuration. The time-dependent configuration obtained from harmonic motion on the surface in Eq. (5.1), with initial conditions $\underset{\sim}{r}(0) = \underset{\sim}{r}$ and $d\underset{\sim}{r}(0)/dt = \underset{\sim}{p}$, is given by

$$\underset{\sim}{r}(t) = \underset{\sim}{r}_0 + \underset{\approx}{U}^{\mathrm{T}} \cdot \cos(\underset{\approx}{\omega} t) \cdot \underset{\approx}{U} \cdot (\underset{\sim}{r} - \underset{\sim}{r}_0) + \underset{\approx}{U}^{\mathrm{T}} \cdot \underset{\approx}{\omega}^{-1} \cdot \sin(\underset{\approx}{\omega} t) \cdot \underset{\approx}{U} \cdot \underset{\sim}{p} \tag{5.3}$$

The point $\underset{\sim}{r}_0 = \underset{\sim}{r} - (\underset{\sim}{\nabla}\underset{\sim}{\nabla}^{\mathrm{T}} V_e)^{-1} \cdot \underset{\sim}{\nabla} V_e$ is the minimum (or maximum or saddle) point of the harmonic approximation to the excited-state potential, and depends on $\underset{\sim}{r}$ unless the true potential is harmonic.

The instantaneous-normal-mode prediction for the average electronic transition energy is obtained by using Eq. (5.3) in Eq. (3.24). But the equilibrium average over initial $\underset{\sim}{r}$ and $\underset{\sim}{p}$ can be made easier by using a linear approximation to the difference potential,

$$\Delta V(\underset{\sim}{r}(t)) = \Delta V(\underset{\sim}{r}) + (\underset{\sim}{\nabla}^{\mathrm{T}} \Delta V(\underset{\sim}{r})) \cdot (\underset{\sim}{r}(t) - \underset{\sim}{r}) \tag{5.4}$$

Because $\langle \underset{\sim}{p} \rangle_g = 0$, the time-dependent electronic transition energy reduces to an average over initial configurations alone,

$$\Delta E_{\mathrm{IMN}}(t) = \int d^N r \sigma_g(\underset{\sim}{r}) \{ \Delta V + (\underset{\sim}{\nabla}^{\mathrm{T}} \Delta V) \cdot \underset{\approx}{U}^{\mathrm{T}} \cdot [\cos(\underset{\approx}{\omega} t) - \underset{\approx}{1}] \cdot \underset{\approx}{\omega}^{-2} \cdot \underset{\approx}{U} \cdot (\underset{\sim}{\nabla} V_e) \} \tag{5.5}$$

where the equilibrium spatial distribution function, $\sigma_g(\underset{\sim}{r})$, is obtained from the full phase space distribution by integrating over momentum.

Equation (5.5) is more easily calculated as a weighted integration over normal mode frequencies. Some of the eigenvalues ω_α^2 of the force constant matrix may be negative, corresponding to the presence of local maxima or saddle points in $V_e(\underset{\sim}{r})$ [63]. In practice, the resulting imaginary frequencies are simply excluded from the calculation, and the integration is performed over the remaining positive frequencies

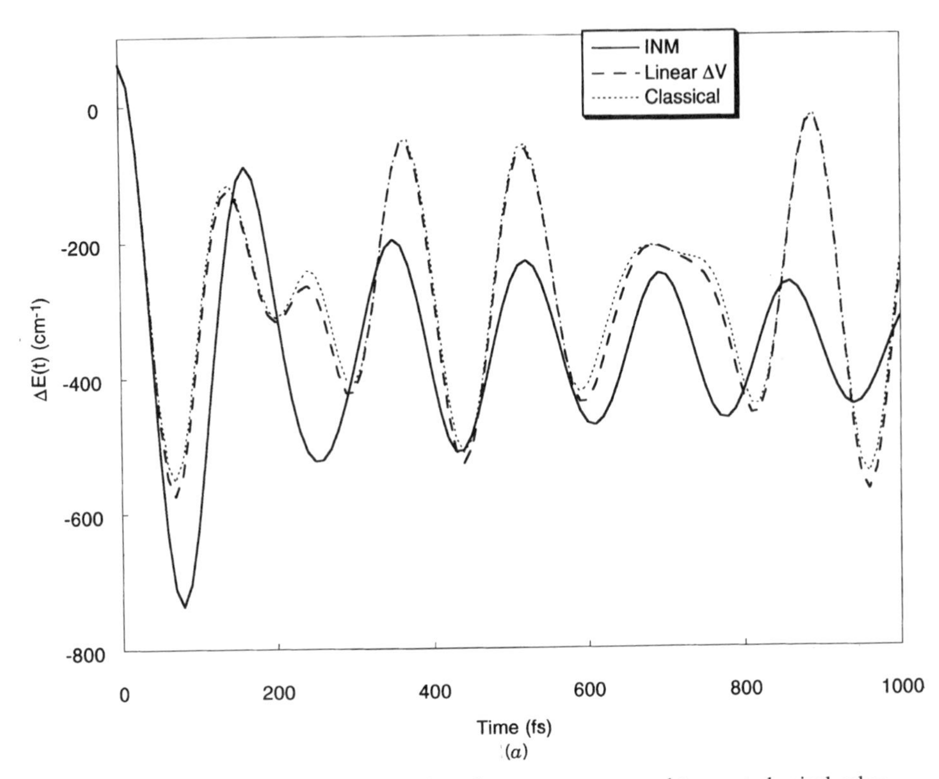

Figure 13. (*a*) Instantaneous normal mode treatment compared to exact classical calculations of the electronic difference potential and a linearly expanded difference potential (see text). Shortcomings of the instantaneous normal mode scheme result from oversimplification of the excited-state dynamics. Small additional errors introduced by linearly expanding the difference potential are most noticable as discrepancies between the two exact dynamics calculations when $\Delta E(t)$ is small, and the nuclear distribution is far from the Franck–Condon region. (*b*) Normalized distribution, $f(\overline{\omega})$, of real instantaneous normal mode frequencies of the model system at room temperature. The model system shows no significant spectral density for imaginary values of $\overline{\omega}$.

$$\Delta E_{\mathrm{INM}}(t) = \langle \Delta V \rangle_g + \int_0^\infty d\overline{\omega} f(\overline{\omega})(\cos \overline{\omega} t - 1) \tag{5.6}$$

where the weighting function in Eq. (5.6) is defined as

$$f(\overline{\omega}) = \int d^N r \sigma_g(\underset{\sim}{r})(\underset{\sim}{\nabla}^{\mathrm{T}} \Delta V) \cdot \underset{\approx}{U}^{\mathrm{T}} \cdot \delta(\overline{\omega} - \underset{\approx}{\omega}) \cdot \underset{\approx}{\omega}^{-2} \cdot \underset{\approx}{U} \cdot (\underset{\sim}{\nabla} V_e) \tag{5.7}$$

The time-dependent electronic transition energy of the three-mode system in the instantaneous normal mode approximation is plotted in Fig. 13. It

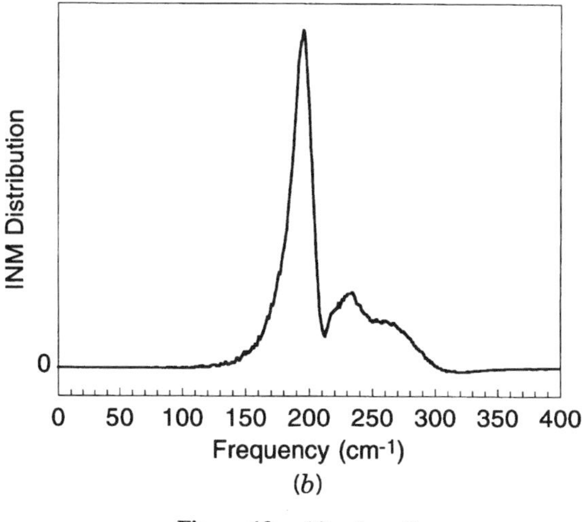

Figure 13. (*Continued*)

is compared with the time-dependent average of the linearly approximated difference potential in Eq. (5.4) resulting from ordinary classical dynamics, and with the classical $\Delta E(t)$ obtained without the additional linear expansion. Like the ground-state linear response approximation of Fig. 12, the instantaneous normal mode treatment fares well at times much less than the characteristic librational period, but founders when the excited-state nuclear distribution leaves the Franck–Condon region.

Like ground-state linear response, the instantaneous normal mode scheme employs the correct initial (classical) distribution, but propagates it in an approximate potential. Essentially, the INM treatment replaces the excited electronic surface with an inhomogeneous collection of multidimensional harmonic surfaces (whereas linear-response substitutes near-equilibrium ground-state dynamics). This added inhomogeneity accounts for the excessive decay rate of the oscillations seen in Fig. 13. The INM propagation, like the semiclassical treatments, uses a locally quadratic approximation; but while the semiclassical wave-packet treatment expands the true potential about the time-dependent center of the moving distribution, the INM scheme permanently assigns each portion of the *initial* nuclear distribution an approximate harmonic potential energy surface.

VI. CONCLUDING DISCUSSION

We have presented the basic theory of ultrafast fluorescence upconversion in a compact way that unifies the description of its diverse applications, captures the physical origin of some of its generic features—especially those at very short times, and explains on a physical level the success of a widely adopted mathematical procedure for deconvolving the signal. In order to investigate how information about the system is filtered through experimental results and numerical simulations, we performed rigorous calculations of time-resolved fluorescence signals on a simple model system and compared the results with those of various approximation schemes.

Our results have been numerous, in keeping with the complexity of time-resolved fluorescence spectroscopy and associated dynamics simulations, and some concluding comments are in order. We began by setting forth the basic theory of time- and frequency-resolved fluorescence from a thermal system with an arbitrary number of nuclear modes and a single, energetically isolated electronic transition. The time-dependent perturbation approximation taken as our starting point is valid for weak fields and for fluorescence slow enough that radiation damping may be neglected; it therefore does not include inhomogeneous relaxation kinetics, a Stokes shift due to selective depopulation via radiative or nonradiative decay [64, 65]. Analytical expressions were obtained for both the observed time-dependent fluorescence spectrum and a more informative deconvolved spectrum. Systematic investigation of the deconvolved time-resolved fluorescence spectrum was motivated by the widespread (and successful) application of numerical deconvolution to experimental and simulated ultrafast emission data in order to reveal otherwise elusive short-time dynamics. In contrast to nonresonant time-resolved spectroscopies such as the optical Kerr effect (OKE) [66], however, the deconvolution does not, even in principle, remove all dependence on the pulses.

Our basic expression, Eq. (2.18), for the observed time-dependent fluorescence spectrum $F(\omega, t_d)$ is similar to the corresponding expression for the stimulated emission contribution to an ultrafast pump-probe measurement obtained previously by Yan and Mukamel [33]. However, in the time-resolved fluorescence signal, excitation-pulse action must precede gate-pulse action. In the pump-probe signal, the probe pulse can act on the system before the pump at short delays. For well-separated pulses the expressions for fluorescence upconversion and pump-probe stimulated emission signals are equivalent (apart from frequency-dependent prefactors). (See Appendix A.) In pump-probe measurements, moreover, the stimulated emission contribution to the signal is accompanied, in general, by additional contributions due to excited-state absorption and ground-state bleaching.

When the excitation and gate pulses cease to overlap, both the observed and the deconvolved spectra can be expressed in terms of excited-state density matrix increments generated by the excitation pulse and window operators that describe the gate-pulse-mediated measurement of the probability density in a narrow region of nuclear configuration space. The doorway-window treatment does not make explicit recourse to the third-order time-dependent response functions, which are perhaps more necessary to describe ultrafast experiments, such as femtosecond photon echoes,[6] that focus on electronic rather than nuclear coherence [69]. The doorway and window operators take particularly simple forms under the classical Franck approximation, which is valid for short pulses.

A fundamental distinction was identified between the excited-state density matrix increment (or doorway operator) appropriate to the observed time-dependent spectrum and the one appropriate to the deconvolved spectrum. The former is a positive-definite operator, which describes the nuclear distribution generated by the excitation pulse. The density matrix increment of the deconvolved spectrum is *not* positive-definite in general; it is a composite quantity constructed by the deconvolution procedure from a range of interpulse delays.

The shape of the excited-state density matrix increment for the deconvolved spectrum provides physical justification for the success of deconvolution in uncovering fine features of the short-time dynamics. Such a justification is needed because the doorway-window formulas obtained here for the observed and deconvolved time-dependent fluorescence spectra [Eqs. (B.1) and (2.24), respectively] have the same formal structure. The two window operators are equivalent under the classical Franck approximation, but a subtle difference in the classical Franck expressions for the excited-state density matrix increments is important. As demonstrated for a Gaussian distribution, the momentum width of the initial increment for the deconvolved spectrum can be narrower than the spatial width and the uncertainty principle would allow. This nonphysical composite wave packet spreads less rapidly as it explores the excited potential surface. The reduced uncertainty accounts for the ability of the deconvolved spectrum to reveal subtle features of the dynamics that would otherwise be missed, unless shorter excitation and detection pulses could be used.

Our semiclassical treatment adds to the existing wave-packet formalism by providing some concrete details specific to ultrafast time-resolved fluorescence or pump-probe spectroscopy. In Eq. (3.12), we obtained a semiclassical expression for the deconvolved peak frequency that can be evaluated (in principle analytically) at short times by using Eqs. (4.3) and (4.4). The initial values of the moments were found in Eqs. (3.13–3.17) and Eq. (3.19), which define the Gaussian doorway functions for the observed and deconvolved spectra under the classical Franck approximation. Gausian wave packets have been useful in describing low-resolution continuous-wave spec-

[6]Photon echo peak shifts [67] and time-gated photon echoes [68] have recently been applied to the study of short-time solvation dynamics.

tra (e.g., absorption) and in situations where the relevant nuclear motion is brief (e.g., direct photodissociation cross sections and resonance Raman spectra of dissociating species).[7] The importance of short timescales, which justifies a semiclassical wave-packet approach, is equally true for ultrafast nonlinear optical measurements, including time-resolved fluorescence and femtosecond pump-probe spectroscopy. Some features that are not describable by single-wave-packet methods may well be adequately handled by more advanced semiclassical treatments based on cellular dynamics [71–73], path-integrals [74, 75] or an $\hbar$-expansion of the Wigner distribution function [76].

The semiclassical closed-form expression for the peak emission frequency in terms of the time-dependent moments of the excited-state density matrix increment, and the formulas for the initial values of those moments, provide what is perhaps the most rudimentary and general explanation possible for the Gaussian behavior of the short-time Stokes shift. Although the excited-state density matrix increment is distorted from the equilibrium distribution, it is born standing still. Hence the motion and spreading of the excited-state distribution, which govern the departure of the peak frequency from its initial value, occur only at second order in time.

As framed here for fluorescence, and elsewhere for pump-probe spectroscopy, the doorway-window and semiclassical wave-packet treatments are most useful when the pulses are well separated. In order to compare the short-time Stokes shift dynamics during the excitation pulse and after pulse overlap, we expanded the respective quantum mechanical expressions for the time-resolved spectrum in powers of the excited-state propagation time. There were interesting differences in the two resulting peak-frequency expressions through second order [Eqs. (4.2) and (4.5)]. The initial peak frequency during the excitation pulse is the excitation-pulse center frequency, but the zeroth-order term of the expansion for short, non-overlapping pulses is the equilibrium value of the electronic difference potential. Neither of the peak-frequency expressions has a linear term, but the quadratic coefficients, which describe the initial Stokes shift, are not the same. The quadratic coefficient during $T' < \tau_{\mathrm{exc}}$ depends on both the resonance offset and the squared gradient of the difference potential. The quadratic coefficient for non-overlapping pulses depends only on the squared-gradient of the difference potential, but with a prefactor half again as large as during pulse overlap. These differences should be taken into account in quantitative interpretations of experimental data on inertial solvation dynamics.

[7]Moment expansions of time-dependent distribution functions have proven useful in other areas of physical chemistry as well; see, for example, ref. [70].

It is worth noting an interesting consequence of the differing inti-tial values—Ω_{exc} and $\langle \Delta V \rangle_g$, respectively—for $\omega_{\text{peak}}(0)$ and $\Delta E(0)$. At very short times, with short pulses, $\omega_{\text{peak}}(T')$ tends quadratically toward $\Delta E(0)$—upward or downward—before the Stokes shift due to nuclear motion commences. Appendix C documents this very short time process in simulations on our model system. It occurs only during the excitation pulse, and is not to be confused with the anti-Stokes shift exhibited in Fig. 8.

We found a potentially significant distinction between two models of the short-time dynamics of the average electronic transition energy that have been regarded as more-or-less equivalent. These are the rigid-cage approximation put forward by Maroncelli [20] and the difference-potential-only model advanced by Perera and Berkowitz [37]. Careful examination of the rigid-cage approximation revealed that it agrees exactly with the quantum mechanical average transition energy through second order in time, when the electronic difference potential is a sum of single-solvent interactions with the solute. In contrast, the difference-potential-only approximation, which can be defined for an arbitrary difference potential, contains an additional second order term that can be nonzero if the difference potential is a sensibly nonlinear function of nuclear coordinates.

Numerical results presented here for a three-mode model system suggest that interpretation of fluorescence upconversion results can be complex, and approximate simulations can be inaccurate. Excitation and gate pulses with bandwidth not much wider than the range of system absorption or emission frequencies—or, equivalently, duration not much shorter than system vibrational periods—distorted the fluorescence signal, causing a marked loss of information in the time-dependent fluorescence spectrum. In particular, the initial peak emission frequency of the observed spectrum, which must be accurately determined in order to uncover the short-time dynamics, was found to red-shift toward the steady state value as either of the pulse bandwidths was reduced. Deconvolution restored some, but not all, of the lost information.

Additional effects of nonzero pulse duration seen at short interpulse delay times are due to overlap of the excitation and gate pulses. In calculations neglecting the effects of pulse overlap, the "observed" peak emission frequency from our model system initially is too high—the signal exhibits an initial peak frequency artificially close to the ideal unrelaxed value. By neglecting pulse overlap, some nonphysical negative propagation times in the excited electronic state are introduced, which tend to postpone the Stokes shift.

Some previous calculations on pump-probe spectroscopy have incorporated finite-bandwidth effects from pulses of nonzero duration. Examples include Stock and Domcke's simulations of ultrafast pump-dispersed probe measurements on a three-state quantum mechanical model of pyrazine [77] and recent simulations of ultrafast pump-probe measurements on tri-iodide [78]; see also ref. [79]. A recent paper by Li and co-workers [80] puts forward an approximate method for incorporating finite excitation-pulse bandwidth effects in classical calculations of pump-probe signals. That method pre-selects ground-state trajectories through a resonance filter akin to the exponential factors in our Eq. (B.5), and thereby avoids further (excited-state) propagation of trajectories with nonresonant initial coordinates. None of these studies takes account of the pulse overlap between the finite duration pump and probe pulses.

Investigations have also recently been made of excited-state vibrational coherence induced by curve crossing, rather than by short-pulse electronic excitation [19]. Jean's calculations focused on the possible occurrence of time-dependent vibrations in the final state of an electron transfer reaction, and considered the manifestation of these coherences in time-dependent emission signals.

Our few-mode simulations incorporate the effects of optical dephasing only insofar as that process is driven by the dynamical fluctuations of those optically active nuclear degrees of freedom. As the absorption spectrum of the model system comprises discrete electronic-librational lines, the optical dephasing time is evidently longer than the characteristic periods of nuclear motion. (See Appendix C for a plot of the peak emission frequency at long times of the three-mode model system.) If the dephasing time were shorter than the librational timescale (and shorter than the pulse widths), perhaps because a large number of modes were coupled to the electronic transition, then the effects of pulse overlap could be somewhat less prominent than those in Fig. 7.

In practice it can be rather difficult to identify precisely the vertical electronic transition energy for large dye molecules in solution [81]. In the absence of such information, molecular dynamics simulations often have assumed vertical excitation. Both our analytical formulas for the short-time Stokes shift during pulse overlap and a numerical test on our model system show that excitation-pulse center-frequency effects can significantly influence the form of the signal, including an *anti*-Stokes shift under sub-resonant excitation. Again, however, sizable excited-state displacements (compared to kT) in many modes would militate against an anti-Stokes shift by preventing the requisite significant absorbance in the red tail of the absorption band, below the frequency of maximum steady-state fluorescence.

While pulse overlap and finite pulse bandwidth effects were most easily incorporated in a fully quantum mechanical treatment, approximate methods were tested in calculations of the time-dependent average electronic transition energy, to which the observed peak emission frequency was shown to reduce when using ideally short excitation and gate pulses. The excited-state dynamics of our model system were fairly well reproduced by both classical and semiclassical methods; in more anharmonic systems (not shown) the restricted semiclassical waveforms caused large, incorrect values of the transition energy. This finding is no surprise, as the characteristic librational frequencies of the model system are somewhat lower than room temperature.

Two further approximations to classical excited-state dynamics were found to be problematic in the model system. These were the linear response treatments in terms of classical equilibrium correlation functions in the electronic ground and excited states, and a nonequilibrium treatment using instantaneous normal modes. Both the ground-state linear response approximation and the instantaneous normal mode approximation failed at all but the shortest excited-state propagation times. The excited-state linear response treatment, which performed adequately at longer time, gave an incorrect initial value of the average transition frequency. These results from a single model system may not be widely generalizable. Nonetheless, failure in the seemingly innocent system considered, with its small excited-state displacements in a few low-frequency modes, suggests caution in applying linear-response methods (and by extension cumulant expansion methods) and instantaneous normal mode approximations to ultrafast nonlinear optical response properties.

There is good reason to believe that satisfactory theoretical treatment of what is arguably the simplest nonlinear ultrafast measurement can be extended beyond few-mode systems of the kind simulated here. For instance, there is ample evidence from simulation and experiment that the short-time dynamics of bulk polar solvation is dominated by the separable motions of molecules in the first few solvent shells. It is possible that a suitable decomposition can be made, which incorporates the strongly optically active collective solvent modes into a rigorously treated "system," and relegates the remaining solvent modes to a "bath," which is described more crudely, perhaps by using Redfield theory [82]. A preliminary step in this direction has already been taken [61, 83] and will be reported elsewhere.

ACKNOWLEDGMENTS

The authors thank Stephen Bradforth, Mei Du, Graham Fleming, Bob Harris, Eric Hiller, David Jonas, Craig Martens, Alex Matro, David Oxtoby, Mike Raymer, Sandy Rosenthal, and Tim Smith for helpful conversations. Portions of this work contributed to LWU's Ph.D. dissertation at the University of Chicago. This research was supported by grants from the National Science Foundation and the Camille and Henry Dreyfus Teacher-Scholar Award Program.

APPENDIX A: STIMULATED EMISSION AMPLITUDE FOR A PUMP-PROBE SIGNAL

The simulated-emission contribution to the pump-probe signal can be derived from the amplitude to find the system back in its electronic ground state after interacting with the pump and probe fields each once. That amplitude is given by

$$(-i)^2\int_{-\infty}^{\infty} d\tau\int_0^{\infty} d\tau'\langle g|\tilde{V}_{\rm pr}(\tau)\tilde{V}_{\rm pu}(\tau-\tau')+\tilde{V}_{\rm pu}(\tau)\tilde{V}_{\rm pr}(\tau-\tau')|\psi_n|g\rangle$$

$$= -\frac{\mu_s^2 E_{\rm pr}E_{\rm pu}}{4}\int_{-\infty}^{\infty} d\tau\int_0^{\infty} d\tau'[f_{\rm pr}(\tau-t_d)f_{\rm pu}(\tau-\tau')$$

$$\cdot\, e^{i\Omega_{\rm pr}\tau}e^{i\Omega_{\rm pu}(\tau'-\tau)} + f_{\rm pu}(\tau-t_d)f_{\rm pr}(\tau-\tau')e^{i\Omega_{\rm pu}\tau}e^{i\Omega_{\rm pr}(\tau'-\tau)}]$$

$$\cdot\, e^{iH_g\tau}e^{-iH_e\tau'}e^{iH_g(\tau'-\tau)}|\psi_n\rangle \tag{A.1}$$

In Eq. (A.1), the first term in square brackets is directly analogous to Eq. (2.16), with pump and probe fields replacing excitation and gate pulse fields, respectively. The second term in square brackets accounts for the fact that the pump and probe pulses may act in reverse order at short delay times; this additional term vanishes when the pulses cease to overlap in time.

APPENDIX B: DOORWAY-WINDOW PICTURE OF THE OBSERVED FLUORESCENCE SPECTRUM

The time-resolved emission signal is equivalent to the stimulated emission contribution to the pump-probe signal when the pulses do not overlap (apart from prefactors; see Appendix A). The doorway-window treatment of the observed fluorescence spectrum under this condition is formally equivalent to that put forward by Yan and Mukamel [33] for the stimulated emission contribution to pump-probe spectroscopy.

Let us return to Eq. (2.18) for the observed time-dependent spectrum. Following specialization to the case of non-overlapping pulses (and some straightforward changes in variables of integration), that expression for $F(\omega, t_d)$ can be rewritten as

$$F(\omega, t_d) = \mathrm{Tr}_{\rm nuc}[W(\omega)\exp(-iH_e t_d)D(\Omega_{\rm exc})\exp(iH_e t_d)] \tag{B.1}$$

in which the excited-state density matrix increment, or doorway operator, is

$$D(\Omega_{\text{exc}}) = \int_{-\infty}^{\infty} d\tau \int_{-\infty}^{\infty} d\tau' f_{\text{exc}}(\tau) f_{\text{exc}}(\tau') \exp[i\Omega_{\text{exc}}(\tau - \tau')] U(\tau) \rho_g U^{\dagger}(\tau') \tag{B.2}$$

and the window operator has the form

$$W(\omega) = \int_{-\infty}^{\infty} d\tau \int_{-\infty}^{\infty} d\tau' f_{\text{up}}(\tau) f_{\text{up}}(\tau') \exp[-i\omega(\tau - \tau')] U(\tau) U^{\dagger}(\tau') \tag{B.3}$$

The operator $U(t)$ is defined in Eq. (2.23).

In the classical Franck approximation as described in Section III.A, $U(t) \cong \exp(-i\Delta V t)$, the window operator (B.3) reduces to

$$W(\omega) \cong 2\pi\tau_{\text{up}}^2 \exp[-(\omega - \Delta V)^2/\tau_{\text{up}}^2] \cong 2\pi\tau_{\text{up}}^2[1 - \tau_{\text{up}}^2(\omega - \Delta V)^2] \tag{B.4}$$

which differs from Eq. (3.2) only by a constant factor. On the other hand, the excited-state density matrix increment that is actually prepared by the excitation pulse does differ, under the classical Franck approximation, from the one that gives rise to the deconvolved spectrum. In the classical Franck approximation, Eq. (B.2) reduces to

$$\begin{aligned} D(\Omega_{\text{exc}}) &\cong 2\pi\tau_{\text{exc}}^2 \exp[-(\Omega_{\text{exc}} - \Delta V)^2\tau_{\text{exc}}^2/2]\rho_g \exp[-(\Omega_{\text{exc}} - \Delta V)^2\tau_{\text{exc}}^2/2] \\ &\cong 2\pi\tau_{\text{exc}}^2 \left\{\rho_g - \frac{\tau_{\text{exc}}^2}{2}\,[(\Omega_{\text{exc}} - \Delta V)^2\rho_g + \rho_g(\Omega_{\text{exc}} - \Delta V)^2]\right\} \end{aligned} \tag{B.5}$$

Section III.B above explains how the subtle difference of $\tilde{D}(\Omega_{\text{exc}})$ in Eq. (3.1) from the state in Eq. (B.5) makes the former better able to track the short-time evolution of the average electronic transition energy.

APPENDIX C: VERY SHORT TIME AND VERY LONG TIME BEHAVIOR OF THE PEAK EMISSION FREQUENCY

It is of some interest to detail the transition between the short-time evolution of the peak emission frequency during the excitation pulse in Eq. (4.2) and that following pulse overlap [given approximately by Eq. (4.5)]. Figure 14 shows the peak frequency $\omega_{\text{peak}}(T')$ of the deconvolved fluorescence spectrum of the three-mode system for vertically resonant excitation pulses ($\Omega_{\text{exc}} = 0$) of durations $\tau_{\text{exc}} = 0.1$, 1.0, and 10.0 fs. In all three cases, the

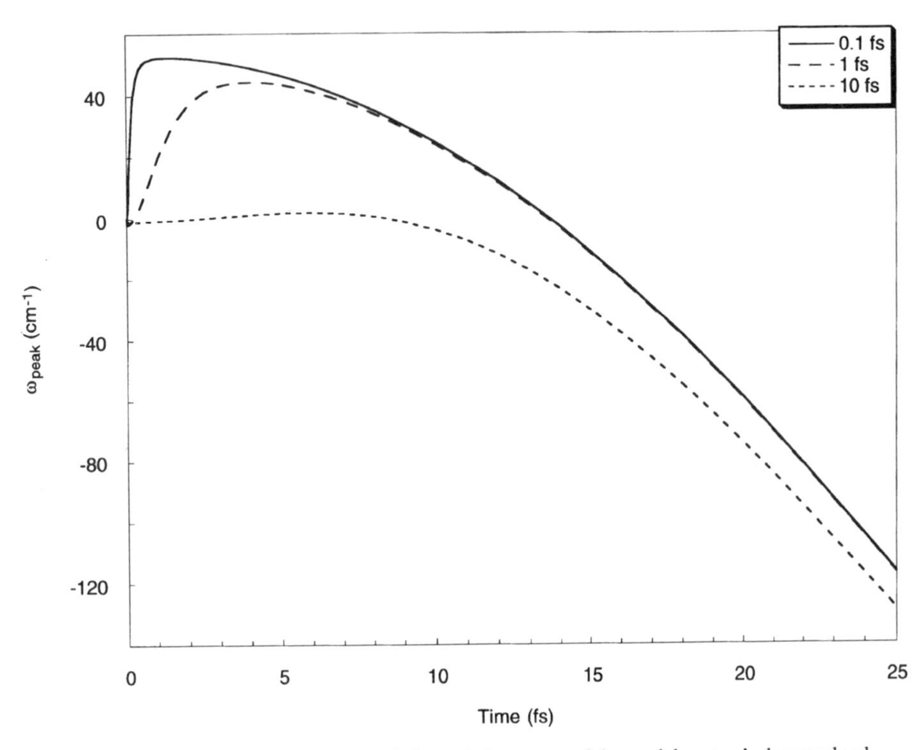

Figure 14. Short-time evolution of the peak frequency of the model system's deconvolved fluorescence spectrum. Results are shown for three different excitation pulse durations—see text of Appendix C for details.

gate pulse is $\tau_{up} = 10$ fs long. The initial change is from the excitation-pulse center frequency toward $\Delta E(0)$. As $\langle \Delta V - \Omega_{exc} \rangle_g = \Delta E(0) = 50.7\ \text{cm}^{-1}$ in the cases shown, the anti-Stokes shift during very short excitation pulses is in accord with the prediction of Eq. (4.2). The shift is larger and more rapid for shorter pulses, as expected from the $1/\tau_{exc}^2$ dependence in Eq. (4.2). At longer propagation times, the dynamical Stokes shift overwhelms this initial behavior, and the peak frequency decreases as T'^2.

For the sake of completeness, we also wish to document the peak emission frequency of our model system at longer times than those shown in Section V. Figure 15 shows the long-time evolution of the peak frequency of the "observed" fluorescence spectrum for $\Omega_{exc} = 0$, $\tau_{exc} = 0.1$ fs, and $\tau_{up} = 10$ fs. Although the system does not approach a steady state, there is a signficant decrease in the amplitude of peak-frequency oscillations about the "equilibrium" value, due to wave-packet spreading, during the first 10 ps. At later times there is partial rephasing in this three-coordinate system.

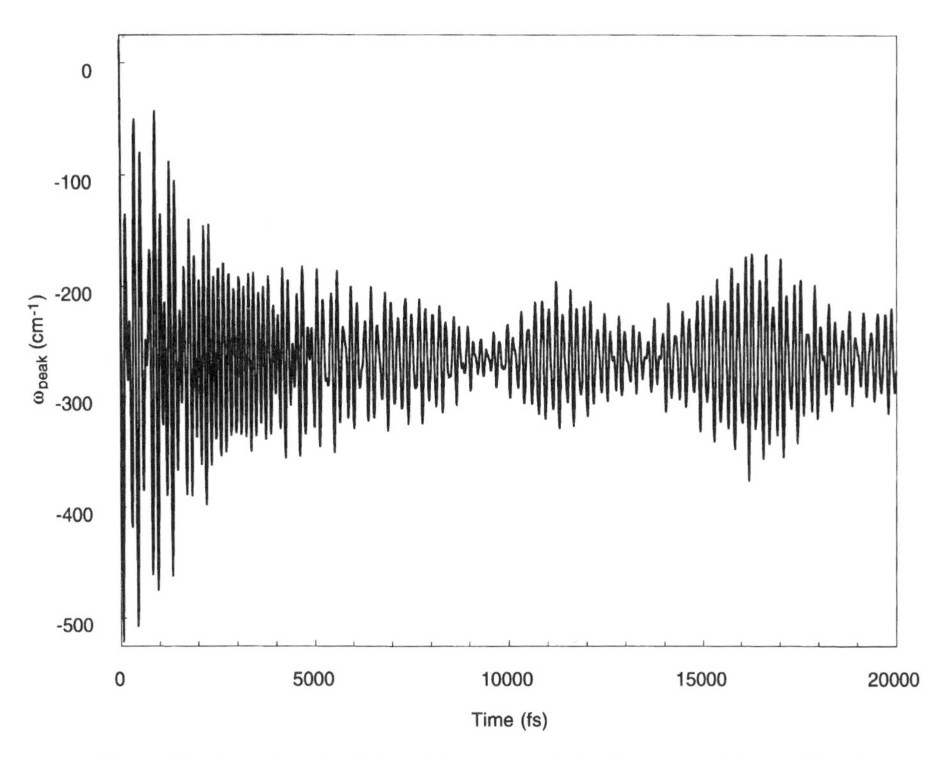

Figure 15. Long-time simulation of the peak emission frequency of the model system under nearly ideal excitation and detection conditions. See text of Appendix C.

REFERENCES

1. R. Jimenez, G. R. Fleming, P. V. Kumar, and M. Maroncelli, *Nature*, **369,** 471 (1994).
2. T. J. Dunn, J. N. Sweetser, I. A. Walmsley, and C. Radzewicz, *Phys. Rev. Lett.*, **70,** 3388 (1993).
3. H. Wang, J. Shah, T. C. Damen, and L. Pfeiffer, *Phys. Rev. Lett.*, **74,** 3065 (1995).
4. R. J. Stanley and S. G. Boxer, *J. Phys. Chem.*, **99,** 859 (1995).
5. P. F. Barbara and W. Jarzeba, *Adv. Photochem.*, **15,** 1 (1990).
6. M. Maroncelli, *J. Mol. Liquids*, **57,** 1 (1993).
7. M. L. Horng, J. A. Gardecki, A. Papazyan, and M. Maroncelli, *J. Phys. Chem.*, **99,** 17311 (1995).
8. P. J. Rossky and J. D. Simon, *Nature*, **370,** 263 (1994).
9. J. Ma, D. Vanden Bout, and M. Berg, *J. Chem. Phys.*, **103,** 9146 (1995).
10. J. T. Fourkas and M. Berg, *J. Chem. Phys.*, **98,** 7773 (1993).
11. S. E. Bradforth, R. Jimenez, F. von Mourik, R. van Grondelle, and G. R. Fleming, *J. Phys. Chem.*, **99,** 16179 (1995).

12. A. Matro and J. A. Cina, *J. Phys. Chem.*, **99,** 2568 (1995).
13. J. H. Eberly and K. Wódkiewicz, *J. Opt. Soc. Am.*, **67,** 1252 (1977).
14. K.-H. Brenner and K. Wódkiewicz, *Optics Comm.*, **43,** 103 (1982).
15. G. Nienhuis, *Physica*, **96C,** 391 (1979).
16. J. S. Melinger and A. C. Albrecht, *J. Chem. Phys.*, **84,** 1247 (1986).
17. R. F. Loring, Y.-J. Yan, and S. Mukamel, *J. Chem. Phys.*, **87,** 5840 (1987).
18. M. G. Raymer, J. Cooper, H. J. Carmichael, M. Beck, and D. T. Smithey, *J. Opt. Soc. Am. B*, **12,** 1801 (1995), determine the full probability distribution of intensity and phase of a transient source (including fluorescence) relevant to a given spatial-temporal field mode.
19. J. M. Jean, *J. Chem. Phys.*, **104,** 5638 (1996); *J. Chem. Phys.*, **101,** 10464 (1994).
20. M. Maroncelli, *J. Chem. Phys.*, **94,** 2084 (1991).
21. J. S. Bader and D. Chandler, *Chem. Phys. Lett.*, **157,** 501 (1989); E. A. Carter and J. T. Hynes, *J. Chem. Phys.*, **94,** 5961 (1991).
22. L. Perera and M. L. Berkowitz, *J. Chem. Phys.*, **96,** 3092 (1992).
23. E. Neria and A. Nitzan, *J. Chem. Phys.*, **96,** 5433 (1992).
24. R. Brown, *J. Chem. Phys.*, **102,** 9059 (1995).
25. B. Bagchi, *Annu. Rev. Phys. Chem.*, **40,** 115 (1989).
26. B. Bagchi and A. Chandra, *Adv. Chem. Phys.*, **80,** 1 (1991).
27. F. O. Raineri, H. Resat, B. C. Perng, F. Hirata, and H. L. Freidman, *J. Chem. Phys.*, **100,** 1477 (1994), and references therein.
28. S. J. Rosenthal, X. Xie, M. Du, and G. R. Fleming, *J. Chem. Phys.*, **95,** 4715 (1991).
29. T. Fonseca and B. M. Ladanyi, *J. Phys. Chem.*, **95,** 2116 (1991), and references therein.
30. R. M. Stratt and M. Cho, *J. Chem. Phys.*, **100,** 6700 (1994).
31. B. M. Ladanyi and R. M. Stratt, *J. Phys. Chem.*, **99,** 2502 (1995); **100,** 1266 (1996).
32. M. Buchner, B. M. Ladanyi, and R. M. Stratt, *J. Chem. Phys.*, **97,** 8522 (1992). This paper focuses on short-term dynamics of neat liquids, rather than solvation dynamics *per se.* See also, R. M. Stratt, *Acct. Chem. Res.*, **28,** 201 (1995).
33. Y.-J. Yan and S. Mukamel, *Phys. Rev. A*, **41** 6485 (1990).
34. D. M. Jonas, S. E. Bradforth, S. A. Passino, and G. R. Fleming, *J. Phys. Chem.*, **99,** 2594 (1995).
35. U. Banin, A. Bartana, S. Ruhman, and R. Kosloff, *J. Chem. Phys.*, **101,** 8461 (1994).
36. T. J. Smith, L. W. Ungar, and J. A. Cina, *J. Lumin*, **58,** 66 (1994).
37. L. Perera and M. L. Berkowitz, *J. Chem. Phys.*, **97,** 5253 (1992).
38. M. Maroncelli and G. R. Fleming, *J. Chem. Phys.*, **86,** 6221 (1987); *J. Chem. Phys.*, **92,** 3251 (1990).
39. S. J. Rosenthal, R. Jimenez, and G. R. Fleming, *J. Mol. Liquids*, **60,** 25 (1994).
40. B. J. Schwartz and P. J. Rossky, *J. Phys. Chem.*, **99,** 2953 (1995); see also *J. Chem. Phys.*, **101,** 6902 (1994); see also **101,** 6917 (1994).
41. P. K. Walhout, J. C. Alfano, Y. Kimura, C. Silva, P. J. Reid, and P. F. Barbara, *Chem. Phys. Lett.*, **232,** 135 (1995).
42. M. Braun, C. Meier, and V. Engel, *J. Chem. Phys.*. **103,** 7907 (1995).
43. A. Messiah, *Quantum Mechanics*, John Wiley & Sons, New York, 1976. Vol. 1, pp. 299–301.

44. R. G. Littlejohn, *Physics Reports*, **138,** 193 (1986), contains a comprehensive mathematical treatment of Gaussian wave packet dynamics.
45. E. J. Heller, *J. Chem. Phys.*, **62,** 1544 (1975); **68,** 2066 (1978).
46. E. J. Heller, R. L. Sundberg, and D. Tannor, *J. Phys. Chem.*, **86,** 1822 (1982).
47. S. Mukamel, *J. Phys. Chem.*, **88,** 3185 (1984).
48. Y. J. Yan and S. Mukamel, *J. Chem. Phys.*, **88,** 5735 (1988).
49. R. D. Coalson and M. Karplus, *Chem. Phys. Lett.*, **90,** 301 (1982); *J. Chem. Phys.*, **93,** 3919 (1990).
50. R. Heather and H. Metiu, *J. Chem. Phys.*, **84,** 3250 (1986).
51. L. W. Ungar and J. A. Cina, *J. Lumin.*, **58,** 89 (1994); **63,** 345 (1995).
52. D. A. McQuarrie, *Statistical Mechanics*, Harper & Row, New York, NY, 1976.
53. J. G. Saven and J. L. Skinner, *J. Chem. Phys.*, **99,** 4391 (2993).
54. G. T. Evans, *J. Chem. Phys.*, **103,** 8980 (1995).
55. T. Kalbfleish, R. Fan, J. Roebber, P. Moore, E. Jacobsen, and L. D. Ziegler, *J. Chem. Phys.*, **103,** 7673 (1995).
56. D. Chandler, unpublished.
57. L. W. Ungar, University of Chicago, Ph.D. dissertation (1994), Chapter 7.
58. S. Roy and B. Bagchi, *J. Chem. Phys.*, **99,** 1310 (1993).
59. D. M. F. Edwards, P. A. Madden, and I. R. McDonald, *Molec. Phys.*, **51,** 1141 (1984).
60. M. Messina, B. C. Garret, and G. K. Schenter, *J. Chem. Phys.*, **100,** 6570 (1994).
61. L. W. Ungar, University of Chicago Ph.D. Dissertation (1994), Chapter 9.
62. G. Seeley and T. Keyes, *J. Chem. Phys.*, **91,** 5581 (1989).
63. M. Cho, G. R. Fleming, S. Saito, I. Ohmine, and R. M. Stratt, *J. Chem. Phys.*, **100,** 6672 (1994).
64. G. J. Blanchard, *J. Chem. Phys.*, **95,** 6317 (1991).
65. R. S. Fee, J. A. Milsom, and M. Maroncelli, *J. Phys. Chem.*, **95,** 5170 (1991).
66. M. Cho, M. Du, N. F. Scherer, G. R. Fleming, and S. Mukamel, *J. Chem. Phys.*, **99,** 2410 (1993).
67. T. Joo, Y. Jia, J. Y. Yu, M. J. Lang, and G. R. Fleming, *J. Chem. Phys.*, **104,** 6089 (1996).
68. W. P. de Boeij, M. S. Pshenichnikov, and D. A. Wiersma, *J. Phys. Chem.*, **100,** 11806 (1996).
69. S. Mukamel, *Principles of Nonlinear Optical Spectroscopy,*, Oxford University Press, New York, 1995. This comprehensive monograph provides a unified description of most nonlinear optical experiments in terms of time-dependent response functions. The connection between that approach and the doorway-window picture in application to pump-probe spectroscopy is outlined in Chapters 11 and 12.
70. M. Guenza and K. F. Freed, *J. Chem. Phys.*, **105,** 3823 (1996).
71. E. J. Heller, *J. Chem. Phys.*, **94,** 2723 (1991).
72. M. A. Sepúlveda, S. Tomsovic, and E. J. Heller, *Phys. Rev. Lett.*, **69,** 402 (1992).
73. M. A. Sepúlveda and F. Grossmann, *Adv. Chem. Phys.*, **96,** 191 (1996).
74. J. Cao and G. A. Voth, *J. Chem. Phys.*, **104,** 273 (1996).
75. M. Topaler and N. Makri, *J. Chem. Phys.*, **101,** 7500 (1994).
76. N. E. Shemetulskis and R. F. Loring, *J. Chem. Phys.*, **97,** 1217 (1992).

77. G. Stock and W. Domcke, *Phys. Rev. A*, **45,** 3032 (1992).
78. G. Ashkenazi, U. Banin, A. Bartana, R. Kosloff, and S. Ruhman, *Adv. Chem. Phys.*, submitted.
79. A. E. Johnson and A. B. Myers, *J. Chem. Phys.*, **104,** 2497 (1996).
80. Z. Li, J.-Y. Fang, and C. C. Martens, *J. Chem. Phys.*, **104,** 6919 (1996).
81. R. S. Fee and M. Maroncelli, *Chem. Phys.*, **183,** 235 (1994).
82. W. T. Pollard and R. A. Friesner, *J. Chem. Phys.*, **100,** 5054 (1994).
83. J. S. Bader and B. J. Berne, *J. Chem. Phys.*, **100,** 8359 (1994).

QUANTUM DESCRIPTION OF THE IMPULSIVE PHOTODISSOCIATION DYNAMICS OF I_3^- IN SOLUTION

GUY ASHKENAZI, URI BANIN, ALLON BARTANA, RONNIE KOSLOFF, AND SANFORD RUHMAN

Department of Physical Chemistry
and the Fritz Haber Center
for Molecular Dynamics,
The Hebrew University, Jerusalem, 91904 Israel

CONTENTS

I. Introduction
  A. Experimental Background
  B. Theoretical Background
  C. Objective
  D. Outline for the Paper
II. Methods
  A. Statics
    1. Describing the State of the System
    2. Visualizing the State of the System
    3. Initial States
  B. Dynamics
    1. The Hamiltonian Operator
    2. The Dissipative Super-Operators
    3. The Evolution Operator
  C. Interpretation
    1. The Impulsive Excitation Picture
    2. Power Absorption of a Laser Pulse
    3. Absorption Spectrum
    4. Raman Spectrum
  D. Method Summery
III. Application
  A. Electronic Potential Energy Surfaces
  B. Photodissociation of I_3^-: The "Pump" Pulse
    1. Statics

Advances in Chemical Physics, Volume 100, Edited by I. Prigogine and Stuart A. Rice.
ISBN 0-471-17458-0

2. Dynamics
3. Interpretation
C. Dynamics of the photo-induced "hole" in I_3^-
1. Statics
2. Dynamics
3. Interpretation
D. Dynamics of Nascent I_2^-: The "Probe" Pulse
1. Statics
2. Dynamics
3. Interpretation
E. Vibrational Excitation of Relaxed I_2^-: The "Push" Pulse
1. Statics
2. Dynamics
3. Interpretation
F. Application Summary
IV. Summary
A. Critical Evaluation
B. Conclusions
APPENDIX A. Numerical Methods
1. Approximating Functions of Operators
2. Newton's Interpolation Method
3. Leja's Interpolation Points
4. Application to Operators
APPENDIX B. System Parameters
References

I. INTRODUCTION

The study of the dynamics and relaxation of simple chemical and photochemical reactions in solution is essential for the understanding of solution phase chemistry in its entirety. This study is part of an extensive experimental and theoretical effort aimed at gaining insight into the mechanism of photochemical reactions in solution. The strategy followed is to reduce the complexity of the mechanism by following the photodissociation dynamics in real time, so that the elementary steps can be studied sequentially. The photodissociation dynamics of I_3^- in different solvents has been chosen as a test case for this study. The significance of this system rises from the ability to observe coherent motion in the condensed phase on a sub-picosecond time scale, utilizing ultrafast pump-probe spectroscopy. This temporal resolution enables the separation in time of major events, which are basic to the understanding of condensed phase chemical dynamics, such as the intramolecular motion leading to bond cleavage, and energy flow between solute and solvent. In particular, direct access to the temporal evolution of photoproducts is made possible by these techniques allowing reconstruction of the time-dependent electronic and nuclear density operator. In turn, this extremely detailed description of dynamics provides a stringent testing ground of various theoretical descriptions of dissipative phenomena of highly excited molecules in solution.

A. Experimental Background

The photoinduced dynamics of I_3^- are inferred from a combination of CW and pump pulse spectroscopic measurements. The parent triiodide ion is linear and symmetric in solution, and stable in many polar solvents. The absorption spectra of I_3^- in solution consists of two bands, peaked at ~290 and ~350 nm, whose spectral assignment has been the subject of debate. The most common explanation is that excitation leads to dissociative states correlating with $I_2^-(^2\Sigma_u^+)$ in the ground state, and I $(^2P_{1/2,3/2})$ in either of its lowest spin-orbit states. The photoproduct's diiodide electronic absorption spectrum is composed of two bands—one in the near UV, partly overlapping the I_3^- absorption spectra, and the other in the near IR, centered at ~740 nm. The separate spectral bands for reactant and product enable a direct interpretation of the pump-probe spectroscopy. The absorption spectrum of I_3^- and I_2^- are shown in Fig. 1.

In the experimental system, solvated I_3^- is subjected to an intense, short UV pump pulse, which dissociates it into $I_2^- + I^*$. A weak probe pulse in the visible region, in resonance with the nascent I_2^- product absorption, interrogates the photoproduct. A probe pulse in the UV region in resonance with the I_3^- absorption interrogates the photoinduced dynamics on the ground electronic surface of I_3^- (Fig. 2). At the application of the pump pulse, a sud-

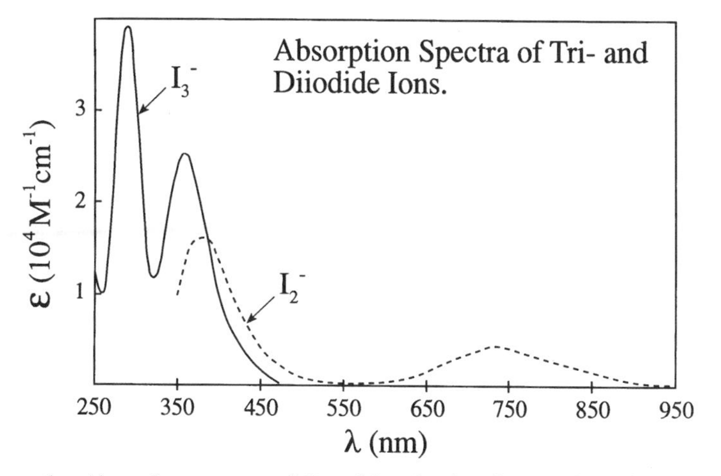

Figure 1. Absorption spectrum of I_3^- and I_2^-, showing the two absorption bands of I_3^- and of I_2^-.

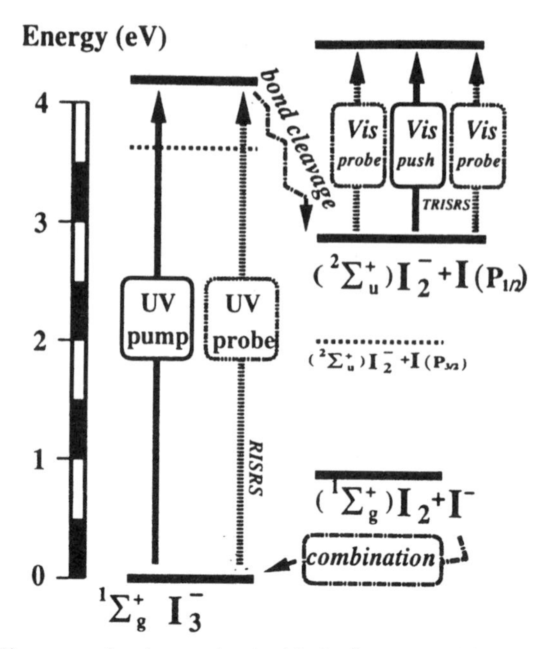

Figure 2. The energy bands associated with the I_3^- system. The UV pump initializes two separate processes. The first is photodissociation to I_2^-, which is probed in the visible region. The second is the RISRS process, which is probed in the UV region. In the TRISRS experiment, the I_2^- photoproduct can be invigorated by a push pulse and probed again in the visible region. The dashed levels show other possible routes to pump probe experiments.

den rise in absorption of the visible probe is observed, followed by a rapid reduction in absorption that lasts for nearly 300 fsec. During the following stages a slight increase of absorption is observed, which is accompanied by weak damped oscillations in the optical density [1]. The parameters of these oscillations strongly depend on the solvent (Fig. 3). The first instantaneous appearance of absorption is associated with the excited state of I_3^-. The stage of rapid reduction in the optical density is assigned to the process of bond fission, and emergence of the nascent fragments. The spectral modulations are attributed to the product state, in which a coherent population oscillates in and out of resonance with the probe, meaning that the I_2^- vibration is synchronized with the bond cleavage. Different probe wavelengths will correspond to different phases in the vibration [Fig. 12(*c*)]. Experimentally this is evident through a π phase shift of the spectral modulations, when probing with a blue and red shifted pulses (Fig. 4) [1].

Probing in the UV range reveals a similar phenomena, assigned to resonant impulsive stimulated raman scattering (RISRS) of the ground state population. This process can be visualized as a coherent "hole" created in

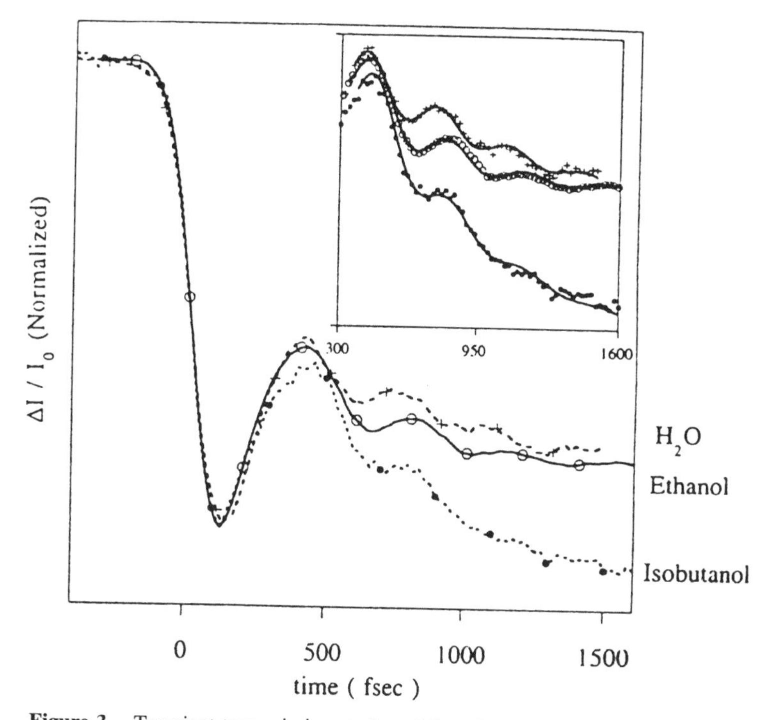

Figure 3. Transient transmission at short delays for three molecular solvents.

the reactant by lost product. Since the "hole" is not stationary it oscillates with the ground surface frequencies (Fig. 5). The spectral modulations of the "hole" shed light on the excitation stage [2]. This vibration can be correlated to the Raman spectrum of I_3^- (Fig. 6) [3].

The energetics of the photodissociation of I_3^- are such that 1.4 eV of excess energy has to be dissipated by the solvent. In an isolated system, kinematic considerations for a linear homonuclear triatomic molecule suggest that $\frac{1}{3}$ of the excess energy appears as vibration. The observed frequency of the modulations suggests a much lower vibrational excitation in the product, so part of this energy must be dissipated into the solvent during the dissociation stage. The rest of the excess energy will be dissipated by the product until thermal equilibrium is reached.

In order to study the evolving vibrational product distribution, at different delay times after the pump pulse, a second intense "push" pulse is applied to the system. The push pulse is in resonance with the nascent I_2^- product absorption, and so initiates a transient RISRS (or TRISRS) process, which sets a coherent motion in the I_2^- ground state population [4]. This motion is observed by modulations in the absorption of a third, weak probe pulse. The

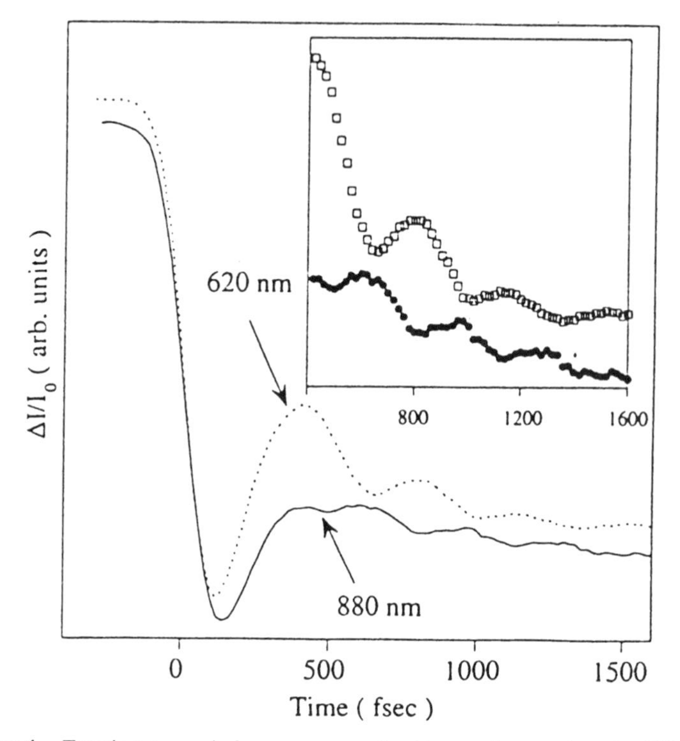

Figure 4. Transient transmission measurements at two probe wavelengths shifted to the blue (620 nm) and to the red (880 nm) of the maximum absorption wavelength for the near IR band of I_2^-. The solvent is water. There is a π radians phase shift of the oscillations between the two probe wavelengths.

parameters of the modulations (frequency, decay time) depend on the delay period between the pump and push pulses (Fig. 7), and so contain information about the vibrational dynamics of the transient species at the push instant. A schematic illustration of the possible pump-probe experiments is depicted in Fig. 2.

The experimental investigation of the I_3^- system is an ongoing story. Dissociation of I_3^- to I_2^- has recently been induced on the lower absorption band of 350 nm showing similar qualitative behavior [5]. To study the influence of initial symmetry, the photodissociation of I_2Br^- to I_2^- + Br [6] has been investigated and shows strong spectral modulations. The analysis presented here can also be applied to other photodissociation processes in solution, such as the dynamics of HgI_2, which has been studied both in the gas phase [7, 8] and in solution [9, 10]. Even at this stage, the degree of detail in the experiments requires an expansion of our theoretical descriptions and visual-

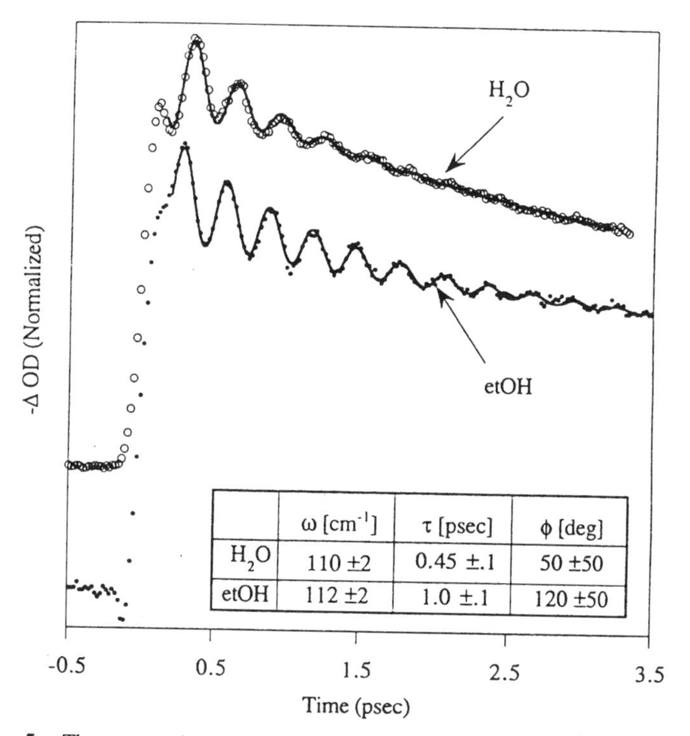

	ω [cm^{-1}]	τ [psec]	φ [deg]
H_2O	110 ±2	0.45 ±.1	50 ±50
etOH	112 ±2	1.0 ±.1	120 ±50

Figure 5. The resonant impulsive stimulated Raman scattering (RISRS) on I_3^- using a UV pump and probe pulse. The spectral modulation corresponds to the symmetric stretch motion of the ground surface.

ization of the underlying chemical physics. This report concentrates on recent developments in this direction.

B. Theoretical Background

The dynamical process describing photodissociation in solution involves many degrees of freedom of the ion-molecule and the surrounding solvent. A full quantum mechanical simulation of the process is therefore prohibitively expensive. For this reason there are two types of approaches to a theoretical investigation of the process:

- Including all relevant degrees of freedom in an approximate fashion.
- Treating the problem exactly within a reduced dimensional model.

To date the theoretical analysis of such systems has proceeded along both these paths. The first approach is usually formulated via the classical mechanics molecular dynamics (MD) setup [11]. If nonadiabatic processes

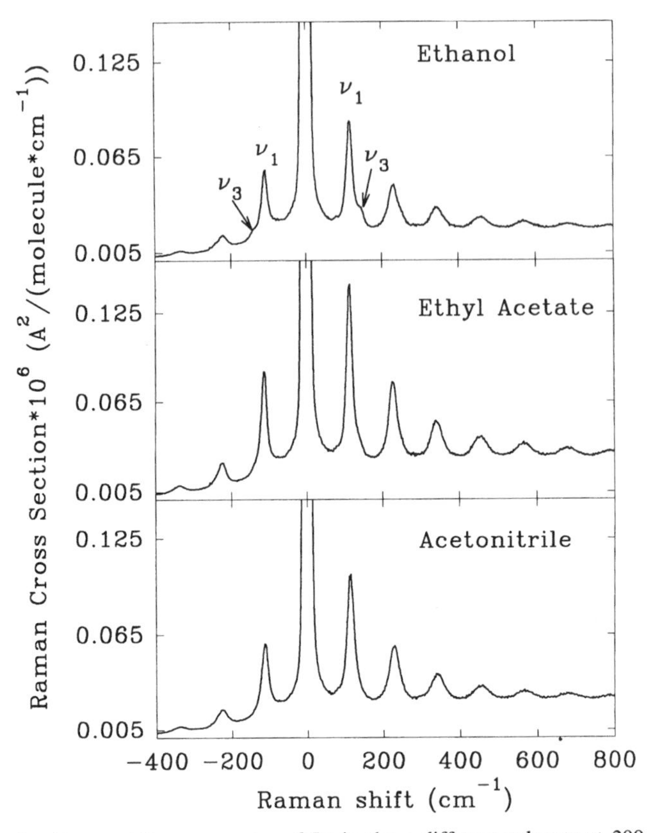

Figure 6. Resonant Raman spectra of I_3^- in three different solvents at 299 nm. ν_1 is the symmetric stretch frequency; ν_3 is the anti-symmetric stretch frequency, visible only in ethanol.

are involved or nuclear quantum phenomena are important, different formulations of quantum-classical simulations have been developed [12, 13, 14, 15, 16, 17].

In this study, the second approach is followed. Even though the participating iodine atoms are heavy and their behavior is expected to be classical, the ultrashort nature of the matter–radiation interaction induces coherent superpositions of quantum states, which should be dealt with in a quantum mechanical framework. An effort was made to include the entire process in this framework. Especially important is the inclusion of the solvent as a quantum species (even if this is only achieved phenomenologically), as it has a profound effect both as a stabilizer and as a destroyer of this quantum coherence.

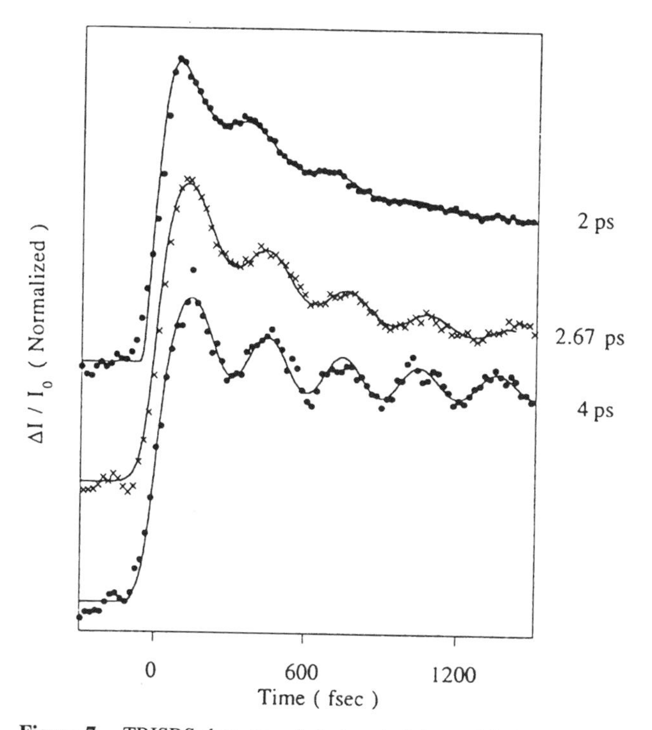

Figure 7. TRISRS data recorded at push delays of 2, 2.7 and 4 ps.

The current work summarizes and expands previous efforts along this direction. The purpose is to supply a complete picture of the theoretical considerations that have been developed in analyzing the I_3^- photodissociation process. The previous theoretical efforts included quantum wave-packet simulations of the photodissociation dynamics [1, 18], analysis of the dynamical "hole" in connection to the RISRS experiments [2, 19], and analysis of the TRISRS experiment utilizing a master equation simulation of the relaxation dynamics [4]. These results are now connected with solvent-induced relaxation dynamics with the purpose of supplying a complete picture.

C. Objective

The main purpose of this work is twofold:

- To develop a complete and general method for a fully quantum mechanical simulation of photodissociation processes in solution.
- To apply this method to a concrete experimental example, to unravel the underlying chemical physics of the processes in the reaction

$$I_3^- \xrightarrow{h\nu} I_2^- + I$$

in solution.

Some additional objectives are:

- Development of a simple uniformly convergent global propagation technique applicable to both Hermitian and non-Hermitian operators. This makes it possible to elegantly calculate all the different types of dynamics involved using a single algorithm.
- Visualization of the quantum state of the system in a way that makes the different components of the dynamics obvious and intuitive.
- Explaining the influence of different dissipation mechanisms on the observed modulations of a nonharmonic oscillator.

D. Outline of the Paper

The guiding principles of the quantum simulation and analysis are the following:

- The system can be represented by a state vector (in the pure case) or a density matrix (when dealing with a statistical ensemble).
- Forces acting on the system are formulated as operators, which induce a change in the state of the system as it evolves in infinitesimal time. These operators generate the dynamics of the system.
- Physical observables are formulated as operators, whose average values (with respect to the state of the system) are the observed measurements.

In Section II the theoretical framework for the quantum mechanical simulation of an impulsive photodissociation process in solution is developed. First, the system should be reduced to a numerically feasible quantum mechanical model, and its initial state must be defined (Section II.A). Second, the operator that generates the dynamics is constructed from the different forces acting within the system. By recursively applying the generator to the initial state, all the dynamical behavior can be reconstructed (Section II.B). Last, the experimentally measurable operators (observables) are constructed and their average value calculated (Section II.C).

In Section III the above framework is applied to an experimental system: the photochemical decomposition of tri-iodide.

$$I_3^- \xrightarrow{h\nu} I_2^- + I$$

The simulation is broken into four parts. In the first, the system is described by a state vector and follows dissociation induced by the pump pulse (Section III.B). In the second and third, the system is described by a density matrix, demonstrating the influence of the solvent on the spectral modulations observed by two probe pulses: one in the UV (Section III.C), the other in the visible (Section III.D). In the last part, the effect of a delayed push pulse is studied (Section III.E).

Finally, in Section IV, the main results are summarized, and potential extensions and limitations are indicated.

II. METHODS

In this section the theoretical framework for a quantum mechanical simulation and analysis of an impulsive photodissociation process in solution is described.

A. Statics

The representation of a complex quantum mechanical system is discussed. The state of the system should be described in a way that can be processed by computers and visualized by humans. Once this is achieved, the initial state of the system can be cast in this way, as the starting point for the next section.

1. *Describing the State of the System*

To describe a quantum mechanical state, the Hilbert space of all possible states should be identified by defining the relevant degrees of freedom. Then a basis for this space is chosen. By expanding the state as a linear combination of the basis elements, it can be written as a vector (for a pure state) or a matrix (for a statistical ensemble). A discrete Fourier transform is used to change representations between conjugated basis sets.

a. Degrees of Freedom. The full description of a photodissociating molecule in solution contains the following degrees of freedom (DOF):

- Radiation field (photons) DOF
- Nuclear and electronic DOF for every solvent atom in interaction with the photodissociating species
- Molecular DOF:
  - Electronic
  - Nuclear—translation, rotations, vibrations.

Unlike classical mechanics, which is a local theory and therefore scales linearly with the number of DOF, quantum mechanics is a global theory and scales exponentially with this number. It becomes prohibitively time-consuming to make a full quantum description of all DOF, and several approximations must be made to reduce the dimensionality of the problem.

The first approximation is to separate the radiation field from the system, and treat it as an external time-dependent variation. The radiation field is taken to be a classical electromagnetic wave, using the rotating wave approximation (RWA). This is a good approximation considering the intensities usually used in photodissociation experiments.

The most drastic reduction in dimensions is gained by separating the DOF of the solvent molecules from the DOF of the photodissociating molecular system, by changing the definition of the system from a closed one (solvent + solute) to an open system (solute only) surrounded by a bath (solvent only). This is done formally by taking a partial trace over the solvent DOF. If no correlations exist between the solvent and solute, the reduced system would remain a pure state, but if such correlations exist, the partial trace will transform the reduced system into a mixed state, and measurements made on the reduced system alone will prove it to be a statistical ensemble. The approximation is not in the trace operation (which is exact, formally), but in the neglect of the correlations with the bath in the dynamics. The bath can be incorporated into the dynamics of the open system by the Liouville von Neumann equation, as a dissipative part formulated within the dynamical semigroup approach [20, 21]. This approximation is good when the "memory" of the bath for correlations with the reduced system is shorter than the timescale of interest. The partition between the system and bath modes is arbitrary. The quality of the approximation improves if more degrees of freedom are included in the primary system. In particular, in solution the "cage" degrees of freedom should be included [22]. In practice, the number of degrees of freedom in the primary system is dictated by the computer resources available.

In the molecular system, the Born–Oppenheimer approximation is used to separate electronic from nuclear DOF, again on the basis of different timescales of the dynamics. The much faster electronic movement creates an effective potential for each configuration of the nuclei, and can be treated as a potential surface in which the nuclei move. Photodissociation experiments involve more than one electronic state simultaneously. In this study a simplifying assumption is used in which only two electronic states are coupled by the pulse at one time, so the electronic DOF can be reduced to the formalism of a two-level system.[1] This approximation is good as long as the electronic potential surfaces do not cross.

[1]Using pseudo-spin notation, $|-\rangle$ and $|+\rangle$ are the lower and upper states of the system, $\hat{\mathbf{P}}_- = |-\rangle\langle-|$ and $\hat{\mathbf{P}}_+ = |+\rangle\langle+|$ are the projection operators on those states, $\hat{\mathbf{S}}_x$, $\hat{\mathbf{S}}_y$, $\hat{\mathbf{S}}_z$ are the angular momentum operators, and $\hat{\mathbf{S}}_-$, $\hat{\mathbf{S}}_+$ are the lowering and raising operators, respectively.

Finally, only the molecular nuclear DOF remain. Here, each system should be studied to obtain the relevant DOF by comparing the timescales of the different motions (for example, rotations are usually much slower than vibrations, and can be neglected from the calculations).

b. Grid and Eigenstate Representations. Two different representations for the nuclear DOF are used throughout this work. The first representation is based on the Fourier method [23]. The state is taken to be a wave function in coordinate space ($\mathbf{r}$), its values sampled on an evenly spaced grid. The spatial extent of the grid is chosen so that the wave function decays exponentially to zero outside its boundaries (finite support). Its density ($\Delta\mathbf{r}$) is chosen to be such that the Fourier transform of the wave function decays exponentially to zero for absolute momentum values $|\mathbf{p}|$ greater than the Nyquist frequency $\mathbf{p}_{max} = h/2\Delta\mathbf{r}$ (band limited). If these conditions are met, the grid representation is equivalent to expansion on a continuous basis of δ-functions in coordinate space $\{|\mathbf{r}\rangle\}$. This basis is a set of eigenstates of the coordinate operator $\hat{\mathbf{R}}$.

The second representation is a discrete basis expansion. The basis is taken as a set of eigenstates of the molecular Hamiltonian operator for the bound ground state, denoted by $|v\rangle$. Since v has the meaning of the energy level of a periodic motion (vibration or rotation), its conjugated variable (via Fourier transform) is the phase of the motion ϕ_v. This pair is known in analytical classical mechanics as *action angle* variables.

A unitary transformation matrix $\mathbf{T}: \mathbf{r} \rightarrow v$ is constructed by finding the grid representation of the eigenstate wavefunctions, using the relaxation method (Eq. 2.42):

$$\mathbf{T}(v, \mathbf{r}) = v(r) = \langle v|\mathbf{r}\rangle \tag{2.1}$$

The transformations $\mathbf{T}$ and $\mathbf{T}^\dagger$ are used to convert states and operators between the two representations.

c. State Vector. The state of the system is represented simultaneously on two electronic surfaces: the initial ground surface and the electromagnetically coupled excited surface. When the system is a pure state $\mathbf{\Psi}$, it can be represented as a pseudo-spin vector

$$\mathbf{\Psi} = |\psi_g\rangle \otimes |-\rangle + |\psi_e\rangle \otimes |+\rangle = \begin{pmatrix} \psi_e \\ \psi_g \end{pmatrix} \tag{2.2}$$

where ψ_g, ψ_e are the nuclear wave functions on the ground and excited surfaces, respectively. These wave functions can be expanded as a linear combination of the aforementioned basis sets, and represented as a vector of complex numbers, which are the coefficients of this combination. If there are n degrees of freedom in the system, this vector will be n-dimensional. When expanded in coordinate representation, the coefficients will be denoted by $\psi(\mathbf{r}) = \langle \mathbf{r}|\psi\rangle$, and in eigenstate space by $\psi(v) = \langle v|\psi\rangle$.

d. Density Matrix. When the system is a mixed state, it is no longer possible to represent it as a single state vector. The most general representation is by a density matrix $\hat{\rho}$

$$\hat{\rho} = \hat{\rho}_g \otimes \hat{\mathbf{P}}_- + \hat{\rho}_e \otimes \hat{\mathbf{P}}_+ + \hat{\rho}_c \otimes \hat{\mathbf{S}}_+ + \hat{\rho}_c^\dagger \otimes \hat{\mathbf{S}}_- = \begin{pmatrix} \hat{\rho}_e & \hat{\rho}_c \\ \hat{\rho}_c^\dagger & \hat{\rho}_g \end{pmatrix} \tag{2.3}$$

where $\hat{\rho}_g$, $\hat{\rho}_e$ are the nuclear populations on the ground and excited surfaces, respectively, and $\hat{\rho}_c$ is the nuclear coherence between these surfaces. These nuclear terms can be expanded as a linear combination of operators of the form $|b\rangle\langle b'|$, where $\{|b\rangle\}$ is a basis set. When expanded in coordinate representation, the coefficients form a matrix $\rho(\mathbf{r}, \mathbf{r}') = \langle \mathbf{r}|\hat{\rho}|\mathbf{r}'\rangle$, and in eigenstate space $\rho(v, v') = \langle v|\hat{\rho}|v'\rangle$. If there are n degrees of freedom in the system, those matrixes will be n^2-dimensional.

e. Conjugated Basis Sets. Given a representation of a state vector $|\psi\rangle$ or a density matrix $\hat{\rho}$ in a basis set, it is easy to transform it to its representation in the conjugated basis and back by using a discrete Fourier transform. For the $(\mathbf{r}, \mathbf{p})$ pair

$$\psi(\mathbf{p}) = \frac{1}{\sqrt{2\pi}} \sum_{\mathbf{r}} \psi(\mathbf{r})\, e^{-\frac{i}{\hbar}\mathbf{p}\cdot\mathbf{r}}\, \Delta\mathbf{r} \tag{2.4}$$

$$\psi(\mathbf{r}) = \frac{1}{\sqrt{2\pi}} \sum_{\mathbf{p}} \psi(\mathbf{p})\, e^{\frac{i}{\hbar}\mathbf{p}\cdot\mathbf{r}}\, \Delta\mathbf{p} \tag{2.5}$$

$$\rho(\mathbf{p}, \mathbf{p}') = \frac{1}{2\pi} \sum_{\mathbf{r},\mathbf{r}'} e^{-\frac{i}{\hbar}\mathbf{p}\cdot\mathbf{r}}\, \rho(\mathbf{r}, \mathbf{r}')\, e^{\frac{i}{\hbar}\mathbf{p}'\cdot\mathbf{r}'}\, \Delta\mathbf{r}\Delta\mathbf{r}' \tag{2.6}$$

$$\rho(\mathbf{r}, \mathbf{r}') = \frac{1}{2\pi} \sum_{\mathbf{p},\mathbf{p}'} e^{\frac{i}{\hbar}\mathbf{p}\cdot\mathbf{r}}\, \rho(\mathbf{p}, \mathbf{p}')\, e^{-\frac{i}{\hbar}\mathbf{p}'\cdot\mathbf{r}'}\, \Delta\mathbf{p}\Delta\mathbf{p}' \tag{2.7}$$

and similarly for the (v, ϕ_v) pair. These transformations can be calculated very efficiently using parallelized fast Fourier transform (FFT) algorithms.

2. *Visualizing the State of the System*

To gain insight on the processes under study, it is important to be able to visualize the quantum mechanical state of the system. Two ways are presented to extract graphical information from the nuclear parts of the system $(\psi_g, \psi_e, \rho_g, \rho_e)$.

A Single Basis Picture. Each basis is a set of eigenstates $\{|b\rangle\}$ of an operator $\hat{\mathbf{O}}$. Once a basis set is selected, and the state $|\psi\rangle$ is expanded in this basis, the probability of finding the result associated with $|b\rangle$ in a measurement of $\hat{\mathbf{O}}$ is $|\psi(b)|.^2$ For a density matrix $\hat{\rho}$, this probability is the diagonal element $\rho(b, b)$. Plotting these probabilities gives a picture of the distribution with respect to a single basis: coordinate distribution for the $\{|\mathbf{r}\rangle\}$ representation, momentum distribution for $\{|\mathbf{p}\rangle\}$, energy distribution for $\{|v\rangle\}$, and phase distribution for $\{|\phi_v\rangle\}$.

Phase Space Picture. A broader picture can be constructed by crossing the information gained from two conjugated representations, revealing the correlations between the conjugated properties. This is done by constructing a phase-space picture of the density matrix,[2] using the Wigner distribution function [24]

$$W(r,p) = \frac{1}{2\pi} \int e^{ipy} \rho\left(r - \frac{1}{2}y, r + \frac{1}{2}y\right) dy \tag{2.8}$$

$$W(v,\phi_v) = \frac{1}{2\pi} \int e^{i\phi_v y} \rho\left(v - \frac{1}{2}y, v + \frac{1}{2}y\right) dy \tag{2.9}$$

The naive interpretation of $W(r,p)[W(v,\phi_v)]$ is the probability of finding the system simultaneously at position r and momentum p [energy v and phase ϕ_v]. Care must be taken when using this interpretation, as there is no meaning for an area smaller than h in phase space, because of Heisenberg's uncertainty principle. Therefore the meaning of probability is associated with integration on an area, not with a specific value (which can be negative!). In particular,

$$\int_r W(r,p)\, dr = \rho(p,p) \tag{2.10}$$

[2]In the case of a state vector $|\psi\rangle$, it can easily be transformed into its associated pure state density matrix $|\psi\rangle\langle\psi|$.

$$\int_p W(r,p)\,dp = \rho(r,r) \tag{2.11}$$

$$\int_v W(v,\phi_v)\,dv = \rho(\phi_v,\phi_v) \tag{2.12}$$

$$\int_{\phi_v} W(v,\phi_v)\,d\phi_v = \rho(v,v) \tag{2.13}$$

i.e., the projections of the Wigner distribution function on a basis give back the single basis picture.

The two phase space pictures $W(r,p)$ and $W(v,\phi_v)$ are complimentary, and reveal different aspects of the dynamics, so both are used side by side in this work. Fig. 8 shows the relation between the two coordinate sets.

3. *Initial States*

Before the application of the laser pulse, the system is in thermal equilibrium with the bath, which is a mixed state $\hat{\rho}$. Because of the large energy gap between electronic surfaces, at room temperature it resides entirely on the ground electronic surface:

$$\hat{\rho}_e = 0 \tag{2.14}$$

$$\hat{\rho}_c = 0 \tag{2.15}$$

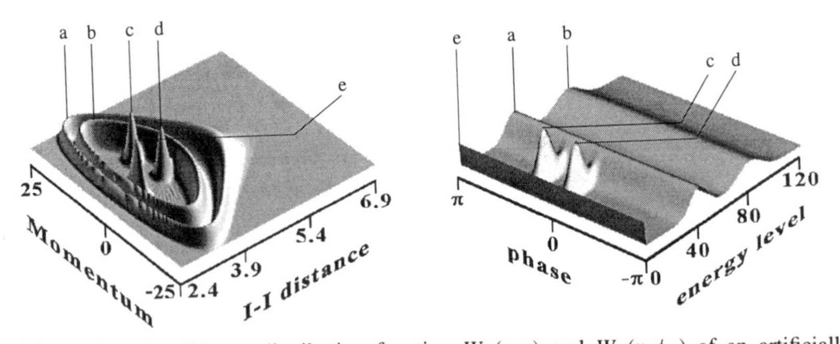

Figure 8. The Wigner distribution function $W_g(r,p)$ and $W_g(v,\phi_v)$ of an artificially mixed state of I_2^- (this distribution is only for illustration, and bears no physical significance). Depicted are two totally dephased states (a, b centered at $v = 80, 40$), two coherent states (c, d centered at $v = 20$ and $\phi_v = \pi/3, 0$) and the ground state (e) which is compact in (r,p) but phase independent in (v,ϕ_v). The ridges of the two dephased states in (r,p) mark the energy iso-lines, which in (v,ϕ_v) are straight lines through the respective ridges at $v = 40, 80$. The angle between the peaks of the two coherent states and the ground state in (r,p) is the distance between their peaks in (v,ϕ_v). Distances are in Å, and the momentum units are Å/ps.

which also means that no correlation exists between the ground and excited electronic surfaces. The statistical mechanical postulate that at equilibrium the phase of all quantum states is random, means that $\rho_g(v, v')$ is diagonal.[3] In an open system, the diagonal elements obey the Boltzmann distribution:

$$\rho_g(v, v) = \frac{e^{-E_v/k_b T}}{\sum_v e^{-E_v/k_b T}} \tag{2.16}$$

where E_v is the energy of the eigenstate $|v\rangle$.

B. Dynamics

The dynamics of a quantum system is governed by one of the following equations: The Schrödinger equation, Eq. (2.17), for a state vector, and the Liouville von Neumann equation, Eq. (2.18), for a density matrix.

$$i\hbar \frac{\partial}{\partial t} |\psi\rangle = \hat{\mathbf{H}} |\psi\rangle \tag{2.17}$$

$$i\hbar \frac{\partial}{\partial t} \hat{\rho} = \hat{\mathcal{L}}(\hat{\rho}) \tag{2.18}$$

where $\hat{\mathbf{H}}$ is called the Hamiltonian operator, and $\hat{\mathcal{L}}$ is the Liouvillian super-operator. In the case of a closed system, the dynamics are purely Hamiltonian, and $\hat{\mathcal{L}}$ is derived directly from $\hat{\mathbf{H}}$

$$\hat{\mathcal{L}} = \hat{\mathcal{L}}_H(\hat{\rho}) = [\hat{\mathbf{H}}, \hat{\rho}] \tag{2.19}$$

In an open system, a dissipative term is added to the Liouvillian which describes the interactions with the bath

$$\hat{\mathcal{L}} = \hat{\mathcal{L}}_H + i\hbar\hat{\mathcal{L}}_D(\hat{\rho}) \tag{2.20}$$

This section deals with the dynamics originating from the above equations. The structure of the Hamiltonian and the dissipative super-operators of the Liouvillian will be presented, followed by a numerical algorithm for propagating a state in time.

[3]This can be seen by calculating the Wigner distribution function of a diagonal matrix $\rho(v, v)$. This gives $W(v_0, \phi_v) = const$, i.e., for a given eigenstate $|v_0\rangle$, all phases are equally probable.

1. *The Hamiltonian Operator*

The evolution of a state can be divided into periods that are influenced by the radiation field, and of field-free evolution. In the field-free periods, there is no coupling between the electronic components of the state, therefore each nuclear component can be propagated independently. In this case, two separate Hamiltonians are used for the propagation of the ground and excited surface components. These are simply a sum of the kinetic and potential energy operators

$$\hat{\mathbf{H}}_{g/e} = \hat{\mathbf{T}} + \hat{\mathbf{V}}_{g/e} = \frac{\hat{\mathbf{P}}^2}{2m} + V_{g/e}(\mathbf{r}) \tag{2.21}$$

where $V_g(\mathbf{r})$ and $V_e(\mathbf{r})$ are the ground and excited state potential surfaces, respectively.

In the presence of the electromagnetic field, the two surfaces are coupled with the interaction of the field with the dipole moment operator

$$\begin{aligned}\hat{\mathbf{H}} &= \hat{\mathbf{H}}_g \otimes \hat{\mathbf{P}}_- + \hat{\mathbf{H}}_e \otimes \hat{\mathbf{P}}_+ + \epsilon(t)\hat{\mu} \otimes \hat{\mathbf{S}}_+ + \epsilon(t)^*\hat{\mu} \otimes \hat{\mathbf{S}}_- \\ &= \begin{pmatrix} \hat{\mathbf{H}}_e & \epsilon(t)\hat{\mu} \\ \epsilon(t)^*\hat{\mu} & \hat{\mathbf{H}}_g \end{pmatrix}\end{aligned} \tag{2.22}$$

where $\hat{\mathbf{H}}_{e/g}$ are the surface Hamiltonians, $\epsilon(t)$ is the time-dependent field, and $\hat{\mu}$ is the electronic transition dipole. Using the rotating-wave approximation, the field can be written as:

$$\epsilon(t) = \bar{\epsilon}(t)e^{-i\omega_L t} \tag{2.23}$$

where ω_L is the carrier frequency of the laser, and $\bar{\epsilon}(t)$ is the envelope of the pulse. To avoid the fast oscillating carrier frequency, a rotating-frame approach is used. A rotation operator is defined to transform the state

$$\hat{\mathbf{R}}(\omega_L t) = e^{i\omega_L t\hat{\mathbf{S}}_z} \tag{2.24}$$

$$|\tilde{\psi}\rangle = \hat{\mathbf{R}}(\omega_L t)|\psi\rangle \tag{2.25}$$

$$\hat{\tilde{\rho}} = \hat{\mathbf{R}}(\omega_L t)\hat{\rho}\hat{\mathbf{R}}^\dagger(\omega_L t) \tag{2.26}$$

In this representation, the equation of motion for a state vector is[4]

$$
\begin{aligned}
i\hbar \frac{\partial}{\partial t} |\tilde{\psi}\rangle &= i\hbar \frac{\partial}{\partial t} (\hat{\mathbf{R}}(\omega_L t)|\psi\rangle) \\
&= i\hbar \frac{\partial \hat{\mathbf{R}}(\omega_L t)}{\partial t} |\psi\rangle + i\hbar\, \hat{\mathbf{R}}(\omega_L t) \frac{\partial}{\partial t} |\psi\rangle \\
&= -\hbar\omega_L \hat{\mathbf{S}}_z \hat{\mathbf{R}}(\omega_L t)|\psi\rangle + \hat{\mathbf{R}}(\omega_L t)\hat{\mathbf{H}}|\psi\rangle \\
&= -\hbar\omega_L \hat{\mathbf{S}}_z \hat{\mathbf{R}}(\omega_L t)|\psi\rangle \\
&\quad + (\hat{\mathbf{H}} - \bar{\epsilon}(t)e^{-i\omega_L t}\hat{\mu}\hat{\mathbf{S}}_+ - \bar{\epsilon}(t)e^{i\omega_L t}\hat{\mu}\hat{\mathbf{S}}_- + 2\bar{\epsilon}(t)\hat{\mu}\hat{\mathbf{S}}_x)\hat{\mathbf{R}}(\omega_L t)|\psi\rangle \\
&= ((\hat{\mathbf{H}}_g + \hbar\omega_L/2) \otimes \hat{\mathbf{P}}_- + (\hat{\mathbf{H}}_e - \hbar\omega_L/2) \otimes \hat{\mathbf{P}}_+ + 2\bar{\epsilon}(t)\hat{\mu} \otimes \hat{\mathbf{S}}_x)|\tilde{\psi}\rangle \\
&= \tilde{\mathbf{H}}|\tilde{\psi}\rangle
\end{aligned}
\tag{2.27}
$$

Thus, working in the rotating frame, we can substitute the Hamiltonian of the system $\hat{\mathbf{H}}$ with an effective Hamiltonian

$$
\tilde{\mathbf{H}} = \begin{pmatrix} \hat{\mathbf{H}}_e - \hbar\omega_L/2 & \bar{\epsilon}(t)\hat{\mu} \\ \bar{\epsilon}(t)\hat{\mu} & \hat{\mathbf{H}}_g + \hbar\omega_L/2 \end{pmatrix} \tag{2.28}
$$

In the effective Hamiltonian the potential surfaces are closer by an amount equal to the photon energy $\hbar\omega_L$, and the time dependence is only in the slowly varying pulse envelope.

Calculating the Hamiltonian in Coordinate Representation. An efficient way for numerically calculating the operation of the Hamiltonian for a state in the coordinate representation is the Fourier method [23]. It is based on two facts: First, the Hamiltonian is composed of operators that are local in either the coordinate or momentum representations as in Eq. (2.21); second, there is an efficient way to transform a state between the two representations as in Eqs. (2.4–2.7). A local operator means that its representation in the appropriate basis reduces to a diagonal matrix. Specifically, applying the kinetic energy operator to a state in the coordinate representation requires:

[4]It is helpful to note that $[\hat{\mathbf{S}}_+, \hat{\mathbf{R}}(\theta)] = (1 - e^{i\theta})\hat{\mathbf{S}}_+\hat{\mathbf{R}}(\theta)$ and $[\hat{\mathbf{S}}_-, \hat{\mathbf{R}}(\theta)] = (1 - e^{-i\theta})\hat{\mathbf{S}}_-\hat{\mathbf{R}}(\theta)$, so that $[\hat{\mathbf{H}}, \hat{\mathbf{R}}(\theta)] = \bar{\epsilon}(t)\hat{\mu}(e^{-i\theta}\hat{\mathbf{S}}_+ + e^{i\theta}\hat{\mathbf{S}}_- - 2\hat{\mathbf{S}}_x)\hat{\mathbf{R}}(\theta)$.

1. Transforming the state into momentum representation as in Eq. (2.4).
2. Multiplying each momentum grid point p by $p^2/2m$.
3. Transforming the state back into coordinate representation as in Eq. (2.5).

Applying the potential energy operator to a state in the coordinate representation is simpler: just multiply each grid point by the value of the potential at that point.

Calculating the Hamiltonian in Eigenstate Representation. In eigenstate representation the Hamiltonian takes the form a matrix. Its operation on a state vector or a density matrix is simply done by matrix–vector or matrix–matrix multiplication (for which optimized parallel algorithms exist). The Hamiltonian matrix is constructed from the respective matrix elements

$$\mathbf{H}(v', v) = \langle v' | \hat{\mathbf{H}} | v \rangle \tag{2.29}$$

where the operation $\hat{\mathbf{H}}|v\rangle$ is calculated in coordinate representation using the Fourier algorithm.

2. *The Dissipative Super-Operators*

The relaxation dynamics under study involves the dynamical behavior of a quantum mechanical molecular system in a solvent. On a purely Hamiltonian level, the overall system can be described by the Hamiltonian

$$\hat{\mathbf{H}} = \hat{\mathbf{H}}_S + \hat{\mathbf{H}}_B + \hat{\mathbf{H}}_{SB} \tag{2.30}$$

Here $\hat{\mathbf{H}}_S$, $\hat{\mathbf{H}}_B$, and $\hat{\mathbf{H}}_{SB}$ stand, respectively, for system, bath and system–bath interaction. To obtain a traceable computation scheme the full dynamics is replaced with an appropriate reduced dynamics within the subspace of the quantum mechanical system. The interaction of the system with the bath can be represented in the form

$$\hat{\mathbf{H}}_{SB} = \sum_i \hat{\mathbf{V}}_i \hat{\mathbf{B}}_i \boldsymbol{\Gamma}_i \tag{2.31}$$

where $\boldsymbol{\Gamma}_i$ is the interaction strength, and $\hat{\mathbf{V}}_i$, $\hat{\mathbf{B}}_i$ are operators of the system and of the bath, respectively. By reducing these interaction terms to the system coordinates alone, the influence of the bath can be described by a dissipative term $\mathcal{L}_D$. Using general arguments based on positivity and causality, the semi-group analysis derives for the dissipative term $\mathcal{L}_D$ in the form [20, 21, 25]

$$\mathcal{L}_D(\hat{\rho}) = \sum_i \gamma_i \left(\hat{\mathbf{V}}_i \hat{\rho} \hat{\mathbf{V}}_i^\dagger - \frac{1}{2} \{\hat{\mathbf{V}}_i \hat{\mathbf{V}}_i^\dagger, \hat{\rho}\}_+ \right) \tag{2.32}$$

where the parameters γ_i are amplitudes describing the relaxation. This is a general formulation, and the parameters γ_i and the operators $\hat{\mathbf{V}}_i$ characterize the dissipation process.

In analyzing dissipative processes in solution, there are three specific, important cases of Eq. (2.32) that should be considered:

1. If $\mathbf{V}$ is unitary, Eq. (2.32) collapses to

$$\hat{\mathcal{L}}_D(\hat{\rho}) = \sum_i \gamma_i \, (\hat{\mathbf{V}}_i \hat{\rho} \hat{\mathbf{V}}_i^\dagger - \hat{\rho}) \tag{2.33}$$

   This form is appropriate for what might be referred to as Poisson processes: i.e., for processes in which the system suffers isolated changes due to sudden interactions with the environment—the isolated binary collision model is an obvious example. Under these conditions, the operator $\hat{\mathbf{V}}_i$ becomes the S matrix of a collision with particles of the environment. The process is characterized by a sum of independent uncorrelated scattering events acting on the molecular subsystem at a rate γ_i.

2. By setting $\hat{\mathbf{V}}_i$ to be Hermitian, Eq. (2.32) becomes

$$\hat{\mathcal{L}}_D(\hat{\rho}) = -\frac{1}{2} \sum_i \gamma_i [\hat{\mathbf{V}}_i, [\hat{\mathbf{V}}_i, \hat{\rho}]] \tag{2.34}$$

   The form is characteristic of a system strongly driven by a Gaussian random process. It can be derived directly for a quantum system coupled to a stochastic process [26], or driven by a δ-correlated random noise [27].

3. The third common case of Eq. (2.32) is of energy pooling. This case describes a relaxation through resonant energy transfer between the system and bath. It is best described by choosing as the operator $\hat{\mathbf{V}}_i$ in Eq. (2.32) as the raising or lowering operators of the systems manifold. Figure 9(a, b) demonstrates the energy relaxation of a harmonic oscillator starting from a coherent state. It should be noticed that the shape of the probability density stays compact throughout the process. The dephasing accompanying the energy relaxation can be interpreted as a geometric effect of increase of solid angle as the density moves toward the origin.

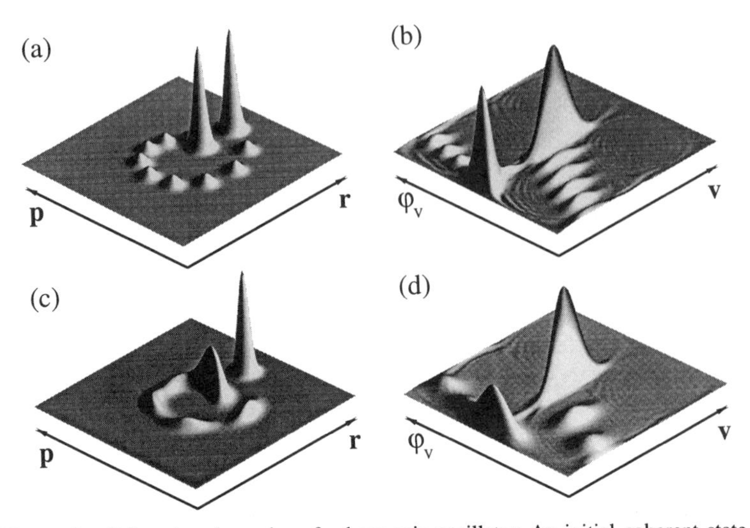

Figure 9. Relaxation dynamics of a harmonic oscillator. An initial coherent state positioned at the outer turning point is propagated for one period. A stroboscopic display is shown, in which the first and last states are enhanced by a factor of 10. The upper panels show an energy relaxation process in Eq. (2.32), where the operator $\hat{\mathbf{V}} = \hat{\mathbf{a}}$ describes relaxation to zero temperature. The relaxation parameter $\gamma = 0.3\omega$, the oscillator frequency. A time interval of $1/10\omega^{-1}$ between snapshots is used. The left panels show the Wigner distribution $W(r, p)$, while the right panels show the Wigner distribution $W(v, \phi_v)$. Notice in panel (*b*) that energy relaxation is accompanied by broadening in the phase distribution due to the approach of the distribution to the zero energy origin. The lower panels shows a combined dephasing, as in Eq. (2.35), and energy relaxation process, $\gamma_d = 0.015\omega^2$. A time interval of $1/5\omega^{-1}$ between shapshots is used.

A pure dephasing process is defined by an energy conserving dissipation process. Based on the Poisson and Gaussian dissipation, a pure dephasing process can be constructed. In a Gaussian process this amounts to choosing $\hat{\mathbf{V}}_i$ to be a function of the Hamiltonian operator $\hat{\mathbf{V}}_i = f(\hat{\mathbf{H}})$. Keeping only the linear term leads to

$$\hat{\mathcal{L}}_D(\hat{\rho}) = -\tfrac{1}{2}\gamma_d[\hat{\mathbf{H}}, [\hat{\mathbf{H}}, \hat{\rho}]] \tag{2.35}$$

Figure 9(*c*, *d*) demonstrates a Gaussian dephasing process of the harmonic oscillator. This dissipative process can be derived from δ-correlated random fluctuations in the oscillator's frequency.

A pure dephasing Poisson process can be constructed from the unitary operator $\hat{\mathbf{V}}_i = \exp(-\frac{i}{\hbar}\hat{\mathbf{H}}\tau)$, leading to

$$\hat{\mathcal{L}}_D(\hat{\rho}) = \gamma_d(e^{-\frac{i}{\hbar}\hat{\mathbf{H}}\tau}\hat{\rho}e^{\frac{i}{\hbar}\hat{\mathbf{H}}\tau} - \hat{\rho}) \tag{2.36}$$

where τ is a characteristic time parameter. The origin of this process can be elastic collisions of the system with solvent particles. Considering τ to be from a Gaussian distribution, $p(\tau) = 1/\sqrt{2\pi\sigma_\tau} \exp(-\tau^2/2\sigma_\tau^2)$, the dephasing generator becomes

$$\hat{\mathcal{L}}_D(\hat{\rho}) = \gamma_d(e^{-\frac{\sigma_\tau^2}{\hbar^2}[\hat{\mathbf{H}}, [\hat{\mathbf{H}}, \ ]]}\hat{\rho} - \hat{\rho}) \tag{2.37}$$

where the exponent is interpreted as a power expansion. The leading term of Eq. (2.37) is equivalent to Eq. (2.35); therefore only Gaussian dephasing processes will be considered in this study.

The three dissipative processes are the building blocks for constructing the dissipative superoperator for the photo-induced dynamics of I_3^-.

A phase space approach to dissipative dynamics in solution and its relation to classical mechanics has been the subject of a recent effort by Kohen and Tannor [28].

An alternative to the present approach is to construct the dissipative superoperator through a second-order perturbation expansion in the system–bath coupling constant. This expansion relates the dissipative constants γ_i to Fourier transforms of bath correlation functions [29, 30, 21, 31, 32]. This approach has been employed by Pugliano et al. [10] to calculate the relaxation coefficients in a master equation calculation for the vibrational relaxation of IHgI. The correlation functions are then calculated by a MD simulation of the solvent. For the I_3^- system, detailed knowledge of the bath correlation functions and the system–bath interactions is lacking. This justifies, at this stage, the phenomenological approach; in the future, the results can be related to detailed information on the solvent.

3. *The Evolution Operator*

From the differential equations in Eq. (2.17) and (2.18) an integral evolution operator and super-operator are constructed

$$|\psi\rangle(t') = \hat{\mathbf{U}}(t, t')|\psi\rangle(t) \tag{2.38}$$

$$\hat{\rho}(t') = \hat{\mathcal{U}}(t, t')\hat{\rho}(t) \tag{2.39}$$

In the case of a time-independent Hamiltonian, the integration is simple

$$\hat{\mathbf{U}}(t, t + \Delta t) = e^{-\frac{i}{\hbar}\hat{\mathbf{H}}\Delta t} \tag{2.40}$$

$$\hat{\mathcal{U}}(t, t + \Delta t) = e^{-\frac{i}{\hbar}\hat{\mathcal{L}}\Delta t} \tag{2.41}$$

and is exact for any Δt. A small variation on this equation is used to find the eigenstates of the ground state Hamiltonian $\hat{\mathbf{H}}_g|\phi_n\rangle = E_n|\phi_n\rangle$. Since $\{|\phi_n\rangle\}$ is a basis set, an arbitrary state can be written as $|\psi\rangle(0) = \Sigma_n a_n|\phi_n\rangle$. Using an imaginary time step in Eq. (2.41) results in

$$|\psi\rangle(t) = \hat{\mathbf{U}}(0, -i\Delta t)|\psi\rangle(0) = e^{-\frac{1}{\hbar}\hat{\mathbf{H}}\Delta t} = \sum_n a_n|\phi_n\rangle$$

$$= \sum_n a_n e^{-\frac{1}{\hbar}E_n\Delta t}|\phi_n\rangle \tag{2.42}$$

The decay of contributions from eigenstates with higher energy is exponentially faster than for ones with a lower energy, so Eq. (2.42) will quickly converge to

$$|\psi\rangle(t) \propto |\phi_0\rangle \tag{2.43}$$

which is the lowest eigenstate. Taking $|\psi'\rangle = |\psi\rangle - \langle\phi_0|\psi\rangle|\phi_0\rangle$ as the initial state will result in the next eigenstate $|\phi_1\rangle$, and so on.

When the evolution is induced by the radiation field, the Hamiltonian becomes time-dependent and Eq. (2.41) has to be amended by time ordering. The first-order Magnus expansion was employed [33] to approximate the evolution operator

$$\hat{\mathbf{U}}(t, t + \Delta t) = \exp\left[-\frac{i}{\hbar}\left(\int_t^{t+\Delta t} \frac{\hat{\mathbf{H}}(t')\,dt'}{\Delta t}\right) \cdot \Delta t\right] \tag{2.44}$$

$$\hat{\mathcal{U}}(t, t + \Delta t) = \exp\left[-\frac{i}{\hbar}\left(\int_t^{t+\Delta t} \frac{\hat{\mathcal{L}}_H(t')\,dt'}{\Delta t} + \hat{\mathcal{L}}_D\right)\Delta t\right] \tag{2.45}$$

This amounts to averaging the Hamiltonian over each time step, so Δt should be small in comparison with the rate of change in the pulse amplitude.

All the equations for the evolution operator and super-operator have the same functional form

$$f(\hat{\mathbf{O}}) = e^{-\frac{i}{\hbar}\hat{\mathbf{O}}\Delta t} \tag{2.46}$$

The basic idea underlying the propagation algorithm is that approximating a function of an operator is equivalent to approximating a scalar function in the domain of eigenvalues of this operator [34]. The scalar function is approximated by the Newton interpolation polynomial:

$$f(z) \approx \mathcal{P}_{N-1}(z) \equiv \sum_{n=0}^{N-1} a_n \prod_{j=0}^{n-1} (z - x_j) \tag{2.47}$$

where the x_j's are interpolation points in which the value of $f(x_j)$ is sampled, and the a_n's are the expansion coefficients calculated to give $f(x_j) \equiv \mathcal{P}(x_j)$.

If the approximation is good for all the scalars that are equal to eigenvalues of an operator, then it is a good approximation for the function of this operator. For this procedure to work, a uniformly convergent method has to be applied to sum the power series into the different final results. The Chebychev polynomial expansion [35] has uniform convergence properties and therefore has been utilized in both time-dependent and time-independent calculations, but its interpolation domain lies on the real axis, so it has a very limited support for non-Hermitian operators. Since the dissipative terms in the Liouville von Neumann equation are anti-Hermitian, a different method had to be developed, in which the interpolation points reside in the complex plane. This method is the Newton interpolation with Leja interpolation points [18], which is detailed in the Appendix.

Once the interpolation points are chosen and the coefficients calculated, the function of the operator can be calculated by recursively applying the operator to an initial state, and summing the resulting polynomial. For example, if $\hat{\mathcal{L}}$ is the time-independent Liouvillian of the system, and

$$f(z) = e^{-\frac{i}{\hbar} z \Delta t} \approx \mathcal{P}_{N-1}(z) \tag{2.48}$$

for all z residing in the complex eigenvalue range of $\hat{\mathcal{L}}$, then

$$\hat{\rho}(t + \Delta t) = \hat{\mathcal{U}}(t, t + \Delta t)\hat{\rho}(t) \approx \mathcal{P}(\hat{\mathcal{L}})\hat{\rho}(t) \tag{2.49}$$

C. Interpretation

The primary interrogation tool of the photodissociation dynamics is light–matter interaction. A combination of CW and pump-probe spectroscopic experiments are then to be used. Theoretically, physical measurements applied to the system can be calculated through the use of observables, which are operators associated with the measurement process. The ability to separate the dynamics from the observation is based on the use of weak electromagnetic interactions, for which the perturbation of the system is limited. This section describes the construction of the relevant observables for a pump-probe photodissociation experiment.

1. *The Impulsive Excitation Picture*

When the pulse is short on the timescale of nuclear motion, and long compared to the timescale defined by the electronic excitation, the construction of observables associated with this pulse can be simplified using the impulsive two-level coordinate-dependent approximation.

a. The Coordinate-Dependent Two-Level Approximation. The starting point is a rearrangement of the effective Hamiltonian [Eq. (2.28)] defining the difference potential

$$2\hat{\mathbf{\Delta}}(\mathbf{r}) = \hat{\mathbf{V}}_e(\mathbf{r}) - \hat{\mathbf{V}}_g(\mathbf{r}) - \hbar\omega_L \qquad (2.50)$$

This leads to the modified Hamiltonian

$$\tilde{\mathbf{H}} = \begin{pmatrix} \hat{\mathbf{\Delta}} & \hat{\mathbf{W}} \\ \hat{\mathbf{W}} & -\hat{\mathbf{\Delta}} \end{pmatrix} + \begin{pmatrix} \hat{\mathbf{\Delta}} & 0 \\ 0 & \hat{\mathbf{\Delta}} \end{pmatrix} + \begin{pmatrix} \hat{\mathbf{H}}_g & 0 \\ 0 & \hat{\mathbf{H}}_g \end{pmatrix} \qquad (2.51)$$

where the time-dependent transition operator $\hat{\mathbf{W}} = -\bar{\epsilon}\hat{\mu}$ is also defined. This partition of the effective Hamiltonian is the basis for the definition

$$\tilde{\mathbf{H}} = \hat{\mathbf{H}}_3 + \hat{\mathbf{H}}_2 + \hat{\mathbf{H}}_1 \qquad (2.52)$$

using spin notation

$$\hat{\mathbf{H}}_1 = \hat{\mathbf{H}}_g \otimes \hat{\mathbf{I}} \qquad \hat{\mathbf{H}}_2 = \hat{\Delta}(\mathbf{r}) \otimes \hat{\mathbf{I}} \qquad \hat{\mathbf{H}}_3 = 2\hat{\mathbf{W}}(\mathbf{r},\mathbf{t}) \otimes \tilde{\mathbf{S}}_x + 2\hat{\Delta}(\mathbf{r}) \otimes \hat{\mathbf{S}}_z \qquad (2.52a)$$

The original Hamiltonian has been partitioned into a direct product of spatially dependent operators and spin operators.

The partitioning of the Hamiltonian is the basis for approximating the evolution operator that propagates the wave function for a pulse duration t_f, as a product of terms

$$\hat{\mathbf{U}}(t_f) = e^{-\frac{i}{\hbar}\tilde{\mathbf{H}}t_f} \approx e^{-\frac{i}{\hbar}\hat{\mathbf{H}}_3 t_f} e^{-\frac{i}{\hbar}\hat{\mathbf{H}}_2 t_f} e^{-\frac{i}{\hbar}\hat{\mathbf{H}}_1 t_f} \qquad (2.53)$$

The approximation is based on the Trotter formula [36]. Its use is well known in the split operator propagation method for the time-dependent Schrödinger equation introduced by Fiet and Fleck [37, 38]. The approximate evolution operator in Eq. (2.53) is used to propagate an initial wave function that is exclusively on the ground surface, for a pulse duration t_f

$$\Psi(\mathbf{r}, 0) = \begin{pmatrix} 0 \\ \psi_g(\mathbf{r}, 0) \end{pmatrix} \tag{2.54}$$

A simple case is of an initially stationary wave function at $t = 0$, i.e., $\hat{\mathbf{H}}_\mathbf{g}\psi_g(\mathbf{r}, 0) = E_g\psi_g(\mathbf{r}, 0)$. The operation of the propagator $e^{-\hat{\mathbf{H}}_1 t}$ on $\Psi(0)$ will lead to a global phase shift

$$\Psi_1 = e^{-\frac{i}{\hbar}\hat{\mathbf{H}}_1 t_f}\Psi(0) = e^{-\frac{i}{\hbar}E_g t_f}\Psi(0) \tag{2.55}$$

The next step is to propagate with $e^{-\frac{i}{\hbar}\hat{\mathbf{H}}_2 t_f}$

$$\Psi_2 = e^{-\frac{i}{\hbar}\hat{\mathbf{H}}_2 t_f}\Psi_1 = e^{-\frac{i}{\hbar}E_g t_f}\begin{pmatrix} 0 \\ e^{-\frac{i}{\hbar}\hat{\Delta} t_f}\psi_g(\mathbf{r}, 0) \end{pmatrix} \tag{2.56}$$

The operator $\exp(-\frac{i}{\hbar}\hat{\boldsymbol{\Delta}}(\mathbf{r})t_f)$ causes a momentum change in ψ_g, since $\hat{\boldsymbol{\Delta}}(\mathbf{r})$ is a function of $\mathbf{r}$.

The final step is to operate with $\exp(-\frac{i}{\hbar}\hat{\mathbf{H}}_3 t_f)$ on Ψ_2. Since $\hat{\mathbf{H}}_3$ is not diagonal in the g/e representation, $\hat{\mathbf{H}}_3$ is diagonalized for each position $\mathbf{r}$. At this point it is assumed that the field envelope $\bar{\epsilon}$ is constant. This amounts to a square pulse shape. For a nonsquare pulse, $\hat{\mathbf{W}}(r, t)$ is approximated as [39]:

$$\hat{\mathbf{W}} \approx \hat{\mu}\,\frac{1}{t_f}\int_0^{t_f} \bar{\epsilon}(t')\,dt' \tag{2.56a}$$

Closed form expressions can be obtained also for $\epsilon(t) = \bar{\epsilon}\,\mathrm{sech}(t)/\tau$ [40]. The eigenvalue equation of the Hamiltonian $\mathbf{H}_3$ becomes $(\lambda + \Delta)(\lambda - \Delta) - W^2 = 0$, leading to

$$\lambda_{1,2} = \pm\sqrt{W^2 + \Delta^2} \tag{2.57}$$

The diagonalization matrixes can be described as a rotation matrix for each position $\mathbf{r}$ [41]

$$\hat{\mathbf{S}}(\mathbf{r}) = e^{-i\theta(\mathbf{r})\tilde{\mathbf{S}}_\mathbf{y}} = \begin{pmatrix} \cos\theta/2 & -\sin\theta/2 \\ \sin\theta/2 & \cos\theta/2 \end{pmatrix} \tag{2.58}$$

where: $\tan\theta(\mathbf{r}) = \frac{|\hat{\mathbf{W}}|}{\hat{\Delta}}$. With this form the propagator for each spatial point $\mathbf{r}$

becomes

$$e^{-\frac{i}{\hbar}\hat{\mathbf{H}}_3 t_f} = \mathbf{S}^{-1}(\mathbf{r}) \begin{pmatrix} e^{-\frac{i}{\hbar}\lambda_1 t_f} & 0 \\ 0 & e^{-\frac{i}{\hbar}\lambda_2 t_f} \end{pmatrix} \mathbf{S}(\mathbf{r}) \tag{2.59}$$

For convenience the spatial dependent Rabi frequency is defined as

$$\Omega(\mathbf{r}) = \frac{1}{\hbar}\sqrt{W^2 + \Delta(\mathbf{r})^2} \tag{2.60}$$

and also the angles

$$\sin\,\theta = \frac{|W|}{\sqrt{W^2 + \Delta^2}}, \qquad \cos\,\theta = \frac{\Delta}{\sqrt{W^2 + \Delta^2}} \tag{2.61}$$

Combining the successive contributions of the propagator in Eq. (2.53) onto the initial wave function leads to the wave function at $t = t_f$ after the pulse. The ground surface wave function becomes

$$\psi_g(\mathbf{r}, t_f) = e^{-\frac{i}{\hbar}E_g t_f} e^{-\frac{i}{\hbar}\Delta t_f}[\cos(\Omega t_f) + i\cos\,\theta\,\sin(\Omega t_f)]\psi_g(\mathbf{r}, 0) \tag{2.62}$$

and the excited surface wave function becomes

$$\psi_e(\mathbf{r}, t_f) = e^{-i\phi} e^{-\frac{i}{\hbar}\Delta t_f}[\sin\,\theta\,\sin(\Omega t_f)]\psi_g(\mathbf{r}, 0) \tag{2.63}$$

where

$$\phi = \left(\omega + \frac{E_g}{\hbar}\right) t_f + \frac{\pi}{2} \tag{2.63a}$$

Equations (2.63) and (2.62) constitute extremely useful interpretive tools for analyzing impulsive pump-probe experiments.

b. The Ground Surface Dynamical "Hole" and the Excited Surface Density. We now consider the amplitude of the wave function in coordinate space. The amplitude transferred to the excited surface, $A^2(\mathbf{r}) = \sin^2\,\theta\,\sin^2(\hat{\boldsymbol{\Omega}} t_f)$, is missing in the ground surface creating a dynamical "hole." Once created, the "hole," which is not stationary, evolves, due to ground surface dynamics. The RISRS experiment monitors by a probe pulse this photo-induced dynamics.

In momentum space the picture is more involved. Due to cycling of amplitude between the ground and excited surfaces the wave functions gain a coordinate dependent phase that induces a momentum kick

$$\Phi_g = -\frac{1}{\hbar} E_g t_f - \frac{1}{\hbar} \hat{\mathbf{\Delta}} t_f + \arctan[\cos\ \theta(\mathbf{r}) \tan(\Omega(\mathbf{r}) t_f)] \tag{2.64}$$

and

$$\Phi_e = -\phi t_f - \frac{1}{\hbar} \hat{\mathbf{\Delta}} t_f \tag{2.65}$$

On the excited surface, the wave function gains momentum linearly in time due to the coordinate-dependent phase $\Phi_e(\mathbf{r})$. Since for most photodissociation situations $\partial\hat{\mathbf{\Delta}}/\partial r$ is negative, the momentum kick $\delta\hat{\mathbf{P}} \approx \hbar\,\partial\Phi/\partial r$ is positive.

A more involved situation exists on the ground surface. For short times the change in momentum is zero, since the positive momentum change of the term that is linear in $\hat{\mathbf{\Delta}}$ in Eq. (2.64) is exactly compensated by the next term. This is in accordance with the Frank–Condon picture, in which the momentum of the ground surface is not affected by transfer of population to the excited surface. Far from resonance, the coordinate-dependent Rabi frequency Ω becomes very large, leading to rapid cycling. This rapidly changes the direction of the momentum shift from negative to positive, leading to the average asymptotic momentum shift of

$$\delta P = -\frac{W^2 t_f}{2\Delta^2} \frac{\partial\Delta}{\partial r} \tag{2.66}$$

This expression has also been obtained by Cina and Smith [42], using the classical Frank approximation. Detuned from resonance, this momentum shift provides the excitation interaction for impulsive stimulated Raman scattering (ISRS) [43]. On resonance ($\mathbf{r} = \mathbf{r}_h$), the cycling is slower and can lead first to a significant negative momentum shift. A semiclassical estimation of the momentum kick becomes

$$\delta\mathbf{P} \approx \hbar \left.\frac{\partial\Phi}{\partial r}\right|_{r_h} = \left[\frac{\tan(W t_f)}{W} - t_f\right] \left.\frac{\partial\Delta(r)}{\partial r}\right|_{r_h} \tag{2.67}$$

This negative momentum kick causes the "hole" to move in the positive direction of bond extension. For more intense pulses where $\Omega(\mathbf{r}_h) \cdot t_f$ exceeds π, the momentum kick changes sign and becomes positive. An interesting effect is obtained for a 2π pulse, where the "hole" in coordinate space fills up, and a significant positive momentum change is acquired by the ground surface wave function.

A more rigorous definition of the dynamical "hole" is obtained by an orthogonal decomposition of the ground surface density $\hat{\rho}(t_f)$ into a static part and a dynamical part

$$\hat{\rho}(t_f) = \hat{\rho}_d + c^2 \hat{\rho}_s \tag{2.68}$$

where $\hat{\rho}_s = Z^{-1} \exp(-\beta \hat{\mathbf{H}}_g)$ is the equilibrium stationary density operator, and the scalar product between operators is defined as $(\hat{\mathbf{A}} \cdot \hat{\mathbf{B}}) = tr\{\hat{\mathbf{A}}^{\dagger} \hat{\mathbf{B}}\}$. This definition casts the full dynamics of the density operator, including the dissipative relaxation, and leads to full thermal equilibrium. For the RISRS experiment this is usually the initial density operator $\hat{\rho}_0$. The decomposition can be done by requiring that the dynamical part $\hat{\rho}_d$ be orthogonal to the stationary density $\hat{\rho}_s$. This leads to the overlap functional

$$c^2 = tr\{\hat{\rho}(t_f) \cdot \hat{\rho}_s\} / tr\{\hat{\rho}_s^2\} \tag{2.69}$$

Since the decomposition of Eq. (2.68) is linear, all the dynamical observations depend only on $\hat{\rho}_d$ the "hole" density. It is important to notice that the dynamical "hole" is not a pure state, even when pure wave function dynamics can be applied [2].

A quantitative measure of the created coherence is an important tool in the analysis. The integrated square density of the dynamical "hole" creates a natural definition of this measure

$$C^2 \equiv tr\{\hat{\rho}_d^2\} \tag{2.70}$$

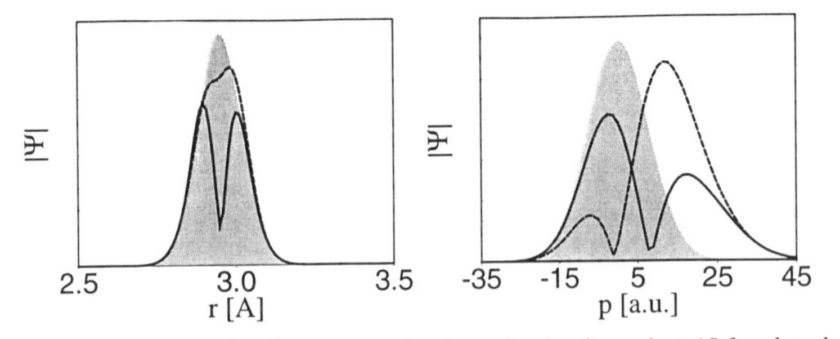

Figure 10. The ground surface wave packet immediately after a short 15-fs pulse, showing the creation of the "hole." The left panel shows the absolute value of the wave function in position space for π (solid) and 2π (dashed) pulses. The right panel shows the same wave function in momentum space. The gray background shows the wave function prior to the application of the pulse.

The coherence measure can be interpreted as the self-dynamical expectation of the dynamical "hole" $\hat{\rho}_d$.

2. *Power Absorption of a Laser Pulse*

The main observable in a pump-probe experiment is the power absorbed from the pulse. With the time-dependent Hamiltonian, Eq. (2.22), the power, or change in energy becomes [44]

$$\mathcal{P} = \frac{dE}{dt} = -2\text{Re}\left(\langle\psi_e|\hat{\mu}|\psi_g\rangle\frac{d\epsilon}{dt}\right) = -2\text{Re}\left(\langle\hat{\mu}\otimes\hat{\mathbf{S}}_+\rangle\frac{d\epsilon}{dt}\right) \tag{2.71}$$

where the last line is appropriate for a density operator description.

Under the conditions of the RWA, the power is linearly proportional to the change in population [44, 45]

$$\mathcal{P} = -\hbar\omega_0\frac{dN_g}{dt} \tag{2.72}$$

The integrated power per pulse, which is the total energy absorbed, can be related to the change in population

$$\Delta E = \int_0^t \mathcal{P}\, dt' = -\hbar\omega\Delta N_g \tag{2.73}$$

a. Transient Absorption of a Weak Pulse. The power absorbed by a probe pulse can be calculated using the linear relation between the integrated power and the change in population Eq. (2.73), using the impulsive two-level coordinate dependent approximation (Section II.C.1). In most cases a weak probe

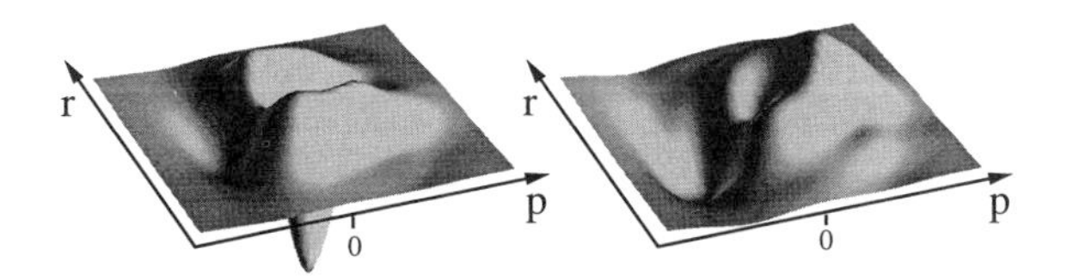

Figure 11. The dynamical "hole" $\hat{\rho}_d$ in I_3^--shown as a Wigner distribution in phase space. The left panel shows the dynamical "hole" created by a π pulse. The point of resonance is in middle of the absorption band. Qualitatively the shape of the "hole" does not change up to intensities of π. Notice that the hole is centered at $p = 0$. The right panel shows the dynamical "hole" for 2π pulse. Notice the rotation in phase space in the "hole's" shape from position to momentum.

pulse is used ($W \ll \Delta(r)$), so the population transfer at a position $\mathbf{r}$ after a time of τ_{pr} becomes

$$P(\mathbf{r}) = [\sin\,\theta \cdot \sin(\Omega\,(\mathbf{r})\tau_{\mathrm{pr}})]^2 \approx \left(\frac{W\sin(\Delta(\mathbf{r})\tau_{\mathrm{pr}}/\hbar)}{\Delta(\mathbf{r})}\right)^2 \tag{2.74}$$

This expression can be rearranged and written as a sinc window function $\hat{\mathbf{W}}(r,r) = P(\mathbf{r})$, which is a diagonal operator in coordinate space:

$$\hat{\mathbf{W}}(r,r) = (\hbar W\tau_{\mathrm{pr}})^2\left(\frac{\sin(\Delta(r)\tau_{\mathrm{pr}}/\hbar)}{\Delta(r)\tau_{\mathrm{pr}}/\hbar}\right)^2 \tag{2.75}$$

To obtain the absorption signal at time t, the local population at each r should be multiplied by the transfer probability and integrated over all r, to give the total population transfer $\Delta N_g(t)$. This amounts to using the window function as the observable which measures the total power absorption of the pulse:

$$\begin{aligned}\Delta E(t) &= -\hbar\omega_{\mathrm{pr}}\Delta N_g(t) = -\hbar\omega_{\mathrm{pr}}\langle\hat{\mathbf{W}}\rangle(t)\\ &= -\hbar\omega_{\mathrm{pr}}\langle\psi(t)|\hat{\mathbf{W}}|\psi(t)\rangle = -\hbar\omega_{\mathrm{pr}}tr(\rho\cdot\hat{\mathbf{W}})\end{aligned} \tag{2.76}$$

For a short probe pulse within the impulsive limit, the observation window is influenced by electronic dephasing. A modified window operator can be calculated by simulating the absorption process starting from $\hat{\rho}_g = \hat{\mathbf{I}}$ using the full dynamics, including dissipation. Projecting out the excited surface density operator $\hat{\rho}_e$ and propagating backward in time to the instant of the peak of the pulse defines the modified window operator $\hat{\mathbf{W}}$.

b. Other Transient Spectroscopies. A different type of probe is obtained, if, after the pulse propagates through the medium, it is dispersed and its spectrum is compared to the spectrum of a reference pulse [46, 47]. In general, this problem is quite difficult, since it requires the solution of both Maxwell's and Schrödinger's equations simultaneously. The difficulty is lessened if the absorption process takes place in a thin sample and the pulse is only slightly depleted. In addition, due to fast electronic dephasing in solution, it is assumed that the dipole created by the pump pulse has decayed by the time the probe pulse has arrived. In order to relate the observation to a molecular property, the absorbed power represented by Eq. (2.73) is integrated. The total amount of energy absorbed from the field becomes

$$\Delta E_f = \int_{-\infty}^{\infty}\mathcal{P}\,dt = 2\mathrm{Re}\left(\int_{-\infty}^{\infty}\langle\hat{\mu}\otimes\hat{\mathbf{S}}_+\rangle\cdot\frac{\partial\epsilon}{\partial t}\,dt\right) \tag{2.77}$$

Defining the Fourier transforms of the instantaneous dipole expectation

$$\langle \hat{\mu} \otimes \hat{\mathbf{S}}_+ \rangle(\omega) = \int_{-\infty}^{\infty} \langle \hat{\mu} \otimes \hat{\mathbf{S}}_+ \rangle(t) e^{i\omega t} \, dt \tag{2.78}$$

and using the Fourier transform of the field $\epsilon(\omega)$, the energy absorption can be written as [48]

$$\Delta E_f = -2\mathrm{Re}\left(\int_{-\infty}^{\infty} i\omega \langle \hat{\mu} \otimes \hat{\mathbf{S}}_+ \rangle(\omega) \cdot \epsilon^*(\omega) \, d\omega \right) \tag{2.79}$$

This suggests the decomposition of energy to frequency components

$$\Delta E_f = \int_{-\infty}^{\infty} \Delta E_f(\omega) \, d\omega \tag{2.79a}$$

Normalizing each frequency component to the energy density of the pulse leads to the expression

$$\sigma_a(\omega) = \frac{\Delta E_f(\omega)}{|\epsilon(\omega)|^2} = -2\omega \mathrm{Im}\left[\frac{\langle \hat{\mu} \otimes \hat{\mathbf{S}}_+ \rangle(\omega)}{\epsilon(\omega)} \right] \tag{2.80}$$

This expression resembles the one presented by Pollard and Mathies [46], and Yan and Mukamel [49], but its derivation is not based on a perturbation expansion, and therefore it is correct for strong fields. The main ingredient of the calculation, the instantaneous dipole $\langle \hat{\mu} \otimes \hat{\mathbf{S}}_+ \rangle(t)$, is extracted directly from the numerical integration of the Liouville von Neumann equation.

Another option for probing the transient motion is to collect the dispersed emission after the probe has propagated through the medium. The emission is a consequence of the induced dipole or polarization created by the probe pulse, which continues to radiate even after the pulse is over. The emission is proportional to the acceleration of the dipole observable $\partial^2 \langle \hat{\mu} \otimes \hat{\mathbf{S}}_x \rangle / \partial t^2$ [48]. This leads to the expression for the dispersed emission

$$\sigma_e(\omega) \propto |\omega^2 (\langle \hat{\mu} \otimes \hat{\mathbf{S}}_x \rangle(\omega))|^2 \tag{2.81}$$

Since interest is in the transient features, the static dispersed emission is subtracted from the transient one.

The same molecular quantity, the instantaneous dipole, governs all light–matter interactions. This is also true if the instantaneous dipole maintains coherence up to the time of the probe pulse [50, 51, 52, 45, 53]. The analysis of this quantity makes it possible to interpret coherent pump-probe spectroscopies. The key point is that any full numerical simulation of a pump-probe sequence automatically imposes a phase-locked pulse sequence. As a result, the instantaneous dipole induced by the first pulse can interfere with the second pulse. This is the basis of the heterodyne experiment [54, 55]. In a recent paper, Domke [56] has shown how to derive the observables of four-wave mixing using a nonperturbative approach based on the calculation of the instantaneous dipole. Due to fast electronic dephasing of the molecular I_3^- in solution, this type of spectroscopy has not been applied.

3. *Absorption Spectrum*

The initial conditions for the absorption spectrum are that all amplitude is in an eigenstate of the ground electronic surface. The main assumption is that the field is weak and that only a very small fraction is transferred to the excited surface. Under these conditions the absorption spectrum can be derived directly from the transient power expression Eq. (2.71) [57]. With these assumptions, the evolution of the wave function on the ground surface is not altered by the field

$$|\psi_g(t)\rangle = e^{-\frac{i}{\hbar}E_g t}|\psi_g(0)\rangle \tag{2.82}$$

where E_g is the ground state energy. The excited surface wave function is obtained using the time-dependent first-order perturbation theory

$$|\psi_e(t)\rangle = \frac{i}{\hbar}\int_0 d\tau e^{-\frac{i}{\hbar}\hat{\mathbf{H}}_e(t-\tau)}\hat{\mu}\epsilon^*(\tau)e^{-\frac{i}{\hbar}E_g\tau}|\psi_g(0)\rangle \tag{2.83}$$

Using the RWA-CW field $\epsilon(t) = E_0 e^{-i\omega_L t}$, Eqs. (2.82) and (2.83) are inserted into the power expression Eq. (2.71)

$$\begin{aligned}\mathcal{P} &= -2\mathrm{Re}\left(\langle\psi_g(t)|\hat{\mu}|\psi_e(t)\rangle\frac{d\epsilon}{dt}\right)\\ &= -2\omega_L\mathrm{Re}\left(\frac{E_0^2}{\hbar}\int_0^t d\tau\langle\psi_g|\hat{\mu}e^{-\frac{i}{\hbar}(\hat{\mathbf{H}}_e - E_g)(t-\tau)}\hat{\mu}|\psi_g\rangle e^{-i\omega_L(t-\tau)}\right)\end{aligned} \tag{2.84}$$

By changing the integation variable from $t-\tau$ to t, extending the integration to $t \rightarrow \infty$, and using the time reversal symmetry the well-known formula obtained by Heller [58] from the time-independent Golden-Rule expression, Eq. (2.85) is obtained the expression

$$\sigma_A(\omega_L) \propto \int_{-\infty}^{\infty} \langle \theta(0)|\theta(t)\rangle e^{i(\omega_L+\omega_i)t}\, dt \tag{2.85}$$

where $|\theta(t)\rangle = \hat{\mu}|\psi_g(t)\rangle$ is the wave function propagated on the excited surface, ω_L is the laser frequency, and $\hbar\omega_i$ is the energy of $|\psi_g\rangle$. The absorption cross section is proportional to the Fourier transform of the autocorrelation function. The steeper the excited surface is, the faster this function decays to zero, and the wider will be its Fourier transform. This correlates the slope of the excited surface at the Frank–Condon region with the bandwidth of the electronic spectrum.

Equation 2.85 can also be written as

$$\sigma_A(\omega_L) \propto \langle \psi_i|\hat{\mathbf{A}}|\psi_i\rangle \tag{2.86}$$

$$\hat{\mathbf{A}} = \hat{\mu}\left\{\int_{-\infty}^{\infty} e^{\frac{i}{\hbar}[\hbar(\omega_L+\omega_i)-\hat{\mathbf{H}}_e]t}\, dt\right\}\hat{\mu} \tag{2.87}$$

The observable $\hat{\mathbf{A}}$ is a function of the Hamiltonian of the excited surface, and is interpolated using the Newtonian interpolation method described in Appendix A.

4. *Raman Spectrum*

In a similar way, the Raman cross section for an induced transition from an initial state $|\psi_i\rangle$ to a final state $|\psi_f\rangle$ is

$$\sigma_{if}(\omega_L) \propto |\langle R_i(\omega_L)|\psi_f\rangle|^2 \tag{2.88}$$

where ω_L is the laser frequency, and $|R_i(\omega_L)\rangle$ is called the Raman wave function.

$$|R_i(\omega_L)\rangle = \left[\int_0^{\infty} e^{\frac{i}{\hbar}(\hbar(\omega_L+\omega_i)-\hat{\mathbf{H}}_e)t} e^{-\Gamma(t)}\, dt\right]|\psi_i\rangle \tag{2.89}$$

where $\Gamma(t)$ is a phenomenological lifetime. If the energies of $|\psi_i\rangle$ and $|\psi_f\rangle$ are

$\hbar\omega_i$ and $\hbar\omega_f$, respectively, the frequency of the scattered light resulting from this transition is $\omega_s = \omega_L + \omega_i - \omega_f$. The operator that creates the Raman wave function from $|\psi_i\rangle$, is a function of the Hamiltonian of the excited surface, and is interpolated using the Newtonian interpolation method described in Appendix A.

D. Method Summary

The methods described follow orthodox quantum mechanics and include three parts: the state vector, the dynamical evolution, and the observables represented by operators. The computational methods supply a phase space picture of the state, based on a discrete representation. The state is followed in time by constructing the evolution operator approximated as a polynomial. The dynamics evolution includes the radiation field as a time-dependent influence, as well as the dissipative influence of the solvent. Finally the relevant spectroscopic observables are cast into the form of observables that can be directly calculated from the state of the system.

III. APPLICATION

To gain insight into the photodissociation event of I_3^-, the calculation follows the sequence of events in time, from the initial thermal state on the ground surface to the final product $I_2^- + I$. Due to the heavy mass of iodine, quantum calculations are computationally intensive; therefore the computational model has to be carefully constructed. The strategy followed is to first use wave-packet simulations. These calculations serve as a guide to Liouville dynamics, which include the dissipative solvent effects. Following the methods developed in the previous section, a model of the encounter is set, involving a limited number of degrees of freedom. Within this model a fully converged quantum mechanical calculation is carried out by the following sequence:

- An initial state is prepared.
- Evolution in time is simulated.
- Experimental measurements are predicted.

The simulation is broken into four parts. In the first, the act of excitation of I_3^- by the pump pulse leading to dissociation is studied. The same pump pulse photo-induced ground surface dynamics is studied in the second part. In the third part, the system studied is the hot and coherently vibrating nascent I_2^-, interrogated by a delayed ultrafast probe pulse. In the last part, the dissipation of the excess vibrational energy from the system is investigated by a delayed push pulse, acting on the spectroscopically decayed I_2^- population. Insight into the photodissociation process can be obtained by comparing the calculated observables with the experiment. A single set of potential energy surfaces unifies all calculations.

All the parameters used in the simulations can be found in Appendix B.

A. Electronic Potential Energy Surfaces

Common to all types of simulations are the electronic potential energy surfaces and transition dipole functions, which interpolate smoothly from reactant I_3^- to product I_2^- and to the excited I_2^- used for interrogation.

The solvent free ground surface potential of I_3^- can be calculated to reasonable accuracy by *ab initio* methods [Fig. 12(*a*)]. The colinear structure of the molecule is reconstructed and the three vibrational frequencies calculated are within 1% of the experimental values [59]. In the calculation, the potential was fit to a harmonic potential in the symmetric (111 cm^{-1}) and antisymmetric (143 cm^{-1}) stretching modes. This description is appropriate for low vibrational excitation energies, as in a thermal distribution at room temperature. If recombination events are studied, where high excitation of the ground surface vibration is anticipated, the present description should be replaced with the full *ab initio* potential.

The experimental change in these vibrational frequencies for different solvents is also within 1%. Nevertheless the antisymmetric stretch vibration is accompanied by a charge redistribution of the molecular ion. In polar solvents this charge redistribution may lead to loss of symmetry, which is represented by a double well potential.

The excited electronic energy surfaces of I_3^- are very difficult to calculate. Preliminary calculations show a multitude of excited surfaces [60], which split further due to spin–orbit interactions. As an interim solution, the upper excited potential energy was fit to a LEPS functional form using the absorp-

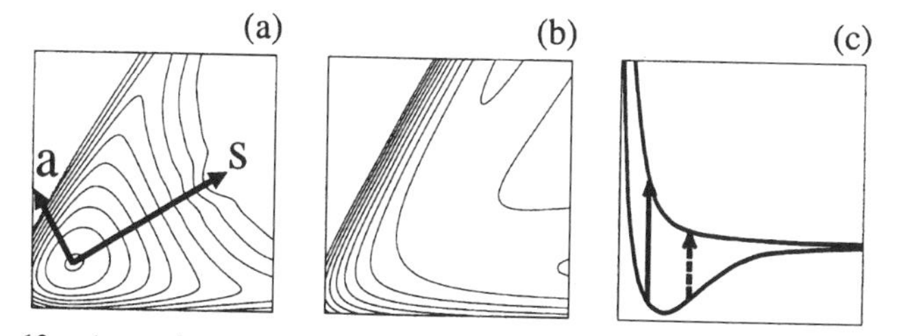

Figure 12. Approximate potential surfaces for the $I_3^- \rightarrow I_2^-$ reaction. (*a*) A contour map of the ground binding state of I_3^- (From Yamashita et al. [60]). The symmetric and antisymmetric stretching mode coordinates are shown. (*b*) The excited dissociative state of I_3^-, with two symmetrical exit channels leading to $I_2^- + I^*$ (the LEPS potential is shown). (*c*) The potential curves for the ground and excited states of I_2^-. The arrows represent the probe resonance, which is blue- (solid arrow) and red- (dashed) shifted from the center of absorption. (*a*) and (*b*) are drawn in Jacoby mass-scaled coordinates (Eq. 3.4).

tion spectra and the asymptotic I_2^- potentials, [1] [Fig. 12(*b*)]. The absorption spectrum of I_3^- shows only a small variation in different solvents, in accordance with *ab initio* calculations, which show a very small change in the charge distribution of I_3^- upon excitation.

The ground and excited surface potentials of I_2^- were taken from the gas phase experiments [61]. The potential energy surfaces have been calculated extensively and show good agreement with the experimental results [62]. The solvent can induce a charge distribution symmetry breaking of I_2^- that can modify these potentials significantly. Nevertheless, there is only a small observed solvent dependence of the absorption spectrum of I_2^-. A more elaborate treatment would include a solvation coordinate also [63, 64]. The ground and excited potential energy surfaces of I_2^- used in the present calculation are shown in Fig. 12(*c*).

B. Photodissociation of I_3^-: The "Pump" Pulse

The initial stage of the photolysis is now studied, when the electromagnetic perturbation initiates the photodissociation journey of the molecule. The impulsive nature of the excitation creates conditions at the very beginning of the trip, whose consequences are still evident much later down the road.

1. *Statics*

Degrees of Freedom. The model is based on the fact that I_3^- is a linear triatomic molecule. The calculation therefore concentrates on the symmetric and antisymmetric vibrational modes of the molecule, denoted r_s and r_a, respectively. As no bending mode overtones appear in the experimental resonance Raman spectrum, it is concluded that bending is not a major component at the early stages of the reaction coordinate, and the bending modes are excluded from the calculation. The rotation of the molecule is also neglected since a separation of timescales exists between the rotational and vibrational motion. These two global coordinates are transformed into atomic coordinates

$$\begin{aligned} r_{I_A-I_B} &= (r_s + 2r_a)/2 \\ r_{I_B-I_C} &= (r_s - 2r_a)/2 \end{aligned} \tag{3.1}$$

These are further transformed into mass-weighted Jacoby coordinates

$$x = r_{I_A-I_BI_C} = a \cdot r_{I_A-I_B} + b \cdot \cos\beta \cdot r_{I_B-I_C} \tag{3.2}$$

$$y = r_{I_B-I_C} = b \cdot \sin\beta \cdot r_{I_B-I_C} \tag{3.3}$$

$$a = \sqrt{2/3} \qquad b = \sqrt{2/3} \qquad \beta = \pi/3$$

This last set of coordinates is asymptotically equivalent to I_2^- vibration and relative translation of the photofragments. In these coordinates, the equations of motion for the three atoms reduce to a single equation for a pseudoparticle, with the mass of an iodine atom, moving in a two-dimensional world. In the Jacoby representation, the symmetric stretching mode is a line in the direction of 30° from the x axis, and the antisymmetric stretching mode is perpendicular to it (see Fig. 12).

Initial State. A density matrix in a two-dimensional world is a four-dimensional entity. The amount of CPU time and storage required for a four-dimensional simulation prohibit such calculations. To keep the calculation on a two-dimensional scale, the interaction with the solvent is neglected, so the dynamics do not mix pure states, and a state vector representation is adequate. This approximation is reasonable only for very short times, before the photofragments hit the walls of the solvent cage.

At room temperature, many of the excited vibrational states of the I_3^- ground electronic surface are populated. The initial density matrix in Eq. (2.16) is divided into its pure state components, which are the eigenstates of the ground surface, each multiplied by a Boltzmann weight according to its energy. The simulation starts with an initial set of these wave functions, represented on a two-dimensional grid and propagated independently. At the end of the simulation, each wave function is transformed back to a density matrix representation of a pure state, and the sum of these matrixes gives the final mixed state.

The ground electronic surface is considered to be harmonic in both vibrational modes, so the grid representation of the initial wave functions was calculated analytically. These wave functions will be denoted by their symmetric and antisymmetric excitations, respectively, i.e., $|v_{1,0}\rangle$ means a state with one quantum in the symmetric stretching mode and at ground state with respect to the antisymmetric motion. The initial states are shown in Fig. 13.

All vibrational levels up to an energy of 300 cm^{-1} were calculated. This means that at room temperature 61% of the population was accounted for in the simulation. It will be shown that this subset of the population is dominant in the experimental observations (see Tables I and II).

2. *Dynamics*

a. Electromagnetic Field. The laser pulse plays a major part in setting the dynamics in motion, by coupling the two surfaces, thus creating a population on the excited surface and leaving a population vacancy on the ground surface (to be studied in the next section). The excitation is tuned to the $I_3^- \rightarrow I_2^- + I^*$ electronic transition, slightly red-shifted from the center of absorption at 308 nm. It is taken to be circularly polarized, according to the rotating-wave approximation. In the resulting effective Hamiltonian of

Eq. (2.28) the excited state potential crosses the ground potential surface at the line of resonance, and the coupling between the shifted surfaces $\bar{\epsilon}(t)\hat{\mu}$ becomes a slowly varying Gaussian field of 60 fs FWHM (full width half maximum).

b. Propagation Results. The initial states (Fig. 13) were propagated using the time-dependent evolution operator in Eq. (2.45) with a time step of $\Delta t = 5$ fs for a total time of $t_f = 240$ fs. Under the experimental conditions, the intensity is large enough that a significant fraction of the population is transferred to the excited surface. Considering individual initial vibrational states, the fraction transferred depends primarily on the symmetric stretch excitation. This effect is summarized in Table I. The maximum quantum efficiency with a 308-nm pump is for $v = 1$ in the symmetric stretching mode. The pulse is not exactly in resonance with the maximum thermal absorption, which occurs at the peak of the $|v_{0,0}\rangle$ eigenstate [Fig. 13(*a*)], but coincides with the upper right lobe of the $|v_{1,0}\rangle$ state [Fig. 13(*b*)]. Further excitation of the symmetric stretching mode will move the main population away from the line of resonance [Fig. 13(*d*)]. The Boltzmann weights of more excited states further lowers their contribution in the final product (Table II), so the initial states used are sufficient to describe the experimental results.

The pump pulse used in the experiment was long in comparison to dynamics in the symmetric stretching mode on the excited surface. Significant coupling is present for a duration of ~180 fs [Fig. 14(*a–d*)]. This means that pho-

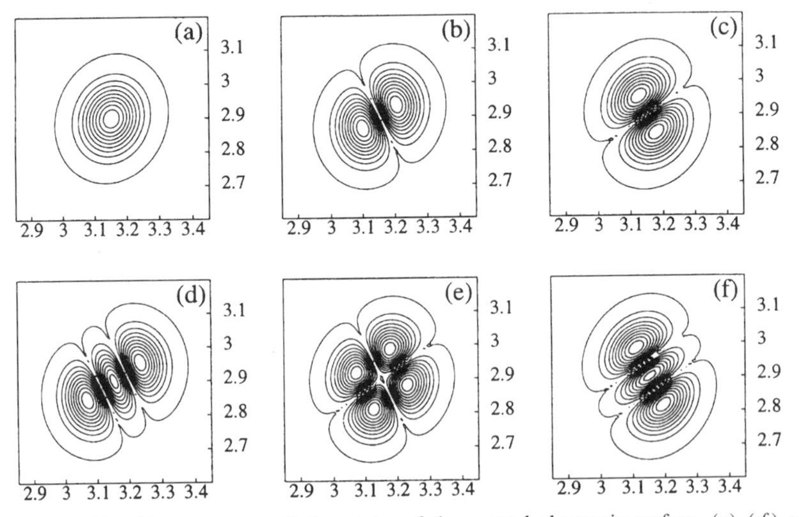

Figure 13. Contour maps of eigenstates of the ground electronic surface. (*a*)–(*f*) are $|v_{0,0}\rangle$, $|v_{1,0}\rangle$, $|v_{0,1}\rangle$, $|v_{2,0}\rangle$, $|v_{1,1}\rangle$, $|v_{0,2}\rangle$, respectively. The scaling is in angstroms.

TABLE I
The Influence of the Initial Vibrational Excitation on the Population Transferred to the Excited Surface[a]

A \ S	0	1	2	3
0	18.6%	41.6%	27.3%	5.0%
1	17.1%	40.4%		
2	15.8%			

S = excitation in the symmetric stretch mode.
A = excitation in the anti-symmetric stretch mode.

todissociation dynamics proceeds while amplitude is still fed to the excited surface. For this pulse duration, the impulsive approximation (Section II.C.1) has only qualitative meaning. Most of this motion is along the symmetric stretching mode as opposed to dynamics leading toward bond cleavage, i.e., motion along the antisymmetric stretch, which is small (see Figure 12a).

Once the pump pulse is over, the dynamics on the ground and excited surfaces decouple. Since the focus of the calculation is shifted to the products, free propagation of the wave function is carried out on the excited surface alone. The excited wave functions $|\psi_e^{s,a}\rangle(t_f)$ induced by the excitation pulse from the initial state $|\psi_g\rangle(0) = |v_{s,a}\rangle$, were used as the initial states. The propagation was carried out by the time-independent evolution operator in Eq. (2.41) generated by a time-independent Hamiltonian Eq. (2.21) with a time step of $\Delta t = 2000$ a.u. Intermediate 50-fs snapshots of the evolving wave functions were stored. Since both potentials are symmetric along the atomic coordinates [Eq. (3.2)], so are all the wave functions, throughout the propagation. This means that it is enough to follow the dynamics of just one of the product channels, leading to $I_A + (I_BI_C)^-$. To save memory and com-

TABLE II
The Boltzmann Distribution of the Initial Vibrational States[a]

A \ S	0	1	2	3
0	20.5%	12.0%	7.1%	4.1%
1	10.3%	6.0%	3.5%	2.1%
2	5.2%	3.0%	1.8%	1.0%
3	2.6%	1.5%	0.9%	0.5%

S = excitation in the symmetric stretch mode.
A = excitation in the anti-symmetric stretch mode.

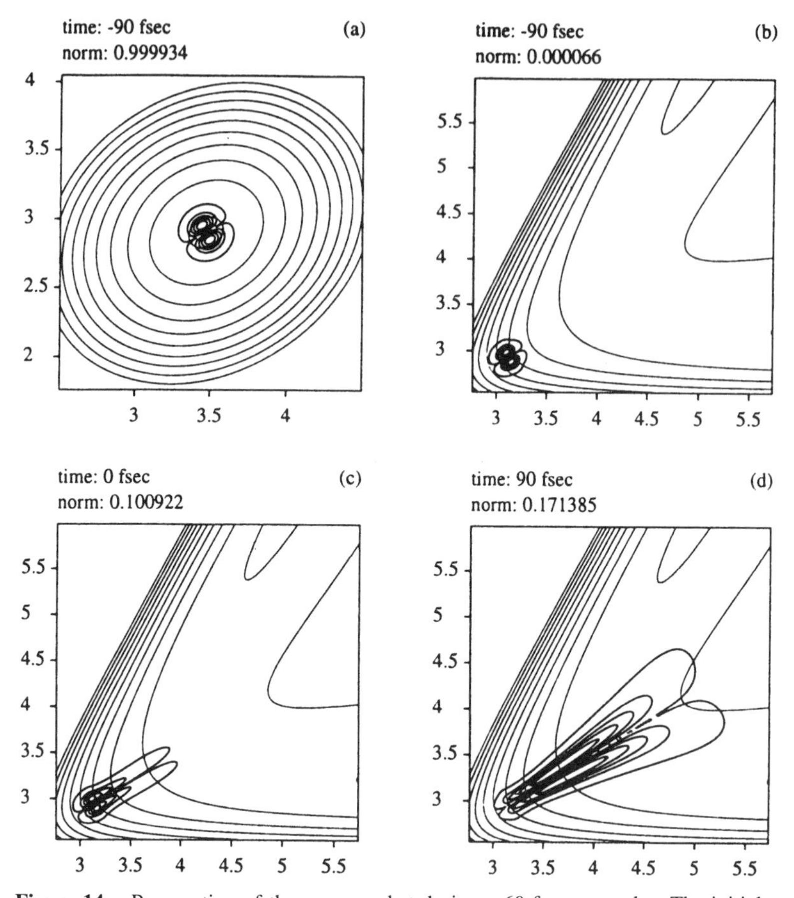

Figure 14. Propagation of the wave packet during a 60-fs pump pulse. The initial state shown is $|v_{0,1}\rangle$. (*a*) The wave function on the ground state at the very beginning of the pulse. (*b*) The excited surface wavefunction. (*c*) The wave function at the peak of the pulse. (*d*) The wave function after most of the pulse is over. The scaling is in angstroms.

putation time, and this exit channel is fully described by the grid. The other channel is blocked by an absorbing boundary condition[5] [65], which makes it possible to take the grid twice as far in the x direction as in the y direction. Figures 15 and 16 show several snapshots starting from two different initial wave functions.

[5]Absorbing boundary condition is a pseudopotential $\hat{\mathbf{V}}_{absorbing}$ added to the Hamiltonian. This potential is zero on most of the grid, and negative imaginary on the boundary. Any part of a wavefunction propagating under the influence of this anti-Hermitian potential will decay exponentially to zero before it can overflow the grid boundary.

At the early stage of the dynamics the motion is along the direction of the symmetric stretch, down the potential slope, across the saddle point, and "up hill" in the direction of the three-body dissociation. Only when the wave function crosses the saddle point, does motion in the direction of the antisymmetric stretching become significant. This motion, is an essential component of the reaction coordinate leading to bond cleavage.

The wave function that was initiated as an antisymmetric eigenstate on the ground surface is "better suited" for such motion, and enters the second stage earlier, falling rapidly into the exit channels without ascending deep into three-body dissociation [Fig. 16(*a*)]. On the other hand, the wave function that started as a symmetric eigenstate has most of its population centered along the symmetric stretching mode, so it continues ascending much further before bifurcating into the exit channels (Fig. 15(*a*)].

The wave function enters the exit channels with much excess energy, which translates into two asymptotic modes: vibration of the I_2^- fragment, and translation of the I atom away from the I_2^- molecule. Coherent vibrational motion of the antisymmetric wave function is seen clearly in Fig. 16(*b*), and less evidently in Fig. 15(*b*).

After longer propagation, the wave function is no longer compact because of the large distribution in translational energy, which smears the wave packet into a long "snake" Figs. 15(*c*) and 16(*c*). The total excess energy is relatively well defined by the photon energy dispersion in the pump (0.04 eV). This imposes an obvious correlation between vibrational and translational degrees of freedom in the evolving product. The long serpentine appearance of the wave packet initially in $|v_{1,0}\rangle$ indicates an extremely large dispersion of the wave packet, whose initial dynamics is primarily along the symmetric stretch. The shape of the wave packet at long delays also indicates that the portion entering earlier into the exit channel is characteristic of high translational and low vibrational energy content; whereas the more delayed portions are poor in translation and highly excited vibrationally. This dispersion is much less apparent in the case of I_3^- initially excited in the antisymmetric stretching mode. This initial displacement of density away from the ridge of perfect symmetry tends to direct the population into the exit channels in a much more compact and uniform fashion, limiting the degree of dispersion in vibration and translation. One should keep in mind that, in a solvent, the translational motion will be rapidly dissipated due to hitting with the solvent molecules.

3. *Interpretation*

It is evident that the shape of the wave packet at very long times still bears a memory of its early history, at times shorter than a few hundred femtoseconds, in which the dynamics are governed by the electromagnetic pulse

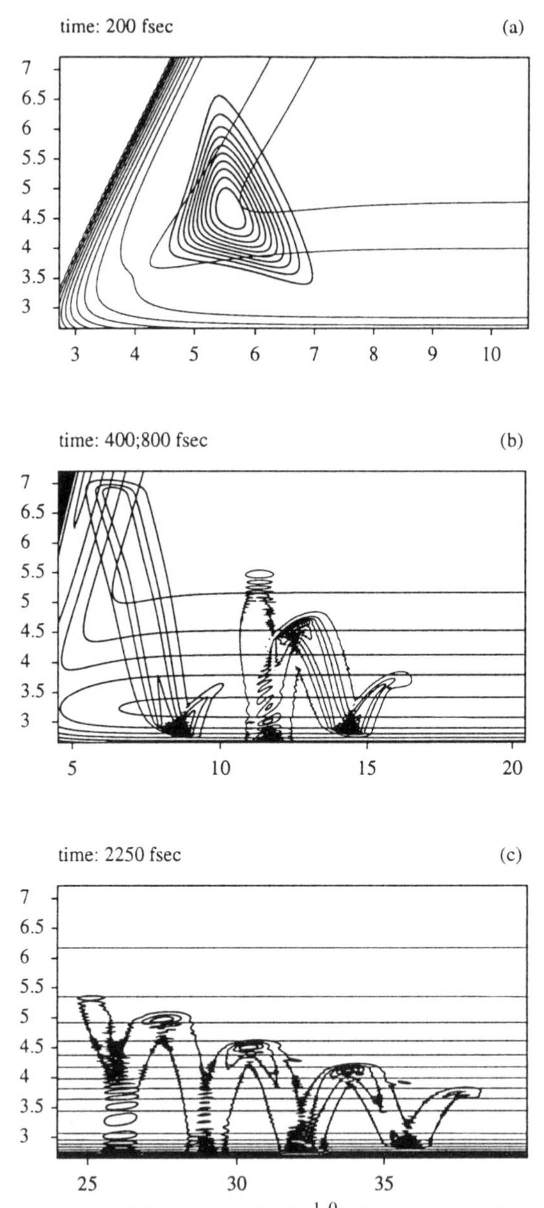

Figure 15. Propagation of the wave packet $|\psi_e^{1,0}\rangle(t)$ generated by the pump pulse, on the excited surface. The initial state $|\psi_g\rangle(0)$ is $|v_{1,0}\rangle$. The first wave function in (*b*) is "chopped" at the top of the frame by an absorbing potential. The scaling is in angstroms.

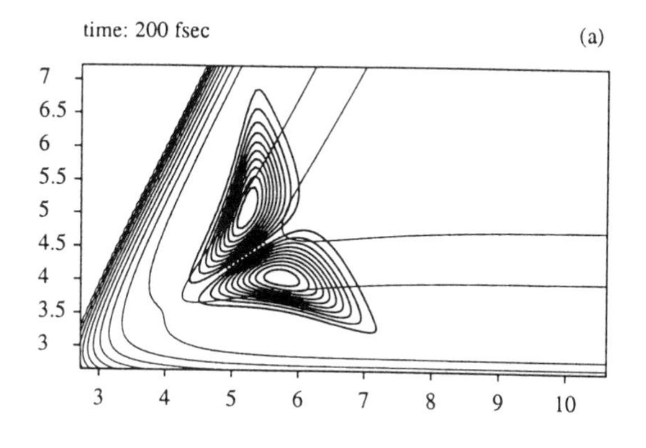

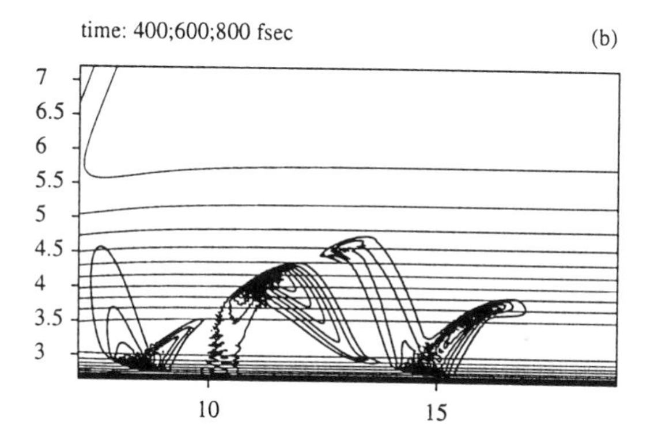

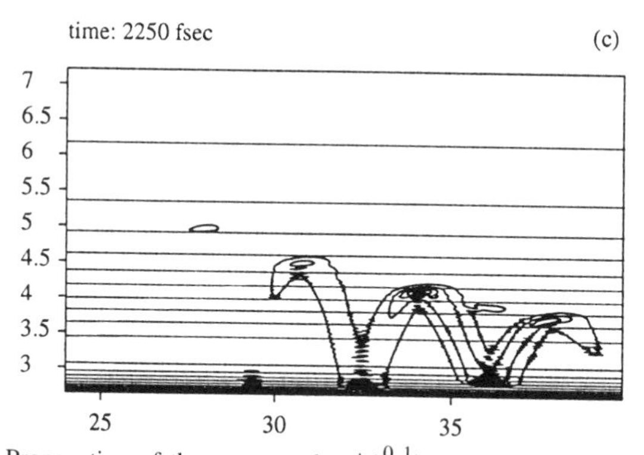

Figure 16. Propagation of the wave packet $|\psi_e^{0,1}\rangle(t)$ generated by the pump pulse, on the excited surface. The initial state $|\psi_g\rangle(0)$ is $|v_{0,1}\rangle$. The scaling is in angstroms.

and the Franck–Condon region of the excited potential. There are two other observables that correlate with the same part of the potential: the electronic absorption spectrum and the Raman spectrum.

a. Absorption Spectrum. The slopes of the excited potential surface in the Frank–Condon region have a major influence on the width of the electronic absorption spectrum. Calculation of the absorption spectrum, using Eq. (2.87), then gives the calculated measure for this slope. Fitting the calculated spectra with the experimental one has served to calibrate the parameters of the LEPS potential.

b. Raman Spectrum. A complementary probe of the first stages of the dissociation dynamics is recovered from the Raman spectra. Raman wave functions were calculated for all initial vibrational states using Eq. (2.89), integrating to a time of 75 fs. The excitation energy E_L is on resonance with the transition.

The Raman wave function shown in Fig. 17 is very similar to the wave packet created by the pump pulse [Fig. 14(*c*)]. Thus the Raman spectra and the RISRS experiment carry complimentary information on the initial stage of the photodissociation. Individual Raman spectra were extracted from the Raman wave functions using Eq. (2.88), and then Boltzmann averaged according to the initial vibrational energies. Figure 18 shows the resulting Raman spectrum at $T = 300$ K.

The main features in the spectrum correspond to harmonics of the symmetric stretching frequency, and so are attributed to the population of highly excited states of the symmetric stretching mode by the Raman process. In the time domain, this correlates to the long propagation along this mode.

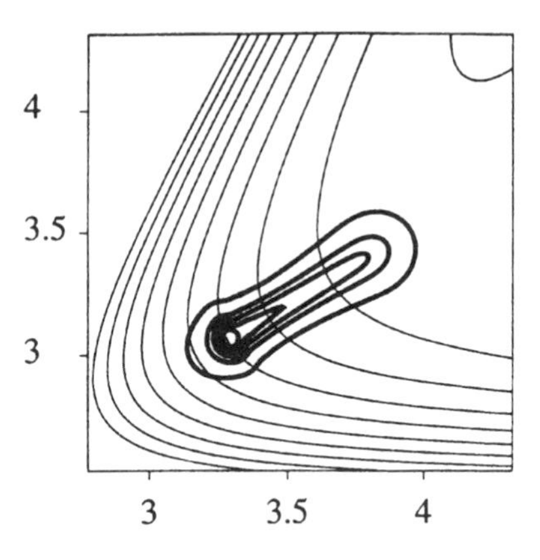

Figure 17. The Raman wavefunction excited from $|v_{1,0}\rangle$ on the LEPS potential. The scaling is in angstroms.

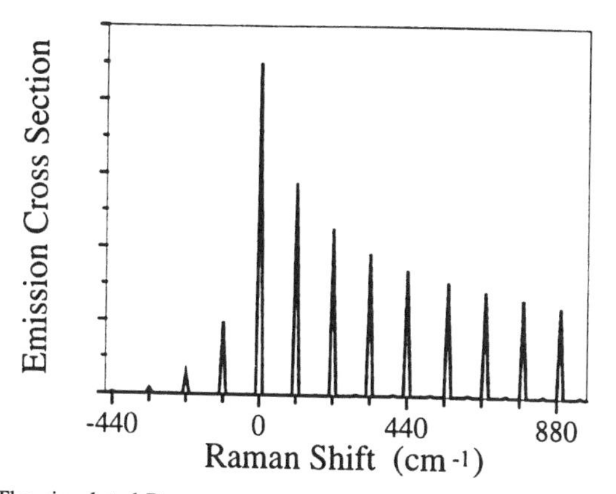

Figure 18. The simulated Raman spectrum at $T = 300$ K of I_3^- calculated with the LEPS potential.

The odd harmonics of the antisymmetric stretching frequency are symmetry forbidden, so only the second harmonic is expected in the spectrum, as the propagation along the antisymmetric mode is very small. On examining the spectrum, it can be seen that the second harmonic of the antisymmetric stretch has, indeed, only a very small signature.

The Raman spectrum is sensitive to the shape of the excited potential, as can be seen by comparing the above results with preliminary work on the *ab initio* potential. This potential induces spreading along the antisymmetric stretching mode as well (Fig. 19). The resulting spectrum (Fig. 20) differs in

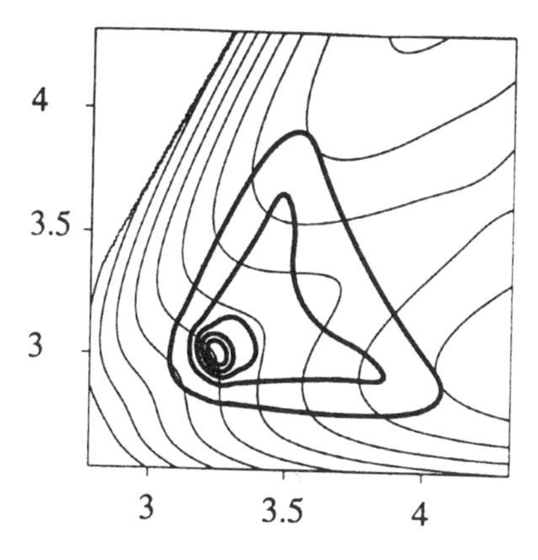

Figure 19. The Raman wave function excited from $|v_{1,0}\rangle$ on the *ab initio* potential. The scaling is in angstroms.

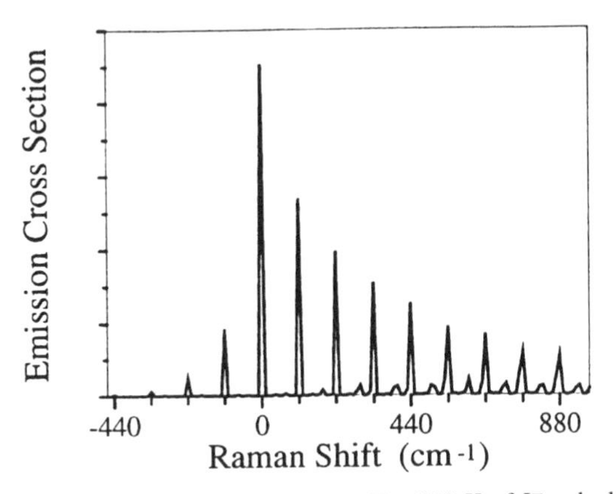

Figure 20. The simulated Raman spectrum at $T = 300$ K of I_3^- calculated with the *ab initio* potential.

two ways from the previous one. First, the decay of the signal corresponding to the higher harmonics of the symmetric stretch is more rapid. Second, the second harmonic of the antisymmetric stretch has a more pronounced signature, which increases for higher Raman shifts (corresponding to longer propagation times on the excited surface). This is consistent with the shape of the Raman wave function, in which the motion along the antisymmetric stretch mode is larger at later times.

C. Dynamics of the Photo-Induced "Hole" in I_3^-

The photoinduced density on the excited surface leading to photodissociation is missing on the ground surface. The missing density or the coherent "hole" is nonstationary and will therefore continue to ring long after the pump pulse has turned off. Observation of this ringing sheds light on the interaction of the reactant I_3^- with its solvent.

1. *Statics*

a. Degrees of Freedom. The initial position of the dynamical "hole" is determined by the ground–excited difference potential $2\hat{\mathbf{\Delta}} = \hat{\mathbf{V}}_e(\mathbf{r}) - \hat{\mathbf{V}}_g(\mathbf{r}) - \hbar\omega_L$ in Eq. (2.50). The gradient of the difference potential along the antisymmetric stretch is almost zero; therefore this mode can be excluded from the calculations. The simulation of the excitation stage in Section III.B.2 has verified this assumption. For this reason a one-dimensional representation in the symmetric stretching coordinate is an adequate description.

These consideration should be reevaluated for a shorter pump pulse and the sharper *ab initio* potential of Fig. 20 is used. In this case the difference potential $\hat{\mathbf{\Delta}}$ will include a contribution from the antisymmetric stretching mode. The induced "hole" in this direction will have inversion symmetry, in accordance with the Raman spectrum; therefore the fundamental frequency will not be excited. A modulation with a period of the first overtone ≈ 100 fs will be a signature of this motion.

b. Initial State. The dynamical "hole" density is recreated by solving the Liouville dynamics for the symmetric stretching mode. The initial state was a thermal density on the ground surface at 300 K. The pump field amplitude used in the simulation corresponds to 1.1π pulse conditions on resonance for the 60-fs pulse. The total depletion of the ground surface population in these conditions is 8%. Electronic dephasing has a profound influence on the amount of coherence generated on the ground surface and, as a result, on the depth of the observed modulations. This dephasing process can be thought off as limiting the time during which the two surfaces are coupled by the radiation. In accordance with Section II, Eq. (2.35), a Gaussian dissipative model is used, leading to

$$\hat{\mathcal{L}}_D(\hat{\rho}) = -\tfrac{1}{2}\gamma_z[\hat{\mathbf{I}} \otimes \hat{\mathbf{S}}_\mathbf{z}, [\hat{\mathbf{I}} \otimes \hat{\mathbf{S}}_\mathbf{z}, \hat{\rho}]] \tag{3.4}$$

The electronic coherence lifetime γ_z^{-1} is estimated to be in the range of 10–100 fs. Finally a Gaussian vibrational dephasing process was also added

$$\hat{\mathcal{L}}_D(\hat{\rho}) = -\tfrac{1}{2}\gamma_g[\hat{\mathbf{H}}_\mathbf{g} \otimes \hat{\mathbf{I}}, [\hat{\mathbf{H}}_\mathbf{g} \otimes \hat{\mathbf{I}}, \hat{\rho}]] \tag{3.5}$$

The dephasing characteristic time was set at $\tau_{dg} = (\omega_g^2\gamma_g)^{-1} = 960$ fs. With this timescale the vibrational dephasing has only a small effect on the creation of the dynamical "hole." Fig. 21(*a*) displays the dynamical "hole" immediately after the pump pulse. Under these conditions the resulting dynamical "hole" is mainly a shift of density along the position coordinate, with relatively weak contribution of direct shift along the momentum direction.

2. *Dynamics*

The dynamics of the "hole" $\hat{\rho}_d$ are studied directly by propagating $\hat{\rho}_g$ obtained above after the excitation pulse has turned off. The evolution operator is generated by the ground surface Hamiltonian and the vibrational dephasing term in Eq. (3.5). Figure 21 shows phase space snapshots of the dynamical "hole" at different time delays after the pump pulse. Note that the phase space distribution consists of balanced positive and negative parts, since the integrated phase space volume of the "hole" is zero ($\hat{\rho}_d^e$ is traceless). The dynamics is influenced by the ground surface Hamiltonian and the vibrational dissipation. The ground surface Hamiltonian rotates the distribution in phase space—a full cycle corresponds to one vibrational period

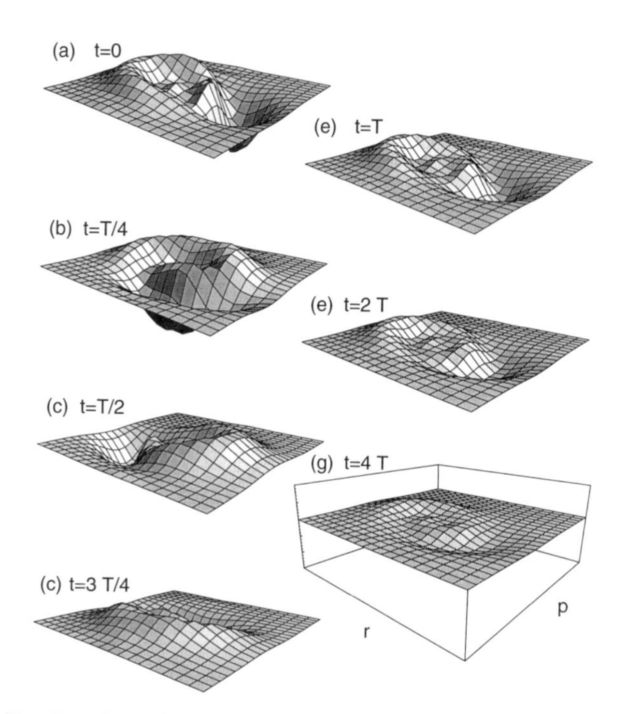

Figure 21. Snapshots of the Wigner distribution function $W_d(r,p)$ of the dynamical "hole." The initial density is a Boltzmann distribution at 300 K, pumped by a pulse of 60 fs FWHM and wavelength of 308 nm. T is the period of the vibration.

[Fig. 21(*a–e*)]. Since the "hole" in this case is created in an initial thermal distribution at the bottom of the I_3^- potential there is no anharmonic contribution to the dynamics. Therefore only the dissipative terms eliminate the "hole" by smoothing it out. In the present case, the major effect of dissipation is assumed to be vibrational dephasing, which manifests itself in smearing the lobe and anti-lobe of the dynamical "hole" while they undergo circular motion in phase space. Energy relaxation processes are assumed not to contribute in view of the longer timescale of vibrational relaxation found for the I_2^- in Section III.E.

In a pump-probe experiment, many experimental measurements are taken at different delays between the pulses. The idea is that the strong pump pulse prepares the molecules in a nonstationary state, and the weak probe pulse serves as a tool for measuring the evolved state at a certain point in time.

After the measurement takes place, these molecules are regarded as gone (they either dissociate to give a different molecule, or recombine and cool back to their initial state). The next pulse pair encounters the same initial conditions as the last, prepares the molecules at the same nonstationary state, and measures the evolved state at a different point in time. Only the power absorbed by the probe is calculated; thus, the state of the system on the excited surface after its interaction with the probe pulse has no experimental significance, and is therefore eliminated.

3. *Interpretation*

A full simulation of the probe signal in a pump-probe absorption experiment is shown in Fig. 22. The modulations are fit to an exponentially decaying sin(t) function. The decay time, 960 fs, agrees with the decay time of the coherence measure. It is important to stress that the ability to directly measure the rate of vibrational dephasing from the decay of oscillations in optical density is not at all trivial.

Figure 23 compares the results of the full simulation of the pump-probe experiment (Fig. 22) with the experimental results in ethanol. The expo-

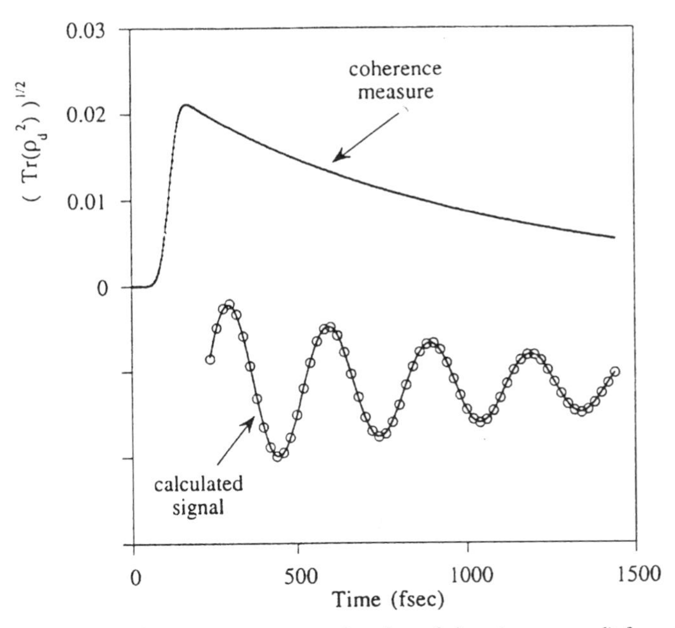

Figure 22. The coherence measure as a function of time (upper panel) for a full simulation of the pump-probe experiment. An electronic dephasing time of 5 fs and vibrational dephasing time of 960 fs have been used. The calculated signal is shown in the bottom panel. The solid line is a fit to an exponentially decaying sin function of $\omega_v = 112\ \mathrm{cm}^{-1}$ and decay time $\tau_{dg} = 960$ fs.

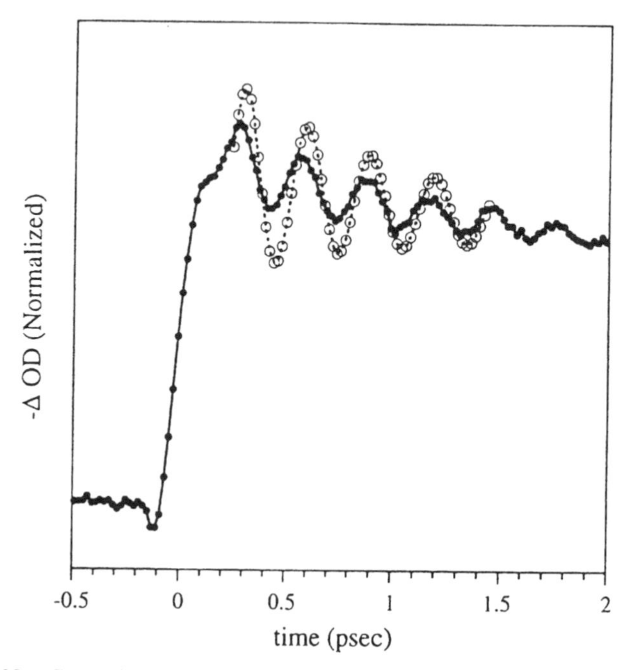

Figure 23. Comparison of the experimental results in ethanol (solid points) with the simulation of the pump-probe experiment (open circles). The baseline is identical for both signals. The slow background exponential decay, obtained from the fit to the experimental data, was added also to the simulated result. The frequency, phase, and decay time of the simulated signal agrees with the experimental data.

nentially decaying sine function affords a good fit to both the measured and calculated signals, allowing the extraction of various molecular and solvent-dependent parameters. The frequency of the experimental and simulated spectral modulation matches the ground state symmetric stretching frequency. The absolute phase of the spectral modulations reflects the position in phase space of the coherent "hole" and is identical, within experimental error, to the simulations. Finally the decay rate of the spectral modulation allows the extraction of a dephasing constant, assumed in the simulation to be due to purely vibrational dephasing. Identifying the decay rate with changes of the coherence measure in the molecular system has been verified in Fig. 22.

The major discrepancy between the calculated and experimental signal is the depth of modulation. This might be due to propagation effects of the radiation through a dense medium; these are not considered in the simulation.

The RISRS experiment is able to measure the ground surface vibrational frequency of the solvated system and the vibrational dephasing rate. With a shorter pump and probe higher harmonics can be observed.

D. Dynamics of Nascent I_2^-: The "Probe" Pulse

The role of the probe laser pulse in this section is that of a silent observer, quietly recording the details of the trip of hot I_2^- to equilibrium.

1. Statics

a. Degrees of Freedom. The system at hand is the nascent I_2^- molecule. The model for I_2^- is one-dimensional, including the single vibrational mode of the diatomic molecule $(I_B I_C)^-$. This mode can be represented in coordinate space by the y Jacoby coordinate in Eq. (3.4), or expanded in eigenstate space $\{|v_i\rangle\}$, where $\hat{\mathbf{H}}_g|v_i\rangle = E_i|v_i\rangle$. The relative translation of the photofragments (the x Jacoby coordinate), will be shortly discarded as it bears no effect on the experimental observables. As for I_3^-, the rotations are once again neglected due to the divergence of timescales.

Since the system has just one dimension, it can be represented as a density matrix, and the influence of the solvent can be added to the dynamics through the open system formalism developed at the previous section.

b. Initial State. The final states of Section III.B.2 serve as the starting point for the current calculations. Each final state $\psi_e^{s,a}(x, y)$ is first transformed to momentum representation along the x coordinate by an FFT as in Eq. (2.4), and then to the I_2^- vibrational eigenstate representation along the y coordinate, using a unitary basis transformation from $\{|y\rangle\}$ to $\{|v_i\rangle\}$ as in Eq. (2.1), resulting in $\psi_e^{s,a}(p, v)$. A density matrix was constructed from each of these wave functions

$$\rho_g^{s,a}(p, p', v, v') = |\psi_e^{s,a}(p, v)\rangle\langle\psi_e^{s,a}(p, v)| \tag{3.6}$$

A partial trace over the momentum degree of freedom (p) was taken, resulting in the desired one-dimensional representation in v space $\rho_g^{s,a}(v, v')$. The effect of this reduction on the measured observables was checked by comparing the transient absorption of $\psi_e^{s,a}(x, y)$ and $\rho_g^{s,a}(v, v')$, and the difference was found to be negligible. This can be attributed to the strong correlation between v and p, as seen in Fig. 16(c) and Fig. 15(c), meaning that each v collects contributions from a very narrow band of p's in the partial trace process, and has no coherence with p's outside this band.

The initial state for the I_2^- relaxation study was created by Boltzmann averaging over all initial wave functions:

$$\rho_g(v, v') = \sum_{s,a} e^{-E_v^{s,a}/k_B T} \rho_g^{s,a}(v, v') \tag{3.7}$$

where $E_v^{s,a}$ is the energy of the eigenstate $|v_{s,a}\rangle$ of the ground surface of I_3^-.

The time instant chosen for the reduction was set at $t = 680$ fs in which the dissociation process already separated the wave function into the asymptotic channels. In order to compensate for the dissipation at earlier times the initial density was propagated backwards in time under dissipation-free evolution to a time of $t = 255$ fs. This time is estimated as the time at which the molecular ion I_2^- has separated from the I atom and can be considered as a separate motion influenced by the solvent. The state $\rho_g(v, v')$ at $t = 255$ fs was then used as the initial state for the dissipation studies. Figure 24 shows the Wigner plots of the initial distribution. The left panel shows the Wigner function $W_g(r, p)$ and its projections on the coordinate and momentum coordinates. On examining the figure it is clear that a very broad initial distribution is created. The right panel shows the initial Wigner function $W_g(v, \phi_v)$, which reveals that the I_2^- photoproduct is created in a vibrationally hot state, with a mean energy of 0.55 eV. The projection on the vibrational state displays a very wide vibrational distribution peaked at 0.4 eV ($v = 34$). The dissipation of this energy by the solvent will be the main subject of the next section.

2. *Dynamics*

The probe pulse is not to be incorporated into the Hamiltonian, and only its experimental outcome, the total power absorbed, is calculated in Section III.D.3.

a. Potential and Interaction with the Solvent. The ground and excited potential curve of I_2^- has been described in Section III.A. The excited surface does not take part in the dynamics, as there is no coupling between the

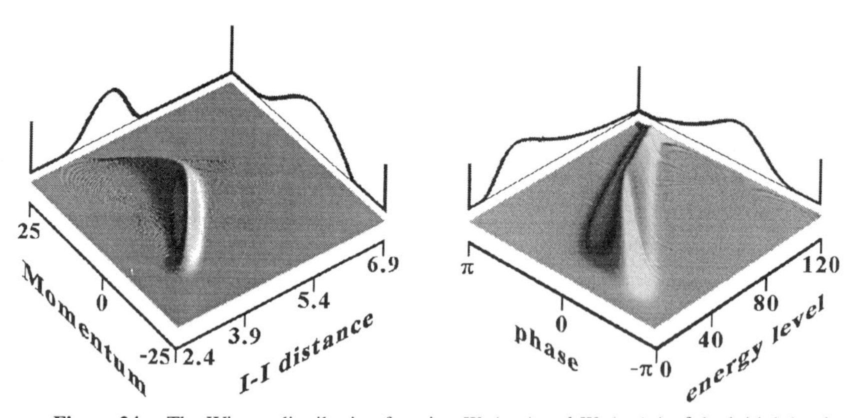

Figure 24. The Wigner distribution function $W_g(r, p)$ and $W_g(v, \phi_v)$ of the initial density at time $t = 255$ fs. The projections on the variables are also shown, displaying the initial position, momentum, vibrational energy, and phase distributions. Distances are in angstroms, and the momentum units are Å/ps.

surfaces (the electromagnetic field of the probe is excluded from the Hamiltonian), so only $\hat{\rho}_g$ will be considered in the dynamics.

During its nuclear motion the nascent I_2^- molecule is constantly influenced by the bath. This influence can be caused by direct impact on the nuclei due to collisions with the bath molecules, or by solvent dipoles interacting with the electrons of the molecule. The solvent's influence can be divided into a static part, described by a modification of the effective potential, and a dynamical part, which, viewed by the reduced molecular system, has a stochastic nature, so each I_2^- molecule "feels" random collisions and dipole fluctuations. To describe these processes, the phenomenological framework of Section II.B.2 is used. The dissipation process are divided in two categories:

- Relaxation: Processes that allow energy flow between the system and the bath. They will be referred to as T_1 processes.
- Dephasing: Elastic processes that conserve the energy of the system, but erase the coherence between the different populations in $\hat{\rho}_g$. These will be termed T_2 processes.

The dissipative part of the Liouvillian is likewise divided into two categories

$$\hat{\mathcal{L}}_D = \hat{\mathcal{L}}_{D_{nr}} + \hat{\mathcal{L}}_{D_{nd}} \tag{3.8}$$

where $\hat{\mathcal{L}}_{D_{nr}}$ is the nuclear energy relaxation term and $\hat{\mathcal{L}}_{D_{nd}}$ is the nuclear dephasing term.

In general the dissipative energy exchange process is described by

$$\hat{\mathcal{L}}_{D_{nr}}(\hat{\rho}_{\mathbf{g}}) = \sum_{ij} \gamma_{ij} \left[\hat{\mathbf{P}}_{ij}\hat{\rho}_{\mathbf{g}}\hat{\mathbf{P}}_{ij}^{\dagger} - \frac{1}{2}\{\hat{\mathbf{P}}_{ij}\hat{\mathbf{P}}_{ij}^{\dagger}, \hat{\rho}_{\mathbf{g}}\}_{+} \right] \tag{3.9}$$

where $\hat{\mathbf{P}}_{ij}$ is a projection from state $\langle v_i|$ to state $|v_j\rangle$, $\hat{\mathbf{P}}_{ij} = |v_j\rangle\langle v_i|$, and γ_{ij} is a rate constant Detailed balance is imposed when the rate constants obey the relation $\gamma_{ij}/\gamma_{ji} = e^{-(E_j - E_i)/k_BT}$. In this study the source of the energy relaxation is the cage effect of solvent molecules pushing against the I_2^- ion. It is therefore reasonable to assume that the relaxation rate is proportional to the vibrational amplitude. A linear relation of the rate γ_{ij} with vibrational amplitude was therefore assumed. Since the oscillator is anharmonic, the rate was chosen to be proportional to the classical vibrational amplitude of the energy of level i. Specifically, only nearest-neighbor transitions were included in the calculation. Finally, the rate coefficients were all proportional to a global parameter Γ that empirically determines the relaxation rate τ_{nr}. For low excitation energies the relaxation model matches the energy relaxation rate of a harmonic oscillator. The energy relaxation parameters were adjusted to the harmonic model used previously with τ_{nr} = 3ps [2].

Random elastic collisions of the solvent with the molecule, which are fast on the vibrational time scale, are described by a Gaussian random process. This process leads to a dephasing of the vibrational motion. By choosing $\hat{\mathbf{V}}_i = \hat{\mathbf{H}}_g$ in Eq. (2.34) the nuclear dephasing term is obtained

$$\hat{\mathcal{L}}_{D_{nd}}(\hat{\rho}_{\mathbf{g}}) = -\gamma_{nd}[\hat{\mathbf{H}}_g, [\hat{\mathbf{H}}_g, \hat{\rho}]] \tag{3.10}$$

$$\tau_{nd}(v) = 1/(\omega(v)^2 \gamma_{nd}) \tag{3.11}$$

where $\tau_{nd}(v)$ is the dephasing time constant of the vibrational level v, and $\omega(v)$ is the vibrational frequency of that level. The value of $\tau_{nd}(0)$ was taken to be 960 fs. In the Heisenberg picture $\hat{\mathbf{H}}_g$ is a constant of motion ($\hat{\mathcal{L}}^{\dagger}_{D_{nd}}(\hat{\mathbf{H}}_g) = 0$), so this term represents a process where the system on the ground surface does not exchange energy with the bath. A similar dephasing term was used to describe the dissipative dynamics of the "hole" in the reactant I_3^-.

Since the nuclear relaxation term is easily constructed in terms of the ground surface eigenstates, the eigenstate representation was used for the Hamiltonian as well, which is diagonal in its own basis representation

$$\mathbf{H}_g(v', v) = \delta_{v',v} E_v \tag{3.12}$$

b. Propagation Results. To obtain insight concerning the different mechanisms of dissipation, different types of dynamics were studied and compared. As a benchmark, the dissipation-free Hamiltonian dynamics were simulated. The free dynamics were contrasted with dissipative evolution that includes energy relaxation processes [T_1, Eq. (3.9)]. Finally dephasing processes were also included [$T_1 + T_2$, Eq. (3.10)].

Figure 25 compares snapshots of the Wigner function $W_g(r,p)$ of the different dynamical processes. The free evolution is described in the left panel. The motion observed is a winding motion, since the outer energy shells have a lower vibrational frequency due to anharmonicity. As a result the free dynamics stretches the initial distribution around the zero energy point $(r - r_{\text{eq}}), p = 0$. The anharmonic effect is more readily understood by examining Fig. 26, which displays the Wigner function in action angle variables. The phase velocity of the higher quantum numbers (higher action) is slower than the lower ones, because of the anharmonicity of the I_2^- potential. As a result the dynamics can be visualized as a winding motion of the distribution around a cylinder whose axis represents the action. The observable consequence of the anharmonic motion is scrambling of the phase compactness. This leads to a reduction of the amplitude of the spectral modulations of the probe pulse.

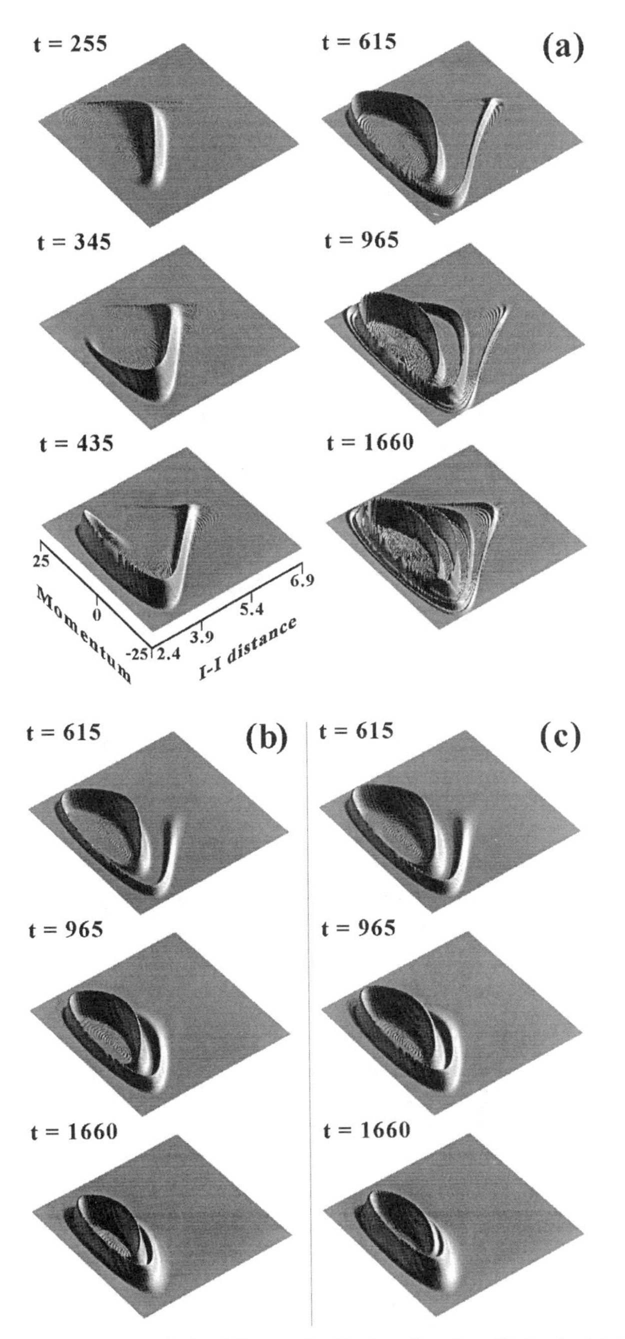

Figure 25. Snapshots of the Wigner distribution function $W_g(r, p)$: (*a*) Hamiltonian dynamics only; (*b*) Dynamics with energy relaxation (T_1); (*c*) Dynamics with energy relaxation and dephasing ($T_1 + T_2$). Times are in femtoseconds where the panels correspond to approximately 0, $\frac{1}{4}$, $\frac{1}{2}$, 1, 2, 4 vibrational periods. Distances are in angstroms, and the momentum units are Å/ps.

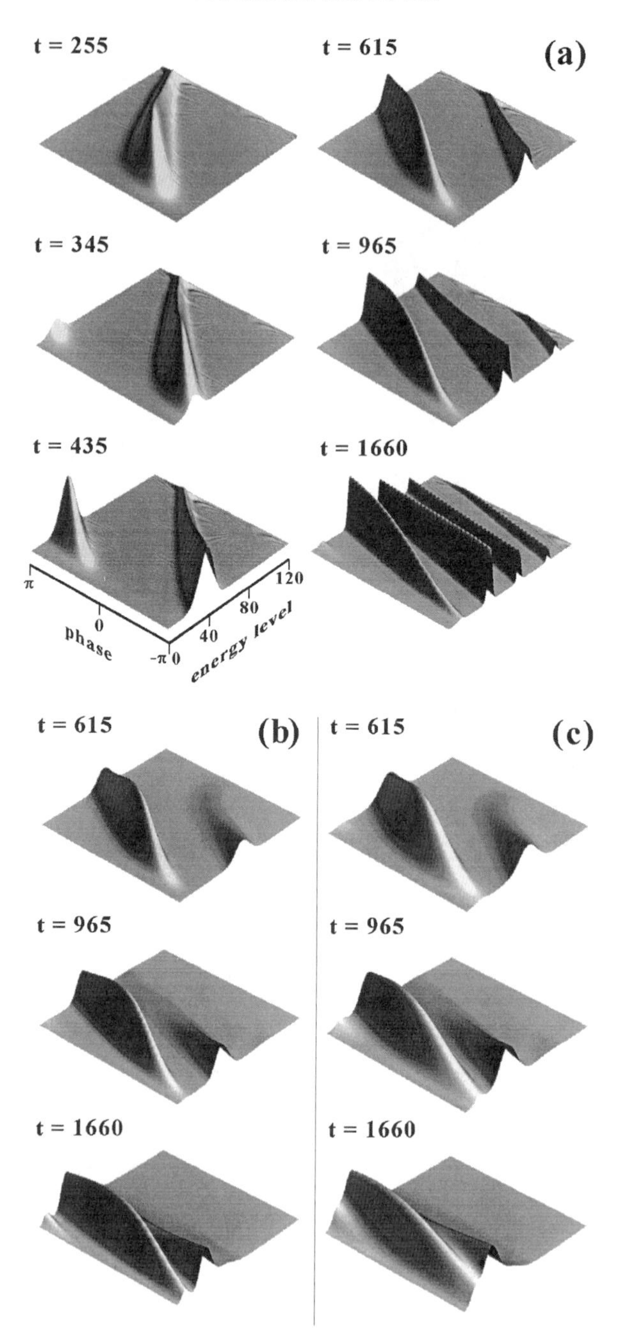

Figure 26. Snapshots of the Wigner distribution function in action angle coordinates, $W_g(v, \phi_v)$: (*a*) Hamiltonian dynamics only; (*b*) Dynamics with energy relaxation (T_1); (*c*) Dynamics with energy relaxation and dephasing ($T_1 + T_2$).

The energy relaxation process is described in Figs. 25(*b*) and 26(*b*). In the (r,p) phase space, energy relaxation is manifested by the distribution $W_g(r,p)$ approaching the origin. In the action angle representation, the relaxation moves the distribution $W_g(v,\phi_v)$ towards lower action values, thus decreasing the loss of phase due to anharmonicity. The effect of removing energy is enhancement of the modulation in phase, as can be seen by comparing panels (*a*) and (*b*) of Fig. 26 at $t = 1660$ fs.

Finally the right panel (*c*) in Figs. 25 and 26 shows the added influence of dephasing on the dynamics. As can be seen in Fig. 26 the features along the phase coordinate at a time of $t = 1660$ fs are smoother in panel (*c*), so that the observed modulation would be smaller.

3. *Interpretation*

The different types of dynamics have a profound effect on the transient observables, specifically on the measured transient spectrum of a weak probe pulse. Two different wavelengths were used, one blue-(at 620 nm) and one red-(at 880 nm) shifted from the center of the absorption band (at 740 nm), which induces a transition to a dissociative state of I_2^-. To obtain the absorption signal of the probe pulse, the impulsive limit in weak fields is employed [Eq. (2.76)], where $\tau_{pr} = 60$ fs. A phase space image of the window functions associated with the two probe pulses is shown in Fig. 27.

The window function is momentum independent, but it is almost δ correlated in the r coordinate (actually, it is sinc correlated), with a sharp peak at the resonance point (3.1 Å for 620 nm and 3.4 Å for 880 nm). These points are equal to the classical inner (620 nm) and outer (880 nm) turning points of $v = 2$. This correlation is reflected in the phase coordinate at low v values, where the window is centered around $\pm\pi$ for 620 nm and around 0 for 880 nm, which are the phases of the inner and outer turning points, respectively. For $v > 2$ the lines at $r = 3.1$ Å and at $r = 3.4$ Å cross the iso-energy lines at two points each (compare Fig. 27 and Fig. 8), corresponding to inward and outward motion at the resonance point, which explains the bifurcation of the phase dependency for larger energy values. Because of the oscillatory shape of the eigenstates, the window in the v representation is a wide oscillatory function, with a large peak at $v = 3$, but on the average the window is more sensitive to the lower vibrational states.

a. Transient Spectrum of Individual States. The effect of the initial vibrational state on the resulting spectral modulations was studied first. The transient absorption for the different $\rho_g^{s,a}(v,v')$ was calculated with a dissipation-free Hamiltonian. Examining Fig. 28 it is clear that the modulations of the probe signal increase upon excitation of the antisymmetric stretching mode. The phase of the signal is also shifted to earlier times because the wave packet enters the exit channel earlier.

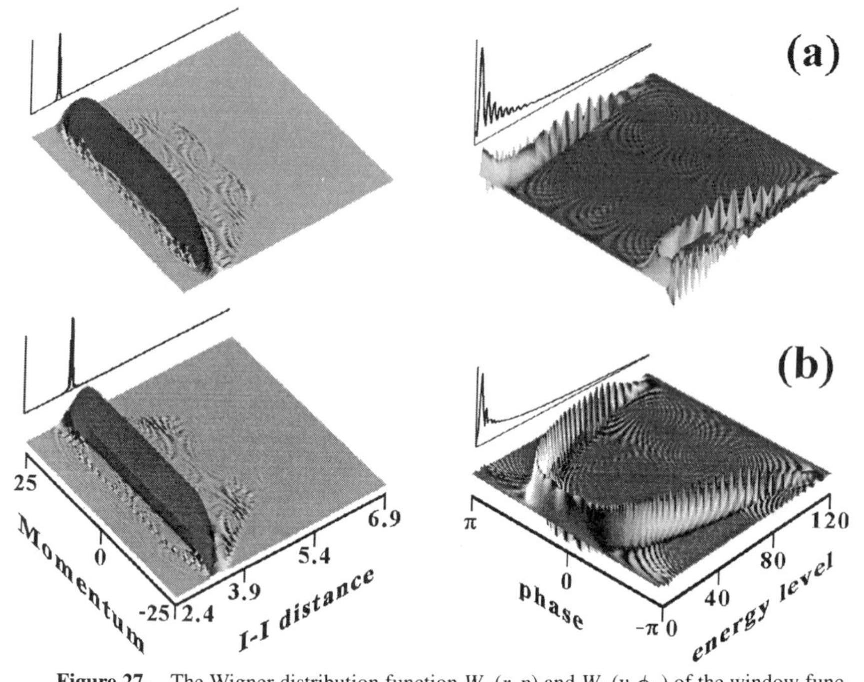

Figure 27. The Wigner distribution function $W_g(r,p)$ and $W_g(v,\phi_v)$ of the window function in Eq. (2.75) of a 60-fs pulse with central frequency of (*a*) 620 nm and (*b*) 880 nm. The projection on the position and vibrational energy coordinates is also shown.

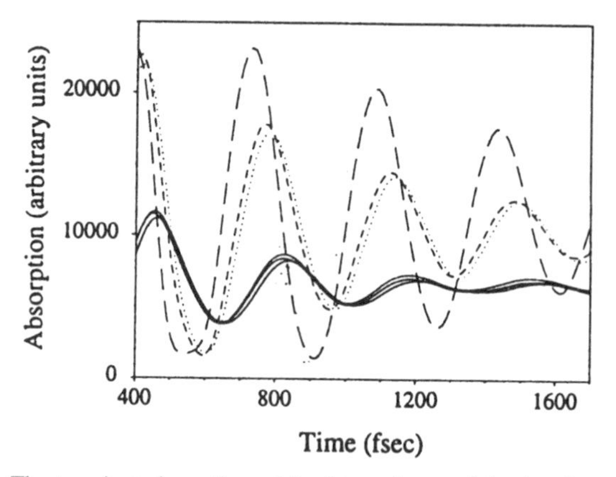

Figure 28. The transient absorption of the I_2^- products originating from different initial wave packets. The evolution was carried out with a dissipation-free Hamiltonian, and the spectrum calculated for a probe pulse of 60-fs duration and a wavelength of 620 nm. All signals are normalized with respect to the excited state population ($tr(\hat{\rho}_g^{s,a})$). The solid lines represent the pure symmetric stretch excitations $|v_{0,0}\rangle$, $|v_{1,0}\rangle$ and $|v_{2,0}\rangle$, which all bunch together. The antisymmetric stretch excitations are represented as broken lines: $|v_{0,1}\rangle$ (dotted), $|v_{1,1}\rangle$ (dashed line) and $|v_{0,2}\rangle$ (long dashed line).

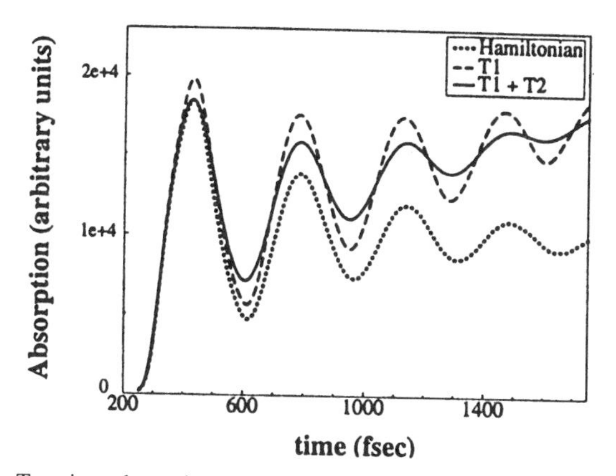

Figure 29. Transient absorption of a 60-fs probe pulse at 620 nm as a function of the time delay between the photodissociation pump and the probe. (*a*) Free Hamiltonian dynamics; (*b*) dynamics including energy relaxation processes (T_1); (*c*) dynamics including vibrational dephasing processes ($T_1 + T_2$).

b. Transient Spectrum Under Dissipation Conditions. To see the effect of the different dissipation mechanisms on the observed spectrum, the total population $\rho_g(v, v')$ was probed during propagation under the influence of the bath operators. Observing Fig. 29, the spectral modulations in the signal of the free evolution dynamics decay, due to the anharmonic nature of the potential, causing the winding motion in Fig. 26(*a*). The suprising observation is that addition of energy relaxation causes the modulations to become more persistent. The reason is that the high vibrational part of the distribution, which is prone to anharmonic motion, is pushed to the lower more harmonic part of the potential and refocused [Figure 26(*b*)]. This transfer is obtained without wiping out the vibrational coherence. The baseline of the modulation also increases, due to the flow of amplitude from the high vibrations into the observation window. As expected when dephasing is added, the spectral modulations decay [Fig. 26(*c*)]. This observation makes it possible to assign the decay of spectral oscillation primarily to dephasing processes.

c. Transient Spectrum at Different Probe Wavelengths. The last calculation (transient spectrum including T_1 and T_2 processes) was repeated for the red-shifted probe, expecting an anti-phased modulation as in the experiment (see Fig. 4). The resulting signal is shown, along with that resulting from the blue-shifted probe, in Fig. 30.

There are two deviations from the experimental scan. The first is the appearance of a double frequency second harmonic part for the red-shifted probe, which disappears after the second modulation (a dephasing process timescale). The second is the phase difference, which only approaches π at the end of the scan (a relaxation process time scale). Examining the window

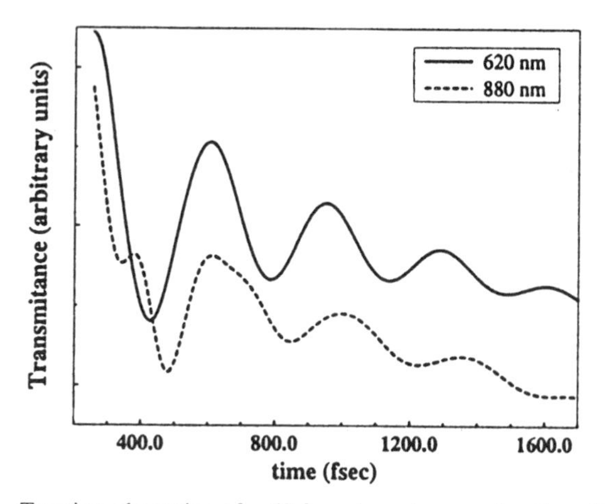

Figure 30. Transient absorption of a 60-fs probe pulse as a function of the time delay between the photodissociating pump and the probe. (*a*) Probe wavelength at 620 nm. (*b*) Probe wavelength at 880 nm.

functions of the probe pulses (Fig. 27), the source of the double frequency is obviously the large bifurcation of the phase dependency at 880 nm. As long as the phase distribution of the population is narrower than this bifurcation [this is true at short times—Fig. 26(*a*)], the distribution crosses the window twice at each cycle of its winding motion. This behavior vanishes as the distribution broadens, due to anharmonicity and dephasing processes [Fig. 26(*c*)]. The absence of this feature from the experimental scan might indicate the existence of earlier dephasing mechanisms not accounted for in the calculation, such as electronic dephasing during the pump pulse, or collisions with the cage prior to bond cleavage. It is also possible that this behavior is obscured by the sharp fall in absorption at the beginning of the scan, attributed to a transition to a higher electronic surface, which is not included in the model. The discrepancy of the phase difference between the different wavelengths is also explained by the shape of the window functions, in which a π phase difference is only present for very low values of v. Only when the distribution cools enough so that a significant portion of the population occupies the lower energy region would an anti-phased behavior appear. This suggests that the initial experimental distribution is cooler than the one obtained from the calculations of the pump pulse.

E. Vibrational Excitation of Relaxed I_2^-: The "Push" Pulse

Typically the dephasing rate is faster, or even much faster, than the energy relaxation rate. The observable consequence is that the spectral modulations have decayed long before the vibrational distribution reaches equilibrium. In order to regain the observable, the spectral modulations are reinvigorated by adding a push pulse that transfers amplitude to the dissociative excited surface while leaving a dynamical "hole" in the ground surface. Once created, the dynamical "hole" will oscillate, creating new spectral modulations. This phenomena is analogous to the RISRS experiment in the parent I_3^- molecule. The transient frequency of these oscillations can be associated with the energy distribution of the system, and the decay of the modulations can be associated with the dephasing rate. Since both the energy distribution and the phase coherence manifest themselves in the frequency and decay rate of the spectral modulations of the same observable, this experiment makes it possible to measure T_1 and T_2 simultaneously.

1. Statics

As in the previous section, the system under study is the I_2^- molecule, so the same degrees of freedom apply here as well. The difference is that the ground and excited electronic states are coupled by the push pulse, so the full dual-surface representation of the density matrix is needed to represent the system.

The initial state for the dynamics is $\hat{\rho}_g(t)$ of the previous section, where t is taken to be 2, 4, 6, 8, and 10 ps before the application of the pulse, $\hat{\rho}_e = \hat{\rho}_c = 0$.

Figure 31(a) demonstrates the state of the system after the dephasing process has smoothed out the initial phase dependence (as seen from the flat projection on ϕ_v), and with that eliminated the spectral modulations. At the same time, the relaxation process is still far from completion, as seen from the projection on v which peaks at $v = 10$ and has almost no population at $v = 0$.

The energy distribution of the population is slowly decaying toward a Boltzmann distribution due to T_1 processes, but even after as much as 10 ps thermal equilibrium is still not reached. Figure 32 shows the diagonal elements of the density $\langle v|\hat{\rho}|v\rangle$, which are the relative populations at each energy level, for all initial states.

2. Dynamics

a. Interaction with the Solvent and Electromagnetic Field. In view of the two electronic surfaces involved the dissipation mechanisms are divided into electronic and nuclear dissipative processes: $\hat{\mathcal{L}}_D = \hat{\mathcal{L}}_{D_{el}} + \hat{\mathcal{L}}_{D_{nd}}$, where $\hat{\mathcal{L}}_{D_{el}}$ is the electronic dephasing term and $\hat{\mathcal{L}}_{D_{nd}}$ is the nuclear dephasing. Nuclear relaxation processes are neglected for the duration of the pulse, to simplify the calculations. This is justified by the larger time scale of these processes ($\approx$ 3 ps) in comparison with the pulse's duration.

The main electronic dissipative process is dephasing. The mechanism involved originates from fast fluctuations of the dipoles in the solvent, which induce fluctuations in the energy gap between the ground and excited surface. This results in fluctuations of the transition frequencies between the surfaces [compare Eq. (2.60)], and the wave packet looses the coherence ($\hat{\rho}_c$) between its lower and upper parts ($\hat{\rho}_g$ and $\hat{\rho}_e$). By choosing $\hat{\mathbf{V}}_i = \hat{\mathbf{I}} \otimes \hat{\mathbf{S}}_z$ in Eq. (2.34) the electronic dephasing term is obtained

$$\hat{\mathcal{L}}_{D_{el}}(\hat{\rho}) = -\gamma_{el}[\hat{\mathbf{I}} \otimes \hat{\mathbf{S}}_z, [\hat{\mathbf{I}} \otimes \hat{\mathbf{S}}_z, \hat{\rho}]] \tag{3.13}$$

The relaxation coefficient can be related to the electronic dephasing time $\gamma_{el} = 1/\tau_{el}$. The electronic dephasing time τ_{el} was taken to be 30 fs. On the timescale of the experiment nonradiative decay process are negligible, therefore no other electronic dissipative terms are included in the calculation. For simplicity, the nuclear dephasing term is set as a pure dephasing term of the ground surface [Eq. (3.10)]

$$\hat{\mathcal{L}}_{D_{nd}}(\hat{\rho}) = -\gamma_{nd}[\hat{\mathbf{H}}_g \otimes \hat{\mathbf{I}}, [\hat{\mathbf{H}}_g \otimes \hat{\mathbf{I}}, \hat{\rho}]] \tag{3.14}$$

The Hamiltonian is constructed in the eigenstate representation. The operator $\hat{\mathbf{H}}_g$ is the same diagonal matrix as in the previous section, but since $\{|v_i\rangle\}$ are not eigenstates of the upper surface Hamiltonian, $\hat{\mathbf{H}}_e$ is not diagonal and its matrix elements are calculated using Eq. (2.29).

The push pulse has the same characteristics as the pump pulse in Section III.B, except for the carrier frequency, which is attuned to the $I_2^- \rightarrow I^- + I$ electronic transition, slightly blue-shifted from the center of absorption at 616 nm. When the electronic surfaces are coupled by the pulse, population is fed to the upper repulsive potential, and quickly propagated very far from the Frank–Condon region. This "runaway" population is the dissociated I + I^-, which is not probed by the experiment. It will be discarded at the end of the pulse, when the focus of attention is back on the ground I_2^- state. To avoid the need to represent these long propagations, an absorbing boundary condition was imposed on the excited surface.[6] The boundary was set in the eigenstate representation for large values of v, since the highest vibrational states of the ground surface correspond to large propagation distance when projected on the excited surface.

Once the pulse is over and the electronic surfaces decouple, the system can be again treated as in Section III.D, considering only $\hat{\rho}_g$ and setting $\hat{\mathcal{L}}_D = \hat{\mathcal{L}}_{D_{nr}} + \hat{\mathcal{L}}_{D_{nd}}$.

[6]The absorbing boundary condition for a density matrix is a super-operator of the form $\hat{\mathcal{L}}_A(\hat{\rho}) = \{\hat{\rho}, \hat{\mathbf{V}}_{absorbing}\}_+$ added to the Hamiltonian. The value of $\hat{\mathbf{V}}_{absorbing}$ is zero on most of the grid and negative imaginary on the boundary. Any part of the density propagating under the influence of this anti-Hermitian potential will decay exponentially to zero before it can overflow the grid boundary.

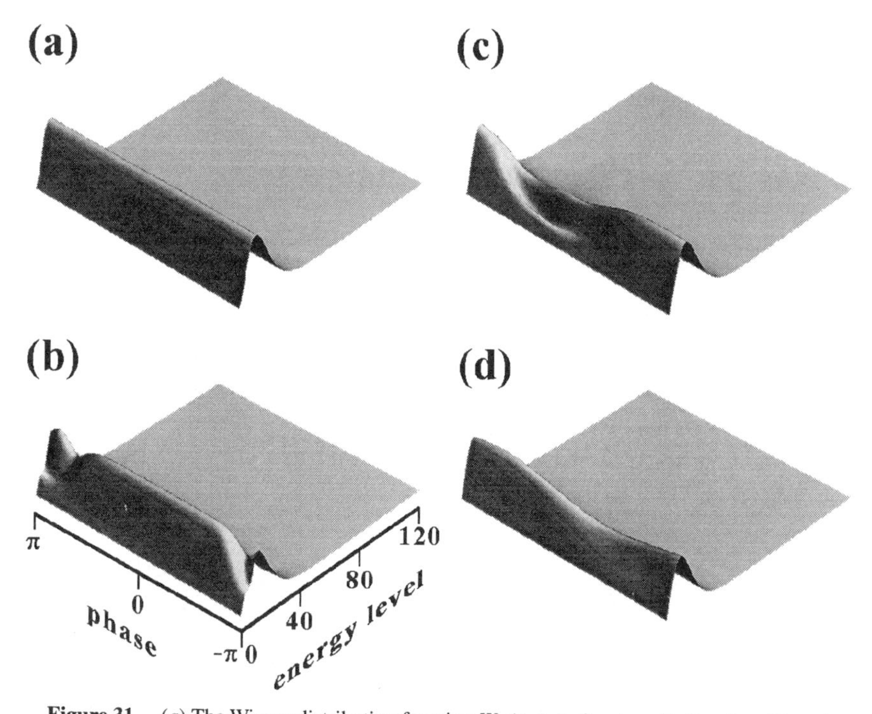

Figure 31. (*a*) The Wigner distribution function $W_g(v, \phi_v)$ after 4 ps. Notice that although the vibrational energy has not reached equilibrium the distribution is phase independent. (*b*) The same distribution function after the application of a push pulse of 60 fs at 616 nm, which photodissociated 20% of the I_2^- population. (*c, d*) The dynamics of the hole after 100,200 fs.

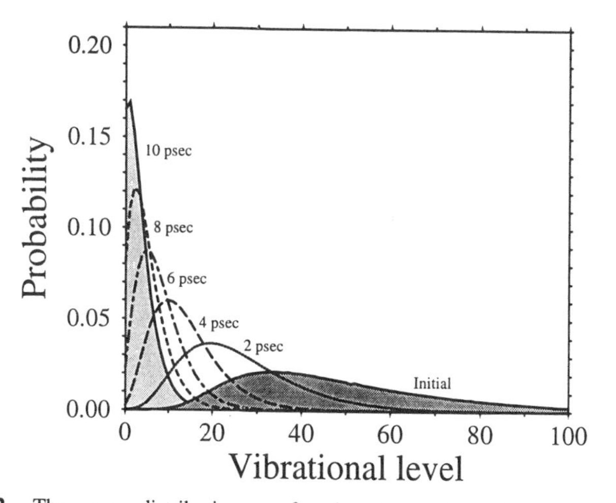

Figure 32. The energy distribution as a function of time delay from the pump pulse.

b. Propagation Results. The application of the push pulse in Fig. 31(*b*) has "drilled a hole" in a specific area in the phase coordinate (around $\pm\pi$), which will travel across ϕ_v [Fig. 31(*c*)] until smoothed out by the dephasing [Fig. 31(*d*)]. The winding motion along the phase coordinate renews the spectral modulations. As seen earlier (Fig. 26), the velocity of this motion depends on v, because of the anharmonicity of the vibrational manifold. Thus, the frequency of observed spectral modulations will change with the initial energy distribution at the onset of the push pulse. This phenomena was observed in experiment [4], and recreated in simulation by applying the push pulse at different time delays from the pump, and observing the resulting transient spectrum. The (r, p) picture of the same process is shown in Fig. 33. Notice that the push pulse has drilled two holes in the distribution [Fig. 33(*b*)], at the crossing of the energy iso-line and the resonance line at 3.1 Å (cf. Fig. 27). The hole travels clockwise along the iso-energy line [Fig. 33(*c*)], until smoothed out by the dephasing [Fig. 33(*d*)].

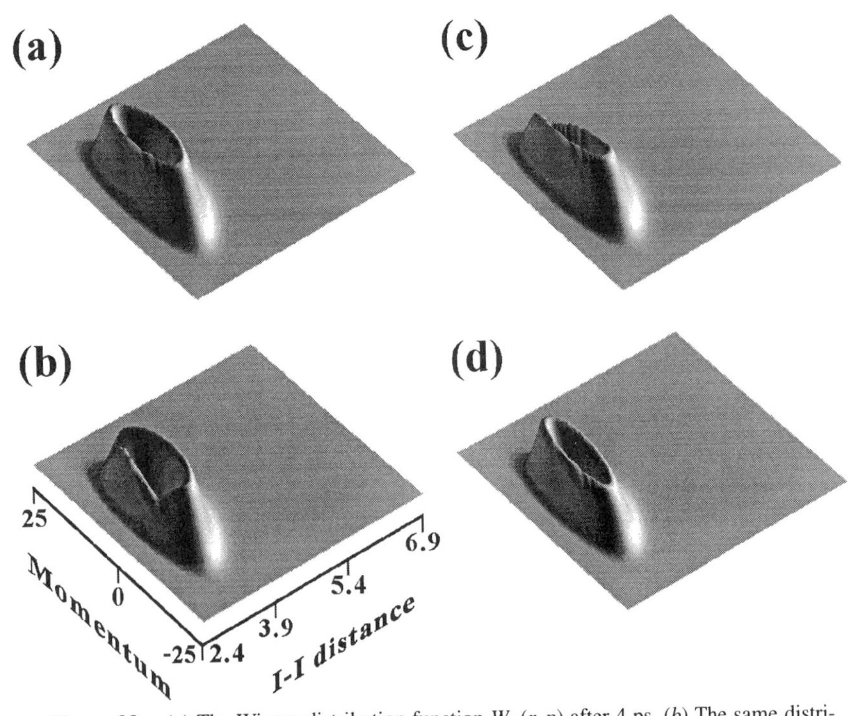

Figure 33. (*a*) The Wigner distribution function $W_g(r, p)$ after 4 ps. (*b*) The same distribution function after the application of a push pulse of 60 fs at 616 nm, which photo dissociated 20% of the I_2^- population. (*c, d*) The dynamics of the hole after 100,200 fs. Distances are in angstroms, and the momentum units are Å/ps.

3 *Interpretation*

The transient spectra for different pump-push delays are shown in Fig. 34. As in the experiment, there is an initial decrease in the absorption during the application of the pulse (due to loss of I_2^- population via photodissociation), followed by a decaying spectral modulation. The percent of photodissociated population falling within the pushing window function increases with time as the population cools and enters the center of the absorbing part of the push window function at lower vibrational levels (cf. Fig. 27 and Fig. 32). The same cooling process is also responsible for the observed increment in the frequencies of the spectral modulations. The modulations were fitted with a decaying harmonic behavior (Fig. 35), and the fit results are compared with the experiment in Table III.

The observed decay times of the modulations are very short (fast decay) for the hot populations measured at short delays, and saturate at a much longer value at longer delays. This trend is seen both in experiment and in simulation. The decay time of the modulations (τ_t) is smaller at high energies because there are contributions to the loss of coherence from the anharmonicity effect (cf. Fig. 29) as well as from the dephasing processes. The slight decrease in τ_t at longer delays can be attributed to the fact that the rate of the dephasing process varies as the square of the oscillator frequency in Eq. (3.11), which is larger at lower vibrational energies. It is satisfying to note that this trend will eventually lead to the chosen parameter $\tau_{nd}(0) \approx 1$ ps,

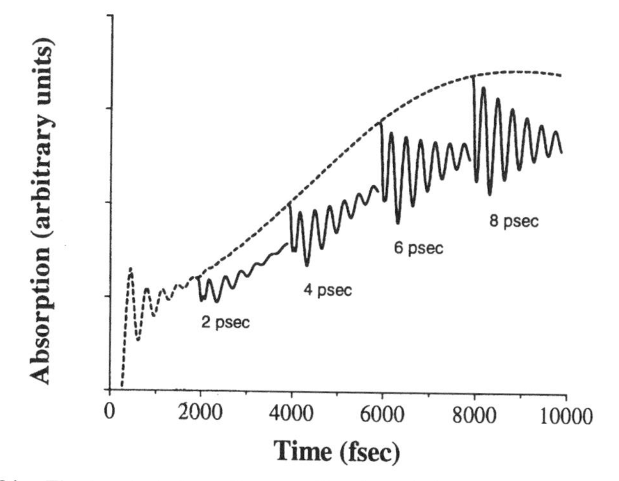

Figure 34. The transient absorption as a function of time showing the perturbation of the push pulse at different times on the dynamics. The dashed line shows the unperturbed dynamics. The pump pulse is at 308 nm, the push and probe at 616 nm.

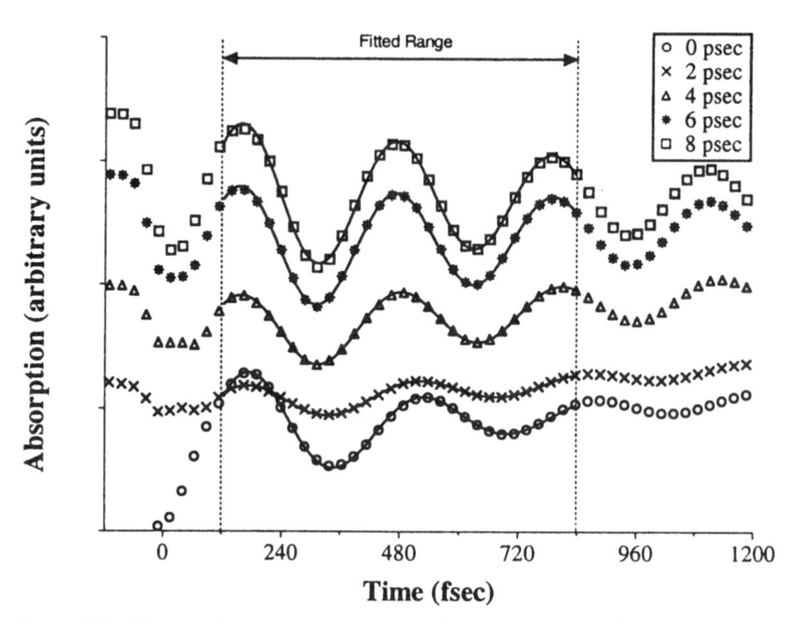

Figure 35. The transient absorption as a function of time delay after the push pulse for different push delays. The middle section was fitted to the function $C + A \cdot e^{-t/\tau_t} \sin(\omega_t t + \phi) + B \cdot t$.

meaning that the asymptotic damping rate of the modulations can be associated with a pure dephasing mechanism.

The values for the frequencies are harder to interpret, as their associated energies do not follow an exponential decay law (as should be expected from a T_1 process), in both experiment and theory. This can be explained by exam-

TABLE III
Fitted Parameters for the TRISRS Spectral Modulations

	Experimental [4]		Calculated	
Push Delay (ps)	ω_t (cm^{-1})	τ_t (ps)	ω_t (cm^{-1})	τ_t (ps)
No push			97 ± .2	0.4 ± .01
2.0	102 ± 4	0.4 ± .2	99 ± .2	0.7 ± .02
2.67	105 ± 3	0.9 ± .2		
4.0	112 ± 2	1.2 ± .3	106 ± .1	1.1 ± .02
6.0	112 ± 2	1.7 ± .4	108 ± .2	1.3 ± .05
8.0			109 ± .3	1.2 ± .07
10.0			110 ± .4	1.1 ± .08
13.4	113 ± 2	1.2 ± .2		

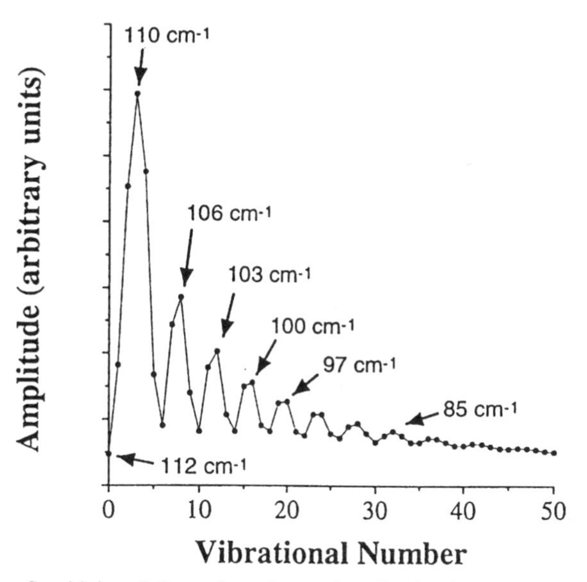

Figure 36. Sensitivity of the probe pulse to the vibrational quantum number, obtained from the diagonal elements of the window function in the v, v' representation. Some typical vibrational frequencies of different vibrational levels are also shown.

ining the sensitivity of the probe pulse to different vibrational levels. As was seen in Fig. 27, the sensitivity is not a smooth function of the vibrational level, but rather oscillatory, with large peaks at low vibrational levels. The behavior at the lower end of the spectrum is magnified in Fig. 36.

By comparing Fig. 31, Fig. 36, and Table III, it is clear that the observed modulation frequency is associated not with the mean energy of the distribution, but rather with its lower energy tail, because of the preference of the window function for lower energies. While the average energy obeys an exponential decay law, its lower energy tail quickly reaches the highest peak in the window at $v = 3$, and stays there throughout the relaxation process, thus masking the continuing decay of the rest of the distribution.

F. Application Summary

The key point in understanding the results of this section is that photoinduced events in the past propagate and influence the outcome of observables at later times. These events have a finite memory span due to electronic and nuclear dissipative processes induced by the solvent. The decay of the spectral modulations is due to a combination of dephasing and loss of compactness due to unharmonic dynamics. The energy relaxation processes are able to enhance the spectral modulations by transferring the population to lower more harmonic regions of the potential without losing the vibrational coherence. When the dephasing rate is larger than the energy relaxation rate, spectral modulations are invigorated by a push pulse, shedding light on the full scope of vibrational relaxation in solution.

IV. SUMMARY

A. Critical Evaluation

1. Theoretical Framework

The calculation methods developed and applied to the photodissociation dynamics of I_3^- are based on the strict rules of quantum mechanics. In Wigner's terminology "orthodox" quantum mechanics is followed, starting from state vectors to describe the system and operators to describe measurable observables. The evolution of the system is described by differential generators either $\hat{\mathbf{H}}$ for a closed system or $\hat{\mathcal{L}}$ for an open quantum system. All the methods applied are numerically exact, meaning that within the model, the measurable observables converge exponentially with respect to the numerical parameters. Moreover, the interaction with the radiation field is included explicitly and not perturbatively. Employment of these methods assures that the finite precision of the computation can be ruled out as a source of discrepancy with the experiment. The main difficulty in following such a program lies in the exponential scaling of quantum calculations with the number of degrees of freedom. This fact severely limits the scope of the calculation to a small number of modes. Under these conditions, one may wonder, can converged quantum calculation become relevant to an extremely complex scenario such as the photoreaction of I_3^- in solution?

2. Relevance to Experiment

The relevance of the simulation is closely tied to the experimental conditions under study. The photodissociation of I_3^- is characterized by a series of timescale separations. This situation enables the use of the impulsive limit in which the pulse duration is long compared to the electronic transition time, but short compared to the vibrational and rotational periods (see Table IV). Under these circumstances a carefully constructed reduced dimensional quantum description is able to capture the essential dynamics.

The main asset of the computational methods developed in this study are their flexibility in addressing different types of experimental probes. These can be time-dependent, such as the spectral modulations of the product; or time-independent, such as the absorption and Raman spectrum. Each probe gives information on different aspects of the dynamics: The absorption spectrum is associated with the Frank–Condon area of the excited surface; the Raman spectrum covers further ground down the dissociative slope, and can

TABLE IV
Timescales Involved in the I_3^- Photodissociation

Physical Entity		Timescale
Electronic transition	$1/\omega_0$	1 fs
Electronic dephasing	$1/\gamma_z$	5–60 fs
Pulse fluctuations	$1/\gamma_\phi$	100 fs
Pulse duration	t_f	30–60 fs
Pump Probe Delay time	t_{pr}	100–8000 fs
Bond cleavage	τ_{bond}	250 fs
Vibrational period	$1/\omega_v$	300 fs
Vibrational dephasing	$1/\omega_v^2\gamma_v$	400–1000 fs
Vibrational relaxation	T_1	3–10 ps
Reorientation	T_R	10–100 ps

distinguish between motions along the different coordinates; the transient spectrum has an accumulated memory of the entire potential all the way to dissociation, and may serve as a probe for those parts not covered by absorption and Raman spectra. Crossing the information gained from all probes is made possible because they are bundled together as the outcome of a unified framework.

The current simulations are able to point to the experimental levers that are the most relevant to understand the dynamics. The effects of the initial vibrational excitation in the reactants were found to be important. The extremely large effect that the excitation in the antisymmetric stretching mode has on every aspect of the photochemical outcome, be it vibrational excitation in the products, the emergence time of isolated fragments, or the depth of modulation in the transient spectrum, underscores this clearly. The antisymmetric excited population propagates in the antisymmetric direction from the very beginning of the dynamics. This leads to faster bond cleavage and lower vibrational excitation in the products, and helps to keep a compact wave packet for longer times, which enhances the observed spectral modulations. Of special interest here is the strong effect of symmetry breaking on the chemical evolution of the system. This is a particular example of a more general phenomena. The intimate nature of the mutual interaction between the system and the solvent in the present case underlines the prominent role that asymmetric solvation must play in determining the outcome of the reaction.

The calculated Raman spectrum associates the appearance of overtones of the antisymmetric stretch with early motion along the antisymmetric direction, which was shown to have a major part in the creation of spectral modulations. Such overtones, as well as the appearance of the fundamental antisymmetric frequency, were measured experimentally by Anne Meyers and coworkers [3] in resonance Raman spectra of I_3^- in several organic solvents, and proved to be solvent dependent. This indicates that the solvent has a significant influence on the early stages of the dynamics, and can encourage motion in the antisymmetric direction, by inducing nonsymmetric dipole fluctuations, thus breaking the potential symmetry. In other words, the solvent "helps" the molecule to decide which nuclei are primed to pair up to give the diatomic product. The appearance of the fundamental antisymmetric frequency (which is symmetry forbidden) in protonated solvents (alcohols and water) suggest that these solvents prefer an asymmetric configuration of the ion. Such help from the solvent was shown to be as valuable to the creation of coherent modulations as the impulsive nature of the pulse.

After its emergence, the photoproduct is subjected to various forces which influence the observed spectral modulations. There are energy transfer and dephasing processes originating from interaction with the bath, and there is the anharmonic nature of the I_2^- potential well. The latter makes different components of the wave packet travel at different phase velocities, thus loosing its compact shape, causing a damping of the modulations. Considering only the anharmonic effect, the wave packet is still fully coherent, and if no dissipation processes were taking place as well, a full revival [66] of the compact shape would be expected at a later time. Since dephasing processes are present, destroying the coherence between different populations, such recurrence is impossible. This makes the anharmonic effect practically (experimentally) indistinguishable from the damping caused by the dephasing process, and both contribute to the total decay of the oscillating signal. The TRISRS technique is the key for separating the two components: the anharmonic contribution decreases as the population cools, bringing the measured total decay rate to an asymptotic value equal to the pure dephasing rate.

The role of the energy relaxation in this process is somewhat surprising, as it enhances, rather than decreases, the observed modulations. This is because the relaxation narrows the energy band in which the product population exists, and does so in a coherent fashion, preserving the existing phase dependence and only diminishing the anharmonic effect. The stronger the coupling between the molecular system and the solvent bath, the more efficient this process. It might be that this mechanism is responsible for the more persistent modulations observed experimentally in the highly polar water, in contrast with the rapidly decaying modulations measured in the less polar isobutanol.

The present analysis of the TRISRS signals contains for the first time a complete representation of all dissipative mechanisms influencing the observables. The results confirm the conclusion of earlier analysis and demonstrate that this spectroscopy unveils the evolving vibrational density. As pointed out there, this method is especially well adapted for probing highly excited vibrational ensembles, complementing information obtained from the time-dependence of fragment absorption spectra. It is important to point out, however, that due to the strong bias any single probe frequency has toward detecting a specific region of the vibration ladder, it is highly recommended that this spectroscopy be implemented with multipole frequencies both of "push" and of probe pulses [4].

3. *Imagery*

Following events in time naturally submits itself to the use of visual tools. The phase space snapshots are chosen as the primary display. Without losing the full quantum picture one is able to directly compare classical and semiclassical calculations. The extension of the phase space image to action-angle representation supplies a new viewpoint at these complex dissipative processes. In particular it allows one to follow the role of phase and phase relaxation directly. This imagery further underlines the limited information content of any single spectroscopic probing method in capturing the full dynamic picture. It is often the case that the phase space picture shows dynamic richness and complexity, yet a given spectroscopic probe shows only a marginal evolution. Unraveling this richness is an immense experimental challenge and requires a multitude of probing techniques.

4. *Theoretical Alternatives*

The fully quantum approach applied to the I_3^- photoreaction is in no way exclusive. The exponential scaling of quantum calculations has served as the primary motivation for the development of a multitude of approximate methods. The most common of which is classical molecular dynamics [11]. Quantum effects are then introduced by employing semiclassical methods or a series of approximate mixed quantum-classical methods [12, 13, 14, 15, 16, 17]. These methods can be related on a mean field approach, usually formulated within the TDSCF approximation [67, 68, 69]. Although these methods are usually applied under uncontrollable approximations conditions, they are able to incorporate many degrees of freedom. Bearing in mind that the purpose of the calculation is to gain insight on the system under study, these methods can become extremely useful. The present approach limits the scope of the calculation in order to obtain a controllable model. Under these circumstances specific effects and their influence on the observables can be studied.

5. *Dissipative Model*

The dissipative quantum dynamics in this study attempts to reach beyond the well-studied two-level system coupled to a bath [32, 70, 71, 72]. The experimental conditions are such that a vast amount of energy has to be dissipated from the highly excited vibrational manifold, thus multilevel excitation is observed. Under these conditions phase coherence is maintained between many levels simultaneously.

The present approach to vibrational relaxation proves to be conceptually more straightforward than alternative treatments based on an eigenvalue decomposition of the density operator and its interactions [73]. In particular, conservation of coherence despite ongoing energy relaxation arises naturally from following the change in observables in time. This is a particular example of the advantages of an explicit time-dependent approach in studying transient phenomena.

The current dissipative model is based on a phenomenological approach that classifies and parameterizes a small class of dissipative mechanisms. The drawback of this approach is that the solvent properties are only implicitly addressed. It is possible by using the connection between the semigroup approach and the weak coupling limit [21, 30, 73, 74] to relate the semigroup parameters with Fourier transforms of bath correlation functions. This approach has been used by Pugliano et al. [9] to obtain Master equation parameters for the relaxation of IHgI in solution.

A key point in applying a reduced dynamical approach is where to position the partition between the system and bath. This partition determines which degrees of freedom are treated explicitly and which implicitly. The choice in this study, based primarily on computational considerations, was to follow the molecular identities, therefore the primary degrees of freedom are the ones associated with iodine. The large mass mismatch and the similarities of many spectroscopic observables in different solvents partly justify this choice. There is growing tendency to include in the primary zone at least one collective solvation mode. This approach is used in the Brownian oscillator model [75] to couple the electronic and nuclear dissipative dynamics. A collective solvent mode has also been used in the description of the recombination dynamics of I_2^- [64, 76, 77]. The addition of a cage or solvent degree of freedom to the present calculation seems a natural extension. Its inclusion would also correct for inhomogeneous effects not considered presently.

6. *Potentials*

Another drawback of current computation model is the inability to calculate good *ab initio* potentials for this electronic complex system. The hypothesis underlying all the computations in this work, namely that the photodissociation reaction only involves two electronic surfaces, is naive. Preliminary calculations show that the large number of electrons and the strong spin–orbit coupling create a multitude of intercrossing surfaces and recombination paths, which should be incorporated in the simulation. The addition of more electronic surfaces is easily achieved within the developed framework, and is not very costly: the computation scales as the square of the number of surfaces, and not exponentially with it (as is the case with additional degrees of freedom). The effect of the solvent on the potential energy surfaces also has to be addressed. The charge distribution in the ions is the key quantity. Progress in this direction is on its way.

B. Conclusions

The causal approach of following events in time is the key in gaining insight into a complex dynamical encounter of the photodissociation of I_3^- in solution. Therefore the combination of ultrafast spectroscopic methods and time-dependent quantum mechanical methods naturally fit together to study a process that is characterized by a series of timescale separations.

Acknowledgments

We thank David Tannor and S. Rice for stimulating discussions. This research was supported by Moked, Israel Science Foundation. The Frarkas and the Fritz Haber Research Centers are supported by the Minerva Gesellschaft für die Forschung, GmbH München, FRG.

APPENDIX A: NUMERICAL METHODS

1. Approximating Functions of Operators

Iterative propagation schemes have become the methods of choice in quantum dynamical modeling and simulations. The reason is their superior efficiency when the size of the problem increases. These schemes are based on the ability to perform numerically the elementary step of mapping a state vector by an operator (e.g., $|\phi\rangle = \hat{\mathbf{H}}|\psi\rangle$), or a density matrix by a super-operator (e.g., $\hat{\sigma} = \hat{\mathcal{L}}\hat{\rho}$).[7] The propagators are defined by their recursive application of the elementary mapping step [34]. This amounts to approximating a function of the Hamiltonian as a polynomial.

The original propagation scheme [78], was developed to solve the time-dependent Schrödinger equation. The evolution operator $\hat{\mathbf{U}}(t) = e^{-i\hat{\mathbf{H}}t/\hbar}$ was approximated as a polynomial. A spectral expansion based on the Chebychev orthogonal polynomial was used leading to

$$\psi(t) = e^{-i/\hbar\hat{\mathbf{H}}t}\psi(0)$$

$$\approx e^{-i/\hbar(\Delta E/2 + E_{min})t} \sum_{n=0}^{N_{ch}-1} a_n \left(\frac{\Delta E t}{2\hbar}\right) T_n(\hat{\mathbf{H}}_{\mathbf{norm}})\psi(0) \qquad (A.1)$$

[7]Since the algebra of vectors and operators is isomorphic to the algebra of matrixes and superoperators, a general notation will be used from here on. ξ will represent a generalized state, and O a generalized operator, so $|\xi\rangle$ and $\hat{\mathbf{O}}$ might as well be read as $\hat{\xi}$ and $\hat{O}$.

where the expansion coefficients become $a_n(\alpha) = i^n(2 - \delta_{n0})J_n(\alpha)$ and $T_n(x)$ are the Chebychev polynomials: $T_n(\cos\theta) = \cos(n\theta)$ [34]. For stability in Eq. (A.1) the Hamiltonian is normalized: $\hat{\mathbf{H}}_{\mathbf{norm}} = 2(\hat{\mathbf{H}} - \overline{H})/\Delta E$ where $\overline{H} = (E_{max} + E_{min})/2$ is the center of the eigenvalue spectrum and $\Delta E = (E_{max} - E_{min})$ is the eigenvalue range of $\hat{\mathbf{H}}$. The normalized Hamiltonian has its eigenvalues distributed on the real axis between -1 and 1.

Examining Eq. (A.1) it can be noticed that the time variable only appears in the expansion coefficients a_n. The computationally intensive part, which is the evaluation of the mapping induced by the Chebychev polynomial: $\phi_n = T_n(\hat{\mathbf{H}}_{norm})\psi(0)$ is time-independent. This observation has led to the development of propagation schemes for other functions of the Hamiltonian. Examples include the Green function, allowing the calculation of Raman spectra [35], or reactive scattering cross sections [79, 80]; the delta function, allowing the calculation of absorption spectra [81] and density of states [82, 83]; propagation in imaginary time [84] and filter diagonalization [85, 86, 87], allowing the calculation of eigenstates. The method can be classified as a spectral expansion of a function of the Hamiltonian operator.

For approximating an analytic function $f(z)$ the spectral expansion possess exponential convergence [78]. Comparison of the Chebychev propagator with other propagation schemes has shown that the Chebychev expansion is usually superior in both accuracy and efficiency to other methods [88].

These findings have led to a prolification of the use of the algorithm and to the exploration of its range of validity. It was found that the original Chebychev algorithm can become unstable when the Hamiltonian operator $\hat{\mathbf{H}}$ is not Hermitian. A complex non-Hermitian Hamiltonian arises naturally when absorbing boundary conditions are introduced [65, 89, 90, 91, 92, 93, 94]. The reason for the instability is that support for the Chebychev polynomials is on the real axis. Although the Chebychev method can tolerate some complex character [90], large complex eigenvalues of the Hamiltonian cause severe instability. Complex eigenvalues are also obtained for the Liouville super-operator in a dissipative environment [20, 21]. Solving the Liouville von Neumann equation for dissipative open systems was the motivation for developing an alternative propagation scheme that could tolerate complex eigenvalues [95].

The new propagator was based on the Newtonian interpolation polynomial. The support points or interpolation points were located on a polygon in the complex plane, and therefore tolerated complex eigenvalues of the Liouville super-operator that are contained within the polygon. Using the theory of interpolation in the complex plane it can be shown that a uniform converging interpolation in a domain D is obtained when the interpolation points are located on the circumference of the domain.

The choice and ordering of the points is crucial to the stability of the algorithm. Evenly distributed points on the exterior of the polygon were obtained by a conformal mapping of the polygon onto a circle, where evenly distributed points are easily obtained by symmetry considerations. An inverse transform distributed the points back onto the polygon. Ordering the points was found to be crucial to the stability of the algorithm and to directly influence the calculation of the divided difference coefficients. It was found that to obtain stability a complete staggering of points was required [95]. The scheme was applied to the evolution operator in simulating photoinduced processes in solution [45, 2] and photoinduced desorption from metal surfaces [96]. Another application has been the calculating of the S matrix in reactive scattering using absorbing boundaries [97, 98].

In practical applications the Newtonian-based propagator was found to be hard to use. The difficulty could be traced to the Schwartz–Christoffel conformal mapping algorithm [99], which is required to obtain the uniformly distributed points on the circumference of the interpolation domain. The mapping algorithm severely limited the order of the Newtonian interpolation polynomial in the complex plane. This is in contrast to Newtonian interpolation on the real axis, where no limit to the order of the polynomial was found [34, 100]. In this work a new approach to defining the interpolation points (termed Leja points [101]) is used. This method is able to overcome the difficulty in locating the interpolation points.

In parallel with these developments, the original Chebychev expansion was generalized, first by shifting the support from the real axis to a line shifted into the complex plane [83]. This shift greatly enhances the stability of the method. The domain of stability becomes an ellipse in the complex plane. Another alternative is to modify the recursion relation of the Chebychev polynomial by adding a damping term [102]. A more rigorous fix to the problem is to define a spectral expansion in the complex plane. The Faber polynomials that are a generalization of the Chebychev polynomials constitute such a set. With the use of the Faber polynomial it has recently been shown that a stable uniform approximation in the complex plane is possible [103].

Before continuing, a brief comparison of the two methods is appropriate. If for the Newtonian propagator the zeros of the Chebychev polynomial are chosen as sampling points, the two methods are numerically equivalent [34]. Formally, if the expansion coefficients in the Chebychev series are calculated by a Gaussian–Chebychev quadrature rule, then the expansion becomes an interpolation formula mathematically equivalent to the Newtonian interpolation formula [34]. For the practitioner it will be shown that the Newtonian method is more flexible when different functions of the Hamiltonian are required simultaneously.

2. Newton's Interpolation Method

The propagation method is based on the Newtonian interpolation formula in which an analytic function $f(z)$ is approximated as a polynomial

$$f(z) \approx \mathcal{P}_{N-1}(z) \equiv \sum_{n=0}^{N-1} a_n \prod_{j=0}^{n-1} (z - x_j) \tag{A.2}$$

By definition, at the sampling points x_j, $f(x_j) \equiv \mathcal{P}(x_j)$. The coefficient a_n is the nth divided difference coefficient [104] defined as

$$a_0 = f[x_0] = f(x_0)$$

$$a_1 = f[x_0, x_1] = \frac{f(x_1) - f(x_0)}{x_1 - x_0} \tag{A.3}$$

$$a_k = f[x_0, x_1, \ldots, x_k] = \frac{f(x_k) - \mathcal{P}_{k-1}(x_k)}{\prod_{j=0}^{k-1} (x_k - x_j)} \tag{A.4}$$

The advantage of this method is the complete freedom in choosing the function $f(z)$. The only demand is the ability to calculate its value at the sampling points so it is even possible to interpolate an integral of a function that can be only integrated numerically, e.g., Eq. (2.89), which is not solvable in the Chebychev method.

When interpolating the function of an operator, rather than of a scalar, the same divided difference coefficients are used in the expansion

$$f(\hat{\mathbf{O}}) \approx \mathcal{P}_{N-1}(\hat{\mathbf{O}}) = \sum_{n=0}^{N-1} a_n \prod_{j=0}^{n-1} (\hat{\mathbf{O}} - x_j \hat{\mathbf{I}}) \tag{A.5}$$

The choice and order of the interpolation points, x_j is the crucial step in the algorithm.

3. Leja's Interpolation Points

The first step is to establish the domain D of eigenvalues of $\hat{\mathbf{O}}$,[8] and shift it on the real axis so it is symmetric with respect to the imaginary axis. Once

[8]For the Hamiltonian operator $\hat{\mathbf{H}}$, the extent of D on the real axis is from $\hat{\mathbf{H}}_{\min} = \hat{\mathbf{V}}_{\min}$ to $\hat{\mathbf{H}}_{\max} = \hat{\mathbf{V}}_{\max} + \hat{\mathbf{P}}^2_{\max}/2m$. The extent on the imaginary axis is from 0 to $-\hat{\mathbf{V}}^{\max}_{\text{absorbing}}$. For the Liouville superoperator $\hat{\mathcal{L}}$, the real extent is from $\hat{\mathcal{L}}_H^{\min} = \hat{\mathbf{H}}_{\min} - \hat{\mathbf{H}}_{\max}$ to $\hat{\mathcal{L}}_H^{\max} = -\hat{\mathcal{L}}_H^{\min}$, and the imaginary from 0 to $\hat{\mathcal{L}}_D^{\max}$, which is negative.

the domain is defined the algorithm used to generate the interpolation points can begin:

1. A line encircling the domain D in the complex plain is defined. For practical purposes it will be chosen as a polygon. The domain is scaled in size, without changing its shape, to make the interpolation process stable. While its exact size will be fixed in step 4, initial coordinates should be of the order of 1.
2. Trial points $\{y_i\}$ are calculated to be equally distributed on the circumference contour of the domain D. The number of trial points is 1.5 to 3 times the number of requested interpolation points.
3. The interpolation points $\{x_i\}_{i=0}^{N-1}$ are chosen from $\{y_i\}$. The first interpolation point can be chosen arbitrarily

$$x_0 = y_0 \tag{A.6}$$

Other interpolation points are chosen so they maximize the denominator of Eq. (A.4). After choosing n such points, the product

$$\mathcal{I}(y_i) = \prod_{j=0}^{n-1} (y_i - x_j) \tag{A.7}$$

is calculated for each trial point y_i. The trial point for which $\mathcal{I}$, Eq. (A.7), is maximal becomes x_n. If Eq. (A.7) goes to zero or infinity for large n's, the size of D should be adjusted to correct that [e.g., scaled down if Eq. (A.7) overflows].
4. The optimal interpolation points are calculated by normalizing the size of D. A point z in the center of the domain is chosen arbitrarily, and a normalizing factor is calculated by

$$\rho = \prod_{j=0}^{N-1} (z - x_j)^{1/N} \tag{A.8}$$

Each of the x_j's is then divided by ρ, to yield $\tilde{z}_j$. The result is N sampling points on the contour of a scaled domain $\tilde{D}$. The normalization is essential to keep Eq. (A.2) stable. If $\tilde{D}$ is too small this will result in divergence of the divided differences (the a_k's), while if it is too large it will diverge the product term in Eq. (A.2).

4. Application to Operators

After choosing the interpolation points, the operator is shifted and scaled so all of its eigenvalues reside inside the domain $\tilde{D}$

$$\tilde{\mathbf{O}} = (\hat{\mathbf{O}} - \overline{O}) \cdot \frac{1}{\sigma} \tag{A.9}$$

Here, $\overline{O}$ is the shift,[9] and σ is a scaling factor. To compensate for the change from $\hat{\mathbf{O}}$ to $\tilde{\mathbf{O}}$, the interpolation polynomial is used to approximate a scaled function $\tilde{f}(z) = f(z\sigma + \overline{O})$

$$\begin{aligned} f(\hat{\mathbf{O}})|\xi\rangle &\equiv \tilde{f}(\tilde{\mathbf{O}})|\xi\rangle \approx \tilde{\mathcal{P}}_{N-1}(\tilde{\mathbf{O}})|\xi\rangle \\ &\equiv a_0|\xi\rangle + a_1(\tilde{\mathbf{O}} - \tilde{z}_0)|\xi\rangle + a_2(\tilde{\mathbf{O}} - \tilde{z}_1)(\tilde{\mathbf{O}} - \tilde{z}_0)|\xi\rangle \end{aligned} \tag{A.10}$$

with the $\tilde{z}_k$'s residing on the contour of $\tilde{D}$, and the a_k's calculated by

$$a_0 = \tilde{f}(\tilde{z}_0) \tag{A.11}$$

$$a_k = \frac{\tilde{f}(\tilde{z}_k) - a_0 - \sum_{l=1}^{k-1} a_l\,(\tilde{z}_k - \tilde{z}_0)\cdots(\tilde{z}_k - \tilde{z}_{l-1})}{(\tilde{z}_k - \tilde{z}_0)\cdots(\tilde{z}_k - \tilde{z}_{k-1})} \tag{A.12}$$

To calculate the product terms in Eq. (A.10) a recursive relation is used

$$\begin{aligned} |\phi_0\rangle &= |\xi\rangle \\ |\phi_1\rangle &= (\tilde{\mathbf{O}} - \tilde{z}_0\hat{\mathbf{I}})|\phi_0\rangle \\ |\phi_{n+1}\rangle &= (\tilde{\mathbf{O}} - \tilde{z}_n\hat{\mathbf{I}})|\phi_n\rangle \end{aligned} \tag{A.13}$$

The final result is obtained by accumulating the sum

$$|\phi\rangle = \sum_{n=0}^{N-1} a_n|\phi_n\rangle \tag{A.14}$$

The sum is truncated when the residuum $a_n|||\phi\rangle_n||$ is smaller than a prespecified tolerance. Since the quality of the approximation of the function $f(\hat{\mathbf{O}})$ is equivalent to a scalar function in the domain D, before performing the actual calculation the accuracy can be checked on the scalar function. Figure 37 shows contour maps of the accuracy of the interpolation for some test cases. A few guidelines for choosing the interpolation points can be deduced from experience and from these figures:

1. It becomes obvious that when using interpolation points residing only on the real axis (as in the original Chebychev algorithm) the domain

[9]The real extent of the eigenvalues of $\hat{\mathcal{L}}$ is symmetric with respect to the imaginary axis, so $\overline{O}$ is identically zero in this case.

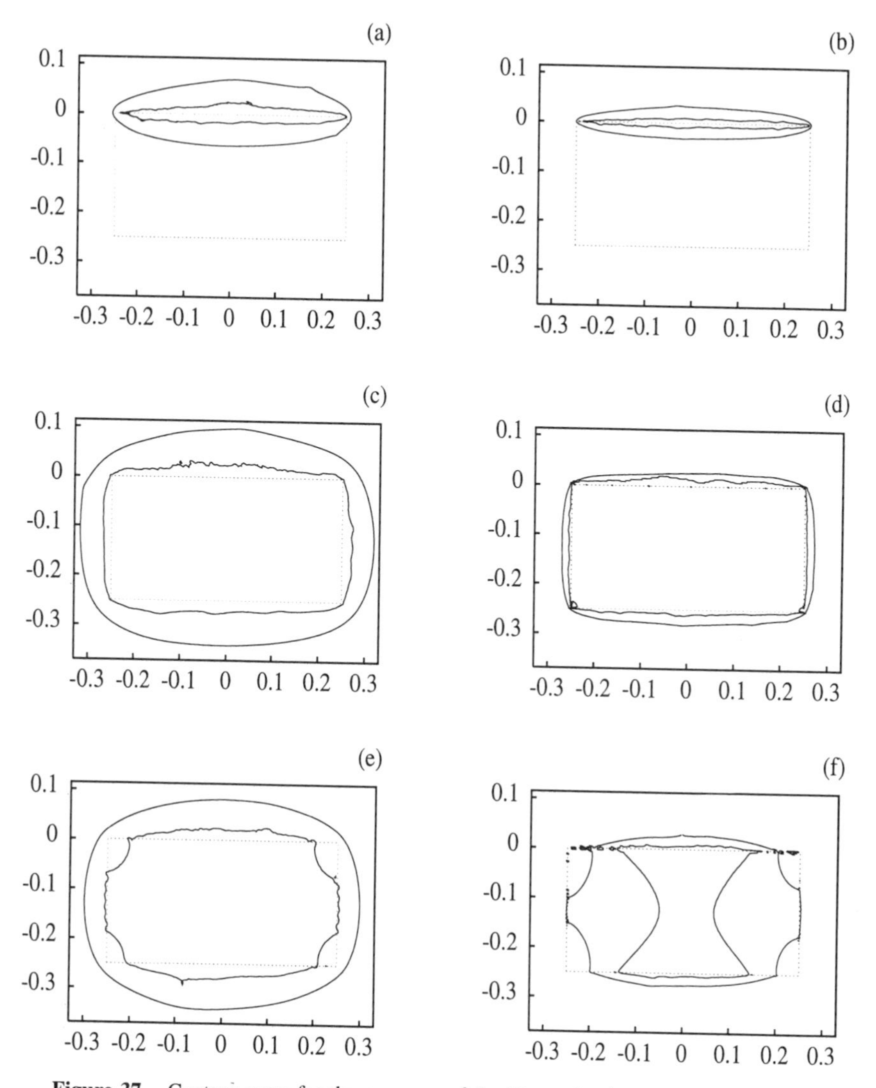

Figure 37. Contour maps for the accuracy of the Newtonian interpolation. The approximated function is an evolution operator as in [Eq. (2.41)], with a time step of 100 a.u. and a Hamiltonian with $\Delta E = \hat{\mathbf{H}}_{\max} - \hat{\mathbf{H}}_{\min} = 0.5$ a.u. The inner line is the boundary of the domain in which the relative error of the interpolation is less than 10^{-5}, the outer line is for relative error larger than 1 (stability boundary). (*a*) and (*b*) were calculated with the Chebychev algorithm, with interpolation points on the real axis; (*c*)–(*f*) were calculated with Newtonian interpolation, with interpolation points on the dotted rectangle. The number of points used in each map is: (*a*) 64; (*b*) 128; (*c*) 64 chosen from 96; (*d*) 200 chosen from 300; (*e*) 64 chosen from 67; (*f*) 200 chosen from 210.

of stability is a small region around the real axis. Choosing the same number of points on the circumference of a rectangular domain leads to a much better coverage in the complex plain. If the domain contains all of the eigenvalues of the interpolated operator, stability is assured for very long time steps. Figure 37(*a*) and Fig. 37(*b*) show the domain of convergence of the Chebychev scheme.

2. The number of trial points has to exceed the number of actual interpolation points. If too few trial points are used, the interpolation becomes inaccurate, especially in the vicinity of the sharp corners of the domain. Higher order interpolation polynomials require more trial points, since the density of the points increases, making the divided difference terms more sensitive to the choice of the interpolation points. The actual number of trial points needed for a low order ($N < 100$ terms) polynomial is 1.5 times the number of interpolation points, while for higher order polynomials the ratio will be bigger (for an 800-term polynomial a ratio of 1 : 3 was needed). Too many trial points will slow the calculation, but this calculation is performed only once before the propagation cycle. Comparison of Fig. 37(*c*) with Fig. 37(*e*) and of Fig. 37(*d*) with Fig. 37(*f*) shows the effect of not choosing enough trial points on the convergence domain.
3. Employing more interpolation points than is required for obtaining the desired accuracy inside the domain hampers the calculation. As a result the stability area shrinks and if some eigenvalues reside outside, but close to, the domain the accuracy is degraded. This is the reason why the Chebychev algorithm is stable only for short time steps when an absorbing potential is employed, but diverges when using larger time steps that require higher order polynomials. For comparison, the results in Section III.B.2 were calculated using 2000 a.u. (atomic units) time steps (700 terms in the polynomial). The same calculations carried out with the Chebychev algorithm diverged for time steps larger than 100 a.u. (64 terms). This effect can be seen by comparing Fig. 37(*a*) with Fig. 37(*b*), or Fig. 37(*c*) with Fig. 37(*d*).

APPENDIX B: SYSTEM PARAMETERS

TABLE B.I
Potential Surfaces Parameters in Atomic Units

Ground Potential Surface for I_3^- [11]

$$V_g(r_{ab}, r_{bc}) = \frac{1}{2}k(r_{ab} - r^{eq})^2 + \frac{1}{2}k(r_{bc} - r^{eq})^2 + \chi(r_{ab} - r^{eq})(r_{bc} - r^{eq})$$

$$k = 0.04598 \qquad \chi = 0.01323 \qquad r^{eq} = 5.480$$

Excited Potential Surface for I_3^- [11]

$$V_e(r_{ab}, r_{bc}, r_{ac}) = \Delta E_0 + Q_1 + Q_2 + Q_3 - (J_1^2 + J_2^2 + J_3^2 - J_1 J_2 - J_2 J_3 - J_3 J_1)^{1/2}$$

$$Q_i = \tfrac{1}{2}[{}^1E(r_i) + {}^3E(r_i)] \qquad {}^1E(r_i) = {}^1D[1 - e^{-{}^1\beta(r_i - {}^1r^{eq})}]^2 - {}^1D$$

$$J_i = \tfrac{1}{2}[{}^1E(r_i) - {}^3E(r_i)] \qquad {}^3E(r_i) = {}^3D[1 + e^{-{}^3\beta(r_i - {}^3r^{eq})}]^2 - {}^3D$$

$${}^1r^{eq} = 6.104 \quad {}^1\beta = 0.6138 \quad {}^1D = 0.0404$$

$${}^3r^{eq} = 5.637 \quad {}^3\beta = 0.5292 \quad {}^3D = 0.0371 \quad \Delta E_0 = 0.1361$$

Ground Potential for I_2^- [4]

$$V_g r(r) = D[1 - e^{-\beta(r_i - r^{eq})}]^2$$

$$r^{eq} = 6.104 \quad \beta = 0.6138 \quad D = 4.04 \cdot 10^{-2}$$

Excited Potential for I_2^- [4]

$$V_e x(r) = D + V e^{-\beta(r - r^{eq})}$$

$$r^{eq} = 6.104 \quad \beta = 1.852 \quad D = 4.04 \cdot 10^{-2} \quad V = 2.168 \cdot 10^{-3}$$

Absorbing Potential [65]

$$V(\bar{y}) = -i \cdot A \cdot N \cdot e^{-2/\bar{y}} \qquad \bar{y} = (y - y_i)/(y_f - y_i)$$

$$A = 0.018 \qquad N = 13.22 \quad y_i = 8.94 \qquad y_f = 9.69$$

$$V(\bar{v}) = -i \cdot A \cdot \bar{y}^2 \qquad \bar{v} = (v - v_i)/(v_f - v_i)$$

$$A = 0.05 \qquad v_i = 100 \quad v_f = 120$$

TABLE B.II
Electromagnetic Field Parameters in Atomic Units

$$\epsilon(t) = \bar{\epsilon}(t) e^{i\omega t} \qquad \bar{\epsilon}(t) = A e^{\frac{-2 \ln 2}{\tau^2}(t - t_0)^2}$$

Pump Pulse [1]

$\omega = 0.1479$ $\quad A = 2.5 \cdot 10^{-4}$ $\quad \tau = 2400$ $\quad t_0 = 4800$

Dipole function $\quad \hat{\mu} = a \cdot \hat{\mathbf{I}}$ $\quad a = 3.7$

Push Pulse [4]

$\omega = 7.396 \cdot 10^{-2}$	$A = 5.9 \cdot 10^{-4}$	$\tau = 2400$	$t_0 = 4800$
Dipole function	$\hat{\mu} = a \cdot \hat{\mathbf{I}}$	$a = 2.24$	

TABLE B.III
Typical Parameters of Propagation in Atomic Units

Mass Scaled Grid

$\Delta x = 1.2\ 10^{-2}$	$N_x = 1024$	$x_{\min} = 6.24$
$\Delta y = 1.2\ 10^{-2}$	$N_y = 512$	$y_{\min} = 3.42$
Mass	$2.315 \cdot 10^5$	

Propagation

With pulse	$\Delta t = 200$	Order = 70
Without pulse	$\Delta t = 1000$	Order = 200

REFERENCES

1. U. Banin, R. Kosloff, and S. Ruhman, *Israel. J. Chem.*, **33,** 141 (1993).
2. U. Banin, A. Bartana, S. Ruhman, and R. Kosloff, *J. Chem. Phys.* **101,** 8461 (1994).
3. A. E. Johnson and A. B. Myers, *J. Chem. Phys.*, **102,** 2519 (1995).
4. U. Banin, R. Kosloff, and S. Ruhman, *Chem. Phys.*, **183,** 289–307 (1994).
5. T. Kühne and P. Vöhringer, *Ultrafast Phenomena*, 1996.
6. S. Ruhman, E. Gordon and E. Gershgoren, *Ultrafast Phenomena*, 1996.
7. M. Gruebele, M. Dantus, R. M. Bowman, and A. H. Zewail, *J. Chem. Phys.*, **91,** 7437 (1989).
8. L. R. Khundkar and A. H. Zewail, *Annu. Rev. Phys. Chem.*, **41,** 15 (1990).
9. N. Pugliano, D. K. Palit, A. Z. Szarka, and R. M. Hochstarsser, *J. Chem. Phys.*, **99,** 7273 (1993).
10. N. Pugliano, A. Z. Szarka, S. Gnanakaran, M. Triechel, and R. M. Hochstarsser, *J. Chem. Phys.*, **103,** 6498 (1995).
11. I. Benjamin, U. Banin, and S. Ruhman, *J. Chem. Phys.*, **98,** 8337 (1993).
12. T. J. Martinez, M. Ben-Nun, and G. Ashkenazi, *J. Chem. Phys.*, **104,** 1996.
13. G. D. Billing, in *Numerical Grid Methods and Their Applications to Schrödinger's Equation*, C. Cerjan, Ed., Kluwer Academic Publishers, The Netherlands, 1993, p. 121.
14. R. Kosloff and A. D. Hammerich, *Faraday Discuss. Chem. Soc.*, **91,** 239–247 (1991).
15. F. Webster, E. Tang, P. Rossky, and R. Friesner, *J. Chem. Phys.*, **100,** 4835 (1994).
16. E. R. Bittner and P. Rossky, *J. Chem. Phys.*, **103,** (1995).
17. Z. L. Zadoyan, V. Apkarian, and C. C. Martens, *J. Phys. Chem.*, **99,** 7453 (1995).

18. G. Ashkenazi, R. Kosloff, S. Ruhman, and H. Tal-Ezer, *J. Chem. Phys.*, **103,** 10005 (1995).
19. A. Bartana, U. Banin, S. Ruhman, and R. Kosloff, *Chem. Phys. Lett.*, **229,** 211 (1994).
20. G. Lindblad, *Commun. Math. Phys.*, **48,** 119 (1976).
21. R. Alicka and K. Landi, *Quantum Dynamical Semigroups and Applications*, Springer-Verlag, 1987.
22. L. Liu and H. Guo, *J. Chem. Phys.*, **103,** 8541, 1996.
23. R. Kosloff, in *Numerical Grid Methods and Their Applications to Schrödinger's Equation.* C. Cerjan, Ed., Kluwer Academic Publishers, The Netherlands, 1993.
24. E. Moyal, *J. Phys. Chem.*, **45,** 99 (1949).
25. V. Gorini, A. Kossokowski, and E. C. G. Sudarshan, *J. Math. Phys.*, **17,** 821 (1976).
26. R. Kosloff, *Physica*, **110A,** 346 (1982).
27. A. Frigerio and V. Gorini, *J. Math. Phys.*, **17,** 2123 (1976).
28. D. Kohen and D. J. Tannor, *J. Chem. Phys.*, (1996).
29. A. G. Redfield, *IBM Jr.*, **1,** 19 (1957).
30. E. B. Davies, *Commun. Math. Phys.*, **39,** 91 (1974).
31. J. L. Skinner, *Annu. Rev. Phys. Chem.*, **39,** 463 (1988).
32. J. L. Skinner, B. B. Laird, and L. Root, *J. Lumin.*, **45,** 6 (1990).
33. H. Tal Ezer, R. Kosloff, and C. Cerjan, *J. Comp. Phys.*, **100,** 179 (1992).
34. R. Kosloff, *Annu. Rev. Phys. Chem.*, **45,** 145 (1994).
35. R. Kosloff, *J. Phys. Chem.*, **92,** 2087 (1988).
36. W. Magnus, *Comm. Pure and Appl. Math.*, **7,** 659 (1954).
37. M. D. Feit, J. A. Fleck, Jr., and A. Steiger, *J. Comm. Phys.*, **47,** 412 (1982).
38. M. D. Feit and J. A. Fleck, Jr., *J. Chem. Phys.*, **78,** 301 (1983).
39. L. Allen and J. H. Eberly, *Optical Resonance and Two-Level Atoms*, Dover, 1987.
40. B. M. Garraway, K.-A. Suominen, and S. Stenholm, *Phys. Rev. A*, **45,** 3060 (1992).
41. C. Cohen-Tanoundji and J. Allen, *Quantum Mechanics*, Wiley, New York, 1977.
42. T. J. Smith, L. W. Ungar, and J. A. Cina, *J. Lumin.*, **58,** 66 (1994).
43. Y. X. Yan, L. T. Cheng, K. A. Nelson, in *Advances in Nonlinear Spectroscopy*, R. G. H. Clarke, and R. E. Hester, Eds., Wiley, New York, 1987.
44. R. Kosloff, A. Dell Hammerich, and D. Tannor, *Phys. Rev. Lett.*, **69,** 2172 (1992).
45. A. Bartana, R. Kosloff, and D. Tannor, *J. Chem. Phys.*, **99,** 196 (1993).
46. W. T. Pollard and R. A. Mathies, *Annu. Rev. Phys. Chem.*, **43,** 497 (1992).
47. W. T. Pollard, S. L. Dexheimer, Q. Wang, L. A. Peteanu, C. V. Shank, and R. A. Mathies, *J. Phys. Chem.*, **96,** 6147 (1992).
48. J. D. Jackson, *Classical Electrodynamics*, Wiley, 1975.
49. Y. J. Yan and S. Mukamel, *Phys. Rev. A*, **41,** 6485 (1990).
50. N. F. Scherer, R. J. Carlson, A. Matro, M. Du, A. J. Ruggiero, V. Romero-Rochin, J. A. China, G. R. Fleming, and S. A. Rice, *J. Chem. Phys.*, **95,** 1487 (1991).
51. N. F. Scherer, L. D. Ziegler, and G. R. Fleming, *J. Chem. Phys.*, **96,** 5544 (1992).
52. N. F. Scherer, D. M. Jonas, and G. R. Fleming, *J. Chem. Phys.*, **99,** 153 (1993).
53. S. Rice, H. Tang, and R. Kosloff, *J. Chem. Phys.*, **104,** 5457 (1996).
54. M. Cho, N. F. Scherer, G. R. Fleming, and S. Mukamel, *J. Chem. Phys.*, **77,** 202 (1982).

55. W. P. de Boeiji, M. S. Pshenichnikov, and D. A. Weirsma, *Chem. Phys. Lett.*, **238,** 1 (1995).
56. G. Stock, L. Seidner, and W. Domcke, *J. Chem. Phys.*, **103,** 3998 (1995).
57. R. Baer and R. Kosloff, *Chem. Phys. Lett.*, **200,** 183 (1992).
58. D. J. Tannor and E. J. Heller, *J. Chem. Phys.*, **77,** 202 (1982).
59. D. Danovich, J. Hrusak, and S. Shaik, *Chem. Phys. Lett.*, **233,** 249 (1995).
60. G. Asahkenazi, K. Yamashita, R. Kosloff, and S. Ruhman, in preparation (1996).
61. E. C. M. Chen and W. E. Wentworth, *J. Phys. Chem.*, **89,** 4099 (1985).
62. P. E. Maslen, J. M. Papanikolas, J. Faeder, R. Parson, and S. V. ONeil, *J. Chem. Phys.*, **101,** 5731 (1994).
63. P. K. Walhout, J. C. Alfano, K. A. M. Thakur, and P. Barbara, *J. Phys. Chem.*, **99,** 7568 (1995).
64. I. Benjamin, P. Barbara, B. J. Gertner, and J. T. Hynes, *J. Phys. Chem.*, **99,** 7557 (1995).
65. A. Vibok and G. G. Balint-Kurti, *J. Phys. Chem.*, **96,** 7615 (1992).
66. I. Sh. Averbuck and N. F. Perelman, *Phys. Lett.*, **139A,** 449 (1989).
67. R. B. Gerber, A. Garcia-Vela, *J. Chem. Phys.*, **98,** 427 (1993).
68. P. Jungwirth and R. B. Gerber, *J. Chem. Phys.*, **102,** 6046 (1995).
69. H. Stock, *J. Chem. Phys.*, **103,** 2888 (1995).
70. F. Bloch, *Phys. Rev.*, **102,** 104 (1956).
71. M. Toplar and N. Mackri, *J. Chem. Phys.*, **101,** 7500 (1994).
72. P. Hänggi, P. Talkner, and M. Borkovec, *Rev. Mod. Phys.*, **62,** 251 (1990).
73. W. T. Pollard and R. A. Freisner, *J. Chem. Phys.*, **100,** 5054 (1994); *Adv. Chem. Phys.*, **93,** 77 (1996).
74. E. B. Davies, *Quantum Theory of Open Systems*, Academic Press, 1976.
75. S. Mukamel, *Annu. Rev. Phys. Chem.*, **41,** 647 (1990).
76. R. Bianco and J. T. Hynes, *J. Chem. Phys.*, **102,** 7885 (1995).
77. R. Bianco and J. T. Hynes, *J. Chem. Phys.*, **102,** 7864 (1995).
78. H. Tal-Ezer and R. Kosloff, *J. Chem. Phys.*, **96,** 5618 (1992).
79. D. J. Kouri, M. Arnold and D. J. Hoffman, *Chem. Phys. Lett.*, **203,** 96 (1993).
80. W. Zhu, Y. Huang, D. J. Kouri, C. Chandler and D. J. Hoffman, *Chem. Phys. Lett.*, **217,** 73 (1994).
81. B. Hartke, R. Kosloff, and S. Ruhman, *Chem. Phys. Lett.*, **158,** 238 (1989).
82. Y. Huang, W. Zhu, D. J. Kouri, and D. J. Hoffman, *Chem. Phys. Lett.*, **214,** 451 (1993).
83. D. J. Kouri, W. Zhu, Y. Hoang, and D. J. Hoffman, *Chem. Phys. Lett.*, **220,** 312 (1994).
84. R. Kosloff and H. Tal-Ezer, *Chem. Phys. Lett.*, **127,** 223 (1986).
85. D. Neuhauser, *J. Chem. Phys.*, **93,** 2611 (1990).
86. D. Neuhauser, *J. Chem. Phys.*, **100,** 5076 (1994).
87. M. R. Wall and D. Neuhauser, *J. Chem. Phys.*, **102,** 8011 (1995).
88. C. Leforestier, R. Bisseling, C. Cerjan, M. Feit, R. Freisner, A. Guldberg, A. D. Hammerich, G. Julicard, W. Karrlein, H. Dieter Meyer, N. Lipkin, O. Roncero, and R. Kosloff, *J. Comp. Phys.*, **94,** 59 (1991).
89. C. Leforestier and R. E. Wyatt, *Chem. Phys. Lett.*, **78,** 2334 (1983).
90. R. Kosloff and D. Kosloff, *J. Comp. Phys.*, **63,** 363 (1986).

91. D. Neuhauser and M. Baer, *J. Chem. Phys.*, **90,** 4351 (1989).
92. M. S. Child, *Mol. Phys.*, **72,** 89 (1991).
93. G. G. Balint-Kurti and A. Vibok A, in *Numerical Grid Methods and Their Application to Schrödinger's Equation*, C. Cerjan, Ed., Kluwer Academic Publishers, The Netherlands, 1993, p. 412.
94. C. W. MacCurdy, and C. K. Stround, *Comp. Phys. Comm.*, **63,** 323 (1991).
95. M. Berman, R. Kosloff, and H. Tal-Ezer, *J. Phys. A*, **25,** 1283 (1992).
96. P. Saalfrank, R. Baer, and R. Kosloff, *Chem. Phys. Lett.*, **230,** 463 (1994).
97. S. Aurbach and C. Leforestier, *Comp. Phys. Comm.*, **78,** 55 (1994).
98. S. M. Auerbach and W. H. Miller, *J. Chem. Phys.*, **100,** 1103 (1994).
99. L. M. Trefethen, *J. Sci. Stat. Comput.*, **1,** 82 (1980).
100. U. Peskin, R. Kosloff, and N. Moiseyev, *J. Chem. Phys.*, **100,** 8849 (1994).
101. Lotharreichel, *Bite*, **30,** 332 (1990).
102. V. A. Mandelshtam and H. S. Taylor, *J. Chem. Phys.*, **102** (1995).
103. Y. Hoang, D. J. Kouri, and D. J. Hoffman, *J. Chem. Phys.*, **101,** 10493 (1994).
104. M. Abramowitz and I. A. Stegun, *Handbook of Mathematical Functions*, Dover, 1972.

MICROSCOPIC SIMULATIONS OF COMPLEX FLOWS

MICHEL MARESCHAL

Centre for Non-Linear Phenomena and Complex Systems,
Université Libre de Bruxelles,
1050 Brussels, Belgium

CONTENTS

I. Introduction
II. The Techniques
  A. Molecular Dynamics
  B. Nonequilibrium Molecular Dynamics
  C. Lattice Gases
    1. Definitions
    2. Properties of Lattice Gases
    3. Applications
    4. Conclusions
  D. Direct Simulation Monte Carlo Method
  E. Other Types of Simulations Using Particles
III. Rheology of Fluids
  A. Simple Fluids
  B. Towards More Complex Fluids
IV. Nonequilibrium Steady States
V. Flows and Instabilities
  A. Flows
  B. Rayleigh-Bénard Instability
VI. Conclusions

I. INTRODUCTION

One of the major contributions of microscopic simulations to our understanding of nonequilibrium fluid behavior has been the discovery made by Alder and Wainwright [1] that the velocity–time autocorrelation function had a long-time tail. This long-time memory effect appeared counterintuitive at the time, although kinetic theory computations had already given some indi-

Advances in Chemical Physics, Volume 100, Edited by I. Prigogine and Stuart A. Rice.
ISBN 0-471-17458-0

cations that the simple ideas borrowed from the description of dilute systems could not be straightforwardly applied to dense fluids and liquids. Twenty-five years later, this result appears as the most clear-cut evidence that hydrodynamics remains a valid explanation of fluid behavior at atomic time and space scales. The idea which quite naturally came to explain why the velocity of a particle would keep such a long-time correlation was at the hydrodynamical level of description: The particle would very quickly exchange momentum with its direct environment, which, in turn, would slowly, in a diffusive (hydrodynamic) way, transfer it to the remaining part of the fluid. This intuitive explanation was supported by a direct inspection of fluid motion and velocity correlations computed in the fluid at equilibrium. It was also supported by a lot of further theoretical computations ranging from phenomenological fluctuating hydrodynamics to full kinetic theory treatments. (For an excellent and still up-to-date review on mode-coupling theories, see ref. [2].) Later work has provided further evidence that equilibrium thermal fluctuations would indeed behave according to the fluctuating hydrodynamic theory, even at wavelengths corresponding to a few atomic distances.

In this chapter, we will survey recent work supporting the idea that the same mechanisms remain when fluids are maintained out of equilibrium. Simulations using the molecular dynamics technique can reproduce complex collective behaviors such as the ones appearing during a hydrodynamic instability. The Rayleigh–Bénard transition to ordered convection in a fluid layer heated from below has been reproduced with a few thousand interacting particles, the trajectory of which follows Newton's equations of motion. Time scales, space scales, and amplitudes of the various forcings present are orders of magnitude different from what they are in laboratory experiments. Nevertheless, the continuum macroscopic description seems to remain valid and can be tested quantitatively. These microscopic simulations give modern scientists a direct visualization of Laplace's demon, linking the deterministic motion of microscopic particles to the macroscopic behavior, which has been shown to be, in many circumstances, chaotic and unpredictable.

The macroscopic modeling of flows rests on the balance equations for the mass, momentum, and energy densities, as functions of time and space. These are precisely the densities of the conserved quantities that evolve in time and space only because of macroscopic inhomogeneities. The conservation equations need to be supplemented with phenomenological relations linking macroscopic fluxes with forces. They are subsequently solved once initial conditions and boundary values of the hydrodynamic field variables are given.

Typically, the equations obtained are nonlinear and their solutions not known in generality. Numerical techniques are thus required in order to study complex time-dependent phenomena. Solving the equations on a computer, however, leads to time and space discretization, and the algorithms to be used are not so different from those that would be directly based on solving the dynamical trajectories of the particles that constitute the fluid. One can imagine that instead of solving the equations for the field variables one could replace these by real particles. Molecular dynamics modeling is doing precisely that. Now a simulation with as many particles as a fluid contains in a laboratory experiment would certainly be too costly, but the idea is that, since in many flow problems there is no intrinsic length scale, one could set up a simulation at much smaller scales and study whether these simulations would still reproduce macroscopic hydrodynamic behavior.

The idea of replacing field variables by particles appeared computationally interesting when it was realized that it could be done with extremely simplified particle models: point particles, with no mass or interaction energy, which were allowed to move on a lattice at discrete intervals of time. This caricature of a fluid is a cellular automaton. Its programming is simple and fast, and it has been proved that, with proper conditions and limits, its average evolution would be the same as the one given by the macroscopic incompressible Navier–Stokes equation. These models have even led to computationally efficient algorithms, the so-called lattice Boltzmann approach. One must recognize, however, that the microscopic status of this algorithm is not fully obvious. Actually, the lattice gas models have received considerable attention within the statistical mechanics community, not because of their efficiency with respect to traditional Navier–Stokes solvers (as a matter of fact, they do not compare well), but because they have obtained the status of "nonequilibrium Ising model." Their simplicity permits the study of larger system sizes, with longer timescales, and also, in some cases, the generation of better statistics than do more sophisticated molecular dynamical models. There is a price to pay, however, in that in many circumstances these models miss essential ingredients of microscopic fluid models so that it is worth developing a more realistic modeling. This is, for instance, the case in the study of the effect of fluctuations on the behavior of fluid near bifurcation points, and, more generally, in nonequilibrium steady states, a recurrent theme of this chapter.

This article will be divided into two main parts. First, the techniques used for microscopic flow simulations will be presented. The focus will be on the molecular dynamics method since this is the method which, although computationally expensive, has the firmest theoretical foundations. We will, however, also present two similar but different and more efficient techniques used in very specific contexts: the lattice gas cellular automata and the direct simulation Monte Carlo methods.

In a second part, some of the most significant results obtained will be sketched, or at least reference to the original literature will be given. The physics presented here will be concerned mainly with two aspects that are specific to the microscopic simulations: first, the characterization of nonequilibrium states, and in particular the long-ranged correlations that appear in the fluctuations of hydrodynamic densities in constrained fluids. This is an important effect showing the coherence of nonequilibrium states which, we believe, is essential to explain the appearance of order and the emergence of structures once a mechanism for an instability exists. In addition, they constitute still another illustration of the mode-coupling mechanism that was alluded to at the beginning of this introduction. The second aspect that will be discussed is a series of examples of flows, mainly two-dimensional, for which comparison with macroscopic approaches has been carried out: among these are flow past an obstacle, channel flow, Couette flow, Rayleigh–Bénard transition to convection. In some of these examples, microscopic modeling can provide information missing in the macroscopic approach: this is the case in the boundary conditions of two-phase flows in channels or in shock fronts at high Mach numbers.

There are a few collective works dealing with the subject of this article: the proceedings of a conference held in Boulder in 1982 [3] are the first to provide an idea of the possibilities offered by molecular simulations to study nonequilibrium or/and nonlinear fluid behavior. Very recently, a conference was held at Lyon on a similar subject, the proceedings of which are to be published in 1997 [4]. In the meantime, there have been the proceedings of the Brussels conference in 1989 and those of a summer school held at Alghero in 1991 [5]. Recent summer school proceedings concerned with molecular dynamics and Monte Carlo simulation techniques have also been published: Alghero in 1992 [6], Varenna in 1993 [7] and Como in 1995 [8]. The most recent reference concerning the lattice gas technique concerned the 1994 conference in Princeton [9]. General textbooks about the techniques [10], the statistical mechanics of equilibrium [11] and nonequilibrium [12–14] fluids are also available. Let us end this list of general references with significant review articles dealing with flow simulations [15], Monte Carlo [16] and molecular dynamics [17] techniques, and also a recent review of chaotic dynamical systems related to the present approach [18]. In this article, however, we shall refer mostly to the original articles rather than to review papers or to collective works.

II. THE TECHNIQUES

A. Molecular Dynamics

Molecular dynamics (MD) is the most direct simulation technique used in statistical mechanics. It consists of numerically solving on a computer the classical equations of motion of a system of interacting particles, very much like a straightforward representation of what is really happening. Equilibrium thermodynamic quantities are then obtained by a time-averaging of dynamical observables. Historically, it soon became clear that considering a few hundred particles in a periodic geometry would allow an accurate computation of the equation of state of simple models.

Early on, in the late 1950s and early 1960s, systems of hard spheres and hard disks were considered. At the 1956 IUPAP conference on statistical mechanics, held in Brussels, Alder and Wainwright [19] reported computations on systems of a few hundred particles that were followed during a few thousand collisions. Although results providing strong evidence of the existence of the solid–fluid transition in these systems were only communicated a year later in Varenna [20], it was already apparent that machine computations would be extremely useful in the study of both equilibrium and nonequilibrium states of matter. (The application of "electronic machine" computation to aerodynamics is explicitly mentioned in the discussion following Alder's report in the proceedings of the 1956 conference in Brussels, ref. [19].)

The later extension of the method to models with regular potentials [21], the subsequent progress made by Verlet and his group [22–24] concerning the algorithm and the programming, and also the growing availability of computers have helped to spread the use of the molecular dynamics technique among the physics and chemistry communities. Nowadays, software packages are available on Internet servers so that anyone can reproduce early molecular dynamics computations on a personal computer much more rapidly and with much more powerful visualization facilities than during the early developments.[1] At the large computational facilities, on the other hand, larger and larger systems have been integrated for larger and larger times. Recently, Brad Holian from Los Alamos National Laboratory performed simulations of a fracture occurring in a solid under stress on a highly parallel computer, the CM-5: the integration over 10,000 time steps (around 100 ps) of a three-dimensional system made of 100 million atoms required 150 computer hours [25]. This is far from being a routine computation, even at Los Alamos, but these numbers help to provide an idea of the limitations both of the sizes and of the timescales involved nowadays in atomic and molecular simulations.

In Fig. 1, the phase diagrams of hard-sphere and Lennard–Jones model

[1]There are a few servers from which standard Fortran source codes can be downloaded. The most useful one is certainly the CCP5 program library which also distributes the codes described in the book by Allen and Tildesley [10]; on the web: www.dl.ac.uk/CCP/CCP5/main.html. The CCP5 server makes also available as a MacIntosh application built by D. J. Evans, which illustrates some parts of his book [13]. Another application for book illustrating that is relevant to the domain covered here, and which is available in Mac and Unix formats, is provided by D. C. Rapaport (email: rapaport@phys8.ph.biu.ac.il). The corresponding book is due to appear shortly from Cambridge University Press. Last but not least, D. W. Herrmann has a few programs available on a server located at Heidelberg (on the web: wwwcp.tphys.uni-heidelberg.de). These illustrate the exercises of his book.

systems are qualitatively reproduced. Despite its simplicity, the hard-sphere fluid has many advantages that make it worthwhile to study. First of all, the numerical integration of the trajectory of the system is not approximate: the dynamics is a succession of binary encounters that are determined exactly—at least the only limitation is the precision of the computer.[2] At collisions, the projection of the relative velocity on the line joining the centers is reversed instantaneously, and, between collisions, particles are otherwise moving freely. The programming consists of computing the times for the next collisions, which amounts to solving a second-order algebraic equation for every pair of particles, selecting the shortest of these times, updating the particles' positions up to that time, performing the collision, and then restarting the procedure. Hard-sphere transport properties have all the important features of simple atomic, dense fluids and liquids. This is due to the fact that the longer-range attractive Van der Waals forces do not modify, in an appreciable way, the values of the transport coefficients [27], and, thanks to the Enskog equation (which generalizes the Boltzmann equation for hard spheres to the moderately dense regime), it is understood that the main physical process needed for a transport theory in dense fluids is the collisional transfer mechanism, giving rise to the potential part of the thermodynamic fluxes [28].

The soft-sphere model is very similar to the hard-sphere system: the potential is short-ranged and repulsive but it is regular, and therefore it is more adapted to computations on vector processors. Actually "soft-sphere" is a generic name for all potentials that are regular, short-ranged, and purely repulsive. The so-called WCA potential, named after Weeks, Chandler, and Anderson, is a particular "soft-sphere" model, which consists of a Lennard–Jones potential cut at the distance where the potential is a minimum and shifted upwards by the amount of the potential at that distance. Both the potential energy and its derivative are smooth and continuous functions of the interparticle distance.

Apparent from Fig. 1(*a*) is the fact that, for hard spheres, the temperature does not play any role. This is easily understood once it is realized that hard-sphere systems have no potential energy and that an increase of the temperature is then equivalent to a rescaling of the velocities of all the particles. Therefore, two systems in the same configuration but set at different temperatures will follow identical trajectories in phase space, giving rise to

[2]The computation itself is faster than equivalent regular potential models on a scalar processor but the programming is such that it cannot take much advantage of modern vector and parallel architectures: the reason is intrinsic to the model. One has to perform the next collision before any further computing. A good description of the algorithms can be found in ref. [26].

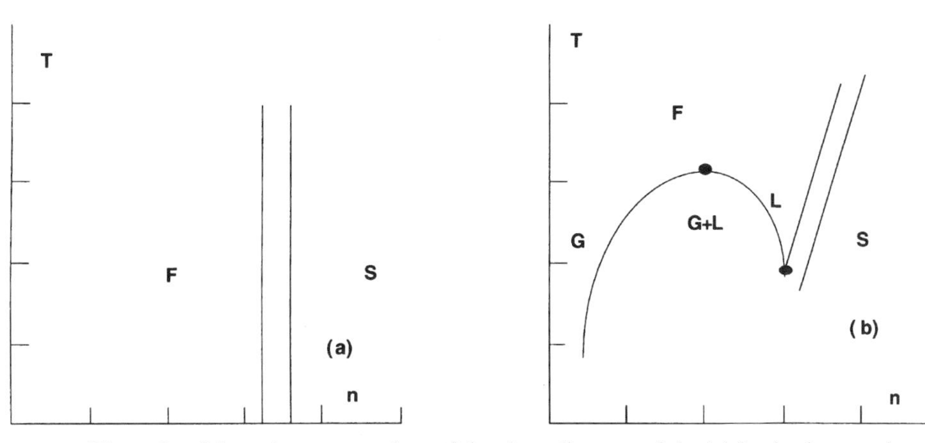

Figure 1. Schematic representations of the phase diagrams of the (*a*) hard spheres and (*b*) Lennard–Jones models.

identical time-averaging of static quantities and trivial rescaling of dynamic ones. Transport coefficients have thus a trivial dependence on the temperature: they simply scale as its square root. It has to be noted, however, that such a property does not extend to nonequilibrium states: two hard-sphere systems submitted to two different temperature gradients will not be related by a scaling relation.

In Fig. 1(*b*), the phase diagram of the Lennard–Jones model is shown qualitatively. It contains a critical point and a triple point. Recently the precise location of these points has been made possible thanks to the Monte Carlo–Gibbs ensemble technique [29, 30]. In particular, the influence of the dimensionality and of the cutoff distance of the potential are important. In nonequilibrium studies, for the simulation of flows, one tries to avoid any influence of the critical point or of possible phase transitions so that the cutoff distance taken for the potential, for instance, has no real influence on the developments of flows. At first sight, it might appear that the dimensionality could be of more relevance, particularly since, due to the long-time tails of the Green–Kubo integrands, the finiteness of the transport coefficients in two dimensions is still a question of debate. However, since simulations are done with finite systems during finite times, it is to be expected that possible long-time divergence will produce only a (very) small indeterminacy over the value of the transport coefficient of interest. Let us mention also that an adequate choice of state properties and, therefore, of the corresponding models, is in order when properties specific to the neighborhood of the critical point, such as a large compressibility, are needed [31].

The MD technique has been widely developed in the last few years in order to compute thermodynamic properties of various materials. Quite often, as already stressed, nonequilibrium studies focus on properties or behaviors that do not much depend on the specific model chosen. On a hydrodynamic scale, the choice of the model will result in setting values for the transport coefficients or prescribing the equation of state. These values, in turn, will influence the evolution only through the dimensionless numbers that characterize flow properties, for example, the Reynolds or Rayleigh numbers. Thus, most of the flow properties will not depend on the model chosen, except for various specific cases where one is precisely looking at the departure from macroscopic behavior. Microscopic modeling can be directly related to large-scale behavior, near a solid boundary, at an interface, or in a strong shock front. Regarding the computation of flows, one is, therefore, often led to consider for example, simple models, like the ones with purely repulsive interactions that have been described above, or even simpler ones, and to study their properties when the fluid is contained within complex boundaries or when it is submitted to various constraints.

B. Nonequilibrium Molecular Dynamics

Early studies of nonequilibrium states took place at the beginning of the 1970s. These first attempts are seen retrospectively as a test of whether it could be done. Independently the three groups—Hoover and Ashurst [32]; Gosling, McDonald, and Singer [33]; and Lees and Edwards [34]—were among the first to simulate a fluid system in a steady shear (see Fig. 2). The transport coefficient is obtained either from the "macroscopic" velocity profile or, more directly, from the ratio between the imposed velocity gradient and the computed response of the system, namely the shear stress. (For a microscopic definition of fluxes see ref. [35].) The original motivation was to produce an alternative to the Green–Kubo way of computing linear transport coefficients. Indeed, once integrated in time, the fluctuations of the equilibrium time-correlation functions of the fluxes directly lead to the value of the transport coefficients [36, 37]. However, since the statistics needed to obtain a satisfactory signal-to-noise ratio is hard to get, it appeared that a direct computation in a constrained fluid could be a useful shortcut.

Figure 2 shows three different ways of computing the viscosity by shearing the fluid. In Fig. 2(*a*), the fluid is maintained between two moving boundaries. In the method developed by Hoover and Ashurst [32], boundary atoms are in a fluid state (artificially maintained); this choice was made after many trials (moving a flat surface, moving rigid rows of fixed particles, wall potential, and so forth) because it seemed to perturb less significantly the fluid near the boundary itself. In particular, a solid wall does generate density oscillations over a range of a few atomic distances from the boundary. A recent simulation by Liem, Brown, and Clarke [38] of a channel flow made with 37,500 fluid particles in a pipe, and simulating in addition 5,610 solid wall particles,

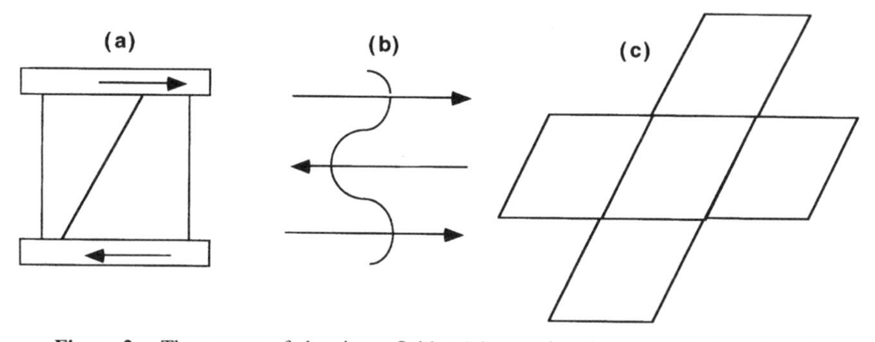

Figure 2. Three ways of shearing a fluid: (*a*) by moving the top and bottom boundaries in opposite directions; (*b*) by an external periodic forcing; and (*c*) by a continuous deforming of the periodic boundary conditions (Lees–Edwards).

does indeed show a long-range ordering well inside the fluid region. A similar result is displayed in the channel flow simulation of Koplick, Banavar, and Willemsen [38].

In Fig. 2(*b*), a periodic forcing is imposed on the fluid. This method was first used by Gosling et al. [33] since it is compatible with periodic boundary conditions. In two dimensions, this flow has since been studied by other methods and is sometimes referred to as the "Kolmogorov flow." In the original work, the method appeared unable to produce precise results. Indeed, as in the other cases, the forcing exerted on the fluid needs to be fairly large, resulting in a sizeable dissipation of mechanical into thermal energy. The temperature is therefore not constant during the simulation time, and some adjustments are necessary. This is true for all three methods described, but even more so in the case shown in Fig. 2(*b*), since the shear, and hence the dissipation is not even homogeneous. Progress in the thermostatting technique has made it possible to readdress the simulation of this flow by molecular dynamics. It appears to be one of the simplest cases where an instability can be studied in detail at the microscopic level [39]. Actually, the latter model displays successive instabilities while increasing the amplitude of the external forcing, with period-doubling bifurcations towards fully-developed turbulence, relaminarization, and so forth. The detailed study of these bifurcations by molecular dynamics appears feasible in the near future, that is, with today's computing capacity.

Last, and certainly not least, Lees and Edwards [34] presented a method that permits combining the advantages of periodic boundary conditions and homogeneous Couette shear flow: The boundary cells are moving according to the imposed homogeneous (and of infinite extension) shear. This geometry avoids many difficulties of the simulations, since the system is homogeneous and has no physical boundaries but it represents, as a matter of fact a simulation of a somewhat unphysical state. It has allowed the most precise MD studies of the viscosity transport coefficients in fluid systems, a problem that will be described in more detail in the third section of this article. In the absence of any boundary effects, the problems to face are basically the unusually large amplitude of the forcing, which is still required to produce a good signal to noise ratio, and the precise control of the temperature in these highly dissipative models.

Concerning this last point, it should be noted that early elementary techniques used an "*ad hoc* velocity rescaling," a revealing name for a technique based on the idea that one could control the temperature by constraining the mean kinetic energy for the fluid system simply by a (small) adjustment at each time step of the updating of particle positions. A similar but more sophisticated procedure has been proposed by Denis Evans, inspired by Gauss's principle at least constraint [40]. It amounts to adding a friction term in the equations of motion for the particles. The friction, with no definite sign, is treated as a Lagrange multiplier, the value of which is thus determined by the isokinetic (or isoenergetic) constraint. Since then, the technique has been tested in various different geometries, with various model fluids, and is known to produce values of thermodynamic properties in agreement with other methods. In particular, it has been shown that equilibrium time-correlation functions generated under these Gaussian thermostatted equations of motion are identical, in the thermodynamic limit, to those obtained from a Newtonian (i.e., without friction) dynamics [41]. The method of determining, at every time, the friction coefficient can be derived from the general technique of constrained dynamics developed by Ciccotti et al. [42]. The main difference in the application of the latter technique to rigidly bonded molecules is that, in the case of constraining the energy or part of the energy, the constraint is not holonomic.

Interesting developments of the thermostatting techniques were also done independently and around the same time by Nosé [43] and Hoover [44]. A time-reversible friction coefficient was proposed, first by Hoover, Ladd, and Moran [44], and later by Evans. At first sight, the two techniques appear different, but they have been shown to lead to the same modified equations of motion for the particles, to which we will refer to as the Nosé–Hoover equations. The friction, ζ_{NH}, introduced by Hoover, is a new dynamical variable of the system representing the action of the thermostat. It has also an equation of motion based on the idea of integral feedback: the instantaneous kinetic energy, $K(t)$, fluctuates around the average value that the controlled temperature does impose, K_0.

$$\zeta_{NH} = \frac{1}{\tau^2} \int dt \left[\frac{\hat{K}(t) - K_0}{K_0} \right] \tag{2.1}$$

where

$$\hat{K}(t) = \sum_{i \in \alpha} \frac{(\mathbf{p}_i)^2}{2m}$$

and

$$K_0 = \frac{3(N-1)}{2} k_B T_{imposed}$$

The $N - 1$ factor in the last equation comes from the fact that the total momentum is usually conserved and therefore three degrees of freedom are frozen (the relation between temperature and kinetic energy is straightforwardly generalized to the case where there is another number of degrees of freedom frozen).

The equations of motion can be derived from this prescription of integral feedback. The important point is now that the friction coefficient can be negative, and this will happen whenever the kinetic energy falls below the imposed kinetic energy for a sufficiently long time. Strictly speaking, these equations are reversible. For every trajectory that evolves forward in time, one has a reversed one that can be obtained by reversing all velocities and by changing the sign of the friction and of its time-derivative. The system of equations is not conservative, however, so that the volume in phase space is not conserved. Since the friction is fluctuating around zero, the volume in phase space is also oscillating in time, with alternating expansions (negative friction) and contractions (positive friction). The equations of motion for the dynamical variables read

$$\frac{d\mathbf{p}_i}{dt} = \mathbf{F}_i + \mathbf{F}_e - \zeta_{NH}\mathbf{p}_i$$

$$\frac{d\zeta_{NH}}{dt} = \frac{1}{\tau^2}\left(\frac{\hat{K} - K_0}{K_0}\right) \tag{2.2}$$

where $\mathbf{p}_i$ is the ith particle momentum, $\mathbf{F}_i$ is the force exerted by the interactions with the other particles, $\mathbf{F}_e$ is an external field and the time τ is a free parameter, related to the strength of the coupling of the system with the thermostat. Some pathologies [45] can result from a poor choice for this parameter but these cases are usually well understood, and ergodic behavior for the system's trajectories is in general observed. Some judicious choices of this parameter have been proposed, following a discussion of the physical processes involved in the coupling mechanism with the thermostat [46]. An alternative, consisting of adding a coupling to a chain of thermostats, has also been suggested, an idea which may prove useful when the model lacks the expected ergodic behavior [47].

It is worth mentioning the change of phase space dimension D is directly related to the friction through the following equation: $\Delta D = (\zeta_{NH}/\nu)D$, with ν the collision frequency. The flow in phase space can display behaviors that are typical of dissipative systems, rather than of conservative ones: attractors as well as repellors and fractal invariant measures have been reported in some simple systems studied by simulations [48–51] or even by more mathematical approaches [52]. It is still an open question to understand whether these changes are related to physical properties typical of nonequilibrium stationary states or if they are linked to the technique which is used. The same physical problem, but simulated by using stochastic boundary probability distribution in phase space remains smooth and does not behave as a fractal object. (See the article by G. L. Eyink and J. L. Lebowitz [5] and the discussions of the round-table on Lyapounov spectra and irreversibility in ref. [5].) The proof of a smooth probability density for non-equilibrium states has been given in ref. [53]. Whether or not these effects are real, the related discussions have nevertheless lead to many new and original works, of mathematical as well as numerical nature, on the properties of stationary states ensemble densities (e.g., ref. [54] and references cited therein) and on the relation between chaotic and irreversible properties. It is to be expected that further progress will be achieved in this field in the very near future.

The Nosé–Hoover equations can be derived as the Hamilton's equations of motion, the Hamiltonian being the system's total energy plus an extra term containing the kinetic and potential energy of an extra variable, s. This Hamiltonian reads:

$$H_{\text{total}} = H_{\text{system}} + \frac{p_s^2}{2Q} + k_B T \ln(s) \tag{2.3}$$

Using this formulation, it is possible to show that a microcanonical ensemble average of the total Hamiltonian (system plus the extra variables s and its conjugate momentum) will result in a canonical average of the system's Hamiltonian (see ref. [19]). Actually, this seems to be the major theoretical foundation justifying the use of Nosé–Hoover's equations of motion. These equations produce trajectories, the time-average of which leads to canonical averages. Provided ergodic behavior is observed, one is then sure to generate sets of states sampling the equilibrium distribution at the imposed temperature. The technique does not imply, of course, that real microscopic systems would obey equations that possess properties typical of macroscopic behavior (like irreversibility and nonconservative flows), but it provides an extremely useful way of introducing the temperature into the equations of motion! (Let us mention that pressure can also be controlled by a similar technique, see the chapter by S. Nosé in ref. [19].)

The equations can be generalized to cases where the system is not in equilibrium. Suppose for instance that one wants to thermalize a fluid sheared by a periodic external force like in the Kolmogorov flow [Fig. 2(*b*)]. The local temperature is then defined in terms of the comoving kinetic energy of the particles, the kinetic energy obtained from the so-called peculiar velocities. One has, indeed, to remove the kinetic energy relative to the flow motion in order to relate the local temperature to the velocity distribution function of the particles. This can be achieved by introducing a local thermostatting, and therefore several local friction coefficients. The system is divided into computational cells, labelled by the index α. In each cell, one needs to compute, at every time step, local values for the number of particles, N_α, the local momentum, u_α, and the local peculiar kinetic energy $K_\alpha(t)$. In the case of a stationary nonequilibrium state, these quantities can be also averaged on an intermediate timescale, but in full generality they can be taken as instantaneous values, only smoothed by the requirement that there should be a sufficient number of particles in every cell. The equations then read as follows

$$\zeta_{NH}(\alpha) = \frac{1}{\tau^2} \int dt \left[\frac{\hat{K}_\alpha(t) - K_0}{K_0} \right]$$

with

$$\hat{K}_\alpha(t) = \sum_{i \in \alpha} \frac{(\mathbf{p}_i - \mathbf{u}_\alpha)^2}{2m}$$

and

$$K_0 = \frac{3(N_\alpha - 1)}{2} k_B T_{imposed} \tag{2.4}$$

Two remarks are in order: first, the particles move from cell to cell during the time evolution. Once they change cell, they contribute to the averages of their new local environment and they are thermalized by the new local Nosé–Hoover friction. This may cause discontinuities in the evolution. Actually the discrete time step already introduces discontinuities, and it is important that those should be small. This is achieved by having a sufficiently high number of particles per cell. With experience, known macroscopic profiles are recovered once a few tens of particles per cells are considered. Second, it is important that the local momentum is calculated from the actual system's values and not from the expected macroscopic profile. This has indeed been shown to introduce some artificial effects, which are not always reproducible once a so-called unbiased thermostat [55] is used.

Other straightforward generalizations have been considered. A fluid or a solid submitted to a temperature gradient can be simply simulated by sandwiching a system of model particles between two slices considered as local thermostat set at different temperatures [56] (see Figure 3). The system will then develop a heat flux from the hot reservoir to the cold one. The value of this flux divided by the temperature gradient that establishes in the system leads to a straightforward computation of the heat conductivity (and to a checking of Fourier's law).

The problem to face in this kind of simulation is really that the temperature difference between the two reservoirs has to lead to a computable heat flux, much larger than typical fluctuations. This usually leads to temperature gradients that are unusually large (at least compared to gradients of laboratory experiments). An increase of the temperature by 10% for argon near the triple point in a three-dimensional 1,000-particle system produces a gradient of 3×10^9 K/m. It is important to realize that such gradients can provoke departures from the usual phenomenological relations linking fluxes and forces linearly, i.e., the Navier–Stokes and Fourier relations. Molecular dynamics is therefore able to study the limits of validity of these linear relations between fluxes and forces. Results from very precise and detailed studies are available for the viscosity (see Section III), and interesting behaviors have been obtained for the heat conductivity. In general, one has then to test whether or not the values obtained do depend on the amplitude of the forcing. Surprisingly, the departures from linear relations occur at extremely large values *for simple atomic models* [57]. For argon, shear rates of 10^{12} s^{-1} are needed in order to compute effects of more than a few percent. The

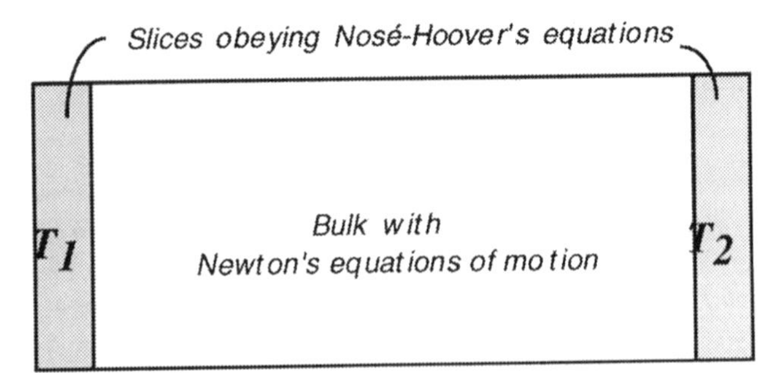

Figure 3. System with a temperature gradient.

situation is similar for the heat conductivity, bulk viscosity, and interspecies diffusion [58]. Along the same lines, shock-wave studies by MD both in dense solid [59] and dense fluid [60–62] have also led to the conclusion that extremely large values of the constraints are needed in order to witness deviations from Navier–Stokes laws. It is worth mentioning that this robustness of the linear laws can only be guessed from theoretical studies where the macroscopic laws are derived by statistical approaches. In the Chapmann–Enskog method for solving the Boltzmann equation, the smallness parameter on which the expansion of the solution is based is the ratio of the mean free path to the inhomogeneity length [63]. In the case of the temperature gradient alluded to above, this parameter is only around 1% since the mean free path is smaller than the interatomic distance. Similarly, nonlinear shear occurs when the shear rate becomes comparable to the collision frequency. This kind of behavior makes it plausible that the assumption of hydrodynamic behavior starts to be valid very early when one increases the scale of observation from the atomic scale. This was also observed in the seminal work of Alder on the long-time tails mentioned in the introduction [1]. All observations of nonequilibrium behavior in atomic fluids confirm this understanding, but deviations from linear laws occur at much smaller values of shear rates once molecular internal motion is considered, or for more complex systems like colloids and polymers. We will come back to this point in the Section III dealing with rheological fluid properties.

Important progress was also made once it was realized that the nonequilibrium forcing could be achieved by an extra term in the Hamiltonian. The most famous artificial constraints are the so-called slodd [64] equations to simulate an homogeneous Couette shear and the Evans–Gillan [65, 66] method to simulate a heat flow. In both cases, the forcing is homogeneous and is compatible with periodic boundary conditions. In the case of a homogeneous shear, Lees–Edwards boundary conditions are used and the constraint amounts to an instantaneous increment of the initial velocities of the particles. Thus the linear velocity profile is forced ($\langle v_x \rangle = \gamma y$ for example, at the initial time), and it is maintained thereafter only by the boundary conditions. In the case of the temperature gradient, the external forcing is achieved by a force that depends on the relative value of the particle's velocity compared to the average thermal velocity. The proof that the system's response is identical to the response to a real temperature gradient is, however, limited to the linear domain, and the response is probably not identical once nonlinear terms in the thermodynamic flux-force relation can no longer be neglected.

Let us also mention a technique where the perturbation to the system is made much smaller than in the methods described above. This is the so-called subtraction technique in which one repeatedly computes the difference in the response of an unperturbed and a perturbed system along one equilibrium trajectory [67]. The method is restricted to short-time responses, since, due to the Lyapounov instability of trajectories in phase space, perturbed and unperturbed trajectories will very quickly become uncorrelated [68]. The advantage of the method is that it allows small perturbation studies, and hence it lies well within the domain of validity of the linear regime. Its drawback is that the response can only be computed for short times, before Lyapunov instability actually decorrelates the perturbed and unperturbed trajectories. It does not give access to the long-time behavior of the response.

Several algorithms to compute the system's trajectory have been proposed and their relative efficiencies compared in various works (e.g., ref. [69]). The following one, which we give in the equation below, amounts to replacing a second-order derivative by a centered difference form

$$r_\alpha(t) = r_\alpha(t - \delta t) + v_\alpha\left(t - \frac{\delta t}{2}\right)\delta t + O(\delta t^3) \qquad \alpha = x, y, z$$

$$\zeta(t) = \zeta(t - \delta t) + \frac{1}{\tau^2}\left[\frac{K\left(t - \frac{\delta t}{2}\right)}{K_0} - 1\right]\delta t + O(\delta t^3)$$

$$v_\alpha\left(t + \frac{\delta t}{2}\right) = \frac{v_\alpha\left(t - \frac{\delta t}{2}\right)\left[1 - \frac{1}{\tau^2}\,\zeta(t)\,\delta t\right] + \frac{F_\alpha(t)}{m}\,\delta t}{1 + \frac{1}{\tau^2}\,\zeta(t)\,\delta t} + O(\delta t^3) \qquad (2.5)$$

where $r_\alpha(t)$ is the αth component of the particle's position, v_α its velocity, and δt is the time step. The force $F(t)$ is computed at time t from all forces exerted by all other particles; velocities are computed at time $t - \delta t/2$, as is the instantaneous kinetic energy. This set of finite-difference equations is known as the velocity version of the Verlet algorithm, coupled with the numerical solution of the equation for the friction coefficient obtained by an Euler method. Methods have been developed in order to speed up the computation. Most of the processing time is spent in computing the forces, because the number of operations to be performed is proportional to the number of pairs of particles. For short-range potentials (or potentials with a cutoff), it proves extremely useful to either make and regularly update a list of neighbors or to introduce a "multiple time steps" algorithm [70].

To end this section, let us mention the explicit form taken by the relevant observables in transport studies. The number of particles density is defined as follows

$$n(\mathbf{r}, t) = \sum_{i=1}^{N} \delta(\mathbf{r} - \mathbf{r}_i(t)) \tag{2.6}$$

whereas the momentum density reads

$$n(\mathbf{r}, t)\mathbf{u}(\mathbf{r}, t) = \sum_{i=1}^{N} \mathbf{v}_i \delta(\mathbf{r} - \mathbf{r}_i(t)) \tag{2.7}$$

and the energy density,

$$n(\mathbf{r}, t)e(\mathbf{r}, t) = \sum_{i=1}^{N} e_i \delta(\mathbf{r} - \mathbf{r}_i(t))$$

where

$$e_i = \frac{p_i^2}{2m} + \frac{1}{2} \sum_{i \neq j} V(r_{ij}) \tag{2.8}$$

Using the definitions given above, and Newton's equations of motion for the particles, it is possible to derive the five basic equations of mass, momentum, and energy density conservation.

$$\begin{aligned} \frac{\partial n}{\partial t} &= -\nabla \cdot n\mathbf{u} \\ \frac{\partial n\mathbf{u}}{\partial t} &= -\nabla \cdot \mathbf{P} \\ \frac{\partial ne}{\partial t} &= -\nabla \cdot \mathbf{J}_e \end{aligned} \tag{2.9}$$

These equations, in turn, permit the identification of the microscopic expressions for the stress tensor and the heat flux

$$P_{\alpha\beta} = \sum_{i=1}^{N} \left[m v_{i,\alpha} v_{i,\beta} - \frac{1}{2} \sum_{j \neq i} r_{ji,\alpha} \frac{\partial V(r_{ji})}{\partial r_{ji,\beta}} \right] \delta(\mathbf{r} - \mathbf{r}_i(t)) \tag{2.10}$$

$$\mathbf{J}_Q(\mathbf{r},t) = \sum_{i=1}^{N} \left[\mathbf{v}_i e_i - \frac{1}{2} \sum_{j \neq i} \mathbf{r}_{ji} \mathbf{v}_i \cdot \frac{\partial V(r_{ji})}{\partial \mathbf{r}_{ji}} \right] \delta(\mathbf{r} - \mathbf{r}_i(t)) \qquad (2.11)$$

Let us stress that these expressions are valid for small enough gradients (small with respect to interatomic distance) and for regular potentials. The case of hard potentials has to be treated with a little care since the microscopic expressions contain a singular part due to the instantaneous nature of the collision. The extension of these definitions to the case of hard spheres can be found in the article by J. J. Erpenbeck in Ref. [3].

One is often led to consider cases where there is an inhomogeneous velocity field. In these cases, the expressions given above can still be used, replacing particle velocities by peculiar velocities, which are the velocities obtained by subtracting the local velocity field value. These local values are obtained from a spatial average; one has to divide the system into cells in which one averages the values of the observables. The gradients need to be sufficiently small for the variations of the densities over a cell size to be negligible. In Eqs. (2.10) and (2.11), the microscopic fluxes are obviously the sum of kinetic and potential terms. The first are fluxes directly related to the motion of the particles, while the second are momentum and energy transfer through the interactions at distance.

It is also worth mentioning that specific programming techniques have to be introduced when dealing with large system sizes [71, 72], and that these techniques can be optimized with respect to the computer architecture. When using parallel [73] or vector [74] processing one needs a programming strategy that takes advantage of the way the computing is done.

Before we turn to specific computations that have been made using the molecular dynamics technique, we shall describe simplified simulation methods that turn out to be more efficient than molecular dynamics, namely the lattice gas and the Direct Simulation Monte Carlo (DSMC) methods. Both of these strictly describe the evolution of a set of particles with no interaction energy. Their use in the present context is justified by the fact that, whatever the microscopic model consists of, the macroscopic motion, on the largest time and space scales, is hydrodynamical.

C. Lattice Gases

1. *Definitions*

It has already been argued that the interaction potential is not important for the study of large-scale properties that are specific to nonequilibrium states. In the case of lattice gases, modelization is reduced to the minimum (and probably less than that) required to produce hydrodynamic behavior on the large time- and space-scales of the system. This technique became popular after Frisch, Pomeau, and Hasslacher published their letter [75] showing that particles in an hexagonal lattice with simple collision rules would, in the appropriate limit, evolve according to the incompressible Navier–Stokes equations. This family of models was named cellular automata, since it appeared that they were a particular case of a class of models that could display very complex behavior from simple local deterministic rules [76]. Actually, it is worth noting that a similar model had been proposed a little bit earlier by Wolfram [77], in the context of his general study of cellular automata. Because the model failed to conform to hydrodynamic behavior on large scales, it has since been forgotten, but many of the ingredients of the lattice gas models for hydrodynamics are already present in Wolfram's study.

A cellular automaton is a lattice where the nodes can be in any of a finite set of states. Simultaneous updates of the node states take place at discrete time intervals, and the rules are such that the state of a node at time $t + 1$ depends on the states of its neighbors at time t. These models were developed first by John von Neumann in the late 1940s as a practical way of investigating the question of knowing what is the minimum set of rules needed to account for the replication of biological cells [78]. They became popular in the 1970s, in part thanks to the so-called "game of life" [79], an example of a Turing machine capable of universal computations. These models have rules governing their evolution and can sometimes display behaviors or properties similar to those of physical systems, like time-irreversible entropy, unpredictability, fractals and so forth. However, they are not meant to model real physical systems, they have no energy and there seems to be no basic principles giving keys as to their evolution.

The numerical solution by a finite-difference method on a lattice of fluid dynamical equations is a kind of cellular automaton, except for the fact that the node states are continuous rather than discrete (as a matter of fact, on the computer they *are* discrete but this is not really the point). Would one consider only a discrete set of states to describe the fluid at lattice nodes, such a numerical scheme would become a cellular automaton. This is precisely what is achieved by the lattice gas modeling.

Another way of considering the introduction of lattice gases to simulate hydrodynamic flows is, instead, to consider these models as truly microscopic. Space is discretized, and particles move on a lattice at discrete time intervals, jumping from node to neighboring node, velocity being thus also discrete. There is no interaction between the particles, except at collisions, which occur at the lattice nodes when several particles meet. The collision rules are such that mass and momentum are conserved. This conservation property is known to be crucial for the existence of hydrodynamic equations for mass and momentum densities.

The first such model studied in the literature is known as HPP (Hardy,

Pomeau, de Pazzis) model and it was proposed in 1972 [80] to study time correlation function on a simplified model that would eventually allow for an exact investigation of long-time behavior, thus making it possible to obtain the asymptotic long-time tails of the stress–stress time-correlation functions independently of any approximation. The model consists of a two-dimensional square lattice. On each lattice site there are at most four particles, which may have four velocities, all equal in absolute value but pointing in the four possible directions to the nearest neighbors. Configurations are excluded in which more than one particle is in a given state. This exclusion principle, propagated by the evolution rules, is imposed for computational efficiency and has no physical meaning. The evolution is as follows: At discrete time intervals, particles are moved from the sites they occupy to the sites they are pointing to. Then, collisions are performed that preserve the mass and momentum of the incoming configuration at each node. The only efficient collisions to be considered in this simple case are head-on collisions. When two particles (and only two) arrive at a node, the postcollision momenta are the incoming ones rotated by 90° (see Fig. 4).

Let $s_i(r,t)$ be the dynamical variable: it takes values 0 or 1 depending if the node r at time t is occupied, that is if there is a particle with velocity c_i pointing in direction i ($i = 0$, 1, 2 or 3 and $i = i + 4$). The equation of motion reads

$$s_i(r + c_i, t + 1) = s_i + s_{i+1}s_{i+3}(1 - s_i)(1 - s_{i+2}) - s_i s_{i+2}(1 - s_{i+1})(1 - s_{i+3}) \tag{2.12}$$

on the right-hand side of equation (12), the s_i are evaluated at node r and time t. This equation is the automaton rule, and it is easily verified that it agrees with the definition given at the beginning of this section. Besides, the actual computation can be performed by the construction of a table linking input states to output states. The state of a node is represented by a 4-bit word in the computer, there are 16 different states at each node, and the collision table

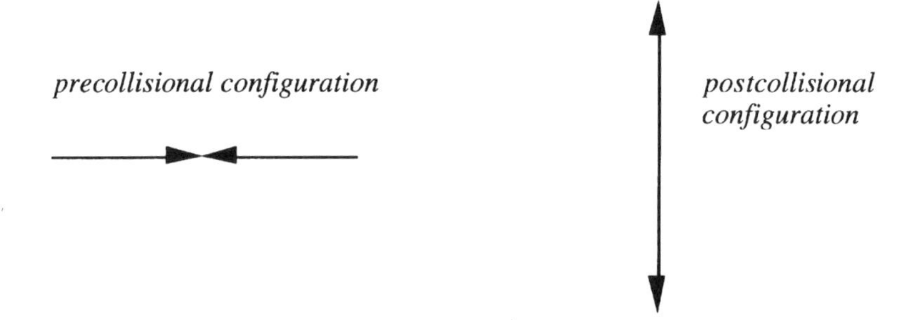

Figure 4. The only momentum-exchange collisions considered in the HPP model.

consists of a vector of 16 elements, each of which is a 4-bit word. An alternative is to compute the state at time $t + 1$ by performing logical operations on the s_i at time t. This procedure is of comparable computational efficiency in the HPP model but becomes more efficient in the case of more complex models in higher dimensions: for instance, in three dimensions, states are represented by 24-bits words and collision tables would involve a 16-million component vector. For efficiency reasons, the state of the lattice must also be stored in the main memory, which also provides a direct bound on the lattice sizes that can be considered.

The computation of a flow from such a model requires in addition the determination of an initial condition and the specification of what happens at the boundaries. This being done, the evolution can be computed according to the rules just given. The flow properties will be obtained as averages, both in time and space, of the momentum and mass densities of the exact trajectories computed. This indicates that the theoretical analysis of the model needs the introduction of a probability measure on the lattice, the evolution of which is governed by the equivalent of the Liouville equation for this specific model. The problem is then to prove that the average densities obey equations, in the proper limits, which are the macroscopic hydrodynamic equations. In the case of the model considered here, collisions only affect momentum and mass, but energy is trivially conserved. One should expect therefore that the models simulate the equations for mass and momentum densities, not the energy equation, which would require more intricate collision rules. The hydrodynamic limit to be considered is the limit where one lets frequency and wave vectors (inhomogeneities) go to zero in such a way that frequencies divided by wave vector (sound speed) and by their squares (sound damping and momentum diffusion) are finite.

2. *Properties of Lattice Gases*

The connection between, on one side, the various "microscopic" lattice-gas modelings and, on the other side, macroscopic or large-scale behavior has been extensively studied in the last years by several authors. Below are listed a few of the most significant results obtained, which somehow make more precise the range of applicability of lattice-gas models to reproduce incompressible hydrodynamical flows.

(a) The HPP model correctly reproduces sound propagation, and the sound speed is $c_s = c/2^{1/2}$, c being the particle constant velocity. Sound propagation is isotropic despite the square lattice; however, the sound damping is not isotropic as one would expect. Actually, the computation of the stress tensor involves a fourth-rank tensor that has the symmetries of the lattice. On a square lattice, the fourth-rank tensor can be expressed with three independent scalar quantities. This implies that the stress tensor has three viscosities, and not two, as one would expect for an isotropic fluid. Thus, sound adsorption is not isotropic in the HPP model, which is a quite unphysical feature. The requirement of symmetry can be fulfilled in two dimensions by the triangular lattice and, in the three-dimensional case, by a projection in three dimensions of the four-dimensional face-centered cubic hyperlattice [81]. The corresponding lattice gases are known, respectively, as the FHP and FCHC models.

(b) The equilibrium measure is a Fermi-like distribution function of the velocities. The origin of this is the exclusion principle, which is imposed for convenience. As a consequence, hydrodynamic equations can be recovered only in the small-velocity regime. The flow velocity divided by the particle velocity, u/c, is a smallness parameter which permits expansion of the equilibrium distribution. Terms of higher order appear in the equations and are also unphysical. For instance, the hydrostatic pressure has correction terms proportional to the square of this smallness parameter. A thermodynamic pressure can also be calculated from the partition function of the model. Kinetic and thermodynamic pressure are equal in the small-velocity regime. Let us note that equilibrium distributions exist only for those models that obey detailed (or, at least, semidetailed) balance. Extensions of the FHP or FCHC models that do not obey semi-detailed balance have been proposed. No equilibrium state then exists, and consequently no partition function and no thermodynamics can be straightforwardly calculated [82].

(c) The model is not Galilean-invariant because the particles move with a constant velocity. As a consequence, the inertial term in the hydrodynamic equations is multiplied by a factor $g(d)$ that depends on the local density per node. It is only in the incompressible limit that this g-factor can be eliminated by a rescaling of the time variable. The incompressible limit consists of letting the flow velocity, u/c, go to zero (compared to the particle velocity) and the wave vector (or the inhomogeneity) go to zero while keeping their ratio fixed. The density variations then evolve only because of the fluid motion (divergence of u is zero), and the density gradients disappear from the equations, except in the reversible pressure-gradient term. In that limit, the incompressible Navier–Stokes equations are obtained as the long-wavelength, long-time solutions of the lattice-gas model [83].

(d) The original FHP model consists of particles on a triangular lattice with six possible velocities, one in each direction pointing to the nearest neighbor. The collision rules also consider head-on collisions, with a probabilistic rotation of the two incoming velocities after collision, either by $+60^\circ$ or by -60°. The probabilities for each event should be equal, otherwise the chiral symmetry would be lost. The same

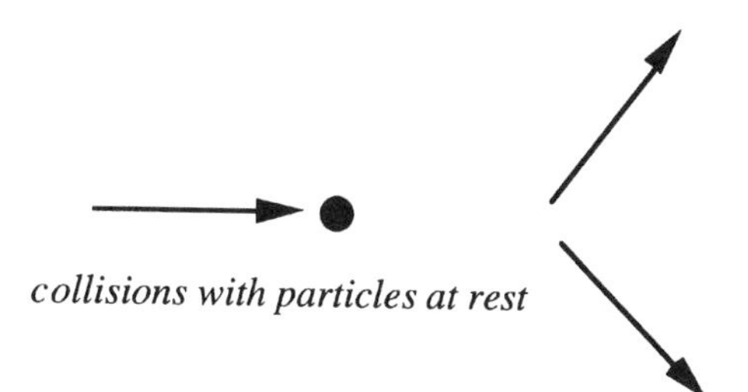

Figure 5. On an hexagonal lattice, collisions with particles at rest are to be considered in order to have a positive bulk viscosity. Here is an example of a collision with conservation of mass and momentum, but not of the kinetic energy. The inverse collision should be allowed with an equal probability, otherwise detailed balance would be lost and spontaneous deviations from equilibrium would be possible (and indeed observed).

is true for quadrupole collisions, the velocities being turned either by $+60^\circ$ or by -60° with an equal probability. The model has since been generalized to a seven-velocity model, by adding the possibility of an immobile particle, energy not being conserved at collision (or, equivalently, considering transfer from kinetic to some kind of internal energy, see Figure 5). The models with particles at rest have a nonvanishing bulk viscosity, contrary to the original FHP model. The introduction of a nonvanishing bulk viscosity is necessary because the original six-velocity model behaves effectively, as if the bulk viscosity is negative.

Actually, various collision rules have been considered for the triangular lattice and it has been realized that the shear viscosity decreases when more collisions exchanging momentum among the particles are allowed. This is important since a low viscosity leads to a more turbulent regime, which is the domain where the lattice gas could prove its robustness with respect to finite difference methods. Actually, the dimensionless number, characteristic of how turbulent the flow is, is the Reynolds number, $\mathrm{Re} = L \cdot u/\nu$, with L the inhomogeneity length, u the typical flow velocity, and ν the kinematic viscosity. High Reynolds number values can be achieved by setting large values to L and u and a small value to the viscosity. Now L is controlled by the computational cost. Usually the CPU time scales as the square of L in two dimensions—proportional in fact to the number of nodes. The flow velocity is bounded by the incompressibility condition (a small fraction of the sound speed) so that adding collision rules to decrease the viscosity is meaningful. In three dimensions, the search for optimal collision rules has been done in a subtle way by M. Hénon [85]. With the arguments given above, one also finds that the CPU time scales like Re^3 in two dimensions, which implies that lattice gases will not perform more efficiently than finite difference Navier–Stokes solvers.

TABLE I
Values of Lattice-Gas Parameters in Some Real Systems

System	FHP Model	Water	Air
Viscosity	$0.7\ l^2/\tau$	$0.01\ \mathrm{cm}^2/\mathrm{s}$	$0.15\ \mathrm{cm}^2/\mathrm{s}$
Sound speed	$1/2^{1/2} l/\tau$	$10^5\ \mathrm{cm/s}$	$3.10^4\ \mathrm{cm/s}$
Lattice spacing		$10^{-7}\ \mathrm{cm}$	$5.10^{-6}\ \mathrm{cm}$
Time step		$10^{-12}\ \mathrm{s}$	$10^{-10}\ \mathrm{sec}$

What are the intrinsic scales of the model? There is no particle size, no microscopic parameters, so that the basic scales introduced are not obvious to determine. In Table I, we have estimated the values of the lattice spacing and of the time steps for two different fluids, water and air, under normal pressure and temperature conditions. One needs at least two quantities in order to do this and we have naturally chosen the sound speed and the kinematic viscosity. The viscosity and the sound speed in the FHP model are given functions of the lattice spacing and of the time step; therefore, assigning the values for air and water, this permits solution of the two equations to determine the corresponding parameters for those fluids. The values obtained are of the order of magnitude of the interparticle distances and of the relaxation time for both fluids, an order of magnitude larger than corresponding values in typical MD simulations. This is to say that the modeling is not truly microscopic. Of course, in hydrodynamic flows, where there is no intrinsic length scale, the flow properties are functions of Re only. In that case, one is led to consider that the main limitation to achieving flows with high Reynolds numbers is really the number of lattice points one is able to consider. This is very much like the situation of traditional Navier–Stokes solvers. In that respect, lattice gases do not show any advantage, as was already noticed by Orszag [86], and the scaling of the CPU time with respect to the Reynolds number is not advantageous either. However, the great advantage of the particle method is that no numerical instability has to be feared. Another advantage that is often cited is the ease with which boundary conditions can be incorporated (and programmed) in the model.

3. *Applications*

Extensions and applications taking advantage of the simplicity of the modeling and of the exact nature of the computing have been proposed in the last years. Let us mention:

(a) *The lattice Boltzmann method.* The lattice-gas simulations made are of course noisy since the microscopic, or at least mesoscopic, fluctuations are incorporated in the model. This is an advantage for problems where one looks for the possible effects of fluctuations on macroscopic behavior. Near a bifurcation point, for instance, one would like to know the sensitivity of the system to fluctuations. However, in the problem of computing a flow, fluctuations appear as an unnecessary source of statistical noise. In order to smooth out these effects of the underlying microscopic world, a formulation of the lattice gas model in terms of densities has been proposed. Actually, it amounts to solving the Boltzmann equation for discrete velocities and on a lattice

$$f_i(r + c_i, t + 1) = f_i(r, t) + \sum_j A_{ij} \left[f_j(r, t) - f_{\mathrm{eq}_j}(r, t) \right] \tag{2.13}$$

with A_{ij} being a transition probability between densities of velocity directions i and j at node r; it is proportional to the collision frequency and it can be tuned to control the viscosity. With a proper choice of the transition matrix A_{ij}, Eq. (2.13) is, as a matter of fact, a discretized version of the BGK (Bhatnagar, Gross, Krook) model for the Boltzmann equation, that is a model that simplifies the relaxation process while keeping the nonlinearity of the collision operator [87]. This is not to be confused with the Boltzmann approximation, which is usually done in the theoretical computation of the lattice gas properties—and which by the way, provides very good values.

The lattice Boltzmann method [88] presents many appealing features. It permits lowering the viscosity in a more efficient way; it does not suffer from the non-Galilean invariance disease; and lattices different from the triangular one may be considered. However, it has also been realized that the algorithms used to perform the simulations were identical to some algorithms developed in the framework of numerical solutions to Navier–Stokes equations. Somehow the modeling has shifted from the micro- or mesoscopic level to the macroscopic one.

(b) *The existence of diffusion* in two and three-dimensional lattice gases has also received much interest in the recent years. This is the kind of "fundamental" problem that is recurrent in statistical mechanics. As a matter of fact such problems were at the very origin of the HPP model and it is only justice that lattice gases can provide some useful information on the existence of diffusion in two-dimensional fluids [89]. The question is whether the Green–Kubo integrand for the self-diffusion coefficient, that is the velocity–velocity time-correlation function, decreases sufficiently fast at long times in order that its time-integral remain finite. As it has already been mentioned in the introduction, in two dimensions, the long-time behavior was computed by Alder and Wainwright in hard-disk MD simulations to behave as $1/t$, leading to a logarithmic divergence for the transport coefficient. The origin of the behavior has been traced to the mechanism by which a particle would transfer its momentum to the surrounding fluid. Being a hydrodynamic diffusive process, it leads to a power-law decay of the same type as the time-correlation decay. It is, however, argued that the use of bare transport coefficients in the estimated decay is not consistent, since these are eventually not well-defined in two dimensions. As a consequence, arguments are given to renormalize the transport coefficients in the estimate of the long-time behavior of the autocorrelation function, leading then to a $1/[t \ln t]$ decay and so to a finite time-integral.

It was not easy to check this conjecture from MD simulations and the test provided by the simulations made in lattice gases was made possible thanks to a clever trick found by Frenkel and his coworkers [90]. In the computation of the velocity autocorrelation function, one is interested in the correlation in time of a particle velocity with itself at a later time. Now, in a lattice gas model, particles have no identity and the postcollision configuration does not say which particle is which. First attempts modified the model in order to distinguish between particles, giving a color to one of the colliding particles and tracing it through a colored trajectory. The idea that permitted an increase in the statistics to a tremendously higher level (by a factor of 10^6!) is rather to take advantage of this indiscernability and to sum over different trajectories where an incoming particle is, with equal probabilities, traced in the different possible outgoing states. So the scalar product of a particle velocity with itself at a later time is in fact to be taken with the average velocity of all outgoing particles, which can possibly be found to result from the original particle trajectory. It is easily understood that the number of trajectories considered grows exponentially with the number of collisions performed, leading to a much better statistics. It has also to be realized that the coding of the computation is not that difficult in the sense that one has to count possible particle paths and then average all possible outgoing velocities with the proper weight.

Evidence provided by the simulations indeed confirmed the theoretical arguments [91]. The algebraic slow decay is followed in time by a slightly more rapid decay, on an extremely long-time scale.

Let us mention also that recently the lattice Boltzmann approach has been used to check the theoretical mode-coupling predictions for the stress–stress time autocorrelation function [92]. Since the model does not contain any thermal fluctuation, the authors had to study rather the decay of an ensemble of initial disturbances. The excellent agreement found with the mode-coupling theories cannot however be taken as a test of the theory. Indeed, this is an illustration of the fact that the model is not truly microscopic since one studies densities rather than particles (how a Boltzmann fluid particle would exhibit long time correlations since it is supposed to obey the molecular chaos hypothesis!). This work is to be understood as a confirmation that mode-coupling theories correctly estimate, within a hydrodynamical framework, the slow decay at long times of the stress fluctuations [93], within a macroscopic framework. A similar test of the so-called mode-coupling theories has been made possible through the simulation of a simple flow at various sizes differing by an order of magnitude each; such testing is still impossible by a direct MD simulation [94], requiring too large a difference in size and therefore too much computer time. Concerning these last two cases, it has to be stressed that the lattice gas model is limited to the computation of the kinetic part of the fluxes, since there is no potential energy and no momentum transfer. Therefore, this type of model does not permit the investigation of the apparent discrepancy between mode-coupling predictions and the computations made on various fluid models for the potential and cross terms of the response [95].

(c) *The generalization to temperature-dependent problems* has been proposed by several authors [96]. The nontrivial energy conservation at collisions requires the consideration of particles with many velocities. Lattice gases with four- and five-velocity amplitudes have been proposed, leading to the so-called 19-bit and 24-bit models, in two dimensions. As for the pressure, the temperature that appears in the equations is different from the mean kinetic energy: Thermodynamic and kinetic temperatures differ by terms of the order of the square of the flow velocity divided by the sound speed. Of course this discrepancy is, as for the pressure, removed if one linearizes the equations around equilibrium so that the thermal fluctuation spectrum of the model is in agreement with the well-known Landau–Placzek formula.

(d) Colloidal suspensions: A very elegant application of the LGCA (Lattice Gas Cellular Automata) technique has been proposed by Ladd [97] in the case of *colloidal suspensions*. The important question is the computation of the hydrodynamic interactions between the Brownian particles in the suspensions. Models were proposed in which the solvent is treated as a lattice gas that exchanges momentum and angular momentum at the moving boundaries of the large colloidal particles. The technical problems are important, especially the treatment of the (moving) boundaries between fluid and Brownian particles. A recent formulation of a lattice Boltzmann approach for the fluid and a "Newtonian" integration for the large-size particles seems to provide a powerful way of performing simulations on relatively large scales. The technique has been applied to basic questions concerning the behavior of Brownian particles, to the computation of the viscosity as function of the concentration of the colloidal particles, and to the computation of sedimentation speeds. Note that fluctuations have to be artificially added to the lattice Boltzmann model to provide a source of thermal noise [98]. These results have been also discussed by Cichocki and Felderhof, who question their range of validity [99].

(e) *Two-phase flows and spinodal decomposition* kinetics have been also studied by the mean of lattice gas models. Rothman has proposed a generalization of an FHP model where particles are given a color (e.g., blue or red) and whenever there is a collision and a choice for the outgoing directions of the blue (or red) particles, particles would be chosen to move preferably towards neighbors of the same color, a kind of majority rule. This type of evolution rule does not obey detailed balance so that the corresponding models do not give rise to any true equilibrium state. However the red and blue particles do separate into two different phases and the interfacial tension seems to obey Laplace's equation. (For this model and many other references on lattice-gas applications, see the review article in ref. [100].) A more microscopic model has been proposed for a kind of liquid-vapor phase separation kinetics. It consists of allowing particles moving away from each other and distant by r nodes to flip their momentum so as to make them move towards each other. The total momentum is conserved on the average but not locally, and the model is again not obeying detailed balance; however, the emphasis is on the kinetics of phase separation. The rules contain some stochastic part meant to mimic an attractive interaction. The existence of the equivalent of a phase transition for a sufficiently large value of r has been observed, and critical behavior can be studied.

4. *Conclusions*

To conclude this section, let us stress that the lattice gas cellular automata models have been shown to produce flows that are solutions of the Navier–Stokes equations in the incompressible fluid limit. Only in this case, with the proper lattice symmetry (triangular or hypercubic), rigorous statements are available. The diseases displayed by the lattice gases can be cured by generalizing the original ideas to lattice Boltzmann models, but at the cost of the loss of the microscopic scale description.

Interesting applications have been provided by these models, ranging from the study of the existence of diffusion to the computation of the viscosity of colloidal suspensions. Some care has to be taken, however, with models which do not obey detailed balance. These models do not possess equilibrium states from which thermodynamical properties can be computed; instead, they have irreversibility built-in, and their range of validity is not well defined.

D. Direct Simulation Monte Carlo Method

The community of rarefied gas dynamics has been also interested in the solution of flow problems starting from a microscopic description. In the case of dilute and very dilute systems, the fundamental description is often taken to be the Boltzmann equation rather than the microscopic dynamical equations of the particles. Except for the artificial cases where velocities of the particles are reversed and lead to a temporary antithermodynamic behavior [101], it is hard to imagine states for dilute systems where the Boltzmann equation could be incorrect. It has therefore been used quite extensively to study flows, shock profiles, and boundary problems in dilute and very dilute systems. It is not surprising either that numerical schemes have been proposed to solve the equation. The method that has proved to be the most efficient has been developed by Bird [102, 103] since 1964 and has been since popularized in engineering as well as statistical mechanical contexts [104]. It consists in a probabilistic treatment of representative sets of particles and is based on the same assumptions as those necessary to derive the Boltzmann equation. A Monte Carlo procedure is then constructed, leading to the name "direct simulation Monte Carlo" method, which we now briefly describe. The method has also received some theoretical justification based on the theory of Markov processes [105].

The volume of the gas to be modeled is divided into cells, so that each cell can be considered as homogeneous. Initially, in every cell is put a number of particles sufficient to represent the state of the gas at that point. Actually, only the local number density and the second moment of the velocity distribution function are needed so that this number of particles need not be very large. From experience, it appears that a minimum of 20 particles per cell is required for ordinary flow problems. The time evolution is then performed in a succession of two alternate steps: the free motion of the particles during the time step and then a representative set of collisions among the particles belonging to the same cell. The rules for the collision process are deduced from the scattering properties and depend on the interaction potential as well as the molecular chaos assumption, which is known to be a good approximation in a dilute system. A pair of particles is chosen at random in a cell, and the acceptance-rejectance method is used to ascertain that there is a good distribution of relative velocity at collision. For hard spheres, the collision frequency is proportional to the relative velocity. The pair is accepted if their relative velocity is larger than a random number times the maximum relative velocity sampled up to then. This procedure is a practical implementation of the rejectance-acceptance method to generate a specified distribution. When accepted, the collision is performed by choosing a random impact parameter, and the velocities of the colliding particles are changed according to the

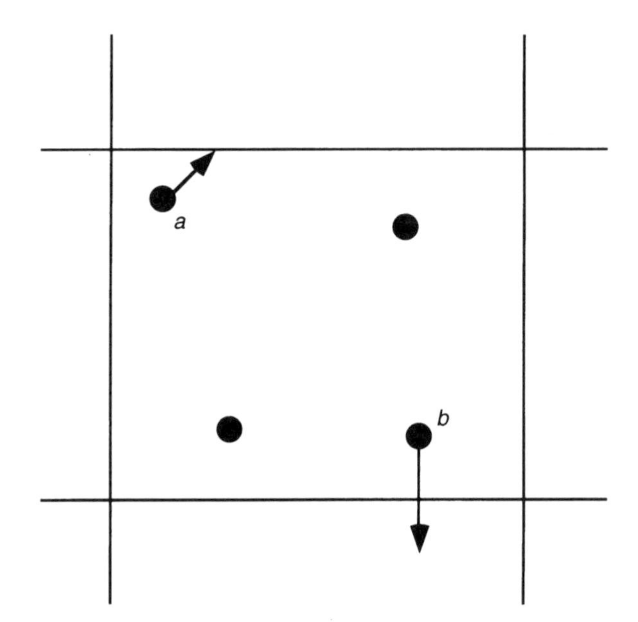

Figure 6. Particles a and b may collide as if they were in a precollisional configuration. Their relative velocity would then be randomly rotated.

collision law. The process is repeated until the correct number of collisions corresponding to the time step, given the collision frequency which depends on the local density, has been done.

Only particles belonging to the same cell are allowed to collide. However, as can be seen in Fig. 6, some unphysical collisions can take place. Most of the time, this is unimportant for the evolution of the flow, but when the vortex size is not much larger than the cell size, this may result in unphysical evolution. Meiburg [106] has shown by comparing molecular dynamics simulations of a two-dimensional flow past an obstacle and a DSMC simulation, that occasionally differences can occur. Collisions of the type shown in Fig. 6 do not conserve the angular momentum, and this may interfere with the flow evolution. The vorticity of the DSMC solution could not be the same as the one emerging from a full dynamical treatment. Bird [107] has subsequently proposed a modified version of his algorithm, curing this pathology by dividing the cell into subcells in order to choose particles for collisions. Candidate particles for collisions are taken in the same part of the computational cell, a procedure that then reduces their distance at collision.

The method has been applied in many different contexts and has received considerable attention from engineers[3] as well as from chemists and physi-

[3]See the various Rarefied Gas Dynamics Conference Proceedings, the most recent of which is ref. [108].

cists. The foundation of the method is quite intuitive, and some more "rigorous" approaches, leading to slightly different numerical algorithms, have been proposed [109]. However, it is fair to say that the proposed modifications are not very different and certainly not more efficient. Besides the comparison in some known test cases has always been in favor of the original scheme.

Recently an interesting use of the method that is comparable to the approach in lattice gas models has also been proposed. The idea is again that one is not obliged to be realistic in the modeling, but rather one can tune the free parameters of the model in order to study flows, and eventually flows with high Reynolds numbers. One can eventually modify the collision frequency of a given molecular model in order to decrease the viscosity and therefore increase the Reynolds number. It has been suggested that one can forget the origin of the method and use it to simulate hydrodynamics with adjustable transport coefficients. This large-scale behavior should be independent of the fact that it is realized with a dilute system, except for the fact that the bulk viscosity will remain equal to zero, and that potential fluxes are absent. This has been indeed done and it has lead to interesting results [110].

Let us also mention a recent and important work that extends the method to the moderately dense regime. This extension has been recently proposed by Alder and coworkers and named the CBA method (consistent Boltzmann algorithm) [111]. The Enskog kinetic equation for hard spheres generalizes the Boltzmann treatment by taking into account two effects that become essential at higher densities: the increase of the collision frequency due to the increase of the number density and the collisional transfer mechanism. The increase of the collision frequency can be accounted for by multiplying the number of collisions to be performed in a time step by the (local equilibrium) pair correlation function at contact. This is a well-known equilibrium property that contains all the information needed for the nonideal part of the equation of state. Its dependence on the local density is well accounted for by a variety of fitting formulas [112]. The collisional transfer is more difficult to treat. It is also more important, since it is responsible for the existence of potential fluxes in dense fluids, the transport mechanism that becomes dominant at high density. In the DSMC technique, collision between hard spheres amounts to a random rotation of their relative velocity. In the CBA algorithm, at every collision, after the random rotation of the relative velocity, the positions of the two colliding hard spheres are in addition displaced by an atomic distance along the line of the transferred momentum and in opposite directions. This produces an instantaneous transfer of momentum leading to the correct equation of state. The results obtained for the transport properties are in remarkable agreement with those computed by MD in a large range of densities. Very recently, another direct simulation method to solve the Enskog equation has been proposed by Santos [113]. The method seems to produce exactly the solution of the Enskog equation in the limit of large densities. Let us also mention that one can instead solve directly the Enskog equation by a Monte Carlo procedure. This method has also been developed recently and the results of its application to shock front computations compare very well with MD [114, 115].

The DSMC method has also been carefully tested by looking at the equilibrium and nonequilibrium fluctuation properties of model systems. In the case of equilibrium systems, the question raised is actually the following: Using the DSMC method to generate the particle's trajectories rather than full MD, and computing the same average values of observables in the various conditions considered, are we going to obtain similar results? We already know that the DSMC treatment leads to a correct computation of the equilibrium and nonequilibrium flow patterns, at least on scales larger than the mean free path and larger than the size of the computational cell. This has also been shown in the case of shock waves that are accurately described by the DSMC method—at least when compared to MD for dilute gases [116]. Is the same still true when computing fluctuations?

Let us recall that the description of fluctuations is not included in the Boltzmann equation. It is only because the simulation of the Boltzmann equation is made with particles rather than densities that the question of fluctuations arises. Lattice Boltzmann simulation methods, for instance, do not produce thermal fluctuations. In some sense, computing the fluctuation spectrum out of the DSMC simulations amounts to testing whether the DSMC method simulates a *fluctuating* Boltzmann equation [117, 118] or the Boltzmann equation.

In Fig. 7, we show the results of an equilibrium computation of the dynamical structure factor, a quantity that is central in the characterization of the time evolution of density fluctuations (see Section IV for more details on this function). The computation was made with 3,000 hard spheres in 60 computational cells: 1 million collisions per particle have been performed. The system is one-dimensional and its length is around 30 mean free paths. The dots are the results of the simulation, while the curve is the hydrodynamic form of the dynamic structure factor. The area under the central (Rayleigh) line is $C_p/C_v - 1$ (ratio of heat capacities); the width of the finite-frequency (Brillouin) peak is v, the kinematic viscosity, to a very good accuracy, showing indeed the correct description of density fluctuations by the DSMC method at large scales, at least in equilibrium. This figure is shown in order to illustrate the fact that the DSMC method is truly mesoscopic, providing good agreement with the fluctuating hydrodynamics approach. A more detailed discussion about the nonequilibrium density fluctuations will be given in Section IV. It is also to be noted that, in contrast to the compu-

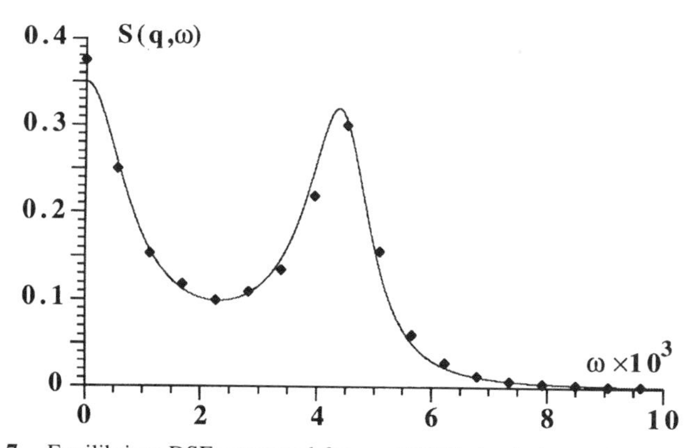

Figure 7. Equilibrium DSF computed from a DSMC simulation and compared with the corresponding Landau–Placzek formula.

tation of flows, the computation of fluctuations is only valid on the scale set in the simulation.

Similar results had been previously obtained on nonequilibrium fluids. Correlation function of the fluctuations of temperature had been computed by a DSMC simulation of a fluid submitted to a temperature gradient [119]. The comparison made with the fluctuating hydrodynamics theory gives very good agreement between the two approaches. The nonequilibrium simulation was made by Garcia and coworkers and involved a computation with 20,000 particles in a system of length around 50 mean free paths. Temperatures are set to $T_L = 1$ and $T_R = 3$ in system's units. The simulation has run for approximately 200,000 collisions per particle. Similar results were also obtained for a fluid under a shear [120]. The shear was realized by moving boundaries in the vertical direction, keeping periodicity in the other two directions. (Actually, in the DSMC method, periodicity in the other two directions can be ignored, since the geometry can be taken to be one-dimensional in the case considered here. Only the components of the positions in the inhomogeneity directions need to be computed, the other can be left out.) Again 20,000 particles were used and a total of 1 billion collisions were performed in order to average the data.

Detailed comparisons between the DSMC method and MD have been done for shock waves and complete agreement has been found, except at densities where the nonideality of the fluid starts to be important [121, 122]. The peculiar property of dilute systems is that one can easily realize a stationary simulation and average the particles' properties with as much precision as desired. Indeed, a stable shock front appears if one injects a dilute gas from one side as shown in Fig. 8, and expel it, in the denser hotter state, from the

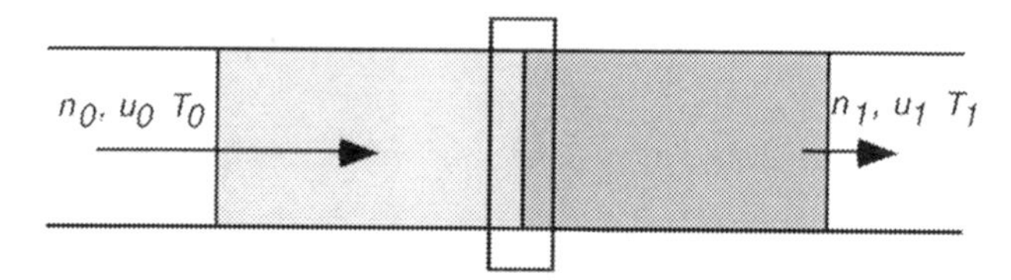

Figure 8. A shock front is simulated in the comoving frame. The unshocked gas is injected at left with a velocity that is $u_0 = -u_s$, while the shocked fluid is moving toward right with velocity $u_1 = u_p - u_s$, u_p being the velocity of a virtual piston compressing the fluid. A stable shock front is obtained and its statistical properties can be computed by the DSMC as well as by MC simulations.

other side, once it has crossed the discontinuity of the shock front. Injection boundary conditions are easy to implement in a dilute system, since equilibrium configurations in boundary cells can be reconstructed at each time step. Such a procedure is clearly more difficult in dense systems where the techniques used only produce transients.

The shocked and unshocked states far from the front are related through the three conservation equations so that the six parameters appearing in Fig. 8 are not independent and reduce to three; fixing the Mach number, defined as the ratio of the front velocity to the sound speed in the unshocked region, further reduces the number of parameters to two. The profiles shown in Fig. 9 are for a Mach number of 4.5, and they correspond to the following values of the gas states (in SI units): $\rho_0 = 4.42$ kg/m^3 ($\rho_1 = 1.54$), $u_0 = 1410$ m/s ($u_1 = 403$), $T_0 = 270\ K$ ($T_1 = 2000$).

The comparison in Fig. 9 shows that MD and DSMC results are clearly indistinguishable and that both of them disagree with the predictions based on Navier–Stokes equations, which are solved numerically using a local equation of state and a local value for the heat conductivity and the viscosity obtained from a kinetic theory computation for the hard-sphere model are dependent on the horizontal coordinate, through the value of the local density and temperature. The continuum predictions compare more favorably with the microscopic simulation results if one follows a recipe proposed by Holian, namely to use the second-moment of the velocity distribution function only in the direction of the propagation in order to determine the temperature at which transport coefficients have to be computed in the Navier–Stokes description [123].

To conclude this section, let us stress that the probabilistic method described here is much more efficient than MD. Actually, probabilistic treatments are more efficient than exact dynamical treatment in the dilute regime, but their computational costs become comparable once the density increases.

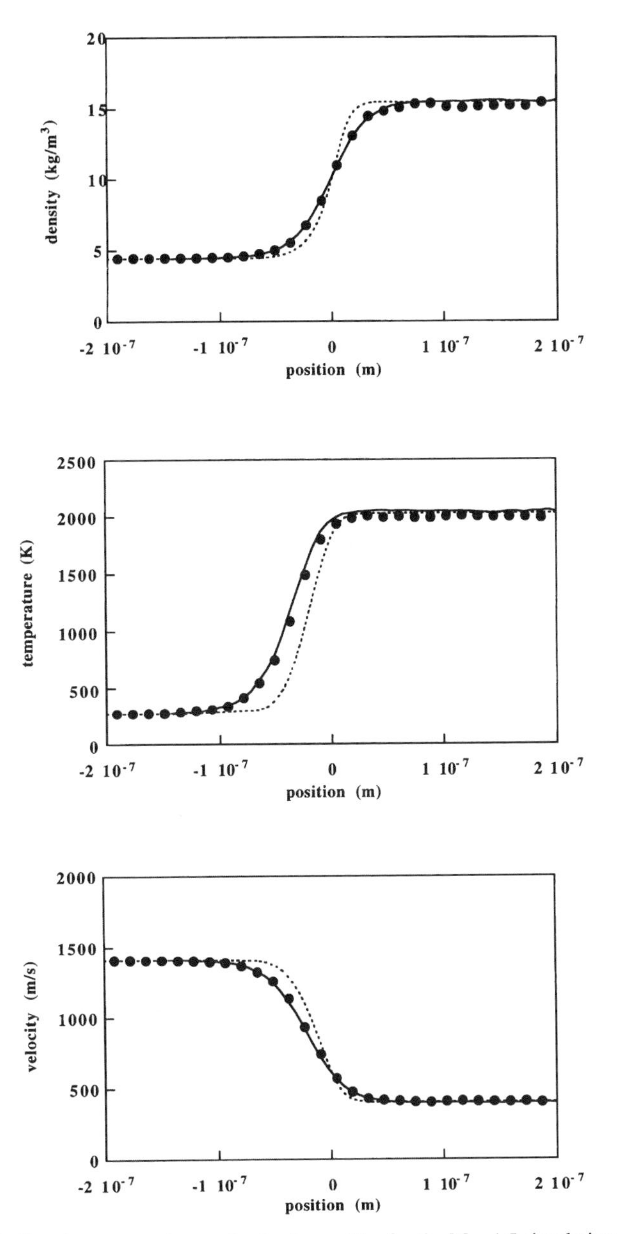

Figure 9. Density, temperature and velocity profiles for the $M = 4.5$ simulations. NS results are given by the dotted line, DSMC by the continuous curve, while MD results are shown by dots. SI units are used.

This is linked to the fact that the stochastic method has a length scale, which is the mean free path, and when the density becomes large, this length scale becomes smaller than the atomic diameter. The number of particles needed to describe an inhomogeneity becomes then more important than in molecular dynamics, leading in turn to a higher cost (and also probably to a non-exact solution!). However, the DSMC technique can be safely used to describe compressible flows with a zero bulk viscosity in the same way as the lattice gas method is used for incompressible Naver–Stokes hydrodynamics.

E. Other Types of Simulations Using Particles

There are many other cases where one needs to treat problems containing two or three different length- or timescales. In these cases a fully dynamical treatment still looks prohibitively expensive. Approximate methods have been developed that are based on mesoscopic equations.

Brownian dynamics is certainly the most widely kwown and widely used of these methods. A Brownian suspension of particles is modeled by a set of interacting particles on which a force deriving from a Gaussian white noise is exerted. The solvent is characterized by its (bare) viscosity and its thermal energy. The particles have a definite diffusion coefficient, or mobility, and one then tries to compute the viscosity of the suspension. Ionic solutions and polyelectrolytes [124] are simulated successfully along these lines. When the density of the suspension becomes larger, one can no longer neglect any hydrodynamic interactions between large colloidal particles. The diffusion coefficient itself becomes a tensor that depends on the configuration of the other particles. In order to take these collective interactions into effect, Stokesian dynamics [125] has been developed. All the short-range effects, called the lubrication, as well as the long-ranged effects, are evaluated to compute an effective diffusion coefficient.

Let us note that fluctuating hydrodynamics itself can be simulated on computers. The fluctuating hydrodynamics equations were postulated by Landau and Lifshitz in the 1940s. These amount to a Langevin treatment of hydrodynamics and consist of adding a Gaussian white noise to the linearized hydrodynamic equations for the velocity and energy equations. More precisely, the hydrodynamic fluxes are divided into two parts, a deterministic part, depending on the local gradients, and a fluctuating part which is modeled as a Gaussian white noise. The amplitude of the noise is determined from the requirement that the equilibrium static fluctuations should coincide with known Gaussian distributions derived from thermodynamics. Although the method has proved very useful for understanding the spectra obtained from light-scattering experiments [126], it is only recently that robust numerical implementations have been proposed. The added noise contains derivatives of a "delta-correlated" force, in time and space, and therefore the treatment of the delta-function singularities has to be done with care, once a finite difference scheme is adopted [127].

The Langevin equation (or Brownian approach) and the Landau-Lifschitz equations are mathemtically well-posed within the framework of linear equations. Adding noise to nonlinear equations, however, require a description at a more fundamental level. This has been achieved by the Master Equation, which describes the system by a probability distribution, in the appropriate space, whose evolution is governed by birth-and-death processes. Considerable theoretical works have been done concerning chemical as well as hydrodynamical systems in far-from-equilibrium states [128]. Numerical simulation of the master equations has been proposed by Gillespie [129]. The problem of a Master Equation description of hydrodynamical flows seems to be difficult to solve, however, the reason being that one has to treat differently the effects of the free motion of particles (which actually does not give rise to any noise since it is exact and fully taken into account at the macroscopic level) and the effects of the collisions. Indeed, the noise is linked to the non-conservation of the momentum and energy fluxes. Let us mention here an interesting solution given by Breuer and Petruccione who presented shock-like solutions of Burger's equation [130] as a test case, before attacking more complex flow problems [131].

There have been also many other attempts in which simulations of macroscopic equations were performed by direct representations of particles obeying dissipative dynamics. The methods developed in the late 1970s in the context of astronomy [132, 133] were recently revisited by Hoover and coworkers to simulate the Bénard instability [134]. A more ambitious, but somewhat less successful, approach was taken by Hoogerbrugge [135] in a kind of mixed method between molecular dynamics, lattice gas dynamics dissipative dynamics, and the stochastic approach.

Foams were also simulated by a prescription derived from the minimization of a local a free-energy density functional [136–138]. Similar approaches are developed within the framework of the description of the growth of grain in metallurgy [139]. These are again examples where a full dynamical treatment is made impossible due to the differences to be considered in the timescales.

III. RHEOLOGY OF FLUIDS

A. Simple Fluids

At first sight, it may appear that the molecular dynamics simulations of Couette flow could directly provide extremely useful insights between the microscopic behavior of a fluid, including its structural response, and its rheological properties, such as those measured in laboratory experiments [140, 141].

A fairly large number of studies have been performed in recent years and a lot of information and understanding has been gained. However, it is fair to say that one is only at the beginning of the stage where computer simulations and experiments can compare in a quantitative as well as in a qualitative way.

The basic law to investigate is the relation between, on one hand, the imposed shear stress and the velocity gradient that develops inside the fluid, and, on the other hand, the law of Newton and its possible extensions. When there is a constant gradient of the α component of the velocity along the β direction, Newton's law reduces to

$$P_{\alpha\beta} = -\eta(\gamma)\gamma$$

where

$$\gamma = \frac{dv_\alpha}{dr_\beta} \tag{3.1}$$

The variable γ is called the strain rate. In general this is a tensor, and in the case of the Couette shear flow described above, the shear rate tensor has only one nonvanishing element. For clarity, in Eq. (3.1), we have dropped the tensorial notation. We have, in that same equation, displayed explicitly the dependence of the viscosity η on the shear rate itself, although, in the linear regime, for small enough shear rates, one expects that the viscosity is independent of the shear rate. Nonlinear behavior is, however, quite common for the viscosity. In particular, complex structured fluids like polymers display a variety of behaviors that can be explained by a dependence of the viscosity on the shear rate. These behaviors, for polymers as for other fluids composed of complex molecules, are not easy to study, either from a kinetic theory approach or even from direct simulation, since the molecular relaxations involve very different timescales. The characteristic times for polymer molecular motion can be as long as several seconds, making a full microscopic description impossible.

The first molecular studies have, in fact, tried to characterize the nonlinear behavior in the much simpler case of atomic dense fluids modeled by hard or soft spheres or Lennard–Jones particles [142]. The first surprise is that the nonlinear behavior does not show up before the shear rate gets to very high values, 10^{12} sec^{-1} for liquid argon! Actually, this value corresponds to the situation where the shear rate becomes comparable to the characteristic frequency which, in a liquid, corresponds to the frequency of an harmonic oscillator approximation. Most of the time, in a liquid, particles oscillate in the minimum of the effective potential of their neighbors, which can be roughly approximated by a harmonic well. The corresponding frequency is of the order of the inverse of a picosecond, that is the shear rate value at which deviations from the linear regime begin to be important. This line of argument also explains why the phenomena that are reported in the simulations of atomic fluids can be compared with experimental data from much larger particles, like colloidal particles, for which the relaxation time can be milliseconds, so that the shear rate values at which nonlinearities develop can be smaller by several orders of magnitude. The dimensionless number that drives the flow is the ratio of the two timescales: the inverse of the shear rate and the atomic relaxation time, a number which has been named the Deborah number. Once the Deborah number is of order one, nonlinear behavior can be expected.

The nonlinear effects that have been observed in these simulations revealed many features like shear thinning [143] (decrease of the effective viscosity at high shear rate), shear dilatancy (a change in the density of the fluid under the influence of the shear, a phenomenon linked to pressure differences induced by the stress), shear-induced melting [144], and spatial ordering of the particles [145]. Ordering transitions, and subsequent shear thickening [146, 147] after an order-to-disorder transition [148] at still higher shear rate, have been also reported by several authors [149], although the phenomenon may be linked with the way the system is thermalized. A variant of the thermalization procedure, very inelegantly named "profile-unbiased thermostat," which amounts to using peculiar velocities to compute the temperature, has lead to simulations where this ordering transition was not found. Simulations have, however, been performed confirming this ordering transition [150]. In addition, simulations made with Brownian particles and for which there is no need to thermalize, have displayed the same phenomenon [151], which occurs at high density. Moreover, there seems to be also some experimental support for the idea that the transition is real [152]. In any case, the isotropy of the pair correlation function is lost in the nonlinear regime [153], a phenomenon which can be easily visualized on a MacIntosh microcomputer by using the software provided by Denis Evans (see footnote 1, p. 321).

Early studies concerning the computation of the viscosity of simple atomic liquids concentrated on checking the possible nonanalytic dependence of the viscosity on the shear rate. A nonanalytic dependence appeared plausible after mode-coupling theories, which were developed in order to explain the long-time behavior of the velocity time autocorrelation function, were generalized to the other Green–Kubo integrands as well. These are equilibrium theories but they aim to predict a frequency and a wave vector dependence for the transport coefficients. The general idea is that is that one can generalize the relations between the stress tensor and the strain rate to smaller time and space scales (scales at which hydrodynamics is not expected to be valid) by adding a possible time and wavelength dependence of the proportionality coefficient [154]. This was not really a new idea. Maxwell had already made a model for a frequency dependence of the viscosity, and the possible dependence of the viscosity on the frequency is called the viscoelasticity. The name originates from the fact that, at short times, before the decay of the stress–stress time correlation function in fact, the liquid can react to a stress by an elastic response, equilibrating a tangential force very much like the reaction of a solid and, only later, flow under the applied stress. Mode-coupling theories [155] were then applied to the computation of these Green–Kubo integrands. They extended the predictions of the long-time slow decay of the time-correlation functions, which were seen for the velocity, to other transport coefficients as well. In three dimensions, the $t^{-3/2}$ long-time behavior, corresponds, once Fourier-transformed into frequency, to a nonanalytic dependence on the frequency, namely $\omega^{1/2}$. Similarly, a possible nonanalytic wave-vector dependence was also predicted, and similar arguments lead to another prediction of a nonanalytic dependence on the shear rate for the viscosity itself, in far-from-equilibrium regimes [157].

$$\begin{aligned}\eta(\gamma) &= \eta(0) + A\gamma^{1/2}\\ \eta(\omega) &= \eta(0) + A'\omega^{1/2}\\ \eta(k) &= \eta(0) + A''k^{3/2}\end{aligned} \tag{3.2}$$

The first computations made on fluid models undergoing a shear with Lees–Edwards boundary conditions and with a Gaussian or Nosé-Hoover thermostat, seemed to confirm these predictions. Since then, larger system sizes have been considered, alternative perturbation techniques have allowed smaller shear rate, and other molecular models have been considered as well. There seems to be a persisting disagreement between the theoretical (mode-coupling) predictions and the molecular dynamics result, concerning the amplitude and even the existence of the long-time tail for the potential part of the Green–Kubo integrand. More recently, fairly accurate computations have been reported that suggest that the dependence of the viscosity on the shear rate is analytic [157]. While there is no discussion that the kinetic part of the stress–stress time autocorrelation function has a long-time tail (in fact, the behavior of the kinetic part is very similar to the velocity autocorrelation function), the persistent disagreement between theoretical predictions and simulation estimates have led to new studies that have shown how difficult it is to obtain a definite answer from the numerical results concerning the nonanalyticities suggested by Eqs. (3.2). Much work has still to be done in this domain. The dependence of the stresses developing inside the fluid on the system sizes is a possible way to explore this problem, in a new perspective, and is similar, in fact, to what has been tested for lattice gas models [158]; however, it requires considering fluid models differing in sizes by orders of magnitude, a requirement that appears very costly on the computational side.

Let us finally mention that attention has recently been brought to an error [159, 160] in the book by McQuarrie [161] in a formula given without proof and expressing the viscosity in terms of an average of one-body phase space observables.

B. Towards More Complex Fluids

Although of much importance, simple atomic fluids do not give rise directly to effects that can be quantitatively compared to experiments. Some of the nonlinear effects predicted by the simulations have found an experimental correspondence in studies of colloidal suspensions. Spherical particles with radii of at least several microns can be chemically tailored in such a way that their interaction forces are spherically symmetric and repulsive over short ranges [162]. Such systems have been the first to display the phenomenon of hard-sphere solidification, providing experimental support to the simulation prediction [163]. Quite recently, both simulations and experiments have shown many order–disorder transitions [164] that confirm the validity of a simple modeling. The timescales involved are, however, much larger than the inverse frequency of liquid argon, as has been mentioned earlier. This suggests that the Deborah number can be of order one for shear rates smaller than for atomic liquids by several orders of magnitude so that they can be realized in laboratory experiments. This has indeed been the case.

Besides using this "primitive" scaling argument, a few pioneering studies have been performed to explore the effects of the complexity of the geometry of the molecules. Of course, this kind of research is not limited to transport properties; many new features appear in the phase diagrams of more complex molecules. Actually many of the order–disorder transitions that are known in liquid crystals have been reproduced with elongated molecules such as hard ellipsoids [165]. Great progress has also been achieved in the modeling of these static properties by the development of techniques permitting the simple expression of the (holonomic) constraints that occur in the equations of motions for rigid bodies [166, 167]. Alkanes have been modeled as a chain of point particles interacting through Lennard–Jones potentials but rigidly linked to two neighboring sites. Rotation is allowed in the chain, but all the very rapid vibrational motion is neglected.

Even with such simplified models, it is still quite challenging to deal well with systems having different timescales. As an example, a study of liquid chlorine in various flows [168] has been performed in which it was found difficult to get clear stress responses out of the noise in the low-strain-rate regions. It is much easier to get clear signals at intermediate strain rates (that is when the longer relaxation time determining the onset of nonlinearity is the rotational time, which is about one order of magnitude larger than the diffusion time). As expected, the first and largest nonlinearity is associated with molecular reorientation, while the redistribution of the center of mass, which leads to nonlinear behavior in atomic fluids (the pair correlation function becoming anisotropic), only plays a secondary role. Along the same lines, there have also been a few studies concerning the rheology of *n*-alkane molecules in shear flow [169].

The general picture that emerges from these studies is that such fluids display shear thinning for small to intermediate shear rates, while for higher strength of the shear rate the behavior depends, as in the atomic case, on the technique used for thermostatting. As for simpler systems, the nonlinearity appears for strain rates that are still huge compared to those found in experiments (10^8 sec^{-1}). This is about the lower limit that can be realized with modern supercomputers if one wants to remain at the level of an atomic description. Using the Green–Kubo approach would be of no help in this context since the computational cost appears comparable to that of a nonequilibrium simulation.

An alternative approach to simulate polymer systems is to start from a coarse-grained model such as the bead-spring or the bead-rod chain model [170]. In such an approach, one disregards the small-scale details of the system (the chemistry) in favor of more global properties (the physics). It relies on the existence of well-known scaling laws for static and dynamic properties of polymers [171], which provide the link between the results obtained for short chains in the simulations and the experimental data for longer realistic chains. More recent works include the study of the dynamics of a many-chain system and of one chain in a solvent system [172]. Successively, the same model was used to study the deformation and the orientation of one chain in a solvent undergoing steady shear flow [173, 174]. This system has been experimentally studied by several techniques, such as electric birefringence, light scattering, and small-angle neutron scattering [175]. The simulation results were found to be in very good agreement (in the rescaled sense) with experiments, and this suggests the existence of an underlying scaling for such nonequilibrium states. This scaling is partially supported by the simulation results but it remains somewhat conjectural, and clearly one needs further experimental and theoretical investigations.

It is important to keep in mind that the success of the latter approach relies on the large timescale separation between the solvent and the polymer dynamics. The ratio of these two timescales increases (roughly) as N^2, where N is the number of monomers in the chain. Such a separation allows one to work at values of the shear rate such that the chain is strongly deformed, while the solvent is still Newtonian. This corresponds to typical experimental setups. It has not yet been possible to focus on genuine rheological properties, such as the intrinsic viscosity, because of the signal-to-noise problem. The latter properties can probably be studied for models of a single chain in solvent, as well as for a melt of chains [176] in simple shear flow.

It is also important to realize that many new properties exist once the isotropy of the molecular model is lost. As a consequence of this symmetry breaking, for example, new hydrodynamic modes can appear [177]. The nature of these modes is not always clear, but anyway, for an anisotropic liquid, with no particular symmetries, there are, in general, five shear plus two bulk viscosities. The computation of these and their relations to experimentally observable quantities have been performed in recent studies of liquid crystals and ferrofluids [179, 180].

IV. NONEQUILIBRIUM STEADY STATES

Besides computing the actual value for the viscosity coefficient, simulations have also been used to obtain results that permit the characterization of nonequilibrium fluid states. Kinetic theories have been developed for nonequilibrium states, but they are unnecessarily complex, and their validity is restricted to the dilute density regime. It has appeared that more phenomenological approaches like fluctuating hydrodynamics would be equally revealing and easier to develop. We present in this section a brief description of this approach and of the simulations performed in order to confirm theoretical predictions.

A central quantity in the study of fluctuations is the density–density space-time correlation function known as the Van Hove function [181] and defined by the following equation

$$G(r,t) = \frac{1}{n_0} \langle \delta n(r,t) \delta n(0,0) \rangle \tag{4.1}$$

Here, n_0 is the average equilibrium number density, $\delta n(r,t)$ is the number density fluctuation at time t and location r, and the average is taken over an equilibrium ensemble. The fluctuating hydrodynamics computation of the Van Hove function, known as the Landau-Placzek formula [182, 183], has been compared with values obtained from direct computations on assemblies of a few hundred to a few thousand particles in a periodic geometry. Quite remarkably, the Landau-Placzek formula proves to be already valid at the largest scales of space and time for these computer experiments [184, 185]. The validity of the hydrodynamic description of equilibrium fluctuations is certainly one of the most significant results of computer simulations.

The out-of-equilibrium fluid states are more difficult to handle, both theoretically and computationally. Indeed, since the microscopic distribution in phase space is in general not known, the theoretical prediction, thus far, has been essentially based on an extension of the fluctuating hydrodynamics description to nonequilibrium stationary states. These approaches led to the following predictions for the Van Hove function [186–188]: First, the function loses its symmetries; due to the presence of external constraints, it is neither isotropic nor time-reversible. Second, equal-time fluctuations are correlated over distances comparable to the system size. This behavior is reminiscent of the critical Ornstein–Zernike equilibrium fluctuations. Due to these long-range correlations, the nonequilibrium systems can be seen as more coherent than equilibrium ones, as soon as an external forcing is switched on.

Computing nonequilibrium Van Hove functions is difficult since the systems need to be simulated in nonhomogeneous states. The simulation of a temperature gradient or a shear stress might lead to boundary effects that significantly affect the system (see, for example ref. [4]). Besides, the required precision is already very demanding computationally for equilibrium states. This is, of course, even more so in nonequilibrium situations, where predicted deviations from equilibrium values are generally small.

Besides the Van Hove function, defined in Eq. (4.1), we shall, in the equations below, also mention the intermediate scattering function (ISF), $F(\boldsymbol{q}, t)$ which is the Fourier transform, with respect to the space variable, of $G(\boldsymbol{r}, t)$; and the dynamic structure factor (DSF), $S(\boldsymbol{q}, \omega)$, where the Fourier transformation is also performed on the time variable. The traditional way of addressing the (equilibrium) fluctuation problem is the following: First, the $t = 0$ value of the Van Hove function is fixed by the Einstein theory of equilibrium fluctuations [189]

$$G(|\mathbf{r} - \mathbf{r}'|, 0) = \frac{1}{n_0} \langle \delta n(\mathbf{r}, t) \delta n(\mathbf{r}', t) \rangle = n_0 k_B T \chi_T \delta(\mathbf{r} - \mathbf{r}') \tag{4.2}$$

where χ_T is the isothermal compressibility. Second, it is assumed that local fluctuations of the conserved variables (number of particles, velocity, and energy densities) obey the linearized hydrodynamic equations when they relax to equilibrium. For example, the ISF of a fluid made of spherical molecules can be written [190]

$$\frac{F(q,t)}{F(q,0)} = \frac{\gamma - 1}{\gamma} e^{-D_T q^2 t} + \frac{1}{\gamma} e^{-\Gamma q^2 t} [\cos(c_s q t) + b(q) \sin(c_s q t)] \tag{4.3}$$

In this formula, the symbols have their usual meaning: γ is the ratio of specific heats, D_T the thermal diffusivity, c_s the isentropic sound speed, Γ the sound attenuation coefficient, and q the amplitude of the wave vector. In Eq. (4.3), the first term gives the contribution of the entropy mode to the number density fluctuations, while the second term corresponds to the two propagating sound modes. Similar expressions are known for the velocity and temperature fluctuations.

As an alternative, and perhaps simpler, way to derive these formulas, one can write down the linearized hydrodynamic equations for $\delta n(\boldsymbol{r}, t)$, $\delta u(\boldsymbol{r}, t)$ and, for instance, $\delta T(\boldsymbol{r}, t)$ and then add a noise. The noise is produced by the rapid local variation of the purely dissipative parts of the (nonconserved) heat and momentum fluxes. Therefore, to the linearized Navier–Stokes equation for δu and to the linearized Fourier equation for δT, is added a term that has the form of the divergence of a fluctuating flux. The fluctuating fluxes are Gaussian white noises with amplitudes such that the thermodynamic values, as given in Eq. (4.2), are recovered. For instance, for the fluctuating heat flux, $J_Q^f(r, t)$

$$\langle J_{Q\alpha}^f(\mathbf{r}, t) J_{Q\beta}^f(\mathbf{r}', t') \rangle = 2\lambda k_B T \delta(\mathbf{r} - \mathbf{r}') \delta(t - t') \delta_{\alpha,\beta}^{Kr} \qquad \alpha, \beta = x, y, z \tag{4.4}$$

where λ is the thermal conductivity.

The extension of this procedure to stationary nonequilibrium states is straightforward. In an inhomogeneous system, the amplitude of the noise, see Eq. (4.4), may vary in space mainly because of the variations of the local temperature, and possibly of the thermal conductivity, so the noise sources are not everywhere the same. The basic assumption is then to give Eq. (4.4) a local meaning. The rapidly varying fluctuating parts of the fluxes are still Gaussian white noises, but the amplitudes are fixed by the local equilibrium values of the more slowly varying quantities, like the temperature and the transport coefficients (the latter, because of their dependence on the density or the temperature). Note that the value of the temperature and the transport coefficient appearing in the right-hand side of Eq. (4.4) are not fluctuating quantities, but can be understood as the time-averages of fluctuating quantities ("*deterministic*").

In the geometry shown in Fig. 10, where a fluid is under the influence of a stationary thermal gradient, the theoretical prediction of (local) fluctuating hydrodynamics (the wave vector being perpendicular to the heat flux) reads [191]

$$F(q, t)^{Non-eq} = F(q, t)^{eq} [(1 + A_T) e^{-D_T q^2 t} - A_v e^{-v q^2 t}] \tag{4.5}$$

$$A_v^* = \frac{c_p}{T(v^2 - D_T^2)} \qquad A_T^* = A_v^* \left(\frac{v}{D_T} \right) \tag{4.6}$$

$$A = A^* \frac{(\mathbf{q}_\perp \cdot \nabla T)^2}{q^4} \tag{4.7}$$

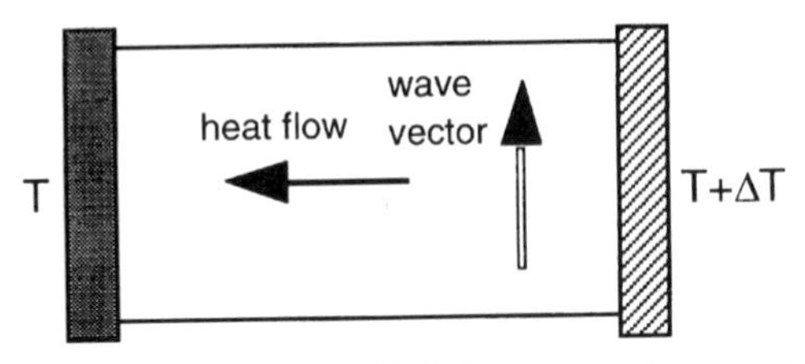

Figure 10. Geometry of the system and of the wave vector chosen perpendicular with respect to the heat flux.

Here, υ is the kinematic viscosity and $\mathbf{q}_\perp$ a unit vector perpendicular to the wave vector. The effect is seen to affect essentially the (slow) diffusive mode and is proportional to the square of the temperature gradient. When the wave vector is directed along the heat flux, the effect is first-order in the gradient, so that Eqs. (4.5–4.7) are no longer valid. Actually Eqs. (4.5–4.7) can be derived by simply considering an infinitely fast relaxation by the sound modes. Considering, however, a wave vector along the gradient, this is no longer so, and the density fluctuations are more efficiently relaxed by sound waves moving along the heat flux than by those moving in the opposite direction. As a result, the dynamic structure factor becomes anisotropic; one Brillouin peaks increases while the other decreases by the same amount. Neglecting boundary effects, the Brillouin part of the dynamic structure factor can be shown to change according to the following equation (see ref. [7])

$$S(q,\omega)^{\text{Brillouin}} = \frac{\Gamma q^2}{(\omega \pm c_S q)^2 + \Gamma q^2}\left(1 \pm \frac{c_S}{\Gamma q^2}\,\frac{\hat{\mathbf{q}}\cdot\nabla T}{T}\right) \tag{4.8}$$

where $\hat{\mathbf{q}}$ is a unit vector in the direction of the heat flux.

These are the basic theoretical predictions to leading order in the nonequilibrium constraint's amplitude. Other correlation functions have been computed involving the other densities. A striking property of these correlation functions is the appearance of long-range correlation in the equal-time value of these functions [192]. Temperature fluctuations, for example, are correlated over the size of the system. Similar computations have been made by various authors, using different sophistication levels in their approaches and examining the effects of possible interfering phenomena: reflecting boundaries, variation of thermodynamic coefficients, non-linearities, etc. The experimental tests, on the other hand, have been extremely difficult to perform, and it is only recently that satisfactory agreement concerning the Rayleigh peak predictions has been reported [193]. We shall see, in the following discussion, how molecular simulations permit study of these fluctuation properties.

The macroscopic predictions just mentioned can also be directly computed in a microscopic simulation. To each thermodynamic quantity corresponds an average over the values taken by an observable along its trajectory. The intermediate scattering function is defined as the time autocorrelation function

$$F(\mathbf{q}, t) = \langle \delta n_{\mathbf{q}}(t) \delta n_{-\mathbf{q}}(0) \rangle \tag{4.9}$$

$$\delta n_{\mathbf{q}}(t) = \frac{1}{\sqrt{N}} \sum_{a=1}^{N} \exp[i\mathbf{q} \cdot \mathbf{r}_a(t)] \tag{4.10}$$

In Eqs. (4.9–4.10), $\boldsymbol{r}_a(t)$ is the ath particle position at time t, and the wave vector $\mathbf{q}$ can only be given discrete values since the system is finite. In a cube of side L, one has for example (n_α being an integer)

$$\mathbf{q} = \frac{2\pi}{L}\mathbf{n} \qquad \mathbf{n} \equiv (n_x, n_y, n_z) \tag{4.11}$$

A hard-sphere fluid model has been used in order to compute the intermediate scattering function, both at equilibrium and under a temperature gradient [194]. The simulation has been performed with a thousand particles enclosed in a parallelepiped, with the largest side, $L_x = 43.68$ ($L_y = L_z = L_x/5$) in sphere diameter units. In these units, the chosen number density was 0.3. The system was made periodic in the y and z direction while thermal walls formed the boundary in the x direction. Any encounter of a sphere with the wall resulted in a diffuse reflection with outgoing velocities sampled from a Maxwellian at the desired temperature. (A Nosé–Hoover procedure for a hard-sphere fluid is not adequate since the energy is purely kinetic.) The result of the nonequilibrium computation is displayed in Fig. 11, where a comparison is made with the linearized hydrodynamic fluctuating equations. In system units, the left and right temperatures of the nonequilibrium runs were chosen to be $T_L = 0.5$ and $T_R = 1.5$ (in system's units). Since there is a temperature drop at the wall, the boundary temperatures for the macroscopic simulation were chosen to be 0.52 and 1.46, in agreement with the extrapolation of the temperatures computed from inside the system, by a local (slice) averaging of the kinetic energy density.

The comparison shown in Fig. 11 is quite satisfactory. It would become less so if we choose a larger wave vector. The presence of a second peak in the DSF spectrum is believed to be due to the reflection of the thermal sound waves at the boundary. The extra peak is centered at about half the value of the first peak ($c_S q$). This effect is also present in the macroscopic computation and would only disappear in the limit of a large system, sound

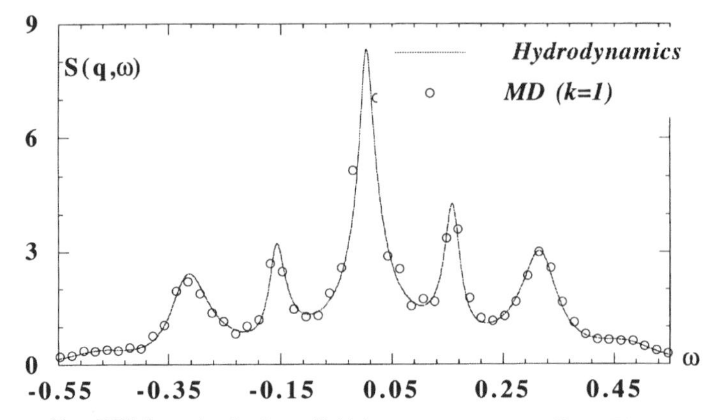

Figure 11. DSF for a hard-sphere fluid in a temperature gradient. The comparison is between MD and linearized fluctuating hydrodynamics.

being damped before reaching the boundary. The continuous curve in Fig. 11 has been generated by a numerical solution [195] of the fluctuating hydrodynamic equations with appropriate boundary conditions and appropriate values for the equation of state and for the transport coefficients (namely the Enskog values).

A slightly different system has been chosen in order to compute the ISF with a wave vector perpendicular to the temperature gradient. The interaction potential, in this case, was chosen to be the WCA (Weeks, Chandler, Anderson) potential. Periodic boundary conditions were also set in the x direction, with the geometry displayed in Fig. 12. In that geometry, it is possible to make an average over the entire simulation cell, since the effect to be computed is proportional to the square of the temperature gradient, it does not change sign when the gradient is reversed in the other half of the computational cell.

Equations of motion for the particles that belong to the shaded area are of the Nosé–Hoover type. On the contrary, particles in the "bulk" obey the

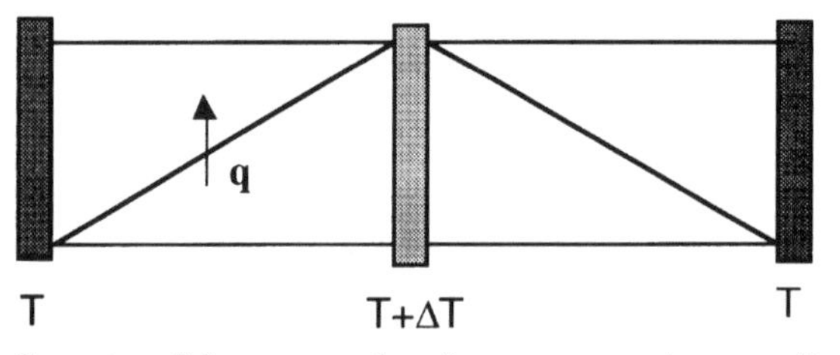

Figure 12. Geometry of the system when the wave vector is perpendicular to the temperature gradient: the simulation cell is periodic in the three directions of space.

usual Newton's equations of motion. These equations are integrated according to the Verlet algorithm [see Eqs. (2.5)] and several million time steps are necessary to generate the time correlation functions with an accuracy of around 5%.

In Fig. 13, the intermediate scattering functions computed for different temperature gradients are displayed. For all the cases considered, the average temperature was set to 1, in system units, and the thermostat-shaded regions were tuned so as to increase the temperature gradient: 0.9–1.1, 0.8–1.2, 0.7–1.3, 0.6–1.4, and 0.5–1.5. A remarkable feature of the results displayed is that the frequencies and the dampings that appear are not affected by the temperature gradient. The amplitude of the noise is affected to a considerable extent, increasing with the gradient in agreement with the theoretical predictions, [Eqs. (4.5–4.7)].

A more quantitative test of the theory is shown in Fig. 14. There, the gradient-dependence of the zero-frequency signal is plotted for a soft-sphere system of 2,000 particles at a reduced mean number density of 0.7. The linearity of the plot as well as the value for the angular coefficient are in good agreement (5%) with the theory.

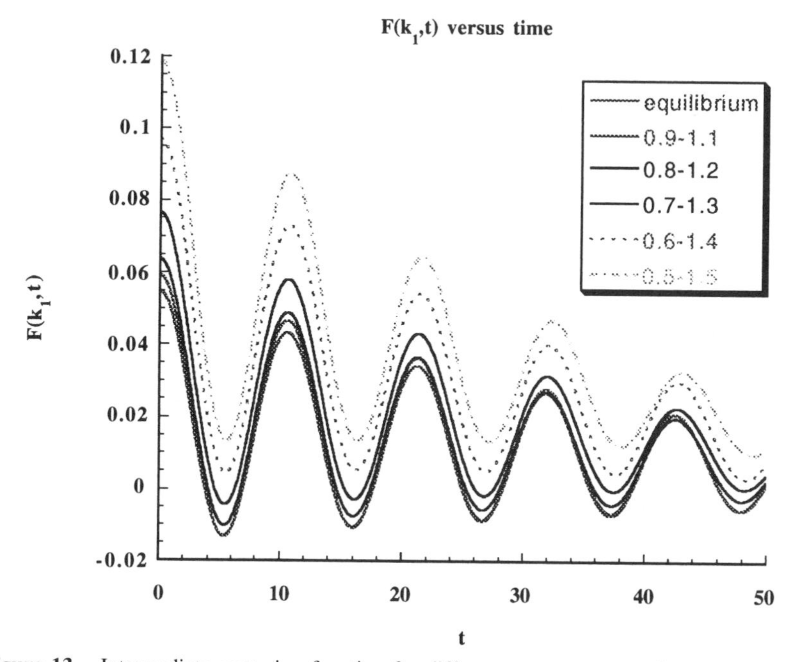

Figure 13. Intermediate scattering function for different temperature gradient. The square-root dependence is quantitatively obeyed. The dependence in k^{-4} is more difficult to check since the range of accessible wave vectors is more restricted.

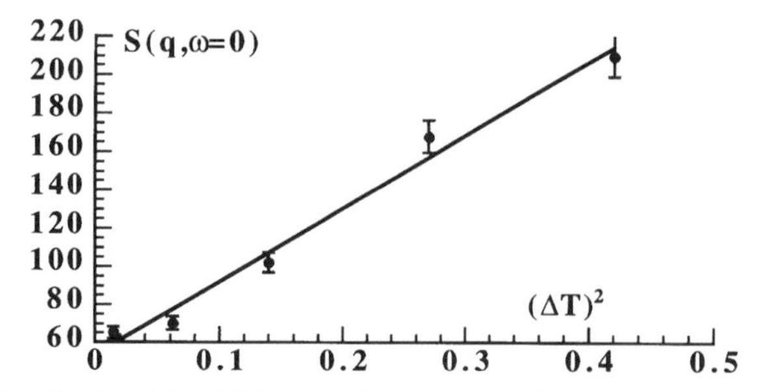

Figure 14. Amplitude of the DSF at zero frequency as a function of the gradient squared.

To summarize the results reported in this section, MD simulations of dense fluids show very good agreement with theoretical predictions based on a local fluctuating hydrodynamic hypothesis for the deviations from the stationary state. Wave vectors considered correspond to wavelengths of a few atomic distances, actually to the largest wavelengths that can be considered. This is in order to approach as closely as possible the hydrodynamic regime behavior and to have also well-separated sound and entropy peaks in the structure factor. Their typical timescales are a few picoseconds. Although the constraints are large when extrapolated to macroscopic objects, they have been chosen so as to remain within the range of validity of the linear force-flux phenomenological relations. This can be understood on the basis that, at the chosen density, the mean free path is much smaller than the atomic diameter.

V. FLOWS AND INSTABILITIES

A. Flows

The first results concerning the computations of velocity fields for fluid flows were done in the late 1980s thanks to the availability of computers capable of handling a few tens of thousand particles for several thousand time steps. Nowadays, such computations appear routine due to the development of so-called supercomputers. The first example is the simulation made by Meiburg [196] in his doctoral thesis. He used 40,000 particles (hard spheres) in a three-dimensional box, and "pulled" a plate inside—a moving boundary condition in fact, impulsively started in an equilibrated fluid—watching the vorticity which forms behind (see Fig. 15). The aim of the computation was to compare the MD computation, taken as a reference, to a DSMC simulation with 210,000 particles for this case study. The origin of the differences, which were observed between the developments of the flows in the two simulations was traced to a nonconservation of the angular momentum in the DSMC treatment of collisions. The problem was later readdressed by

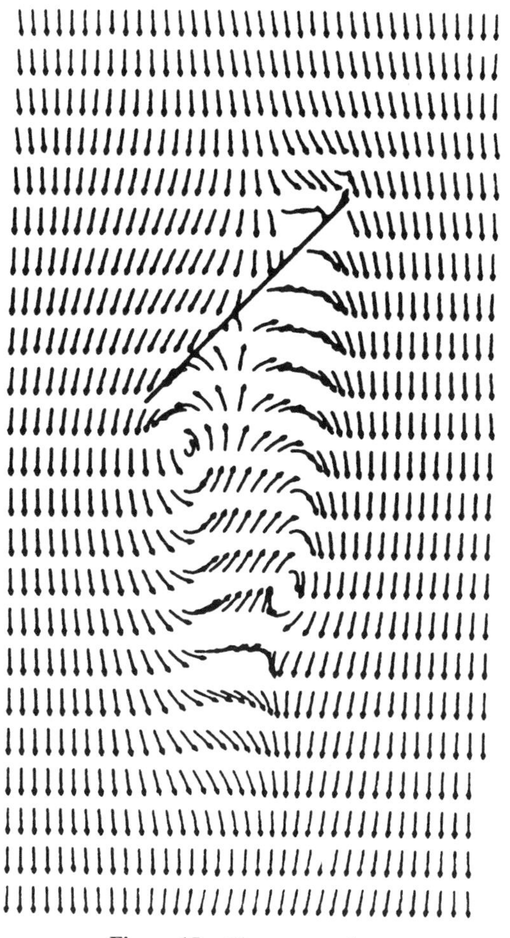

Figure 15. Flow past a plate.

Graeme Bird [197], who improved his technique in order to avoid spurious effects when the size of the computational cell is not sufficiently large compared to the size of the vortices that are formed behind the plate, and in general in a highly turbulent fluid.

Another simulation was done a little bit later by Rapaport [198, 199], who used an experimental computing facility, a loosely-coupled array of processors built by IBM (New York). The processors were array processors arranged in such a way that the computing could be efficiently distributed among them, at least in that particular computation. The number of particles was 160,000 two-dimensional soft spheres, in a rectangular geometry of size 1,500 Å, and in which a circular reflecting boundary, of size 250 Å, would model a two-dimensional projection of a cylindrical obstacle. The fluid would move due to an injection condition that would accelerate particles against the obstacle. The eddies formed behind the cylinder were very much like what is observed in fluid flows, the so-called von Karman streets, with eddies just behind the obstacle growing in time and eventually moving along the flow. Also some oscillatory behavior was observed. A similar simulation was reported more recently with a similar molecular model, using a few tricks to give a more precise control over the inlet flow [200].

The difference between these results and previous ones obtained in order to compute the viscosity by NEMD (Non-Equilibrium MD) was precisely that one was looking for the instability, the transition from laminar to turbulent regime in the fluid flow. The Couette flow is known to be stable whatever the amplitude of the flow, at least within the limits of validity of the linear theory. Besides the work mentioned in Section III, Couette flows were also simulated in dilute fluid systems [201, 202], in order to compare simulation results with known analytic velocity profiles that eventually depart from the linear domain. Now, in order to really describe a fluid undergoing a transition to roll patterns and developing vortices in its velocity field, one has to use a much larger system, essentially because it is necessary to average over a large number of molecules in order to see regular fluid–velocity patterns emerging. Even in the cases reported, spatial averages have to be supplemented with time averages, so that several successive configurations contribute to the computation of transients macroscopic values for the velocity. This is easy in the case of a stationary state, but may become impossible for very complex, rapidly evolving patterns.

Channel flow [203, 204], also named Poiseuille flow, is another example where comparison has been made between microscopic realizations, made with several thousand particles, and macroscopic predictions. In this case, the boundaries are to be simulated as well. Usually diffuse reflecting boundary conditions can be used. Now, even if the diffuse [205, 206] nature of the reflection should lead to a zero-velocity boundary condition for the flow (no-slip), there still remains a small slip at the boundary leading to a microscopic layer where local equilibrium is violated. This is, however, a well-understood problem within the framework of kinetic theory (see ref. [63]). For a macroscopic boundary condition however, the no-slip condition is usually taken, and it is deduced from experimental evidence as well as from arguments valid for dilute systems. The simulation of flows by molecular dynamics has made it possible to make a direct inspection of the state of the fluid near a solid boundary, the latter being modeled in various ways. It always appears that, on average, the fluid velocity vanishes near the wall. There is still a microscopic slip, as mentioned above, but this effect is insignificant for macroscopic developments. The ordering that occurs near a solid boundary does not affect this behavior, and is even a key factor in explaining the transmission of the shear stress across the fluid–solid interface [207].

In a channel flow, heat is usually dissipated at the boundaries through the reflecting boundary conditions, so that the temperature is not uniform in the system. Using a DSMC technique, this was carefully investigated by Garcia and Mansour [4], who found a temperature profile which is quartic, with one minimum at the center and two maxima, symmetric with respect to the center of the channel and very close to it. This profile is reminiscent of a theoretical computation by Santos [208] based on a perturbation analysis of a simple Boltzmann model for the channel flow. These effects are more important for the dilute systems that were used than for dense liquid-like fluids where the mean free path, or its equivalent, tends to zero.

A very interesting problem, the contact angle of a two-phase flow in a channel, has been studied by molecular computation. The problem has been posed within a macroscopic context [209]. The no-slip boundary condition that is imposed in continuum descriptions leads to divergences for the shear stress in the case of two nonmixing phases flowing along a solid boundary (at least with Newtonian behavior). A moving contact-line is quite compatible with the no-slip condition; in fact, a rolling motion is permissible [210]. There have been several suggestions made in order to clarify the point. Adopting a slip boundary condition would indeed remove the divergence, as would unsteady motion; but what exactly is happening at the fluid–fluid–solid interface? The problem has been looked at by molecular dynamics simulation by two groups [211–214]. The simulation involved non-mixing Lennard–Jones model particles. This is a model where one uses an artificial trick, in which two species have different intra- and interspecies interactions. There would be a normal Lennard–Jones interaction among the members of the same species, but the attractive part of the potential would be significantly reduced for the interspecies forces. When multiplying this attractive part by a parameter ranging from 0 to 1, it has been known that nonmixing conditions already appear when the parameter is around 0.5, leading to an interfacial tension (adopting other parameter values so as to correspond to liquid argon near its triple point) of the order of the water–hexane tension at normal temperature [215]. The interface structure is similar to the one that has been studied in detail for the liquid–vapor equilibrium of Lennard–Jones [216, 217].[4] The interface is well located, extending only over a few molecular diameters. Similar models have been used to study

[4]Note, however, that a full microscopic treatment of the surface tension requires a more general definition in terms of microscopic variables than the one found in most textbooks (including ref. [196]). A recent effort along this direction, providing microscopic expressions for curved interfaces as well, can be found in the work of M. Baus and R. Lovett [218].

amphiphilic adsorption at liquid–liquid interfaces [219], surfactant behavior [220], and demixing kinetics [221].

These simulations have involved a few thousand particles, the solid wall being modeled by particles similar to that of the fluid but fixed at their position by an artificially large mass. The direct inspection by MD has shown that the no-slip condition is still valid except very near the moving contact-line, in a region extending over two or three molecular diameters. There, the fluid would try to stick to the solid—unsuccessfully, since it would require an infinite force to do so—and the result is a compromise with a small nonvanishing velocity for the fluid at the surface. The exact boundary condition has not been yet resolved by direct inspection and it is expected that further work, involving larger system sizes, is needed to achieve progress on this problem. The local high stress in the fluid near the contact line leads to deviations from purely Newtonian behavior. Also, the angle of the fluid–fluid interface [222] with respect to the solid surface is observed to depend on the fluid velocity, in agreement with experimental observation. In this field, more precise and quantitative studies, implying larger system sizes and larger integration times, are required in order to confront experimental data and theoretical predictions.

The general phenomenon of wetting has been, of course, the object of many different studies by molecular dynamics. Usually artificial modeling is adopted [223], like the one just described for the study of the moving contact-line. However, a realistic study of a water droplet on a solid substrate has been done recently [224]. Such studies have shown that a droplet could be modeled with a few thousand particles, in the sense that a meaningful computation of a static quantity like the contact angle at rest can be performed on these systems. The extension to dynamical problems presents some interest, since the dynamics of droplet spreading is still a matter of intense experimental [225, 226] as well as theoretical work [227].

The studies performed by MD [228] have used the artificial modeling described above for the liquid–liquid interface. By an artificial tuning of the attractive part of the Lennard–Jones interaction potential one can scan situations ranging from no-wetting to complete wetting. Actually, when the attractive part between solid substrate and the liquid droplet is increased, becoming even larger than the intraspecies attractive interaction, a spreading in the form of a monomolecular layer which grows on the solid substrate has been observed. This is indeed in agreement with observations made in experiments performed with polymers [229]. In the layer, molecules do diffuse, although the motion is affected by the solid structure. The layer growth rate computed from MD simulations is slow, fitting a logarithmic law rather than the expected and experimentally observed square-root dependence. It is to be noted, however, that a square root growth rate has been computed from a simulation [230] using a Nosé–Hoover thermostatting and another modeling of a smooth solid surface. Monte Carlo simulations have also reproduced this phenomenon, but lack any possible connection to real time [231].

Coalescence and rupture of liquid droplets have also been modeled in MD simulations using around 10,000 Lennard–Jones particles undergoing the phenomena. Two drops of around 400 atoms are placed in a solvent undergoing a steady shear flow; the solvent is made of around 10,000 particles. The general picture of the coalescence mechanism is as follows [232]: The two drops will be attracted to each other by the so-called disjoining pressure, which is known to exist at close separation. A macroscopic description would require adding this force to the Navier–Stokes equations, together with a criterion saying when coalescence is achieved. This information should be obtained from microscopic modeling. The simulations performed have not yet provided them because they were fully three-dimensional and too limited in size to smooth out the thermal fluctuations. Larger simulations in the future should probably lead to better understanding. Similar rupture simulations [233] were done in various geometries, various flow patterns, and with different system sizes. The general picture seems to be that rupture occurs roughly as a time-reverse sequence of the coalescence dynamics.

B. Rayleigh-Bénard Instability

The Rayleigh–Bénard instability was discovered more than 100 years ago experimentally by Bénard, and its mechanisms were later explained theoretically by Lord Rayleigh at the beginning of this century [234]. It is one of the most fascinating examples of transitions leading, in nonequilibrium conditions, to a dissipative structure where part of the dissipation energy would be used to build an ordered state rather than increase the entropy production. This instability has been used in many different contexts in order to illustrate general nonequilibrium concepts and theories, and a possible microscopic realization would have a pedagogical interest for the reason that order appears, on large scales, out of the molecular chaos which prevails on the small scale. This instability is moreover a case study where quantitative studies have been carefully performed so as to assess the exact validity of microscopic simulations in their reproduction of macroscopic phenomena on a small scale. A third motivation for its presentation here is in the fact that, once its validity has been established, it can be used in order to produce "experimental" data in a general investigation of nonequilibrium states. We shall, in the following discussion, show that one can study the enhancement of molecular diffusion in an ordered state by the MD techniques, and confront the theoretical predictions in a manner similar to laboratory experiments.

The phenomenon to be understood can sometimes be reduced to a two-dimensional geometry. Suppose that we have a fluid layer that extends to infinity in the horizontal direction. If the layer is heated from below, the reference state (convective) loses its stability, and an ordered structure develops once the Rayleigh number is made larger than a critical value (between 600 and 2,000 depending on the boundary conditions). This dimensionless number is defined as

$$\mathrm{Ra} = \frac{\alpha g \Delta T l^3}{\upsilon D_T} \tag{5.1}$$

In Eq. (5.1), α is the thermal expansion coefficient, υ the kinematic viscosity, D_T the thermal diffusivity, g the acceleration of gravity, and ΔT the temperature difference between the top and bottom of the layer. After the instability, the fluid develops a pattern that, in this infinite layer geometry, is made of parallel rolls. This is in fact a two-dimensional structure, in the plane perpendicular to the roll axis, which can be modeled by a two-dimensional microscopic simulation. Although the fluid layer height that can be achieved in a molecular simulation is fairly small, critical values of Ra can be realized by an artificial increase of the forcing parameters, g and ΔT. Any further increase of Ra can only be obtained by a corresponding increase of the layer height, which scales with the number of particles used in the simulation. In dimension 2, Ra increases linearly with N.

The first simulations of the Bénard transition were made precisely in a two-dimensional geometry [235]. Hard disks were enclosed in a rectangle whose horizontal sides were thermal diffusive reflecting boundaries and vertical sides were specularly reflecting. The disks would be accelerated downwards in between their collisions. Actually, this external force does not modify the cost of the computation to be made. Systems of 5,000 disks, in a geometry of aspect ratio equal to 2.82, were integrated during several thousand relaxation times (the relaxation time is defined as the mean time of collision, each particle has collided once on the average). The Rayleigh number computed from the values of the external forcing and from the values of the thermodynamic parameters of the model was overcritical, around 2,000. The velocity field which then emerged was indeed showing an ordered roll structure, characteristic of the roll pattern that was predicted on the basis of the hydrodynamics equation (actually within the incompressible-fluid Boussinesq approximation). The strange value of the aspect ratio was the consequence of a bad interpretation of a theoretical result by one of the authors; however, it resulted in a series of simulations [236] where one would eventually see the influence of this parameter. In particular, for some values of the aspect ratio, no unique stationary solution exists, and the final state appears to be dependent on the initial conditions. Some velocity patterns show up and then disappear very quickly; for example, a three-roll structure appeared and then quickly disappeared before a seemingly more stable two-roll pattern emerged. This unstable behavior is actually very much like the observed behavior of fluids, or the computed solutions of the macroscopic Boussinesq equations after the instability threshold. A similar behavior was also reported in a simulation made later with a larger number of particles and with periodic boundary conditions in the horizontal direction [237]. The aspect ratio was set equal to 4, and a six-roll pattern lived for some time before two rolls merged and finally disappeared.

Of course, this very unstable pattern could be thought of as resulting from large amplitude fluctuations of the system, due to the small number of particles involved. If it is indeed true that large fluctuations are produced in these simulations, one should be aware that the solutions of the hydrodynamic equations themselves can show similar unstable behavior. Small perturbations can lead to completely different behaviors, and different initial conditions can lead to different final states. For instance, the solutions of the Boussinesq equations for aspect ratios near 2 have shown a behavior similar to that of the MD simulations. The question of a quantitative comparison between the two descriptions was then raised. How do the solutions of the microscopic model compare to the solutions of the Boussinesq and Navier–Stokes equations, with, as an input, the transport coefficients and the equation of state of the model considered, namely the hard-disk fluid? The answer is that the two approaches are essentially indistinguishable [238]. Numerical analyses of both sets of equations have led to the following conclusions:

1. Profiles of density, pressure, and temperature computed in the microscopic fluid are, within their statistical uncertainties, in agreement with those of the continuum approaches. The accuracy of the computation was around 5% for quantities depending on two coordinates. Averaging over the horizontal coordinate would eventually lead to an accuracy with less than 1% uncertainty (see for instance the density and temperature profiles in Fig. 18).
2. The effects of compressibility are very limited, even if the velocities of the rolls are not small compared to the thermal speeds–around one fifth in the simulations with the largest Rayleigh number. Of course, at the densities chosen for the simulation, the sound speed is larger than the thermal speed, but not by much.
3. The phenomena of unstable roll patterns seem to be independent of the various settings of the simulations. Other simulations, made with other interaction potentials and different boundary conditions, and even different density regimes, have displayed the same behavior, which has also been reproduced by simulating the hydrodynamic equations.

4. The instability point cannot be precisely located since the fluctuations are huge, due to the small number of particles involved. Larger simulations have been made in order to characterize the fluctuations when approaching the instability point, and qualitative agreement with theory has been reported [239]. Also some oscillatory behavior was found for a two-roll structure in a small geometry, which would normally lead to a one-roll structure if there were no periodic boundary conditions in the horizontal direction.
5. The Bénard instability has also been simulated using the DSMC technique [240]; however the latter requires a much larger number of particles, because of the decrease of the transport coefficients in the denominator of the Rayleigh number. Thus, although DSMC performs more efficiently with an equal number of particles, it is not really attractive in the present case [241].

We now turn to the problem of self-diffusion in a Rayleigh–Bénard convective flow. In this problem, the interplay between the small-scale random walk of the tracer particles, giving rise to the molecular diffusion processes, and the large-scale hydrodynamical flow-field of the fluid is by no means trivial, as was shown for the first time by Taylor [242] in 1954. (See also ref. [243] for a recent review.) Passive mass transport is of much interest in many complex situations such as, for example, a fluid subject to turbulent motion, or in flows in which spatiotemporal patterns have set in. These fluid states are themselves difficult to describe so that the modelization of the mass transport cannot be fully understood from the analysis of the basic starting equations: the Navier–Stokes equations for the fluid flow and the diffusion equation (in a fluid at rest) for the tracer. Phenomenological treatments, however, make it possible to obtain most of the qualitative behavior of the scalar transport [244].

More recently, a lot of attention has been devoted to the understanding of passive mass transport when the fluid is undergoing a steady convective flow in a one-dimensional geometry. This is one of the simplest examples where the interaction between diffusion and a steady flow structure can be examined in detail. Indeed, an analytical form for the fluid flow field can be obtained from a perturbation expansion [245]. The non-dimensional parameter characterizing the relative importance of the flow field mass transport and the molecular diffusion is given by the Péclet number, Pe. This number is defined as the ratio $\mu l/D\pi$ with u, a characteristic flow field velocity, l the typical length (in Rayleigh–Bénard convection experiments, l is taken to be the fluid layer height) and D, the molecular diffusion coefficient. It represents the ratio of the diffusion-over-a-roll time over the inverse shear rate in the flow field. MD simulations, in this case, permit the exploration of a range of values for the Péclet number that can hardly be reached in an experiment—between one and a few tens. They also give access to the full range of timescales from the molecular times up to the effective diffusion times.

Sagues and Horsthemke [246] have derived a diffusion equation for the tracer in the vicinity of the convective instability threshold, with an effective diffusion coefficient, D^*. This result has been later extended to the large Péclet number domain so that the scaling of D^* with Pe has been derived in the whole Pe range [247], going from a Pe^2 behavior, in the low Pe regime, to a $Pe^{1/2}$ behavior at high Pe. In the limit of large Pe, Shraiman [248] and, independently, Rosenbluth [249] have also found an enhancement of D^* proportional to $Pe^{1/2}$.

At large Pe, a crossover between two behaviors was predicted by Young [250] to depend on the timescale investigated. For times smaller than the time needed to diffuse over a distance equal to a roll width, no linear increase of the mean square displacement in time was found. Rather the number of convective cells invaded by the tracer was predicted to increase in time according to $t^{1/3}$ for rigid and $t^{1/4}$ for stress-free boundary conditions [251, 252]. For larger times, on the other hand, the diffusion law again applied, and the number of invaded cells increased according to $t^{1/2}$.

Experimental results have been reported by Solomon and Gollub [253]. They have found a diffusive behavior for impurities injected into a fluid layer with convective rolls in the direction perpendicular to the roll axis. Over the range of Pe inspected—between a few hundred and a million—they could confirm the square-root scaling behavior. Anomalous diffusion predictions were also tested and verified experimentally [254]. Numerous experiments have been reported concerning effective diffusion coefficients in turbulent media [255].

In order to perform the simulations, the fluid particles are contained between two horizontal layers, the top ($y = L_y$) and bottom ($y = 0$) boundaries, by an external potential that increases to infinity with the inverse of the 12th power of the distance. Thus, the effect of this constraining force is felt by the fluid particles over a distance of the order of σ. There is also another external force, $\mathbf{g}$, accelerating the fluid particles downwards in order to mimic a gravitational force. The system is periodic in the horizontal (x-axis) direction so that the central cell is indefinitely repeated as shown in Fig. 16.

The fluid is thermalized inside two slices adjacent to the top and bottom boundaries. For reasons explained below, the thermalization mechanism is performed through the local Nosé–Hoover technique: for the particles which belong to these slices, the equations of motion are modified by a coupling with a reservoir at temperature T_{top} or T_{bottom}. These two horizontal slices are divided into cells. At each time step the instantaneous number of particles N_α and the instantaneous cell velocity, $\mathbf{u}_\alpha$, of cell α are computed from the

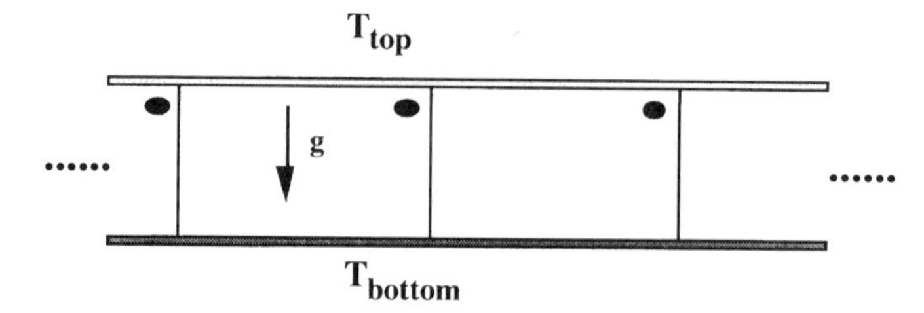

Figure 16. Geometry of the simulation cell. The shaded regions are Nosé–Hoover thermostats, periodicity is imposed horizontally, and there is a vertical acceleration.

values of the velocities of the particles belonging to that cell. For particles belonging to cell α, the equations of motion are those of Eq. (2.1). For the particles which belong to the bulk, that is those which do not belong to any of these boundary layer cells, their equation of motion is simply Eq. (2.1) with a vanishing Nosé–Hoover friction constant.

Other simulation parameters are fixed by the physical problem being solved. For instance, when the boundary conditions to be simulated are stress-free, critical values values of the Rayleigh number are lower and therefore easier to realize. The Rayleigh number that can be achieved in a MD simulation is proportional to the number of particles; this implies that the CPU time required is proportional to the square of the number of particles (at best), or, equivalently, to the square of Ra. Therefore, in order to keep the CPU time within reasonable limits, it is preferable to choose a low-Ra critical value. Thus the tangential fluid velocity at the boundary is not constrained.

Since the fluid velocity can vary along the boundary, a local thermalization has to be set in: the boundary layers have been divided into cells with as few particles as possible. In every boundary cell, a different reservoir coupling is switched on. However, a sufficient number of particles has to be within a boundary cell. Indeed, every time a particle enters (or leaves) one of these cells, it produces a discontinuity in the local density and the local fluid velocity, and therefore also in the equations of motion. The effect of this discontinuity decreases with the number of particles contributing to N_α and to $\mathbf{u}_\alpha$. We found empirically that a number of around forty particles per cell was large enough to avoid any strong perturbation. The cell size extends therefore over 5 σ in the vertical direction and 20 σ in the horizontal direction.

The mean square displacements of the fluid particles in the horizontal direction are computed. Adopting periodic boundary conditions in this direction, the minimum number of rolls required is therefore 2 and the aspect ratio (L_x/L_y) has been set to exactly 2.

As already mentioned, the fluid model we use resembles a fluid made of hard spheres. The thermodynamic properties of a hard-sphere fluid are well understood [256]. For instance, the transport coefficients computed from the Enskog equation are in good agreement with those computed from molecular simulations, even at large densities (see ref. [18]). We have performed a few equilibrium simulations in order to validate this Enskog model. The results are listed in Table II. The self-diffusion coefficient has been computed from the slope of the linear increase of the mean square displacement of the particles, while viscosity and thermal conductivity have been obtained through their Green–Kubo expressions. The fluid was thermalized by a Nosé–Hoover thermostat at $k_B T = 1$ (in system's units) and the number density $n\sigma^2 = 0.7$, well within the fluid region.

The choice of the temperature (or the mean vertical temperature in nonequilibrium cases) is rather arbitrary, as it is for a hard-sphere fluid where the temperature simply rescales the time variable. The choice of the (mean) density, on the contrary, is determined by the following two requirements: it should be in a domain where the transport coefficients are small, and the fluid should be considered as incompressible.

The first requirement follows from the form of the Rayleigh number. Replacing the coefficients of Eq. (5.1) by their values for the model used (Carnahan–Stirling equation of state and Enskog values of the viscosity and thermal diffusivity), and keeping in mind that $mgl \approx k_B \Delta T$, Ra can be put in the form of a function of the density and temperature times N (see ref. [15] for more details). With N and $k_B \Delta T$ fixed, Ra is a maximum for densities between $n\sigma^2 = 0.2$ and $n\sigma^2 = 0.4$. In order to simulate an incompressible fluid, on the other hand, the speed of sound should be made as large as possible so that the largest possible density (around 0.4) seems the most appropriate. This leads to an estimate of Pe $\approx$ 10. Other realistic choices of the density and of the number of particles permit simulation of fluids in the range between Pe $\approx$ 1 and 100.

TABLE II
Computed and Reference Values of the Transport Coefficients in the Model Fluid[a]

	Number Density $n\tau^2$	Computed Value	Enskog Value
Viscosity, ν	0.7	9.33	10.63
Thermal conductivity λ	0.7	0.91	0.98
Molecular diffusion, D	0.4	0.09	0.055

[a]The reference value is computed from the Enskog equation (first-order Chapmann–Enskog expression in two dimensions, see ref. [21]). (Units are chosen so that the diameter is one, the particle mass is one, and the mean temperature is one.)

TABLE III
Parameters of the Simulations Performed[a]

N	T_{bot}	T_{top}	l	Ra	t_c	t_{diff}	Pe*
9,800	1.6	0.4	110.7	2,635	6,500	136,161	21
9,800	1.65	0.35	110.7	3.092	4,700	136,161	29
9,800	1.7	0.3	110.7	3,586	4.300	136,161	32
9,800	1.8	0.2	110.7	4,684	3,600	136,161	38
5,000	2.0	0.1	79.1	3,056	2,800	69,520	25

[a] N is the number of particles for the simulation, T_{top} and T_{bottom} are the fixed temperatures of the top and bottom thermostats, l is the fluid total layer height, Ra is the estimated Rayleigh number, t_c and t_{diff}, are the times required for a particle to move over a roll-distance by advection and by diffusion respectively. (Units are chosen so that the particle diameter, the mass, and the system mean temperature are one.) *Péclet number.

We have performed [257] five different simulations whose parameters are listed in Table III. The first four have been performed with 9,800 particles at different Rayleigh and Péclet numbers, and a fifth one has been made with 5,000 particles in order to check any size dependence of the results. For convenience, in Table III we also list the values of the diffusion time obtained from the values of the computed equilibrium transport coefficients. The advective time, t_c, has been computed from an estimate of the typical fluid velocity.

The system was first started with a uniform density and temperature fixed to the average values we want to have, $n\sigma^2 = 0.4$ and $k_B T = 1$. We follow the evolution of the velocity, density, and temperature fields. These are computed from both a time and space average in statistical cells. We have 40 times 20 of these cells, and values recorded inside these cells are averaged over a few thousand time steps. This time average is done in order to eliminate the thermal noise that would otherwise be dominant. It is only after this averaging procedure that a smooth velocity pattern emerges. An interval of 50,000 time steps seems to be the minimum required to generate a regular field.

The velocity field changes during many transients. The stable pattern does not emerge immediately and many intermediate states appear and disappear before a stationary state is reached. This stage takes around 500,000 time steps, which is computationally costly. On a Cray YMP this took around 50,000 seconds for the 9,800-particle system. The other states were then started with an initial condition obtained from the final configuration of the first simulation. Transients were also observed with this procedure, but the time necessary to reach stability was somewhat reduced.

The fluid velocity is not constant along the streamlines. We found two reasons for this. First, the number density increases as the top boundary is approached. Because of mass conservation, this implies a reduced mean

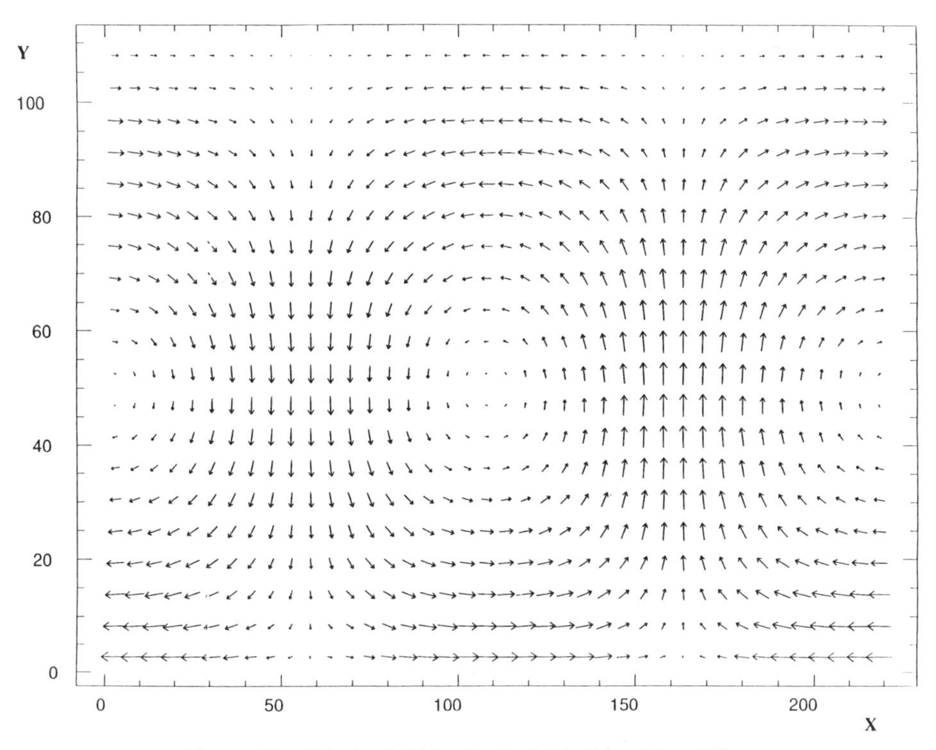

Figure 17. Velocity field for the Rayleigh–Bénard instability.

fluid velocity, which, in turn, induces the vertical asymmetry of the roll. A second reason becomes apparent when the velocity field is observed more closely (see Fig. 17). The two rolls do not have the same extent. The roll displayed in the center of the cell has a radius that is 20% smaller than the other roll, which is split between the two sides of the simulation cell. This is an important fluctuation that lasts for the entire simulation time. This asymmetry leads to an upstream mass flow taking place within a narrower pipe than the downstream flow. Therefore the upstream velocity is somewhat larger than the downstream one. The fluctuations are intrinsic in the MD system (as in a real fluid); moreover, the importance of the effect is fairly large and its reduction by further averaging, either in time or in space, cannot be easily imagined because of the huge computational cost.

Figure 18 displays the profiles of density and temperature averaged over the x direction as a function of the height. These profiles are very similar to those obtained in hard-sphere fluid models and they perfectly match the solutions obtained from the full two-dimensional Navier–Stokes equations, completed by the transport coefficients and equation of state of our model. Even the density variation near the upper boundary matches quantitatively

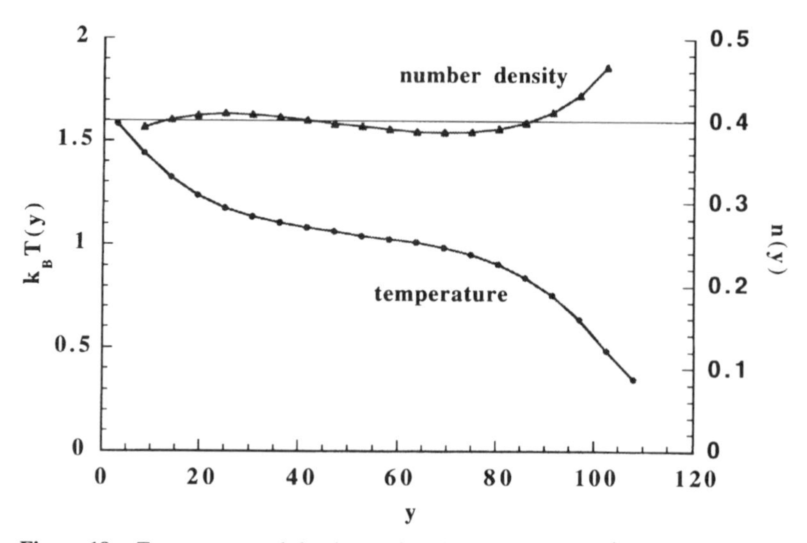

Figure 18. Temperature and density profiles for the Rayleigh-Bénard simulation. Averages are performed horizontally.

the macroscopic profile, once a local value for the thermodynamic parameters is fed into the Navier–Stokes equations. It is worth mentioning that the effect of compressibility is negligible and that the profiles look unchanged when the Boussinesq equations are used.

The mean square displacements in the x direction are shown in Fig. 19. The computations are performed with the real displacements along the x direction of every particle initially located in the unit cell of the simulation. The positions of the particles are recorded every 5,000th step, and the total durations of the runs were 750,000 steps. The correlations are kept over 500,000 time steps and they are displayed in Fig. 19.

One can distinguish three timescales corresponding to different behaviors:

1. The first, shortest, timescale is evident when molecular self-diffusion is dominant. This happens for very short times on the scale used for the plot ($t \approx 100$ in system's units). The diffusion coefficients computed from the Einstein formula

$$\langle \Delta x^2(t) \rangle = 2Dt \tag{5.2}$$

compare well with their values computed from a periodic equilibrium system at the mean temperature and density.

2. The mean square displacement then starts increasing much more than would be expected from molecular self-diffusion. For t of the order of

3,000, there is a saturation and the rate of increase diminishes somewhat. This is the intermediate timescale when particles are accelerated by the flow field. Later on, most of them come back near their original positions.

3. Finally, on the largest timescale, the mean square displacement tends to again become linear and the slope computed from the plot can be identified with twice an effective diffusion coefficient, D^*. Note, however, that the larger the time, the more linear the increase of the mean square displacement appears. Also the higher the Rayleigh number, the sooner the linear regime starts.

A log–log plot of the mean square displacement (MSD) versus time during the last of these three periods makes it possible to compute the exponent of the time dependence of the MSD. The linear fit of these plots produces an angular coefficient equal to 1 within 3% if it is performed with the part of the graph between $t = 12{,}000$ and $t = 16{,}000$ (in system's units). Performing the fit from a time interval between 8,000 and 16,000 increases the dispersion of the angular coefficients to around 10%. Quite clearly, the mean square displacement reaches the asymptotic regime of enhanced diffusion. The evaluation of D^* has been carried out using Eq. (5.2) in the time interval between 12,000 and 16,000.

We used the four points corresponding to the $N = 9{,}800$ simulations of Fig. 19 to make a power-law fit. The result of the estimate reads

$$D^*/D = 1.02\ Pe^{0.51} \tag{5.3}$$

which is in extremely good agreement with the theoretical predictions. The square-root dependence is well reproduced in the MD computation. The value of the prefactor, on the other hand, necessitates a much better precision for the D^* computation than can be done in such simulations. Indeed, we have not shown any error bars in Fig. 19, since these are difficult to estimate. The only way to perform such an estimate would be to repeat similar independent computations and to observe dispersion of the results, since the basic fluctuations are the very slow changes of the velocity patterns, which may take fairly long times to average out.

The $N = 5{,}000$ results are qualitatively similar to those of the larger system ($N = 9{,}800$). The behavior of the mean square displacement, its time-derivative, and the diffusion enhancement are all equivalent to the behavior in the system with $N = 9{,}800$ particles. However, the square that appears in Fig. 19 is a little bit away from the curve-fit line. Although the mean values computed in this smaller system are also in quantitative agreement with the macroscopic description, the effect of the fluctuations can be expected to be

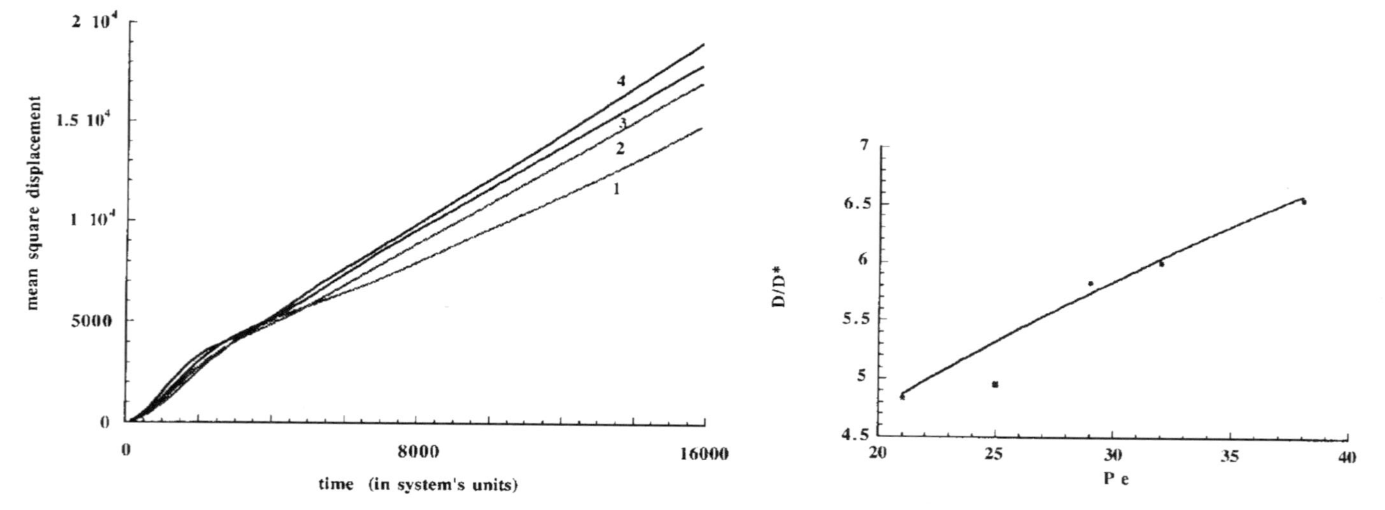

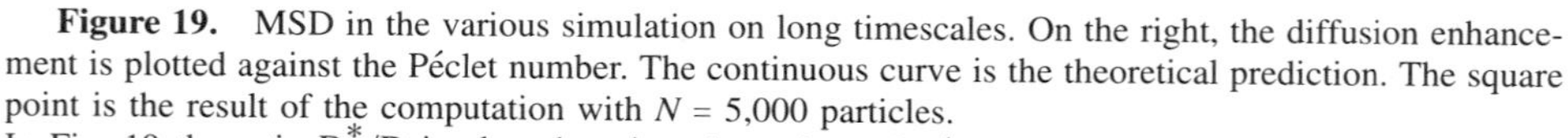

Figure 19. MSD in the various simulation on long timescales. On the right, the diffusion enhancement is plotted against the Péclet number. The continuous curve is the theoretical prediction. The square point is the result of the computation with $N = 5{,}000$ particles.

In Fig. 19 the ratio D^*/D is plotted against the estimated Péclet number. The Péclet number can be written as the ratio of the diffusion time over the advective time t_{diff}/t_C. The diffusion times are listed in Table III for each simulation performed and have been obtained as the height squared, divided by the molecular diffusion coefficient. The advective times, on the other hand, have been computed from the oscillations occurring in the time-derivatives of the mean square displacements. They are also listed in Table III. Indeed, the period of oscillation represents the average time needed for the fluid particles to perform one roll rotation on streamlines. Another estimate of Pe has been done by computing the fluid velocity on the separatrix between rolls. The result of this estimate is very similar to the one given in Table III and used in Fig. 19.

more important. A smaller number of particles requires a larger time average to wipe out the effects of the thermal fluctuations. Even if the time required to display the phenomenon is smaller in the system with $N = 5{,}000$ particles, the computational time required to reach the same precision of the diffusion coefficient might be larger.

VI. CONCLUSIONS

We have surveyed various techniques used in the microscopic modeling of flows, as well as a few significant results obtained by using microscopic modeling in recent years. To conclude this chapter, we shall list a few outstanding and open problems in nonequilibrium fluid systems that could fruitfully be addressed by large-scale computer simulations.

1. The behavior of the viscosity in a sheared atomic fluid has not yet received a complete and satisfactory understanding, at least at a microscopic level. The existence of a long-time tail in the Green–Kubo integrand is certain, but its importance is not yet fully estimated. The first results were obtained more than 10 years ago; since then, new simulations on larger systems have led to different conclusions. The problem could be readdressed now, since present computing power permits the treatment of fluid systems on different length scales. A systematic study of the behavior as a function of the size could produce useful information, very much like what has been done for lattice gases (unfortunately the latter do not have potential fluxes). Also, the viscosity at large density increases very much near solidification. Understanding the mechanisms responsible for this behavior is essential for an understanding of stress relaxation in liquids.
2. The possibility of studying more complex systems, like alkanes and small polymers, will undoubtedly generate further quantitative studies on rheological behavior as function of the molecular geometry. Relaxation mechanisms of polymer motions in various flows (shear, elongational, or four rollers) are within the range of computation. Comparisons with experiments can even be done, provided scaling properties in nonequilibrium conditions and at relatively small scales are confirmed.
3. The stationary nonequilibrium steady states seem to obey the fluctuating hydrodynamical descriptions very well, even far from equilibrium, but testing the theories when one is very near an instability remains to be done. Are the fluctuations still of the same type, or do new phenomena appear? How much is the great sensitivity of molecular dynamics systems due to their small sizes and how much is it due to the proximity of the critical point? Computations leading to good statistics can now be done in order to obtain *quantitative* results.

4. Nonequilibrium dynamical systems have much in common with turbulent fluids. Techniques borrowed from the study of chaotic dynamical systems have been used in this context. Could one then find more connections between these two fields? Let us mention the recent study by Gallavotti and Cohen [54] along these lines, in which they attempt to formulate ideas developed by Ruelle to obtain statistics of turbulent flows within the context of nonequilibrium stationary states. Many of the hypotheses made could be directly tested against large-scale simulations.
5. The usage of Lyapunov exponents, fractal dimensions, and fractal measures are already familiar to the community of nonequilibrium molecular dynamics. Although the models in nonequilibrium molecular dynamics are made explicitly irreversible, can they be used to explore the connection between chaotic properties and transport? Various attempts have been made to link Lyapunov exponents and transport coefficients [258, 259].[5] Quantitative tests have also been made for diffusion and should be extended to the other transport coefficients in the near future.
6. The question of the nature of the stationary states is still open. Are nonequilibrium states unique in the sense that they are independent of their preparation? Is the fractal nature of the distributions of any physical relevance, or is it simply due to the techniques used?
7. The development of techniques in itself could provide useful tools to investigate larger scale phenomena. Also there is the possibility of using stochastic methods, or hybrid codes, and mixing different approaches (depending on the local nature of the flows). These could be very promising and of much use to more applied problems.

ACKNOWLEDGMENTS

I have greatly benefited from many discussions and much encouragement: Giovanni Ciccotti, Eddy Kestemont, Brad Holian, Mamad Malek Mansour, and Grégoire Nicolis have been extremely helpful during all the recent years.

Financial support of the Pole d'Attraction Interuniversitaire of the Belgian Government is gratefully acknowledged.

REFERENCES

1. B. J. Alder and T. E. Wainwright, *Phys. Rev. A*, **1,** 18 (1970).
2. Y. Pomeau and P. Résibois, *Phys. Rep.*, **19,** 63 (1975).

[5] A very good introduction to this field is given in ref. [260].

3. H. J. M. Hanley, Ed., *Nonlinear Fluid Behavior, Physica*, **118A,** 1–454 (1983): proceedings of the 1982 conference held at Boulder, Colorado.
4. Proceedings of the conference "The Microscopic Approach to Complexity by Non-equilibrium Molecular Simulations," Centre Européen de Calcul Atomique et Moléculaire (CECAM), Lyon, 1996, *Physica A*, M. Mareschal Ed., special issue to appear in 1997.
5. M. Mareschal, Ed., *Microscopic Simulations of Complex Films:* Proceedings of the Brussels conference held in 1989, NATO Advanced Science Institute Series in Physics, **B236,** Plenum Press, New York, 1990; M. Mareschal and B. L. Holian, Eds., *Microscopic Simulations of Complex Hydrodynamic Phenomena:* 1991 Alghero summer school proceedings, NATO Advanced Science Institute Series in Physics, **B292,** Plenum Press, New York, 1992.
6. M. P. Allen and D. J. Tildesley, *Computer Simulation in Chemical Physics*, NATO Advanced Science Institute Series, **C397,** Kluwer, Dordrecht, 1993.
7. M. Baus, L. F. Rull, and J. P. Ryckaert, Eds., *Observation, Prediction and Simulation of Phase Transitions in Complex Liquids*, NATO Advanced Science Institute Series, **C460,** Kluwer, Dordrecht, 1994.
8. Proceedings of the conference "Lattice Gas 94," Princeton, June, 1994, *J. Stat. Phys.*, **81,** special issue (1995).
9. K. Binder anda G. Ciccotti, Eds., *Monte Carlo and Molecular Dynamics of Condensed Matter Systems*, Societa Italiana de Fisica, Bologna, 1995.
10. M. P. Allen and D. J. Tildesley, *Computer Simulation of Liquids*, Clarendon Press, Oxford, 1987.
11. J. P. Hansen and I. R. McDonald, *Theory of Simple Liquids*, 2nd ed., Academic Press, London, 1986.
12. W. G. Hoover, *Computational Statistical Mechanics*, Elsevier, Amsterdam, 1992.
13. D. J. Evans and G. P. Morriss, *Statistical Mechanics of Nonequilibrium Liquids*, Academic Press, New York, 1990.
14. D. C. Rapaport, *The Art of Molecular Dynamics Simulation*, Cambridge University Press, Cambridge, 1995.
15. J. Koplick and J. R. Banavar, *Ann. Rev. Fluid Mech.*, **27,** 257 (1995).
16. K. Binder Ed., *Monte Carlo Methods in Statistical Physics*, 2nd ed., Springer-Verlag, Berlin, 1986.
17. G. Ciccotti, D. Frenkel, and I. R. McDonald, *Simulations of Liquids and Solids*, North-Holland, Amsterdam, 1987 (collection of reprints of historical papers with a few comments made by the editors).
18. P. Gaspard and X.-J. Wang, *Phys. Rep.* **235,** 291 (1993).
19. B. J. Alder and T. E. Wainwright in *Transport Processes in Statistical Mechanics*, (1956 Brussels Conference Proceedings) I. Prigogine, Ed., Wiley Interscience, New York, 1958.
20. B. J. Alder and T. E. Wainwright, *J. Chem. Phys.* **27,** 1208 (1957).
21. A. Rahman, *Phys. Rev.*, **136,** 405 (1964).
22. L. Verlet, *Phys. Rev.*, **159,** 98 (1967).
23. L. Verlet, *Phys. Rev.*, **165,** 201 (1968).
24. D. Levesque and L. Verlet, *Phys. Rev. A*, **2,** 2514 (1970).
25. Brad Holian, in the round-table discussions ref. [9].
26. J. J. Erpenbeck and W. W. Wood in *Modern Theoretical Chemistry*, Vol. 6, Part B: *Time-Dependent Processes*, B. J. Berne, Ed., Plenum Press, New York, 1977.

27. J. Piasecki and P. Resibois, *J. Math. Phys.*, **14,** 1984 (1973); P. Résibois, Y. Pomeau and J. Piasecki, *J. Math. Phys.*, **15,** 1238 (1974).
28. P. Résibois and M. Deleener, *Kinetic Theory of Classical Fluids*, J. Wiley, New York, 1976.
29. B. Smit, Ph. deSmedt, and D. Frenkel, *Molec. Phys.*, **68,** 931 (1989).
30. B. Smit and D. Frenkel, *Molec. Phys.*, **68,** 951 (1989).
31. H. Luo, G. Ciccotti, M. Mareschal, M. Meyer, and B. Zappoli, *Phys. Rev. E*, **51,** 2013 (1995).
32. W. T. Ashurst and W. G. Hoover, *Phys. Rev. A*, **11,** 658 (1975); W. G. Hoover and W. T. Ashurst, in *Theoretical Chemistry*, Vol. I. H. Eyring and D. Henderson, Eds., Academic Press, New York, 1975, p. 1.
33. E. M. Gosling, I. R. McDonald, and K. Singer, *Molec. Phys.*, **26,** 1475 (1973).
34. A. W. Lees and S. F. Edwards, *J. Phys. C*, **5,** 1921 (1972).
35. J. H. Irving and J. G. Kirkwood, *J. Chem. Phys.*, **18,** 817 (1950).
36. D. Levesque, L. Verlet, and J. Kurkijarvi, *Phys. Rev. A*, **7,** 1690 (1973).
37. C. Hoheisel and R. Vogelsang, *Comp. Phys. Rep.*, **8,** 1–70 (1988).
38. S. Y. Liem, D. Brown, and J. H. R. Clarke, *Phys. Rev. A*, **45,** 3706 (1992); J. Koplick, J. R. Banavar, and J. F. Willemsen, *Phys. Fluids A*, **1,** 781 (1989).
39. P. Peeters, Ph.D. thesis, University of Brussels, 1992 (unpublished).
40. D. J. Evans and G. P. Morriss, *J. Chem. Phys.*, **77,** 63 (1983).
41. D. J. Evans and G. P. Morriss, *J. Chem. Phys.*, **77,** 451 (1984).
42. G. Ciccotti and J. P. Ryckaert, *Comp. Phys. Rep.*, **4,** 345 (1986).
43. S. Nosé, *J. Chem. Phys.*, **81,** 511 (1984); S. Nosé, *Molec. Phys.*, **52,** 255 (1984).
44. W. G. Hoover, *Phys. Rev. A*, **31,** 1695 (1985); W. G. Hoover, A. J. C. Ladd, and B. Moran, *Phys. Rev. Lett.*, **48,** 1818 (1982); J. Evans, *J. Chem. Phys.*, **78,** 3297 (1983).
45. B. L. Holian, A. F. Voter, and R. Ravelo, *Phys. Rev. E*, **52,** 2338 (1995).
46. S. Nosé, in M. Meyer and V. Pontikis, Eds., *Computer Simulations Techniques in Materials Sciences*, Kluwer, Dordrecht, 1991.
47. D. J. Tobias, G. J. Martyna, and M. L. Klein, *J. Chem. Phys.*, 97, **12** 959 (1993).
48. W. G. Hoover, H. A. Posch, B. L. Holian, M. J. Gillan, M. Mareschal, and C. Massobrio, *Molec. Sim.*, **1,** 79 (1987).
49. A. J. C. Ladd and W. G. Hoover, *J. Stat. Phys.*, **38,** 973 (1985).
50. B. Moran and W. G. Hoover, *J. Stat. Phys.*, **48,** 709 (1987).
51. W. G. Hoover and B. Moran, *Phys. Rev. A*, **40,** 5319 (1989).
52. N. I. Chernov, G. L. Eyink, J. L. Lebowitz, and Ya. G. Sinai, *Comm. Math. Phys.*, **154,** 569 (1993); N. I. Chernov, G. L. Eyink, J. L. Lebowitz, and Ya. G. Sinai, *Phys. Rev. Lett.*, **70,** 2209 (1993).
53. S. Goldstein, C. Kipnis, and N. Ianiro, *J. Stat. Phys.*, **41,** 915 (1985).
54. G. Gallavotti and E. G. D. Cohen, *J. Stat. Phys.*, **80,** 931 (1995).
55. D. J. Evans and G. P. Morriss, *Phys. Rev. Lett.*, **56,** 2172 (1986).
56. A. Tenenbaum, G. Ciccotti, and R. Gallico, *Phys. Rev. A*, **25,** 2778 (1982); A. Tenenbaum, *Phys. Rev. A*, **28,** 3132 (1983).
57. C. Pierleoni and J. P. Ryckaert, *Phys. Rev. A*, **44,** 5314 (1991).
58. D. J. Evans, *Phys. Rev. A*, **34,** 1449 (1986); C. Massobrio and G. Ciccotti, *Phys. Rev. A*, **30,** 3191 (1984).

59. A. Paskin and G. J. Dienes, *J. Appl. Phys.*, **43,** 1605 (1972).
60. D. H. Tsai and R. A. McDonald, *High Pressures*, **8,** 403 (1976).
61. B. L. Holian and G. K. Straub, *Phys. Rev. Lett.*, **43,** 1598 (1979).
62. B. L. Holian, W. G. Hoover, B. Moran, and G. K. Straub, *Phys. Rev. A*, **22,** 2798 (1980).
63. P. J. Clause and M. Mareschal, *Phys. Rev. A*, **38,** 4241 (1988).
64. A. J. C. Ladd, *Mol. Phys.*, **53,** 459 (1984).
65. M. J. Gillan and M. Dixon, *J. Phys. C*, **16,** 869 (1983).
66. D. J. Evans, *Phys. Lett. A*, **91,** 457 (1982).
67. G. Ciccotti, G. Iacucci, and I. R. McDonald, *J. Stat. Phys.*, **21,** 1 (1979).
68. J. P. Ryckaert, A. Bellemans, G. Ciccotti, and G. V. Paolini, *Phys. Rev. A*, **39,** 259 (1989).
69. H. J. C. Berendsen and W. F. van Gunsteren, in *Molecular Dynamics Simulations of Statistical Mechanical Systems*, G. Ciccotti and W. G. Hoover, Eds., North-Holland, Amsterdam, 1986.
70. M. Tuckerman, B. J. Berne, and G. J. Martyna, *J. Chem. Phys.*, **97,** 1990 (1992).
71. G. S. Grest, B. Dunnweg, and K. Kremer, *Comp. Phys. Commun.*, **55,** 269 (1989).
72. D. C. Rapaport, *Comp. Phys. Rep.*, **9,** 1 (1988).
73. D. C. Rapaport, *Comp. Phys. Comm.*, **62,** 217 (1991).
74. D. C. Rapaport, *Comp. Phys. Comm.*, **62,** 198 (1991).
75. U. Frisch, B. Haaslacher, and Y. Pomeau, *Phys. Rev. Lett.*, **56,** 1505 (1986).
76. N. H. Packard and S. Wolfram, *J. Stat. Phys.*, **38,** 901 (1985); S. Wolfram, *Nature*, **311,** 419 (1984).
77. S. Wolfram, *J. Stat. Phys.*, **45,** 471 (1986).
78. J. Von Neumann, in *Theory of Self-Reproducing Automata*, A. W. Burks, Ed., University of Illinois Press, Urbana, 1966.
79. M. Gardner, *Wheels, Life and other Mathematical Amusements*, W. H. Freeman, San Francisco, 1983.
80. J. Hardy, Y. Pomeau, and O. de Pazzis, *Phys. Rev. Lett.*, **31,** 276 (1973).
81. J. P. Rivet, M. Hénon, U. Frisch, and D. d'Humières, *Europhys. Lett.*, **7,** 231 (1988).
82. M. H. Ernst, in *Fundamental Problems in Statistical Mechanics VII*, H. Van Beijeren, Ed., Elsevier Science, Amsterdam, 1990.
83. U. Frisch, B. Haaslacher, and Y. Pomeau, *Phys. Rev. Lett.*, **56,** 1505 (1986).
84. H. Bussemaker and M. H. Ernst, *Physica A*, **194,** 258 (1993).
85. B. Dubrulle, U. Frisch, M. Hénon, and J. P. Rivet, *J. Stat. Phys.*, **59,** 1187 (1990).
86. S. Orszag and V. Yakhot, *Phys. Rev. Lett.*, **56,** 1691 (1986).
87. P. J. Bhatnagar, E. P. Gross, and M. Krook, *Phys. Rev.*, **94,** 511 (1954); S. Harris, *An Introduction to the Boltzmann Equation*, Holt, Rhinehart & Winston, New York, 1971.
88. R. Benzi, S. Succi, and M. Vergassola, *Phys. Rep.*, **222,** 145 (1992).
89. E. G. D. Cohen in *Microscopic Simulations of Complex Hydrodynamic Phenomena*, M. Mareschal and B. L. Holian, Eds., NATO Advanced Science Institute Series, **B292,** Plenum Press, New York, 1992.
90. M. A. Van der Hoef and D. Frenkel, *Phys. Rev. Lett.*, **66,** 1591 (1991).
91. M. A. van der Hoef and D. Frenkel, *Trans. Theory Stat. Phys.*, **24,** 1227 (1995).
92. M. J. H. Hagen, C. P. Lowe, and D. Frenkel, *Phys. Rev. E*, **51,** 4287 (1995).
93. M. H. Ernst, E. H. Hauge, and J. M. J. van Leeuwen, *Phys. Rev. Lett.*, **25,** 1254 (1970).

94. L. P. Kadanoff, G. McNamara, and G. Zanetti, *Phys. Rev. A*, **40,** 4527 (1989).
95. J. J. Erpenbeck and W. W. Wood, *J. Stat. Phys.*, **24,** 455 (1981).
96. P. Grosfils, J. P. Boon, and P. Lallemand, *Phys. Rev. Lett.*, **68,** 1077 (1992).
97. A. J. C. Ladd, *J. Fluid Mech.*, **271,** 285 (1994); A. J. C. Ladd and D. Frenkel, *Phys. Fluids A*, **2,** 1921 (1990).
98. A. J. C. Ladd, *Phys. Rev. Lett.*, **70,** 1339 (1993).
99. B. Cichocki and B. U. Felderhof, *Phys. Rev. E*, **51,** 5549 (1995).
100. D. H. Rothman and S. Zaleski, *Rev. Mod. Phys.*, **66,** 1418 (1994).
101. J. Orban and A. Bellemans, *Phys. Lett. A*, **24,** 620 (1967).
102. G. A. Bird, *Phys. Fluids*, **6,** 1518 (1963).
103. G. A. Bird, *Molecular Gas Dynamics and Direct Simulation of Gas Flows*, Clarendon Press, Oxford, 1994.
104. A. L. Garcia, M. Malek Mansour, G. C. Lie, M. Mareschal, E. Clementi, *Phys. Rev. A*, **36,** 4348 (1987).
105. W. Wagner, *J. Stat. Phys.*, **66,** 1011 (1992).
106. E. Meiburg, *Phys. Fluids*, **29,** 3107 (1986).
107. G. A. Bird in *Microscopic Simulations of Complex Hydrodynamic Phenomena*, M. Mareschal and B. L. Holian, Eds., NATO Advanced Science Institute Series in Physics, **B292,** Plenum Press, New York, 1992.
108. J. K. Harvey and G. Lord, *Rarefied Gas Dynamics 19*, Vol. 1 and 2, Clarendon Press, Oxford, 1995.
109. K. Nanbu, in *Rarefied Gas Dynamics*, V. Boffi and C. Cercignani, Eds., B. G. Tuebner, Stuttgart, 1986.
110. M. Malek Mansour, A. L. Garcia, and M. Mareschal, *J. Comp. Phys.*, **119,** 94 (1995).
111. F. J. Alexander, A. L. Garcia, and B. J. Alder, *Phys. Rev. Lett.*, **74,** 5212 (1995).
112. J. A. Barker and D. Henderson, *Rev. Mod. Phys.*, **48,** 587 (1976).
113. J. M. Montanero and A. Santos, *Phys. Rev. E*, **54,** 438 (1996).
114. A. Frezzotti and C. Sgarra, *J. Stat. Phys.*, **73,** 193 (1993).
115. A. Frezzotti, in *Rarefied Gas Dynamics*, J. Harvey and G. Lord, Eds., Oxford University Press, Oxford, 1995.
116. E. Salomons and M. Mareschal, *Phys. Rev. Lett.*, **69,** 269 (1992).
117. M. Kac, *Probability Theory and Related Topics in Physical Science*, Wiley Interscience, New York, 1959.
118. J. Logan and M. Kac, *Phys. Rev. A*, **13,** 458 (1976); A. Onuki, *J. Stat. Phys.*, **18,** 475 (1978).
119. M. Malek Mansour, A. L. Garcia, G. C. Lie, and E. Clementi, *Phys. Rev. Lett.*, **58,** 874 (1987).
120. A. L. Garcia, M. Malek Mansour, G. C. Lie, M. Mareschal, and E. Clementi, *Phys. Rev. A*, **36,** 4348 (1987).
121. E. Salomons and M. Mareschal, *Phys. Rev. Lett.*, **69,** 269 (1992).
122. M. Mareschal and E. Salomons, *Transl. Theory Stat. Phys.*, **23,** 281 (1994).
123. B. L. Holian, C. W. Patterson, M. Mareschal, and E. Salomons, *Phys. Rev. E*, **47,** R24 (1993).
124. P. Turq, F. Lantelme, and H. L. Friedman, *J. Chem. Phys.*, **66,** 3039 (1977).

125. J. F. Brady and G. Bossis, *Ann. Rev. Fluid Mech.*, **20,** 111 (1988).
126. B. J. Berne and R. Pecora, *Dynamic Light Scattering*, J. Wiley & Sons, New York, 1976.
127. A. L. Garcia, M. Malek Mansour, G. C. Lie, and E. Clementi, *J. Stat. Phys.*, **47,** 209 (1987).
128. G. Nicolis and M. Malek Mansour, *Phys. Rev. A*, **29,** 2845 (1984).
129. D. T. Gillespie, *J. Phys. Chem.*, **81,** 2340 (1977); D. T. Gillespie, *Markov Processes*, Academic Press, New York, 1992.
130. H. P. Breuer and F. Petruccione, *Phys. Lett. A*, **172,** 49 (1992); H. P. Breuer and F. Petruccione, *J. Phys. A*, **25,** L661 (1992).
131. H. P. Breuer and F. Petruccione, *Phys. Rev. E*, **50,** 2795 (1994).
132. L. B. Lucy, *Astronomical Journal*, **82,** 1013 (1977).
133. J. J. Monaghan, *Ann. Rev. Astron. Astrophy.*, **30,** 543 (1992).
134. H. A. Posch, W. G. Hoover, and O. Kum, *Phys. Rev. E*, **52,** 1711 (1995); O. Kum, W. G. Hoover, and H. Posch, *Phys. Rev. E*, **52,** 4899 (1995).
135. P. J. Hoogerbrugge and J. M. V. A. Koelman, *Europhys. Lett.*, **19,** 155 (1992).
136. K. Kawazaki, T. Nagai, K. Nakashima, *Phil. Mag. B*, **60,** 759 (1989).
137. T. Okuzono, E. Kawazaki, *Phys. Rev. E*, **51,** 1246 (1995).
138. X.-F. Yuan and R. C. Ball, *J. Chem. Phys.*, **101,** 9016 (1994).
139. G. Martin and L. Kubin, Eds., *Non-Linear Phenomena in Materials Science II*, Trans-Tech Publications, Zurich, 1992.
140. R. B. Bird, R. C. Armstrong, and O. Hassager, *Dynamics of Polymeric Liquids*, Vol. 1, Wiley Interscience, New York, 1987.
141. R. Byron Bird and C. F. Curtiss, *Physics Today*, 36 (1984) (January, special issue dedicated to non-linear fluids).
142. C. Trozzi and G. Ciccotti, *Phys. Rev. A*, **29,** 916 (1984).
143. W. Loose and S. Hess, *Rheol. Acta*, **28,** 91 (1989).
144. M. J. Stevens and M. O. Robbins, *Phys. Rev. F*, **48,** 3778 (1993).
145. O. Hess, W. Loose, T. Weider, and S. Hess, *Physica B*, 156, 505 (1989).
146. T. Weider, U. Stottut, W. Loose and S. Hess, *Physica A*, **174,** 1 (1991).
147. D. J. Evans and G. P. Morriss, *Phys. Rev. Lett.*, **56,** 2172 (1986).
148. S. Hess, *Intl. J. Thermophys.*, **6,** 657 (1985).
149. J. J. Erpenbeck, *Phys. Rev. Lett.*, **52,** 1333 (1984).
150. W. Loose and S. Hess, *Phys. Rev. A*, **37,** 2099 (1988).
151. W. Xue and G. S. Grest, *Phys. Rev. Lett.*, **64,** 419 (1990).
152. B. J. Ackerson and P. N. Pusey, *Phys. Rev. Lett.*, **61,** 1033 (1988).
153. S. Hess and H. M. Koo, *J. Non-Equil. Thermodyn.*, **14,** 159 (1989).
154. J. W. Dufty, *Phys. Rev. A*, **30,** 622 (1984).
155. W. E. Alley, B. J. Alder and S. Yip, *Phys. Rev. A*, **27,** 3174 (1983).
156. C. S. Kim, J. W. Dufty, A. Santos, and J. J. Brey, *Phys. Rev. A*, **40,** 7165 (1989).
157. M. Ferrario, G. Ciccotti, B. L. Holian, and J. P. Ryckaert, *Phys. Rev. A*, **44,** 6936 (1991).
158. L. P. Kadanoff, G. McNamara, and G. Zanetti, *Phys. Rev. B*, **40,** 4527 (1989).
159. M. P. Allen, D. Brown, and A. J. Masters, *Phys. Rev. E*, **49,** 2488 (1994).
160. M. P. Allen, *Phys. Rev. E*, **50,** 3275 (1994).

161. D. McQuarrie, *Statistical Mechanics*, J. Wiley & Sons, New York, 1971.
162. J. N. Israelachvili, *Intermolecular and Surface Forces*, Academic Press, New York, 1991.
163. P. N. Pusey and W. van Megen, *Nature*, **320,** 340 (1986).
164. E. J. Meijer and D. Frenkel, *Phys. Rev. Lett.*, **67,** 1110 (1991); M. Dijkstra, D. Frenkel, and H. N. W. Lekkerkerker, *Physica A*, **193,** 374 (1993); C. Smits, W. J. Briels, J. K. G. Dhont, and H. N. W. Lekkerkerker, *Progr. Colloid Polym. Sci.*, **79,** 287 (1989).
165. M. P. Allen, G. T. Evans, D. Frenkel, and B. M. Mulder, *Adv. Chem. Phys.*, **86,** 1–66 (1993).
166. J. P. Ryckaert, G. Ciccotti, and H. J. C. Berendsen, *J. Comp. Phys.*, **23,** 327 (1977).
167. G. Ciccotti and J. P. Ryckaert, *Comp. Phys. Rep.*, **4,** 345 (1986).
168. M. N. Hounkonnou, C. Pierleoni, and J. P. Ryckaert, *J. Chem. Phys.*, **97,** 9335 (1992).
169. G. P. Morriss, P. J. Davis, and D. J. Evans, *J. Chem. Phys.*, **94,** 7420 (1991); A. Berker, S. Chynoweth, U. C. Klomp, and Y. Michopoulos, *J. Chem. Soc. Faraday Trans.*, **88,** 1719 (1992).
170. B. Dunweg and K. Kremer, *J. Chem. Phys.*, **99,** 6983 (1993); K. Kremer, in *Monte Carlo and Molecular Dynamics of Condensed Matter Systems*, K. Binder and G. Ciccotti, Eds., Editrice Compositori, Bologna, 1995.
171. P. G. de Gennes, *Scaling Concepts in Polymer Physics*, Cornell University Press, Ithaca, 1979.
172. C. Pierleoni and J. P. Ryckaert, *J. Chem. Phys.*, **96,** 8539 (1992).
173. C. Pierleoni and J. P. Ryckaert, *Phys. Rev. Lett.*, **71,** 1724 (1993).
174. C. Pierleoni and J. P. Ryckaert, *Macromolecules*, **28,** 5097 (1995).
175. P. Lindner and R. C. Oberthur, *Physica B*, **156,** 410 (1989); A. Link and J. Springer, *Macromolecules*, **26,** 464 (1993).
176. Z. Xu, J. J. de Pablo and S. Kim, *J. Chem. Phys.*, **102,** 5836 (1995).
177. D. Forster, *Hydrodynamic Fluctuations, Broken Symmetry and Correlation Functions*, W. A. Benjamin, Reading, MA, 1975.
178. P. D. Fleming III and C. Cohen, *Phys. Rev. B*, **13,** 500 (1976).
179. A. Kilian and S. Hess, *Liquid Crystals*, **8,** 465 (1990); S. Hess, J. F. Schwrzl, and D. Baalss, *J. Phys. Condens. Matt.*, **SA279,** 2 (1990).
180. H. Ehrentraut and S. Hess, *Phys. Rev. E*, **51,** 2203 (1995).
181. L. Van Hove, *Phys. Rev.*, **95,** 249 (1954).
182. L. P. Kadanoff and P. C. Martin, *Ann. Phys.*, **24,** 419 (1963).
183. P. Résibois, *Physica*, **49,** 591 (1970).
184. W. A. Alley and B. J. Alder, *Phys. Rev. A*, **27,** 3158 (1983).
185. A. Rahman, in Proceedings of the 1971 IUPAP Conference (Chicago): *Statistical Mechanics*, New concepts, New problems, New applications, S. A. Rice, R. C. Freed, and J. C. Light, Eds., University of Chicago Press, Chicago, 1972; M. Schoen and C. Hoheisel, *Mol. Phys.*, **57,** 445 (1986).
186. I. Procaccia, D. Ronis, and I. Oppenheim, *Phys. Rev. Lett.*, **42,** 287 (1979).
187. T. R. Kirkpatrick, E. G. D. Cohen, and J. R. Dorfmann, *Phys. Rev. Lett.*, **42,** 862 (1979).
188. A. M. S. Tremblay, M. Arai, and E. D. Siggia, *Phys. Rev. A*, **23,** 1451 (1980).
189. L. Landau and E. Lifshitz, *Fluid Mechanics*, Pergamon Press, London, 1963.
190. J. P. Boon and S. Yip, *Molecular Hydrodynamics*, Dover, New York, 1991.
191. B. M. Law and J. V. Sengers, *J. Stat. Phys.*, **57,** 531 (1989).

192. G. Nicolis and M. Malek Mansour, *Phys. Rev. A*, **29,** 2845 (1984).
193. W. B. Li, P. N. Segré, J. V. Sengers, and R. W. Gammon, *J. Phys. A*, **23,** 119 (1994).
194. M. Mareschal, M. Malek Mansour, G. Sonnino, and E. Kestemont, *Phys. Rev. A*, **45,** 7180 (1992).
195. A. L. Garcia, M. Malek Mansour, G. Lie, and E. Clementi, *J. Stat. Phys.*, **47,** 209 (1987).
196. E. Meiburg, *Phys. Fluids*, **29,** 3107 (1986).
197. G. A. Bird, *Phys. Fluids*, **30,** 364 (1987).
198. D. C. Rapaport and E. Clementi, *Phys. Rev. Lett.*, **57,** 695 (1987).
199. D. C. Rapaport, *Phys. Rev. A*, **36,** 3288 (1987).
200. S. T. Cui and D. J. Evans, *Mol. Sim.*, **9,** 179 (1992).
201. D. K. Bhattacharya and G. C. Lie, *Phys. Rev. Lett.*, **62,** 897 (1989).
202. D. K. Bhattacharya and G. C. Lie, *Phys. Rev. A*, **43,** 761 (1991).
203. L. Hannon, G. C. Lie, and E. Clementi, *Phys. Lett. A*, **119,** 174 (1986).
204. M. Sun and C. Ebner, *Phys. Rev. Lett.*, **69,** 3491 (1992).
205. G. Mo and R. Rosenberger, *Phys. Rev. A*, **42,** 4688 (1990); **44,** 4978 (1991).
206. G. Mo and R. Rosenberger, *Phys. Rev. A*, **46,** 4813 (1992).
207. P. A. Thompson and M. O. Robbins, *Phys. Rev. A*, **41,** 6830 (1990).
208. M. Tij and A. Santos, *J. Stat. Phys.*, **76,** 1399 (1994); M. Alaoui and A. Santos, *Phys. Fluids A*, **4,** 1273 (1992).
209. P. G. de Gennes, *Rev. Mod. Phys.*, **57,** 827 (1985).
210. E. B. Dussan, *Ann. Rev. Fluid Mech.*, **11,** 371 (1979).
211. J. Koplik and J. R. Banavar, and J. F. Willemsen, *Phys. Rev. Lett.*, **60** 1282 (1988).
212. J. Koplik and J. R. Banavar, and J. F. Willemsen, *Phys. Fluids A*, **1,** 781 (1989).
213. M. O. Robbins and P. A. Thompson, *Science*, **253,** 916 (1991).
214. P. A. Thompson and M. O. Robbins, *Phys. Rev. Lett.*, **63,** 766 (1989); P. A. Thompson and M. O. Robbins, *Phys. Rev. A*, **41,** 6830 (1990).
215. M. Meyer, M. Mareschal, and M. Hayoun, *J. Chem. Phys.*, **89,** 1067 (1988).
216. J. S. Rowlinson and B. Widom, *Molecular Theory of Capillarity*, Clarendon Press, Oxford, 1982.
217. E. Salomons and M. Mareschal, *Europhys. Lett.*, **16,** 85 (1991).
218. M. Baus and R. Lovett, *J. Chem. Phys.*, in press.
219. M. Mareschal, M. Meyer, and P. Turq, *J. Phys. Chem.*, **95,** 10723 (1991).
220. B. Smit, A. G. Schlijper, L. A. M. Rupert, and N. M. van Os, *J. Phys. Chem.*, **94,** 6933 (1990).
221. S. Toxvaerd, preprint.
222. J. Koplik and J. R. Banavar, *Phys. Fluids A*, **6,** 480 (1994).
223. J. H. Sikkenk, J. O. Indekeu, J. M. J. van Leeuwen, E. O. Vossnack, A. F. Bakker, *J. Stat. Phys.*, **52,** 23 (1988).
224. J. Hautman and M. L. Klein, *Phys. Rev. Lett.*, **67,** 1763 (1991).
225. F. Heslot, N. Fraysse and A. M. Cazabat, *Nature*, **338,** 640 (1989).
226. P. G. de Gennes and A. M. Cazabat, *C. R. Acad. Sci. Paris*, **310,** 1601 (1990).
227. L. Leger and J. F. Joanny, *Rep. Prog. Phys.*, **55,** 431 (1992).
228. J. X. Yang, J. Koplik, and J. R. Banavar, *Phys. Rev. Lett.*, **67,** 3539 (1991); J. X. Yang, J. Koplik, and J. R. Banavar, *Phys. Rev. A*, **46,** 7738 (1992).

229. N. Fraysse, M. P. Valignat, A. M. Cazabat, F. Heslot, and P. Levinson, *J. Colloid Int. Sci.*, **158,** 27 (1993).
230. J. A. Nieminen, D. B. Abraham, M. Karttunen, and K. Kaski, *Phys. Rev. Lett.*, **69,** 124 (1992).
231. J. Deconinck, S. Hoorelbeke, M. P. Valignat, and A. M. Cazabat, *Phys. Rev. E*, **48,** 4549 (1993).
232. J. Koplik and J. R. Banavar, *Science*, **257,** 1664 (1992).
233. J. Koplik and J. R. Banavar, *Phys. Fluids A*, **5,** 521 (1993).
234. S. Chandrasekhar, *Hydrodynamic and Hydromagnetic Stability*, Clarendon Press, Oxford, 1961.
235. M. Mareschal and E. Kestemont, *Nature*, **329,** 427 (1987).
236. M. Mareschal and E. Kestemont, *J. Stat. Phys.*, **48,** 1187 (1987).
237. D. C. Rapaport, *Phys. Rev. Lett.*, **60,** 2480 (1988).
238. A. Puhl, M. Malek Mansour, and M. Mareschal, *Phys. Rev. A*, **40,** 1999 (1989).
239. D. C. Rapaport, *Phys. Rev. A*, **46,** 1971 (1992).
240. F. J. Alexander, A. L. Garcia, and B. J. Alder, in the proceedings of the Conference "The Microscopic Approach to Complexity by Molecular Simulations," Centre Européen de Calcul Atomique de Moléculaire, Lyon 1996, *Physica A*, M. Mareschal Ed., special issue to appear in 1997.
241. T. Watanabe, H. Kaburaki, and M. Yokokawa, *Phys. Rev. E*, **49,** 4060 (1994); A. L. Garcia, F. Baras, and M. Malek Mansour, *Phys. Rev. E*, **51,** 3784 (1995).
242. G. I. Taylor, *Proc. R. Soc. London*, **219,** 186 (1953); **223,** 446 (1954); **A225,** 473 (91954).
243. C. Van den Broeck, *Physica A*, **168,** 677 (1990).
244. H. K. Moffat, *Rep. Prog. Phys.*, **46,** 621 (1983).
245. P. Manneville, *Dissipative Structures and Weak Turbulence*, Academic Press, San Diego, 1990.
246. F. Sagues and W. Horsthemke, *Phys. Rev. A*, **34,** 4136 (1987).
247. P. McCarty and W. Horsthemke, *Phys. Rev. A*, **37,** 2112 (1988).
248. B. I. Shraiman, *Phys. Rev. A*, **36,** 261 (1987).
249. M. N. Rosenbluth, H. L. Berk, I. Doxas, and W. Horton, *Phys. Fluids*, **30,** 2636 (1987).
250. W. Young, A. Pumir, and Y. Pomeau, *Phys. Fluids A*, **1,** 462 (1989).
251. Y. Pomeau, A. Pumir, and W. R. Young, *C. R. Acad. Sci. (Paris)*, **306,** 741 (1988).
252. E. Guyon, J. P. Hulin, C. Baudet, and Y. Pomeau, *Nucl. Phys. B*, **2,** 271 (1987).
253. T. H. Solomon and J. P. Gollub, *Phys. Fluids*, **31,** 1372 (1988).
254. O. Cardoso and P. Tabeling, *Europhys. Lett.*, **7,** 225 (1988).
255. W. Y. Tam and H. L. Swinney, *Phys. Rev. A*, **36,** 1374 (1987).
256. J. A. Barker and D. Henderson, *Rev. Mod. Phys.*, **48,** 587 (1976); D. M. Gass, *J. Chem. Phys.*, **54** 1898 (1971).
257. S. Vannitsem and M. Mareschal, *Phys. Rev. E*, **51,** 5564 (1995).
258. P. Gaspard and G. Nicolis, *Phys. Rev. Lett.*, **65,** 1693 (1990).
259. J. R. Dorfman and P. Gaspard, *Phys. Rev. E*, **51,** 28 (1995).
260. J. R. Dorfman, From Molecular Chaos to Dynamical Chaos, (Lecture notes, University of Utrecht, 1995).

MICROSCOPIC SIMULATIONS OF CHEMICAL INSTABILITIES

F. BARAS AND M. MALEK MANSOUR

Centre for Nonlinear Phenomena and Complex Systems,
Université Libre de Bruxelles,
1050 Brussels, Belgium

CONTENTS

I. Introduction
II. Stochastic Theory of Reactive Fluids
  A. Global Description
    1. Fluctuation Near a Pitchfork Bifurcation
    2. Hopf Bifurcation
  B. Reaction-Diffusion Master Equation
    1. The Langevin Limit
    2. Onset of Spatial Correlations
    3. Spatial Correlations Near a Pitchfork Bifurcation
    4. Critical Fluctuations in One-Dimensional Systems
III. Microscopic Simulations
  A. Microscopic Simulations of Reactive Fluids
  B. Bird Algorithm for Boltzmann Dynamics
  C. Hard-Sphere Chemistry
  D. Microscopic Simulation of an Exothermic Chemical Reaction
  E. Microscopic Simulation of Continuously Stirred Tank Reactor (CSTR)
    1. Multiple Steady-State Transitions
    2. Hopf Bifurcation
  F. Microscopic Simulation of Spatially Extended Systems
IV. Concluding Remarks
Acknowledgments
References

I. INTRODUCTION

The macroscopic theory of chemical kinetics is based on the idea of random collisions. Consider for instance a gaseous mixture or a dilute solution

Advances in Chemical Physics, Volume 100, Edited by I. Prigogine and Stuart A. Rice.
ISBN 0-471-17458-0

of chemically active molecules. Each of them undergoes random thermal motion and, from time to time, suffers an encounter with one of its partners in the mixture that is itself a random event. Depending on the geometry and the relative kinetic energy of colliding molecules, some of these encounters can be reactive, giving rise to the formation of new chemical species. It is a matter of observation that, despite the enormous complexity associated with motion at the molecular level, the time course of a reaction can be expressed in terms of a limited number of macroscopic observables. The founders of chemical kinetics were quick to realize this point. As early as 1879, Guldberg and Waage stated, in their basic paper, that the rate of a chemical reaction is proportional to the average number of collisions between the reacting molecules. The latter is in turn proportional to the product of appropriate powers of the concentrations $c_1, c_2, \ldots, c_n$ of the constituents of the mixture. This leads to the well-known law of *mass action*. Later advances by Arrhenius, Bodenstein, Polanyi, and Eyring, among others, have followed this idea. In addition, they have produced workable models for the coefficient of proportionality—*the rate constant*—and its dependence on the temperature and other physicochemical parameters.

One of the merits of these ideas was to lead to a simple quantitative formulation. Let W_α stand for the rate of reaction α, and k_α for the corresponding rate constant. Adding up the effect of all reactions, one obtains the time evolution of the composition variables in the form

$$\left(\frac{d}{dt} c_i\right)_{\text{reaction}} = \sum_\alpha \nu_{i\alpha} W_\alpha(k_\alpha; c_1, c_2, \ldots, c_n) \tag{1.1}$$

The *stoichiometric* coefficient $\nu_{i\alpha}$ describes the numbers of molecules of the constituent i participating in reaction α and can be positive, negative, or zero.

In the real world, nonuniformities in the composition are inevitable. Once present, they give rise to the phenomenon of diffusion, which tends to reestablish homogeneity. Diffusion is a complex process of molecular origin. Again, however, in the absence of heterogeneities like interfaces or membranes, its rate is amenable to a simple expression in terms of the composition gradients

$$\left(\frac{d}{dt} c_i\right)_{\text{diffusion}} = \nabla \cdot \sum_j D_{ij} \nabla c_j \tag{1.2}$$

where the diffusion matrix **D** depends in general on state variables. In their general form, Eqs. (1.1) and (1.2) represent a set of coupled nonlinear partial differential equations for the composition variables that are quite difficult to handle.

Two major simplifications are commonly introduced to allow further progress. The first one rests on an important result of kinetic theory, which shows that the transport coefficients, such as viscosity, thermal conductivity, and diffusion coefficients, depend only weakly on state variables. The state dependence of **D** can therefore be neglected if one limits oneself to systems exhibiting "moderate" inhomogeneities, evolving away from equilibrium phase transition regions, and made up of simple molecules that do not undergo conformational changes. A second, and perhaps less obvious simplification, consists in neglecting the cross-diffusional processes related to the nondiagonal elements of **D**. This second simplification implies that the diffusional motion of each molecule is totally independent of the presence of the other molecules. Obviously, this can only be justified in dilute solution where the collision frequency between chemically active molecules is negligible as compared to the one with inert solvent molecules. Fortunately the above restrictions are easily fulfilled in common laboratory experiments, so that for most practical purposes we can write

$$\left(\frac{d}{dt}\,c_i\right)_{\text{diffusion}} = D_i \nabla^2 c_i \tag{1.3}$$

where D_i is simply the diffusion coefficient of the component i. Combining Eqs. (1.1) and (1.3) and limiting ourselves to isothermal systems in mechanical equilibrium (no convection), we arrive at a set of n coupled nonlinear partial differential equations, known as *reaction-diffusion equations*:

$$\frac{\partial}{\partial t}\,c_i(\boldsymbol{r},t) = f_i(c_1,\ \ldots\ ,c_n;\lambda_1,\ \ldots\ ,\lambda_m) + D_i \nabla^2 c_i \tag{1.4}$$

where $\lambda_1, \ldots, \lambda_m$ represent a set of *control parameters* related to the values of the rate constants and to the imposed external constraints. Depending on their values, the reaction-diffusion equations can exhibit a variety of complex phenomena such as a multiplicity of states, oscillations, chaos, and pattern formation [1–3].

There is at present ample experimental evidence corroborating these features [4–8]. Ideas about self-organization and complexity can be found in the literature well before the current massive experimental evidence became available. In the mid 1950s thermodynamic criteria predicting the existence of oscillations in dissipative systems far from equilibrium were clearly formulated [9, 10]. Furthermore, in the mid 1960s the basic work of Smale and of the Soviet school drew attention to the fact that stationary states and multiperiodic orbits are far from being the only possible modes of behavior of dynamical systems [11].

The first analyses performed on the model systems were based on numerical simulations. The diversity of the behaviors was such that the need for more fundamental understanding quickly became obvious. Stability theory of stationary states was a first useful tool. It was, however, the extensive use of bifurcation analysis and dynamical systems theory that allowed one to connect the main features of self-organization phenomena to the system's parameters and to the nonequilibrium constraints [12–14].

There exists a great variety of bifurcation phenomena. Stationary, time-periodic, quasi-periodic solutions can be generated in agreement with experimental findings. The system may remain homogeneous in space or develop spontaneous symmetry breaking, leading to spatial patterns. There may be a single bifurcation, or a cascade of bifurcations to solutions of increasing complexity, which sometimes culminate into chaos. Bifurcation is, therefore, the basic mechanism of diversification of the behavior of physicochemical systems [15].

Obviously, the statistical properties of nonequilibrium systems must show a profound change when a bifurcation point is crossed. In fact, the very existence of macroscopic structures implies strong correlations between various regions of the system, even though they are far apart from each other. This is a rather remarkable point, in view of the fact that self-organization phenomena can also occur in systems dominated by short-range interactions or even in ideal systems.

The occurrence of multistability automatically raises some basic questions related to the relative importance of each of the available states and of the kinetics of the transitions between them. The nature of the problem becomes especially clear in systems of large spatial extension, for which the thresholds of successive instabilities are squeezed in a narrow region of parameter space. As a result, the multiplicity associated to different orientations or "phases" of the various states accessible to the system is greatly enhanced. The question can then legitimately arise as to whether transitions induced by the microscopic dynamics cannot give rise to a qualitative change by wiping out, through destructive interference, any systematic behavior, thus compromising the very existence of bifurcations. Such a phenomenon, if true, would imply an intricate coupling between macroscopic and microscopic behaviors, and would thus mark the limits of validity of the purely macroscopic description of instabilities and bifurcation phenomena, on which most of our predictions are based.

Similar arguments can be advanced in connection with the phenomena taking place in the time domain, such as excitation through ultrashort light pulses [16], explosive behavior arising in chemical [17, 18] or thermal combustion [19], and chaotic dynamics [20–22].

An analysis of the above-mentioned problems is clearly outside the strict realm of macroscopic description. Our principal goal in the present review is to explore the properties of nonequilibrium chemical systems by means of an enlarged description in which information pertaining to microscopic behavior is incorporated. Such an enlarged description is also the only one to address the fundamental problem of how order, on a macroscopic scale, may emerge from the highly disordered phenomena associated with dynamical chaos at the molecular level.

These considerations and the fact that nonequilibrium transitions may also occur in ideally dilute and perfectly homogeneous reactive systems attracted a considerable amount of theoretical work in the early 1970s. The main effort was oriented in the setting up of a *stochastic* description that constitutes a convenient intermediate between purely microscopic and macroscopic theories and which accounts for microscopic dynamics through the fluctuations of the macrovariables [23–25]. The most successful approach was the so called *master equation* formulation of chemical reactions, which gives a "mechanistic" point of view of what is going on at the molecular level [1, 26–28]. Many exotic properties of fluctuations were discovered in this way. In particular, for simple situations involving a single composition variable, it was shown that in the close vicinity of a pitchfork (cusp) bifurcation point the solution of the master equation can be cast into the exponential of a "stochastic potential," which turns out to be the Landau–Ginzburg potential familiar in equilibrium critical phenomena [29]. In more complex situations, such as transitions to time periodic behavior, the very existence of stochastic potentials was questioned [30–32]. This last point actually highlights the profound difference between nonequilibrium bifurcations and equilibrium phase transitions that can be formulated entirely in terms of thermodynamic potentials.

Despite the interest and the novelty of the above results, stochastic approaches to chemical systems suffer from a paucity of experimental studies [33]. For this reason, the amount of theoretical work devoted to this subject slowed down significantly in the early 1980s. A similar situation arose in the 1950s when a kinetic theoretical approach based on the Boltzmann equation was proposed to deal with gas phase chemical reactions [34]. Deviations from the Arrhenius law, sometimes up to 10% for exothermic reactions [35], were predicted but no experimental confirmation has yet been reported.

With the ever-growing power of computers, a new way of investigating nonequilibrium phenomena became available in the mid 1980s. This method, known as *molecular dynamics* (MD), had been pioneered by Alder in the 1950s [36]. It consists in solving numerically Newton's equations of motion for an assembly of particles interacting through a given potential. An appropriate average over space and time of the basic mechanical variables, i.e., mass, momentum, and energy, gives access to macroscopic quantities such as pressure, temperature, etc. Regarded at first as a curiosity by the scientific community, the method quickly became extremely widespread thanks to its ability to bridge the gap between the phenomenological analysis of large-scale macroscopic phenomena and modeling of these phenomena at the microscopic level.

For a long time the main application of MD has been in the context of equilibrium statistical physics and liquid theory (see, for example, ref. [37]). More recently, MD has been successfully extended to the study of nonequilibrium systems as well [38–40], including a variety of complex phenomena arising in simple fluids such as shock waves [41], flows past an obstacle [42] or Rayleigh-Bénard instability [43, 44]. Today, MD is considered by physicists as a sort of ideal laboratory where they can design a desired "experiment" to check, for example, the validity of a theory in extreme conditions where real experiments are difficult or even impossible to perform (see, for example, ref. [45]). It is precisely in this spirit that we shall use MD throughout this review. In doing so, one of our main objectives will be to clarify the status of the stochastic theory of isothermal chemical reactions based on the master equation. The progress achieved during recent years in the microscopic simulation techniques of chemical systems offers, in this respect, additional possibilities of interesting cross-fertilization with theoretical developments.

The master equation formalism in perfectly homogeneous media will be laid down in the next section. We review the main assumptions that form the basis of this description, the formal properties of its solutions, and some results established in the early literature on this subject in connection with bifurcations leading to steady-state (Section II.A.1) and to time-periodic solutions (Section II.A.2). We subsequently turn to the local formulation of the master equation (Section II.B) which allows one to clarify the status of the Langevin formulation of reaction-diffusion equations (Section II.B.1) and the onset of long-range spatial correlations in nonequilibrium systems (Sections II.B.2 and II.B.3). We end Section II.B by reviewing some of the results obtained recently for the case of one-dimensional systems (Section II.B.4).

Section III is devoted to the survey of the microscopic simulation of reactive fluids. The basics of molecular dynamic techniques and its generalization to reactive media will be introduced in Section III.A. As it turns out, the very nature of chemical dynamics requires high-performance computational algorithms, so that we are forced to limit the simulation to dilute hard-sphere mixtures in the Boltzmann limit. The basis of such simulations is discussed in Section III.B. In order to take advantage of the relative simplicity of hard-sphere dynamics, one has to give a meaning to reactive hard-sphere collisions. The principles of "hard-sphere chemistry" are described in Section III.C. In Section III.D we consider exothermic chemical systems, where the limitations of equilibrium rate theory are illustrated on a representative example. The microscopic simulation of isothermal chemical systems in perfectly homogeneous media is discussed in Section III.E. A comparison with the stochastic theory of chemical systems is conducted on a number of case studies, including the pitchfork (Section III.E.1) and Hopf bifurcation (Section III.E.2). The microscopic simulation of spatially extended reactive systems is discussed in section III.F, along with the validity of the reaction-diffusion master equation. Finally the main conclusions and perspectives are presented in Section IV.

II. STOCHASTIC THEORY OF REACTIVE FLUIDS

The phenomenological theory of reaction-diffusion equations, Eq. (1.4), rests on the fundamental assumption of a clear-cut separation between macroscopic behavior, as described by the equations of chemical kinetics coupled to mass transfer, and dynamical processes at the microscopic level. Each event is considered to result from an average over all microscopic characteristics. Everything happens as if in each sub-volume ΔV of the system, the molecules feel mainly the average effect of all their partners therein. These very molecules are thus at the origin of the effect they themselves feel. We call this the *mean field* description, a term that originated in early treatments of equilibrium phase transitions in fluids and in magnetic systems pioneered by Van der Waals and by Weiss, and since extended by Landau [46].

This lumping of all but the macroscopic degrees of freedom ignores spontaneous deviations from average behavior, the *fluctuations*. These deviations are always present in a macroscopic system, which, because of the complexity of molecular motions, can be viewed as a spontaneous generator of noise. So far there is no workable purely microscopic theory for reaction-diffusion systems, so in what follows we approach the problem of fluctuations in terms of the theory of stochastic processes [47]. As we have a fairly simple picture of what is going on at the molecular level, we can easily construct a stochastic process that mimics as closely as possible the molecular aspects of reaction-diffusion systems.

A. Global Description

Consider an ideal isothermal chemical system at mechanical equilibrium. Postponing until later the discussion of the role of the diffusion, we focus here on the description of chemical processes alone. The composition of the system can only change through reactive collisions, which, because of the existence of activation energies, are typically rare events compared to nonreactive ones. This suggests that one can lump all the microscopic (position-momenta) degrees of freedom and view the evolution as a succession of jumps corresponding to the change of composition through chemical reactions, interrupted by *long waiting-time* intervals. During these intervals the large number of nonreactive collisions will give rise to a randomization and a loss of memory. These remarks strongly suggest that the evolution can be represented by a *jump Markov process* in an appropriate phase space. The evolution of the probability distribution for the total number of particles U_i of species i is then governed by the master equation [26]

$$\frac{d}{dt} P(\{U_i\};t) = \sum_{\{U'_i\}} W(\{U'_i\}|\{U_i\})P(\{U'_i\};t) \tag{2.1}$$

where the W's are transition probabilities per unit time. They are proportional to the frequency of collisions between molecules of the constituents involved in each reaction, the proportionality factor being the macroscopic rate constant. The latter determine the fraction of collisions that are reactive. These constants are the only explicit reference to macroscopic properties introduced in the stochastic description.

An important property of the transition probabilities is their extensivity, which expresses the physically obvious fact that the rate of a chemical process in a volume V must be proportional to V times a suitable function of *intensive* variables: the concentrations. Alternatively, one may use the "total number of particles," $\mathcal{N}$, as the extensivity parameter, in which case the associated intensive variables would be mole fractions. Some authors prefer using the latter instead of the former, mainly because mole fractions are dimensionless quantities, so that rate constants appearing in the master equation always have the same dimension of a frequency, no matter what the molecularity of the reaction. Apart from the above consideration, there is no fundamental difference between using either of the extensivity parameters, or even adopting a new one, such as the number of particles of a given chemical whose concentration is known to remain constant. For the sake of generality, in what follows we shall use the symbol Ω for the extensivity parameter and require that

$$W(\{U'_i\}|\{U_i\}) = \Omega w(\{U'_i/\Omega\}, \{U_i/\Omega\}, 1/\Omega) \equiv \Omega w(\{u'_i\}, \{u_i\}, 1/\Omega) \tag{2.2}$$

where $\{u_i\}$ represent the "densities" (concentrations or mole fractions) of chemical components $\{i\}$.

In ideal isothermal systems, the transition probabilities can be easily constructed through combinatorial arguments [1]. As an example, for the reaction

$$U + U \xrightarrow{k} A \tag{2.3a}$$

one has

$$W(U+2, A-1 \mid U, A) = \frac{\tilde{k}}{\Omega} \frac{(U+2)(U+1)}{2!} \tag{2.3b}$$

where $\tilde{k}$ is proportional to the frequency of reactive collisions between U particles. It would be exactly equal to the frequency of reactive collisions, if one had used $\mathcal{N}$ instead of Ω as the extensivity parameter. Traditionally the combinatorial factor, 2!, is incorporated into the definition of the rate constant k appearing in Eq. (2.3a), so that in terms of the latter one has

$$W(U+2, A-1 \mid U, A) = \frac{k}{\Omega} (U+2)(U+1) \tag{2.3c}$$

Equation (2.1) for this single step thus reads

$$\frac{d}{dt} P(U, A; t) = \frac{k}{\Omega} [(U+2)(U+1)P(U+2, A-1; t) - U(U-1)P(U, A; t)] \tag{2.3d}$$

Note the presence of the factor Ω in the denominator of the above relations, reflecting the extensivity of the transition probabilities.

The exact solution of the master equation, Eq. (2.1), is known only for some very particular cases. The extensivity property of the transition probabilities proves to be the clue for setting up approximation schemes. Before discussing this point, it is important to note that the master equation can also be studied numerically. Here, the evolution of the system is viewed as a random walk in a discrete phase space (space of "numbers of particles" of different species) for which transitions occur at randomly spaced time intervals. The Markovianity of the process leads to an exponential distribution of waiting times [48]. From this distribution and the transition probabilities corresponding to each elementary chemical step, explicit realizations of the process can be constructed, along the lines of a Monte Carlo type of simulation first developed by Gillespie [49, 50]. Similar techniques are described in refs. [51, 52].

In thermodynamic equilibrium, the master equation obeys *detailed balance* and can be solved exactly. The result is a multi-Poissonian distribution (multinomial in closed system), in perfect agreement with the predictions of equilibrium statistical mechanics of ideal systems [1]. Moreover, it can be shown that the construction of transition probabilities, sketched in relations (2.3), is the only one compatible with the requirement of having multi-Poissonian distribution at equilibrium [53]. In more complex situations, such as non-ideal or non-isothermal systems, one can use the explicit form of the equilibrium partition function as a guideline to possibly derive the correct form of the transition probabilities [54–56]. For instance, in the previous example [Eq. (2.3)] taking place in a dense nonideal mixture, the combinatorial factor $U(U-1)/2!$ in W must be replaced by an expression that accounts for the pair correlation function of the medium.[1] We shall not comment more on this problem, since throughout this review we will restrict ourselves to ideal mixtures in the Boltzmann limit.

Equation (2.1) constitutes a generalization of the macroscopic description, as it is based on less restrictive assumptions than the latter. As we shall see later on, in certain situations it will give rise to predictions that cannot be inferred from the macroscopic kinetic laws. Nevertheless, it is of interest to look into the properties of the master equation when the macroscopic laws are expected to be valid, i.e., in the thermodynamic limit $\Omega \to \infty$.

This problem has been investigated extensively by Van Kampen, who used the extensivity properties of the transition probabilities to set up an expansion of the master equation in powers of the inverse system size Ω [57, 58]. An alternative approach is due to Kubo et al. [59] who recast the original master equation into a Hamilton–Jacobi formulation using the *ansatz*

$$P \approx \exp[\Omega\, \Phi(c_1, \ldots, c_n; t)] \tag{2.4}$$

where c_i represents the density (concentration or mole fraction) of the compound i. The function Φ is usually called the *stochastic potential*, by analogy to thermodynamic potentials of equilibrium statistical mechanics. A similar approach was used by Nicolis et al. [1, 60] at the level of the generating function corresponding to the master equation. An alternative approximation scheme in complex plane was proposed by Gardiner et al. [61, 62]. Finally a rigorous mathematical study of the connection between the jump Markov process and the corresponding continuous (diffusion) process, in the limit of large system size, has been performed by Kurtz [63]. The main results of these investigations are summarized below.

In the thermodynamic limit, $\Omega \to \infty$, the probability distribution becomes sharp around the deterministic path, with probability one, provided that it is initially so. Strictly speaking, this result holds only for a finite time, i.e., the

[1] M. Moreau, private communication.

limits $\Omega \to \infty$ and $t \to \infty$ do not commute, unless the deterministic equations have a unique globally asymptotically stable stationary solution [29]. In any case, we see from Eq. (2.4) that the deterministic path is the one for which the stochastic potential Φ is maximum. The deterministic path must therefore be associated with the most probable path of the stochastic process and not necessary with its average path. Again this result is in perfect agreement with Gibb's theory of equilibrium statistical mechanics, where macroscopic observables are the set of particular values of state variables for which the appropriate thermodynamic potential is an extremum.

As to the behavior of fluctuations around the deterministic path, it can be shown that for large but finite system size, the jump Markov process whose associated probability distribution obeys the master Eq. (2.1) converges, with probability one, to a continuous Markov process (diffusion process) [63] described through the following set of Langevin equations [29]

$$\frac{d}{dt} c_i(t) = f_i(c_1, \ldots, c_n) + \varepsilon^{1/2} F_i(t) \qquad \varepsilon \equiv \frac{1}{\Omega} \tag{2.5}$$

where $\{F_i(t)\}$ are Gaussian random noises, with

$$\langle F_i(t) \rangle = 0 \qquad \langle F_i(t) F_j(t') \rangle = Q_{ij}(\bar{c}_1, \ldots, \bar{c}_n)\delta(t - t') \tag{2.6}$$

The associated probability density obeys the following *Fokker–Planck* equation

$$\frac{\partial}{\partial t} P(\{c_i\}; t) = \sum_i \left(-\frac{\partial}{\partial c_i} f_i + \frac{\varepsilon}{2} \sum_j Q_{ij} \frac{\partial^2}{\partial c_i \partial c_j} \right) P(\{c_i\}; t) \tag{2.7}$$

The correlation matrix Q_{ij} depends on the deterministic paths $\bar{c}_1, \ldots, \bar{c}_n$, which are solutions of the deterministic equations

$$\frac{d}{dt} \bar{c}_i(t) = f_i(\bar{c}_1, \ldots, \bar{c}_n) \tag{2.8}$$

A remarkable property of Q_{ij} is that it obeys a fluctuation-dissipation theorem and it is identical to the one obtained through thermodynamic principles using the Landau–Lifshitz derivation [54, 64]. Again, the validity of the above results can only be guaranteed for some finite time, $t < \tau(\varepsilon)$, unless the deterministic equations have a unique globally asymptotically stable stationary solution [29].

We first note that for $\varepsilon \to 0$, the Langevin equations, Eqs. (2.5), reduce to the original deterministic equations, Eqs. (2.8). For $\varepsilon \neq 0$, but small, the evolution of the system will still be dominated by the macroscopic law, provided the rate functions $\{f_i\}$ remain much larger than the noise amplitude, i.e., away from possible turning points of the macroscopic equations. As a result, the associated probability density reduces to a propagating multi-Gaussian, whose width remains of the order of ε, at least for some finite time that depends itself on ε. During this time interval, the fluctuations of intensive variables are negligibly small and behave as

$$\langle (c_i - \langle c_i \rangle)(c_j - \langle c_j \rangle) \rangle \equiv \langle \delta c_i \delta c_i \rangle \approx O(\varepsilon) \tag{2.9}$$

As a consequence, the rate functions $f_i(c_1, \ldots, c_n)$ appearing in the Langevin equations, Eqs. (2.5), can be linearized around the deterministic path so that Eq. (2.7) reduces to a *linear* Fokker–Planck equation that was first derived by Van Kampen [57].

Contrary to the case with Eq. (2.7), the validity of this linear Fokker–Planck equation and the resulting Gaussian behavior of the fluctuations cannot be guaranteed for an arbitrarily long time, even if the deterministic equations have a unique globally asymptotically stable stationary solution. An example is the case of systems exhibiting explosive behavior [17, 19]. These systems are characterized by a long induction time, followed by an abrupt transition to the final state. During the induction time, the rate functions remain small, and fluctuations will have time to grow, so that near the ignition time some of sample paths would have reached the final state, whereas some others will still be in the preexplosive regime (see also ref. [65]). As a consequence, the probability distribution will exhibit *transient bimodality* near the ignition time, showing the failure of the Gaussian behavior of the fluctuations [66]. The Fokker–Planck equations, Eq. (2.7), however, remains applicable. Some aspects of this problem will be discussed in more detail in Section III.D, which is devoted to microscopic simulations of exothermic reactions.

The above discussion leaves open the problem of the validity of Langevin formulation of chemical systems in the long-time limit, i.e., near a stable attractor of the deterministic equations. Before discussing this important point, it is instructive to first review the behavior of the stationary (time-independent) solution of the master equation. Despite the fact that the deterministic equations may have multiple stationary or nonstationary (such as time periodic or even chaotic) solutions, the associated master equation has *always* a unique stationary solution. The discussion of the necessary mathematical conditions for the uniqueness of the solution of the master equation is beyond the scope of the present review. Suffice to say that they are routinely fulfilled for any realistic chemical system of finite size [67].

If the macroscopic equations admit a unique globally asymptotically stable stationary state then, for large Ω, the stationary solution of the master equation reduces to a Gaussian distribution around this macroscopic stationary state. The Gaussian behavior gradually breaks down as one approaches a bifurcation point, resulting in a dramatic increase of fluctuations. Two different cases have been studied so far in the literature, namely the case of a pitchfork (or cusp) bifurcation and the case of a Hopf bifurcation. We shall now examine these cases separately with some detail.

1. Fluctuation Near a Pitchfork Bifurcation

The pitchfork bifurcation occurs when a simple eigenvalue of the linearized deterministic equations crosses zero and some additional symmetry properties are fulfilled. Near the bifurcation point, the associated mode exhibits a "critical slowing down." The dynamics of the system is then governed by this slow mode so that, without loss of generality, one can concentrate on a single-variable problem that is best illustrated through the so-called Schlögl model [68]

$$A + 2U \underset{k_2}{\overset{k_1}{\rightleftarrows}} 3U$$

$$U \underset{k_4}{\overset{k_3}{\rightleftarrows}} B \tag{2.10}$$

where the concentrations of A and B are controlled from outside and are used as control parameters. In practice, this can be achieved in an isothermal continuously stirred tank reactor (CSTR). The associated master equation reduces to a simple *birth-and-death* process

$$\frac{d}{dt} P(U;t) = \upsilon(U-1)P(U-1;t) - \upsilon(U)P(U;t)$$

$$+ \mu(U+1)P(U+1;t) - \mu(U)P(U;t) \tag{2.11a}$$

where the birth rate υ and the death rate μ read

$$\upsilon(U) = \frac{3}{\Omega} U(U-1) + \Omega(1+\lambda')$$

$$\mu(U) = \frac{1}{\Omega^2} U(U-1)(U-2) + (3+\lambda)U \tag{2.11b}$$

In writing the above relations, we have used the following notations:

$$k_1 A = \frac{3}{\Omega} \qquad k_2 = 1/\Omega^2$$

$$k_3 = (3 + \lambda) \qquad k_4 B = \Omega(1 + \lambda') \tag{2.11c}$$

The evolution equation for the macroscopic density $\overline{u} = \overline{U}/\Omega$ reads

$$\frac{d\overline{u}}{dt} = -\overline{u}^3 + 3\overline{u}^2 - \overline{u}(3 + \lambda) + (1 + \lambda') \tag{2.12}$$

which, at the stationary state, leads to the following cubic equation

$$(\overline{u}_s - 1)^3 + \lambda(\overline{u}_s - 1) + (\lambda - \lambda') = 0 \tag{2.13}$$

We note that this equation has the same form as the van der Waals equation near the critical point. This analogy becomes more evident if one introduces the so-called *kinetic potential* $\mathcal{V}(\overline{u})$, through

$$\frac{d\overline{u}}{dt} = -\frac{\partial \mathcal{V}(\overline{u})}{\partial \overline{u}} \tag{2.14}$$

For positive values of λ or λ', this potential has a single minimum that corresponds to the unique real solution of Eq. (2.13), which is stable. There exists however a region, located at the bottom left corner of the (λ, λ') plane, within which the potential exhibits two minima separated by a maximum (see Fig. 1). The location of these extrema coincides with the three real roots of Eq. (2.13), one unstable and two stable. These correspond respectively to the maximum and the minima of $\mathcal{V}(\overline{u})$. The curve limiting the region of three real roots has a singular point at the origin $\lambda = \lambda' = 0$, known as *cusp*. The system has at this point a triple root $\overline{u}_s = 1$. Assuming that one moves in the λ, λ' plane along the line $\lambda = \lambda'$, one has

- a root $\overline{u}_s = 1$ for all λ
- and two roots $\overline{u}_\pm = 1 \pm \sqrt{-\lambda}$, for $-1 \leq \lambda < 0$ (2.15)

This is depicted in Fig. 2, which displays the phenomenon of *cusp bifurcation*. As can be seen, the "upper" and "lower" branches are symmetrical, a property shared by the kinetic potential $\mathcal{V}(\overline{u})$, which exhibits two minima

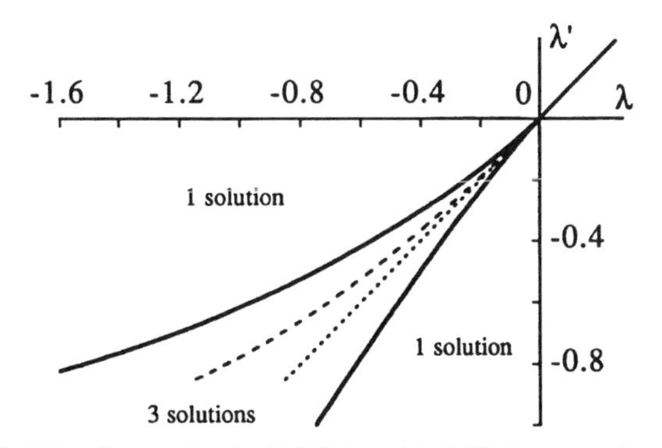

Figure 1. Stability diagram for the Schlögl model (2.10) in space of parameters (λ, λ'). ······ Macroscopic coexistence line, as given by the macroscopic equation, Eq. (2.12), - - - - Stochastic coexistence line, as given by the master equation, Eq. (2.21).

with equal depth for $\lambda = \lambda' < 0$. For this reason, the line $\lambda = \lambda'$ is referred to as *macroscopic coexistence* line.

If one moves in the multiple steady state region away from the line $\lambda = \lambda'$, the "upper" and "lower" branches are no longer symmetrical. Moreover, depending on where the system is initially, different paths will be traced during the time evolution to a final steady state. This is the phenomenon of *hysteresis*, which is one of the most characteristic features of systems involving multiple steady-state transitions. Again, this is strikingly similar to the liquid–vapor transition as described by the classical van der Waals theory. These analogies prompts us to speak of *nonequilibrium phase transitions* in connection with the chemical model (2.10) [69]. An important question is whether the one-to-one correspondence with equilibrium phase transitions remains true in the presence of fluctuations.

To clarify this issue, we consider the master equation, Eq. (2.11), which

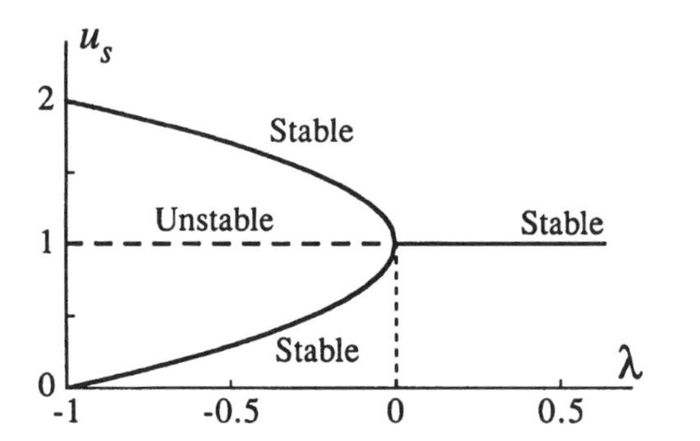

Figure 2. Stability diagram for the Schlögl model, Eq. (2.10), for $\lambda = \lambda'$.

can be solved exactly at the stationary state [70]. For large Ω, its solution can be cast into a form very much like Eq. (2.4)

$$P_{st}(u) = N\phi(u)\exp[\Omega\,\Phi(u)][1 + O(1/\Omega)] \tag{2.16a}$$

where N stands for the normalization constant and

$$\Phi(u) = u[1 - \ln(u/3)] + \mathrm{Re}[(u - i\eta)\ln(u - i\eta) - (u - i\xi)\ln(u - i\xi)]$$

$$\phi(u) = \frac{1}{\sqrt{u}}\,\frac{u^2 + 3 + \lambda}{3u^2 + 1 + \lambda'}$$

$$\eta^2 = \frac{1 + \lambda'}{3} \qquad \xi^2 = 3 + \lambda \qquad u = \frac{U}{\Omega} \tag{2.16b}$$

The form of this stationary density is particularly well suited for asymptotic analysis. Using the steepest descent, a number of properties can be easily established.

Let us first consider the case of $\lambda = \lambda'$. Away from the bifurcation point, $\lambda = 0$, $P_{st}(u)$ reduces to a Gaussian with the following properties [71]:

$$\lim_{\Omega \to \infty} \frac{\langle U \rangle}{\Omega} = \begin{cases} 1 & \text{for} \quad \lambda < 0 \\ 1 - \sqrt{-\lambda} & \text{for} \quad -1 \leq \lambda < 0 \end{cases} \tag{2.17}$$

and

$$\lim_{\Omega \to \infty} \frac{\langle (\delta U)^2 \rangle}{\Omega} = \begin{cases} \dfrac{4}{\lambda} + 1 & \text{for} \quad \lambda > 0 \\ -\dfrac{2}{\lambda} - \dfrac{3}{\sqrt{-\lambda}} + 1 & \text{for} \quad -1 \leq \lambda < 0 \end{cases} \tag{2.18}$$

This last result shows that, away from the bifurcation point, $\langle (\delta U)^2 \rangle$ is proportional to Ω. As one approaches the bifurcation point from either side, the fluctuations tend to diverge as the inverse power of the "distance" from the bifurcation point, which is known as *classical law* of divergence. This in turn implies the breakdown of Gaussian behavior in the close vicinity of the bifurcation point. In fact, right at the bifurcation point, $\lambda = 0$, the probability density becomes quartic

$$P_{st}(u) = N\exp\left(-\Omega\,\frac{u^4}{16}\right)[1 + O(1/\Omega)] \tag{2.19}$$

with

$$\lim_{\Omega \to \infty} \frac{\langle (\delta U)^2 \rangle}{\Omega^{3/2}} = 4\,\frac{\Gamma(3/4)}{\Gamma(1/4)} \tag{2.20}$$

where Γ stands for the gamma function [70]. The enhancement of fluctuations and the change of the probability law at the bifurcation point deserves some further comments. A basic law of probability theory, known as the central limit theorem (see, for example ref. [72]), states that the sum of n independent random variables with common distributions approaches a Gaussian distribution as $n \to \infty$. Intuitively, for the type of ideal system considered here, one would expect that this property should apply. Indeed, an extensive quantity U, such as the total number of particles of a certain constituent, can always be represented as the sum of random variables U_r, each of them pertaining to a small subvolume of the system. If the size of these subvolumes is chosen to be larger than the range of intermolecular forces, then the random variables $\{U_r\}$ are almost statistically independent and the central limit theorem applies. Hence, in the thermodynamic limit one would expect U to be distributed according to a Gaussian distribution. That this is no longer so at the bifurcation point means that the system can in no way be subdivided into uncorrelated subsets. This of course reflects the spatial coherence associated with bifurcation. We shall come back to this problem when we discuss the local properties of fluctuations.

Of special interest are predictions of the master equation concerning the asymptotic behavior of the average value $\langle U \rangle$ in the multiple steady-state region. The relation in Eq. (2.17) shows that the average density of U coincides exactly with the "lower" macroscopic branch $\bar{u}_- = 1 - \sqrt{-\lambda}$, ignoring completely the existence of the "upper" branch $\bar{u}_+ = 1 + \sqrt{-\lambda}$. This result totally contradicts the macroscopic prediction according to which the two branches are equally probable along the line $\lambda = \lambda'$. It implies that the "stochastic coexistence line," based on the master equation, differs from the macroscopic coexistence line. In fact, the probability density in Eq. (2.16) admits two equally probable states along a line for which the stochastic potential exhibits two equal height maxima, i.e.,

$$\Phi(\bar{u}_+(\lambda, \lambda')) = \Phi(\bar{u}_-(\lambda, \lambda')) \tag{2.21}$$

where $\bar{u}_\pm$ are stable stationary solutions of the macroscopic equation, Eq. (2.13). This relation defines a curve on the (λ, λ') plane that represents the stochastic coexistence line (see Fig. 1). Along this line, one has

$$\lim_{\Omega \to \infty} \begin{cases} \dfrac{\langle U \rangle}{\Omega} = \dfrac{c_+ \overline{u}_+ + c_- \overline{u}_-}{c_+ + c_-} \\ \dfrac{\langle (\delta U)^2 \rangle}{\Omega^2} = \dfrac{c_+ c_-}{(c_+ + c_-)^2} (\overline{u}_+ - \overline{u}_-)^2 \\ \dfrac{\langle (\delta U)^3 \rangle}{\Omega^3} = \dfrac{c_+ c_- (c_- - c_+)}{(c_+ + c-)^3} (\overline{u}_+ - \overline{u}_-)^3 \end{cases} \tag{2.22a}$$

with

$$c_\pm = \phi(\overline{u}_\pm)\sqrt{-2/\Phi''(\overline{u}_\pm)} \tag{2.22b}$$

We first note that the average density $\langle U \rangle$ lies somewhere in between the most probable values, which coincide with the two stable stationary solutions of the macroscopic Eq. (2.13). Next, the fluctuations are of the order of $O(\Omega^2)$, which shows that there is no way to distinguish between fluctuations and average values. Finally, the probability density is not symmetrical along the coexistence line, since $\langle (\delta U)^3 \rangle$ is of the order of $O(\Omega^3)$.

Let us now consider the Langevin limit of the master equation, Eq. (2.5). For $\lambda = \lambda'$, the associated Fokker–Planck equation reads [73, 74]

$$\frac{\partial}{\partial t} P(u;t) = \left\{ -\frac{\partial}{\partial u} [-(u-1)^3 - \lambda(u-1)] + \frac{\varepsilon Q(\overline{u}_s)}{2} \frac{\partial^2}{\partial u^2} \right\} P(u;t) \tag{2.23a}$$

with

$$Q(\overline{u}_s) = (\overline{u}_s + 1)^3 + \lambda(\overline{u}_s + 1) \tag{2.23b}$$

where $\overline{u}_s$ obeys Eq. (2.13). The stationary solution of this equation reads

$$P_s(u) = N \exp\left[\frac{-2\varepsilon^{-1}}{Q(\overline{u}_s)} \left(\lambda \frac{(u-1)^2}{2} + \frac{(u-1)^4}{4} \right) \right] \tag{2.24}$$

For $\lambda \geq 0$ and to dominant order in ε, this result is in quantitative agreement with the stationary solution of the master equation. In particular, for $\lambda \gg 0$, the fluctuations around the macroscopic stationary state $\overline{u}_s = 1$ are of the order of $\varepsilon^{1/2}$, whereas at the bifurcation point, $\lambda = 0$, they are of the order of $\varepsilon^{1/4}$. There exists, therefore, a critical range of the bifurcation parameter λ_c, for which both the quadratic and the quartic terms in Eq. (2.24) are of the same order in ε. A close inspection of Eq. (2.24) gives $\lambda_c \approx O(\varepsilon^{1/2})$. We note that for $\lambda \approx \lambda_c$ the chemical relaxation toward the stationary state exhibits a critical slowing down [29, 75]. In particular, the chemical relaxation time τ_{ch} behaves as

$$\tau_{\text{ch}} = \frac{1}{\lambda_c} \approx O(\varepsilon^{-1/2}) \tag{2.25}$$

The above results establish the validity of the Langevin approach before, and up the bifurcation point. This, however, is no longer the case in the multiple steady-state region since the probability density, Eq. (2.24), remains symmetrical along the line $\lambda = \lambda' < 0$, implying that the latter is indeed the coexistence line for the Langevin approach. It is important to realize that this is not simply a minor quantitative correction of one description as compared to the other. For instance, for any region located between the two existence lines in the (λ, λ') plane (Fig. 1), the most probable state for the master equation is the least probable one for the Langevin approach, and vice versa [76]. Similarly, one can always find particular values of λ and λ' for which the first passage times between different states can differ by many orders of magnitude depending on which approach is used [77].

The origin of this major discrepancy lies at the heart of Langevin-type approaches to nonequilibrium phenomena. Indeed, a Langevin formalism is always characterized by a macroscopic law of evolution to which a noise term is added. The amplitude of the noise is directly related to the macroscopic law through a fluctuation-dissipation theorem, which guarantees that at equilibrium the resulting probability density becomes equivalent to one of the familiar Gibbs ensembles of equilibrium statistical physics [78]. This can be achieved provided the macroscopic equations have a unique stationary solution, a condition that is fulfilled up to a bifurcation point. Beyond the bifurcation point, however, the system typically possesses more than one stable attractor, i.e., the probability density is multimodal, each local maximum characterizing a possible stable solution of the corresponding macroscopic equations. As a result, the noise amplitude would have different values for each of the coexisting macroscopic states. While a Langevin approach might be suitable to describe the behavior of the system around a given maximum, it seems highly unlikely that it would correctly describe the global behavior of the system.

One exception to the above conclusions is the case of the Langevin approach to equilibrium phase transition phenomena, such as the Glauber equation for spin systems [79]. In equilibrium systems the noise has a "thermal" origin, i.e., its amplitude is a function of the temperature, which is uniform. For example, a liquid droplet and its surrounding vapor phase are at the same temperature and therefore are driven by the same "noise." That this is not the case in chemical systems shows simply the limits of any analogy one could expect between equilibrium and nonequilibrium transitions. Blindly applying methods developed for the former to the latter might lead to results that are qualitatively wrong. Unfortunately, so far there does not exist experimental data to clarify the situation. Computer experiments remain, therefore, the most promising tool to shed some light on this important issue. We shall therefore come back to this problem later, when we discuss the microscopic simulations of isothermal systems.

The above analysis raises an important question, namely the possibility of finding a stochastic process with continuous realization (diffusion process) that would correctly approximate the master equation in the multiple steady-state region. As we have pointed out, a traditional Langevin approach has to be excluded, mainly because the amplitude of the driven noise is independent of actual values of the process and depends only on the "average" behavior of the system. Another possibility is obviously a diffusion process with *process-dependent* noise. In this approach, the noise amplitude would be a function of the random variable u and therefore would automatically take different values around each of the deterministic stationary solutions. Extending the basic results of Kurtz, Horsthemke and Brenig succeeded in deriving such a noise-dependent diffusion process whose probability density obeys the following *nonlinear* Fokker-Planck equation [80]

$$\frac{\partial}{\partial t} P(u;t) = \left\{ -\frac{\partial}{\partial u}[-(u-1)^3 - \lambda u + \lambda'] + \frac{\varepsilon}{2} \frac{\partial^2}{\partial u^2} Q(u) \right\} P(u;t) \quad (2.26a)$$

where the noise amplitude Q is now a function of the random variable u

$$Q(u) = (u+1)^3 + \lambda u + \lambda' \quad (2.26b)$$

Before, and up to the bifurcation point, the stationary solution of this equation reduces asymptotically to the one we obtained previously, Eq. (2.24) [81]. Beyond, but close to the bifurcation point, it gives much better agreement with the master equation than the corresponding noise-independent process [82, 83]. It comes, however, as a surprise that the coexistence line based on Eq. (2.26) does not match completely the one based on the master equation as one moves on to the multiple steady-state region [84]. Although the deviations remain extremely small (see Table I), there exists, nevertheless, a narrow region between these coexistence lines where the two approaches lead to significantly different results (see also ref. [85]). This suggests that in

TABLE I
"Phase Coexistence" Lines, as Given by the Master Equation (λ_M) and by the Nonlinear Fokker–Planck Equation (λ_F)

λ'	−.25	−.50	−.90	−.95	−.99
λ_M	−.26437	−.56882	−1.278	−1.438	−1.649
λ_F	−.26437	−.56881	−1.276	−1.433	−1.627

the region of multiple steady states, the continuous process obtained as an asymptotic limit of the master equation might be non-Markovian. To our knowledge, this problem has not yet received any solution and remains open.

Note that it is of course always possible to construct an *ad hoc* Fokker–Planck equation whose stationary density matches asymptotically the stationary solution of the master equation [86]. Such a construction, however, remains artificial, since it is not based on a systematic expansion of the master equation. Furthermore, it requires the explicit form of the stationary solution of the master equation, which it is not always possible to obtain.

A satisfactory analysis of fluctuations must clearly go beyond the static properties to which most of this section was limited. Indeed, the mechanisms by which a system evolves in a spontaneous fashion, away from a reference state, are essential for the elucidation of nonequilibrium transition phenomena. We have already mentioned the case of systems undergoing explosive behavior. We end this section by recalling briefly the "initial value problem" for stochastic processes undergoing pitchfork bifurcation, as well as their spectral properties.

As far as the master equation is concerned, Matsuo has developed an asymptotic evaluation of its eigenvalues using a WKB type of method [87]. Most of the results, however, refer to the eigenvalue problem of the Fokker–Planck equation, Eq. (2.23) [88]. In discussing this problem in the multiple steady-state region, Caroli et al. applied a WKB approximation [89], whereas Van Kampen was able to develop a model amenable to a solvable equation of the Schrödinger type [90] (see also ref. [91] for a survey). An interesting result of these investigations is the justification of Kramers theory, which describes the transition between the two stable states, viewed as minima of a double-well potential, over the "potential barrier" constituted by an intermediate unstable state (an interesting reformulation of Kramers theory is given by Gardiner [92]). As it turns out, this theory corresponds to the regime of the first "excited state" above the "ground state" corresponding to the stationary solution of Eq. (2.23). Englund et al. [93] have made a detailed comparison of the Kramers first passage time and WKB approximation with numerical solutions for the case of the absorbing optical bistability. A singular perturbation approach extending the Kramers problem to many coupled variables has been developed by Schuss and Matkowsky [94]. Finally, the "extrema statistics" of the master equation, for instance the evaluation of mean first passage time between probability extrema, is discussed in refs. [95, 96].

Another question of great interest is to know how an initial probability distribution centered on an unstable state is deformed in time, until it reaches the final stationary state [97]. For the case of systems featuring multiple steady states, Suzuki proposed a scaling theory that consists of splitting the evolution into an "initial regime," amenable to a linearized Fokker–Planck equation, and an "intermediate" regime amenable to a macroscopic description [98]. A matching between the two regimes is required. During the first stage, the distribution is continuously broadened, but remains centered on the unstable state. As time goes on, the variance of this distribution ceases to be extensive, and one has to switch to the second stage. A rough estimation of this transition time leads to $t_0 \approx \log(\varepsilon^{-1})$, which becomes exceedingly long in the thermodynamic limit $\varepsilon \to 0$ (for a review, see [99]). The evolution of the system rests then on the dynamics of *inhomogeneous fluctuations*, which may occur in small volumes with appreciable probability, despite the fact that the total volume of the system, V, may become very large.

2. *Hopf Bifurcation*

A Hopf bifurcation occurs when the real part of a pair of complex conjugate eigenvalues of the linearized deterministic equations vanishes. Near the bifurcation point, the associated modes exhibit a "critical slowing down," very much like the case of pitchfork bifurcations, except that now the dynamics of the system is governed by two such modes. Again, without loss of generality, one can concentrate on a two-variable problem that is historically illustrated through the following trimolecular model, also known as "Brusselator" [100, 101],

$$\begin{aligned} A &\xrightarrow{k_1} U \\ B + U &\xrightarrow{k_2} V + C \\ 2U + V &\xrightarrow{k_3} 3U \\ U &\xrightarrow{k_4} E \end{aligned} \tag{2.27}$$

where A, B, C, and E are assumed to remain constant. Ideally, their concentrations are controlled throughout the reaction space to a uniform and time-independent value. In practice, this is achieved in CSTR by pumping A and B at constant rates and by removing C and E as soon as they are produced. The macroscopic evolution equation for the densities $\overline{u}$ and $\overline{v}$ reads

$$\frac{d\overline{u}}{dt} = \overline{u}^2\overline{v} - (\beta + 1)\overline{u} + \alpha \equiv f(\overline{u}, \overline{v})$$

$$\frac{d\overline{v}}{dt} = -\overline{u}^2\overline{v} + \beta\overline{u} \equiv g(\overline{u}, \overline{v}) \tag{2.28a}$$

where we have set

$$\frac{k_1 A}{\Omega} = \alpha \qquad \frac{k_2 B}{\Omega} = \beta \qquad k_3 = \Omega^{-2} \qquad k_4 = 1 \tag{2.28b}$$

The system admits a unique stationary solution

$$\overline{u}_s = \alpha \qquad \overline{v}_s = \frac{\beta}{\alpha} \tag{2.29}$$

which is stable, provided $\beta < \beta_c$ where

$$\beta_c = \alpha^2 + 1 \tag{2.30}$$

For $\beta > \beta_c$ this stationary state becomes unstable and the system evolves to a stable time-periodic solution known as a limit cycle. Unlike the periodic behavior associated with the pendulum and other simple mechanical devices, this solution is reached for any arbitrary initial condition, with an amplitude and period that depend only on the characteristic parameters appearing in the macroscopic equations, i.e., α and β (see Fig. 3). Such *chemical clocks* are, perhaps, the most ubiquitous difference between the behavior of nonequilibrium systems and that of the systems in thermodynamic equilibrium.

The master equation associated with Eq. (2.27) has been analyzed in detail by Turner [102]. The main results can be summarized as follows: In the thermodynamic limit $\Omega \to \infty$, the master equation admits time-dependent solutions $P(t)$, in the form of sharp probability peaks rotating along the limit cycle. In addition to these solutions, it admits also a stationary distribution P_s, in the form of a crater whose edges project on to the limit cycle [103]. The existence of both stationary and time-dependent solutions reflects the symmetry-breaking associated with the formation of a limit cycle. Each individual realization breaks the gauge symmetry, as it has a definite phase. In

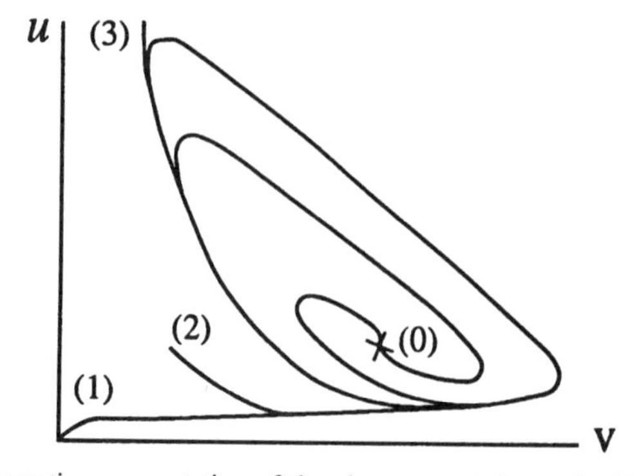

Figure 3. Schematic representation of the phase portrait beyond a Hopf bifurcation, for a variety of initial conditions (0, 1, 2, 3). The cross indicates the location of the unstable stationary state.

a statistical ensemble, however, phases are averaged out and one obtains the stationary distribution P_s. This function P_s turns out to be the envelope of $P(t)$ as it moves in time, and has a number of intriguing properties. Firstly, for any smooth function $S(u, v)$ one has an interesting ergodic theorem: the time average $\overline{S(u,v)}$ of $S(u, v)$ over a long period interval is identical to the statistical average $\langle S(u, v)\rangle$ evaluated with P_s. And, secondly, the height of P_s integrated over the direction normal to the limit cycle, is inversely proportional to the tangential speed along this limit cycle [104]. In other words, the faster the system moves along a portion of the limit cycle, the smaller will be the statistical weight of that portion.

For any arbitrary initial conditions the solution of the evolution equations, Eq. (2.28), ends up in the long time limit on a unique stable attractor. For $\beta \leq \beta_c$, this attractor is a zero dimensional manifold (fixed point), whereas for $\beta > \beta_c$ it is a one-dimensional manifold (limit cycle). In this sense, we can assert that the asymptotic solution of the macroscopic equations is unique. We therefore expect that the Langevin limit of the master equation associated to the model (2.27) should remain valid beyond the bifurcation point. This considerably simplifies the mathematical analysis of the problem and allows one to study in detail the behavior of fluctuations beyond, but close to the bifurcation point [105, 106]. To this end, consider the Langevin limit of the master equation associated to the model (2.27) [29]

$$\frac{du}{dt} = f(u, v) + \varepsilon^{1/2} F(t)$$

$$\frac{dv}{dt} = g(u, v) + \varepsilon^{1/2} G(t) \qquad (2.31a)$$

with

$$\langle F(t)F(t')\rangle = Q_{uu}(t)\delta(t-t') \equiv (\overline{u}^2\overline{v} + (\beta+1)\overline{u} + \alpha)\delta(t-t')$$
$$\langle G(t)G(t')\rangle = Q_{vv}(t)\delta(t-t') \equiv (\overline{u}^2\overline{v} + \beta\overline{u})\delta(t-t')$$
$$\langle F(t)G(t')\rangle = Q_{uv}(t)\delta(t-t') \equiv -(\overline{u}^2\overline{v} + \beta\overline{u})\delta(t-t')$$
$$Q_{uv}(t) = Q_{vu}(t) \tag{2.31b}$$

where the functions $f(u, v)$ and $g(u, v)$ are defined in Eq. (2.28) and ε is the "noise amplitude" (inverse of the extensivity parameter).

As we already mentioned, beyond the bifurcation point the macroscopic path ends up, in the long time limit, on a limit cycle that can be represented as a closed curve Γ in the (u, v) plane, circling the unstable stationary state (see Fig. 3). As one moves close to the bifurcation point, the "amplitude" of Γ decreases. For the case of a Hopf bifurcation, this amplitude must vanish smoothly as $\beta \rightarrow \beta_c$,[2] so that one is allowed to approximate the limit cycle by the lowest possible degree polynomial equation in (u, v). Such a polynomial must, of course, lead to a closed curve in the (u, v) plane which could be an ellipse (second-degree polynomial). Unfortunately, this is not the most common situation, so that, in general, one has to first perform a nonlinear transformation to reduce the limit cycle to an ellipse. The procedure, however, is well known and, as a matter of fact, there exist several alternative techniques to perform the complete transformation [14]. The resulting equation represents the so called "normal form" of a Hopf bifurcation.

$$\frac{dz}{dt} = (i\omega + \gamma)z - (\beta_1 + i\beta_2)|z|^2 z \tag{2.32}$$

The complex "order parameter" $z(t)$, together with its complex conjugate $z^*(t)$, account for both the amplitude and phase of the oscillation; $\gamma(\geq 0)$ represents the "distance" from the bifurcation point; ω is the angular frequency; and β_1, β_2 are two real parameters.

A fundamental property of the normal form, Eq. (2.32), is universality, in the sense that its structure is independent of the details of the underlying system. The specificity of the latter appears only through the various parameters appearing in the equation $(\omega, \gamma, \beta_1$ and $\beta_2)$. For the Brusselator, Eqs.

[2]Note that there exist other types of bifurcations (usually referred to as "sub-critical") where the transition to the limit cycle occurs in an "abrupt" fashion. A well-known example is the case of Van der Pol oscillator whose statistical behavior has been analyzed in ref. [107].

(2.28), they are given by [108]

$$\omega = \alpha \qquad \gamma = (\beta - \beta_c)/2$$

$$\beta_1 = \frac{\alpha^2 + 2}{2\alpha^2} \qquad \beta_2 = \frac{\alpha^4 - 7\alpha^2 + 4}{6\alpha^3} \tag{2.33}$$

A full discussion of the nonlinear transformation, $(u, v) \to (z, z^*)$, is beyond the scope of the present work (see for example ref. [14]). In any case, the same procedure can also be applied to the Langevin equations, Eq. (2.31), from which important statistical properties of the system can be derived [109]. Before presenting the main results, it is important to notice that the "Langevin-normal form" equation, obtained through the above procedure, is not just the macroscopic normal form equation, Eq. (2.32), with an additive noise term, as in Eq. (2.31), but contains some extra terms, usually referred to as "spurious drift" terms. The origin of these terms is directly related to the nonlinear character of the transformation, as can be easily realized by performing the transformation at the level of the Fokker–Planck equation, associated to Eq. (2.31), using the "chain rule." Although in some circumstances the spurious terms may play an important role,[3,4] detailed analysis shows that in the present problem they can only lead to small corrections (of the order of ε) [112].

We stress that the validity of the normal form equation, and the corresponding Langevin formulation, can only be guaranteed in the close vicinity of the bifurcation point, i.e., $\beta - \beta_c$ has to be considered as a small parameter. On the other hand, the noise amplitude ε is also small, so that the statistical properties of the system are determined by the relative importance of these small parameters.

The simplest case arises when the limit $\beta \to \beta_c$ is taken prior to the limit $\epsilon \to 0$, since it leads to the behavior of the system right at the bifurcation point. In this case, the macroscopic equations have a unique stationary solution and, as we already discussed, to dominant order in ε the master equation reduces to a Fokker–Planck equation that is equivalent to the Langevin equations, Eq. (2.31) [102]. Two other regimes have so far been studied:

1. The limit $\beta \to \beta_c$ and the limit $\varepsilon \to 0$ are taken jointly, with $\beta - \beta_c \approx O(\varepsilon^{1/2})$. In this case, the relaxation of the "radial" variable r, defined as $z = re^{i\theta}$, towards its stationary state $r_{st} = \sqrt{\gamma/\beta_1}$ (radius of the limit cycle) exhibits a critical slowing down, with a relaxation time of the order of $O(\varepsilon^{-1/2})$. The corresponding stationary probability density is

[3]This is, for example, the case with "multiplicative noise" processes [110].
[4]The effect of multiplicative noise on Hopf bifurcation has been considered in ref. [111].

quartic, i.e.,

$$P_{st}(r) \approx r \exp\left[\frac{2\varepsilon^{-1}}{Q}\left(\gamma\,\frac{r^2}{2} - \beta_1\,\frac{r^4}{4}\right)\right] \tag{2.34}$$

where Q is an "effective" noise amplitude, whose value is related in a complicated manner to the original noise amplitudes, Eq. (2.31b), evaluated at the unstable macroscopic stationary state. As can be easily verified, the "amplitude fluctuation" behaves as $\langle|\delta r|\rangle \approx O(\varepsilon^{1/4})$. On the other hand, for all finite times the phase variable, θ, follows its macroscopic path, $\theta = \overline{\theta}(t)$, with

$$\overline{\theta}(t) = (\omega - \beta_2\gamma/\beta_1)t \tag{2.35}$$

In the long time limit $t \to \infty$, however, taken prior to the limit $\varepsilon \to 0$, it approaches a uniform distribution

$$\lim_{t \to \infty} P_{st}(\theta|r) = \frac{1}{2\pi} \tag{2.36}$$

where $P_{st}(\theta|r)$ denotes the conditional probability of θ, for a given value of r. The details of the above relaxation mechanism are not completely understood at the present time.

2. The limit $\varepsilon \to 0$ is taken prior to the limit $\beta \to \beta_c$. In this case, the radial variable is Gaussian around its stationary value. As to the phase fluctuation, $\delta\theta = \theta - \overline{\theta}(t)$, in the long time limit it behaves as ($\beta - \beta_c \neq 0$)

$$\langle\delta\theta^2\rangle \approx h\varepsilon t \tag{2.37a}$$

with

$$h = \frac{\beta_1 Q}{\beta - \beta_c} \tag{2.37b}$$

This behavior of the phase fluctuation is usually known as "phase diffusion" [105, 113]. Moreover, in the long time limit, the time correlation function of the order parameter behaves, as [114a]

$$\langle z(t)z^*(0)\rangle \approx e^{i\omega t}e^{-h\varepsilon t} \tag{2.38}$$

This in turn implies that the time correlation function of the original variables (u, v) oscillates in time, with the same period as the limit cycle, but with an amplitude that decreases exponentially with a relaxation time inversely proportional to the extensivity parameter (system size). In particular, the macroscopic behavior is recovered in the limit $\varepsilon \to 0$. Note that in the limit $t \to \infty$, with $\varepsilon \neq 0$, the phase variable becomes uniformly distributed, as in case (1) [cf. Eq. (2.36)]. We will have the opportunity to check the above results in Section III, where we will discuss the microscopic simulation of chemical reactions.

Note finally that there exist a few studies concerning the stochastic analysis of the limit cycle far from bifurcation where it is assumed that the distribution transverse to the limit cycle is Gaussian (equivalent to case b) [105, 114b]. In this case, a suitable change of variables to a moving frame traveling along the macroscopic limit cycle allows us to analyze the behavior of fluctuations.

B. Reaction-Diffusion Master Equation

The stochastic approach presented in the previous sections is particularly well suited in describing the statistical properties of well-stirred chemical systems. The main advantage of such a global description lies in its simplicity, which permits detailed analytical investigations. However, its applicability to non-stirred media remains questionable, even if one limits oneself to macroscopically homogeneous systems. In fact, the global master equation selects the very limited class of exceptionally large fluctuations that appear at the level of the entire system, disregarding important nonequilibrium features originated by local fluctuations. For instance, we know that chemical systems can exhibit coherent behavior that manifests itself in the form of regular spatial patterns or of rhythmic phenomena emerging abruptly in a hitherto homogeneous and stationary system. Both phenomena imply sharp and reproducible correlations between distant parts of the system. How can this ever happen in a dilute mixture of molecules behaving almost like hard spheres and hence incapable of recognizing each other over more than a few angströms? An answer to this question cannot even be formulated without introducing the concept of localized fluctuations, which break spatial homogeneity and introduce in the problem an intrinsic size parameter much smaller than the size of the entire system V.

The master equation formulation of reaction-diffusion systems was first proposed by Kitahara [115]. Its basic lines can be summarized as follows [1, 116–118]: We subdivide the reaction volume into spatial cells $\{\Delta V_r\}$ and consider as variables the numbers of particles $\{U_{ir}\}$ in these cells. We assume, as before, that the set of variables $\{U_{ir}\}$ defines a Markov process. In addition to the intuitive arguments given in the very beginning of Section

II.A, the Markov assumption is motivated by the local character of the equation of evolution, Eq. (1.4), which allows one to describe the instantaneous state of the system by the same variables as at equilibrium (local equilibrium assumption). The random variables $\{U_{ir}\}$ change as a result of two processes: chemical reactions, which will be modeled as before by a jump Markov process; and diffusion, whereby a particle may jump to an adjacent cell; the latter will be assimilated to a random walk. The resulting probability distribution $P(\{U_{ir}\};t)$ obeys the so-called *reaction-diffusion master equation*

$$\frac{d}{dt} P(\{U_{ir}\};t) = \sum_{r,\{U'_{ir}\}} W(\{U'_{ir}\}|\{U_{ir}\})P(\{U'_{ir}\};t)$$
$$+\sum_{i} \frac{\tilde{D}_i}{2d} \sum_{r,l} \{(U_{ir}+1)P(\ldots,U_{ir}+1,U_{ir+l}-1,\ldots;t)$$
$$- U_{ir}P(\{U_{ir}\};t)\} \tag{2.39}$$

The sum l runs over the first nearest neighbors of the cell r, and $\tilde{D}_i$ represents *the mean jump frequency* of species i. It is related to Fick's diffusion coefficient D_i of the species by

$$D_i = \frac{l^2}{2d}\tilde{D}_i \tag{2.40a}$$

where d represents the space dimension and l is the characteristic length of a cell

$$\Delta V = l^d \tag{2.40b}$$

Note that again the transition probabilities are extensive quantities proportional to the volume ΔV of the cells.

Before discussing the general properties of the master equation, Eq. (2.39), it is appropriate to review the conditions under which it is expected to correctly describe realistic reaction-diffusion systems. Beside the Markovianity hypothesis, the stochastic theory of reactive fluids relies on the fundamental assumption that the state of the system can be completely specified in terms of a limited number of macroscopic variables. For isothermal systems, these are just the composition variables. The lumping of all microscopic degrees of freedom except the composition variables can only be justified in systems remaining permanently in a local thermal equilibrium state, which in turn requires a "large" number of molecules per cell. Detailed numerical studies show that a few hundreds of molecules are enough in most practical situations. The local equilibrium assumption is also a necessary condition that allows one to approximate the extremely complex motion of molecules by a simple random walk. This is a reasonable approximation, provided the linear dimensions of a cell remains *at least* of the order of the mean free path, since otherwise the microscopic characteristics of individual molecules, such as their velocity distribution function, must also be incorporated into the theory.[5] The cell size, however, cannot be chosen arbitrarily large, even for macroscopically homogeneous systems. In fact, the reaction-diffusion master equation considers each cell as a perfectly coherent entity, which in turn implies that the linear dimensions of a cell must be smaller than the correlation length. Now, the correlation length is at least equal to the reactive mean free path, defined as the average distance traveled by a particle before it undergoes a reactive collision. We therefore arrive at the conclusion that the linear dimensions of a cell should be typically of the order of the reactive mean free path. We will have the opportunity to check the above intuitive arguments in Section III.E, which is devoted to microscopic simulations of isothermal chemical systems.

1. The Langevin Limit

The analysis of the reaction-diffusion master equation Eq. (2.39), proves to be much more involved than the global master equation, Eq. (2.5). Beside the large number of variables, another technical difficulty arises from the lack of an obvious perturbation parameter. Indeed, as we already mentioned, in the absence of any further information as to the range of spatial correlations, the linear dimensions of ΔV must remain of the order of the reactive mean free path. This could hardly be regarded as a large parameter playing a role analogous to that of the volume V in the global master equation, Eq. (2.5). More importantly, as can be seen from Eqs. (2.40), the mean jump frequency $\tilde{D}$ depends on ΔV in a nontrivial way, indicating that any "regular" expansion of Eq. (2.39) in the inverse power of ΔV is bound to fail. Even a singular perturbation expansion of Eq. (2.39) would not be correct if it does not relate, one way or another, the linear dimension of ΔV to the distance from a bifurcation point where long-range correlations are expected [120]. In fact, expanding Eq. (2.39) blindly in the inverse power of ΔV, although mathematically correct, may lead to unphysical results contradicting the macroscopic reaction-diffusion equations [121]. The question thus arises as to the relation of the reaction-diffusion master equation with

[5] Such a "phase space" formulation of the master equation has been proposed by G. Nicolis and I. Prigogine, ref. [119].

the more phenomenological Langevin-type formulation of reaction-diffusion systems, based on the Landau–Lifshitz theory of fluctuating hydrodynamics (see also ref. [122]).

The macroscopic theory of fluids rests mainly on the local equilibrium assumption. Each spatial "point" is associated to a volume element which, although much smaller than the volume of the system, contains nevertheless enough particles to maintain its state close to thermal equilibrium. As discussed already in the beginning of Section II, the reaction-diffusion master equation is based on the same considerations. In particular, for realistic systems, the total number of particles within a cell is large enough so that the change of the corresponding composition variables can indeed be approximated as a continuous process. For example, in a gas phase reaction taking place under standard atmospheric conditions, the average number of particles per cubic reactive mean free path is at least of the order of 10^6, even for extremely fast reactions. We can therefore use the average total number of particles per cell, $\Delta\mathcal{N}$, as the extensivity parameter, instead of ΔV, and expand the reaction-diffusion master equation in a similar manner as in Section II.A. The reaction-diffusion master equation then reduces to a Fokker–Planck equation whose sample paths obey the following set of Langevin equations [29]

$$\frac{du_r}{dt} = f(u_r) + \frac{\tilde{D}}{2d} \sum_l (u_{r+l} - u_r) + \varepsilon^{1/2} F_r(t) \tag{2.41}$$

$$\varepsilon \equiv 1/\Delta\mathcal{N} \tag{2.42}$$

where $\{u_r\} = \{U_r/\Delta\mathcal{N}\}$ are the local mole fractions, f stands for the macroscopic rate function, and $F_r(t)$ is a "multi-Gaussian white noise," with

$$\langle F_r(t) F_{r'}(t')\rangle = \delta(t-t')\left\{ Q(\bar{u}_r)\delta^{Kr}_{r,r'} + \frac{\tilde{D}}{2d}\sum_l [(\bar{u}_{r+l} - \bar{u}_r)\delta^{kr}_{r,r'} - (\delta^{kr}_{r+l,r'} - \delta^{kr}_{r,r'})(\bar{u}_r + \bar{u}_{r'})] \right\} \tag{2.43}$$

As before, the sum l runs over the first nearest neighbors of the cell r. We note that the strength of both chemical noise Q and diffusional noise are evaluated along the deterministic path $\bar{u}(r,t)$. Remarkably, the expression of the noise correlations in Eq. (2.43) is the discretized version of the well-known results based on Landau–Lifshitz formalism

$$\langle F(r,t)F(r',t')\rangle = \delta(t-t')\{\delta(r-r')Q(\overline{u}(r,t)) + 2D\nabla\cdot\nabla'\overline{u}(r,t)\delta(r-r')\} \tag{2.44a}$$

$$\varepsilon^{1/2}F_r(t) = F(r,t) \tag{2.44b}$$

which can be viewed as an extension of the fluctuation-dissipation theorem to far-from-equilibrium systems [54].

The above results establish the validity of the nonlinear Langevin equation with process-independent noise for finite times $t < \tau^*(\varepsilon)$. Its validity for $t \to \infty$ is only guaranteed if the deterministic equation has a unique stable stationary solution. This is true before and up to the bifurcation point. But even in this case, the analysis of Eq. (2.43) remains extremely complex, mainly because the noise contribution arising from diffusion gives rise to nondiagonal elements (i.e. $\partial^2/\partial u_r \partial u_{r'}$, $r \neq r'$) in the associated Fokker–Planck equation. There are, however, two exceptions: first, far before the bifurcation point where the macroscopic rate law can be linearized [117, 118, 123]; next, close to a pitchfork bifurcation point, where it can be shown that the diffusional noise term becomes negligible as one moves to the bifurcation point [29, 124]. We shall examine these cases separately.

2. *Onset of Spatial Correlations*

One of the most important results of the formalism discussed in Section II.A was that chemical kinetics, if it acted alone, would generate non-Poissonian fluctuations away from thermodynamic equilibrium. One can easily show that such deviations arise even in the linear range of irreversible processes and are generated by terms proportional to the mass flow traversing the system [56]. In contrast to this, if diffusion acted alone it would lead to Poissonian fluctuations within each cell. Indeed, in a system subject to zero flux or to periodic boundary conditions, diffusion conserves the total number of each chemical species $\{U_i\} = \sum_r \{U_{ir}\}$. Thus, any function of these global variables $P_G(\{U_i\})$ is a stationary eigenfunction of the diffusion operator. Moreover, as diffusion simply amounts to a rearrangement of the particles over N_c cells into which the system is divided, we can write the most general steady-state solution arising from diffusion in the form [125]

$$P(U_r) = \frac{U!}{\prod_r U_r!} N_c^{-U} P_G(U) \qquad U = \sum_r U_r \tag{2.45}$$

where, to avoid cumbersome notations, we have limited ourselves to a single variable system. The contraction of this multinomial distribution into one cell, or into a number n of cells such that $n \ll N_c$, leads in the thermodynamic limit ($N_c \to \infty$) to a (multi-) Poissonian distribution in which spatial correlations are nonexistent.

What happens when these two conflicting tendencies coexist, as in the reaction-diffusion master equation, Eq. (2.39)? Intuitively, we expect that diffusion will propagate in space the non-Poissonian fluctuations generated locally by the chemical kinetics, thus giving rise to spatial correlations. In simple situations, the range of the latter will be the average length over which particles can diffuse without undergoing reactive transformations. This is substantiated by detailed calculations on exactly solvable models [126, 127].

As an example, consider the following reactions taking place in an ideal mixture within a large closed vessel

$$\begin{aligned} S + A &\underset{k_{-1}}{\overset{k_1}{\rightleftharpoons}} 2U \\ S + U &\underset{k_{-2}}{\overset{k_2}{\rightleftharpoons}} B + S \end{aligned} \tag{2.46}$$

where S stands for solvent molecules. After some time the concentrations of A, U, and B reach their equilibrium value. There is then no net flux from A to B and every reaction is balanced by its reverse. This property of microreversibility implies the property of detailed balance for the reaction-diffusion master equation, Eq. (2.39), which can then be solved exactly. One finds Poissonian fluctuations and an absence of spatial correlations.

Obviously, the property of microreversibility is lost when the system is subject to nonequilibrium constraints implying matter or energy transport [128]. For instance, if we supply A in large excess and eliminate B continuously, the reaction scheme (2.46) is effectively replaced by

$$\begin{aligned} S + A &\xrightarrow{k_1} 2U \\ S + U &\xrightarrow{k_2} B + S \end{aligned} \tag{2.47}$$

where U is now an intermediate product in the flux of matter from A to B. The reaction-diffusion master equation can again be solved exactly. In the continuum limit of a three-dimensional system, the static correlation function takes a form similar to the well known Ornstein-Zernike result of equilibrium critical phenomena [129]

$$\langle \delta u(r) \delta u(r') \rangle - \langle u_{\mathrm{st}} \rangle \delta(r - r') = \frac{\Phi_{AB}}{|r - r'|} \exp\left(-\frac{|r - r'|}{l_c} \right) \tag{2.48}$$

where $\langle u_{st} \rangle = 2k_1 a/k_2$ is the stationary average concentration of U, Φ_{AB} is proportional to the macroscopic net flux from A to B, and l_c represents the correlation length

$$l_c = \sqrt{\frac{D}{k_2}} \tag{2.49}$$

For typical values of k_2 and the Fick diffusion coefficient D, l_c will be larger than the range of intermolecular interactions by several orders of magnitude. If we let the system approach equilibrium, the amplitude of the correlation function vanishes, being proportional to the flux of matter imposed on the system (Φ_{AB} in Eq. (2.48)). In contrast, the correlation length remains macroscopic and is determined entirely by the system's intrinsic parameters. The existence of an intrinsically determined correlation length leads to the conclusion that the properties of fluctuations *are not universal* but depend on their scale. For short wavelength fluctuations, the dynamical interplay between reaction and diffusion hardly shows up, and one essentially finds a multi-Poissonian distribution (modified, perhaps, by the effect of intermolecular forces); it is only when the wavelength of the fluctuations becomes comparable to l_c that typical nonequilibrium features, such as non-Poissonian fluctuations, can be observed.

The appearance of such a spatial correlation can be explained qualitatively from the following intuitive argument. When, by fluctuation, two U particles are created through the reaction $S + A \rightarrow 2U$, they are obviously correlated. This correlation will spread in space, as they diffuse, until at least one of the particles becomes inactivated through the second step, $U \rightarrow B$. Now, the average distance L through which a particle U diffuses during a time t, is $L \approx \sqrt{Dt}$. Since the average lifetime of a U particle is $1/k_2$, the effective correlation length is then the average distance traveled by the particle through diffusion, before it recollides reactively, i.e., $\sqrt{D/k_2}$. Clearly, this time and hence the correlation length increase as $k_2 \rightarrow 0$. For small k_2 however, the concentration of U will increase and the inverse nonlinear reaction $2U \rightarrow A + S$ will no longer be negligible. As a consequence, nonlinear modes come into play and the problem becomes more involved.

For a general reaction scheme the situation remains very similar. Away from the bifurcation point, the behavior of the correlation function is typically described by an exponential decay such as Eq. (2.48). The correlation length takes also the same form as Eq. (2.49), but now the coefficient k_2 is replaced by the slowest mode describing the decay of fluctuations around the reference state. To be more specific, let us consider once more the Schlögl model (2.10). For $\lambda = \lambda' \gg 0$, i.e. away from the bifurcation point, the evolution equation can be linearized around $u = 1$ (cf. Section II.A.1). The corresponding reaction-diffusion master equation can then be solved, leading to a correlation function similar to Eq. (2.48) with a correlation length

$$l_c = \sqrt{\frac{D}{\lambda}} \tag{2.50}$$

In the vicinity of the bifurcation point ($\lambda \to 0$) the correlation length diverges as $\lambda^{-1/2}$, indicating the possibility of a nonequilibrium transition. We call this a *classical law* of divergence. At this point, however, one is no more allowed to neglect the nonlinear modes, which can modify the law of divergence of the correlation length, or even simply suppress the transition altogether.

3. *Spatial Correlations Near a Pitchfork Bifurcation*

We would now like to see in a more general way how the competition between nonlinear reactions and diffusion affects the properties of spatial correlations in the vicinity of a pitchfork bifurcation point. Before we address the quantitative aspects of this question, it will be instructive to present some general arguments, based mainly on dimensional analysis.

Let V be a finite volume within which both reactions and diffusion are taking place. A systematic perturbation scheme was set up, based on the smallness of the ratio $\tau_D/\tau_{\mathrm{ch}}$, where τ_D and τ_{ch} are the characteristic times for diffusion and reaction, respectively [130]. It was proved that in the limit of this ratio going to zero, the probability distribution assumes a form similar to Eq. (2.45), referred to as *diffusional equilibrium*. The function $P_G(U)$, appearing in Eq. (2.45), turns out to be the probability distribution for the total number of particles obeying the global master equation, Eq. (2.1). The above limit, expressing the predominance of diffusion over chemical reaction, leads thus to *mean field* results. Clearly this conclusion cannot be extended to the thermodynamic limit $V \to \infty$ (with fixed cell size ΔV), since the characteristic diffusion time increases for inhomogeneities of increasing wavelength. In fact, it is easy to verify [130] that

$$\frac{1}{\tau_D} \approx D V^{-2/d} \tag{2.51}$$

where d is the dimensionality of the system. On the other hand, away from the bifurcation point ($\lambda = \lambda' \gg 0$, for the Schlögl model), the characteristic chemical time is clearly independent of the volume ($\tau_{\mathrm{ch}} \approx 1/\lambda$). For some finite volume of the system, the characteristic diffusion and chemical time will therefore be of the same order and, as a result, a solution of the form (2.45) will cease to be valid.

As one moves close to the bifurcation point, however, the chemical process exhibits critical slowing down. In fact, in the previous section we established that for the Schlögl model the inverse chemical relaxation time is given by [cf. Eq. (2.25)]

$$\frac{1}{\tau_{\mathrm{ch}}} \approx O(V^{-1/2}) \tag{2.52}$$

Now, from Eq. (2.51) it is clear that the rate of increase of the characteristic diffusion time τ_D for $V \to \infty$ also slows down as the dimensionality becomes larger. Comparing Eqs. (2.51) and (2.52), we observe that there exists a critical dimensionality, $d_c = 4$, for which the characteristic reaction and diffusion times are of the same order in V. For dimensionalities $d > d_c$, diffusion will be dominant and the probability distribution will relax to the diffusional equilibrium form, Eq. (2.45). Hence, mean field behavior will result. For $d < d_c$, diffusion is not strong enough to correlate the cells perfectly. We expect, therefore, the appearance of spatial correlations showing "nonclassical" behavior, or even the suppression of bifurcation altogether.

The above arguments clearly suggest that for high enough dimensionalities the competition between diffusion and chemical kinetics should result in *macroscopically long-range* spatial correlations. We expect therefore that the properties of the system would still be represented reasonably well by a division of space into large cells, as long as their size does not exceed the range of correlations. This will allow us to use the inverse of the cell volume as a perturbation parameter. Using this *ansatz*, a singular perturbation scheme has been set up to solve the reaction-diffusion master equation for the Schlögl model. One of the main conclusions of this analysis was that the diffusional noise term, appearing in the Langevin equation, Eq. (2.41), becomes negligible as one moves to the bifurcation point. The associated Fokker–Planck equation can then be solved at the stationary state to give

$$P_s(\{u_r\}) \approx \exp\left\{-\frac{2\varepsilon^{-1}}{Q(\bar{u}_s)} \sum_r \left[-\int du_r\, f(u_r) + \frac{\tilde{D}}{8d} \sum_l (u_{r+l} - u_r)^2\right]\right\} \tag{2.53}$$

For the Schlögl model, with $\lambda = \lambda'$, Q is given by Eq. (2.23b) and

$$-\int du_r\, f(u_r) = \lambda\, \frac{(u_r - 1)^2}{2} + \frac{(u_r - 1)^4}{4} \tag{2.54}$$

These results establish the link between the master equation approach and a well-known method developed in the context of equilibrium critical phenomena and widely used since then in analyzing nonequilibrium transitions as well. Indeed, Eq. (2.53) features the exponential of the discretized *Landau–Ginzburg functional*, which can be studied by renormalization group methods [131, 114]. As is well known, the result of this analysis is the existence of a critical dimensionality d_c, in agreement with our scaling arguments, and the concomitant appearance of nonclassical exponents describing the divergence of variances and correlation lengths.

Despite the successful application of this approach to chemical systems, some fundamental questions are left open [132]. Indeed, the microscopic mechanisms responsible for transitions in equilibrium and nonequilibrium systems are fundamentally different. An equilibrium phase transition is the result of a competition between the intermolecular interaction forces, which tend to order the system, and the thermal agitation, which has the opposite effect. In a nonequilibrium transition, however, the essential ingredient is the applied external constraint, which maintains the system in far from equilibrium conditions. The main consequence of this difference is that, beyond the bifurcation point, results based on Langevin–Landau–Ginzburg formalism totally contradict those based on the master equation (cf. Section II.A.1). As has been shown recently, this discrepancy is even more striking in problems involving time symmetry breaking (limit cycle) in the presence of local fluctuations [133].

Equation (2.53) allows one to extract some further properties of reaction-diffusion systems by very simple arguments, such as the law of divergence of the correlation length in the limit $\lambda \to 0$. We have already stressed that the validity of Eq. (2.53) is based on the fact that the cell size can be taken as large as the correlation length l_c

$$\Delta V = l_c^d \approx O(\varepsilon^{-1}) \tag{2.55}$$

As one moves through the bifurcation point, $\lambda \to 0$, the correlation length diverges according to a certain law of the form

$$l_c \approx \lambda^{-\nu} \tag{2.56}$$

On the other hand, relation (2.25) shows that the critical region is characterized by values of λ, such that

$$\lambda = \frac{1}{\tau_{\text{ch}}} \approx O(\varepsilon^{1/2}) \tag{2.57}$$

In fact, one can easily check that for $\lambda \approx O(\varepsilon^{1/2})$ the quartic and the quadratic terms in Eq. (2.53) are of the same order in ε. Combining the above three relations, one gets:

$$\nu d = 2 \tag{2.58}$$

This gives the exact Ising exponent ν in two and four dimensions and a rather good approximation in three dimensions (0.67 instead of 0.64). Given the extreme simplicity of the analysis, the result is quite satisfactory.

4. *Critical Fluctuations in One-Dimensional Systems*

The results obtained in the previous section are based on the hypothesis that the correlation length diverges as $\lambda \to 0$. This is strongly supported by the analogy with equilibrium critical phenomena, at least for $d \geq 2$. The question thus arises as to the behavior of the correlation function and local fluctuations in one-dimensional systems. This question is of special interest, since most of the results of computer experiments available to date refer to this case. There exist two different ways to address the problem, depending on whether or not we accept the validity of the Landau–Ginzburg type of behavior of the probability distribution, Eq. (2.53), in one dimension.

Let us first assume that (Eq. 2.53) remains valid for $d = 1$. It is then easy to check that the probability distribution (2.53) possesses the property of *spatial Markovianity*, in the sense that the conditional probability $P_{1/M}$ of having u_k particles in a cell "k" given that there are $u_{k+1}, u_{k+2}, \ldots, u_{k+M}$ particles in cells "$k + 1$", ... , "$k + M$", respectively, depends only on the value of u that is specified in the nearest neighbor cell: $P_{1/M}(u_k|u_{k+1}, \ldots, u_{k+M}) = P_{1/1}(u_k|u_{k+1}), \forall M > 1$. Here, by definition $P_{1/M} = P_{M+1}/P_M$, in which P_M denotes the reduced M-cell probability distribution. This property also implies that $P_{1/1}$ obeys the *Chapman–Kolmogorov* equation [134] and thus all the information on the statistical properties of the system is contained in the one- and two-cell probabilities $P_1(u_k)$ and $P_2(u_k, u_{k+1})$. The "many-body" problem in Eq. (2.53) is thus reduced, at least formally, to a "two-body" problem. Note that this simplification holds only for a one-dimensional infinite system, in which case the system possesses translational and permutational invariance, i.e., $P_1(u_m = u) = P_1(u_n = u)$ and $P_{1/1}(u_m = u|u_n = w) = P_{1/1}(u_n = u|u_m = w), \forall n, m$.

Using steepest descent and a rather involved singular perturbation expansion, it can be shown that the local (one cell) probability distribution for the Schlögl model, along the line $\lambda = \lambda' \to 0$, takes the following form [135]

$$P_1(\delta u) \approx \exp\left\{-\frac{2}{3\varepsilon}\sqrt{\tilde{D}}\,[2\lambda + (\delta u)^2]^{3/2}\right\} \tag{2.59}$$

where we have set $\delta u = u - \bar{u}_s = u - 1$. In particular, at the critical point $\lambda = 0$, one finds

$$P_1(\delta u) \approx \exp\left(-\frac{2}{3\varepsilon}\sqrt{\tilde{D}}\,|\delta u|^3\right) \tag{2.60}$$

which is a rather unusual distribution. Using this result and the spatial Markovianity, the spatial correlation function can also be computed explicitly. At the critical point, one finds

$$\frac{\langle \delta u(k)\delta u(k+r)\rangle}{\langle(\delta u)^2\rangle} = \frac{2\sqrt{\tilde{D}}}{\varepsilon}\int_0^\infty dx\,\frac{x^2}{1+r|x|\tilde{D}^{-1/2}}\exp\left(-\frac{2}{3\varepsilon}\sqrt{\tilde{D}}\,|x|^3\right) \tag{2.61}$$

The system is thus characterized by long-range spatial correlations decaying as $1/r$. This corresponds to a marginal situation, as far as phase transitions are concerned, since the integral of the correlation function has a weak, logarithmic divergence.

It is important to realize that the above results hold only in the limit $\varepsilon \to 0$, and are therefore extremely difficult to check through direct particle simulations. Moreover, it comes as a surprise that the reaction-diffusion master equation does not share the property of spatial Markovianity, even in one dimension: the property of spatial Markovianity holds only near the critical point and in the limit $\varepsilon \to 0$, where the diffusional noise becomes negligible.

An alternative theory, that holds for any values of ε, was developed some years ago. Although based on rather heuristic arguments, it has the merit of being easily checked through microscopic simulations. For the sake of clarity, we consider a chemical system involving a single composition variable changing through elementary jumps of ± 1, such as the Schlögl model (2.10). Summing the corresponding reaction-diffusion master equation over all cell variables, except one, say U_α, one finds at the steady state the following recurrence relation

$$P(U_\alpha + 1) = P(U_\alpha)\,\frac{\upsilon(U_\alpha) + \tilde{D}(E(\xi|U_\alpha))}{\mu(U_\alpha + 1) + \tilde{D}(U_\alpha + 1)} \tag{2.62}$$

where υ and μ are birth and death rates, respectively, and $E(\xi|U_\alpha)$ represents the "conditional average" of nearest neighbors cells to α

$$E(\xi|U_\alpha) = \sum_{U_{\alpha\pm1}} U_{\alpha\pm1}P_{1/1}(U_{\alpha\pm1}|U_\alpha) \tag{2.63}$$

As before, the function $P_{1/1}(U_{\alpha\pm1}|U_\alpha)$ is the conditional probability of having $U_{\alpha\pm1}$ particles in the cell $\alpha \pm 1$, given that there are U_α particles in the cell α. Following the discussions of Section II.B.2, we expect that the distribution of local fluctuations is close to a Poissonian distribution and that spatial correlations between adjacent cells are thus very small. As a first approximation, one may be tempted to neglect the cell–cell correlation altogether. The conditional average becomes then independent of U_α, $E(\xi|U_\alpha) = \langle U \rangle$, and the reaction-diffusion master equation reduces to a *nonlinear master equation* for the one-cell probability $P_1(U_\alpha)$ [136]. Such a mean field approximation is reasonable, provided the cell size is chosen to be of the order of the correlation length [137]. It has been used in the past to investigate the possibility of a "nucleation" type of mechanism for the onset of bifurcation in nonequilibrium systems [138].

The above approximation of the reaction-diffusion master equation is not totally satisfactory, for it ignores completely the spatial correlations between cells. As we have shown in Section II.B.2, even well before bifurcation, spatial correlations exist and have a macroscopic range. A better approximation consists of assuming a linear dependence of $E(\xi|U_\alpha)$ on the random variable U_α, i.e., $E(\xi|U_\alpha) = c_1 + c_2 U_\alpha$. Using Eq. (2.63), the constants c_1 and c_2 can easily be expressed in terms of the first and the second moment of the probability distribution

$$E(\xi|U_\alpha) = \langle U_\alpha \rangle + \frac{\langle \delta U_\alpha \delta U_{\alpha+1} \rangle}{\langle (\delta U_\alpha)^2 \rangle} (U_\alpha - \langle U_\alpha \rangle) \tag{2.64}$$

Thanks to this assumption, the "many-body" problem described by the reaction-diffusion master equation, Eq. (2.39), is reduced to a "one-body" problem, Eq. (2.62). The explicit form of the solution for the correlation functions is, however, not easy to obtain, mainly because the recurrence relation, Eq. (2.62), is nonlinear (see ref. [139] for details).

We note that the relation in Eq. (2.64) is exact for a multi-Gaussian distribution. It is therefore a consistent approximation whenever the master equation can be linearized around a macroscopic state. The question thus arises as to the validity of the approximation (2.64) as one moves into the vicinity of a bifurcation point, where the Gaussian behavior ceases to be valid. Unfortunately, this question cannot be answered on the basis of a theoretical analysis, since Eq. (2.64) is based mainly on heuristic arguments and not on a systematic derivation. We are therefore forced to rely on numerical analysis.

Extensive analyses have been carried out in the past, both in one- and two-dimensional cases, and a quantitative agreement has been obtained [140, 141]. Figure 4 depicts the results of such a numerical simulation of the full

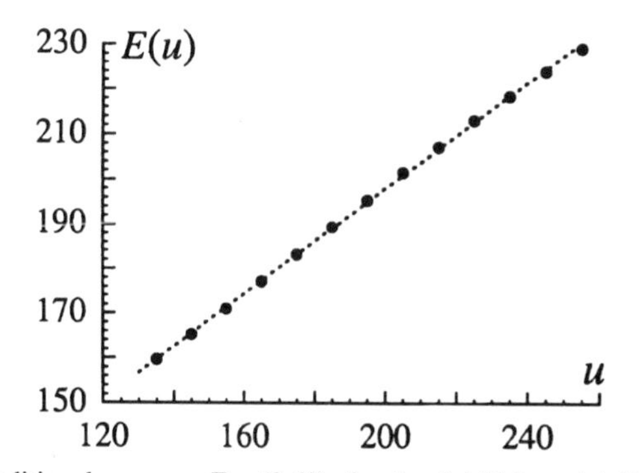

Figure 4. Conditional average, Eq. (2.63), for the Schlögl model (2.10) obtained from a numerical simulation of the master equation (solid circles). The dashed line is a linear fit, with correlation coefficient = 0.9998. Parameters are $\lambda = \lambda' = 0$, $\Omega = 200$, and $\tilde{D} = 1$.

reaction-diffusion master equation for the Schlögl model (2.10), right at the bifurcation point $\lambda = \lambda' = 0$. The system is made up of 101 cells, with periodic boundary conditions, mean jump frequency $\tilde{D} = 1$, and the average total number of particles per cell of 200. Such a small number of particles clearly precludes any asymptotic type of behavior, so that none of the approximation methods presented so far could possibly be applicable. Surprisingly, the numerical results are in quite good agreement with the assumption of Eq. (2.64). Not only is the "measured" conditional average linear, but its slope matches the theoretical predictions to within 5%. This is illustrated further in Fig. 5 where the spatial correlations obtained both theoretically and numerically are compared. To our knowledge there exists so far no theoretical explanation for such a quantitative agreement.

Note that the approximation in Eq. (2.64) is inconsistent with the property of spatial Markovianity. Given the above numerical agreement with the master equation, one can legitimately question the very validity of the Landau–Ginzburg functional, Eq. (2.53) in one-dimensional systems. Similar discrepancy between the reaction-diffusion master equation and Langevin–Landau–Ginzburg approach have been noticed recently in the context of a Hopf bifurcation [133].

Another interesting problem arises whenever a system is prepared initially on an unstable, or marginally stable state, over the entire reaction space, except for a small region whose state is set close to the stable stationary state. After a short transient time, the system develops a wave front that propagates into the unstable region. Macroscopic analysis predicts that such a front, propagating with a constant velocity and a given width, can be stable [142–145]. The question thus arises as to the influence of statistical fluctuations on the stability of the front, and possible deviations from macroscopic

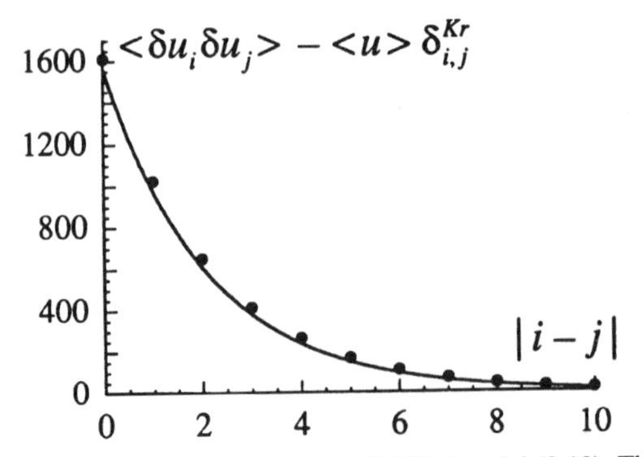

Figure 5. Spatial correlation function for the Schlögl model (2.10). The curve represents the theoretical result based on assumption of Eq. (2.64), whereas the solid circles are from the numerical simulation of the master equation. The estimated statistical errors are of the order of 1% (about the size of the circles). Parameter values as in Fig. 4.

predictions concerning its width and velocity. This problem has been investigated recently by Lemarchand et al., using a variety of different numerical approaches, including microscopic simulations [146, 147]. It has been found that the macroscopic predictions concerning the stability and the shape of the front remain basically correct, although in some limiting cases quantitative deviations up to 25% have been observed [148].

III. MICROSCOPIC SIMULATIONS

As stated in the introduction, molecular dynamics (MD) consists in solving numerically Newton's equations of motion for an assembly of particles interacting through a given potential, and can thus be considered as an "ideal experiment" in bridging the gap between the phenomenological analysis of large-scale macroscopic phenomena and their modeling at the microscopic level. We use the word "ideal experiment" to stress the important fact that in no way can MD replace the real world experiment. No matter how cautious one may be, MD remains a numerical technique that is limited by the finite precision of digital computers, just like any other numerical technique. Is this a serious limitation, as far as the macroscopic behavior of the simulated fluid is concerned? Very likely not, though there does not exist a full mathematical proof for it.

There are other aspects of which one should fully be aware before considering MD as "experiment." For instance, a real fluid is always embedded in an environment that we usually refer to as "thermal reservoir" in the context of equilibrium statistical mechanics, or as "boundary conditions" in nonequilibrium systems. One way or another, we need to give a microscopic prescription as to how the molecules of the simulated fluid are interacting with their environment. Even if we limit ourselves to simulations of equilibrium systems with periodic boundary conditions (microcanonical ensemble), thus avoiding the tricky problem of fluid–boundary interaction, we still need to know the exact interaction laws between the molecules—and we certainly do not. At best, we can give a reasonable estimate of these interaction laws. Again, one can ask the question as to whether this is a relevant factor in a macroscopic context. In some cases the answer is yes and in some others, the answer is no.

It is a matter of observation that in many areas of physics the macroscopic behavior of the simulated fluid depends only very weakly on the details of microscopic interaction potentials. For example, the hydrodynamical behavior of a hard-sphere fluid is essentially the same as a Lennard–Jones fluid, provided one is not too close to an equilibrium phase transition region [149, 150]. In some other areas, such as problems dealing with the motion of macromolecules in an inert solvent, or simply the statistical properties of liquid water, a more complete knowledge of the interaction potentials is generally required (see, for example, ref. [151]). Here, the main difficulty comes from the fact that these interaction potentials are not reducible to pair potentials. Usually, they are evaluated separately through heavy quantum mechanical computations. These approaches are sometimes referred to as "*ab initio*" molecular dynamics [152].

To sum up, MD is indeed a valuable tool with which one can design an "ideal world"; what we measure in that way is nothing but the macroscopic properties of that "ideal world." Keeping this in mind, we shall now summarize the basic ideas underlying the molecular dynamic simulation of reactive fluids.

A. Microscopic Simulations of Reactive Fluids

The MD simulation of reactive fluids rests on the same basic principles as that of nonreactive fluids. How well should we know the details of a reaction mechanism? Is a full quantum mechanical evaluation required, or is a simple "colored" hard-sphere modeling good enough? Again the problem at hand dictates the choice of the microscopic reaction mechanism to be adopted. For example, if the main purpose is the study of statistical properties of complex reactive systems, such as those discussed in the previous section, then one should privilege methods that give the best possible computer performances so that one can get a relevant signal-to-noise ratio. If, on the contrary, one needs to study the detailed dynamics of a single chemical reaction, then a much more complete knowledge of the reactive mechanism is required [153]. One example, among others, is the microscopic simulation of a single exothermic chemical reaction that may occur under the influence of shock waves generated through the very strong collision of solid bodies [154].

In this article, we shall be mainly concerned with the MD simulation of reactive fluids capable of exhibiting nonequilibrium transitions. To do so, we have to face some basic difficulties that are directly related to the very nature of chemical dynamics. A first problem arises in connection with the validity of the macroscopic rate equations describing the time evolution of the composition variables in dilute (ideal) mixtures. This implies that one needs to have "enough" elastic collisions between consecutive reactive collisions in order to ensure mechanical and thermal equilibrium. As a consequence, only a fraction of the computing time will contribute effectively to the evolution of the system, which results in much wasted bookkeeping and a corresponding waste of CPU time.

A second problem is related to the fact that chemical timescales τ_c—for example, the period of a limit cycle in an oscillating system—are frequently in the macroscopic range. To get reliable statistics, one needs to run the corresponding microscopic simulation over an amount of time several times larger than τ_c. This again implies an extremely large amount of running time.

To cope with these difficulties, one is forced to simplify as much as possible both the Newtonian and the chemical dynamics. This can be done by limiting the simulation to hard-sphere or hard-disk dynamics and by considering dilute mixtures in the Boltzmann limit.

Let us consider a system made up of an assembly of N hard spheres confined in a box of volume V with given boundary conditions. The statistical properties of such a fluid, like for instance, its equation of state or the behavior of its pair correlation function and transport coefficients, depend crucially on the ratio between the molecular diameter d, and the mean molecular spacing $\delta = n^{-1/3}$, where $n = N/V$ is the (mean) number density. For example, $d/\delta \approx .6$ corresponds to a moderately dense Enskog gas. As the value of d/δ is decreased, the statistical properties of the system approach those predicted by the Boltzmann equation. Both the laboratory and MD experiments show that the validity of the Boltzmann equation is secured as long as $d/\delta < .14$ (see ref. [155] for more details). Note that $d/\delta \approx .11$ for air at standard atmospheric conditions.

Suppose now that we wish to conduct a MD experiment within the Boltzmann regime. A legitimate question is whether it is possible to set up a simple algorithm, specially designed for Boltzmann dynamics, instead of using the exact Newtonian dynamics. This question was answered positively by Bird [155] who proposed an efficient algorithm known as *direct simulation Monte Carlo* method (DSMC).

B. Bird Algorithm for Boltzmann Dynamics

Some 20 years ago, G. A. Bird proposed his now famous DSMC algorithm to simulate the Boltzmann equation. The original purpose of the method was to deal with problems where the use of the standard hydrodynamic descriptions becomes questionable, such as the computation of high Knudsen number flows of a rarefied gas past an object (e.g., high-altitude flight). Bird's method has become popular since it is in excellent agreement with experimental and numerical data. Its basic steps are summarized below [156].

As with the usual molecular dynamic methods, the state of the system is the set of particle positions and velocities, $\{\mathbf{r}_i, \mathbf{v}_i\}$, $i = 1, \ldots, N$, where N is the total number of particles. The evolution is decomposed in time steps Δt, typically a fraction of the mean collision time for a particle. Within a time step, the free flight motion and the particle interactions (collisions) are assumed to be decoupled. The free flight motion for each particle i is computed as $\mathbf{r}_i(t + \Delta t) = \mathbf{r}_i(t) + \mathbf{v}_i(t)\Delta t$, along with the appropriate boundary conditions. After all the particles have been moved, they are sorted into spatial cells, typically a fraction of a mean free path (m.f.p.) in length. Each cell is assumed to be perfectly homogeneous, i.e., all particles within it are considered to be candidate collision partners, regardless of their exact positions. This major hypothesis allows one to consider, during the time step Δt, a "homogeneous" Boltzmann equation within each cell.

A set of representative collisions, for the time step Δt, is chosen in each cell. For each selected pair, a random impact parameter is generated and the collision is performed. Note that linear momentum and energy are conserved in the evaluation of the collision, whereas the angular momentum is only conserved on average. After the collision process has been completed in all cells, the particles are moved according to their updated velocities and the procedure is repeated as before.

The theory underlying the simulation of the homogeneous Boltzmann equation was first proposed by Kac [157] who assumed that the collisional process can be described as a *jump Markov process*. With this hypothesis, the probability of the collision for a pair of hard-sphere particles is directly proportional to the modulus of their relative speed, and the waiting times between collisions are exponentially distributed [158]. As the computational cost of evaluating waiting times increases very rapidly with the number of particles, Bird proposed an ingenious alternative, which consists of calculating these waiting times by dividing an estimate of the average distance of colliding particles by the modulus of their relative speed. Note that very recently Bird proposed several modifications to DSMC that improve the performance and the flexibility of his original algorithm [159].

To summarize, the main hypothesis in Bird's algorithm is that the cells are assumed to be perfectly homogeneous, i.e., all particles within a cell are considered to be potential collision partners, regardless of their exact positions. This basic hypothesis simplifies considerably the dynamics and allows the algorithm to be up to three orders of magnitude faster than the corresponding exact hard-sphere dynamics. On the other hand, it also raises questions as to the reliability of the algorithm.

From a macroscopic point of view, extensive use of DSMC by Bird and others, in a variety of problems dealing with nonequilibrium gas dynamics, has always shown perfect agreement with experimental data [155, 160, 161]. For example, it correctly yields the density profile of a relatively high Mach number (> 2) shock wave [162]. It also reproduces correctly the data obtained through hard-sphere molecular dynamics in extremely strong shock wave conditions (Mach number > 100), a domain far beyond the validity of Navier–Stokes equations [163].

At the microscopic level, DSMC have also been used to study the behavior of fluctuation spectra in dilute gases subjected to strong nonequilibrium constraints, both for systems under temperature gradient [164, 165] and velocity gradient [166] (shear). The results were shown to be in very good agreement with those obtained by the Landau–Lifshitz fluctuating hydrodynamics [78], whose validity is now well established [129, 167].[6] We can therefore conclude that the Bird algorithm reproduces perfectly both the macroscopic behavior and the fluctuation spectrum of dilute gases, even under severe nonequilibrium regimes.

At this stage one could ask whether it is not worth simplifying the microscopic dynamics even further and consider, for example, the lattice gas cellular automata (LGCA) or the lattice Boltzmann (LB) method, that allow the simulation of relatively high Reynolds number flows [169] and other instabilities [170] in a variety of systems, including the case of reactive fluids [22, 147, 171, 172]. The answer depends on the type of question one wants to clarify through microscopic simulations. As we already stressed, our main interest is in using microscopic simulations as an "experimental" tool to account for cases where high-precision laboratory experiments are difficult or impossible to perform.[7] To this end, we need a method whose validity goes beyond macroscopic formulations, and which is also able to reproduce the correct fluctuation spectra. This last requirement is crucial since one of our main objectives is to clarify the status of stochastic theory

[6]Some recent experimental results concerning systems under temperature gradient can be found in ref. [168].

[7]Measurement of fluctuation spectra in nonreactive fluid mixtures has recently been reported in ref. [173].

of chemical systems. Clearly, the Bird algorithm fully satisfies all the above requirements. On the other hand, so far it is not clear whether the LGCA or LB methods contain more information than the macroscopic equations, at least in nonequilibrium systems. It has only recently been shown that LGCA reproduces correctly the fluctuation spectrum in equilibrium systems [174–176]. (See ref. [177] for an up-to-date account of the field.)

C. Hard-Sphere Chemistry

As we already stressed, the best strategy in setting up a microscopic simulation of chemical systems is through dilute hard-sphere gas dynamics. The principle of hard sphere chemistry was first introduced in 1975 by Portnow [178] and, almost simultaneously, by Ortoleva and Yip [179]. It was subsequently developed further by Boissonade who applied this technique to the study of composition fluctuations [180] and nucleation type of phenomena [181] in isothermal nonequilibrium chemical systems.

We first define what we mean by "reactive hard-sphere collisions." We assign to each species a "color." A reactive collision occurs if the colliding particles have "enough" energy, i.e., if their relative kinetic energy exceeds some threshold related to the activation energy of the reaction. If this is the case, then the colors of the particles are changed, according to the chemical step under consideration. A condition to be met in such a simulation is that the frequency of reactive collisions must remain small compared to the frequency of elastic collisions, since otherwise one could get important nonequilibrium effects, even for isothermal chemical reactions.

To clarify this point, let us consider the following isothermal chemical reaction:

$$A + B \xrightarrow{k} C + D \tag{3.1}$$

where the reaction rate, k, obeys the Arrhenius law,

$$k \sim \exp\left(-\frac{E}{k_B T}\right) \tag{3.2}$$

and E and k_B represent the activation energy and the Boltzmann constant, respectively. Since reactive collisions involve mainly energetic particles, the chemical reaction will give rise to a transfer of translational energy from the reacting particles A and B to the outgoing particles C and D. If the ratio $E/k_B T$ is large enough, the rate of reactive collisions is much smaller than the elastic ones and the high-velocity tail of the Maxwell–Boltzmann distribution is replenished as fast as it is depleted by reactive encounters. There exists, however, a range of activation energy (or temperature) values for which the rate of reaction can become comparable to the rate of thermal equilibration. Consequently, one may expect a slowing down of the reaction speed as compared to the usual macroscopic rate law, based on local equilibrium hypothesis.

Pioneering calculations along these lines were made by Prigogine and Xhrouet [34], who solved the Boltzmann equation for the reaction scheme (3.1) to obtain a first-order correction to the corresponding reaction rate formula. Further work by Present [182] and by Ross and Mazur [183] has led to essentially the same results (for a more advanced theory, see refs. [184–186]). For realistic systems, however, the deviations from equilibrium theory were shown to remain within the uncertainties associated with collision diameters and steric factors. This limitation is overcome in computer experiments, where deviations from equilibrium rate theory have indeed been detected [187].

Consider now reactions in nonisothermal medium. A typical example is the exothermic gas-phase reaction

$$F + S \xrightarrow{k} B + S + heat \tag{3.3}$$

expressing the transformation of reactant (fuel) F to product B through the reactive collisions with "solvent" particles S. In addition to the above-mentioned effects, the heat released by the reaction will further increase the energy of the S particles, which may well recollide with another F particle before being completely thermalized. If the reaction is fast enough, this continuous flow of energy from F to S particles may cause a non-negligible perturbation of the equilibrium distribution, which in turn can modify significantly the phenomenological rate equations. This problem was analyzed in detail by Prigogine and Mahieu in 1950, using a Chapman–Enskog type of approach [35]. In addition to the deformation of the Maxwell distribution occurring already in the isothermal case, it was found that further deviations from the usual rate law are now scaled by the value of heat of reaction relative to the thermal energy. It is reasonable to expect that such effects can become important, particularly in explosive systems where the reaction is strongly accelerated.

To our knowledge there exists so far no experimental evidence for these effects. Indeed, combustion in exothermic chemical systems is often accompanied with the appearance of flame fronts or of hydrodynamic effects. Furthermore, the precise knowledge of the underlying mechanism is very difficult to establish. A deviation from a given macroscopic theory can thus always be attributed to the incomplete knowledge of the reaction mechanism or, perhaps, to the intervention of transport or hydrodynamic effects. One way of overcoming these difficulties is to perform microscopic computer experiments. We shall discuss this problem in the next section.

D. Microscopic Simulation of an Exothermic Chemical Reaction

When an exothermic reaction proceeds in a closed vessel in contact with a thermal reservoir, a *thermal instability* may occur. This phenomenon results from the sudden impossibility of balance between the chemical heat production and the heat loss to the walls. Early studies to determine the conditions for the onset of thermal explosion are due to Semenov for well-stirred systems and to Frank–Kamenetskii for inhomogeneous systems [188]. Their theory is restricted to time-independent situations and rests on the assumptions that the consumption of reactant can be neglected and the fluid is motionless. Heat is thus transported only by conduction.

In the framework of this theory, the critical behavior of a self-heating system is discussed in terms of the stationary solutions of the heat conduction equation with an Arrhenius dependence of the reaction rate

$$\frac{d}{dx}\,\kappa\,\frac{d}{dx}\,T + qf = 0 \tag{3.4}$$

where κ is the thermal conductivity, q the heat of reaction, and f the reaction rate, which is proportional to both the collision frequency ν and the Arrhenius factor, Eq. (3.2). In what follows, we shall consider the simplest geometry of a rectangular box confined between two thermal reservoirs located at $x = 0$ and $x = 2a$, with periodic boundary conditions in the other directions. The boundary conditions for Eq. (3.4) thus read

$$T(x = 0) = T(x = 2a) = T_R \tag{3.5}$$

where T_R represents the wall temperature.

One can show that the behavior of the solutions of Eq. (3.4) can be entirely specified by means of two parameters δ and ε, measuring the heat generation and the inverse activation energy, respectively

$$\delta = \frac{f(T_R)qa^2}{\varepsilon\kappa T_R} \tag{3.6a}$$

$$\varepsilon = \frac{k_B T_R}{E} \tag{3.6b}$$

Most of the work reported in the literature concerns the case of constant thermal conductivity and collision frequency. Even in this case, no closed analytic solution of Eq. (3.4) is known, although useful approximations in the limit of large activation energy ($\varepsilon \ll 1$) have been reported [189, 190].

A first general result is that the temperature profile presents a maximum in the middle of the system ($x = a$). This maximum temperature T_s may be computed numerically for any $\varepsilon \neq 0$. The stationary state diagram describing the behavior of T_s, as a function of the second control parameter δ, is shown by the solid curve in Fig. 6. Provided ε is smaller than a critical threshold and for a certain range of values δ, the solution is triple-valued. The region of multistability terminates at the ignition and extinction points δ_i and δ_e, respectively, the lower and upper branches being stable and the middle one unstable. For a system initially at the reservoir temperature T_R, the temperature will reach a cool stationary state close to T_R for $\delta < \delta_i$ whereas for δ slightly higher than δ_i, the temperature jumps to a much higher value. This upper branch does not describe adequately the explosive state, since in this case the assumption of no consumption of the reactant is no more applicable.

Although it is customary to neglect the temperature-dependence of thermal conductivity κ and collision frequency ν in thermal explosion theory, there are certain cases where this is not appropriate. This is especially true for dilute hard-sphere fluids, where κ and ν have the form

$$\kappa = \frac{75}{64d^2}\left(\frac{k_B^3 T}{\pi m}\right)^{1/2} \equiv \kappa_0 T^{1/2} \tag{3.7a}$$

$$\nu = 4nd^2\left(\frac{\pi k_B T}{m}\right)^{1/2} \equiv \nu_0 T^{1/2} \tag{3.7b}$$

The effects of this temperature-dependence on criticality have been studied numerically in different papers [191, 192]. Figure 6 depicts the diagram of stationary states (dashed line) when the corrections given by expressions (3.7) are incorporated.

The first MD simulation of an exothermic chemical system was performed by Chou and Yip [193] for a hard-disk fluid in contact with thermal reservoirs and undergoing the reaction

$$A + A \xrightarrow{k} A + A + heat \tag{3.8}$$

where the reaction rate, k, obeys the Arrhenius law, Eq. (3.2). The above scheme simulates optimally the Semenov model of combustion since the consumption of reacting particles is neglected. Although the observed temperature profiles were in qualitative agreement with the macroscopic predictions,

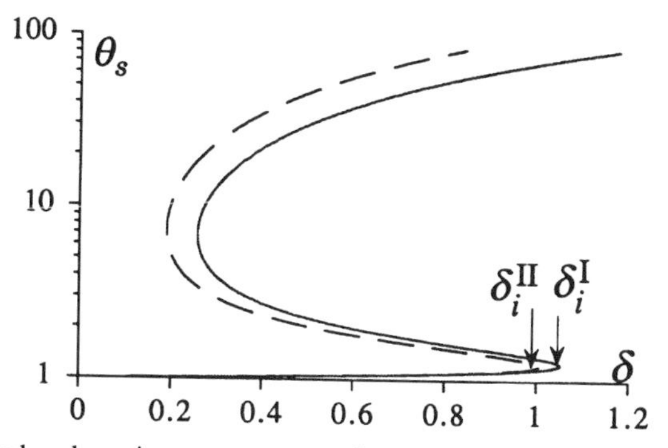

Figure 6. Reduced maximum temperature $\theta_s = k_B T_s/E$ (logarithmic scale) versus the Frank–Kamenetskii parameter δ for $\varepsilon = 0.143$. The solid line represents the solution of Eq. (3.4) with κ and ν constant; the dashed line denotes the solution in gaseous media, where κ and ν are given by Eqs. (3.7). δ_i is the critical value for ignition.

important quantitative discrepancy was noted by the authors. A possible origin for this discrepancy was thought to be the non-negligible temperature slip at the thermal walls. We shall show presently that, in addition, the deviations from macroscopic law observed by Chou and Yip arise from the deformation of the local equilibrium distribution [194]. We shall subsequently discuss the incidence of this deformation on the onset of thermal instability [195].

We consider a system made of an assembly of $N = 4000$ hard spheres of diameter d, confined in a rectangular box with $L_x = 20\,\lambda$ long in the X direction, and with a cross section of $(1 \times 1)\,\lambda^2$ in the Y and Z directions, where λ denotes the mean free path (m.f.p.). The number density is thus $n = 200$ particles per cubic m.f.p. The walls located at $x = 0$ and $x = L_x$ act as thermal reservoirs: each time a particle hits one of these walls, it is re-injected into the system, after having its velocity sampled from a Maxwellian distribution at the reservoir temperature. Periodic boundary conditions are assumed in the other directions. In what follows, the energy will be scaled by the equilibrium (reservoir) thermal energy, so that the reservoir temperature $T_R = 1$. The system is further divided into 40 "statistical" cells in the X direction, over each of which a space averaging is performed to give instantaneous values of the basic mechanical quantities. A time averaging then leads to the local macroscopic quantities, such as temperature, pressure, etc.

The "energy of collision" can be defined in many different ways. For our microscopic simulation, we adopt the following definition due to Present [196]:

$$E_{\text{col}} = \frac{m_i m_j}{2(m_i + m_j)} \, \frac{|(\mathbf{v}_i - \mathbf{v}_j) \cdot (\mathbf{r}_i - \mathbf{r}_j)|^2}{|\mathbf{r}_i - \mathbf{r}_j|^2} \tag{3.9}$$

where m_i represents the mass of the particle "i". If now we assume that the collision is reactive whenever $E_{col} > E$, then it can be shown that for dilute systems the reaction rate indeed obeys the Arrhenius law, Eq. (3. 2) (For details see ref. [197]).

We start our simulation by considering a heat of reaction $q = 2k_BT_R$. The activation energy will always be set to $E = 7\ k_BT_R$. It can be shown from macroscopic analysis of the Semenov model that for the above value of the parameters the system exhibits two well-separated stable states corresponding to "cool" and "hot" (explosive) behavior ($\delta < \delta_i$). We first concentrate on the properties of the cool state, which can be reached by setting the initial value of the system's temperature close to the reservoir temperature. (MD simulations dealing with the transient behavior in an adiabatic explosive chemical systems have been reported [198].)

A time average over 10^5 collisions per particle (CPP) was performed, after the stationary state have been reached (about 10^4 CPP). The measured temperature profile is presented in Fig. 7, where the macroscopic result is also depicted. The statistical error, estimated from successive runs of 10^4 CPP, does not exceed 0.2%. As can be seen, there exists a large temperature slip at the boundaries and a significant difference with the macroscopic profile, all in qualitative agreement with the results of ref. [193]. A simple way to deal with the temperature slip is to set the boundary value problem in the macroscopic equation according to the actual value of the temperature measured in the extreme cells. As can be seen in Fig. 8, much better results are obtained by this procedure, which partly confirm the conclusions presented in ref. [193]. However, the improvement is not enough to resolve the discrepancy with the simulation data. This is particularly clear in Fig. 9, where

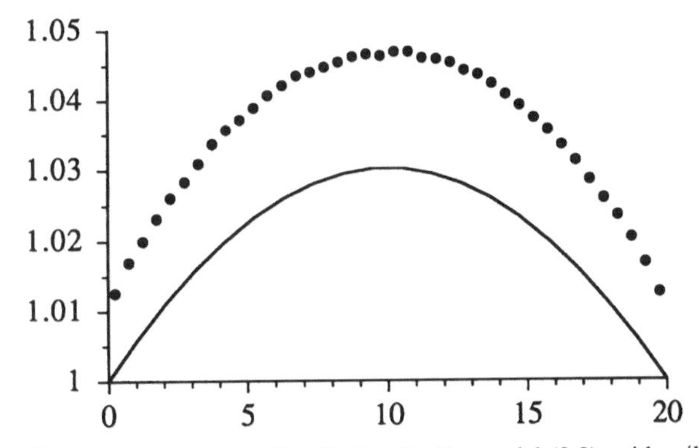

Figure 7. Stationary temperature distribution for the model (3.8), with $q/k_BT_R = 2$, and $L_x = 20\ \lambda$. The dots denote the MD data, while the full line represents the numerical solution of the macroscopic equation, Eq. (4), with $f = f_0$ [see Eq. (3.11)].

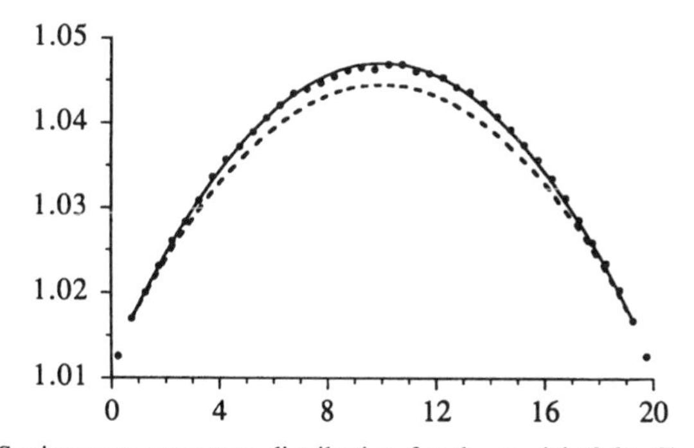

Figure 8. Stationary temperature distribution for the model (3.8). The dots denote the MD data, while the full line represents the numerical solution of the macroscopic equation, Eq. (3.4), with the correction, Eq. (3.10), and the dashed lines depict the numerical solution without the correction. The parameters are the same as in Fig. 7.

the results obtained for a larger value of the heat of reaction $q = 8k_BT_R$ are depicted. Note that in order to remain in the "nonexplosive" state in this latter case, we had to reduce the system size to $L_x = 10\lambda(N = 2000)$.

One possible explanation of the simulation results is the interference with hydrodynamical effects. To check this, we first note that, although microscopically we are dealing with a three-dimensional system, the measured macroscopic data concern effectively a one-dimensional system, since $L_x \gg L_y, L_z$. A transport by advection will therefore affect necessarily the pressure profile in the X direction. The latter is found to be constant in all our computer experiments, thus eliminating the possibility of the presence of pressure waves across the system. Besides, one of the principal factors at the origin of thermal convection, namely the gravitational field, is absent in our simulations. To be complete, we have also measured the static velocity–temperature

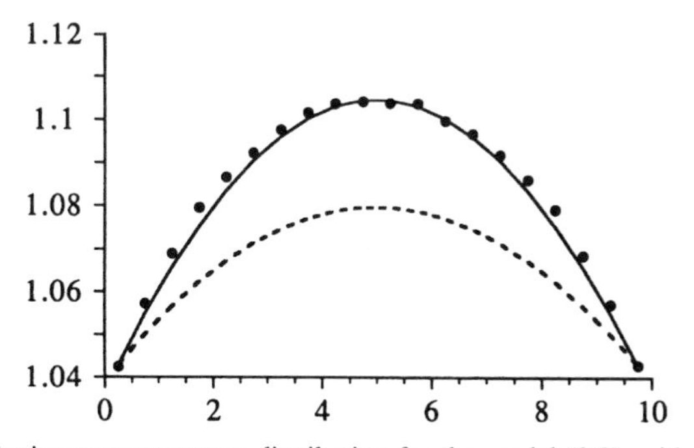

Figure 9. Stationary temperature distribution for the model (3.8), with $q/k_BT_R = 8$ and $L_x = 10\ \lambda$. See captions of Figs. 7 and 8 for details.

correlation functions in the X direction and found it to be zero. We therefore conclude that the hydrodynamical effects, if any, could not exceed the statistical error (0.2%), which is far below the observed discrepancy, especially for $q = 8k_BT_R$ (see Fig. 9).

Next, we compute the deviation of the velocity distribution from local equilibrium, using the method described in ref. [35]. The calculations are tedious and lengthy so that here we report only the final result for the corrected reaction rate

$$f = f_0\left[1 + \frac{1}{8}X_A^2\left(\frac{E}{k_BT}\right)^2\frac{q}{k_BT}\left(\frac{E}{k_BT} + \frac{1}{2}\frac{q}{k_BT} - \frac{3}{2}\right)\exp\left(-\frac{E}{k_BT}\right)\right] \tag{3.10}$$

where X_A denotes the mole-fraction of the reacting particles A (since we are dealing with a one component system, $X_A = 1$) and f_0 represents the usual macroscopic rate law

$$f_0 = \frac{1}{2}\nu n \exp\left(-\frac{E}{k_BT}\right) \tag{3.11}$$

ν being the collision frequency given by Eq. 3.7b. The factor $\frac{1}{2}$ takes account to the fact that the reaction involves a pair of the same particles. Note that the correction to the reaction rate increases with the heat of reaction, as expected from our earlier discussion of model (3.3). Indeed, the correction is about 1% for $q = 2k_BT_R$ and increases to 6% for $q = 8k_BT_R$. The macroscopic temperature profile computed from the corrected rate law, Eq. (3.10), is now in very good agreement with the simulation result (see Figs. 8 and 9).

Our result clearly establishes the deformation of the Maxwellian distribution and, as a corollary, the limitations in the validity of the usual macroscopic rate law. This raises the further question as to what extent this deviation from local equilibrium distribution will also alter the onset of thermal explosion [195]. To answer this question, we have performed a series of runs to locate the critical value of the Frank–Kamenetskii parameter δ_i, Eq. (3.6a). The system is initially set at the wall temperature T_R by assigning to each particle a velocity sampled from a Maxwellian distribution at T_R. When the reaction proceeds, the maximum local temperature is monitored. From the temporal evolution of this temperature one is able to determine whether the system is subcritical or supercritical for a given value of δ. We then alternate simulations between sub- and supercritical conditions until a fairly precise value δ_i for the transition is obtained.

Table II shows the measured critical values δ_i and the predictions based

TABLE II
Critical Value of the Frank–Kamenetskii Parameters, Eq. 3.6[a]

E[a]	6 k_BT_R	7 k_BT_R	8 k_BT_R
δ_i^{exp}	.50	.38	.25
δ_i^{I}	1.09	1.05	1.02
δ_i^{II}	1.04	1.01	1.00
δ_i^{NE}	.62	.50	.39
δ_i^{*}	.44	.37	.30

δ_i^{exp}, from the DSMC simulation; δ_i^{I}, from Eq. (3.4) with no temperature dependence of κ and ν; δ_i^{II}, from Eq. (3.4) for a gas media with κ and ν given by Eqs. (3.7); δ_i^{NE}, from Eq. (3.4) for a reaction rate corrected by nonequilibrium effects [Eq. (3.10)]; δ_i^*, from Eq. (3.4) with the boundary conditions given by Eqs. 3.12 and the corrected reaction rate [Eq. (3.10)].

on the macroscopic theory, Eq. (3.4), with boundary conditions given by Eq. (3.5), for three different choices of the activation energy. As can be seen, the discrepancy with the macroscopic theory is very important, always exceeding 50% in relative values. The discrepancy is reduced to about 20% when nonequilibrium corrections to the reaction rate in Eq. (3.10) are taken into account, but it is still far from being satisfactory.

A plausible explanation for this lack of quantitative agreement can be found by noticing that the boundary conditions used for the calculation of the critical parameter, δ_i, do not allow any temperature slip near the walls. Now, this temperature slip is always present and, in principle, can be incorporated into the calculations, as in our previous analysis of the "cool" state. To this end, however, we need to have a fairly good measure of the temperature profile at the critical point δ_i, which proves to be almost impossible to obtain. The main reason for this is related to the fact that the real part of one of the eigenvalues of the linear operator, associated to Eq. (3.4), approaches zero when the system is moved toward the critical point δ_i. As a consequence, the system exhibits long-range large-amplitude fluctuations that preclude any reliable statistics in a reasonable amount of computing time.

Another alternative is provided by kinetic theory. The discontinuity of temperature at a wall bounding an unequally heated gas can be expressed by the following type of equation

$$g\left(\frac{dT}{dx}\right)_{\text{wall}} = T - T_R \tag{3.12a}$$

where g is called the temperature jump distance. For dilute hard-sphere gases and in the case of a purely diffusive wall, elementary kinetic theory [199, 200] gives us an approximate theoretical expression for the jump distance

$$g \approx \left(\frac{\pi}{2k_B} \right)^{1/2} \frac{\kappa_0}{2nk_B} \tag{3.12b}$$

where n stands for the unperturbed (global) number density and κ_0 is defined in Eq. (3.7a).

First, we have checked the validity of the above approximation by simulating a system submitted to a thermal gradient without chemical reaction. As shown in Fig. 10, we observe a very good agreement between simulation data and the solution of the heat conduction equation when the boundary conditions given by Eqs. (3.12) are used. Next, we have solved Eq. (3.4) to obtain the stationary temperature profile for the cool state (Figs. 7–9), using these new boundary conditions. The agreement with our previous results remains excellent. Finally, we have repeated our calculations with the boundary conditions, Eq. (3.12), to obtain a new estimation of the critical parameter δ_i, hereafter denoted as δ_i^*. These estimated values are also reported in Table II. The agreement with the simulation results is now considerably improved.

To our knowledge, none of the theoretical predictions so far reported in the literature has been checked quantitatively through experimental work, mainly for reasons underlined previously. As we have shown here, thanks to microscopic simulations of exothermic chemical systems, some aspects of the problem are now clarified.

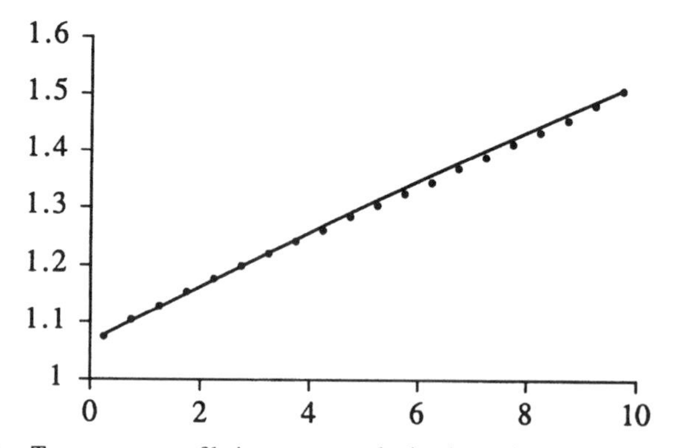

Figure 10. Temperature profile in a system submitted to a thermal gradient $\Delta T/T_R = 0.6$. The dots denote the DSMC data for a system 10 λ long. The full line represents the solution of the heat conduction equation with boundary conditions (3.12). The thermal conductivity coefficient is given by Eq. (3.7).

E. Microscopic Simulation of Continuously Stirred Tank Reactor (CSTR)

We now consider the microscopic simulation of isothermal chemical reactions, limiting ourselves to homogeneous systems, as can be achieved in a continuously stirred tank reactor, postponing until Section III.F the study of local fluctuations. Our main objective here will be to clarify the status of the stochastic theory of chemical systems in the presence of bifurcation. The Bird algorithm is particularly well adapted for this case, since it allows the simulation of a homogeneous Boltzmann gas simply by associating the entire system volume to a single collisional cell. The very nature of the problem, however, raises some basic difficulties that can affect considerably the efficiency of the algorithm.

The first problem is related to nonequilibrium effects due to the deformation of the Maxwell–Boltzmann distribution through Arrhenius type of reactive collisions. To avoid these nonequilibrium effects the frequency of reactive collisions must be significantly smaller than the frequency of elastic collisions, entailing important waste of CPU time. One way to overcome this difficulty is to further simplify the reactive collision rules by the following procedure. Consider a typical bimolecular chemical step

$$A + B \xrightarrow{k} C + D \tag{3.13}$$

with

$$k = \nu \exp\left(-\frac{E}{k_B T}\right) \equiv \nu k_A \tag{3.14}$$

where ν is the collision frequency, Eq. (3.7b). After a collision between two particles A and B has occurred, we choose randomly $k_A\%$ of the collisions to be reactive, where k_A stands for the Arrhenius factor defined in Eq. (3.14). Obviously, this procedure avoids the deformation of the Maxwell–Boltzmann distribution, since it does not involve any systematic energy transfer between reactants and products. It is, however, restricted to isothermal chemical systems.

Another problem with the Bird algorithm is that it is restricted to binary collisions only, i.e., to second-order chemical reactions. On the other hand, simple chemical models exhibiting complex behavior, such as the Brusselator or the Schlögl model, rely on trimolecular collisions (cf. Section II). It has been shown that the trimolecular step can be approximated by a pair of bimolecular steps involving different timescales, so that an adiabatic elimination of a fast variable leads to an effective trimolecular step [201]. Nevertheless, such a scheme is inappropriate for MD studies because the species represented by the slow variables undergo far fewer reactive collisions per unit time than those represented by fast variables. This results in much wasted bookkeeping with a corresponding waste of CPU time. Thus, with present-day computers, we are left with two alternatives: either to set up a microscopic model "mimicking" a three-body collision or search for a suitable second-order reaction scheme in which the difference of timescale between intermediate variables is as small as possible. The former alternative has been considered successfully by Mareschal and De Wit for the Brusselator model [202]. Here we choose the second alternative and look for a chemical model satisfying the following three constraints: (1) it consists of binary collisions only; (2) it has no significant separation of timescales; and (3) it involves as few reactants as possible. As was shown in ref. [203], the above requirements are fully satisfied by the following chemical model:

$$U + W \xrightarrow{k_1} V + W \tag{3.15a}$$

$$V + V \underset{k_{-2}}{\overset{k_2}{\rightleftharpoons}} W + S \tag{3.15b}$$

$$V + S \xrightarrow{k_3} S + S \tag{3.15c}$$

where the concentration of the S particles is supposed to remain constant. In what follows, we shall refer to them as "solvent" particles. The macroscopic rate equations corresponding to the above model in a CSTR condition read

$$\frac{du}{dt} = -k_1 uw + \alpha(u_f - u) \tag{3.16a}$$

$$\frac{dv}{dt} = k_1 uw - 2(k_2 v^2 - k_{-2} ws) - k_3 vs + \alpha(v_f - v) \tag{3.16b}$$

$$\frac{dw}{dt} = k_2 v^2 - k_{-2} ws \tag{3.16c}$$

where u, v, w, and s are the mole fractions of U, V, W, and S; $k_{\pm i}$ are the rate constants of the ith reaction; α is the inverse of the residence time, or feed rate; and u_f and v_f are the feed mole fractions of U and V, respectively. The system is assumed to be impermeable to W.

In order to carry out a microscopic modeling of the feed process we introduce two more pairs of reactions:

$$S + S \underset{a_-}{\overset{a_+}{\rightleftharpoons}} S + U \tag{3.17a}$$

$$S + S \underset{b_-}{\overset{b_+}{\rightleftharpoons}} S + V \tag{3.17b}$$

The forward reaction corresponds to inflow and the reverse reaction to outflow; $a_\pm$ and $b_\pm$ are the fractions of reactive collisions for these two reactions. As can be seen, an *S–S* reaction can result in the production of either *U* or *V* molecules. The fraction of "reactive" *S–S* collisions resulting in the production of a *U* molecule is $u_f/(u_f + v_f)$ and the fraction of "reactive" collisions resulting in the production of a *V* molecule is then just $v_f/(u_f + v_f)$. It can be easily checked that the reactions (3.17) lead indeed to the correct macroscopic feed terms, provided we set

$$a_+ = \frac{2\alpha u_f}{s^2} \qquad b_+ = \frac{2\alpha v_f}{s^2} \quad a_- = b_- = \frac{\alpha}{s} \tag{3.18}$$

Note that the factor 2 in the above relations for a_+ and b_+ is related to the fact that the forward reactions, Eqs. (3.17), involve a pair of the same molecules (cf. Eq. (3.11)).

There remains one last problem, which arises whenever the concentration of some of the species have to be kept constant all along the simulation. For our model, this is the case of the solvent molecules that undergo reactive collisions with the other species and participate, in addition, to reactions designed to mimic the feed terms. An elegant procedure for solving this problem has been proposed by Boissonade [180, 181]. The idea is to introduce one more participant, say molecules *A*. Every time a solvent particle is created (destroyed) in a collision with the other species, an *S*(*A*) particle is chosen at random in the same collisional cell and replaced by an *A*(*S*) particle. Since the *A* molecules do not participate in any reaction, they merely constitute a reservoir of particles maintaining the solvent concentration fixed. We note that the simulation procedure is completely specified in terms of kinetic constants, feed rate, and the concentration of *S* particles.

1. *Multiple Steady-State Transitions*

For certain ranges of parameter values, the macroscopic equations, Eq. (3.16), can admit multiple steady states and limit-cycle oscillations. In this section we concentrate on a possible occurrence of a pitchfork bifurcation. We first note that the stationary state mole fractions $\{u_s, v_s, w_s\}$ obey the following relations

$$w_s = \frac{k_2}{sk_m} v_s^2, \qquad u_s = \frac{\alpha}{k_1 w_s + \alpha} u_f \tag{3.19a}$$

$$k_1 k_2 (k+\alpha) v_s^3 - \alpha k_1 k_2 (u_f + v_f) v_s^2 + \alpha k_m (k+\alpha) v_s - \alpha^2 k_m v_f = 0 \tag{3.19b}$$

where we have set

$$k = k_3 s, \qquad k_m = k_{-2} s \tag{3.20}$$

As is well known, the general solution of the cubic equation, Eq. (3.19b), can be entirely described in terms of two parameters only (see, for example, ref. [14]). Thus, without loss of generality, we are allowed to impose a certain number of suitable relations among the various parameters of the problem. One guideline stems from the fact that at a pitchfork bifurcation point the cubic equation, Eq. (3.19b), must admit a triple root. On the other hand, the stationary state mole fractions of the chemically active components, and that of solvent molecules, should not be significantly different from one another for, otherwise, the microscopic simulation of the model will become highly inefficient.

Keeping the above comments in mind, we find after some algebra that if we set

$$k = \frac{1}{3}(u_f + v_f) \sqrt{2\alpha \frac{k_1 k_2}{k_m} \frac{(u_f + v_f)}{(u_f - 2v_f)}} - \alpha \tag{3.21}$$

then Eq. (3.19b) reduces to the following simple form:

$$(v_s - v_s^{(1)})^3 + \frac{4\alpha}{(u_f + v_f)} \frac{k_m}{k_1 k_2} (v_f - u_f/8)(v_s - v_s^{(1)}) = 0 \tag{3.22}$$

where we have introduced the "reference" stationary state

$$v_s^{(1)} = \frac{1}{3}(u_f + v_f) \frac{\alpha}{k+\alpha} \tag{3.23}$$

As can be seen, for $v_f > u_f/8$ the stationary solution is unique, i.e., $v_s = v_s^{(1)}$, whereas for $v_f < u_f/8$ one has three stationary solutions, showing clearly that the system undergoes a bifurcation at $v_f = u_f/8$. More detailed analysis shows that the latter corresponds to a pitchfork bifurcation point. This is illustrated in Fig. 11, where the stability diagram for the variable u is

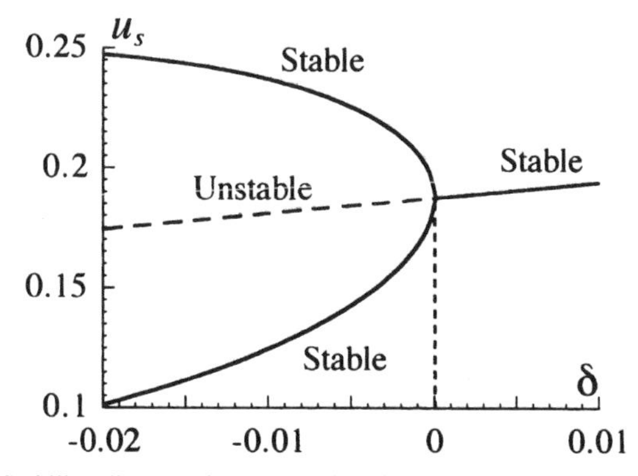

Figure 11. Stability diagram for the model (3.15). Equation (3.24) defines δ and the parameters are given in Eq. (3.25).

depicted. The parameter δ in this figure represents the "distance" from the bifurcation point, defined as

$$v_f = \frac{u_f}{8} + \delta \tag{3.24}$$

the other parameters being set to

$$k_1 = \nu \qquad k_2 = \nu/2 \qquad k_m = \nu/26 \qquad u_f = 1/4 \qquad \alpha = 0.28\nu \tag{3.25}$$

where ν is the collision frequency. Note that in writing the relation in Eq. (3.21) we have implicitly assumed that $u_f - 2v_f > 0$, which is clearly satisfied in the vicinity of the bifurcation point, $\delta \approx 0$.

The main difference between the bifurcation diagram of Fig. 11 and the one for the Schlögl model (Fig. 2), is that the former is no longer symmetrical as one moves from positive to negative values of δ. Numerical simulations of the set of Langevin equations associated to the model (3.15) lead to the conclusion that the lower branch is "more probable" than the upper branch. The master equation, on the other hand, predicts just the opposite. We thus have a good opportunity to confront the above predictions with those obtained through a microscopic simulation of the model (3.15).

For the microscopic simulation, we consider a system made of an assembly of 2,000 hard spheres of diameter d, with a number density $n = 3 \times 10^{-3}$ particles per d^3. The solvent mole fraction is set s to $= 0.35$, i.e., 35% of the system is made up of solvent molecules. We choose $\delta = -0.003$ so the system evolves within the multiple steady state region; the values of the other parameters are given in (3.25). After the system has reached the stationary

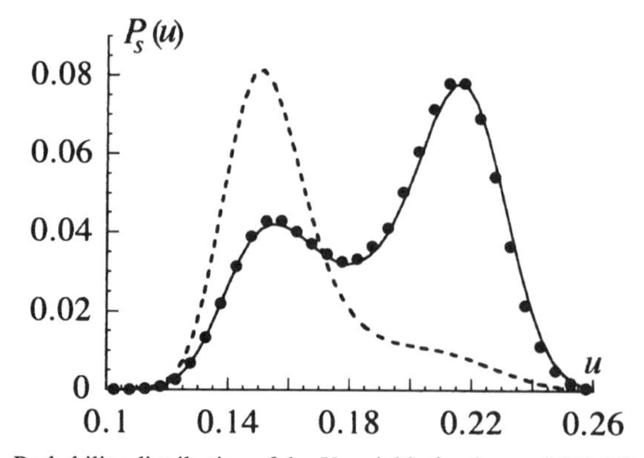

Figure 12. Probability distribution of the U variable for the model (3.15) with $\delta = -0.003$. The solid curve corresponds to results of the master equation, whereas the dashed curve represents those based on the Langevin equation. The circles are obtained through a microscopic simulation based on the Bird algorithm. The other parameters are given in Eq. (3.25).

regime, statistics are taken over 10^6 CPP to construct the probability distribution (histogram) for the reactive components u, v, and w. The statistical error, estimated from 10 successive runs of 10^5 CPP, does not exceed 8%.

The "measured" probability distribution of the U variable is presented in Fig. 12, together with those corresponding to the Langevin equation and the master equation, which were also obtained numerically using standard techniques [49, 204]. The result of the Langevin approach is manifestly inconsistent with both the microscopic and master equation predictions. On the other hand, the agreement between the master equation and the microscopic simulation is excellent, better than the estimated statistical error. Once more, these results show the utility of microscopic simulations in testing the validity of phenomenological theories.

2. *Hopf Bifurcation*

Let us now consider the case of a Hopf bifurcation leading to sustained oscillations. A full bifurcation analysis of this case proves to be much more involved than the previous one, so that here we present only the final result. Figure 13 shows the stability diagram for Eqs. (3.16), where the mole fraction of the V component in the feed stream, v_f, is chosen as the bifurcation parameter, the other parameters being set to

$$k_1 = k_2 = k_{-2}s = \nu \qquad u_f = 1 \qquad k_3 s = 10 \qquad \alpha = 0.033\nu \tag{3.26}$$

As can be seen, the system undergoes a Hopf bifurcation for a value of v_f located at about 0.2045. This is confirmed by numerical integration of dif-

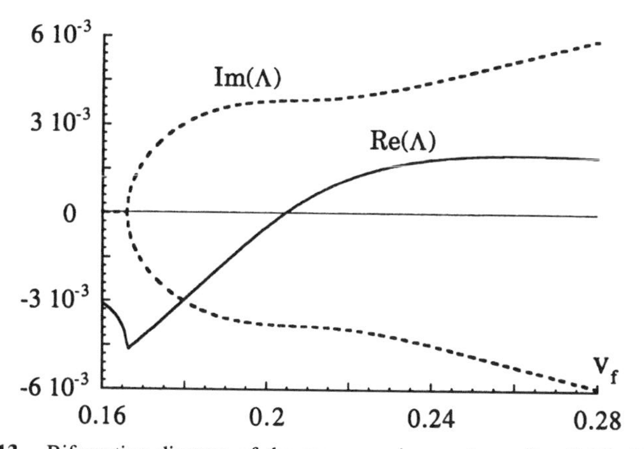

Figure 13. Bifurcation diagram of the macroscopic equations, Eq. (3.16). The full line and the dashed line represent the real part and the imaginary part, respectively, of the largest eigenvalue Λ of the corresponding linearized operator. The parameters are given in Eq. (3.26).

ferential equations, Eq. (3.16), exhibiting sustained oscillations whenever v_f exceeds the above critical value (see Fig. 14).

For our microscopic simulation, we consider a system made of an assembly of 5,000 hard spheres of diameter d, with $s = 0.1$. As before, the number density is set to $n = 3 \times 10^{-3}$ particles per d^3. The results of the simulation, for the parameter values given in Eq. (3.26) with $v_f = 4/15$, reproduce sustained oscillations. We have performed two different types of analysis of the simulation data. First we consider 30 runs, all starting from the same "macroscopic" initial condition, extending over 4 periods of oscillations. A

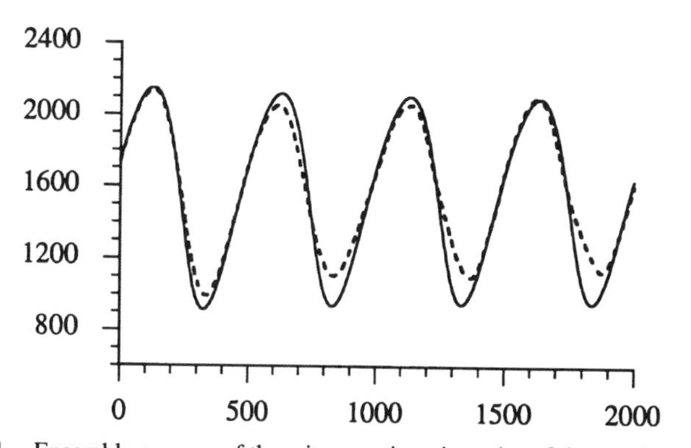

Figure 14. Ensemble average of the microscopic trajectories of the number of U particles (dashed line) compared to the numerical solution of the deterministic equations (solid line), for $v_f = 4/15$.

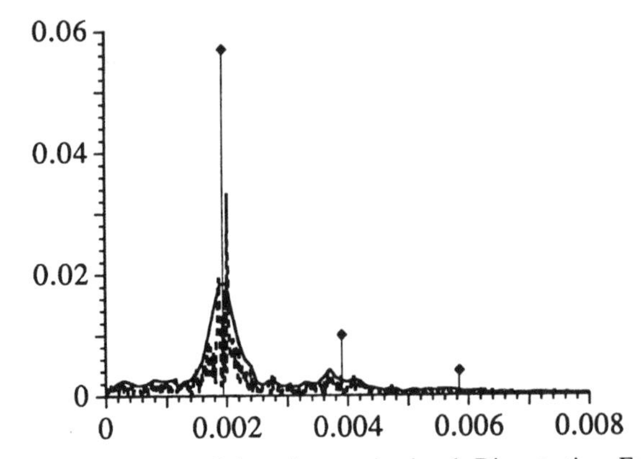

Figure 15. Power spectrum of the microscopic signal. Discrete time Fourier transform on the entire time series of 120 oscillations (dashed curve), average of the power spectrum over 6 times 20 oscillations (solid line), and power spectrum corresponding to the macroscopic equations (vertical solid lines).

single run corresponds to about 50 million collisions. In Fig. 14 the ensemble average of the microscopic trajectories is compared to the numerical solution of the deterministic Eqs. (3.16). Despite the small number of samples, the agreement is quite good. This procedure allows also for an estimation of the statistical errors, which turn out to be of the order of 5%. Next, we consider a single run of a total length of 120 oscillations corresponding to about 10^9 collisions. A discrete time Fourier transform is performed and the power spectrum is computed by considering first the entire time series (dashed line in Fig. 15). We then divide the time series in 6 portions of 20 oscillations each. The solid line in Fig. 15 represents the average of the power spectrum over these 6 parts. In this same figure, we have also plotted the power spectrum corresponding to the macroscopic equations (vertical solid lines). We see that there is very good agreement for the location of the fundamental frequency and reasonable agreement for the first harmonic.

We can also compare the results of our microscopic simulations to the predictions of the stochastic theory of chemical reactions based on the master equation formalism (cf. Section II.A.2). We first consider the static properties of the process, e.g., equal time correlation functions. A comparison with the results of the microscopic simulation is presented in Table III. The deviations never exceed 5%, which is within the expected statistical errors.

Next, we analyze the dynamical properties of the system by computing the time correlation function for different species. In Fig. 16 we plot the time correlation function of the U component, $\langle u(t)u(0)\rangle$, generated by the Bird algorithm and by simulation of the master equation. Again, we observe a

TABLE III
Comparison Between Static Correlation Functions Obtained from Master Equation and Microscopic Simulations[a]

Correlations	C_{uu}	C_{vv}	C_{ww}	C_{uv}	C_{uw}	C_{vw}
Bird	9.10×10^{-2}	1.38×10^{-1}	5.99×10^{-1}	-7.16×10^{-2}	-1.43×10^{-1}	2.74×10^{-1}
Master eq.	8.92×10^{-2}	1.33×10^{-1}	5.70×10^{-1}	-6.96×10^{-2}	-1.39×10^{-1}	2.63×10^{-1}

[a]The parameters are given in Eq. (3.26), with $v_f = 4/15$. The symbols are defined as $C_{ab} \equiv (\langle ab \rangle - \langle a \rangle \langle b \rangle)/\langle a \rangle \langle b \rangle$.

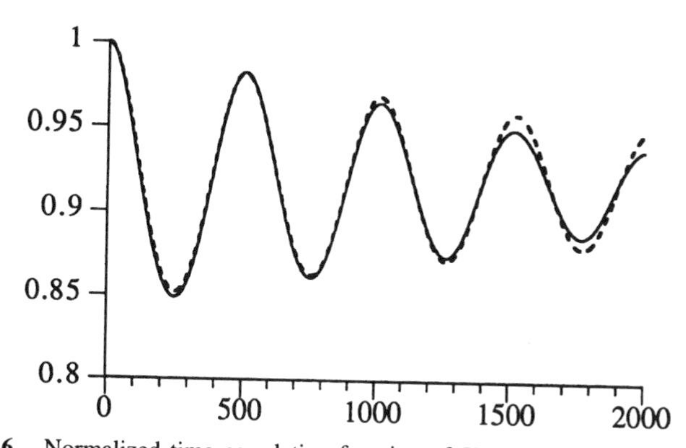

Figure 16. Normalized time correlation function of U species generated by the microscopic simulation (solid line) and by master equation formalism (dashed line).

very good agreement between the two curves. We also note that the amplitude is decaying with time. The stochastic theory of limit cycle, described in Section II.A.2, predicts an exponential decay of this amplitude arising from frequency fluctuations (phase diffusion) (cf. Eq. (2.38); see also ref. [205]). This is confirmed in Fig. 17, where the logarithm of the envelope of the time correlation function $\langle u(t)u(0) \rangle$ versus time is depicted.

F. Microscopic Simulation of Spatially Extended Systems

Our main objective in this section is to check the validity of the reaction-diffusion master equation in the vicinity of a bifurcation point, through a microscopic simulation of a representative chemical model, such as the model (3.15). Testing phenomenological theories through microscopic simulations is not always easy, as it involves a number of pitfalls one should be aware of before drawing any definitive conclusions. We had the occasion to note some of them in the previous section devoted to microscopic simulations of perfectly homogeneous isothermal systems. Probing local properties of reactive fluids reveals some other traps, so that it is instructive to concentrate first on

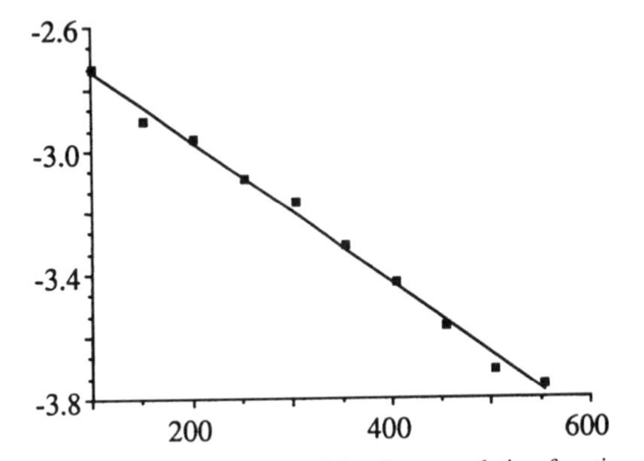

Figure 17. Logarithm of the envelope of the time correlation function versus time.

the following simple model that can be solved exactly [126].

$$\begin{aligned} S + U &\xrightarrow{k_1} U + U \\ S + U &\underset{k_3}{\overset{k_2}{\rightleftharpoons}} S + S \end{aligned} \tag{3.27}$$

where the concentration of S particles (hereafter referred to as "solvent" particles) is supposed to remain constant. As discussed at the end of Section III.E, this can be achieved by introducing one more participant, say molecules A, which constitutes a reservoir of particles maintaining the solvent concentration fixed. We further restrict ourselves to a one-dimensional system with periodic boundary conditions, i.e., a long thin torus.

We first consider the master equation formulation. Dividing the system length L into N_c cells and following the arguments of Section II, one can write

$$\begin{aligned} \frac{d}{dt} P(\{U_i\};t) = \sum_{i=1}^{N_c} \{ & \upsilon(U_i - 1)P(\cdots, U_i - 1, \cdots; t) - \upsilon(U_i)P(\{U_i\};t) \\ & + \mu(U_i + 1)P(\cdots, U_i + 1, \cdots; t) - \mu(U_i)P(\{U_i\};t)\} \\ & + \frac{\tilde{D}}{2} \sum_{i=1}^{N_c} \sum_{l=\pm 1} \{(U_i + 1)P(\cdots, U_i + 1, U_{i+l} - 1, \cdots; t) \\ & - U_i P(\{U_i\};t)\} \end{aligned} \tag{3.28a}$$

with

$$v(U_i) = k_1 s U_i + k_3 \frac{s(S-1)}{2} \qquad \mu(U_i) = k_2 s U_i \tag{3.28b}$$

where s and S represent the mole fraction and the number of solvent particles per cell, respectively. Owing to the periodic boundary conditions,

$$U_{N_c+1} = U_1 \qquad U_0 = U_{N_c} \tag{3.29}$$

The master equation, Eq. (3.28), admits a stationary solution provided $k_2 > k_1$ since otherwise the random variables $\{U_i\}$ become unboundedly large as $t \to \infty$. In this case, the stationary probability distribution possesses translational symmetry, i.e., $P_{\rm st}(U_i = a, U_{i+k} = b) = P_{\rm st}(U_j = a, U_{j+k} = b)$, $\forall i, j, k$. For instance, the average number of U particles per cell reads

$$\langle U_i \rangle = \frac{k_3(S-1)}{2(k_2 - k_1)} \equiv \langle U \rangle \qquad \forall i \tag{3.30}$$

The static correlation function is readily found to obey

$$\frac{\tilde{D}}{2}(g_{i+1,j} + g_{i-1,j} - 2g_{i,j}) - s(k_2 - k_1)g_{i,j} = -k_1 s \langle U \rangle \delta_{i,j}^{Kr} \qquad i,j = 1, 2, \ldots, N_c \tag{3.31}$$

where we have defined

$$g_{i,j} \equiv \langle \delta U_i \delta U_j \rangle - \langle U \rangle \delta_{i,j}^{Kr} \tag{3.32}$$

As this equation is subjected to periodic boundary conditions, Eq. (3.29), it can be solved through lattice Fourier transform. After some calculation, one finds

$$g_{i,j} = \frac{2k_1 s \langle U \rangle \alpha}{\tilde{D}(\alpha^2 - 1)(\alpha^{N_c} - 1)} (\alpha^{|i-j|} + \alpha^{N_c - |i-j|}) \qquad |i-j| = 0, 1, \ldots, N_c - 1 \tag{3.33}$$

where we have set

$$\alpha = (1 + \beta/\tilde{D}) + \sqrt{(1 + \beta/\tilde{D})^2 - 1} \tag{3.34a}$$

with

$$\beta = s(k_2 - k_1) \tag{3.34b}$$

For the microscopic simulation, we consider a system made of an assembly of $N = 42{,}000$ hard spheres of diameter d, confined in a rectangular box of $L = 3{,}780\ d$ long, with a number density $n = 5 \times 10^{-3}$ particles per d^3 (the mean free path λ is about 45 d). The diffusion coefficient $D = 29.92$ and the collision frequency $\nu = 0.025$, in system units. The system is divided into 84 collisional cells, each containing an average of 500 particles (the cell volume is about 1.1 λ^3). The other parameters are chosen as follows:

$$k_1 = 0.1\nu \qquad k_2 = 0.15\nu \qquad k_3 = 0.4\nu \qquad s = 0.1 \tag{3.35}$$

A time average over 10^6 collisions per particle (CPP) was performed to measure the spatial static correlation function, after the stationary state had been reached (about 10^5 CPP). The result is presented in Fig. 18, together with that obtained from the master equation, Eq. (3.33). The agreement is definitely not good. Note that traditional hard-disk and hard-sphere MD simulations have, in the past, led to basically the same type of results [206].

Detailed analysis shows that the origin of the observed discrepancy is closely related to the way the solvent concentration is kept constant in the microscopic simulation. As already underlined, each time a solvent particle S is created (destroyed) in a collision with the other species an $S(A)$ particle is chosen at random in the same collisional cell and replaced by an $A(S)$ particle. This procedure ensures the conservation of solvent particles in reactive collisions, but does not prevent them from moving freely from cell to cell. In other words, the number of solvent particles in a cell fluctuates, but the fluctuations arise only because of diffusion. The effect is negligibly small in macroscopic systems, but not in microscopic simulations where the number

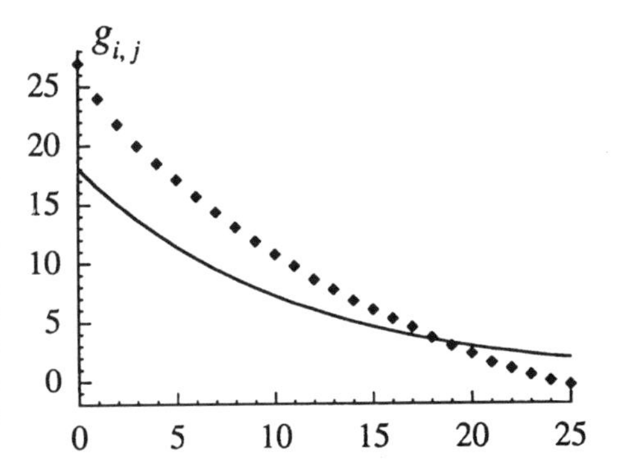

Figure 18. Spatial correlation function g_{ij}, Eq. (3.32), as a function of $|i-j|$. The solid curve corresponds to the solution of the master equation, Eq. (3.33), whereas the squares are obtained through a microscopic simulation based on Bird's algorithm. The estimated statistical errors are less than 4%. The parameters are given in Eq. (3.35).

of particles per cell is generally quite small. For instance, in our case the average number of solvent particles per cell is only 50.

To check the validity of the above arguments, we consider once more the model (3.27), but now we allow the solvent particles to diffuse as well. The corresponding master equation proves to be much more difficult to handle analytically, mainly because the transition probabilities are now nonlinear functions of the state variables $\{U_i, S_i\}$. It can nevertheless be solved numerically and the results are shown in Fig. 19, where a much better agreement with the microscopic results is observed. Still, the agreement is not totally satisfactory. In particular, near the origin the discrepancy is about 9%, well above the expected statistical errors (4%). We note that such a relatively small discrepancy would be quite difficult to detect through hard-sphere MD simulations, since here the statistical errors associated to the measurement of fluctuations can hardly be lowered below 10% within a reasonable CPU time with present-day computers.

The origin of this last discrepancy is deeper and, in a way, more difficult to understand than the previous one. Nevertheless, the simplicity of the model together with the flexibility of the Bird algorithm leads to a complete clarification of the problem. As already stated in Section II.A, the linear dimensions of a cell in the reaction-diffusion master equation cannot be chosen arbitrarily. Too large a cell size violates the cell statistical homogeneity assumptions, whereas too small a cell size may compromise the separability of reaction and diffusion viewed as elementary processes. Intuitive arguments, developed in the beginning of Section II.A, lead to the conclusion that the cell sizes should be typically of the order of the reactive mean free path. In other words, the reaction-diffusion master equation cannot probe processes arising on a scale smaller than the reactive mean free path. For the parameter values we have chosen, the largest kinetic constant $k_3 = 0.4\nu$, so that the reactive mean free path exceeds necessarily 2.5 λ, where λ denotes the usual "elastic" mean free path.

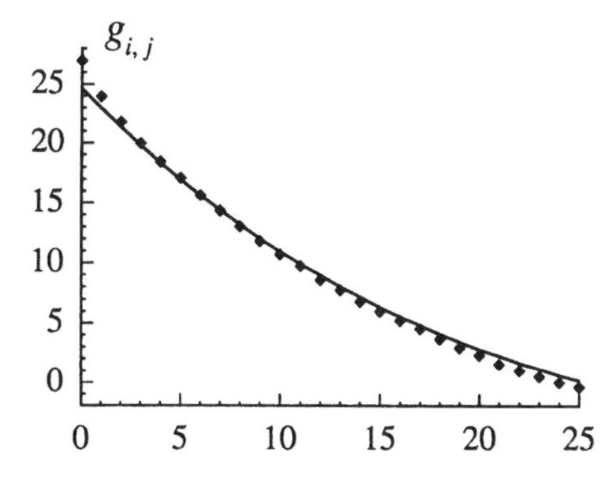

Figure 19. Spatial correlation function g_{ij}, Eq. (3.32), as a function of $|i - j|$. The solid curve corresponds to the numerical solution of the master equation, Eq. (3.28), where the diffusion of solvent particles has been included. The squares are obtained through a microscopic simulation based on Bird's algorithm. The parameters are given in Eq. (3.35).

To check the above intuitive arguments, we consider our microscopic simulation, with the same number of collisional cells as before, but now we divide the system into 28 "statistical" cells, i.e., we group the cells three by three and measure the statistical properties of the system over these enlarged cells. Similarly, we solve numerically the master equation divided also into 28 cells. The results are depicted in Fig. 20 which now shows perfect agreement.

Let us now consider a model giving rise to bifurcation, such as the model (3.15), for which we can write a reaction-diffusion master equation similar to Eq. (3.28), allowing the solvent particles to diffuse as well. Long before a bifurcation point, the dynamics can be linearized around the reference state so that the behavior of the system is basically the same as in the case of the model (3.27). As we move towards a bifurcation point we observe a dramatic increase in the fluctuation lifetime, which is much more pronounced than in the corresponding homogeneous case (cf. Section III.E). As a consequence, statistical errors associated with the measurement of the correlation functions become very important, so that much longer runs are required to get reliable statistics. To avoid this difficulty, we move slightly away from the bifurcation point, but remain close enough so that the fluctuations still keep their non-Gaussian behavior. For the case of the pitchfork bifurcation, a good choice is to set δ to about 10^{-3}, while keeping the other parameters as before (cf. Section III.E.1):

$$k_1 = \nu \qquad k_2 = \nu/2 \qquad k_m = \nu/26 \qquad u_f = 1/4 \qquad \alpha = 0.28\nu \qquad s = 0.35$$

$$\mathbf{v}_f = \frac{u_f}{8} + \delta \tag{3.36}$$

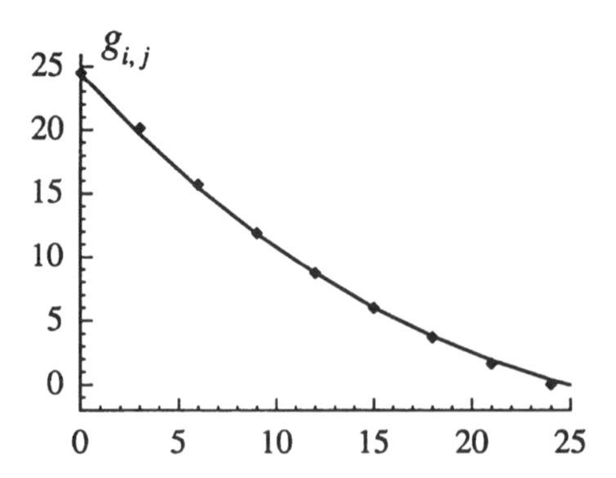

Figure 20. Same as in Fig. 19, except that here the statistics is taken over enlarged cells of about 3 mean free paths long.

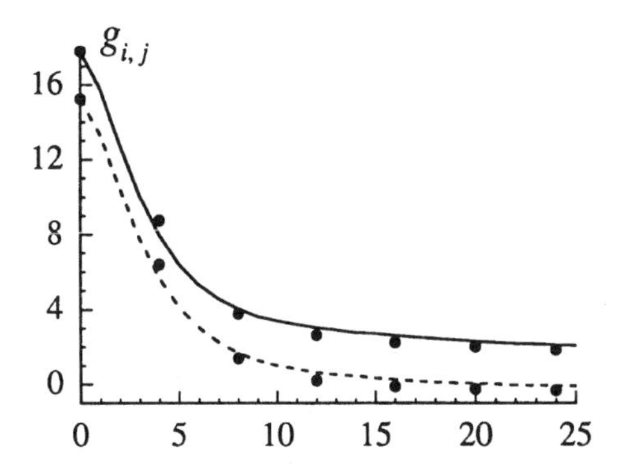

Figure 21. Spatial correlation function g_{ij}, Eq. (3.32), as a function of $|i-j|$ for the model (3.15). The solid and the dashed curves correspond to the numerical solution of the master equation for $\delta = 3 \times 10^{-3}$ and $\delta = 10^{-2}$, respectively. The circles are obtained through the corresponding microscopic simulation based on Bird's algorithm. The statistics are taken over enlarged cells of about 4 mean free paths long. The parameters are given in Eq. (3.36).

For the microscopic simulation, we consider the same basic parameters as those chosen for the simple model (3.27), i.e., $N = 42{,}000$ hard spheres, $L = 3{,}780\ d$ and $n = 5 \times 10^{-3}$ particles per d^3. The system is divided into 84 collisional cells (about 1 mean free path long), and statistics are collected over each such cell as well as over groups of 2, 3, and 4 cells. We run the simulation for two different values of δ: $\delta = 10^{-2}$ and $\delta = 3\ 10^{-3}$. To lower the statistical errors below 8% for this latter case, we had to run the simulation about 5 times longer than in the case of the model (3.27). With the parameter values in Eq. (3.36), the reactive mean free path is close to 3λ. The results, evaluated for groups of 4 cells (about 4λ), are depicted in Fig. 21 and show quantitative agreement with those obtained from the master equation (for more information, see ref. [207]).

IV. CONCLUDING REMARKS

The main purpose of this work was to review the basis and the present status of microscopic simulations of chemical systems. From a practical point of view, the interest of such simulations resides in the fact that they can be used as an "experimental" tool to account for cases where high-precision laboratory experiments are difficult or impossible to perform. Reactive processes, however, are often characterized by relatively long relaxation timescales so that traditional MD simulations are generally too slow for the study of statistical properties of complex chemical systems. With present-day computers, one is forced to search for alternative methods that priviledge computational performances, at the expense of simplifying as much as possible the underlying microscopic dynamics. A basic requirement is that the validity of such a simplified dynamics must go beyond that of macroscopic formulations. In particular, it must be able to reproduce the correct fluctuation spectra, even under strong nonequilibrium conditions.

We have presented strong evidence that the Bird algorithm, extended to reactive gases, is an appropriate way to handle the problem. We have used this simulation technique to analyze the behavior of an exothermic chemical system in contact with thermal reservoirs. As predicted by the phenomenological description, we have observed two distinct behaviors depending on the value of the heat of reaction: a subcritical regime corresponding to the evolution toward a state close to the reservoir temperature, and a supercritical one associated with thermal explosion. In the subcritical regime, a discrepancy is observed between the results of the simulation and predictions based on macroscopic chemical kinetics. The origin of this discrepancy is found to arise from the deformation of the Maxwell distribution through reactive collisions that, in turn, induces a correction to the usual rate law.

The critical threshold for the onset of explosive behavior measured in the simulation differs from the phenomenological predictions. The agreement is significantly improved by taking into account the nonequilibrium corrections and by using appropriate boundary conditions. Nevertheless, some questions remain open. For instance, corrections to the macroscopic description arising from fluctuations might induce a shift of the critical threshold. Moreover, the final value of the pressure in the stationary state is slightly different from the initial one. The time-independent theory of thermal explosions does not take this fact into account. As noted by Zeldovich [207], when ignition occurs, a combustion source, or locus, appears at the hottest point of the system and a flame propagates from this locus. Microscopic simulations can be used to obtain a better understanding of the early stage of development of an explosion [154]. For instance, measurements of space-dependent properties, such as space-time correlation functions, would make it possible to probe the gradual formation of combustion nuclei inducing eventually the explosion phenomenon.

The microscopic simulation of reactive fluids can be further simplified in the case of isothermal chemical systems. This simplification allows the simulation of reactive fluids exhibiting a variety of complex behaviors. We have discussed the case of a three-variable CSTR chemical model giving rise both to sustained oscillations and to multiple steady-state solutions. Here again, we have used the microscopic simulation as an "experimental" tool to check the validity of the stochastic theory of chemical reactions in the homogeneous limit, based on the master equation. In particular, we were able to show that, unlike the master equation, a Langevin formulation of chemical reactions is inappropriate in the multiple steady-state parameters domain. Other questions can be addressed along the same lines. For example, assuming perfect mixing and no hydrodynamic effects, the question has been raised as to whether the chaos observed in the Belousov–Zhabotinsky reaction is of chemical or hydrodynamic origin [208]. Recent simulation results confirm the first proposition [209].

We have also addressed chemical dynamics in spatially extended systems. We have first reviewed in detail the stochastic theory of reaction-diffusion systems. By modeling diffusion as a random walk and reaction as a birth-and-death process, one can derive a reaction-diffusion master equation that complements, in a useful manner, the traditional macroscopic description based on the phenomenological rate equations. We have shown through extensive microscopic simulations that this description leads to the macroscopic reaction-diffusion equations with the correct fluctuation spectrum. The reaction-diffusion master equation can thus be considered as the starting point of a statistical mechanics of nonlinear dynamical systems and provides an interesting view of complex phenomena associated to bifurcations, pattern formation, and chaos.

A basic open question is the possibility of spontaneous symmetry breaking induced by the interplay between reactions and diffusion [210]. Experimentally, time-independent chemical patterns have been observed as the result of externally imposed gradients [211]. Quite recently, evidence for stationary patterns displaying an intrinsic wavelength (Turing patterns) has been reported [212, 213]. Molecular dynamics "experiments" on chemical pattern formation should help in further clarifying this important issue. To do so, however, one has to address a number of important problems. First, to get a stable Turing pattern, the diffusion coefficients of at least two of the constituents must be unequal. This can be achieved easily by choosing different sphere diameters for different reactants. On the other hand, to have an efficient microscopic simulation, the frequency of reactive collisions must be comparable to the frequency of elastic ones. As a consequence, we get non-negligible cross-diffusion effects that considerably increase the complexity of the bifurcation diagram of the corresponding macroscopic equations. Next, Turing structures usually arise in the form of multiplets, all of which are equally stable. From a macroscopic point of view, the system will relax to a particular pattern of the multiplet, depending on the initial conditions. This is not the case in microscopic simulations where one finds that the system randomly switches from one structure to the other, mainly because of the presence of important local composition fluctuations. In principle, the only way to get rid of these fluctuations is to increase the number of particles. Preliminary results for the model given by Eqs. (3.15) indicate that even with 150,000 particles it is not possible to get a completely stable Turing pattern, in the sense that if one waits long enough, the system eventually switches to another pattern of the multiplet. How the above observations are related to experiment is an open question.

It should be realized that there exist other "microscopic" techniques to investigate chemical dynamics. One such technique is based on the "lattice gas cellular automata," where the velocity of the particles is restricted to a finite number of possibilities (typically 4 or 6). Quite recently, this technique has been generalized to deal with complex chemical systems as well [22, 148, 171, 172, 214]. More traditional cellular automata techniques, in a spirit similar to the master equation, have also been extended recently to reactive processes and have proved quite efficient for dealing with diffusion-limited reactions [215–217]. So far these techniques are limited to isothermal reactions.

Last but not least, experimental studies of fluctuation-induced effects in chemistry are long overdue. It may be hoped that with the progress achieved in the theoretical and simulation techniques the time is ripe for the experimentalists to turn to this long-standing problem.

ACKNOWLEDGMENTS

We are grateful to G. Nicolis and J. W. Turner for pertinent comments and fruitful discussions. This work is supported in part by the Belgian Federal Office for Technical, Scientific and Cultural Affairs under the "Pôles d'actions interuniversitaires" program.

REFERENCES

1. G. Nicolis and I. Prigogine, *Self-Organization in Nonequilibrium Systems*, Wiley-Interscience, 1977.
2. H. Haken, *Synergetics: An Introduction*, Springer-Verlag, Berlin, 1977.
3. C. Vidal, G. Dewel, and P. Borckmans, *Au-delà de l'équilibre*, Hermann, Paris, 1994.
4. A. M. Zhabotinski, *Biophysics*, **9,** 329 (1964); A. Winfree, *Sci. Amer.*, **230,** 83 (1974).
5. A. Nitzan and J. Ross, *J. Chem. Phys.*, **59,** 1 (1973); A. Nitzan and J. Ross, *J. Stat. Phys.*, **10,** 379 (1974); A. Nitzan, P. Ortoleva, and J. Ross, *J. Chem. Phys.*, **60,** 8 (1974); C. L. Creel and J. Ross, *J. Chem. Phys.*, **65,** 3779 (1976).
6. C. Vidal and A. Pacault, Eds., *Non-equilibrium Dynamics of Chemical Systems*, Springer-Verlag, Berlin, 1984.
7. H. L. Swinney and V. I. Krinsky, Eds., *Waves and Patterns in Chemical and Biological Media, Physica D*, **49** (1991); Q. Ouyang and H. L. Swinney, in *Chemical Waves and Patterns*, Kluwer, Dordrecht, 1995, p. 269.
8. R. Kapral and K. Showalter, Eds., *Chemical Waves and Patterns*, Kluwer, Dordrecht, 1995.
9. P. Glansdorff and I. Prigogine, *Physica*, **20,** 773 (1954).
10. I. Prigogine and R. Balescu, *Bull. Acad. Roy. Belg.*, **41,** 917 (1955).
11. S. Smale, *Bull. Am. Math. Soc.*, **73,** 747 (1967).
12. Y. Kuramoto, *Chemical Oscillations, Waves and Turbulence*, Springer, Berlin, 1984.
13. P. Manneville, *Dissipative Structures and Weak Turbulence*, Academic Press, New York, 1990.
14. G. Nicolis, *Introduction to Nonlinear Science*, Cambridge University Press, Cambridge, 1995.

15. F. Baras and D. Walgraef, Eds., *Nonequilibrium Chemical Dynamics: From Experiment to Microscopic Simulations*, *Physica A*, **188** (1992).

16. A. H. Zewail, *Science*, **242,** 1645 (1988); L. R. Khund Kar and A. H. Zewail, *Annu. Rev. Phys. Chem.*, **41,** 15 (1990).

17. M. Frankowicz and G. Nicolis, *J. Stat. Phys.*, **33,** 3 (1983); F. Mauricio and S. Velasco, *J. Stat. Phys.*, **43,** 521 (1986); P. Peeters, F. Baras and G. Nicolis, *J. Chem. Phys.*, **93,** 7321 (1990).

18. I. Nagypál and I. R. Epstein, *J. Phys. Chem.*, **90,** 6285 (1986); I. Nagypál and I. R. Epstein, *J. Chem. Phys.*, **89,** 6925 (1988).

19. F. Baras, G. Nicolis, M. Malek Mansour, and J. W. Turner, *J. Stat. Phys.*, **32,** 1 (1983); A. Lemarchand, B. I. Ben Aïn, and G. Nicolis, *Chem. Phys. Lett.*, **162,** 92 (1989).

20. J. E. Keizer and J. Tielsen, *J. Phys. Chem.*, **93,** 2811 (1989); R. F. Fox and J. E. Keizer, *Phys. Rev. Lett.*, **64,** 249 (1990); R. F. Fox and J. E. Keizer, *Phys. Rev. A*, **43,** 1709 (1991); J. E. Keizer, R. F. Fox and J. Wagner, *Phys. Lett. A*, **175,** 17 (1993).

21. G. Nicolis and V. Balakrishnan, *Phys. Rev. A*, **46,** 3569 (1992); P. Peeters and G. Nicolis, *Physica A*, **188,** 426 (1992); P. Geysermans and G. Nicolis, *J. Chem. Phys.*, **99,** 8964 (1993).

22. G. X.-G. Wu and R. Kapral, *Phys. Rev. Lett.*, **70,** 1940 (1993); X.-G. Wu and R. Kapral, *J. Chem. Phys.*, **100,** 5936 (1994).

23. R. Graham and H. Haken, *Z. Physik*, **243,** 289 (1971); R. Graham and H. Haken, *Z. Physik*, **245,** 141 (1971).

24. A. Nitzan, P. Ortoleva, R. Deutch, and J. Ross, *J. Chem. Phys.*, **61,** 1056 (1974).

25. H. Ueyama, *J. Stat. Phys.*, **23,** 463 (1980); P. Hänggi, and H. Thomas, *Phys. Rep.*, **88,** 207 (1982).

26. D. McQuarrie, *Suppl. Rev. Ser. Appl. Prob.*, Methuen, London (1967).

27. N. G. Van Kampen, *Stochastic Processes in Physics and Chemistry*, North-Holland, Amsterdam, 1983.

28. C. W. Gardiner, *Handbook of Stochastic Methods*, Springer-Verlag, Berlin, 1985.

29. M. Malek Mansour, C. Van Den Broeck, G. Nicolis, and J. W. Turner, *Ann. Phys. (USA)*, **131,** 283 (1981).

30. R. Graham and T. Tel, *Phys. Rev. A*, **42,** 4661 (1990); R. Graham, in *Synergetics and Dynamical Instabilities*, G. Gaglioti, H. Haken, and L. Lugiato, Eds., North-Holland, Amsterdam, 1988.

31. E. Sulpice, A. Lemarchand, and H. Lemarchand, *Phys. Lett. A*, **158,** 43 (1991).

32. O. Descalzi and R. Graham, *Z. Physik B*, **93,** 509 (1994); ibid. *Phys. Lett. A*, **170,** 84 (1992).

33. P. H. Richter, I. Procaccia, and J. Ross, *Adv. Chem. Phys.*, **43,** 217 (1980).

34. I. Prigogine and E. Xhrouet, *Physica*, **15,** 913 (1949); C. F. Curtiss, Ph.D. Dissertation, University of Wisconsin 1948 (unpublished).

35. I. Prigogine and M. Mahieu, *Physica*, **16,** 51 (1950).

36. B. J. Alder and T. E. Wainwright, in *Transport Processes in Statistical Mechanics*, I. Prigogine Ed., Interscience, New York (1958); B. J. Alder and T. E. Wainwright, *J. Chem. Phys.*, **31,** 459 (1959); B. J. Alder and T. E. Wainwright, *J. Chem. Phys.*, **33,** 1439 (1960).

37. G. Ciccotti and W. G. Hoover, Eds., *Molecular Dynamics Simulations of Statistical Mechanical Systems*, North-Holland, Amsterdam, 1986.

38. W. G. Hoover, *Computational Statistical Mechanics*, Elsevier Science, Amsterdam (1991); D. J. Evans and G. P. Moriss, Eds., *Statistical Mechanics of Non-Equilibrium Liquids*, Academic Press, New York, 1990.

39. M. Mareschal, Ed., *Microscopic Simulations of Complex Flows*, Nato ASI Series, Vol. 236, Plenum Press, New York, 1990.

40. M. Mareschal and B. Holian, Eds., *Microscopic Simulations of Complex Hydrodynamic Phenomena*, Nato ASI Series, Vol. 292, Plenum Press, New York, 1992.

41. W. G. Hoover, *Phys. Rev. Lett.*, **42,** 1531 (1979); B. L. Holian, W. G. Hoover, B. Moran, and G. K. Straub, *Phys. Rev. A*, **22,** 2798 (1980); B. L. Holian, *Phys. Rev. A*, **37,** 2562 (1988).

42. D. C. Rapaport and E. Clementi, *Phys. Rev. Lett.*, **57,** 695 (1987); L. Hannon, G. Lie, and E. Clementi, *J. Sc. Comp.*, **1,** 145 (1986); D. C. Rapaport, *Phys. Rev. A*, **36,** 3288 (1987); E. Meiburg, *Phys. Fluids*, **29,** 3107 (1986).

43. M. Mareschal and E. Kestemont, *Nature*, **323,** 427 (1987); *J. Stat. Phys.*, **48,** 1187 (1987); D. C. Rapaport, *Phys. Rev. Lett.*, **60,** 2840 (1988).

44. M. Mareschal, M. Malek Mansour, A. Puhl, and E. Kestemont, *Phys. Rev. Lett.*, **61,** 2550 (1988); A. Puhl, M. Malek Mansour, and M. Mareschal, *Phys. Rev. A*, **40,** 1999 (1989).

45. F. Abraham, *Adv. Phys.*, **35,** 1 (1986).

46. L. D. Landau and E. M. Lifshitz, *Statistical Physics*, Part II, Pergamon Press, Oxford, 1981.

47. W. Ebeling and L. Schimansky-Geier, *Statistical Thermodynamics and Stochastic Theory of Nonlinear Systems Far From Equilibrium*, Adv. Series in Stat. Mech., Vol. 8, World Scientific, New York, 1994.

48. S. Karlin and H. Taylor, *A First Course in Stochastic Processes*, Academic Press, 1975.

49. D. T. Gillespie, *J. Comput. Phys.*, **22,** 403 (1976); D. T. Gillespie, *J. Chem. Phys.*, **81,** 2340 (1977).

50. D. T. Gillespie, *Markov Processes: an Introduction for Physical Scientists*, Academic Press, San Diego, 1992.

51. J. S. Turner, *J. Phys. Chem.*, **81,** 237 (1977).

52. P. Hanusse and A. Blanché, *J. Chem. Phys.*, **74,** 6148 (1981).

53. M. Malek Mansour and G. Nicolis, *J. Stat. Phys.*, **13,** 197 (1975).

54. S. Grossman, *J. Chem. Phys.*, **65,** 2007 (1976).

55. J. Keiser, *J. Chem. Phys.*, **67,** 1473 (1977).

56. G. Nicolis and M. Malek Mansour, *Phys. Rev. A*, **29,** 2845 (1984).

57. N. G. Van Kampen, *Canad. J. Phys.*, **39,** 551 (1961).

58. N. G. Van Kampen, *Adv. Chem. Phys.*, **34,** 245 (1976).

59. R. Kubo, K. Matsuo, and K. Kitahara, *J. Stat. Phys.*, **9,** 51 (1975).

60. G. Nicolis and A. Babloyantz, *J. Chem. Phys.*, **51,** 2632 (1969).

61. C. W. Gardiner and S. Chaturvedi, *J. Stat. Phys.*, **17,** 429 (1978).

62. S. Chaturvedi and C. W. Gardiner, *J. Stat. Phys.*, **18,** 501 (1978).

63. T. G. Kurtz, *Math. Progr. Stud.*, **5,** 67 (1976); T. G. Kurtz, *Stoch. Proc. Appl.*, **6,** 223 (1978).

64. A. Van Der Ziel, *Proc. IRE*, **43,** 77 (1955); K. M. Van Vliet, *J. Math. Phys.*, **12,** 1981 (1971).
65. D. K. Kondepudi, in *Noise in Nonlinear Dynamical Systems*, F. Moss and P. V. E. McClintock, Eds., Cambridge University Press, Cambridge, 1989, Vol. 2, pp. 251–270.
66. M. Frankowicz, M. Malek Mansour, and G. Nicolis, *Physica A*, **125,** 237 (1984).
67. S. Karlin and J. McGregor, *Trans. Amer. Math. Soc.*, **85,** 489 (1975).
68. F. Schlögl, *Z. Physik*, **253,** 147 (1972).
69. G. Nicolis, *J. Stat. Phys.*, **6,** 195 (1972); K. J. McNeil and D. F. Walls, *J. Stat. Phys.*, **10,** 439 (1974).
70. G. Nicolis and J. W. Turner, *Physica A*, **89,** 245 (1977).
71. H. K. Jansen, *Z. Physik*, **270,** 67 (1974); I. S. Matheson, D. F. Walls, and C. W. Gardiner, *J. Stat. Phys.*, **12,** 21 (1975).
72. W. Feller, *An Introduction to Probability Theory and Its Applications*, Wiley, New York, 1959, Vol. 2.
73. S. Grossman and R. Schraner, *Z. Physik B*, **30,** 325 (1978).
74. H. Denker, *Physica A*, **103,** 55 (1980).
75. H. Denker and N. G. Van Kampen, *Phys. Lett. A*, **73,** 374 (1979).
76. G. Nicolis and R. Lefever, *Phys. Lett. A*, **62,** 469 (1977); G. Nicolis and J. W. Turner, *Ann. N.Y. Acad. Sci.*, **316,** 251 (1979).
77. B. J. Matkowsky, Z. Schuss, C. Knessel, C. Tier, and M. Mangel, in *Fluctuations and Sensitivity in Nonequilibrium Systems*, W. Horsthemke and D. K. Kondepudi, Eds., Springer-Verlag, New York, 1984.
78. L. D. Landau and E. M. Lifshitz, *Fluid Mechanics*, Pergamon Press, Oxford, 1984.
79. R. G. Glauber, *J. Math. Phys.*, **4,** 294 (1963).
80. W. Horsthemke and L. Brenig, *Z. Physik B*, **27,** 341 (1977).
81. P. Hänggi, *Helv. Phys. Acta*, **51,** 183 (1978).
82. W. Horsthemke, M. Malek Mansour, and L. Brenig, *Z. Physik B*, **28,** 135 (1977).
83. C. Vidal and H. Lemarchand, *La reaction créatice*, Herman, Paris, 1988.
84. Blomberg (preprint).
85. H. Grabert, P. Hänggi, and I. Oppenheim, *Physica A*, **117,** 300 (1983).
86. P. Hänggi, H. Grabert, P. Talkner, and H. Thomas, *Phys. Rev. A*, **29,** 371 (1984).
87. K. Matsuo, *J. Stat. Phys.*, **16,** 169 (1977).
88. H. Risken, *The Fokker-Planck Equation*, Springer-Verlag, Berlin, 1989.
89. B. Caroli, C. Caroli, and B. Roulet, *J. Stat. Phys.*, **21,** 415 (1979); B. Caroli, C. Caroli, and B. Roulet, *Physica A*, **101,** 581 (1980).
90. N. G. Van Kampen, *J. Stat. Phys.*, **17,** 71 (1977).
91. N. G. Van Kampen, *Suppl. Prog. Theor. Phys.*, **64,** 389 (1978).
92. C. Gardiner, in *Stochastic Nonlinear Systems*, L. Arnold and R. Lefever, Eds., Springer-Verlag, Berlin, 1981.
93. J. C. Englund, W. C. Schieve, R. F. Gragg, and W. Zurek, in *Instabilities, Bifurcations and Fluctuations in Chemical Systems*, L. E. Reichl and W. C. Schieve, Eds., University of Texas Press, Austin, 1982; J. C. Englund, W. C. Schieve, and R. F. Gragg, in *Optical Bistability*, C. M. Bowden, Ed., Plenum Press, New York, 1981.
94. Z. Schuss and B. Matskowsky, *SIAM J. Apl. Math.*, **36,** 604 (1979); B. J. Matskowsky, Z.

Schuss, C. Knessel, C. Tier, and M. Mangel, *Phys. Rev. A*, **29,** 3359 (1984); C. Knessel, B. J. Matkowsky, Z. Schuss, and C. Tier, *J. Stat. Phys.*, **42,** 169 (1986).

95. I. Oppenheim, K. E. Shuler, and G. H. Weiss, *Physica A*, **88,** 191 (1977).
96. K. Lindenberg, K. E. Schuler, J. Freeman, and T. J. Lie, *J. Stat. Phys.*, **12,** 217 (1975); K. Lindenberg and V. Seshadri, *J. Chem. Phys.*, **71,** 4075 (1979); V. Seshadri, B. J. West, and K. Lindenberg, *J. Chem. Phys.*, **72,** 1145 (1980).
97. F. de Pasquale and P. Tombesi, *Phys. Lett. A*, **72,** 7 (1979); F. de Pasquale, P. Tartaglia, and P. Tombesi, *Physica A*, **99,** 581 (1979).
98. M. Suzuki, *Prog. Theor. Phys.*, **56,** 77 (1976); M. Suzuki, *Prog. Theor. Phys.*, **56,** 477 (1976).
99. M. Suzuki, *Adv. Chem. Phys.*, **46,** 195 (1981).
100. I. Prigogine and R. Lefever, *J. Chem. Phys.*, **48,** 1695 (1968); R. Lefever and G. Nicolis, *J. Theor. Biol.*, **30,** 267 (1971).
101. R. Lefever, G. Nicolis, and P. Borckmans, *J. Chem. Soc., Faraday Trans. I*, **84,** 1013 (1988).
102. J. W. Turner, in *Proceedings of the International Conference on Synergetics*, H. Haken, Ed., Springer, Berlin, 1980, p. 255.
103. H. Lemarchand, *Physica A*, **101,** 518 (1980).
104. W. Ebeling, *Phys. Lett. A*, **68,** 430 (1978); R. Feistel and W. Ebeling, *Physica A*, **93,** 114 (1978).
105. K. Tomita, T. Ohta, and H. Tomita, *Prog. Theor. Phys.*, **52,** 1744 (1974).
106. R. Schraner, S. Grossman, and P. H. Richter, *Z. Physik B*, **35,** 363 (1979).
107. F. Baras, M. Malek Mansour, and C. Van den Broeck, *J. Stat. Phys.*, **28,** 577 (1982).
108. Y. Kuramoto and T. Tsuzuki, *Progr. Theor. Phys.*, **52,** 1399 (1974).
109. P. H. Coullet, C. Elphick, and E. Tirapegui, *Phys. Lett.*, **111,** 277 (1985).
110. W. Horsthemke and R. Lefever, *Noise Induced Transitions: Theory and Applications in Physics, Chemistry and Biology*, Springer, Berlin, 1984.
111. R. Graham, *Phys. Rev. A*, **25,** 3234 (1982); R. Lefever and J. W. Turner, *Phys. Rev. Lett.*, **56,** 1631 (1986).
112. R. Graham and A. Schenzle, *Phys. Rev. A*, **26,** 1676 (1982).
113. Y. Kuramoto and T. Tsuzuki, *Progr. Theor. Phys.*, **54,** 60 (1975).
114. (a) D. Walgraef, G. Dewel, and P. Borckmans, *J. Chem. Phys.*, **78,** 3043 (1983); (b) M. Dykman, Chu Xiaolin, and J. Ross, *Phys. Rev. E*, **48,** 1646 (1993).
115. K. Kitahara, Ph.D. thesis, Université Libre de Bruxelles, 1974.
116. H. Haken, *Z. Physik B*, **20,** 413 (1975).
117. C. W. Gardiner, K. J. Macneil, D. F. Walls, and I. S. Matheson, *J. Stat. Phys.*, **14,** 307 (1976).
118. H. Lemarchand and G. Nicolis, *Physica A*, **82,** 251 (1976).
119. G. Nicolis and I. Prigogine, *Proc. Natl. Acad. Sci. USA*, **68,** 2102 (1971).
120. G. Nicolis and M. Malek Mansour, *J. Stat. Phys.*, **22,** 495 (1980).
121. L. Arnold and M. Theodosopulu, *Adv. Appl. Prob.*, **12,** 367 (1980).
122. K. Kitahara, H. Metiu, and J. Ross, *J. Chem. Phys.*, **63,** 3156 (1975); H. Metiu, K. Kitahara, and J. Ross, *J. Chem. Phys.*, **64,** 292 (1976).
123. Y. Kuramoto, *Prog. Theor. Phys.*, **52,** 711 (1974).

124. Y. Kuramoto, *Prog. Theor. Phys.*, **51,** 1712 (1974); H. Mori, *Prog. Theor. Phys.*, **53,** 1617 (1975).

125. C. Van den Broeck, W. Horsthemke, and M. Malek Mansour, *Physica A*, **89,** 339 (1977).

126. C. W. Gardiner, K. J. McNeil, and D. F. Walls, *Phys. Lett. A*, **53,** 205 (1975).

127. M. Malek Mansour and C. Van den Broeck, in *Instabilities, Bifurcations and Fluctuations in Chemical Systems*, L. E. Reichl and W. C. Schieve, Eds., University of Texas Press, Austin, 1982.

128. K. Seki and K. Kitahara, *J. Mol. Liq.*, (in press).

129. B. J. Bern and R. Pecora, *Dynamic Light Scattering with Applications to Chemistry, Biology, and Physics*, Wiley, New York, 1976.

130. C. Van den Broeck, J. Houard, and M. Malek Mansour, *Physica A*, **101,** 167 (1980).

131. G. Dewel, D. Walgraef, and P. Borckmans, *Z. Phys. B*, **28,** 235 (1977).

132. D. Walgraef, G. Dewel, and P. Borckmans, *Adv. Chem. Phys.*, **49,** 311 (1982).

133. F. Baras, *Phys. Rev. Lett.*,**77,** 1398 (1996).

134. W. Feller, An Introduction to Probability Theory and Its Applications, Wiley, New York, 1967, Vol. 1.

135. C. Van den Broeck and M. Malek Mansour, in *Instabilities and Nonequilibrium Structures IV*, E. Tirapegui and W. Zellers, Eds., Kluwer, Dordrecht, 1993, p. 149.

136. M. Malek Mansour and G. Nicolis, *J. Stat. Phys.*, **13,** 197 (1975).

137. M. Delledonne and P. Ortoleva, *J. Stat. Phys.*, **18,** 319 (1978).

138. G. Nicolis, M. Malek Mansour, A. Van Nypelseer, and K. Kitahara, *J. Stat. Phys.*, **14,** 417 (1976); I. Prigogine, R. Lefever, J. S. Tumer, and J. W. Turner, *Phys. Lett.*, **51 A,** 317 (1975); C. Y. Mou, G. Nicolis, and R. M. Mazo, *J. Stat. Phys.*, **18,** 19 (1978).

139. M. Malek Mansour and J. Houard, *Phys. Lett.*, **70 A,** 366 (1979).

140. G. Nicolis and M. Malek Mansour, *Progr. Theor. Phys.*, Suppl. No 64, 249 (1978).

141. P. Hanusse (preprint).

142. R. A. Fisher, *Ann. Eugenics*, **7,** 335 (1937); A. Kolmogorov, I. Petrovsky, and N. Piskunov, *Bull. Univ. Moscow. Ser. Intl. Sec A*, 1 (6), 1, (1937). For a review, see P. Fife, "Mathematical Aspects of Reacting and Diffusing Systems," in *Biomathematics*, Springer, New York, 1979, Vol. 28.

143. F. Schlögl, C. Escher, and R. S. Berry, *Phys. Rev. A*, **27,** 2698 (1983).

144. A. S. Mikhailov, L. Schimansky-Geier, and W. Ebeling, *Phys. Lett. A*, **96,** 453 (1983); L. Schimansky-Geier and Ch. Zülicke, *Z. Physik B*, **82,** 157 (1991).

145. B. Chopard and M. Droz, *Europhys. Lett.*, **15,** 459 (1991); S. Cornell, M. Droz, and B. Chopard, *Phys. Rev. A*, **44,** 4826 (1991); B. Chopard, M. Droz, T. Karapiperis, and Z. Racz, *Phys. Rev. E*, **47,** 1140 (1993).

146. A. Lemarchand, H. Lemarchand, E. Sulpice, and M. Mareschal, *Physica A*, **188,** 277 (1992).

147. A. Lemarchand, A. Lesne, A. Perera, M. Moreau, and M. Mareschal, *Phys. Rev. E*, **48,** 1568 (1993); A. Lemarchand, H. Lemarchand, A. Lesne, and M. Mareschal, in *Far From Equilibrium Dynamics of Chemical Systems*, J. Gorecki et al., Eds., World Scientific, Singapore, 1994.

148. A. Lemarchand, A. Lesne, and M. Mareschal, (preprint).

149. J. P. Hansen and I. R. Mc Donald, *Theory of Simple Liquids*, Academic Press, New York, 1986.

150. M. P. Allen and D. J. Tildesley, *Computer Simulations of Liquids*, Oxford Science Publications, Oxford, 1991.
151. *Int. J. Quant. Chem.*, (special issue in honor of E. Clementi), **42** (1992).
152. L. Kubia, Ed., *Computer Simulation in Material Science*, Kluwer, Dordrecht, 1995.
153. W. Ebeling and M. Jenssen, *Physica A*, **188,** 350 (1992); W. Ebeling, V. Yu. Podlipchuk, and A. A. Valuev, *Physica A*, **217,** 22 (1995).
154. C. T. White, D. H. Robertson, and D. W. Brenner, *Physica A*, **188,** 357 (1992).
155. G. A. Bird, *Molecular Gas Dynamics*, Clarendon, Oxford, 1976.
156. A. L. Garcia, *Numerical Methods for Physics*, Prentice-Hall, New Jersey, 1994.
157. M. Kac, in *Probability Theory and Related Topics in Physical Science*, Wiley, New York, 1959; J. Logan and M. Kac, *Phys. Rev. A*, **13,** 458 (1976).
158. A. Onuki, *J. Stat. Phys.*, **18,** 475 (1978); L. Brenig and C. Van den Broeck, *Phys. Rev. A*, **21,** 1039 (1980).
159. G. A. Bird, Molecular Gas Dynamics and the Direct Simulation of Gas Flows, Clarendon, Oxford, 1994.
160. D. R. Chenoweth and S. Paolucci, *Phys. Fluids*, **28,** 2365 (1985); G. A. Bird, *Phys. Fluids*, **30,** 364 (1987).
161. E. P. Muntz, *Ann. Rev. Mech.*, **21,** 387 (1989).
162. R. E. Meyer, *Introduction to Mathematical Fluid Mechanics*, Dover Publications, New York, 1971; G. A. Bird, *Phys. Fluids*, **13,** 1172 (1970).
163. E. Salomons and M. Mareschal, *Phys. Rev. Lett.*, **69,** 269 (1992).
164. M. Malek Mansour, A. Garcia, G. Lie, and E. Clementi, *Phys. Rev. Lett.*, **58,** 874 (1987).
165. M. Malek Mansour, M. Mareschal, G. Sonnino, and E. Kestemont, in M. Mareschal and B. Holian, Eds., *Microscopic Simulations of Complex Hydrodynamics Phenomena*, NATO Advanced Science Institute Series, Vol. 292, Plenum Press, New York, 1992, p. 87.
166. A. Garcia, M. Malek Mansour, G. Lie, and E. Clementi, *Phys. Rev. A*, **36,** 4348 (1987).
167. J. P. Boon and S. Yip, *Molecular Hydrodynamics*, McGraw-Hill, New York, 1980 (reprinted by Dover, New York, 1991).
168. B. M. Law and J. V. Sengers, *J. Stat. Phys.*, **57,** 531 (1989); B. M. Law, P. N. Segrè, R. W. Gammon, and J. V. Sengers, *Phys. Rev. A*, **41,** 816 (1990); W. B. Li, J. V. Sengers, R. W. Gammon, and P. N. Segrè, *Int. J. Thermophysics*, **16,** 23 (1995).
169. S. Succi, R. Benzi, A. Cali, and M. Vergassola, in M. Mareschal and B. Holian, Eds., *Microscopic Simulations of Complex Hydrodynamics Phenomena*, NATO Advanced Science Institute Series, Vol. 292, Plenum Press, New York, 1992, p. 187; D. H. Rothman, *ibid.*, p. 221; C. Burges and S. Zaleski, *Complex Systems*, **1,** 31 (1987).
170. G. McNamara and B. Alder, in M. Mareschal and B. Holian, Eds., *Microscopic Simulations of Complex Hydrodynamics Phenomena*, NATO Advanced Science Institute Series, Vol. 292, Plenum Press, New York, 1992, p. 125.
171. D. Dab, A. Lawniczak, J. P. Boon, and R. Kapral, *Phys. Rev. Lett.*, **64,** 2462 (1990); D. Dab, J. P. Boon, and Y. Li, *Phys. Rev. Lett.*, **66,** 2535 (1991).
172. R. Kapral, A. Lawniczak, and P. Masiar, *Phys. Rev. Lett.*, **66,** 2539 (1991); X.-G. Wu and R. Kapral, in M. Mareschal and B. Holian, Eds., *Microscopic Simulations of Complex Hydrodynamics Phenomena*, NATO Advanced Science Institute Series, Vol. 292, Plenum Press, New York, 1992, p. 284.

173. P. N. Segrè, R. W. Gammon, and J. V. Sengers, *Phys. Rev. E*, **47,** 1026 (1993); W. B. Li, P. N. Segrè, J. V. Sengers, and R. W. Gammon, *J. Phys.*, **6,** A119 (1994); W. B. Li, P. N. Segrè, R. W. Gammon, and J. V. Sengers, *Physica A*, **204,** 399 (1994).

174. P. Grosfils, J. P. Boon, and P. Lallemand, *Phys. Rev. Lett.*, **68,** 1077 (1992); P. Grosfils, J. P. Boon, R. Brito, and M. H. Ernst, *Phys. Rev. E*, **48,** 2665 (1993).

175. J. R. Weimar, D. Dab, J. P. Boon, and S. Succi, *Europhys. Lett.*, **20,** 627 (1992).

176. A. J. C. Ladd, *Phys. Rev. Lett.*, **70,** 1339 (1993).

177. J. P. Boon, D. Dab, R. Kapral, and A. Lawniczak, *Phys. Rep.*,**273,** 55 (1996).

178. J. Portnow, *Phys. Lett. A*, **51,** 370 (1975).

179. P. Ortoleva and S. Yip, *J. Chem. Phys.*, **65,** 2045 (1976).

180. J. Boissonade, *Phys. Lett. A*, **74,** 285 (1979).

181. J. Boissonade, in *Nonlinear Phenomena in Chemical Dynamics*, C. Vidal and A. Pacault Eds., Springer-Verlag, Berlin, 1981; J. Boissonade, *Physica A*, **113,** 607 (1982).

182. R. D. Present, *J. Chem. Phys.*, **31,** 747 (1959).

183. J. Ross and P. Mazur, *J. Chem. Phys.*, **35,** 19 (1961).

184. B. Schizgal and M. Karplus, *J. Chem. Phys.*, **52,** 4262 (1970); B. Schizgal and M. Karplus, *J. Chem. Phys.*, **54,** 4345 (1971).

185. N. Xystris and J. S. Dahler, *J. Chem. Phys.*, **68,** 374 (1978); N. Xystris and J. S. Dahler, *J. Chem. Phys.*, **68,** 387 (1978).

186. B. Schizgal and D. G. Napier, *Physica A*, **223,** 50 (1996).

187. A. S. Cukrowski, S. Fritzsche, and J. Popielawski, in *Far From Equilibrium Dynamics of Chemical Systems*, J. Popielawski and J. Gorecki, Eds., World Scientific, Singapore, 1991; J. Popielawski, A. S. Cukrowski, and S. Fritzsche, *Physica A*, **188,** 344 (1992).

188. D. A. Frank Kamenetskii, *Diffusion and Heat Transfer in Chemical Kinetics*, Plenum Press, New York, 1969.

189. Y. B. Zeldovich, G. I. Barenblatt, V. B. Librovich, and G. M. Makhviladze, *The Mathematical Theory of Combustion and Explosions*, Plenum Publ. Corp., New York, 1985.

190. N. W. Bazley and G. C. Wake, *Combust. Flame*, **33,** 161 (1978); T. Takeno, *Combust. Flame*, **29,** 209 (1977); W. Gill, A. B. Donaldson, and A. R. Shouman, *Combust. Flame*, **36,** 217 (1979).

191. T. Boddington, C.-G. Feng, and P. Gray, *J. Chem. Soc. Faraday*, **79,** 1499 (1983).

192. G. C. Wake, *Combust. Flame*, **39,** 215 (1980).

193. D-P. Chou and S. Yip, *Combust. Flame*, **47,** 215 (1982); D-P. Chou and S. Yip, in *Chemical Instabilities*, G. Nicolis and F. Baras, Eds., Reidel, Dordrecht, 1984, p. 159.

194. F. Baras and M. Malek Mansour, *Phys. Rev. Lett.*, **63,** 2429 (1989); M. Malek Mansour and F. Baras, *Physica A*, **188,** 253 (1992).

195. F. Baras and G. Nicolis, in *Microscopic Simulations of Complex Flows*, M. Mareschal, Ed., Nato ASI Series, 1990, Vol. 236, p. 339.

196. R. D. Present, *Kinetic Theory of Gases*, McGraw-Hill, New York, 1958.

197. W. Stiller, *Arrhenius Equation and Non-Equilibrium Kinetics*, Teubner Texb 21, Teubner-Verlag, Leipzig, 1989.

198. J. Gorecki and J. Gryko, *J. Stat. Phys.*, **48,** 329 (1987).

199. E. H. Kennard, *Kinetic Theory of Gases*, McGraw-Hill, New York, 1938.

200. D. K. Bhattacharya and G. C. Lie, *Phys. Rev. Lett.*, **62,** 897 (1989).

201. P. Gray and S. K. Scott, *Chemical Oscillations and Instabilities*, Oxford University Press, Oxford, 1994.

202. M. Mareschal and A. De Wit, *J. Chem. Phys.*, **96,** 2000 (1992).

203. F. Baras, J. E. Pearson, and M. Malek Mansour, *J. Chem. Phys.*, **93,** 5747 (1990).

204. A. Garcia, M. Malek Mansour, G. Lie, and E. Clementi, *J. Stat. Phys.*, **47,** 209 (1987).

205. R. Kubo, *Fluctuation, Relaxation and Resonance in Magnetic Systems*, D. ter Haar, Ed., Scottish Universities Summer School in Physics, Oliver and Boyd, Edinburgh, 1962; R. Kubo and K. Tomita, *J. Phys. Soc. Jap.*, **9,** 888 (1954).

206. G. Nicolis, A. Amellal, G. Dupont, and M. Mareschal, *J. Mol. Liquids*, **41,** 5 (1989); A. Amellal and M. Mareschal, *J. Mol. Liq.*, **63,** 199 (1995).

207. F. Baras and M. Malek Mansour, *Phys. Rev. E*,**54,** 6139 (1996).

208. F. Argoul, A. Arneodo, P. Richetti, J.-C. Roux, and H. L. Swinney, *Acct. Chem. Res.*, **20,** 436 (1987); L. Gyorgyi and R. Field, *J. Chem. Phys.*, **92,** 7079 (1988).

209. P. Geysermans and F. Baras, *J. Chem. Phys.*, **105,** 1402 (1996).

210. J. E. Pearson and W. Horsthemke, *J. Chem. Phys.*, **90,** 1588 (1989).

211. Q. Ouyang, J. Boissonade, J. C. Roux, and P. deKepper, *Phys. Lett. A*, **134,** 282 (1989).

212. V. Castets, E. Dulos, J. Boissonade, and P. De Kepper, *Phys. Rev. Lett.*, **64,** 295 (1990); cf. J. J. Perraud, K. Agladze, E. Dulos, and P. De Kepper, *Physica A*, **188,** 1 (1992).

213. R. D. Vigil, Q. Ouayng, and H. L. Swinney, *Physica A*, **188,** 17 (1992).

214. X.-G. Wu and R. Kapral, *Physica A*, **188,** 284 (1992).

215. S. Cornell, M. Droz, and B. Chopard, *Physica A*, **188,** 322 (1992).

216. A. Provata, J. W. Turner, and G. Nicolis, *J. Stat. Phys.*, **70,** 1195 (1993); A. Provata and J. W. Turner, in *Statistical Physics and Thermodynamics of Nonlinear Nonequilibrium Systems*, W. Ebeling and W. Muschuk, Eds., World Scientific, London, 1993.

217. A. Tretyakov, A. Provata, and G. Nicolis, *J. Phys. Chem.*, **99,** 2770 (1995); D. Dab, O. Tribel, G. Nicolis, and M. Promel, *Phys. Rev. Lett.*, **74,** 824 (1995).

DIFFERENTIAL RECURRENCE RELATIONS FOR NON–AXIALLY SYMMETRIC ROTATIONAL FOKKER–PLANCK EQUATIONS

L. J. GEOGHEGAN

Department of Applied Mathematics and Theoretical Physics, The Queen's University of Belfast, Belfast BT7 1NN, Northern Ireland

W. T. COFFEY AND B. MULLIGAN

School of Engineering, Department of Electrical and Electronic Engineering, Trinity College, Dublin, Ireland

CONTENTS

List of Major Symbols
I. Introductory Concepts
A. The Purpose of this Review
B. Relaxation of Fine Ferromagnetic Particles
C. The Discrete Orientation Model
D. Néel's Model
E. Brown's Model
F. Calculation of the Correlation Time from the Eigenvalues and Amplitudes of the Fokker–Planck Equation
G. The Case of Axial Symmetry
II. Differential Recurrence Relations for Non–Axially Symmetric Problems
A. Matrix Elements and Differential Recurrence Coefficients of the Fokker–Planck Operator
B. Calculation of the Correlation Time from the Matrix Elements and Differential Recurrence Coefficients
C. Spherical Harmonic Representation Formulae for the Fokker–Planck Operator
III. Application to the Case Where the Free Energy Arises from Uniaxial Anisotropy with Constant Uniform External Magnetic Field at an Angle Oblique to the Easy Axis
A. Calculation of the Matrix Elements and Differential Recurrence Coefficients from the Spherical Harmonic Representation Formulae

Advances in Chemical Physics, Volume 100, Edited by I. Prigogine and Stuart A. Rice.
ISBN 0-471-17458-0

B. Calculation of the Differential Recurrence Relations for the Sharp Values of the Spherical Harmonics from the Gilbert–Langevin Equation
C. Results of Numerical Computation
IV. Application to the Case Where the Free Energy Arises from Cubic Anisotropy
A. Calculation of the Matrix Elements and Differential Recurrence Coefficients Using the Spherical Harmonic Recurrence Relations
B. Calculation of the Matrix Elements and Differential Recurrence Coefficients using the Product Formula for Spherical Harmonics
V. Brown's High Energy Barrier Approximation
A. Introduction
B. Simplification of the Fokker–Planck Equation near a Stationary Point of the Free Energy
C. Behavior in Quasi-Equilibrium near the Anisotropic Minima
D. Behavior near the Saddle Point
E. Brown's High Energy Barrier Approximation Formula
VI. Application to the Case Where the Free Energy Arises from Uniaxial Anisotropy with Constant Uniform External Magnetic Field at an Angle Oblique to the Easy Axis
A. Coordinate System and Stationary Points of the Free Energy
B. Orthogonal Transformation of the Coordinate System and Approximation of the Free Energy near the Stationary Points
C. Bistable Structure of the Free Energy
D. Geometrical Interpretation of the Effect of Changes in h and ψ on the Positions of the Stationary Points
E. High Energy Barrier Approximation for $\psi = \pi/4$
F. High Energy Barrier Approximation for an Arbitrary value of ψ
Appendix A. Associated Legendre Functions and Spherical Harmonics
Appendix B. Derivation of Brown's Fokker–Planck Equation from the General Fokker–Planck Equation and the Gilbert–Langevin Equation
Appendix C. Orthogonal Transformations
Acknowledgments
References

LIST OF MAJOR SYMBOLS

$\mathbf{r}$	Magnetization orientation
$\mathbf{h}$	External magnetic field orientation
$\mathbf{rh}$	Scalar product of the vectors $\mathbf{r}$ and $\mathbf{h}$
M_s	Spontaneous magnetization
H	Magnitude of external magnetic field
K	Anisotropy constant
ξ	External field parameter
σ	Anisotropy parameter
h	Ratio of external field to anisotropy parameter
$(\mathbf{e}_1, \mathbf{e}_2, \mathbf{e}_3)$	Crystal coordinate system, $\mathbf{e}_3$ denotes the direction of the easy axis
θ, ϕ	Angular spherical polar coordinates of $\mathbf{r}$ relative to $(\mathbf{e}_1, \mathbf{e}_2, \mathbf{e}_3)$

ψ	Angle between $\mathbf{h}$ and $\mathbf{e}_3$
$V(\mathbf{r})$	Gibbs free energy per unit volume
V_i	Values of V at its stationary points
$W(\mathbf{r}, t)$	Distribution of magnetization orientations
$\mathbf{J}(\mathbf{r}, t)$	Current density of representative points
v	Magnetic particle volume
β	Ratio of magnetic particle volume to thermal energy kT
n_i	Number of particles in ith orientation
$\nu_{i,j}$	Transition probability
γ	Gyromagnetic constant
η	Phenomenological damping constant from Gilbert's equation
a, b	Brown's constants
α	Damping parameter
$\partial/\partial\mathbf{r}$	Gradient operator with respect to $\mathbf{r}$
Λ^2	Laplacian operator
L_{FP}	Fokker–Planck operator
A_{FP}, G_{FP}	Real and imaginary components of the Fokker–Planck operator
$p_{j,k}$	Coefficients in representation formula for the Fokker–Planck operator
p_n, W_n, A_n	Corresponding eigenvalues, eigenfunctions, and amplitudes of Fokker–Planck equation
λ_n	Eigenvalues of the corresponding Sturm–Liouville equation
p_1, λ_1	Lowest nonvanishing eigenvalues
τ	Relaxation time $(1/p_1)$
$f(t)$	Response function
$C(t)$	Autocorrelation function
T	Correlation time
P_l^m	Associated Legendre functions
α_i	Direction cosines of magnetization orientation relative to $(\mathbf{e}_1, \mathbf{e}_2, \mathbf{e}_3)$
$X_{l,m}$	Spherical harmonics
$Y_{l,m}$	Normalized spherical harmonics
$N_{l,m}$	Normalization constants
$a_{l,m}$	Fourier coefficients
$\langle X_{l,m} \rangle$	Expectation values of spherical harmonics
$c_{l,m}$	Decay modes
$g_{l,m,p,q}$	Coefficients in recurrence relation for Fokker–Planck operator
$d_{l,m,p,q}$	Matrix elements
$e_{l,m,p,q}$	Differential recurrence coefficients
$e_{i,j,k}$	Permutation symbol
$c_k^{(i)}$	Coefficients in Taylor series approximation of the free energy near the ith stationary point

h_c Critical value of h above which the bistable structure disappears
θ_c Polar angle at which maximum of V occurs when $h = h_c$
P_φ, Q_θ Orthogonal transformations given in appendix C

I. INTRODUCTORY CONCEPTS

A. The Purpose of this Review

The Fokker–Planck equation describing the distribution of representative points on the surface of the unit sphere arises in a variety of physical and chemical problems. In particular we mention the Debye theory of dielectric relaxation [1], dielectric relaxation of nematic liquid crystals [2], Kerr effect relaxation [3], nuclear magnetic resonance [4] and superparamagnetic relaxation of fine ferromagnetic particles [5]. In order to calculate the time or frequency behavior of a system governed by the rotational Fokker–Planck equation, the distribution of orientations function is expanded as a Fourier–Laplace series $W(\mathbf{r}, t) = \sum N_{l,m}^2 \langle X_{l,m} \rangle (t) X_{l,m}(\mathbf{r})$. The expectation values of the spherical harmonics $\langle X_{l,m} \rangle$ on being arranged as a column vector $U(t)$ can be shown to satisfy a set of differential recurrence relations that can be approximated by the matrix differential equation $\dot{U} = AU + B$, whence the solution is obtained by taking a sufficiently large number of components so as to ensure convergence. We shall also demonstrate how the correlation time in the linear response approximation may then be calculated.

It is the purpose of this review to show how the differential recurrence relations are generated for a variety of non–axially symmetric potentials such as cubic anisotropy or an oblique magnetic field superimposed on a uniaxial crystal. In addition it will be shown how asymptotic expressions for the lowest nonvanishing eigenvalue of the Fokker–Planck equation, that is the one associated with particle escape over potential barriers [6], is obtained. We shall present our analysis mainly in the context of superparamagnetism, as this is more general than dielectric relaxation because of the presence of gyromagnetic terms. The results for dielectric relaxation of nematic liquid crystals may be obtained from the magnetic case by simply omitting the gyromagnetic terms.

B. Relaxation of Fine Ferromagnetic Particles

A sufficiently fine, internally homogenous *ferromagnetic particle* [7] lacks the domain structure that complicates ordinary magnetic specimens. It is characterized by a uniform vector *magnetization* $\mathbf{M} = M_s \mathbf{r}$. The magnitude M_s is called the *spontaneous magnetization* and is determined by the material and temperature. The unit vector $\mathbf{r}$ is called the *magnetization orientation* and is determined by the *crystalline anisotropy*, *shape anisotropy* (*internal magnetostatic fields*), and by *external magnetic fields*. Besides the directly controlled or *applied field*, the external magnetic fields may include the fields of other particles, i.e., *interparticle interactions*. Throughout this treatment, the applied field [8], when present, will be assumed to be a constant uniform field $\mathbf{H} = H\mathbf{h}$ of magnitude H and direction $\mathbf{h}$. In a system of such particles the time behavior of $\mathbf{r}$ is treated as a *Markov process* [9] and the resulting probability distribution $W(\mathbf{r}, t)$ is called the *distribution of magnetization orientations*.

In *thermal equilibrium* the system is treated as a *Gibbs canonical ensemble*. From statistical mechanics, the *equilibrium distribution* is

$$W(\mathbf{r}) = A_0 e^{-\beta V(\mathbf{r})} \tag{1.1}$$

Here $\beta = v/kT$ where v is the *magnetic particle volume*, k is Boltzmann's constant, and T the absolute temperature. A_0 is a normalization constant and V is the *Gibbs free energy* per unit volume of the system. The *stable magnetization orientations* are then the local maxima of the equilibrium distribution or equivalently the local minima of the free energy. Suppose V is expressed in spherical polar coordinates as shown in Fig. 1; then V is said to be *non–axially symmetric* when there is an explicit dependency on azimuthal angle φ. In contrast V is said to be *axially symmetric* when there is no such dependency on φ. The most important non–axially symmetric cases are:

1. When V arises from *uniaxial anisotropy* [10] in the presence of a constant uniform external magnetic field at an oblique angle $\psi \neq 0, \pi$ to the easy axis then

$$V = K(1 - \mathbf{r}\mathbf{e}_3^2) - HM_s\mathbf{r}\mathbf{h} \tag{1.2}$$

where K is the anisotropy energy per unit volume and $\mathbf{e}_3$ denotes the direction of the *uniaxial axis*. Referring to Fig. 1, we see that

$$1 - \mathbf{r}\mathbf{e}_3^2 = 1 - \cos^2\theta = \sin^2\theta$$

$$\begin{aligned} \mathbf{r}\mathbf{h} &= (\sin\theta\cos\varphi\mathbf{e}_1 + \sin\theta\sin\varphi\mathbf{e}_2 + \cos\theta\mathbf{e}_3)(\sin\psi\mathbf{e}_1 + \cos\psi\mathbf{e}_3) \\ &= \sin\psi\sin\theta\cos\varphi + \cos\psi\cos\theta \end{aligned} \tag{1.3}$$

so that Eq. (1.2) becomes

$$V = K\sin^2\theta - HM_s\sin\psi\sin\theta\cos\varphi - HM_s\cos\psi\cos\theta \tag{1.4}$$

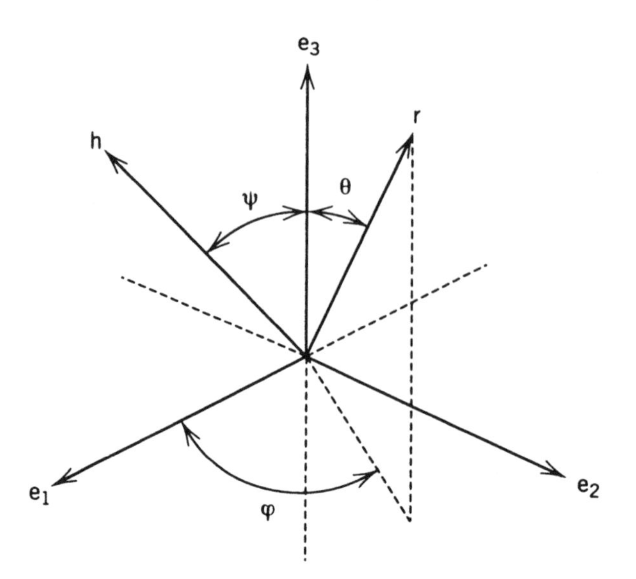

Figure 1. External field and magnetization orientations in terms of spherical polar coordinates.

2. When V arises from *cubic anisotropy* [5, 11, 12]

$$V = K(\mathbf{re}_1^2\mathbf{re}_2^2 + \mathbf{re}_1^2\mathbf{re}_3^2 + \mathbf{re}_2^2\mathbf{re}_3^2) \tag{1.5}$$

Again referring to Fig. 1, we see that

$$\mathbf{re}_1^2 = \sin^2\theta\cos^2\varphi \qquad \mathbf{re}_2^2 = \sin^2\theta\sin^2\varphi \qquad \mathbf{re}_3^2 = \cos^2\theta \tag{1.6}$$

so that Eq. (1.5) becomes

$$V = (K/4)(\sin^4\theta\sin^2 2\varphi + \sin^2 2\theta) \tag{1.7}$$

For a *longitudinally* applied field, $\psi = 0$ so that Eq. (1.4) becomes

$$V = K\sin^2\theta - HM_s\cos\theta \tag{1.8}$$

which is axially symmetric. For a *transversely* applied field, $\psi = \pi/2$ so that Eq. (1.4) becomes

$$V = K\sin^2\theta - HM_s\sin\theta\cos\varphi \tag{1.9}$$

Results of numerical computation in Section III.C suggest that when K and H are sufficiently large, the dependency on φ may be ignored so that Eq. (1.9) can be approximated by

$$V = K \sin^2 \theta - HM_s \sin \theta \tag{1.10}$$

The axially symmetric cases in Eqs. (1.8) and (1.10) will be dealt with in Section I.G. Let V_1 and V_2 denote the values of V at local minima (stable magnetization orientations), which are separated by a local maximum at which the value of V is denoted by V_0. Then we say that the stable magnetization orientations are separated by *energy barriers* having (energy) *barrier heights*

$$\upsilon(V_0 - V_i) \qquad i = 1, 2 \tag{1.11}$$

For a system that is not in thermal equilibrium, the evolution or *relaxation* of the nonequilibrium distribution $W(\mathbf{r}, t)$ into the equilibrium distribution $A_0 e^{-\beta V}$ can be expressed as a sum of decaying exponentials. When the relaxation is dominated by a single exponential $e^{-t/\tau}$, then the corresponding time constant τ is called the *relaxation time*. There are two types of *relaxation mechanisms*:

1. *Debye relaxation* [1] involves physical rotation of the particles and occurs when they are suspended in a liquid carrier or *ferrofluid*. The rate of approach to equilibrium is then determined by the viscosity of the ferrofluid. We shall assume the carrier to be in the solid state throughout so that the Debye relaxation mechanism can be ignored.
2. *Néel relaxation* [13] involves spontaneous Brownian type rotations or jumps in the magnetization orientation within the particle due to *thermal fluctuations*, and occurs when the particles are sufficiently small so that the barrier heights in Eq. (1.11) are comparable to the thermal energy kT. The rate of approach to equilibrium will have an inverse exponential dependence on the ratio of the barrier height to thermal energy $\beta(V_0 - V_i)$.

The magnitudes of the barrier heights give rise to the following cases:

1. *Ferromagnetic behavior* refers to the case when the barrier heights are very large in comparison to kT. This occurs for fine ferromagnetic particles at not too low a temperature and accounts for the apparent stability of the magnetization in certain magnetic storage devices.
2. *Superparamagnetic behavior* [10] or thermal instability of magnetization refers to the case of very small particles where the barrier heights

are small in comparison to kT. Thermal agitation causes continual changes in the magnetization orientation so that the system is constantly in thermal equilibrium. In the presence of a magnetic field the system behaves like an ensemble of paramagnetic atoms with no hysteresis.

3. *Magnetic aftereffect behavior* refers to a narrow range of intermediate barrier heights. The system neither remains in a single state for a long time nor attains thermal equilibrium in a short time so that the relaxation time is of the order of experimental times ($\sim 10^2$ sec).

For a review that serves as an introduction to all aspects of magnetic relaxation theory including both the Néel and Debeye mechanisms, see ref. [14].

C. The Discrete Orientation Model

When the energy barriers are large in comparison with kT but not so large as to prevent changes in the magnetization orientation $\mathbf{r}$ occurring altogether, we may assume [5, 11, 12] that $\mathbf{r}$ is restricted to the stable orientations along the local minima of the free energy. The time behavior of $\mathbf{r}$ is then treated as a discrete Markov process [9] and the distribution W replaced by n_i namely, the number of particles in the ith orientation. For a large number $n = \sum n_i$ of noninteracting particles, n_i changes with time in accordance with the *master equation*

$$\dot{n}_i = \sum_{i \neq j} (\nu_{j,i} n_j - \nu_{i,j} n_i) \tag{1.12}$$

where $\nu_{i,j}$ is the transition probability from orientation i to orientation j, i.e., the probability of a particle in orientation i undergoing a transition to orientation j. The $\nu_{i,j}$ will depend on the absolute temperature T, the anisotropy constant K, and the external field $\mathbf{H}$.

For two orientations as is the case for a uniaxial crystal with constant uniform external magnetic field, we let 1 refer to the positive orientation and 2 to the negative so that Eq. (1.12) reduces to

$$\dot{n}_1 = -\dot{n}_2 = \nu_{2,1} n_2 - \nu_{1,2} n_1 \tag{1.13}$$

Setting $n_2 = n - n_1$ gives

$$\dot{n}_1 = -\dot{n}_2 = \nu_{2,1} n - (\nu_{1,2} + \nu_{2,1}) n_1 \tag{1.14}$$

so that the n_1 approach their final value according to the factor $e^{-(\upsilon_{1,2} + \upsilon_{2,1})t}$, i.e., with time constant or relaxation (Néel) time

$$\tau = (\nu_{1,2} + \nu_{2,1})^{-1} \tag{1.15}$$

It is usually assumed that

$$\nu_{i,j} = \nu_{i,j}^{0} e^{-\beta(V_0 - V_i)} \tag{1.16}$$

where the $\upsilon_{i,j}^{0}$ vary slowly with temperature and thus may be considered constant. When the barrier heights are equal, Eq. (1.15) using the assumption in Eq. (1.16) becomes

$$\tau = Ce^{\beta(V_0 - V_1)} \tag{1.17}$$

where C is a constant.

For more than two orientations, the case of greatest interest is when the free energy per unit volume arises from cubic anisotropy, namely

$$V = K(\alpha_1^2\alpha_2^2 + \alpha_1^2\alpha_3^2 + \alpha_2^2\alpha_3^2) \tag{1.18}$$

where the α_i denotes the direction cosines with respect to the cubic axes. For $K > 0$, V has minima at six orientations of type [100] (i.e. $\alpha_1 = 1$, $\alpha_2 = \alpha_3 = 0$, **r** along a cube edge of the lattice). It has maxima at eight orientations of type [111] (**r** along a body diagonal) and saddle points at twelve orientations of type [110] (**r** along a face diagonal). For $K < 0$, the maxima and minima are interchanged [see Fig. (2)]. The values of V at the orientations [100], [110], and [111] are 0, $K/4$, and $K/3$, respectively.

1. *Positive Cubic Anisotropy K > 0*

Let n_1, n_2 and n_3 denote the numbers of particles at the [100], [010], and [001] orientations, respectively, and n_{-1}, n_{-2} and n_{-3} denote the corresponding numbers in the opposite orientations [see Fig. (3)]. To get from orientation 1 to orientation 2, a particle must surmount an energy barrier whose lowest point is the saddle point at [110]. To get from orientation 1 to orientation -1, it must surmount two successive energy barriers. If the energy barriers are high, it will be unlikely to do this in a single event, and hence we may take

$$\nu_{i,-i} = 0 \qquad \nu_{i,j} = \nu \text{ for } j \neq \pm i \tag{1.19}$$

Eq. (1.12) becomes

$$\dot{n}_{\pm i} = \nu(n_j + n_k + n_{-j} + n_{-k} - 4n_{\pm i}) \quad i \neq j \neq k \quad i, j, k \in \{1, 2, 3\} \tag{1.20}$$

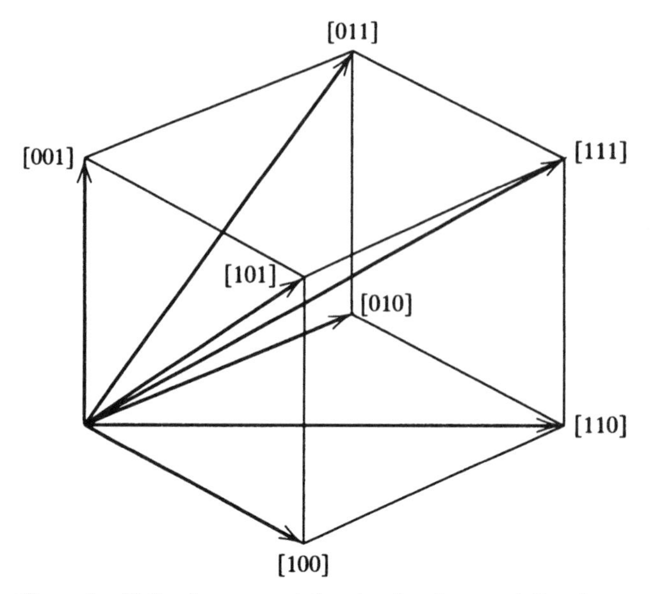

Figure 2. Unit cube representation showing the crystal directions.

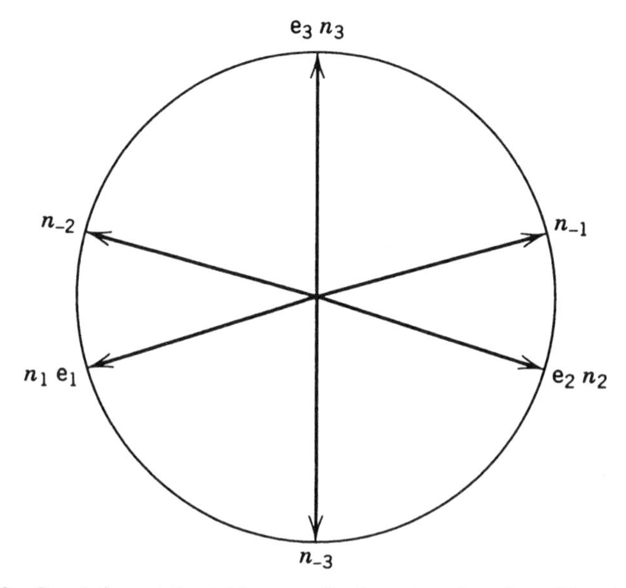

Figure 3. Populations at the stable magnetization orientations for cubic anisotropy with $K > 0$.

Setting $x_i = n_i + n_{-i}$ and $y_i = n_i - n_{-i}$. Then by addition and subtraction, we obtain

$$\dot{x}_i = 2\nu(x_j + x_k - 2x_i) \qquad i \neq j \neq k \qquad i, j, k \in \{1, 2, 3\} \tag{1.21}$$

$$\dot{y}_i = -4\nu y_i \qquad i = 1, 2, 3 \tag{1.22}$$

By symmetry the equilibrium values attained at $t = \infty$ are $n_i = n/6$, and hence $x_i = 1/6$, $y_i = 0$ and the solutions of Eq. (1.22) are

$$y_i = y_i(0)e^{-4\nu t} \qquad i = 1, 2, 3 \tag{1.23}$$

To solve Eq. (1.21), we set $x_j + x_k = n - x_i$ so that

$$\dot{x}_i = 2\nu(n - 3x_i) \qquad i = 1, 2, 3 \tag{1.24}$$

The solutions are then

$$x_i = n/3 + [x_i(0) - n/3]e^{-6\nu t} \qquad i = 1, 2, 3 \tag{1.25}$$

The behavior of the n_i is thus governed by *two* time constants, namely $1/4\nu$ and $1/6\nu$, and ν is given by a formula of the form of Eq. (1.16), namely

$$\nu = \nu^0 e^{-\beta K/4} \tag{1.26}$$

where ν^0 is presumably a function of K and varies slowly with temperature T in comparison to the exponential variation.

2. *Negative Cubic Anisotropy K < 0*

The stable magnetization orientations correspond to the eight corners of a cube. Let n_1, n_2, n_3 and n_4 denote the numbers of particles at the $[111]$, $[11\bar{1}]$, $[1\bar{1}1]$, and $[\bar{1}11]$ orientations respectively and n_{-1}, n_{-2}, n_{-3} and n_{-4} denote the corresponding numbers in the opposite orientations [see Fig. (4)]. As for the $K > 0$ case, we suppose only one barrier at a time can be surmounted. If we let i.ADJ.j mean that the subscripts i and j correspond to adjacent minima and i.NA.j mean the opposite, then

$$\nu_{i,j} = \nu \qquad i.\mathrm{ADJ}.j \qquad \nu_{i,j} = 0 \qquad i.\mathrm{NA}.j \tag{1.27}$$

Eq. (1.12) becomes

$$\dot{n}_i = \nu(n_j + n_k + n_l - 3n_i) \qquad i.\mathrm{ADJ}.j, k, l \qquad i \neq j \neq k \neq l \tag{1.28}$$

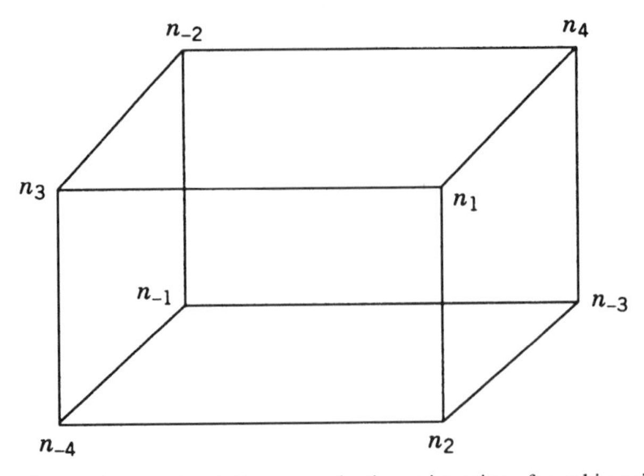

Figure 4. Populations at the stable magnetization orientations for cubic anisotropy with $K < 0$.

On setting $x_i = n_i + n_{-i}$ and $y_i = n_i - n_{-i}$, then by addition and subtraction, we obtain

$$\dot{x}_i = \nu(x_j + x_k + x_l - 3x_i) \qquad i \neq j \neq k \neq l \tag{1.29}$$

and for the y's

$$\begin{aligned}
\dot{y}_1 &= \nu(-3y_1 + y_2 + y_3 + y_4) \\
\dot{y}_2 &= \nu(y_1 - 3y_2 - y_3 - y_4) \\
\dot{y}_3 &= \nu(y_1 - y_2 - 3y_3 - y_4) \\
\dot{y}_4 &= \nu(y_1 - y_2 - y_3 - 3y_4)
\end{aligned} \tag{1.30}$$

Setting $x_j + x_k + x_l = n - x_i$ in Eq. (1.29) gives

$$\dot{x}_i = \nu(n - 4x_i) \tag{1.31}$$

Hence the x_i approach their equilibrium value $n/4$ with time constant $1/4\nu$. Eq. (1.30) can be expressed as

$$\dot{Y} = \nu A Y \tag{1.32}$$

where

$$Y = (y_1, \ldots, y_4)^T \qquad A = \begin{pmatrix} -3 & 1 & 1 & 1 \\ 1 & -3 & -1 & -1 \\ 1 & -1 & -3 & -1 \\ 1 & -1 & -1 & -3 \end{pmatrix} \tag{1.33}$$

Assuming a solution of the form

$$y_i = B_i e^{\lambda \nu t} \tag{1.34}$$

leads to the requirement that λ be an eigenvalue of the system matrix A. The eigenvalues of A are the solutions of

$$\det(A - \lambda I) = 0 \tag{1.35}$$

The *Laplace expansion* is

$$-(\lambda+3)\begin{vmatrix} -(\lambda+3) & -1 & -1 \\ -1 & -(\lambda+3) & -1 \\ -1 & -1 & -(\lambda+3) \end{vmatrix} - \begin{vmatrix} 1 & -1 & -1 \\ 1 & -(\lambda+3) & -1 \\ 1 & -1 & -(\lambda+3) \end{vmatrix}$$

$$+\begin{vmatrix} 1 & -(\lambda+3) & -1 \\ 1 & -1 & -1 \\ 1 & -1 & -(\lambda+3) \end{vmatrix} - \begin{vmatrix} 1 & -(\lambda+3) & -1 \\ 1 & -1 & -(\lambda+3) \\ 1 & -1 & -1 \end{vmatrix} = 0 \tag{1.36}$$

Further application to each minor, followed by simplifying gives the characteristic polynomial

$$(\lambda + 2)^3(\lambda + 6) = 0 \tag{1.37}$$

The eigenvalues are thus $\lambda = -2, -6$. The behavior of the n_i is thus governed by *three* time constants, namely $1/2\nu$, $1/4\nu$, and $1/6\nu$, and ν is given by a formula in the form of Eq. (1.16), namely

$$\nu = \nu^0 e^{-\beta K/12} \tag{1.38}$$

D. Néel's Model

The Néel model [13] was originally applied to a system having uniaxial anisotropy that is initially in thermal equilibrium in the presence of a large magnetic field, which is then switched off at $t = 0$. The assumptions are:

1. *Rigid Coupling Assumption*

The spins of individual magnetic moments within the particle that contribute to the magnetization orientation of the particle are assumed to be rigidly coupled, and hence synchronous rotation of these spins occurs when the magnetization orientation is reversed.

2. *Discrete Orientation Assumption*

When the energy barriers that oppose such reversals are large in comparison to the thermal energy kT, then the magnetization orientation, when treated as a stochastic variable, is assumed to adopt only a finite set of possible orientations, namely the minima of the free energy.

Néel then proposed the following formula for the relaxation time based on the discrete orientation approximation in Eq. (1.17):

$$\tau = \tau_N e^{\beta K} \tag{1.39}$$

The pre-exponential factor is called the *Néel relaxation time* and is given by

$$\tau_N = \frac{M_s}{3K\gamma|\lambda_s|}\sqrt{\frac{\pi}{2\beta G}} \tag{1.40}$$

where γ is the gyromagnetic ratio, λ_s is the empirical magnetization constant, and G is Young's modulus. τ_N is usually taken to be 10^{-10} sec.

E. Brown's Model

Brown's criticism of Néel's calculation is that it does not allow for the particles to spend time in intermediate orientations before jumping, nor does it allow for jumps back to the initial stable orientation. Later he argued that the system must also be explicitly treated as a gyromagnetic one [5, 7].

An uncompensated electron in a ferromagnetic particle [14] possesses a magnetic moment μ and an angular momentum ω which arise from the motion of its charge e and mass m, respectively. They are related by

$$\mu = \gamma\omega \tag{1.41}$$

where for spin electrons $\gamma = -e/mc$ is the gyromagnetic ratio. When the particle is in the presence of an effective magnetic field $\mathbf{H}_{\text{ef}}$, a torque which is given by $\mu \times \mathbf{H}_{\text{ef}}$ acts on each electron. Since this torque must also equal the rate of change of angular momentum of the electron $\dot{\omega}$ then differentiating Eq. (1.41) gives the simple gyromagnetic equation

$$\dot{\mu} = \gamma\mu \times \mathbf{H}_{\text{ef}} \tag{1.42}$$

For a single domain particle all such magnetic moments are assumed to be aligned so that

$$\dot{\mathbf{r}} = \gamma \mathbf{r} \times \mathbf{H}_{\mathrm{ef}} \tag{1.43}$$

This equation represents undamped precession of $\mathbf{r}$ about the axis of $\mathbf{H}_{\mathrm{ef}}$. Due to collisions between precessing electrons, the observable behavior is, however, that of alignment of $\mathbf{r}$ with $\mathbf{H}_{\mathrm{ef}}$. To account for this Gilbert [15] proposed the introduction of an effective damping field term giving

$$\dot{\mathbf{r}} = \gamma \mathbf{r} \times (\mathbf{H}_{\mathrm{ef}} - \eta M_s \dot{\mathbf{r}}) \tag{1.44}$$

where η is a phenomenological damping constant. Taking the scalar product with $\mathbf{r}$ gives

$$\dot{\mathbf{r}}\mathbf{r} = \gamma(\mathbf{r} \times \mathbf{H}_{\mathrm{ef}})\mathbf{r} - \eta\gamma M_s(\mathbf{r} \times \dot{\mathbf{r}})\mathbf{r} = 0 \tag{1.45}$$

Transposing the η in Eq. (1.44) to the left-hand side and taking the vector product with $\mathbf{r}$ gives

$$\dot{\mathbf{r}} \times \mathbf{r} + \gamma\eta M_s(\mathbf{r} \times \dot{\mathbf{r}}) \times \mathbf{r} = \gamma(\mathbf{r} \times \mathbf{H})_{\mathrm{ef}} \times \mathbf{r} \tag{1.46}$$

Expanding the vector triple product on the left-hand side and applying Eq. (1.45) gives

$$(\mathbf{r} \times \dot{\mathbf{r}}) \times \mathbf{r} = -(\dot{\mathbf{r}}\mathbf{r})\mathbf{r} + \dot{\mathbf{r}} = \dot{\mathbf{r}} \tag{1.47}$$

Eq. (1.46) with the aid of Eq. (1.47) becomes

$$\dot{\mathbf{r}} \times \mathbf{r} = -\gamma\eta M_s \dot{\mathbf{r}} + \gamma(\mathbf{r} \times \mathbf{H})_{\mathrm{ef}} \times \mathbf{r} \tag{1.48}$$

Gilberts equation (1.44) can be rearranged to read

$$\dot{\mathbf{r}} \times \mathbf{r} = (\gamma\eta M_s)^{-1}(\dot{\mathbf{r}} - \gamma \mathbf{r} \times \mathbf{H}_{\mathrm{ef}}) \tag{1.49}$$

Equating the right-hand sides of Eqs. (1.48) and (1.49) gives

$$(\gamma\eta M_s)^{-1}(\dot{\mathbf{r}} - \gamma \mathbf{r} \times \mathbf{H}_{\mathrm{ef}}) + \gamma\eta M_s \dot{\mathbf{r}} = \gamma(\mathbf{r} \times \mathbf{H})_{\mathrm{ef}} \times \mathbf{r} \tag{1.50}$$

Solving for $\dot{\mathbf{r}}$ gives the *explicit* form of Gilbert's equation

$$\dot{\mathbf{r}} = aM_s \mathbf{r} \times \mathbf{H}_{\mathrm{ef}} + bM_s(\mathbf{r} \times \mathbf{H}_{\mathrm{ef}}) \times \mathbf{r} \tag{1.51}$$

where

$$a(= g') = \frac{\gamma}{(1 + \alpha^2)M_s} \tag{1.52}$$

$$b(= h') = \frac{\alpha\gamma}{(1 + \alpha^2)M_s} \tag{1.53}$$

$$\alpha = \eta\gamma M_s = b/a \tag{1.54}$$

Equation (1.51) under the approximation $\alpha^2 \cong 0$ reduces to the earlier Landau–Lifshitz equation [16]. The term containing a will be referred to as the gyromagnetic term, and the term containing b as the alignment term. The effective magnetic field can be expressed in terms of the free energy per unit volume V by

$$\mathbf{H}_{\text{ef}} = -\frac{1}{M_s}\frac{\partial V}{\partial \mathbf{r}} \tag{1.55}$$

Brown [7] then supposes that the effective damping (or dissipative) field $\eta M_s \dot{\mathbf{r}}$ describes only the statistical average of the rapidly fluctuating random forces due to thermal agitation, and that for an individual particle this term must become $\eta M_s \dot{\mathbf{r}} + \mathbf{h}(t)$, where

$$\overline{\mathbf{h}(t)} = 0 \qquad \overline{h_i(t_1)h_j(t_2)} = 2\eta\beta^{-1}\delta_{i,j}\delta(t_1 - t_2) \tag{1.56}$$

Here the overbars denote statistical averages over a large number of particles that have all started with the same orientation in the configuration space, and the $h_i(t)$ denote the rectangular components of $\mathbf{h}(t)$. It was assumed that the components obey Isserlis's theorem [14], that is,

$$\begin{aligned} \overline{h(t_1)\ldots h(t_{2n+1})} &= 0 \\ \overline{h(t_1)\ldots h(t_{2n})} &= \sum \prod_{k_i<k_j} \overline{h(t_{k_i})h(t_{k_j})} \end{aligned} \tag{1.57}$$

where the sum is over all such products, each of which is formed by selecting n pairs from $2n$ time points. For example, if $n = 2$, we have

$$\begin{aligned} \overline{h(t_1)h(t_2)h(t_3)h(t_4)} &= \overline{h(t_1)h(t_2)}\,\overline{h(t_3)h(t_4)} \\ &\quad + \overline{h(t_1)h(t_3)}\,\overline{h(t_2)h(t_4)} \\ &\quad + \overline{h(t_1)h(t_4)}\,\overline{h(t_2)h(t_3)} \end{aligned} \tag{1.58}$$

On the basis of this assumption, Brown, using the methods of Wang and Uhlenbeck [17, 18], was able to derive the Fokker–Planck equation for the distribution of magnetization orientations $W(\mathbf{r}, t)$.

This procedure may be circumvented, however, by using an alternative approach given by Brown, which appears to be based on an argument of Einstein [19]. Substituting for the effective magnetic field Eq. (1.55) into the explicit Gilbert equation, Eq. (1.51), gives

$$\dot{\mathbf{r}} = -a\mathbf{r} \times \partial V/\partial \mathbf{r} - b(\mathbf{r} \times \partial V/\partial \mathbf{r}) \times \mathbf{r} \tag{1.59}$$

Since

$$\partial V/\partial \mathbf{r} \perp \mathbf{r} \tag{1.60}$$

we have

$$\mathbf{r}\partial V/\partial \mathbf{r} = 0 \tag{1.61}$$

so that on expanding the triple product in Eq. (1.59) we obtain

$$(\mathbf{r} \times \partial V/\partial \mathbf{r}) \times \mathbf{r} = \partial V/\partial \mathbf{r} - (\mathbf{r}\partial V/\partial \mathbf{r})\mathbf{r} = \partial V/\partial \mathbf{r} \tag{1.62}$$

Eq. (1.59) becomes

$$\dot{\mathbf{r}} = -a\mathbf{r} \times \partial V/\partial \mathbf{r} - b\partial V/\partial \mathbf{r} \tag{1.63}$$

Brown considers the individual magnetization orientations in the system of magnetic particles as a current of representative points moving around the surface of the unit sphere with number density $W(\mathbf{r}, t)$ and current density $\mathbf{J}(\mathbf{r}, t)$. Such representative points are neither created nor destroyed, hence W and $\mathbf{J}$ satisfy the continuity equation

$$\dot{W} = -(\partial/\partial \mathbf{r})\mathbf{J} \tag{1.64}$$

The influence of the random thermal forces is to disperse the concentrations of representative points that occur around the stable magnetization orientations. This can be expressed mathematically by postulating a diffusion term of the form $-k'\partial W/\partial \mathbf{r}$, where $k' > 0$ is constant at a given temperature. Hence

$$\begin{aligned} \mathbf{J} &= W\dot{\mathbf{r}} - k'\partial W/\partial \mathbf{r} \\ &= -aW\mathbf{r} \times \partial V/\partial \mathbf{r} - bW\partial V/\partial \mathbf{r} - k'\partial W/\partial \mathbf{r} \end{aligned} \tag{1.65}$$

and Eq. (1.64) becomes

$$\dot{W} = a \frac{\partial}{\partial \mathbf{r}} \left(W\mathbf{r} \times \frac{\partial V}{\partial \mathbf{r}} \right) + b \frac{\partial}{\partial \mathbf{r}} \left(W \frac{\partial V}{\partial \mathbf{r}} \right) + k' \frac{\partial^2 W}{\partial \mathbf{r}^2} \tag{1.66}$$

Expanding the derivative in the gyroscopic term using the product rule gives

$$\frac{\partial}{\partial \mathbf{r}} \left(W\mathbf{r} \times \frac{\partial V}{\partial \mathbf{r}} \right) = \frac{\partial W}{\partial \mathbf{r}} \left(\mathbf{r} \times \frac{\partial V}{\partial \mathbf{r}} \right) + W \frac{\partial}{\partial \mathbf{r}} \left(\mathbf{r} \times \frac{\partial V}{\partial \mathbf{r}} \right) \tag{1.67}$$

The first term can be written

$$\mathbf{r} \left(\frac{\partial V}{\partial \mathbf{r}} \times \frac{\partial W}{\partial \mathbf{r}} \right) \tag{1.68}$$

Application of the scalar triple product formula to the second term gives

$$\frac{\partial}{\partial \mathbf{r}} \left(\mathbf{r} \times \frac{\partial V}{\partial \mathbf{r}} \right) = \frac{\partial V}{\partial \mathbf{r}} \left(\frac{\partial}{\partial \mathbf{r}} \times \mathbf{r} \right) - \mathbf{r} \left(\frac{\partial}{\partial \mathbf{r}} \times \frac{\partial V}{\partial \mathbf{r}} \right) = 0 \tag{1.69}$$

Eq. (1.66) becomes

$$\dot{W} = a\mathbf{r} \left(\frac{\partial V}{\partial \mathbf{r}} \times \frac{\partial W}{\partial \mathbf{r}} \right) + b \frac{\partial}{\partial \mathbf{r}} \left(W \frac{\partial V}{\partial \mathbf{r}} \right) + k' \frac{\partial^2 W}{\partial \mathbf{r}^2} \tag{1.70}$$

which is the Fokker–Planck equation for the distribution of magnetization orientations. In thermal equilibrium $\dot{W} = 0$, so that W must reduce to the equilibrium distribution

$$W_0 = Ae^{-\beta V} \tag{1.71}$$

where $\beta = \upsilon / KT$ is the ratio of magnetic particle volume to thermal energy. Substituting into Eq. (1.70) gives

$$-\beta a e^{-\beta V} \mathbf{r} \left(\frac{\partial V}{\partial \mathbf{r}} \times \frac{\partial V}{\partial \mathbf{r}} \right) + (b - \beta k') \frac{\partial}{\partial \mathbf{r}} \left(e^{-\beta V} \frac{\partial V}{\partial \mathbf{r}} \right) = 0 \tag{1.72}$$

Clearly the gyroscopic term vanishes and since the prefactor in the final term is nonzero

$$k' = b/\beta \tag{1.73}$$

Hence

$$\dot{W} = a\mathbf{r}\left(\frac{\partial V}{\partial \mathbf{r}} \times \frac{\partial W}{\partial \mathbf{r}}\right) + b\,\frac{\partial}{\partial \mathbf{r}}\left(W\,\frac{\partial V}{\partial \mathbf{r}}\right) + b\beta^{-1}\,\frac{\partial^2 W}{\partial \mathbf{r}^2} \tag{1.74}$$

It is convenient to express Eq. (1.74) as an operator equation

$$\dot{W} = L_{FP}W \tag{1.75}$$

where

$$L_{FP}W = a\mathbf{r}\left(\frac{\partial V}{\partial \mathbf{r}} \times \frac{\partial W}{\partial \mathbf{r}}\right) + b\,\frac{\partial}{\partial \mathbf{r}}\left(W\,\frac{\partial V}{\partial \mathbf{r}}\right) + b\beta^{-1}\,\frac{\partial^2 W}{\partial \mathbf{r}^2} \tag{1.76}$$

is the Fokker–Planck operator [18]. The gradient, divergence, and Laplacian in spherical polar coordinates r, θ, φ are

$$\frac{\partial A}{\partial \mathbf{r}} = \frac{\partial A}{\partial \theta}\,\bar{\mathbf{e}}_1 + \frac{1}{\sin\theta}\,\frac{\partial A}{\partial \varphi}\,\bar{\mathbf{e}}_2 \tag{1.77}$$

$$\frac{\partial \mathbf{A}}{\partial \mathbf{r}} = \frac{1}{\sin\theta}\left(\frac{\partial}{\partial \theta}\,(\sin\theta A_1) + \frac{\partial A_2}{\partial \varphi}\right) \tag{1.78}$$

$$\frac{\partial^2 A}{\partial \mathbf{r}^2} = \Lambda^2 A = \frac{1}{\sin\theta}\,\frac{\partial}{\partial \theta}\left(\sin\theta\,\frac{\partial A}{\partial \theta}\right) + \frac{1}{\sin^2\theta}\,\frac{\partial^2 A}{\partial \varphi^2} \tag{1.79}$$

where $\mathbf{A} = A_1\bar{\mathbf{e}}_1 + A_2\bar{\mathbf{e}}_2$ and $\bar{\mathbf{e}}_1, \bar{\mathbf{e}}_2$ are orthogonal unit vectors on the surface of the unit sphere. The gyroscopic term becomes

$$\begin{aligned} a\mathbf{r}\cdot\left(\frac{\partial V}{\partial \mathbf{r}} \times \frac{\partial W}{\partial \mathbf{r}}\right) &= a\mathbf{r}\cdot\begin{vmatrix} \bar{\mathbf{e}}_1 & \bar{\mathbf{e}}_2 & \bar{\mathbf{e}}_3 \\ \dfrac{\partial V}{\partial \theta} & \dfrac{1}{\sin\theta}\,\dfrac{\partial V}{\partial \varphi} & 0 \\ \dfrac{\partial W}{\partial \theta} & \dfrac{1}{\sin\theta}\,\dfrac{\partial W}{\partial \varphi} & 0 \end{vmatrix} \\ &= \frac{a}{\sin\theta}\left(\frac{\partial V}{\partial \theta}\,\frac{\partial W}{\partial \varphi} - \frac{\partial V}{\partial \varphi}\,\frac{\partial W}{\partial \theta}\right)\mathbf{r}\cdot\bar{\mathbf{e}}_3 \\ &= \frac{a}{\sin\theta}\left(\frac{\partial V}{\partial \theta}\,\frac{\partial W}{\partial \varphi} - \frac{\partial V}{\partial \varphi}\,\frac{\partial W}{\partial \theta}\right) \end{aligned} \tag{1.80}$$

and the alignment term becomes

$$b \frac{\partial}{\partial \mathbf{r}} \left(W \frac{\partial V}{\partial \mathbf{r}} \right) = b \frac{\partial}{\partial \mathbf{r}} \left(W \frac{\partial V}{\partial \theta} \bar{\mathbf{e}}_1 + W \frac{1}{\sin \theta} \frac{\partial V}{\partial \varphi} \bar{\mathbf{e}}_2 \right)$$

$$= \frac{b}{\sin \theta} \frac{\partial}{\partial \theta} \left(\sin \theta W \frac{\partial V}{\partial \theta} \right) + \frac{b}{\sin^2 \theta} \frac{\partial}{\partial \varphi} \left(W \frac{\partial V}{\partial \varphi} \right)$$

$$= bW\Lambda^2 V + b \left(\frac{\partial V}{\partial \theta} \frac{\partial W}{\partial \theta} + \frac{1}{\sin^2 \theta} \frac{\partial V}{\partial \varphi} \frac{\partial W}{\partial \varphi} \right) \tag{1.81}$$

The Fokker–Planck operator Eq. (1.76) is then

$$L_{FP} W = \beta^{-1} b \Lambda^2 W + b W \Lambda^2 V + b \left(\frac{\partial V}{\partial \theta} \frac{\partial W}{\partial \theta} + \frac{1}{\sin^2 \theta} \frac{\partial V}{\partial \varphi} \frac{\partial W}{\partial \varphi} \right) + \frac{a}{\sin \theta} \left(\frac{\partial V}{\partial \theta} \frac{\partial W}{\partial \varphi} - \frac{\partial V}{\partial \varphi} \frac{\partial W}{\partial \theta} \right) \tag{1.82}$$

and hence the Fokker–Planck equation in Eq. (1.75) becomes

$$\dot{W} = \beta^{-1} b \Lambda^2 W + b W \Lambda^2 V + b \left(\frac{\partial V}{\partial \theta} \frac{\partial W}{\partial \theta} + \frac{1}{\sin^2 \theta} \frac{\partial V}{\partial \varphi} \frac{\partial W}{\partial \varphi} \right) + \frac{a}{\sin \theta} \left(\frac{\partial V}{\partial \theta} \frac{\partial W}{\partial \varphi} - \frac{\partial V}{\partial \varphi} \frac{\partial W}{\partial \theta} \right) \tag{1.83}$$

Brown derived the Fokker–Planck equation for the distribution of magnetization orientations in the absence of the discrete orientation assumption, but later applied it to the Fokker–Planck equation to obtain a more effective high energy barrier approximation formula. This enables one to quantify the significance of this assumption.

The major criticism of Brown's Fokker–Planck equation is its failure to address the significance of the rigid coupling assumption. We recall that this assumption was used to generalize the simple gyromagnetic equation for an individual magnetic moment in Eq. (1.42) to a corresponding deterministic equation for the overall magnetization orientation Eq. (1.43).

At the particle surface dipoles exist, and variation of magnetic moment orientation is then possible. Experiments have shown the existence of new micromagnetic structures that lead to nonuniform modes of magnetization reversal in fine ferromagnetic particles [20, 21].

In addition, for large particles, the *curling mode* of magnetization rotation is brought into play due to its being energetically more favorable than the *coherent mode*. Weil [22] tried to apply Néel's theory to granulometry, namely the determination of the size distribution for fine particles whose size could not be measured directly at this time. He measured the remanent magnetization as a function of temperature and analyzed the data by assuming that at a given temperature the particles whose size were below those obtained from the Néel formula become superparamagnetic and do not contribute to the remanence. For a sample of cobalt, he obtained the characteristic bell shape, but a sample of nickel yielded a gap at a certain particle size (of the order of 100–150 Å).

The most plausible explanation is that another mode of magnetization reversal had occurred in particles of that particular size. Direct evaluation of τ from the linewidth of the Mössbauer effect for particles whose size was measured directly [23] showed that τ can sometimes increase with decreasing particle size, a result that was inconsistent with the models of Néel and Brown. To interpret this behavior, Eisenstein and Aharoni defined the functional form of the magnetization curling mode for a sphere away from its nucleation, and thus identified a structure that consisted of a central domain surrounded by a reverse domain that was separated by a thin wall. The calculations were restricted to the case of axial symmetry [24] and later generalized to uniaxial anisotropy [25] and cubic anisotropy [26].

F. Calculation of the Correlation Time from the Eigenvalues and Amplitudes of the Fokker–Planck Equation

By assuming a solution of the Fokker–Planck equation

$$\partial W/\partial t = L_{FP} W \tag{1.84}$$

of the form [7]

$$W(\mathbf{r}, t) = F(\mathbf{r})T(t) \tag{1.85}$$

we can show that the general solution is then of the form

$$W(\mathbf{r}, t) = W_0(\mathbf{r}) + \sum_{n=1}^{\infty} A_n F_n(\mathbf{r}) e^{-p_n t} \tag{1.86}$$

where $F_n(\mathbf{r})$ satisfies Eq. (1.84) with $\partial/\partial t$ replaced by $-p_n$. The eigenvalues p_n and the corresponding eigenfunctions F_n are determined from the requirement of single-valuedness and finiteness. The equilibrium solution $W_0 = Ae^{-\beta V}$ is the eigenfunction corresponding to $p_0 = 0$. The constant A is determined by normalization, and the constants A_n from the initial conditions.

Suppose the constant uniform external magnetic field has magnitude $H + H_1$ for $t < 0$, and H for $t > 0$ where the perturbing field $H_1 \ll H$ is considered to have been "switched on" at $t = -\infty$ and then "switched off" at $t = 0$. The system is therefore assumed to be in thermal equilibrium at $t = 0, \infty$ so that

$$W(\mathbf{r}, 0) = W_1(\mathbf{r}) = A'_0 e^{-\beta V(\mathbf{r}) + \xi_1 \mathbf{rh}}$$

$$W(\mathbf{r}, \infty) = W_0(\mathbf{r}) = A_0 e^{-\beta V(\mathbf{r})} \tag{1.87}$$

where $\xi_1 = \beta H_1 M_s$. The corresponding solution of the Fokker–Planck equation is referred to as the aftereffect solution and can be represented formally as

$$W(\mathbf{r}, t) = e^{-\beta V(\mathbf{r})}[1 + \xi_1 \alpha(t)] + O(\xi_1^2) \tag{1.88}$$

where $\alpha(t)$ decays from an initial value $\alpha(0) = \mathbf{rh}$ to a final value $\alpha(\infty) = 0$. The expectation value of the stochastic variable U is [18]

$$\langle U \rangle = \int UW d\Omega \Big/ \int W d\Omega \tag{1.89}$$

where $d\Omega$ denotes the element of solid angle and the integration is over the unit sphere. In particular let

$$\langle U \rangle_1 = \int Ue^{-\beta V(\mathbf{r}) + \xi_1 \mathbf{rh}} d\Omega \Big/ \int e^{-\beta V(\mathbf{r}) + \xi_1 \mathbf{rh}} d\Omega$$

$$\langle U \rangle_0 = \int Ue^{-\beta V(\mathbf{r})} d\Omega \Big/ \int e^{-\beta V(\mathbf{r})} d\Omega \tag{1.90}$$

denote the expectation values in the presence and absence, respectively, of the perturbing field. The equilibrium expectation value $\langle \mathbf{rh} \rangle_1$ can be expressed in terms of $\langle \mathbf{rh} \rangle_0$ by applying the linear approximation in which terms of $O(\xi_1^2)$ (i.e., containing ξ_1^2 and higher powers) are ignored. Thus

$$\langle \mathbf{rh} \rangle_1 = \frac{\int \mathbf{rh} e^{-\beta V + \xi_1 \mathbf{rh}} d\Omega}{\int e^{-\beta V + \xi_1 \mathbf{rh}} d\Omega} \cong \frac{\int \mathbf{rh} e^{-\beta V}(1 + \xi_1 \mathbf{rh}) d\Omega}{\int e^{-\beta V}(1 + \xi_1 \mathbf{rh}) d\Omega}$$

$$= \frac{\langle \mathbf{rh} \rangle_0 + \xi_1 \langle \mathbf{rh}^2 \rangle_0}{1 + \xi_1 \langle \mathbf{rh} \rangle_0} \cong (1 - \xi_1 \langle \mathbf{rh} \rangle_0)(\langle \mathbf{rh} \rangle_0 + \xi_1 \langle \mathbf{rh} \rangle_0)$$

$$\cong \langle \mathbf{rh} \rangle_0 + \xi_1 (\langle \mathbf{rh}^2 \rangle_0 - \langle \mathbf{rh} \rangle_0^2) \tag{1.91}$$

This is referred to as the *static case*, in that it avoids any considerations concerning the time dependence. The *dynamic case* is accomplished using the formal solution in Eq. (1.88)

$$\langle \mathbf{rh} \rangle \cong \frac{\int \mathbf{rh} e^{-\beta V}[1 + \xi_1 \alpha(t)] d\Omega}{\int e^{-\beta V}[1 + \xi_1 \alpha(t)] d\Omega} = \frac{\langle \mathbf{rh} \rangle_0 + \xi_1 \langle \alpha(t) \mathbf{rh} \rangle_0}{1 + \xi_1 \langle \alpha(t) \rangle_0}$$

$$\cong (\langle \mathbf{rh} \rangle_0 + \xi_1 \langle \alpha(t) \mathbf{rh} \rangle_0)(1 - \xi_1 \langle \alpha(t) \rangle_0)$$

$$\cong \langle \mathbf{rh} \rangle_0 + \xi_1 (\langle \alpha(t) \mathbf{rh} \rangle_0 - \langle \alpha(t) \rangle_0 \langle \mathbf{rh}_0 \rangle) \tag{1.92}$$

The *aftereffect function* is [27]

$$f(t) = \langle \mathbf{rh} \rangle - \langle \mathbf{rh} \rangle_0 = \xi_1 (\langle \alpha(t) \alpha(0) \rangle_0 - \langle \alpha(t) \rangle_0 \langle \alpha(0) \rangle_0)$$

$$= \xi_1 \mathrm{Cov}(\alpha(t), \alpha(0)) \tag{1.93}$$

In particular

$$f(0) = \langle \mathbf{rh} \rangle_1 - \langle \mathbf{rh} \rangle_0 = \xi_1 (\langle \alpha^2(0) \rangle_0 - \langle \alpha(0) \rangle_0^2)$$

$$= \xi_1 \mathrm{Var}(\alpha(0)) \tag{1.94}$$

The *autocorrelation function* is then defined as

$$C(t) = \frac{\mathrm{Cov}(\alpha(t), \alpha(0))}{\mathrm{Var}(\alpha(0))} = \frac{f(t)}{f(0)} \tag{1.95}$$

and the *correlation time* as the area under the curve of the autocorrelation function

$$T = \int_0^\infty C(t)dt \tag{1.96}$$

Assuming Eq. (1.93) to have time dependency

$$f(t) = \sum_{n=0}^{\infty} a_n e^{-p_n t} \tag{1.97}$$

then Eqs. (1.95) and (1.96) give

$$T = \sum_{n=0}^{\infty} p_n^{-1} a_n \bigg/ \sum_{n=0}^{\infty} a_n \tag{1.98}$$

If in addition to Eq. (1.97) the time dependency of Eq. (1.93) is assumed to be

$$f(t) = f(0)\exp(-p_{ef}t) \tag{1.99}$$

then p_{ef} is called the *effective eigenvalue*. Differentiation gives

$$p_{ef} = -\dot{f}(0)/f(0) = \sum_{n=0}^{\infty} p_n a_n \bigg/ \sum_{n=0}^{\infty} a_n \tag{1.100}$$

The *effective relaxation time* is then

$$\tau_{ef} = 1/p_{ef} = \sum_{n=0}^{\infty} a_n \bigg/ \sum_{n=0}^{\infty} p_n a_n \tag{1.101}$$

Assuming that the smallest nonvanishing eigenvalue satisfies $p_1 \ll p_n$, $n \geq 2$, and the corresponding amplitude $A_1 \gg A_n$, $n \geq 2$, then

$$T, \tau_{ef} \cong \tau = 1/p_1 \tag{1.102}$$

The variable τ corresponds to the time constant associated with longest-lived mode in Eq. (1.86) and will be referred to as the *relaxation time*. Suppose the vectors **r** and **h** are represented in terms of a triad of orthogonal unit vectors $\{\mathbf{e}_1, \mathbf{e}_2, \mathbf{e}_3\}$, where $\mathbf{e}_3$ is parallel to the easy axis, then

$$\begin{aligned}\langle \mathbf{rh} \rangle &= \langle (\sin\theta \cos\varphi \mathbf{e}_1 + \sin\theta \sin\varphi \mathbf{e}_2 + \cos\theta \mathbf{e}_3)(\sin\psi \mathbf{e}_1 + \cos\psi \mathbf{e}_3) \rangle \\ &= \cos\psi \langle \cos\theta \rangle + \sin\psi \langle \sin\theta \cos\varphi \rangle \qquad (1.103)\end{aligned}$$

The determination of the dependency of T on ψ is of paramount importance. The values of T in the limiting cases when $\psi = 0, \pi/2$ are denoted by $T_{\|}$ and $T_{\perp}$, respectively, and are referred to as the longitudinal and transverse relaxation times [14, 27, 28].

G. The Case of Axial Symmetry

On assuming that both V and W are axially symmetric, i.e., $\partial V/\partial\varphi = \partial W/\partial\varphi = 0$, then following Brown [7], the Fokker–Planck equation in Eq. (1.83) reduces to

$$\frac{\partial W}{\partial t} = \frac{b}{\sin\theta} \frac{\partial}{\partial\theta} \left[\sin\theta \left(\frac{dV}{d\theta} W + \beta^{-1} \frac{\partial W}{\partial\theta} \right) \right] \qquad (1.104)$$

and the longitudinal component of current density to

$$J_\theta = -b \left(\frac{\partial V}{\partial\theta} W + \beta^{-1} \frac{\partial W}{\partial\theta} \right) \qquad (1.105)$$

Writing $x \equiv \cos\theta$ and

$$W(x,t) = \sum_{n=0}^{\infty} A_n F_n(x) e^{-p_n t} \qquad (1.106)$$

then F_n are the eigenfunctions, A_n the corresponding amplitudes, and

$$\lambda_n = \beta p_n / b \qquad (1.107)$$

the corresponding eigenvalues of the Sturm–Liouville problem

$$\frac{d}{dx} \left[(1 - x^2) e^{-\beta V} \frac{d}{dx} (e^{\beta V} F) \right] + \lambda F = 0 \qquad (1.108)$$

noindent subject to the boundary condition that F must be finite at $x = \pm 1$. For a longitudinally applied field

$$\beta V = -\sigma(x^2 + 2hx) \tag{1.109}$$

where $\sigma = \beta K$ and $h = HM_s/2K$. Brown [7] then derived, using the methods of perturbation theory [30], the following *low energy barrier approximation formula* for the lowest nonvanishing eigenvalue

$$\lambda_1 \cong 2 - \frac{4}{5}\sigma + \frac{96}{875}\sigma^2 + \frac{1}{5}\xi^2 \tag{1.110}$$

where σ and $\xi = \beta HM_s$ are small. The relaxation time is then calculated from

$$\tau_\| = \beta/b\lambda_1 \cong \frac{\beta}{b}\left(2 - \frac{4}{5}\sigma + \frac{96}{875}\sigma^2 + \frac{1}{5}\xi^2\right)^{-1} \tag{1.111}$$

In the opposite, more important limiting case, we assume that the ratios of energy barrier to thermal energy are large. In addition, we impose the restriction that $h < 1$, which insures that the free energy has a bistable structure.

The following calculation is a variation of the *Kramers transition state theory* [31] and is similar to that given by Brown [7] in that we abandon the assumption that requires the stationary points to be at the poles. We will, however, allow for the possibility of this occurring as a special case. This then enables us to treat both the case of a longitudinally applied field, Eq. (1.8), and the axially symmetric approximation to the case of a transversely applied field, Eq. (1.10).

Suppose the free energy has minima at θ_i, $i = 1, 2$, and a maximum at θ_0, with $\theta_1 < \theta_0 < \theta_2$ as shown in Fig. 5. The corresponding values of the free energy per unit volume are V_i, $i = 0, 1, 2$. Hence the high energy barrier assumption can be expressed as $\beta(V_0 - V_i) \gg 1$. The Taylor series of V about θ_i truncated at the $(\theta - \theta_i)^2$ term is

$$V = V_i + (k_i/2!)(\theta - \theta_i)^2 \qquad k_i = (d^2V/d\theta^2)_{\theta=\theta_i} \tag{1.112}$$

Suppose from the outset that *quasi-equilibrium* (see Fig. 5) has been obtained separately within the intervals $R_1 = (0, \alpha_1)$ and $R_2 = (\alpha_2, \pi)$, where α_1, α_2

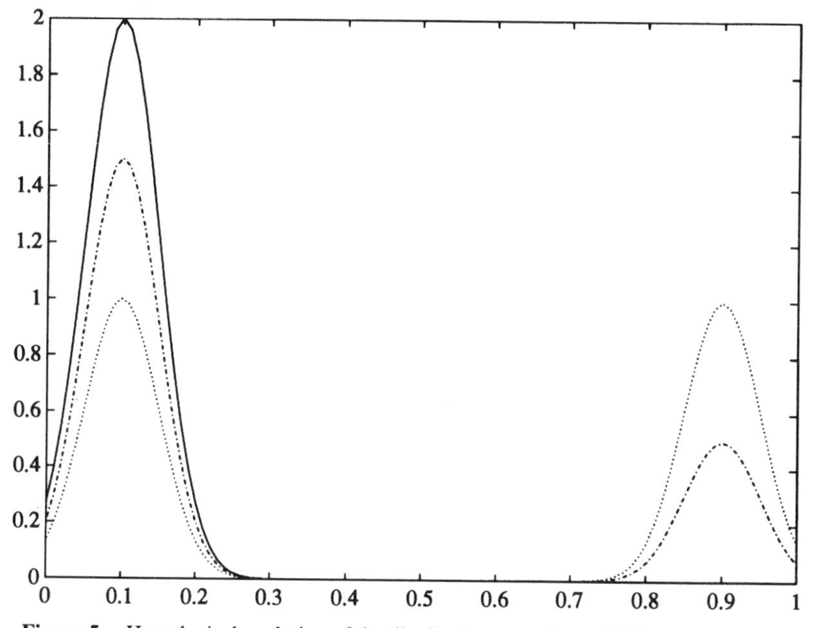

Figure 5. Hypothetical evolution of the distribution towards equilibrium: initially (solid), quasi-equilibrium (dashdots), equilibrium (dots).

are chosen so that

$$0 \leq \theta_1 < \alpha_1 < \theta_0 < \alpha_2 < \theta_2 \leq \pi$$

$$e^{-\beta V_0} \ll e^{-\beta V(\alpha_i)} \ll e^{-\beta V_i} \tag{1.113}$$

and that all but a very small fraction of the particles (treated as representative points) have orientations within the R_i, i.e.,

$$W = W_i e^{-\beta(V - V_i)} \qquad i = 1, 2 \tag{1.114}$$

where $W_i = W(\theta_i)$. The number of representative points in R_i is

$$n_i = 2\pi W_i \int_{\theta \in R_i} e^{-\beta(V - V_i)} \sin\theta d\theta \cong 2\pi W_i I_i \tag{1.115}$$

where from Eq. (1.112)

$$I_i = \int_{\theta \in R_i} e^{-(\beta k_i/2)(\theta - \theta_i)^2} \sin\theta \, d\theta \tag{1.116}$$

The integral I_i is evaluated in two separate limiting cases:

1. When $\theta_i \to 0, \pi$, then $\sin\theta$ is replaced by θ, and the integration is extended over $[0, \infty]$ so that

$$I_i \to \int_0^\infty e^{-(\beta k_i/2)(\theta - \theta_i)^2} \theta \, d\theta = 1/\beta k_i \tag{1.117}$$

2. When θ_i is sufficiently far from the corresponding pole to ensure that $e^{-\beta k_i \theta_i^2/2}$, $e^{-\beta k_i(\pi - \theta_i)^2/2} \gg 1$, then $\sin\theta$ is replaced by $\sin\theta_i$, and the integration extended over $[-\infty, \infty]$ so that

$$I_i \to \sin\theta_i \int_{-\infty}^{\infty} e^{-(\beta k_i/2)(\theta - \theta_i)^2} d\theta = \sin\theta_i \sqrt{2\pi/\beta k_i} \tag{1.118}$$

In the interval (α_1, α_2), W is small, but sufficient to maintain a small net flow of particles from the overpopulated towards the underpopulated minimum. Approximating this flow by a divergenceless current density so that the total current $\dot{n}_1 = -\dot{n}_2 = -2\pi \sin\theta J_\theta$ is independent of θ, then from Eq. (1.105)

$$\frac{\partial W}{\partial \theta} + \beta \frac{\partial V}{\partial \theta} W = \frac{\dot{n}_1}{2\pi b \sin\theta} \tag{1.119}$$

Multiplying by $e^{\beta V}$ and integrating over (θ_1, θ_2) gives

$$W e^{\beta V}\Big|_{\theta_1}^{\theta_2} = \frac{\beta \dot{n}_1}{2\pi b} \int_{\theta_1}^{\theta_2} \frac{e^{\beta V} d\theta}{\sin\theta} \tag{1.120}$$

To evaluate the integral on the right-hand side, $\sin\theta$ is replaced by $\sin\theta_0$, V by Eq. (1.112), and the integration is extended over $(-\infty, \infty)$

$$W e^{\beta V}\Big|_{\theta=\theta_1}^{\theta_2} \cong \frac{\dot{n}_1 e^{\beta V_0}}{2\pi b \sin\theta_0} \sqrt{2\pi\beta/-k_0} \tag{1.121}$$

From Eq. (1.114)

$$We^{\beta V}\Big|_{\theta_1}^{\theta_2} = \frac{1}{2\pi}\left(\frac{n_2 e^{\beta V_2}}{I_2} - \frac{n_1 e^{\beta V_1}}{I_1}\right) \tag{1.122}$$

hence

$$\dot{n}_1 = b\sin\theta_0\sqrt{-k_0/2\pi\beta}\left(\frac{n_2 e^{-\beta(V_0 - V_2)}}{I_2} - \frac{n_1 e^{-\beta(V_0 - V_1)}}{I_1}\right) \tag{1.123}$$

This is equivalent to the discrete orientation approximation Eq. (1.13) namely

$$\dot{n}_1 = -\dot{n}_2 = \nu_{2,1}n_2 - \nu_{1,2}n_1 \tag{1.124}$$

hence the transition probabilities are

$$\begin{aligned}\nu_{i,j} &= b\sin\theta_0 I_i^{-1}\sqrt{-k_0/2\pi\beta}\, e^{-\beta(V_0 - V_i)} \\ &= \begin{cases} bk_i\sin\theta_0\sqrt{-\beta k_0/2\pi}\, e^{-\beta(V_0 - V_i)} & \theta_i \sim 0,\pi \\ (b\sin\theta_0/2\pi\sin\theta_i)\sqrt{-k_0 k_i}\, e^{-\beta(V_0 - V_i)} & \theta_i \text{ otherwise}\end{cases}\end{aligned} \tag{1.125}$$

and the relaxation time is calculated from $\tau = (\nu_{1,2} + \nu_{2,1})^{-1}$. This is the *high energy barrier approximation formula.*

For the case of uniaxial anisotropy with longitudinally applied field (ψ = 0), we use Eq. (1.8), namely,

$$V = K(\sin^2\theta - 2h\cos\theta) \tag{1.126}$$

The stationary points are the solutions of the trigonometric equation

$$V' = 2K\sin\theta(\cos\theta + h) = 0 \tag{1.127}$$

which yields minima at $\theta = 0, \pi$ and a maximum at $\theta = \cos^{-1}(-h)$. Calculating the values of V at the stationary points gives the ratios of energy barrier height to thermal energy

$$\beta(V_0 - V_1) = \sigma(1+h)^2 \qquad \beta(V_0 - V_2) = \sigma(1-h)^2 \tag{1.128}$$

Calculating the values of

$$V'' = 2K(\cos^2\theta - \sin^2\theta + h\cos\theta) \tag{1.129}$$

at the stationary points gives

$$k_0 = -2K(1-h^2) \qquad k_1 = 2K(1+h) \qquad k_2 = 2K(1-h) \tag{1.130}$$

Equation (1.125) becomes

$$\left.\begin{matrix}\nu_{1,2}\\ \nu_{2,1}\end{matrix}\right\} = 2b\beta^{-1}\sigma^{3/2}\pi^{-1/2}(1-h^2)(1\pm h)e^{-\sigma(1\pm h)^2} \tag{1.131}$$

Aharoni [8, 32] showed by numerical solution of the Fokker–Planck equation that $p_1 \cong \nu_{1,2} + \nu_{2,1}$ provides a good approximation to the lowest eigenvalue. The longitudinal relaxation time is given by

$$\tau_1 = \frac{\beta\sigma^{-3/2}\sqrt{\pi}}{2b(1-h^2)[(1+h)e^{-\sigma(1+h)^2} + (1-h)e^{-\sigma(1-h)^2}]} \tag{1.132}$$

Brown [7] also derived Eq. (1.132) using *the method of approximate minimization* [30], which has the advantage of avoiding the assumption of a divergenceless current density in (θ_1, θ_2). A variation of this method was later used to derive Eq. (1.132) by Scully et al. [33]. For zero applied field, $h = 0$; hence, $\nu_{1,2} = \nu_{2,1} = 2b\beta^{-1}\sigma^{3/2}\pi^{-1/2}e^{\sigma}$, and

$$\tau = (\beta\sqrt{\pi}/4b)\sigma^{-3/2}e^{\sigma} \tag{1.133}$$

Recently Coffey et al. [27] have obtained an exact analytic formula for the correlation time from the axially symmetric Fokker–Planck equation, Eq. (1.104), using continued fraction methods, which in the high energy barrier limit reduces to Eq. (1.133).

For a transversely applied field ($\psi = \pi/2$) we use the axially symmetric approximation in Eq. (1.10), namely,

$$V = K \sin\theta(\sin\theta - 2h) \tag{1.134}$$

The stationary points are the solutions of the trigonometric equation

$$V' = 2K\cos\theta(\sin\theta - h) = 0 \tag{1.135}$$

which yields minima at $\theta = \sin^{-1}h$, $\pi - \sin^{-1}h$ and a maximum at $\theta = \pi/2$. Calculating the values of V at the stationary points gives the ratios of energy barrier height to thermal energy

$$\beta(V_0 - V_i) = \sigma(1-h)^2 \qquad i = 1, 2 \tag{1.136}$$

Calculating the values of

$$V'' = 2K(\cos^2\theta - \sin^2\theta + h\sin\theta) \tag{1.137}$$

at the stationary points gives

$$k_0 = -2K(1-h) \qquad k_1 = k_2 = 2K(1-h^2) \tag{1.138}$$

The transition probabilities in Eq. (1.125) become

$$\nu_{1,2} = \nu_{2,1} = (b\sigma/\pi\beta h)\sqrt{(1-h)(1-h)^2}\, e^{-\sigma(1-h)^2} \tag{1.139}$$

hence the longest relaxation time in a transverse field is

$$\tau = (\pi\beta h/2b\sigma)(1-h)^{-1/2}(1-h^2)^{-1/2}e^{\sigma(1-h)^2} \tag{1.140}$$

II. DIFFERENTIAL RECURRENCE RELATIONS FOR NON–AXIALLY SYMMETRIC PROBLEMS

A. Matrix Elements and Differential Recurrence Coefficients of the Fokker–Planck Operator

The *normalized spherical harmonics* [34] are defined

$$Y_{l,m} = N_{l,m}X_{l,m} \qquad |m| \le l \tag{2.1}$$

where $X_{l,m}$ are the spherical harmonics [Appendix A] and

$$N_{l,m} = (-1)^m\sqrt{\frac{(2l+1)(l-m)!}{4\pi(l+m)!}} \tag{2.2}$$

are the normalization constants. The normalized spherical harmonics form a *biorthonormal basis* [34] for $C^2(\Omega)$ (The space of functions defined on Ω having continuous second partial derivatives) where

$$\Omega = \{\mathbf{r} \in R^3, |\mathbf{r}| = 1\} = \{(\theta, \varphi); 0 \le \theta \le \pi, 0 \le \varphi < 2\pi\} \tag{2.3}$$

denotes the unit sphere. In addition they satisfy the *symmetry relation*

$$Y_{l,m}^{*} = (-1)^{m} Y_{l,-m} \tag{2.4}$$

The *matrix elements* of the Fokker–Planck operator L_{FP} w.r.t. normalized spherical harmonics are defined

$$d_{l,m,p,q} = (Y_{l,m}, L_{FP} Y_{p,q}) \tag{2.5}$$

where

$$(A, B) = \int A(\mathbf{r}) B^{*}(\mathbf{r}) d\Omega(\mathbf{r}) = \int_{0}^{2\pi} \int_{0}^{\pi} A(\theta, \varphi) B^{*}(\theta, \varphi) \sin\theta d\theta d\varphi \tag{2.6}$$

denotes the inner product on $C^{2}(\Omega)$. The nonzero matrix elements satisfy the recurrence relation for L_{FP} w.r.t. normalized spherical harmonics

$$L_{FP} Y_{l,m} = \sum_{s=-p}^{p} \sum_{t=-p_s}^{p_s} d_{l+s,m+t,l,m}^{*} Y_{l+s,m+t} \tag{2.7}$$

The *Hilbert adjoint* [18] of the Fokker–Planck operator is defined as

$$(Y_{l,m}, L_{FP} Y_{p,q}) = (L_{FP}^{*} Y_{l,m}, Y_{p,q}) \tag{2.8}$$

and satisfies the recurrence relation

$$L_{FP}^{*} Y_{l,m} = \sum_{s=-p}^{p} \sum_{t=-p_s}^{p_s} d_{l,m,l+s,m+t} Y_{l+s,m+t} \tag{2.9}$$

Suppose the solution of the Fokker–Planck equation has a representation as a *Fourier–Laplace series* [29]

$$W(\mathbf{r}, t) = \sum_{|m| \leq l} a_{l,m}(t) Y_{l,m}(\mathbf{r}) \tag{2.10}$$

then the *Fourier coefficients*

$$a_{l,m} = (W, Y_{l,m}) = \int W Y_{l,m}^{*} d\Omega = \langle Y_{l,m}^{*} \rangle \int W d\Omega = N_{l,m} \langle X_{l,m}^{*} \rangle \int W d\Omega \tag{2.11}$$

satisfy the *differential recurrence relations*

$$\begin{aligned}\dot{a}_{l,m} &= (\dot{W}, Y_{l,m}) = (L_{FP}W, Y_{l,m}) = (W, L_{FP}^{*}Y_{l,m}) \\ &= \left(W, \sum_{s=-p}^{p}\sum_{t=-p_s}^{p_s} d_{l,m,l+s,m+t}Y_{l,+s,m+t}\right) \\ &= \sum_{s,t} d_{l,m,l+s,m+t}^{*}a_{l,+s,m+t}\end{aligned} \tag{2.12}$$

and from Eq. (2.4) the symmetry relation

$$a_{l,m}^{*} = \int WY_{l,m}\,d\Omega = (-1)^m \int WY_{l,-m}^{*}\,d\Omega = (-1)^m a_{l,-m} \tag{2.13}$$

From Eqs. (2.11) and (2.12), the expectation values $x_{l,m} = \langle X_{l,m}\rangle$ satisfy the differential recurrence relations [35]

$$\dot{x}_{l,m} = \sum_{s=-p}^{p}\sum_{t=-p_s}^{p_s} e_{l,m,l+s,m+t}x_{l+s,m+t} \tag{2.14}$$

where the *differential recurrence coefficients* for the expectation values are

$$e_{l,m,l+s,m+t} = N_{l,m}^{-1}N_{l+s,m+t}d_{l,m,l+s,m+t} \tag{2.15}$$

The *decay modes* are defined

$$c_{l,m} = \langle X_{l,m}\rangle - \langle X_{l,m}\rangle_0 \tag{2.16}$$

where, as always, the subscript zero denotes that the expectation value is to be evaluated in the absence of the perturbing field. They satisfy Eq. (2.15), i.e.,

$$\dot{c}_{l,m} = \sum_{s=-p}^{p}\sum_{t=-p_s}^{p_s} e_{l,m,l+s,m+t}c_{l,+s,m+t} \tag{2.17}$$

The expectation values (and hence the decay modes) satisfy the symmetry relations

$$x_{l,-m} = \rho_{l,m}x_{l,m}^{*} \tag{2.18}$$

where

$$\rho_{l,m} = (-1)^m \frac{(l-m)!}{(l+m)!} \tag{2.19}$$

The differential recurrence relations in Eq. (2.14) are obtained as follows: First we apply the recurrence relations for unnormalized spherical harmonics [Appendix A] to obtain the recurrence relation

$$L_{FP} X_{l,m} = \sum_{s=-p}^{p} \sum_{t=-p_s}^{p_s} g_{l+s,m+t,l,m} X_{l+s,m+t} \tag{2.20}$$

The matrix elements and differential recurrence coefficients are then calculated from the formulae

$$\begin{aligned} d_{l,m,l+s,m+t} &= N_{l,m}^{-1} N_{l+s,m+t} \overset{*}{g}_{l,m,l+s,m+t} \\ &= (-1)^t \sqrt{\frac{(2l+2s+1)(l+m)!(l-m+s-t)!}{(2l+1)(l+m+s+t)!(l-m)!}} \\ &\quad \cdot \overset{*}{g}_{l,m,l+s,m+t} \end{aligned} \tag{2.21}$$

$$\begin{aligned} e_{l,m,l+s,m+t} &= N_{l,m}^{-2} N_{l+s,m+t}^{2} \overset{*}{g}_{l,m,l+s,m+t} \\ &= \frac{(2l+2s+1)(l+m)!(l-m+s-t)!}{(2l+1)(l+m+s+t)!(l-m)!} \overset{*}{g}_{l,m,l+s,m+t} \end{aligned} \tag{2.22}$$

The differential recurrence relations in Eq. (2.14) can be expressed as matrix differential recurrence relations as follows. Define the matrices

$$E_{n,m,p} = \{E'_{i,j}\}_{n \le i \le n+p; m \le j \le m+p} = \begin{bmatrix} E'_{n,m} & E'_{n,m+1} & \cdots \\ E'_{n+1,m} & \ddots & \\ \vdots & & E'_{n+p,m+p} \end{bmatrix} \tag{2.23}$$

where the $E'_{n,m}$ are *submatrices* which are defined as follows. Let $n, m \geq 0$ then

$$E'_{-n,-m} = \{e_{n,i-n,m,j-n}\}_{0\le i\le n-1;\,0\le j\le m-1}$$

$$= \begin{bmatrix} e_{n,-n,m,-m} & e_{n,-n,m,1-m} & & \\ e_{n,1-n,m,-m} & \ddots & & \\ & & & e_{n,-1,m,-1} \end{bmatrix} \tag{2.24}$$

$$E'_{-n,m} = \{e_{n,i-n,m,j}\}_{0\le i\le n-1;\,0\le j\le m}$$

$$= \begin{bmatrix} e_{n,-n,m,0} & e_{n,-n,m,1} & & \\ e_{n,1-n,m,0} & \ddots & & \\ & & & e_{n,-1,m,m} \end{bmatrix} \tag{2.25}$$

$$E'_{n,-m} = \{e_{n,i,m,j-m}\}_{0\le i\le n;\,0\le j\le m-1}$$

$$= \begin{bmatrix} e_{n,0,m,-m} & e_{n,0,m,1-m} & & \\ e_{n,1,m,-m} & \ddots & & \\ & & & e_{n,n,m,-1} \end{bmatrix} \tag{2.26}$$

$$E'_{n,m} = \{e_{n,i,m,j}\}_{0\le i\le n;\,0\le j\le m}$$

$$= \begin{bmatrix} e_{n,0,m,0} & e_{n,0,m,1} & & \\ e_{n,1,m,0} & \ddots & & \\ & & & e_{n,n,m,m} \end{bmatrix} \tag{2.27}$$

For $n \geq 1$, define the matrices

$$P_n = \begin{bmatrix} 0 & & 0 & P'_n \\ & 0 & \cdot^{\cdot^{\cdot}} & 0 \\ 0 & P'_2 & 0 & \\ P'_1 & 0 & & 0 \end{bmatrix} \tag{2.28}$$

where the submatrices P'_i are defined

$$P'_i = \begin{bmatrix} 0 & & & 0 & \rho_{i,j} \\ & & & \cdot^{\cdot^{\cdot}} & 0 \\ & 0 & \rho_{i,2} & 0 & \\ 0 & \rho_{i,1} & 0 & & 0 \end{bmatrix} \tag{2.29}$$

Again for $n \geq 1$, define the vectors

$$X_n^T = (x_{1,0}, x_{1,1}, \cdots x_{n,n}) \tag{2.30}$$

$$G_n^T = (e_{1,0,0,0}, e_{1,1,0,0}, \cdots e_{n,n,0,0}) \tag{2.31}$$

Let

$$X_{-n}^T = (x_{n,-n}, x_{n,1-n}, \cdots x_{n,-1}, x_{n-1,-(n-1)}, \cdots x_{-1,-1}) \tag{2.32}$$

then the symmetry relation Eq. (2.18) yields the vector symmetry relation

$$X_{-n} = P_n X_n^* \tag{2.33}$$

For $n \geq 1$, let $E_{-n} = E_{1,-n,n-1}$ and $E_n = E_{1,1,n-1}$; then the differential recurrence relations in Eq. (2.14) are approximated by the complex matrix differential equation

$$\dot{X}_n = E_{-n} X_{-n,-1} + G_n + E_n X_n \tag{2.34}$$

or

$$\dot{X}_n = F_n X_n^* + G_n + E_n X_n \tag{2.35}$$

where

$$F_n = E_{-n} P_n \tag{2.36}$$

Equation (2.35) can then be expressed as a real matrix differential equation

$$\dot{U}_n = A_n U_n + V_n \tag{2.37}$$

where

$$A_n = \begin{bmatrix} \mathrm{Re}E_n + \mathrm{Re}F_n & -\mathrm{Im}E_n + \mathrm{Im}F_n \\ \mathrm{Im}E_n + \mathrm{Im}F_n & \mathrm{Re}E_n - \mathrm{Re}F_n \end{bmatrix} \tag{2.38}$$

is the system matrix and

$$U_n = (\mathrm{Re}X_n, \mathrm{Im}X_n)^T \tag{2.39}$$

$$V_n = (\mathrm{Re}G_n, \mathrm{Im}G_n)^T \tag{2.40}$$

From the equilibrium conditions $\dot{U}_n(0) = \dot{U}_n(\infty) = 0$. The vectors $\dot{U}_n(0)$ and $U_n(\infty)$ are the solutions of

$$A'_n U_n(0) = -V'_n \tag{2.41}$$

$$A_n U_n(\infty) = -V_n \tag{2.42}$$

where the primes denote that H is replaced by $H + H_1$. To solve Eq. (2.37), we take the Laplace transfrom

$$s\tilde{U}_n - U_n(0) = A_n \tilde{U}_n + s^{-1} V_n \tag{2.43}$$

hence

$$\tilde{U}_n = s^{-1}(sI_n - A_n)^{-1} V_n + (sI_n - A_n)^{-1} U_n(0) \tag{2.44}$$

Consider the term

$$s^{-1}(sI_n - A_n)^{-1} V_n \tag{2.45}$$

Applying the method of partial fractions gives

$$\begin{aligned} s^{-1}(sI_n - A_n)^{-1} V_n &= s^{-1} P + (sI_n - A_n)^{-1} Q \\ &= s^{-1}(sI_n - A_n)^{-1}[(sI_n - A_n)P + sQ] \end{aligned} \tag{2.46}$$

where P and Q are undetermined column vectors. Equating the denominators of both sides gives

$$-A_n P + s(P + Q) = V_n \tag{2.47}$$

hence

$$A_n P = -V_n \qquad Q = -P \tag{2.48}$$

From Eq. (2.42)

$$P = U_n(\infty) \qquad Q = -U_n(\infty) \tag{2.49}$$

Equation (2.45) becomes

$$[s^{-1} I_n - (sI_n - A_n)^{-1}] U_n(\infty) \tag{2.50}$$

Substituting into Eq. (2.44) gives

$$\tilde{U}_n(s) - s^{-1}U_n(\infty) = (sI_n - A_n)^{-1}(U_n(0) - U_n(\infty)) \tag{2.51}$$

Let C_n denote the column vector whose components are the real and imaginary parts of the decay modes in Eq. (2.16), i.e.,

$$C_n(t) = U_n(t) - U_n(\infty) \tag{2.52}$$

From Eqs. (2.37) and (2.42), we have

$$\dot{C}_n = A_n C_n \tag{2.53}$$

Taking the Laplace transform of Eq. (2.52) and substituting from Eq. (2.51) gives

$$\begin{aligned}\tilde{C}_n(s) &= \tilde{U}_n(s) - s^{-1}U_n(\infty)\\ &= (sI_n - A_n)^{-1}C_n(0)\end{aligned} \tag{2.54}$$

In particular

$$\tilde{C}_n(0) = A_n^{-1}C_n(0) \tag{2.55}$$

The initial value vector $C_n(0)$ is found as follows. The matrix A'_n and the vector V'_n in Eq. (2.41) can be expressed as

$$A'_n = A_n + H_1 B_n \tag{2.56}$$

$$V'_n = V_n + H_1 W_n \tag{2.57}$$

Equation (2.41) becomes

$$(A_n + H_1 B_n)(C_n(0) + U_n(\infty)) = -V_n - H_1 W_n \tag{2.58}$$

From Eq. (2.42)

$$(A_n + H_1 B_n)(C_n(0) - A_n^{-1}V_n) = -V_n - H_1 W_n \tag{2.59}$$

Hence on ignoring terms of order H_1^2 and higher powers

$$\begin{aligned} C_n(0) &= H_1(A_n + H_1B_n)^{-1}(B_nA_n^{-1}V_n - W_n) \\ &= H_1A_n^{-1}(I + H_1A_n^{-1}B_n)^{-1}(B_nA_n^{-1}V_n - W_n) \\ &\cong H_1A_n^{-1}(I - H_1A_n^{-1}B_n)(B_nA_n^{-1}V_n - W_n) \\ &\cong H_1A_n^{-1}(B_nA_n^{-1}V_n - W_n) \end{aligned} \tag{2.60}$$

Taking the inverse Laplace transform of Eq. (2.51) gives

$$\begin{aligned} U_n(t) &= \exp(A_nt)(C_n(0)) + U_n(\infty) \\ &= R_n \exp(\Lambda_nt)R_n^{-1}(C_n(0)) + U_n(\infty) \end{aligned} \tag{2.61}$$

Here

$$\exp(\Lambda_nt) = \begin{bmatrix} e^{p_1t} & 0 & & 0 \\ 0 & e^{p_2t} & \ddots & \\ & \ddots & \ddots & 0 \\ 0 & & 0 & e^{p_nt} \end{bmatrix} \tag{2.62}$$

where p_i are the eigenvalues and R_n is the matrix whose row vectors are the eigenvectors of A_n. From Eqs. (2.10) and (2.11), the normalized distribution is expressed in terms of the components of $U_n(t)$ as follows

$$\begin{aligned} W\Big/\int W\,d\Omega &= \sum_{|m|\le l} N_{l,m}^2\langle X_{l,m}^*\rangle X_{l,m} \\ &= \sum_{l=0}^{\infty}\Bigg[N_{l,0}^2\langle X_{l,0}\rangle X_{l,0} + \sum_{m=1}^{l}(N_{l,m}^2\langle X_{l,m}^*\rangle X_{l,m} \\ &\quad + N_{l,-m}^2\langle X_{l,-m}^*\rangle X_{l,-m})\Bigg] \\ &= \sum_{l=0}^{\infty}\Bigg[N_{l,0}^2\langle X_{l,0}\rangle X_{l,0} + \sum_{m=1}^{l} N_{l,m}^2(\langle X_{l,m}^*\rangle X_{l,m} + \langle X_{l,m}\rangle X_{l,m}^*)\Bigg] \\ &= \sum_{l=0}^{\infty}\Bigg[N_{l,0}^2\langle X_{l,0}\rangle P_l(\cos\theta) \\ &\quad + 2\sum_{m=1}^{l} N_{l,m}^2 P_l^m(\cos\theta)(\mathrm{Re}\langle X_{l,m}\rangle\cos m\varphi \\ &\quad + \mathrm{Im}\langle X_{l,m}\rangle \sin m\varphi)\Bigg] \end{aligned} \tag{2.63}$$

B. Calculation of the Correlation Time from the Matrix Elements and Differential Recurrence Coefficients

The eigenvalues of the Fokker–Planck equation are precisely the eigenvalues of the infinite matrix representation of the Fokker–Planck operator D whose elements are the matrix elements discussed in the previous section. Let $D_{n,m,p}$ and $E_{n,m,p}$ denote the matrices defined by arranging the matrix elements $d_{l,m,p,q}$ and differential recurrence coefficients $e_{l,m,p,q}$, respectively, as shown in Eqs. (2.23–2.27). From Eq. (2.15)

$$N_{l,m}e_{l,m,p,q} = d_{l,m,p,q}N_{p,q} \tag{2.64}$$

Hence $D_{n,m,p}$ and $E_{n,m,p}$ are related by means of the similarity transformation of the form

$$E = K^{-1}DK \tag{2.65}$$

thus

$$\begin{aligned}\det(E - pI) &= \det(K^{-1}DK - pK^{-1}IK) = \det[K^{-1}(D - pI)K] \\ &= \det\ K^{-1}\det(D - pI)\det\ K = \det(D - pI)\end{aligned} \tag{2.66}$$

i.e., the *characteristic determinants* are equal; hence so too are the eigenvalues. By choosing a suitable value for n, the eigenvalues of D and, in particular, the lowest eigenvalue p_1 can be approximated by the eigenvalues of $D_{-n,-n,2n}$, or equivalently $E_{-n,-n,2n}$. In the case where [cf. Section I.F]

$$p_1 \ll p_k \qquad A_1 \ll A_k \qquad k \geq 2 \tag{2.67}$$

then the correlation time is calculated from Eq. (1.102), namely,

$$T = 1/p_1 \tag{2.68}$$

If Eq. (2.67) is not satisfied, then the following more exact method is used. We recall that the response function Eq. (1.93), with the aid of Eq. (1.103), can be expressed in terms of the decay modes in Eq. (2.16) as

$$f(t) = \langle \mathbf{rh} \rangle - \langle \mathbf{rh} \rangle_0 = \cos\psi c_{1,0}(t) + \sin\psi \operatorname{Re}[c_{1,1}(t)] \tag{2.69}$$

The correlation time Eq. (1.96) can be then expressed as [26, 27]

$$T = \lim_{s \to \infty} \int_0^\infty C(t) e^{-st} dt = \tilde{C}(0) = \tilde{f}(0)/f(0)$$
$$= \frac{\cos \psi \tilde{c}_{1,0}(0) + \sin \psi \mathrm{Re}[\tilde{c}_{1,1}(0)]}{\cos \psi c_{1,0}(0) + \sin \psi \mathrm{Re}[c_{1,1}(0)]} \tag{2.70}$$

Here the initial values of the decay modes are obtained from Eq. (2.60) and the zero frequency values of Laplace transformed decay modes are obtained from Eq. (2.54). For given values of K, H, and ψ, each of these quantities can be expressed as a multiple of the perturbing field magnitude H_1, yielding an expression for T that is independent of H_1.

C. Spherical Harmonic Representation Formulae for the Fokker–Planck Operator

To facilitate easy calculation of expressions for the matrix elements and differential recurrence coefficients, we first obtain a suitable *representation formula for L_{FP}*, [35] i.e.,

$$L_{FP} X_{l,m}(\mathbf{r}) = \sum_{s,t} L_{l+s,m+t,l,m} X_{l+s,m+t}(\mathbf{r}) \tag{2.71}$$

where for each s and t, the operator $L_{l+s,m+t,l,m}$ is one of two types:

1. A product operator that acts on $X_{l+s,m+t}$ by multiplication with an expression that can be expressed as a polynomial in the direction cosines of $\mathbf{r}$ relative to the crystal axes $(\mathbf{e}_1, \mathbf{e}_2, \mathbf{e}_3)$ (see Fig. 1)

$$\alpha_1 = \mathbf{r}\mathbf{e}_1 = \sin\theta \cos\varphi \quad \alpha_2 = \mathbf{r}\mathbf{e}_2 = \sin\theta \sin\varphi \quad \alpha_3 = \mathbf{r}\mathbf{e}_3 = \cos\theta \tag{2.72}$$

2. A composition that consists of such product operators together with the Laplacian operator Λ^2 whose effect on a spherical harmonic is evaluated using relation

$$\Lambda^2 X_{l,m} = -l(l+1) X_{l,m} \tag{2.73}$$

Application of the recurrence relations for spherical harmonics [Appendix A] yields

$$L_{p,q,l,m}X_{p,q} = \sum_{u,v} g_{p,q,l,m,u,v}X_{p+u,q+v} \tag{2.74}$$

where $g_{p,q,l,m,u,v}$ are constants. Evaluation of Eq. (2.74) for each s and t in Eq. (2.71) gives the recurrence relation for L_{FP}

$$L_{FP}X_{l,m} = \sum_{s,t} g_{l+s,m+t,l,m}X_{l+s,m+t} \tag{2.75}$$

Expressions for the nonzero matrix elements $d_{l,m,l+s,m+t}$ and the nonzero differential recurrence coefficients $e_{l,m,l+s,m+t}$ are then calculated using the formulae (cf. Eqs. 2.21, 2.22)

$$d_{l,m,l+s,m+t} = (-1)^t \sqrt{\frac{(2l+2s+1)(l+m)!(l-m+s-t)!}{(2l+1)(l+m+s+t)!(l-m)!}} \cdot g^*_{l,m,l+s,m+t} \tag{2.76}$$

$$e_{l,m,l+s,m+t} = \frac{(2l+2s+1)(l+m)!(l-m+s-t)!}{(2l+1)(l+m+s+t)!(l-m)!} g^*_{l,m,l+s,m+t} \tag{2.77}$$

It is convenient to express the Fokker–Planck operator in Eq. (1.83) as $L_{FP} = A_{FP} + iG_{FP}$, where

$$A_{FP}W = \beta^{-1}b\Lambda^2 W + bW\Lambda^2 V + b\left(\frac{\partial V}{\partial \theta}\frac{\partial W}{\partial \theta} + \frac{1}{\sin^2\theta}\frac{\partial V}{\partial \varphi}\frac{\partial W}{\partial \varphi}\right) \tag{2.78}$$

is the alignment operator and

$$G_{FP}W = \frac{a}{i\sin\theta}\left(\frac{\partial V}{\partial \theta}\frac{\partial W}{\partial \varphi} - \frac{\partial V}{\partial \varphi}\frac{\partial W}{\partial \theta}\right) \tag{2.79}$$

is the gyroscopic operator. Taking the Laplacian of VW gives

$$\Lambda^2 VW = \frac{1}{\sin\theta}\frac{\partial}{\partial\theta}\left(\sin\theta\,\frac{\partial VW}{\partial\theta}\right) + \frac{1}{\sin^2\theta}\frac{\partial^2 VW}{\partial\varphi^2}$$

$$= \frac{1}{\sin\theta}\frac{\partial}{\partial\theta}\left(\sin\theta V\frac{\partial W}{\partial\theta} + \sin\theta W\frac{\partial V}{\partial\theta}\right)$$

$$+ \frac{1}{\sin^2\theta}\frac{\partial}{\partial\varphi}\left(V\frac{\partial W}{\partial\varphi} + W\frac{\partial V}{\partial\varphi}\right)$$

$$= \frac{1}{\sin\theta}\frac{\partial}{\partial\theta}\left(\sin\theta V\frac{\partial W}{\partial\theta}\right) + \frac{1}{\sin\theta}\frac{\partial}{\partial\theta}\left(\sin\theta W\frac{\partial V}{\partial\theta}\right)$$

$$+ \frac{1}{\sin^2\theta}\frac{\partial}{\partial\varphi}\left(V\frac{\partial W}{\partial\varphi}\right) + \frac{1}{\sin^2\theta}\frac{\partial}{\partial\varphi}\left(W\frac{\partial V}{\partial\varphi}\right)$$

$$= V\frac{1}{\sin\theta}\frac{\partial}{\partial\theta}\left(\sin\theta\,\frac{\partial W}{\partial\theta}\right) + 2\,\frac{\partial V}{\partial\theta}\frac{\partial W}{\partial\theta}$$

$$+ W\frac{1}{\sin\theta}\frac{\partial}{\partial\theta}\left(\sin\theta\,\frac{\partial V}{\partial\theta}\right)$$

$$+ V\frac{1}{\sin^2\theta}\frac{\partial^2 W}{\partial\varphi^2} + \frac{2}{\sin^2\theta}\frac{\partial V}{\partial\varphi}\frac{\partial W}{\partial\varphi} + W\frac{1}{\sin^2\theta}\frac{\partial^2 V}{\partial\varphi^2}$$

$$= V\Lambda^2 W + W\Lambda^2 V + 2\left(\frac{\partial V}{\partial\theta}\frac{\partial W}{\partial\theta} + \frac{1}{\sin^2\theta}\frac{\partial V}{\partial\varphi}\frac{\partial W}{\partial\varphi}\right) \tag{2.80}$$

Hence

$$\frac{\partial V}{\partial\theta}\frac{\partial W}{\partial\theta} + \frac{1}{\sin^2\theta}\frac{\partial V}{\partial\varphi}\frac{\partial W}{\partial\varphi} = \frac{1}{2}\,(\Lambda^2 VW - V\Lambda^2 W - W\Lambda^2 V) \tag{2.81}$$

On replacing W with $X_{l,m}$, Eq. (2.78), with the aid of Eqs. (2.73) and (2.81), yields the *alignment representation formula*

$$A_{FP}X_{l,m} = b\left[-l(l+1)\beta^{-1} + \frac{l(l+1)}{2}\,V + \frac{1}{2}\,\Lambda^2 V\right]X_{l,m} + \frac{b}{2}\,\Lambda^2 V X_{l,m} \tag{2.82}$$

It is convenient to define the differential operators

$$\partial_{\pm 1} = \frac{\partial}{\partial \alpha_1} \pm i \frac{\partial}{\partial \alpha_2} \qquad \partial_0 = \frac{\partial}{\partial \alpha_3} \tag{2.83}$$

Applying the chain rule for partial derivatives gives

$$\begin{aligned}\frac{\partial V}{\partial \theta} &= \frac{\partial \alpha_1}{\partial \theta}\frac{\partial V}{\partial \alpha_1} + \frac{\partial \alpha_2}{\partial \theta}\frac{\partial V}{\partial \alpha_2} + \frac{\partial \alpha_3}{\partial \theta}\frac{\partial V}{\partial \alpha_3} \\ &= \cos\theta\left(\cos\varphi\frac{\partial V}{\partial \alpha_1} + \sin\varphi\frac{\partial V}{\partial \alpha_2}\right) - \sin\theta\frac{\partial V}{\partial \alpha_3} \\ &= \frac{\cos\theta}{2}(e^{i\varphi}\partial_{-1}V + e^{-i\varphi}\partial_1 V) - \sin\theta\,\partial_0 V\end{aligned} \tag{2.84}$$

$$\begin{aligned}\frac{1}{\sin\theta}\frac{\partial V}{\partial \varphi} &= \frac{1}{\sin\theta}\left(\frac{\partial \alpha_1}{\partial \varphi}\frac{\partial V}{\partial \alpha_1} + \frac{\partial \alpha_2}{\partial \varphi}\frac{\partial}{\partial \alpha_2} + \frac{\partial \alpha_3}{\partial \varphi}\frac{\partial}{\partial \alpha_3}\right) \\ &= -\sin\varphi\frac{\partial V}{\partial \alpha_1} + \cos\varphi\frac{\partial V}{\partial \alpha_2} \\ &= \frac{i}{2}(e^{i\varphi}\partial_{-1}V - e^{-i\varphi}\partial_1 V)\end{aligned} \tag{2.85}$$

Consider the term in Eq. (2.79)

$$\frac{a}{\sin\theta}\left(\frac{\partial V}{\partial \theta}\frac{\partial X_{l,m}}{\partial \varphi} - \frac{\partial V}{\partial \varphi}\frac{\partial X_{l,m}}{\partial \theta}\right) \tag{2.86}$$

Application of Eqs. (2.84) and (2.85) give

$$\begin{aligned}&\frac{a}{\sin\theta}\left[\frac{\cos\theta}{2}(e^{i\varphi}\partial_{-1}V + e^{-i\varphi}\partial_1 V) - \sin\theta\,\partial_0 V\right]\frac{\partial X_{l,m}}{\partial \varphi} \\ &\quad - \frac{ia}{2}(e^{i\varphi}\partial_{-1}V - e^{-i\varphi}\partial_1 V)\frac{\partial X_{l,m}}{\partial \theta} \\ &= \frac{ae^{-i\varphi}}{2}\partial_1 V\left[\frac{\cos\theta}{\sin\theta}\frac{\partial X_{l,m}}{\partial \varphi} + i\frac{\partial X_{l,m}}{\partial \theta}\right] - a\partial_0 V\frac{\partial X_{l,m}}{\partial \varphi} \\ &\quad + \frac{ae^{i\varphi}}{2}\partial_{-1}V\left[\frac{\cos\theta}{\sin\theta}\frac{\partial X_{l,m}}{\partial \varphi} - i\frac{\partial X_{l,m}}{\partial \theta}\right]\end{aligned} \tag{2.87}$$

The spherical harmonics are expressed in terms of the associated Legendre functions P_l^m by Eq. (A.12), namely,

$$X_{l,m} = e^{im\varphi} P_l^m \tag{2.88}$$

Hence

$$\partial X_{l,m}/\partial\varphi = imX_{l,m} \tag{2.89}$$

The second term on the right-hand side of Eq. (2.87) becomes

$$-a\partial_0 V \frac{\partial X_{l,m}}{\partial\varphi} = -imaX_{l,m}\partial_0 V \tag{2.90}$$

Writing

$$\cos\theta = \alpha^3 \qquad \sin\theta = \sqrt{1-\alpha_3^2} \tag{2.91}$$

the first and third terms on the right-hand side of Eq. (2.87) become

$$\frac{iae^{i(m-1)\varphi}1}{2}\partial_1 V\left(\frac{m\alpha_3}{\sqrt{1-\alpha_3^2}}P_l^m - \sqrt{1-\alpha_3^2}\frac{dP_l^m}{d\alpha_3}\right)$$
$$+\frac{iae^{i(m+1)\varphi}}{2}\partial_{-1} V\left(\frac{m\alpha_3}{\sqrt{1-\alpha_3^2}}P_l^m + \sqrt{1-\alpha_3^2}\frac{dP_l^m}{d\alpha_3}\right) \tag{2.92}$$

Application of the differential relation in Eq. (A.10) namely

$$(1-\alpha_3^2)\frac{dP_l^m}{d\alpha_3} = (l+m)P_{l-1}^m - l\alpha_3 P_l^m \tag{2.93}$$

gives

$$\frac{iae^{i(m-1)\varphi}}{2}\,\partial_1 V\left(\frac{m\alpha_3}{\sqrt{1-\alpha_3^2}}\,P_l^m - \frac{(l+m)P_{l-1}^m - l\alpha_3 P_l^m}{\sqrt{1-\alpha_3^2}}\right)$$

$$+\frac{iae^{i(m+1)\varphi}}{2}\,\partial_{-1} V\left(\frac{m\alpha_3}{\sqrt{1-\alpha_3^2}}\,P_l^m + \frac{(l+m)P_{l-1}^m - l\alpha_3 P_l^m}{\sqrt{1-\alpha_3^2}}\right)$$

$$=\frac{ia(l+m)e^{i(m-1)\varphi}}{2}\,\partial_1 V\left(\frac{\alpha_3 P_l^m - P_{l-1}^m}{\sqrt{1-\alpha_3^2}}\right)$$

$$+\frac{iae^{i(m+1)\varphi}}{2}\,\partial_{-1} V\left(\frac{(l+m)P_{l-1}^m - (l-m)\alpha_3 P_l^m}{\sqrt{1-\alpha_3^2}}\right) \tag{2.94}$$

Multiplying the recurrence relations in Eqs. (A.7) and (A.5) by $-e^{i(m-1)\varphi}/\sqrt{1-\alpha_3^2}$ and $e^{i(m+1)\varphi}/\sqrt{1-\alpha_3^2}$, respectively, gives

$$e^{i(m-1)\varphi}\,\frac{\alpha_3 P_l^m - P_{l-1}^m}{\sqrt{1-\alpha_3^2}} = (l-m+1)X_{l,m-1} \tag{2.95}$$

$$e^{i(m+1)\varphi}\,\frac{(l+m)P_{l-1}^m - (l-m)\alpha_3 P_l^m}{\sqrt{1-\alpha_3^2}} = X_{l,m+1} \tag{2.96}$$

The right-hand side of Eq. (2.94), with the aid of Eqs. (2.95) and (2.96), becomes

$$\frac{ia(l+m)(l-m+1)}{2}\,X_{l,m-1}\partial_1 V + \frac{ia}{2}\,X_{l,m+1}\partial_{-1} V \tag{2.97}$$

The gyroscopic operator in Eq. (2.79), on utilizing the expansion of Eq. (2.86) into Eqs. (2.90) and (2.97), yields the *gyroscopic representation formula*

$$G_{FP}X_{l,m} = \frac{a(l+m)(l-m+1)}{2}\,X_{l,m-1}\partial_1 V - maX_{l,m}\partial_0 V + \frac{a}{2}\,X_{l,m+1}\partial_{-1} V \tag{2.98}$$

$$b\left(\frac{\partial V}{\partial \theta}\frac{\partial X_{l,m}}{\partial \theta}+\frac{1}{\sin^2\theta}\frac{\partial V}{\partial \varphi}\frac{\partial X_{l,m}}{\partial \varphi}\right) \tag{2.99}$$

Applying Eqs. (2.84) and (2.85) gives

$$\begin{aligned}
&b\left[\frac{1}{2}\cos\theta(e^{i\varphi}\partial_{-1}V+e^{-i\varphi}\partial_1 V)-\sin\theta\partial_0 V\right]\frac{\partial X_{l,m}}{\partial\theta}\\
&\quad+\frac{ib}{2\sin\theta}(e^{i\varphi}\partial_{-1}V-e^{-i\varphi}\partial_1 V)\frac{\partial X_{l,m}}{\partial\varphi}\\
&\quad=\frac{be^{-i\varphi}}{2}\partial_1 V\left[\cos\theta\frac{\partial X_{l,m}}{\partial\theta}-\frac{i}{\sin\theta}\frac{\partial X_{l,m}}{\partial\varphi}\right]-b\partial_0 V\sin\theta\frac{\partial X_{l,m}}{\partial\theta}\\
&\quad+\frac{be^{-i\varphi}}{2}\partial_{-1}V\left[\cos\theta\frac{\partial X_{l,m}}{\partial\theta}+\frac{i}{\sin\theta}\frac{\partial X_{l,m}}{\partial\varphi}\right]
\end{aligned} \tag{2.100}$$

From Eqs. (2.88) and (2.90), we have

$$\begin{aligned}
&\frac{be^{i(m-1)\varphi}}{2}\partial_1 V\left[-\alpha_3\sqrt{1-\alpha_3^2}\frac{dP_l^m}{d\alpha_3}+\frac{mP_l^m}{\sqrt{1-\alpha_3^2}}\right]\\
&\quad+b\partial_0 Ve^{im\varphi}(1-\alpha_3^2)\frac{dP_l^m}{d\alpha_3}\\
&\quad-\frac{be^{i(m+1)\varphi}}{2}\partial_{-1}V\left[\alpha_3\sqrt{1-\alpha_3^2}\frac{dP_l^m}{d\alpha_3}+\frac{mP_l^m}{\sqrt{1-\alpha_3^2}}\right]
\end{aligned} \tag{2.101}$$

Application of the differential recurrence relation Eq. (2.93) to the first and third terms, and of the differential relation Eq. (A.10), namely

$$(1-\alpha_3^2)\frac{dP_l^m}{d\alpha_3}=(2l+1)^{-1}[(l+1)(l+m)P_{l-1}^m-l(l-m+1)P_{l+1}^m] \tag{2.102}$$

to the second gives

$$\frac{be^{i(m-1)\varphi}}{2}\partial_1 V\left\{-\frac{\alpha_3}{\sqrt{1-\alpha_3^2}}[(l+m)P_{l-1}^m - l\alpha_3 P_l^m] + \frac{mP_l^m}{\sqrt{1-\alpha_3^2}}\right\}$$

$$+ b\partial_0 V e^{im\varphi}(2l+1)^{-1}[(l+1)(l+m)P_{l-1}^m - l(l-m+1)P_{l+1}^m]$$

$$-\frac{be^{i(m+1)\varphi}}{2}\partial_{-1}V\left[\frac{\alpha_3}{\sqrt{1-\alpha_3^2}}[(l+m)P_{l-1}^m - l\alpha_3 P_l^m] + \frac{mP_l^m}{\sqrt{1-\alpha_3^2}}\right]$$

$$= \frac{be^{i(m-1)\varphi}}{2}\partial_1 V\left[(l+m)\frac{P_l^m - \alpha_3 P_{l-1}^m}{\sqrt{1-\alpha_3^2}} - l\sqrt{1-\alpha_3^2}\,P_l^m\right]$$

$$+ b\partial_0 V e^{im\varphi}(2l+1)^{-1}[(l+1)(l+m)P_{l-1}^m - l(l-m+1)P_{l+1}^m]$$

$$+ b\partial_0 V e^{im\varphi}(2l+1)^{-1}[(l+1)(l+m)P_{l-1}^m - l(l-m+1)P_{l+1}^m]$$

$$-\frac{be^{i(m+1)\varphi}}{2}\partial_{-1}V\left[\frac{(l+m)\alpha_3 P_{l-1}^m - (l-m)P_l^m}{\sqrt{1-\alpha_3^2}}\right.$$

$$\left. + l\sqrt{1-\alpha_3^2}P_l^m\right] \tag{2.103}$$

The recurrence relations in Eqs. (A.8) and (A.4) respectively give

$$(l+m)\frac{P_l^m - \alpha_3 P_{l-1}^m}{\sqrt{1-\alpha_3^2}} = (l+m)(l+m-1)P_{l-1}^{m-1} \tag{2.104}$$

$$-l\sqrt{1-\alpha_3^2}\,P_l^m = -\frac{l(l+m)(l+m-1)}{(2l_1)}P_{l-1}^{m-1}$$

$$+\frac{l(l-m+1)(l-m+2)}{(2l+1)}P_{l+1}^{m-1} \tag{2.105}$$

Addition and multiplication by $e^{i(m-1)\varphi}$ gives

$$e^{i(m-1)\varphi}\left[(l+m)\frac{P_l^m-\alpha_3 P_{l-1}^m}{\sqrt{1-\alpha_3^2}}-l\sqrt{1-\alpha_3^2}\,P_l^m\right]$$
$$=\frac{(l+1)(l+m)(l+m-1)}{(2l+1)}X_{l-1,m-1}$$
$$+\frac{l(l-m+1)(l-m+2)}{(2l+1)}X_{l+1,m-1} \tag{2.106}$$

The recurrence relations Eqs. (A.4) and (A.5) respectively yield

$$l\sqrt{1-\alpha_3^2}\,P_l^m=\frac{l}{(2l+1)}[-P_{l-1}^{m+1}+P_{l+1}^{m+1}] \tag{2.107}$$

$$\frac{(l+m)\alpha_3 P_{l-1}^m-(l-m)P_l^m}{\sqrt{1-\alpha_3^2}}=P_l^{m+1} \tag{2.108}$$

Addition followed by multiplication with $e^{i(m+1)\varphi}$ gives

$$e^{i(m+1)\varphi}\left[\frac{(l+m)\alpha_3 P_{l-1}^m-(l-m)P_l^m}{\sqrt{1-\alpha_3^2}}+l\sqrt{1-\alpha_3^2}\,P_l^m\right]$$
$$=\frac{1}{(2l+1)}[(l+1)P_{l-1}^{m+1}+lX_{l+1,m+1}] \tag{2.109}$$

Application of Eqs. (2.106) and Eq. (2.109) to the first and third terms, respectively, in Eq. (2.103) gives

$$b\left(\frac{\partial V}{\partial\theta}\frac{\partial X_{l,m}}{\partial\theta}+\frac{1}{\sin^2\theta}\frac{\partial V}{\partial\varphi}\frac{\partial X_{l,m}}{\partial\varphi}\right)$$
$$=\frac{b}{2(2l+1)}\partial_1 V[(l+1)(l+m)(l+m-1)X_{l-1,m-1}$$
$$+l(l-m+2)(l-m+1)X_{l+1,m-1}]$$
$$+\frac{b}{(2l+1)}\partial_0 V[(l+1)(l+m)X_{l-1,m}-l(l-m+1)X_{l+1,m}]$$
$$-\frac{b}{2(2l+1)}\partial_{-1}V[(l+1)X_{l-1,m+1}+lX_{l+1,m+1}] \tag{2.110}$$

On replacing W with $X_{l,m}$, Eq. (2.78), with the aid of Eq. (2.110), yields a second alignment representation formula

$$A_{FP}X_{l,m} = -\beta^{-1}bl(l+1)X_{l,m} + bX_{l,m}\Lambda^2 V + b\sum_{k=-1}^{1}(p_{-1,k}X_{l-1,m+k}\partial_{-k}V + p_{1,k}X_{l+1,m+k}\partial_{-k}V) \tag{2.111}$$

where the coefficients are

$$p_{-1,-1} = \frac{(l+1)(l+m)(l+m-1)}{2(2l+1)} \qquad p_{-1,0} = \frac{(l+1)(l+m)}{(2l+1)}$$

$$p_{-1,1} = -\frac{l+1}{2(2l+1)}$$

$$p_{1,-1} = \frac{l(l-m+1)(l-m+2)}{2(2l+1)} \qquad p_{1,0} = \frac{l(l-m+1)}{(2l+1)}$$

$$p_{1,1} = -\frac{l}{2(2l+1)} \tag{2.112}$$

Incorporating the gyroscopic representation formula of Eq. (2.98) gives the representation formula

$$L_{FP}X_{l,m} = -\beta^{-1}bl(l+1)X_{l,m} + bX_{l,m}\Lambda^2 V + b\sum_{j,k=-1}^{1} p_{j,k}X_{l+j,m+k}\partial_{-k}V \tag{2.113}$$

where the additional coefficients are

$$p_{0,-1} = \frac{i(l+m)(l-m+1)}{2\alpha} \qquad p_{0,0} = -i\alpha^{-1}m \qquad p_{0,1} = \frac{i}{2\alpha} \tag{2.114}$$

III. APPLICATION TO THE CASE WHERE THE FREE ENERGY ARISES FROM UNIAXIAL ANISOTROPY WITH CONSTANT UNIFORM EXTERNAL MAGNETIC FIELD AT AN ANGLE OBLIQUE TO THE EASY AXIS

A. Calculation of the Matrix Elements and Differential Recurrence Coefficients from the Spherical Harmonic Representation Formulae

The free energy per unit volume Eq. (1.4) expressed in terms of the direction cosines of the magnetization orientation relative to the crystal axes Eq. (2.72) becomes [35]

$$V(\alpha_1, \alpha_2, \alpha_3) = K(1 - \alpha_3^2) - HM_s \sum_{i=1}^{3} \gamma_i \alpha_i \tag{3.1}$$

The direction cosines of the magnetization are expressed in spherical harmonics using Eq. (A.11)

$$\alpha_1 = \sin\theta\cos\varphi = \tfrac{1}{2}X_{1,1} - X_{1,-1} \tag{3.2}$$

$$\alpha_2 = \sin\theta\sin\varphi = -\tfrac{i}{2}X_{1,1} - iX_{1,-1} \tag{3.3}$$

$$\alpha_3 = \cos\theta = X_{1,0} \tag{3.4}$$

also from Eq. (A.11), we have

$$\alpha_3^2 = \tfrac{1}{3} + \tfrac{2}{3}X_{2,0} \tag{3.5}$$

Equation (3.1) can thus be expressed in spherical harmonics as

$$V = \sum_{s=0}^{2} \sum_{t=1}^{1} \alpha_{s,t} X_{s,t} \tag{3.6}$$

where the nonzero coefficients are

$$\alpha_{0,0} = 2K/3 \qquad \alpha_{2,0} = -2K/3 \qquad \alpha_{1,0} = -HM_s\gamma_3$$

$$\alpha_{1,1} = -\frac{HM_s(\gamma_1 - i\gamma_2)}{2} \qquad \alpha_{1,-1} = HM_s(\gamma_1 + i\gamma_2) \tag{3.7}$$

In order to obtain the recurrence relation for $L_{FP}X_{l,m}$, it will be necessary to obtain an expression for $\Lambda^2 V$ in terms of the direction cosines α_i. To accomplish this we first take the Laplacian of the quantities Eqs. (3.2–3.5) using $\Lambda^2 X_{l,m} = -l(l+1)X_{l,m}$

$$\begin{aligned}\Lambda^2\alpha_i &= -2\alpha_i \\ \Lambda^2\alpha_3^2 &= 2(1 - 3\alpha_3^2)\end{aligned} \tag{3.8}$$

The Laplacian of Eq. (3.1) is then

$$\Lambda^2 V = -K\Lambda^2\alpha_3^2 - HM_s \sum_{i=1}^{3} \gamma_i \Lambda^2 \alpha_i$$

$$= -2K(1 - 3\alpha_3^2) + 2HM_s \sum_{i=1}^{3} \gamma_i \alpha_i \tag{3.9}$$

Hence

$$bX_{l,m}\Lambda^2 V = -2bK(1 - 3\alpha_3^2)X_{l,m} + 2bHM_s \sum_{i=1}^{3} \gamma_i \alpha_i X_{l,m} \tag{3.10}$$

The calculations will be accomplished using the representation formula of Eq. (2.113), namely

$$L_{FP}X_{l,m} = -\beta^{-1}bl(l+1)X_{l,m} + bX_{l,m}\Lambda^2 V + b \sum_{j,k=-1}^{1} p_{j,k}X_{l+j,m+k}\partial_{-k}V \tag{3.11}$$

Differentiating Eq. (3.1) gives

$$\partial_{\pm 1}V = \frac{\partial V}{\partial \alpha_1} \pm i \frac{\partial V}{\partial \alpha_2} = -HM_s(\gamma_1 \pm i\gamma_2) \tag{3.12}$$

$$\partial_0 V = \frac{\partial V}{\partial \alpha_3} = -2K\alpha_3 - HM_s\gamma_3 \tag{3.13}$$

Successive application of Eq. (A.18) gives the recurrence relation

$$\begin{aligned} &-2bK(1 - 3\alpha_3^2)X_{l,m} \\ &\quad = 2bK\{3(l+m)(l+m-1)(2l-1)^{-1}(2l+1)^{-1}X_{l-2,m} \\ &\quad + 2(2l-1)^{-1}(2l+3)^{-1}[l(l+1) - 3m^2]X_{l,m} \\ &\quad + 3(l-m+2)(l-m+1)(2l+1)^{-1}(2l+3)^{-1}X_{l+2,m}\} \end{aligned} \tag{3.14}$$

Application of Eqs. (A.16–A.18) gives the recurrence relations

$$2bHM_s \sum_{i=1}^{3} \gamma_i \alpha_i X_{l,m}$$
$$= bHM_s(\gamma_1 + i\gamma_2)(2l+1)^{-1}[(l+m)(l+m-1)X_{l-1,m-1}$$
$$-(l-m+2)(l-m+1)X_{l+1,m-1}]$$
$$+2bHM_s\gamma_3(2l+1)^{-1}[(l+m)X_{l-1,m} + (l-m+1)X_{l+1,m}]$$
$$+bHM_s(\gamma_1 - i\gamma_2)(2l+1)^{-1}[-X_{l-1,m+1} + X_{l+1,m+1}] \qquad (3.15)$$

Expanding the summation in Eq. (3.11), with the aid of Eq. (3.12) and Eq. (3.13), gives

$$b \sum_{j,k=-1}^{1} p_{j,k} X_{l+j,m+k} \partial_{-k} V$$
$$= -2bK \sum_{j=-1}^{1} p_{j,0}\alpha_3 K_{l+j,m}$$
$$-bHM_s \sum_{j=-1}^{1} [(\gamma_1 + i\gamma_2)p_{j,-1}X_{l+j,m-1}$$
$$+\gamma_3 p_{j,0} X_{l+j,m} + (\gamma_1 - i\gamma_2)p_{j,1}X_{l+j,m+1}] \qquad (3.16)$$

Expanding the first summation on the right-hand side and applying Eq. (A.18) gives

$$-2bK \sum_{j=-1}^{1} p_{j,0}\alpha_3 X_{l+j,m}$$
$$= \frac{2bK}{(2l+1)} \{-(l+1)(l+m)(l+m-1)(2l-1)^{-1}X_{l-2,m}$$
$$+i\alpha^{-1}m(l+m)X_{l-1,m} - (2l+1)(2l-1)^{-1}(2l+3)^{-1}$$
$$\cdot [l(l+1) - 3m^2]X_{l,m} + i\alpha^{-1}m(l-m+1)X_{l+1,m}$$
$$+l(l-m+2)(l-m+1)(2l+3)^{-1}X_{l+2,m}\} \qquad (3.17)$$

where $\alpha = b/a$. Expanding the second summation on the right-hand side of Eq. (3.16) gives

$$
\begin{aligned}
&-bHM_s \sum_{j=-1}^{1} [(\gamma_1 + i\gamma_2)p_{j,-1}X_{l+j,m-1} + \gamma_3 p_{j,0}X_{l+j,m} \\
&\quad + (\gamma_1 - i\gamma_2)p_{j,1}X_{l+j,m+1}] \\
&= \frac{bHM_s(\gamma_1 + i\gamma_2)}{2(2l+1)} [-(l+1)(l+m)(l+m-1)X_{l-1,m-1} \\
&\quad - i\alpha^{-1}(l+m)(l-m+1)(2l+1)X_{l,m-1} \\
&\quad - l(l-m+1)(l-m+2)X_{l+1,m-1}] + \frac{bHM_s\gamma_3}{(2l+1)} \\
&\quad \cdot [-(l+1)(l+m)X_{l-1,m} + i\alpha^{-1}m(2l+1)X_{l,m} \\
&\quad + l(l-m+1)X_{l+1,m}] + \frac{bHM_s(\gamma_1 - i\gamma_2)}{2(2l+1)} \\
&\quad \cdot [(l+1)X_{l-1,m+1} - i\alpha^{-1}(2l+1)X_{l,m+1} + lX_{l+1,m+1}] \qquad (3.18)
\end{aligned}
$$

Equation (3.11), with the aid of Eqs. (3.10) and (3.14–3.18), yields the recurrence relation for $L_{FP}X_{l,m}$, namely,

$$
\begin{aligned}
L_{FP}X_{l,m} &= -\beta^{-1}bl(l+1)X_{l,m} + \frac{2bK}{(2l+1)} \\
&\quad \{-(l-2)(l+m)(l+m-1)(2l-1)^{-1}X_{l-2,m} \\
&\quad + i\alpha^{-1}m(l+m)X_{l-1,m} + (2l+1) \\
&\quad \cdot (2l-1)^{-1}(2l+3)^{-1}[l(l+1) - 3m^2]X_{l,m} \\
&\quad + i\alpha^{-1}m(l-m+1)X_{l+1,m} + (l+3) \\
&\quad \cdot (l-m+2)(l-m+1)(2l+3)^{-1}X_{l+2,m}\} \\
&\quad + \frac{bHM_s(\gamma_1 + i\gamma_2)}{2(2l+1)} [-(l-1)(l+m)(l+m-1)X_{l-1,m-1} \\
&\quad - i\alpha^{-1}(l+m)(l-m+1)(2l+1)X_{l,m-1} - (l+2)(l-m+1) \\
&\quad \cdot (l-m+2)X_{l+1,m-1}] + \frac{bHM_s\gamma_3}{(2l+1)} \\
&\quad \cdot [-(l-1)(l+m)X_{l-1,m} + i\alpha^{-1}m(2l+1)X_{l,m} \\
&\quad + (l+2)(l-m+1)X_{l+1,m}] + \frac{bHM_s(\gamma_1 - i\gamma_2)}{2(2l+1)} \\
&\quad \cdot [(l-1)X_{l-1,m+1} - i\alpha^{-1}(2l+1)X_{l,m+1} + (l+2)X_{l+1,m+1}]
\end{aligned}
\qquad (3.19)
$$

which can be expressed in the form of Eq. (2.75), namely

$$L_{FP}X_{l,m} = \sum_{s=-2}^{2} \sum_{t=-p_s}^{p_s} g_{l+s,m+t,l,m} X_{l+s,m+t} \tag{3.20}$$

where the bounds on t are $p_s = 1$ for $|s| \leq 1$ and $p_s = 0$ for $|s| = 2$. On shifting indices the coefficients become

$$
\begin{aligned}
g_{l,m,l,m-1} &= -\frac{ibHM_s(\gamma_1 - i\gamma_2)}{2\alpha} \\
g_{l,m,l-1,m-1} &= \frac{bHM_s(\gamma_1 - i\gamma_2)(l+1)}{2(2l-1)} \\
g_{l,m,l+1,m-1} &= \frac{bHM_s(\gamma_1 - i\gamma_2)l}{2(2l+3)} \\
g_{l,m,l-2,m} &= \frac{2bK(l+1)(l-m)(l-m-1)}{(2l-1)(2l-3)} \\
g_{l,m,l-1,m} &= \frac{b(l-m)}{(2l-1)} [HM_s\gamma_3(l+1) + 2i\alpha^{-1}Km] \\
g_{l,m,l,m} &= \frac{2bK}{(2l-1)(2l+3)} [l(l+1) - 3m^2] + \frac{ibHM_s\gamma_3 m}{\alpha} \\
g_{l,m,l+1,m} &= \frac{b(l+m+1)}{(2l+3)} [-HM_s\gamma_3 l + 2i\alpha^{-1}Km] \\
g_{l,m,l+2,m} &= -\frac{2bKl(l+m+2)(l+m+1)}{(2l+5)(2l+3)} \\
g_{l,m,l-1,m+1} &= -\frac{bHM_s(\gamma_1 + i\gamma_2)(l+1)(l-m)(l-m-1)}{2(2l-1)} \\
g_{l,m,l,m+1} &= -\frac{ibHM_s(\gamma_1 - i\gamma_2)(l-m)(l+m+1)}{2\alpha} \\
g_{l,m,l+1,m+1} &= -\frac{bHM_s(\gamma_1 - i\gamma_2)l(l+m+2)(l+m+1)}{2(2l+3)}
\end{aligned} \tag{3.21}
$$

The matrix elements are calculated from Eq. (2.76), namely,

$$d_{l,m,l+s,m+t} = (-1)^t \sqrt{\frac{(2l+2s+1)(l+m)!(l-m+s-t)!}{(2l+1)(l+m+s+t)!(l-m)!}}\, g^*_{l,m,l+s,m+t} \tag{3.22}$$

Hence

$$d_{l,m,l-1,m-1} = -\frac{bHM_s(\gamma_1 + i\gamma_2)(l+1)}{2}\sqrt{\frac{(l+m)(l+m-1)}{(2l-1)(2l+1)}}$$

$$d_{l,m,l,m-1} = -\frac{ibHM_s(\gamma_1 + i\gamma_2)\sqrt{(l+m)(l-m+1)}}{2\alpha}$$

$$d_{l,m,l+1,m-1} = -\frac{bHM_s(\gamma_1 + i\gamma_2)l}{2}\sqrt{\frac{(l-m+1)(l-m+2)}{(2l+1)(2l+3)}}$$

$$d_{l,m,l-2,m} = \frac{2bK(l+1)}{(2l-1)}\sqrt{\frac{(l-m)(l-m-1)(l+m)(l+m-1)}{(2l-3)(2l+1)}}$$

$$d_{l,m,l-1,m} = \frac{b}{(2l-1)}[HM_s\gamma_3(l+1) - 2i\alpha^{-1}Km]\sqrt{\frac{(l-m)(l+m)}{(2l-1)(2l+1)}}$$

$$d_{l,m,l,m} = -\beta^{-1}bl(l+1) + \frac{2bK}{(2l-1)(2l+3)} \cdot [l(l+1) - 3m^2] - \frac{ibHM_s\gamma_3 m}{\alpha}$$

$$d_{l,m,l+1,m} = -\frac{b}{(2l+3)}[HM_s\gamma_3 l + 2i\alpha^{-1}Km]\sqrt{\frac{(l-m+1)(l+m+1)}{(2l+1)(2l+3)}}$$

$$d_{l,m,l+2,m} = -\frac{2bKl}{(2l+3)}\sqrt{\frac{(l+m+2)(l+m+1)(l-m+1)(l-m+2)}{(2l+1)(2l+5)}}$$

$$d_{l,m,l-1,m+1} = \frac{bHM_s(\gamma_1 - i\gamma_2)(l+1)}{2}\sqrt{\frac{(l-m)(l-m-1)}{(2l-1)(2l+1)}}$$

$$d_{l,m,l,m+1} = -\frac{ibHM_s(\gamma_1 - i\gamma_2)\sqrt{(l-m)(l+m+1)}}{2\alpha}$$

$$d_{l,m,l+1,m+1} = \frac{bHM_s(\gamma_1 - i\gamma_2)l}{2}\sqrt{\frac{(l+m+2)(l+m+1)}{(2l+1)(2l+3)}} \tag{3.23}$$

The differential recurrence relations for the Fourier coefficients $a_{l,m} = (W, Y_{l,m})$ are [cf. Eq. (2.12)]

$$\dot{a}_{l,m} = \sum_{s=-2}^{2} \sum_{t=-p_s}^{p_s} d^*_{l,m,l+s,m+t} a_{l+s,m+t} \tag{3.24}$$

or explicitly

$$\begin{aligned}
\dot{a}_{l,m} = {} & -\beta^{-1} bl(l+1)a_{l,m} + \frac{bHM_s(\gamma_1 - i\gamma_2)}{2} \\
& \cdot \left[-(l+1)\sqrt{\frac{(l+m)(l+m-1)}{(2l-1)(2l+1)}}\, a_{l-1,m-1} \right. \\
& + i\alpha^{-1}\sqrt{(l+m)(l-m+1)}\, a_{l,m-1} \\
& \left. -1\sqrt{\frac{(l-m+1)(l-m+2)}{(2l+1)(2l+3)}}\, a_{l+1,m-1} \right] \\
& + bHM_s\gamma_3 \left[\frac{(l+1)}{(2l-1)} \sqrt{\frac{(l-m)(l+m)}{(2l-1)(2l+1)}}\, a_{l-1,m} + \frac{im}{\alpha} a_{l,m} \right. \\
& \left. - \frac{l}{(2l+3)} \sqrt{\frac{(l-m+1)(l+m+1)}{(2l+1)(2l+3)}}\, a_{l+1,m} \right] \\
& + \frac{bHM_s(\gamma_1 + i\gamma_2)}{2} \left[(l+1)\sqrt{\frac{(l-m)(l-m-1)}{(2l-1)(2l+1)}}\, a_{l-1,m+1} \right. \\
& \left. + i\alpha^{-1}\sqrt{(l-m)(l+m+1)}\, a_{l,m+1} + l\sqrt{\frac{(l+m+2)(l+m+1)}{(2l+1)(2l+3)}}\, a_{l+1,m+1} \right] \\
& + 2bK \left[\frac{(l+1)}{(2l-1)} \sqrt{\frac{(l-m)(l-m-1)(l+m)(l+m-1)}{(2l-3)(2l+1)}}\, a_{l-2,m} \right. \\
& + \frac{l(l+1) - 3m^2}{(2l-1)(2l+3)} a_{l,m} + \frac{im}{\alpha(2l-1)} \sqrt{\frac{(l-m)(l+m)}{(2l-1)(2l+1)}}\, a_{l-1,m} \\
& + \frac{im}{\alpha(2l+3)} \sqrt{\frac{(l-m+1)(l+m+1)}{(2l+1)(2l+3)}}\, a_{l+1,m} \\
& \left. - \frac{l}{(2l+3)} \sqrt{\frac{(l+m+2)(l+m+1)(l-m+1)(l-m+2)}{(2l+1)(2l+5)}}\, a_{l+2,m} \right]
\end{aligned} \tag{3.25}$$

The differential recurrence coefficients are calculated from Eq. (2.77) namely

$$e_{l,m,l+s,m+1} = \frac{(2l+2s+1)(l+m)!(l-m+s-t)!}{(2l+1)(l+m+s+t)!(l-m)!} g^{*}_{l,m,l+s,m+t} \tag{3.26}$$

hence

$$
\begin{aligned}
e_{l,m,l-1,m-1} &= \frac{bHM_s(\gamma_1 + i\gamma_2)(l+1)(l+m)(l+m-1)}{2(2l+1)} \\
e_{l,m,l,m-1} &= \frac{ibHM_s(\gamma_1 + i\gamma_2)(l+m)(l-m+1)}{2\alpha} \\
e_{l,m,l-1,m-1} &= \frac{bHM_s(\gamma_1 + i\gamma_2)l(l-m+2)(l-m+1)}{2(2l+1)} \\
e_{l,m,l-2,m} &= \frac{2bK(l+1)(l+m)(l+m-1)}{(2l-1)(2l+1)} \\
e_{l,m,l-1,m} &= \frac{b(l+m)}{(2l+1)} [HM_s\gamma_3(l+1) - 2iK\alpha^{-1}m] \\
e_{l,m,l,m} &= -\beta^{-1}bl(l+1) + \frac{2bK}{(2l-1)(2l+3)} [l(l+1) - 3m^2] \\
&\quad - ibHM_s\alpha^{-1}\gamma_3 m \\
e_{l,m,l+1,m} &= -\frac{b(l-m+1)}{(2l+1)} [HM_s\gamma_3 l + 2iK\alpha^{-1}m] \\
e_{l,m,l+2,m} &= -\frac{2bKl(l-m+2)(l-m+1)}{(2l+1)(2l+3)} \\
e_{l,m,l-1,m+1} &= -\frac{bHM_s(\gamma_1 - i\gamma_2)(l+1)}{2(2l+1)} \\
e_{l,m,l,m+1} &= \frac{ibHM_s(\gamma_1 - i\gamma_2)}{2\alpha} \\
e_{l,m,l+1,m+1} &= -\frac{bHM_s(\gamma_1 - i\gamma_2)l}{2(2l+1)}
\end{aligned} \tag{3.27}
$$

The differential recurrence relations for the expectation values $x_{l,m} = \langle X_{l,m} \rangle$ [cf. Eq. (2.14)], and hence the decay modes $c_{l,m}$ [Eq. (2.17)], become

$$\dot{x}_{l,m} = \sum_{s=-2}^{2} \sum_{t=-p_s}^{p_s} e_{l,m,l+s,m+t} x_{l+s,m+t} \tag{3.28}$$

or explicitly [35]

$$\begin{aligned}
\dot{x}_{l,m} = & -\beta^{-1} b l(l+1) x_{l,m} \\
& + \frac{bHM_s(\gamma_1 + i\gamma_2)}{2(2l+1)} [(l+1)(l+m)(l+m-1)x_{l-1,m-1} \\
& + i\alpha^{-1}(2l+1)(l+m)(l-m+1)x_{l,m-1} \\
& + l(l-m+2)(l-m+1)x_{l+1,m-1}] \\
& + bHM_s\gamma_3 \left[\frac{(l+1)(l+m)}{(2l+1)} x_{l-1,m} - i\alpha^{-1} m x_{l,m} - \frac{l(l-m+1)}{(2l+1)} x_{l+1,m} \right] \\
& - \frac{bHM_s(\gamma_1 - i\gamma_2)}{2(2l+1)} [(l+1)x_{l-1,m+1} - i\alpha^{-1}(2l+1)x_{l,m+1} + l x_{l+1,m+1}] \\
& + 2bK \left[\frac{(l+1)(l+m)(l+m-1)}{(2l-1)(2l+1)} x_{l-2,m} - \frac{im(l+m)}{\alpha(2l+1)} x_{l-1,m} \right. \\
& + \frac{l(l+1) - 3m^2}{(2l-1)(2l+3)} x_{l,m} - \frac{im(l-m+1)}{\alpha(2l+1)} x_{l+1,m} \\
& \left. - \frac{l(l-m+2)(l-m+1)}{(2l+1)(2l+3)} x_{l+2,m} \right]
\end{aligned} \tag{3.29}$$

B. Calculation of the Differential Recurrence Relations for the Sharp Values of the Spherical Harmonics from the Gilbert–Langevin Equation

In this section, we present a detailed derivation of the differential recurrence relations for the sharp values of the spherical harmonics in the case of an arbitrary expression for the free energy [37]. The equations that are analogous to the representation formulae of Section II.C are then specialized to the case when the free energy per unit volume is given by Eq. (3.1). This yields a set of differential recurrence relations that are analogous to Eq. (3.29). The derivation constitutes the *Langevin approach*, and its importance lies in the fact that it provides a closer insight into the underlying physics, in contrast to the more abstract *Fokker–Planck approach* [Appendix B], which has until now been implicitly adopted.

The magnetization orientation (Section I.B) is

$$\mathbf{r} = \sum_{i=1}^{3} \alpha_i \mathbf{e}_i \tag{3.30}$$

This satisfies Gilbert's equation, Eq. (1.44), namely,

$$\dot{\mathbf{r}} = a \frac{\partial V}{\partial \mathbf{r}} \times \mathbf{r} + b\left(\mathbf{r} \cdot \frac{\partial V}{\partial \mathbf{r}}\right)\mathbf{r} - b \frac{\partial V}{\partial \mathbf{r}} \tag{3.31}$$

which can be expressed as

$$\dot{\mathbf{r}} = aM_s \mathbf{r} \times \mathbf{H}_{\text{ef}} - bM_s(\mathbf{r} \cdot \mathbf{H}_{\text{ef}})\mathbf{r} + bM_s\mathbf{H}_{\text{ef}} \tag{3.32}$$

where

$$\mathbf{H}_{\text{ef}} = \sum_{i=1}^{3} H_i \mathbf{e}_i = -\frac{1}{M_s} \frac{\partial V}{\partial \mathbf{r}} = -\frac{1}{M_s} \sum_{i=1}^{3} \frac{\partial V}{\partial \alpha_i} \mathbf{e}_i \tag{3.33}$$

is the effective magnetic field. Augmenting $\mathbf{H}_{\text{ef}}$ in Eq. (3.32) by a white noise term (which arises from thermal agitation) gives

$$\lambda = \lambda(t) = \sum_{i=1}^{3} \lambda_i(t)\mathbf{e}_i \tag{3.34}$$

having properties [6]

$$\overline{\lambda_i(t)} = 0 \tag{3.35}$$

$$\overline{\lambda_i(t_1)\lambda_j(t_2)} = 2D\delta_{i,j}\delta(t_1 - t_2) \tag{3.36}$$

where D is a normalization constant. Eq. (3.32) becomes

$$\dot{\mathbf{r}} = aM_s\mathbf{r} \times (\mathbf{H}_{\text{ef}} + \lambda) - bM_s[\mathbf{r} \cdot (\mathbf{H}_{\text{ef}} + \lambda)]\mathbf{r} + bM_s(\mathbf{H}_{\text{ef}} + \lambda) \tag{3.37}$$

which in terms of the components in Eqs. (3.30), (3.33), and (3.34) becomes

$$\sum_{i=1}^{3} \dot{\alpha}_i \mathbf{e}_i = aM_s \sum_{j=1}^{3} \alpha_j \mathbf{e}_j \times \sum_{k=1}^{3} (H_k + \lambda_k)\mathbf{e}_k$$
$$- bM_s \sum_{k=1}^{3} \alpha_k (H_k + \lambda_k) \sum_{i=1}^{3} \alpha_i \mathbf{e}_i$$
$$+ bM_s \sum_{i=1}^{3} (H_i + \lambda_i)\mathbf{e}_i \tag{3.38}$$

hence

$$\dot{\alpha}_i = aM_s \sum_{j,k=1}^{3} e_{i,j,k} \alpha_j (H_k + \lambda_k)$$
$$+ bM_s \sum_{k=1}^{3} (\delta_{i,k} - \alpha_i \alpha_k)(H_k + \lambda_k) \qquad i = 1, 2, 3 \tag{3.39}$$

where $e_{i,j,k}$ is the permutation symbol. This is equivalent to the Langevin equation interpreted in the Stratonovich sense [6, 18, 36]

$$\dot{\xi}_i(t) = h_i(\{\xi_i(t)\}, t) + \sum_{k=1}^{3} g_{i,k}(\{\xi_i(t)\}, t)\lambda_k(t) \qquad i = 1, 2, 3 \tag{3.40}$$

[for the stochastic variables $\xi_i(t)$, i = 1, 2, 3] where

$$h = bM_s \sum_{k=1}^{3} (\delta_{i,k} - \xi_i \xi_k) H_k + aM_s \sum_{j,k=1}^{3} e_{i,j,k} \xi_j H_k \tag{3.41}$$

$$g_{i,k} = bM_s(\delta_{i,k} - \xi_i \xi_k) + aM_s \sum_{j=1}^{3} e_{i,j\,k} \xi_j \tag{3.42}$$

The averaged equation for the *sharp values* α_i, i = 1, 2, 3 is [6]

$$\dot{\alpha}_i = \lim_{t \to 0} \frac{\overline{\xi_i(t+\tau) - \alpha_i(t)}}{\tau} = h_i(\mathbf{r}, t) + D \sum_{j,k=1}^{3} g_{j,k}(\mathbf{r}, t) \frac{\partial}{\partial \alpha_j} g_{i,k}(\mathbf{r}, t) \tag{3.43}$$

In the same way, it can be shown that

$$A_i(\mathbf{r})\dot{\alpha}_i = A_i(\mathbf{r})h_i(\mathbf{r},t) + D\sum_{j,k=1}^{3} g_{j,k}(\mathbf{r},t)\frac{\partial}{\partial\alpha_j}A_i(\mathbf{r})g_{i,k}(\mathbf{r},t) \tag{3.44}$$

Hence expanding the time derivatives of the spherical harmonics [Appendix A] of the magnetization orientation using the chain rule gives

$$\dot{X}_{l,m}\sum_{i=1}^{3}\frac{\partial X_{l,m}}{\partial\alpha_i}\dot{\alpha}_i = \sum_{i=1}^{3}\frac{\partial X_{l,m}}{\partial\alpha_i}h_i(\mathbf{r},t) + D\sum_{i,j,k=1}^{3} g_{j,k}(\mathbf{r},t)\frac{\partial}{\partial\alpha_j}\left[\frac{\partial X_{l,m}}{\partial\alpha_i}g_{i,k}(\mathbf{r},t)\right] \tag{3.45}$$

After a long calculation [37] the second summation on the right-hand side reduces to

$$-b\beta^{-1}l(l+1)X_{l,m} \tag{3.46}$$

Expanding the first term on the right-hand side of Eq. (3.45), using Eq. (3.41), gives

$$M_s\sum_{i=1}^{3}\frac{\partial X_{l,m}}{\partial\alpha_i}\left[b\sum_{k=1}^{3}(\delta_{i,k}-\alpha_i\alpha_k)H_k + a\sum_{j,k=1}^{3}e_{i,j,k}\alpha_j H_k\right] \tag{3.47}$$

The nonzero values of the permutation symbol are

$$\begin{aligned} e_{1,2,3} &= e_{2,3,1} = e_{3,1,2} = 1 \\ e_{1,3,2} &= e_{2,1,3} = e_{3,2,1} = -1 \end{aligned} \tag{3.48}$$

Hence Eq. (3.45) becomes

$$\begin{aligned} \dot{X}_{l,m} = {} & -b\beta^{-1}l(l+1)X_{l,m} \\ & + M_s\frac{\partial X_{l,m}}{\partial\alpha_1}[b(1-\alpha_1^2)H_1 - b\alpha_1\alpha_2H_2 - b\alpha_1\alpha_3H_3 \\ & + a(\alpha_2H_3-\alpha_3H_2)] \\ & + M_s\frac{\partial X_{l,m}}{\partial\alpha_2}[-b\alpha_1\alpha_2H_1 + b(1-\alpha_2^2)H_2 - b\alpha_2\alpha_3H_3 \\ & + a(\alpha_3H_1-\alpha_1H_3)] \\ & + M_s\frac{\partial X_{l,m}}{\partial\alpha_3}[-b\alpha_1\alpha_3H_1 - b\alpha_2\alpha_3H_2 + b(1-\alpha_3^2)H_3 \\ & + a(\alpha_1H_2-\alpha_2H_1)] \end{aligned} \tag{3.49}$$

or

$$\dot{X}_{l,m} = -b\beta^{-1}l(l+1)X_{l,m} + aM_s\left[\left(\alpha_3 \frac{\partial X_{l,m}}{\partial \alpha_2} - \alpha_2 \frac{\partial X_{l,m}}{\partial \alpha_3}\right) H_1\right.$$

$$\left. + \left(i\alpha_3 \frac{\partial X_{l,m}}{\partial \alpha_1} - i\alpha_1 \frac{\partial X_{l,m}}{\partial \alpha_3}\right) iH_2 + \left(\alpha_2 \frac{\partial X_{l,m}}{\partial \alpha_1} - \alpha_1 \frac{\partial X_{l,m}}{\partial \alpha_2}\right) H_3\right]$$

$$+ bM_s\left\{\left[(1-\alpha_1^2) \frac{\partial X_{l,m}}{\partial \alpha_1} - \alpha_1\alpha_2 \frac{\partial X_{l,m}}{\partial \alpha_2} - \alpha_1\alpha_3 \frac{\partial X_{l,m}}{\partial \alpha_3}\right] H_1\right.$$

$$+ \left[i\alpha_1\alpha_2 \frac{\partial X_{l,m}}{\partial \alpha_1} - i(1-\alpha_2^2) \frac{\partial X_{l,m}}{\partial \alpha_2} + i\alpha_2\alpha_3 \frac{\partial X_{l,m}}{\partial \alpha_3}\right] iH_2$$

$$\left. + \left[-\alpha_1\alpha_3 \frac{\partial X_{l,m}}{\partial \alpha_1} - \alpha_2\alpha_3 \frac{\partial X_{l,m}}{\partial \alpha_2} + (1-\alpha_3^2) \frac{\partial X_{l,m}}{\partial \alpha_3}\right] H_3\right\} \tag{3.50}$$

From Eq. (A.15)

$$\frac{\partial X_{l,m}}{\partial \alpha_1} = \frac{m}{2^l l!} (\alpha_1 + i\alpha_2)^{m-1} \frac{d^{l+m}}{d\alpha_3^{l+m}} (\alpha_3^2 - 1)^l \tag{3.51}$$

$$\frac{\partial X_{l,m}}{\partial \alpha_2} = \frac{im}{2^l l!} (\alpha_1 + i\alpha_2)^{m-1} \frac{d^{l+m}}{d\alpha_3^{l+m}} (\alpha_3^2 - 1)^l \tag{3.52}$$

$$\frac{\partial X_{l,m}}{\partial \alpha_3} = \frac{1}{2^l l!} (\alpha_1 + i\alpha_2)^{m} \frac{d^{l+m+1}}{d\alpha^{l+m+1}} (\alpha_3^2 - 1)^l \tag{3.53}$$

hence

$$\frac{\partial X_{l,m}}{\partial \alpha_2} = i\frac{\partial X_{l,m}}{\partial \alpha_1} \tag{3.54}$$

The coefficient of H_3 in the gyroscopic term (i.e., the term containing a) thus simplifies to

$$\alpha_2 \frac{\partial X_{l,m}}{\partial \alpha_1} - \alpha_1 \frac{\partial X_{l,m}}{\partial \alpha_2} = -i(\alpha_1 + i\alpha_2) \frac{\partial X_{l,m}}{\partial \alpha_1} = -imX_{l,m} \tag{3.55}$$

Similarly the coefficient of H_3 in the alignment term (i.e., the term containing b), with the aid of Eqs. (A.18) and (A.20), can be expressed as

$$
\begin{aligned}
& - \alpha_1\alpha_3 \frac{\partial X_{l,m}}{\partial \alpha_1} - \alpha_2\alpha_3 \frac{\partial X_{l,m}}{\partial \alpha_2} + (1 - \alpha_3^2) \frac{\partial X_{l,m}}{\partial \alpha_3} \\
&= -\alpha_3(\alpha_1 + i\alpha_2) \frac{\partial X_{l,m}}{\partial \alpha_1} + (\alpha_1 - i\alpha_2)(\alpha_1 + i\alpha_2) \frac{\partial X_{l,m}}{\partial \alpha_3} \\
&= -m\alpha_3 X_{l,m} + (\alpha_1 - i\alpha_2) X_{l,m+1} \\
&= (2l+1)^{-1}[(l+1)(l+m)X_{l-1,m} - l(l-m+1)X_{l+1,m}] \qquad (3.56)
\end{aligned}
$$

The next recurrence relation follows from Eqs. (A.18), (A.19), and (A.20):

$$
2m\alpha_3 X_{l,m} - (\alpha_1 - i\alpha_2)X_{l,m+1} = (l+m)(l-m+1)(\alpha_1 + i\alpha_2)X_{l,m-1} \qquad (3.57)
$$

From Eqs. (3.51), (3.53), and (3.57),

$$
\begin{aligned}
2\alpha_3 \frac{\partial X_{l,m}}{\partial \alpha_1} - (\alpha_1 - i\alpha_2) \frac{\partial X_{l,m}}{\partial \alpha_3} &= \frac{2m\alpha_3}{(\alpha_1 + i\alpha_2)} X_{l,m} - \frac{(\alpha_1 - i\alpha_2)}{(\alpha_1 + i\alpha_2)} X_{l,m+1} \\
&= (l+m)(l-m+1)X_{l,m-1} \qquad (3.58)
\end{aligned}
$$

Application of the relation

$$
\begin{aligned}
PH_1 + QiH_2 &= \left(\frac{P+Q}{2} + \frac{P-Q}{2} \right) H_1 + \left(\frac{P+Q}{2} - \frac{P-Q}{2} \right) iH_2 \\
&= \frac{P+Q}{2} (H_1 + iH_2) + \frac{P-Q}{2} (H_1 - iH_2) \qquad (3.59)
\end{aligned}
$$

together with Eqs. (3.53), (3.54), and (3.58) to the remaining coefficients in the gyroscopic term gives

$$
\begin{aligned}
& \left(\alpha_3 \frac{\partial X_{l,m}}{\partial \alpha_2} - \alpha_2 \frac{\partial X_{l,m}}{\partial \alpha_3} \right) H_1 + \left(i\alpha_3 \frac{\partial X_{l,m}}{\partial \alpha_1} - i\alpha_1 \frac{\partial X_{l,m}}{\partial \alpha_3} \right) iH_2 \\
&= \frac{i}{2} (H_1 + iH_2) \left[2\alpha_3 \frac{\partial X_{l,m}}{\partial \alpha_1} - (\alpha_1 - i\alpha_2) \frac{\partial X_{l,m}}{\partial \alpha_3} \right] \\
&\quad + \frac{i}{2} (H_1 - iH_2)(\alpha_1 + i\alpha_2) \frac{\partial X_{l,m}}{\partial \alpha_3} \\
&= \frac{i(l+m)(l-m+1)}{2} (H_1 + iH_2) X_{l,m-1} \\
&\quad + \frac{i}{2} (H_1 - iH_2) X_{l,m+1} \qquad (3.60)
\end{aligned}
$$

The recurrence relation Eq. (3.57) can be written

$$\frac{2m\alpha_3}{(\alpha_1 + i\alpha_2)} X_{l,m} - \frac{(\alpha_1 - i\alpha_2)}{\alpha_1 + i\alpha_2)} X_{l,m+1} = (l+m)(l-m+1)X_{l,m-1} \quad (3.61)$$

Multiplication by α_3 gives

$$\frac{2m\alpha_3^2}{(\alpha_1 + i\alpha_2)} X_{l,m} - \alpha_3 \frac{(\alpha_1 - i\alpha_2)}{\alpha_1 + i\alpha_2)} X_{l,m+1} = (l+m)(l-m+1)\alpha_3 X_{l,m-1} \quad (3.62)$$

Since

$$\alpha_3^2 = 1 - \alpha_1^2 - \alpha_2^2 = 1 - (\alpha_1 - i\alpha_2)(\alpha_1 + i\alpha_2) \quad (3.63)$$

Eq. (3.62), with the aid Eqs. (A.18) and (A.20), becomes

$$\begin{aligned}\frac{2m}{(\alpha_1 + i\alpha_2)} X_{l,m} &- m(\alpha_1 - i\alpha_2)X_{l,m} - \alpha_3 \frac{(\alpha_1 - i\alpha_2)}{(\alpha_1 + i\alpha_2)} X_{l,m+1} \\ &= m(\alpha_1 - i\alpha_2)X_{l,m} + (l+m)(l-m+1)\alpha_3 X_{l,m-1} \\ &= \frac{(l+1)(l+m)(l+m-1)}{(2l+1)} X_{l-1,m-1} \\ &\quad - \frac{l(l-m+1)(l-m+2)}{(2l+1)} X_{l+1,m-1} \quad (3.64)\end{aligned}$$

From Eqs. (3.51), (3.53), and (3.64)

$$\begin{aligned}2\frac{\partial X_{l,m}}{\partial \alpha_1} &- (\alpha_1 - i\alpha_2)(\alpha_1 + i\alpha_2)\frac{\partial X_{l,m}}{\partial \alpha_1} - \alpha_3(\alpha_1 - i\alpha_2)\frac{\partial X_{l,m}}{\partial \alpha_3} \\ &= \frac{2m}{(\alpha_1 + i\alpha_2)} X_{l,m} - m(\alpha_1 - i\alpha_2)X_{l,m} - \alpha_3 \frac{\alpha_1 - i\alpha_2)}{(\alpha_1 + i\alpha_2)} X_{l,m+1} \\ &= \frac{(l+1)(l+m)(l+m-1)}{(2l+1)} X_{l-1,m-1} \\ &\quad + \frac{l(l-m+1)(l-m+2)}{(2l+1)} X_{l+1,m-1} \quad (3.65)\end{aligned}$$

From Eqs. (3.51), (3.53), (A.18), and (A.19),

$$
\begin{aligned}
(\alpha_1 + i\alpha_2)^2 \frac{\partial X_{l,m}}{\partial \alpha_1} + \alpha_3(\alpha_1 + i\alpha_2) \frac{\partial X_{l,m}}{\partial \alpha_3} \\
&= m(\alpha_1 + i\alpha_2)X_{l,m} + \alpha_3 X_{l,m+1} \\
&= (2l+1)^{-1}[(l+1)X_{l-1,m+1} + lX_{l+1,m+1}]
\end{aligned}
\tag{3.66}
$$

Application of Eqs. (3.54), (3.59), (3.65), and (3.66) to the remaining coefficients in the alignment term gives

$$
\begin{aligned}
&\left[(1-\alpha_1^2) \frac{\partial X_{l,m}}{\partial \alpha_1} - \alpha_1\alpha_2 \frac{\partial X_{l,m}}{\partial \alpha_2} - \alpha_1\alpha_3 \frac{\partial X_{l,m}}{\partial \alpha_3}\right] H_1 \\
&\quad + \left[i\alpha_1\alpha_2 \frac{\partial X_{l,m}}{\partial \alpha_1} - i(1-\alpha_2^2) \frac{\partial X_{l,m}}{\partial \alpha_2} + i\alpha_2\alpha_3 \frac{\partial X_{l,m}}{\partial \alpha_3}\right] iH_2 \\
&= \frac{1}{2}(H_1 + iH_2)\left[2 \frac{\partial X_{l,m}}{\partial \alpha_1} - (\alpha_1 + i\alpha_2)(\alpha_1 - i\alpha_2) \frac{\partial X_{l,m}}{\partial \alpha_1}\right. \\
&\quad \left. - \alpha_3(\alpha_1 - i\alpha_2) \frac{\partial X_{l,m}}{\partial \alpha_3}\right] \\
&\quad - \frac{1}{2}(H_1 - iH_2)\left[(\alpha_1 + i\alpha_2)^2 \frac{\partial X_{l,m}}{\partial \alpha_1} + \alpha_3(\alpha_1 + i\alpha_2) \frac{\partial X_{l,m}}{\partial \alpha_3}\right] \\
&= \frac{1}{2(2l+1)}(H_1 + iH_2)[(l+1)(l+m)(l+m-1)X_{l-1,m-1} \\
&\quad + l(l-m+1)(l-m+2)X_{l+1,m-1}] \\
&\quad - \frac{1}{2(2l+1)}(H_1 - iH_2)[(l+1)X_{l-1,m+1} + lX_{l+1,m+1}]
\end{aligned}
\tag{3.67}
$$

Equation (3.50) with the aid of Eqs. (3.55), (3.56), (3.60), and (3.67) becomes

$$
\dot{X}_{l,m} = -\beta^{-1} bl(l+1)X_{l,m} - b \sum_{j,k=-1}^{1} p_{j,k} X_{l+j,m+k} \partial_{-k} V \tag{3.68}
$$

where

$$
\begin{aligned}
\partial_{\pm 1} V &= \frac{\partial V}{\partial \alpha_1} \pm i \frac{\partial V}{\partial \alpha_2} = -M_s(H_1 \pm iH_2) \\
\partial_0 V &= \frac{\partial V}{\partial \alpha_3} = -M_s H_3
\end{aligned}
\tag{3.69}
$$

and the $p_{j,k}$ are given by Eqs. (2.112) and (2.114), namely,

$$p_{-1,-1} = \frac{(l+1)(l+m)(l+m-1)}{2(2l+1)} \qquad p_{-1,0} = \frac{(l+1)(l+m)}{(2l+1)}$$

$$p_{-1,1} = -\frac{l+1}{2(2l+1)}$$

$$p_{1,-1} = \frac{l(l-m+1)(l-m+2)}{2(2l+1)} \qquad p_{1,0} = -\frac{l(l-m+1)}{(2l+1)}$$

$$p_{1,1} = -\frac{l}{2(2l+1)}$$

$$p_{0,-1} = \frac{i(l+m)(l-m+1)}{2\alpha} \qquad p_{0,0} = -i\alpha^{-1}m$$

$$p_{0,1} = \frac{i}{2\alpha} \tag{3.70}$$

where $\alpha = b/a$. Equation (3.68) expressed explicitly is [37]

$$\begin{aligned}
\dot{X}_{l,m} = & -\beta^{-1}bl(l+1)X_{l,m} - \frac{b(l+1)(l+m)(l+m-1)}{2(2l+1)} X_{l-1,m-1}\partial_1 V \\
& - \frac{ia(l-m+1)(l+m)}{2} X_{l,m-1}\partial_1 V \\
& - \frac{bl(l-m+1)(l-m+2)}{2(2l+1)} X_{l+1,m-1}\partial_1 V \\
& - \frac{b(l+1)(l+m)}{(2l+1)} X_{l-1,m}\partial_0 V + iamX_{l,m}\partial_0 V \\
& + \frac{bl(l-m+1)}{(2l+1)} X_{l+1,m}\partial_0 V \\
& + \frac{b(l+1)}{2(2l+1)} X_{l-1,m+1}\partial_{-1} V - \frac{ia}{2} X_{l,m+1}\partial_{-1} V \\
& + \frac{bl}{2(2l+1)} X_{l+1,m+1}\partial_{-1} V
\end{aligned} \tag{3.71}$$

We now specialize to the case where the free energy per unit volume is given by Eq. (3.1), namely,

$$V = K(1 - \alpha_3^2) - HM_s \sum_{i=1}^{3} \gamma_i \alpha_i \tag{3.72}$$

Expanding the summation in Eq. (3.68) using Eqs. (3.16), (3.17), and (3.18), gives

$$\begin{aligned}
b \sum_{j,k=-1}^{1} & p_{j,k} X_{l+j,m+k} \partial_{-k} V \\
&= \frac{2bK}{(2l+1)} \{-(l+1)(l+m)(l+m-1)(2l-1)^{-1} X_{l-2,m} \\
&+ i\alpha^{-1} m(l+m) X_{l-1,m} - (2l+1)(2l-1)^{-1}(2l+3)^{-1} \\
&\cdot [l(l+1) - 3m^2] X_{l,m} + i\alpha^{-1} m(l-m+1) X_{l+1,m} \\
&+ l(l-m+2)(l-m+1)(2l+3)^{-1} X_{l+2,m}\} \\
&+ \frac{bHM_s(\gamma_1 + i\gamma_2)}{2(2l+1)} [-(l+1)(l+m)(l+m-1) X_{l-1,m-1} \\
&- i\alpha^{-1}(l+m)(l-m+1)(2l+1) X_{l,m-1} - l(l-m+1)(l-m+2) \\
&\cdot X_{l+1,m-1}] + \frac{bHM_s \gamma_3}{(2l+1)} \\
&\cdot [-(l+1)(l+m) X_{l-1,m} + i\alpha^{-1} m(2l+1) X_{l,m} \\
&+ l(l-m+1) X_{l+1,m}] + \frac{bHM_s(\gamma_1 - i\gamma_2)}{2(2l+1)} \\
&\cdot [(l+1) X_{l-1,m+1} - i\alpha^{-1}(2l+1) X_{l,m+1} + l X_{l+1,m+1}]
\end{aligned} \tag{3.73}$$

Equation (3.68), with the aid of Eq. (3.73), becomes

$$\begin{aligned}
\dot{X}_{l,m} = -\beta^{-1} b l(l+1) X_{l,m} + & \frac{2bK}{(2l+1)} \{(l+1)(l+m)(l+m-1) \\
&\cdot (2l-1)^{-1} X_{l-2,m} - i\alpha^{-1} m(l+m) X_{l-1,m} \\
&- (2l+1)(2l-1)^{-1}(2l+3)^{-1}[l(l+1) - 3m^2] X_{l,m} \\
&- i\alpha^{-1} m(l-m+1) X_{l+1,m} - l(l-m+2)(l-m+1)
\end{aligned}$$

$$\cdot (2l+3)^{-1}X_{l+2,m}\} + \frac{bHM_s(\gamma_1 + i\gamma_2)}{2(2l+1)}$$
$$\cdot [(l+1)(l+m)(l+m-1)X_{l-1,m-1}$$
$$+ i\alpha^{-1}(l+m)(l-m+1)(2l+1)X_{l,m-1} + l(l-m+1)$$
$$\cdot (l-m+2)X_{l+1,m-1}] + \frac{bHM_s\gamma_3}{(2l+1)}$$
$$\cdot [(l+1)(l+m)X_{l-1,m} + i\alpha^{-1}m(2l+1)X_{l,m}$$
$$- l(l-m+1)X_{l+1,m}] + \frac{bHM_s(\gamma_1 - i\gamma_2)}{2(2l+1)}$$
$$\cdot [-(l+1)X_{l-1,m+1} + i\alpha^{-1}(2l+1)X_{l,m+1} - lX_{l+1,m+1}] \quad (3.74)$$

which is identical to Eq. (3.29) with the expectation values of the spherical harmonics replaced by the sharp values of the spherical harmonics hence providing independent confirmation of the result of Eq. (3.29).

C. Results of Numerical Computation

Computation of $\lambda_1 = \beta p_1/b$ in Eq. (1.107), where p_1 is the smallest non-vanishing eigenvalue of the Fokker–Planck equation, was successfully carried out [29] using *Mathematica*© for the case where the gyromagnetic terms (terms containing α^{-1} in Eq. (3.29) and a in the preceding equations (3.74) of Section III.B) are ignored. Assuming p_1 to be real, the characteristic equation was solved using the method of repeated bisection. This method avoided the floating point difficulties that arose from an initial attempt to compute p_1 using the built-in *Eigenvalues* { } function. The results are presented in Figs. 6–10. In particular, Fig. 7 confirms that the Néel relaxation time has an absolute maximum when $\psi = \pi/2$ (transverse relaxation), local maxima at $\psi = 0, \pi$ (longitudinal relaxation), and local minima at $\psi = \pi/4, 3\pi/4$. Figure 10 shows that Kramers approach [5, 7, 11, 14, 20, 31], on being modified as shown in Section I.G to treat the $\psi = \pi/2$ orientation, yields a good approximation to $\lambda_1(\pi/4)$ for $h \geq 0.3$, $\sigma \geq 10$.

IV. APPLICATION TO THE CASE WHERE THE FREE ENERGY ARISES FROM CUBIC ANISOTROPY

A. Calculation of the Matrix Elements and Differential Recurrence Coefficients Using the Spherical Harmonic Recurrence Relations

The free energy per unit volume Eq. (1.5) expressed in terms of the direction cosines of the magnetization orientation relative to the crystal axes Eq. (2.72)

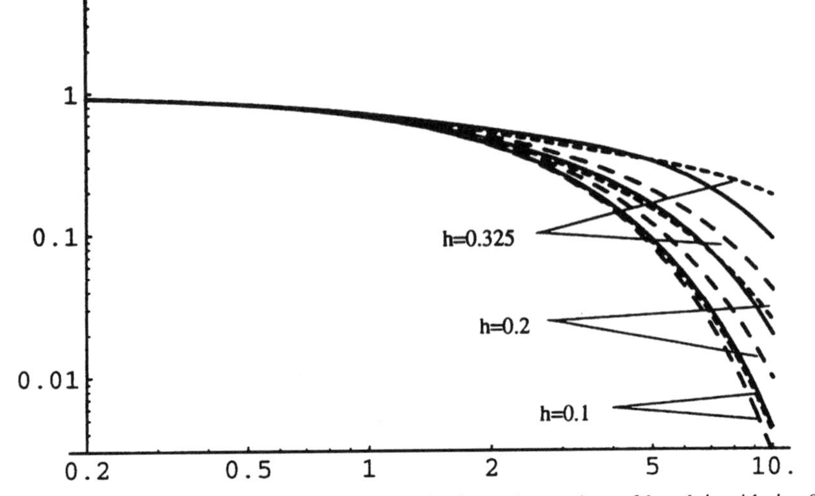

Figure 6. The value of $\lambda_1/2$ as a function of σ for various values of h and ψ, with $\psi = 0$ (solid), $\psi = \pi/4$ (short dashes), $\psi = \pi/2$ (long dashes).

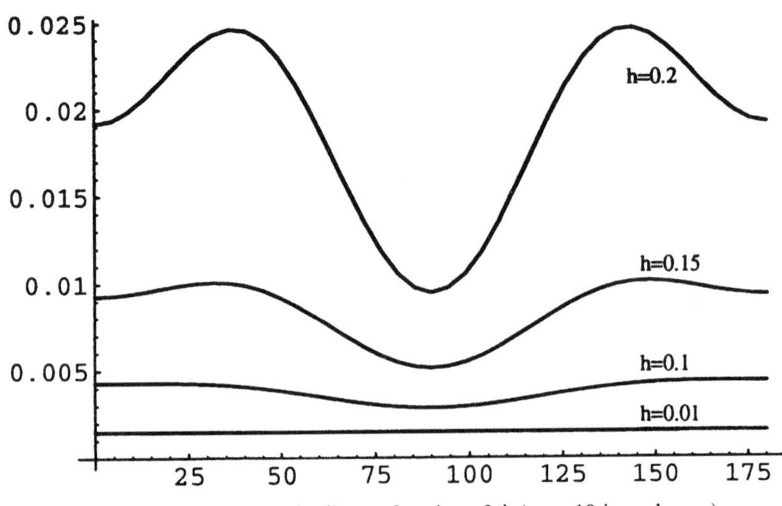

Figure 7. The value of $\lambda_1/2$ as a function of ψ ($\sigma = 10$ in each case).

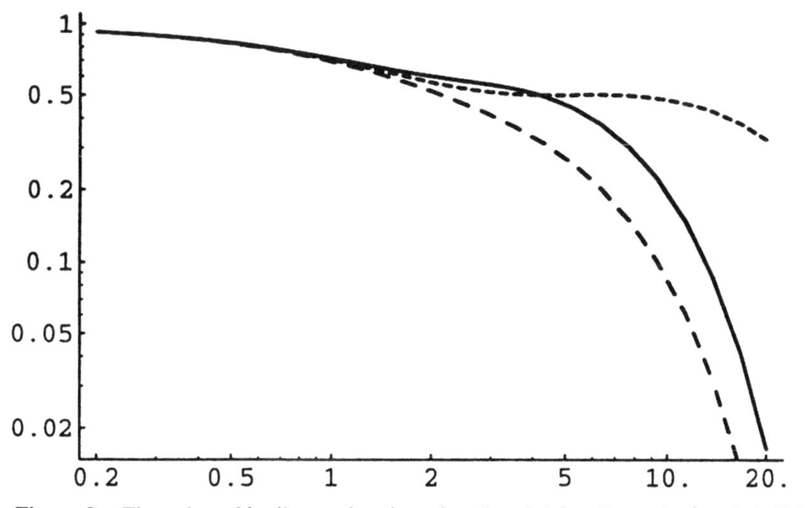

Figure 8. The value of $\lambda_1/2$ as a function of σ (h = 0.4 in all cases); ψ = 0 (solid), $\psi = \pi/4$ (short dashes), $\psi = \pi/2$ (long dashes).

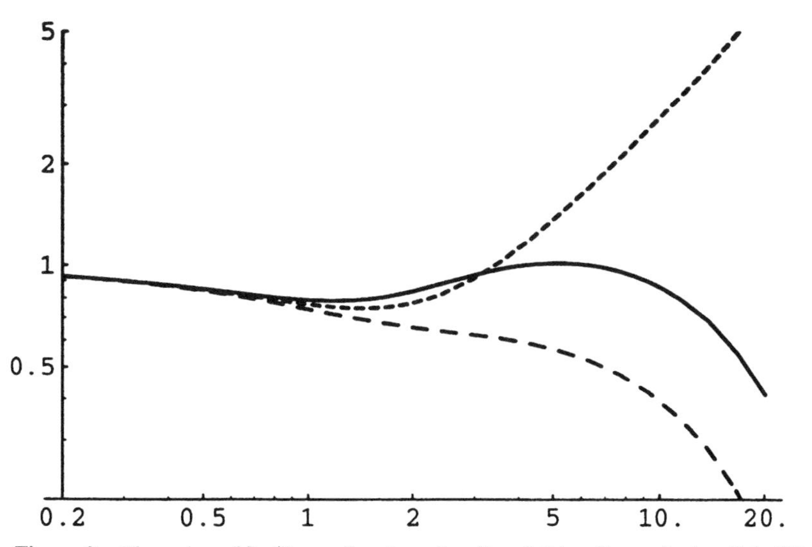

Figure 9. The value of $\lambda_1/2$ as a function of σ (h = 0.6 in all cases); ψ = 0 (solid), $\psi = \pi/4$ (short dashes), $\psi = \pi/2$ (long dashes).

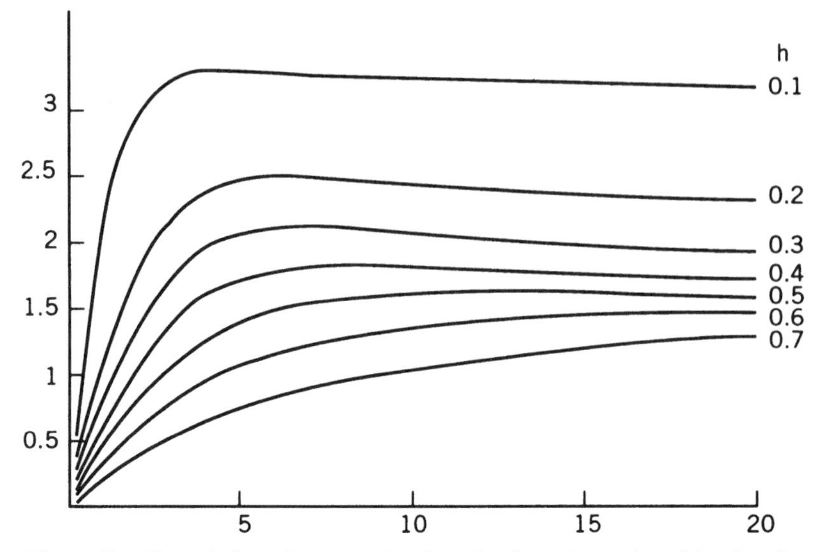

Figure 10. The ratio $\lambda_{\text{num}}/\lambda_{\text{apx}}$ as a function of σ for various valus of h, where λ_{num} denotes the values of $\lambda_1(\pi/2)$ obtained by numerical computation and λ_{apx} denotes the corresponding values obtained from Eq. (1.140). The details of the numerical calculation are given in Ref. 42. The systematic error is due to the use of the axially symmetric formula Eq. (1.140).

becomes

$$V = K(\alpha_1^2\alpha_2^2 + \alpha_2^2\alpha_3^2 + \alpha_1^2\alpha_3^2) \tag{4.1}$$

To obtain an expression for V in terms of spherical harmonics [Appendix A], we rearrange the expression for V in Eq. (1.7) in terms of spherical polar coordinates, as follows

$$\begin{aligned} V &= \frac{K}{4}(\sin^4\theta \sin^2 2\varphi + 4\cos^2\theta \sin^2\theta) \\ &= \frac{K}{8}[\sin^4\theta(1 - \cos 4\varphi) - 8\cos^4\theta + 8\cos^2\theta] \\ &= \frac{K}{8}[-\sin^4\theta \cos 4\varphi + (1 - \cos^2\theta)^2 - 8\cos^4\theta + 8\cos^2\theta] \\ &= \frac{K}{8}(1 - \sin^4\theta \cos 4\varphi - 7\cos^4\theta + 6\cos^2\theta) \end{aligned} \tag{4.2}$$

Using the expansions from Eq. (A.11), namely,

$$P_4^0 = \frac{1}{8}(35\cos^4\theta - 30\cos^2\theta + 3) \tag{4.3}$$

$$P_4^4 = 105\sin^4\theta \tag{4.4}$$

Equation (4.2) becomes

$$V = \frac{K}{5}\left(1 - P_4^0 - \frac{1}{168}P_4^4\cos 4\varphi\right) \tag{4.5}$$

Writing $\cos 4\varphi = \frac{1}{2}(e^{4i\varphi} + e^{-4i\varphi})$ and using the symmetry relation $P_4^4 = 8!P_4^{-4}$ gives

$$\begin{aligned} V &= \frac{K}{5}\left[1 - P_4^0 - \frac{1}{336}(P_4^4 e^{4i\varphi} + 8!P_4^{-4}e^{-4i\varphi})\right] \\ &= \frac{K}{5}\left[1 - X_{4,0} - \frac{1}{336}(X_{4,4} + 8!X_{4,-4})\right] \end{aligned} \tag{4.6}$$

As in Section III.A, an expression for the Laplacian $\Lambda^2 V$ in terms of the direction cosines of the magnetization α_i is required. Taking the Laplacian of Eq. (4.6) using $\Lambda^2 X_{l,m} = -l(l+1)X_{l,m}$, gives

$$\begin{aligned} \Lambda^2 V &= -\frac{K}{5}\left[\Lambda^2 X_{4,0} + \frac{1}{336}(\Lambda^2 X_{4,4} + 8!\Lambda^2 X_{4,-4})\right] \\ &= 4K\left[X_{4,0} + \frac{1}{336}(X_{4,4} + 8!X_{4,-4})\right] \\ &= 4K - 20V \end{aligned} \tag{4.7}$$

The calculation of the recurrence relation for

$$L_{FP}X_{l,m} = A_{FP}X_{l,m} + iG_{FP}X_{l,m} \tag{4.8}$$

will be accomplished using the alignment representation formula Eq. (2.82), which with the aid Eq. (4.7) becomes

$$A_{FP}X_{l,m} = b\left[2K - l(l+1)\beta^{-1} + \frac{l(l+1)-20}{2}V\right]X_{l,m} + \frac{b}{2}\Lambda^2 V X_{l,m} \quad (4.9)$$

and the gyroscopic representation formula Eq. (2.98), namely

$$G_{FP}X_{l,m} = \frac{a(l+m)(l-m+1)}{2}X_{l,m-1}\partial_1 V - amX_{l,m}\partial_0 V + \frac{a}{2}X_{l,m+1}\partial_{-1}V \quad (4.10)$$

The partial derivatives of Eq. (4.1) are

$$\frac{\partial V}{\partial \alpha_1} = 2K\alpha_1(\alpha_2^2 + \alpha_3^2) \quad (4.11)$$

$$\frac{\partial V}{\partial \alpha_2} = 2K\alpha_2(\alpha_1^2 + \alpha_3^2) \quad (4.12)$$

$$\partial_0 V = \frac{\partial V}{\partial \alpha_3} = 2K\alpha_3(\alpha_1^2 + \alpha_2^2) \quad (4.13)$$

Furthermore, Eqs. (4.11) and (4.12) give

$$\begin{aligned}\partial_{\pm 1}V = \frac{\partial V}{\partial \alpha_1} \pm i\frac{\partial V}{\partial \alpha_2} &= 2K(\alpha_1\alpha_2^2 \pm i\alpha_2\alpha_1^2) + 2K\alpha_3^2(\alpha_1 \pm i\alpha_2)\\ &= \pm 2iK\alpha_1\alpha_2(\alpha_1 \mp i\alpha_2) + 2K\alpha_3^2(\alpha_1 \pm i\alpha_2) \quad (4.14)\end{aligned}$$

To develop the recurrence relation for $A_{FP}X_{l,m}$, we start from the recurrence relations in Eqs. (A.16–A.18), namely,

$$\alpha_1 X_{l,m} = \frac{(l+m)(l+m-1)}{2(2l+1)}X_{l-1,m-1} - \frac{(l-m+2)(l-m+1)}{2(2l+1)}X_{l+1,m-1} - \frac{1}{2(2l+1)}X_{l-1,m+1} + \frac{1}{2(2l+1)}X_{l+1,m+1} \quad (4.15)$$

$$i\alpha_2 X_{l,m} = -\frac{(l+m)(l+m-1)}{2(2l+1)}X_{l-1,m-1} + \frac{(l-m+2)(l-m+1)}{2(2l+1)}X_{l+1,m-1} - \frac{1}{2(2l+1)}X_{l-1,m+1} + \frac{1}{2(2l+1)}X_{l+1,m+1} \quad (4.16)$$

$$\alpha_3 X_{l,m} = \frac{(l+m)}{(2l+1)} X_{l-1,m} + \frac{(l-m+1)}{(2l+1)} X_{l+1,m} \tag{4.17}$$

and obtain, from these, recurrence relations for

$$\alpha_3^2 X_{l,m}, \quad \alpha_3^2(\alpha_1^2 + \alpha_2^2) X_{l,m}, \quad (\alpha_1^2 + \alpha_2^2) X_{l,m} = (1 - \alpha_3^2) X_{l,m},$$
$$i\alpha_1\alpha_2 X_{l,m}, \quad -\alpha_1^2\alpha_2^2 X_{l,m}, \quad V X_{l,m} = K[\alpha_1^2\alpha_2^2 + \alpha_3^2(\alpha_1^2 + \alpha_2^2)] X_{l,m} \tag{4.18}$$

To develop the recurrence relation for $G_{FP} X_{l,m}$, we start from the recurrence relations for

$$\alpha_3 X_{l,m}, \quad (\alpha_1 \pm i\alpha_2) X_{l,m}, \quad \alpha_3^2 X_{l,m},$$
$$(\alpha_1^2 + \alpha_2^2) X_{l,m}, \quad i\alpha_1\alpha_2 X_{l,m} \tag{4.19}$$

and obtain from these recurrence relations for

$$X_{l,m}\partial_0 V = 2K\alpha_3(\alpha_1^2 + \alpha_2^2) X_{l,m}$$
$$\pm 2i\alpha_1\alpha_2(\alpha_1 \mp i\alpha_2) X_{l,m}, \quad 2\alpha_3^2(\alpha_1 \pm i\alpha_2) X_{l,m}$$
$$X_{l,m}\partial_{\pm 1} V = \pm 2iK\alpha_1\alpha_2(\alpha_1 \mp i\alpha_2) X_{l,m} + 2K\alpha_3^2(\alpha_1 \pm i\alpha_2) X_{l,m} \tag{4.20}$$

Successive application of Eq. (4.17) gives

$$\alpha_3^2 X_{l,m} = \frac{(l+m)(l+m-1)}{(2l+1)(2l-1)} X_{l-2,m} + \frac{2l(l+1) - 2m^2 - 1}{(2l-1)(2l+3)} X_{l,m} + \frac{(l-m+2)(l-m+1)}{(2l+1)(2l+3)} X_{l+2,m} \tag{4.21}$$

Hence

$$(\alpha_1^2 + \alpha_2^2) X_{l,m} = (1 - \alpha_3^2) X_{l,m} = -\frac{(l+m)(l+m-1)}{(2l+1)(2l-1)} X_{l-2,m} + \frac{2[l(l+1) + m^2 - 1]}{(2l-1)(2l+3)} X_{l,m} - \frac{(l-m+2)(l-m+1)}{(2l+1)(2l+3)} X_{l+2,m} \tag{4.22}$$

Equations (4.21) and (4.22) give

$$
\begin{aligned}
\alpha_3^2(\alpha_1^2+\alpha_2^2)X_{l,m}
&= -\frac{(l+m)(l+m-1)}{(2l-1)(2l+1)}\,\alpha_3^2X_{l-2,m} \\
&\quad + \frac{2[l(l+1)+m^2-1]}{(2l-1)(2l+3)}\,\alpha_3^2X_{l,m} - \frac{(l-m+2)(l-m+1)}{(2l+1)(2l+3)}\,\alpha_3^2X_{l+2,m} \\
&= -\frac{(l+m)(l+m-1)}{(2l-1)(2l+1)}\left[\frac{(l+m-2)(l+m-3)}{(2l-3)(2l-5)}X_{l-4,m}\right. \\
&\quad \left. + \frac{2(l-2)(l-1)-2m^2-1}{(2l-5)(2l-1)}X_{l-2,m} + \frac{(l-m)(l-m-1)}{(2l-3)(2l-1)}X_{l,m}\right] \\
&\quad + \frac{2[l(l+1)+m^2-1]}{(2l-1)(2l+3)}\left[\frac{(l+m)(l+m-1)}{(2l+1)(2l-1)}X_{l-2,m}\right. \\
&\quad \left. + \frac{2l(l+1)-2m^2-1}{(2l-1)(2l+3)}X_{l,m} + \frac{(l-m+2)(l-m+1)}{(2l+1)(2l+3)}X_{l+2,m}\right] \\
&\quad - \frac{(l-m+2)(l-m+1)}{(2l+1)(2l+3)}\left[\frac{(l+m+2)(l+m+1)}{(2l+5)(2l+3)}X_{l,m}\right. \\
&\quad + \frac{2(l+2)(l+3)-2m^2-1}{(2l+3)(2l+7)}X_{l+2,m} \\
&\quad \left. + \frac{(l-m+4)(l-m+3)}{(2l+5)(2l+7)}X_{l+4,m}\right]
\end{aligned}
\tag{4.23}
$$

Rearranging and simplifying gives

$$
\begin{aligned}
\alpha_3^2(\alpha_1^2+\alpha_2^2)X_{l,m} &= -\frac{(l+m)(l+m-1)(l+m-2)(l+m-3)}{(2l-5)(2l-3)(2l-1)(2l+1)}X_{l-4,m} \\
&\quad + \frac{(l+m)(l+m-1)(4m^2-1)}{(2l-5)(2l-1)(2l+1)(2l+3)}X_{l-2,m} \\
&\quad + \frac{(l-m+2)(l-m+1)(4m^2-1)}{(2l-1)(2l+1)(2l+3)(2l+7)}X_{l+2,m}
\end{aligned}
$$

$$
+ \left\{ \frac{2[l(l+1)+m^2-1][2l(l+1)-2m^2-1]}{(2l-1)^2(2l+3)^2} \right.
$$

$$
- \frac{(l+m)(l+m-1)(l-m)(l-m-1)}{(2l-3)(2l-1)^2(2l+1)}
$$

$$
\left. - \frac{(l-m+1)(l-m+2)(l+m+1)(l+m+2)}{(2l+1)(2l+3)^2(2l+5)} \right\} X_{l,m}
$$

$$
- \frac{(l-m+1)(l-m+2)(l-m+3)(l-m+4)}{(2l+1)(2l+3)(2l+5)(2l+7)} X_{l+4,m} \tag{4.24}
$$

From Eqs. (4.15) and (4.16), we have

$$
\begin{aligned}
i\alpha_1\alpha_2 X_{l,m} &= \frac{(l+m)(l+m-1)}{2(2l+1)} i\alpha_2 X_{l-1,\, m-1} \\
&\quad - \frac{(l-m+2)(l-m+1)}{2(2l+1)} i\alpha_2 X_{l+1,m-1} \\
&\quad - \frac{1}{2(2l+1)} i\alpha_2 X_{l-1,m+1} + \frac{1}{2(2l+1)} i\alpha_2 X_{l+1,m+1} \\
&= \frac{(l+m)(l+m-1)}{4(2l-1)(2l+1)} [-(l+m-2)(l+m-3)X_{l-2,m-2} \\
&\quad + (l-m+2)(l-m+1)X_{l,m-2} - X_{l-2,m} + X_{l,m}] \\
&\quad - \frac{(l-m+2)(l-m+1)}{4(2l+1)(2l+3)} [-(l+m)(l+m-1)X_{l,m-2} \\
&\quad + (l-m+4)(l-m+3)X_{l+2,m-2} - X_{l,m} + X_{l+2,m}] \\
&\quad - \frac{1}{4(2l-1)(2l+1)} [-(l+m)(l+m-1)X_{l-2,m} \\
&\quad + (l-m)(l-m-1)X_{l,m} - X_{l-2,m+2} + X_{l,m+2}] \\
&\quad + \frac{1}{4(2l+1)(2l+3)} [-(l+m+2)(l+m+1)X_{l,m} \\
&\quad + (l-m+2)(l-m+1)X_{l+2,m} - X_{l,m+2} + X_{l+2,m+2}]
\end{aligned} \tag{4.25}
$$

The $X_{l,m}$ term vanishes since

$$\frac{(l-m+2)(l-m+1)}{(2l+3)} - \frac{(l+m+2)(l+m+1)}{(2l+3)} + \frac{(l+m)(l+m-1)}{(2l-1)} - \frac{(l-m)(l-m-1)}{(2l-1)} = 0 \tag{4.26}$$

Rearranging and simplifying Eq. (4.25) with the aid of Eq. (4.26) gives

$$\begin{aligned} i\alpha_1\alpha_2 X_{l,m} = &-\frac{(l+m)(l+m-1)(l+m-2)(l+m-3)}{4(2l-1)(2l+1)} X_{l-2,m-2} \\ &+ \frac{(l+m)(l+m-1)(l-m+2)(l-m+1)}{2(2l-1)(2l+3)} X_{l,m-2} \\ &- \frac{(l-m+2)(l-m+1)(l-m+4)(l-m+3)}{4(2l+1)(2l+3)} X_{l+2,m-2} \\ &+ \frac{1}{4(2l+1)(2l-1)} X_{l-2,m+2} - \frac{1}{2(2l-1)(2l+3)} X_{l,m+2} \\ &+ \frac{1}{4(2l+1)(2l+3)} X_{l+2,m+2} \end{aligned} \tag{4.27}$$

Repeated application of Eq. (4.27) gives

$$\begin{aligned} -\alpha_1^2\alpha_2^2 X_{l,m} = &-\frac{(l+m)(l+m-1)(l+m-2)(l+m-3)}{4(2l-1)(2l+1)} i\alpha_1\alpha_2 X_{l-2,m-2} \\ &+ \frac{(l+m)(l+m-1)(l-m+2)(l-m+1)}{2(2l-1)(2l+3)} i\alpha_1\alpha_2 X_{l,m-2} \\ &- \frac{(l-m+2)(l-m+1)(l-m+4)(l-m+3)}{4(2l+1)(2l+3)} i\alpha_1\alpha_2 X_{l+2,m-2} \\ &+ \frac{1}{4(2l+1)(2l-1)} i\alpha_1\alpha_2 X_{l-2,m+2} - \frac{1}{2(2l-1)(2l+3)} \\ &\cdot i\alpha_1\alpha_2 X_{l,m+2} + \frac{1}{4(2l+1)(2l+3)} i\alpha_1\alpha_2 X_{l+2,m+2} \\ = &-\frac{(l+m)(l+m-1)(l+m-2)(l+m-3)}{4(2l-1)(2l+1)} \end{aligned}$$

$$\cdot\left[\frac{1}{4(2l-3)(2l-5)}X_{l-4,m}\right.$$

$$-\frac{(l+m-4)(l+m-5)(l+m-6)(l+m-7)}{4(2l-5)(2l-3)}X_{l-4,m-4}$$

$$-\frac{1}{2(2l-5)(2l-1)}X_{l-2,m}$$

$$+\frac{(l+m-4)(l+m-5)(l-m+2)(l-m+1)}{2(2l-5)(2l-1)}X_{l-2,m-4}$$

$$+\frac{1}{4(2l-3)(2l-1)}X_{l,m}$$

$$\left.-\frac{(l-m+1)(l-m+2)(l-m+3)(l-m+4)}{4(2l-3)(2l-1)}X_{l,m-4}\right]$$

$$+\frac{(l+m)(l+m-1)(l-m+2)(l-m+1)}{2(2l-1)(2l+3)}$$

$$\cdot\left[\frac{1}{4(2l+1)(2l-1)}X_{l-2,m}\right.$$

$$-\frac{(l+m-2)(l+m-3)(l+m-4)(l+m-5)}{4(2l-1)(2l+1)}X_{l-2,m-4}$$

$$-\frac{1}{2(2l-1)(2l+3)}X_{l,m}$$

$$+\frac{(l+m-2)(l+m-3)(l-m+4)(l-m+3)}{2(2l-1)(2l+3)}X_{l,m-4}$$

$$+\frac{1}{4(2l+1)(2l+3)}X_{l+2,m}$$

$$\left.-\frac{(l-m+3)(l-m+4)(l-m+5)(l-m+6)}{4(2l+1)(2l+3)}X_{l+2,m-4}\right]$$

$$-\frac{(l-m+1)(l-m+2)(l-m+3)(l-m+4)}{4(2l+1)(2l+3)}$$

$$\cdot\left[\frac{1}{4(2l+5)(2l+3)}X_{l,m}\right.$$

$$- \frac{(l+m)(l+m-1)(l+m-2)(l+m-3)}{4(2l+3)(2l+5)} X_{l,m-4}$$

$$- \frac{1}{2(2l+3)(2l+7)} X_{l+2,m}$$

$$+ \frac{(l+m)(l+m-1)(l-m+6)(l-m+5)}{2(2l+3)(2l+7)} X_{l+2,m-4}$$

$$+ \frac{1}{4(2l+5)(2l+7)} X_{l+4,m}$$

$$- \left. \frac{(l-m+5)(l-m+6)(l-m+7)(l-m+8)}{4(2l+5)(2l+7)} X_{l+4,m-4} \right]$$

$$+ \frac{1}{4(2l+1)(2l-1)}$$

$$\cdot \left[- \frac{(l+m)(l+m-1)(l+m-2)(l+m-3)}{4(2l-5)(2l-3)} X_{l-4,m} \right.$$

$$+ \frac{(l+m)(l+m-1)(l-m-2)(l-m-3)}{2(2l-5)(2l-1)} X_{l-2,m}$$

$$- \frac{(l-m-3)(l-m-2)(l-m-1)(l-m)}{4(2l-3)(2l-1)} X_{l,m}$$

$$+ \frac{1}{4(2l-3)(2l-5)} X_{l-4,m+4} - \frac{1}{2(2l-5)(2l-1)}$$

$$\left. \cdot X_{l-2,m+4} + \frac{1}{4(2l-3)(2l-1)} X_{l,m+4} \right]$$

$$- \frac{1}{2(2l-1)(2l+3)}$$

$$\cdot \left[- \frac{(l+m+2)(l+m+1)(l+m)(l+m-1)}{4(2l-1)(2l+1)} X_{l-2,m} \right.$$

$$+ \frac{(l+m+2)(l+m+1)(l-m)(l-m-1)}{2(2l-1)(2l+3)} X_{l,m}$$

$$- \frac{(l-m-1)(l-m)(l-m+1)(l-m+2)}{4(2l+1)(2l+3)} X_{l+2,m}$$

$$+ \frac{1}{4(2l+1)(2l-1)} X_{l-2,m+4} - \frac{1}{2(2l-1)(2l+3)} X_{l,m+4}$$

$$\left. + \frac{1}{4(2l+1)(2l+3)} X_{l+2,m+4} \right] + \frac{1}{4(2l+1)(2l+3)}$$

$$\cdot \left[-\frac{(l+m+1)(l+m+2)(l+m+3)(l+m+4)}{4(2l+3)(2l+5)} X_{l,m} \right.$$

$$+ \frac{(l+m+3)(l+m+4)(l-m+1)(l-m+2)}{2(2l+3)(2l+7)} X_{l+2,m}$$

$$- \frac{(l-m+1)(l-m+2)(l-m+3)(l-m+4)}{4(2l+5)(2l+7)} X_{l+4,m}$$

$$+ \frac{1}{4(2l+3)(2l+5)} X_{l,m+4} - \frac{1}{2(2l+3)(2l+7)} X_{l+2,m+4}$$

$$\left. + \frac{1}{4(2l+5)(2l+7)} X_{l+4,m+4} \right] \tag{4.28}$$

Rearranging and simplifying gives

$$-\alpha_1^2\alpha_2^2 X_{l,m} = \frac{(l+m)(l+m-1)\dots(l+m-7)}{16(2l-5)(2l-3)(2l-1)(2l+1)} X_{l-4,m-4}$$

$$- \frac{(l-m+1)(l-m+2)(l+m)(l+m-1)\dots(l+m-5)}{4(2l-5)(2l-1)(2l+1)(2l+3)} X_{l-2,m-4}$$

$$+ \frac{3(l+m)(l+m-1)\dots(l+m-3)(l-m+1)(l-m+2)\dots(l-m+4)}{8(2l-3)(2l-1)(2l+3)(2l+5)} X_{l,m-4}$$

$$- \frac{(l+m)(l+m-1)(l-m+1)(l-m+2)\dots(l-m+6)}{4(2l-1)(2l+1)(2l+3)(2l+7)} X_{l+2,m-4}$$

$$+ \frac{(l-m+1)(l-m+2)\dots(l-m+8)}{16(2l+1)(2l+3)(2l+5)(2l+7)} X_{l+4,m-4}$$

$$- \frac{(l+m)(l+m-1)\dots(l+m-3)}{8(2l-5)(2l-3)(2l-1)(2l+1)} X_{l-4,m}$$

$$+ \frac{(l+m)(l+m-1)(l^2-l-4+m^2)}{2(2l-5)(2l-1)(2l+1)(2l+3)} X_{l-2,m}$$

$$- \left\{ \frac{[(l+m)(l+m-1)\dots(l+m-3)+(l-m)(l-m-1)\dots(l-m-3)]}{16(2l-3)(2l-1)^2(2l+1)} \right.$$

$$+ \frac{[(l+m)(l+m-1)(l-m+1)(l-m+2) + (l+m+1)(l+m+2)(l-m)(l-m-1)]}{4(2l-1)^2(2l+3)^2}$$

$$\left. + \frac{[(l-m+1)(l-m+2)\dots(l-m+4) + (l+m+1)(l+m+2)\dots(l+m+4)]}{16(2l+1)(2l+3)^2(2l+5)} \right\} X_{l,m}$$

$$+ \frac{(l-m+1)(l-m+2)(l^2+3l-2+m^2)}{2(2l-1)(2l+1)(2l+3)(2l+7)} X_{l+2,m}$$

$$- \frac{(l-m+1)(l-m+2)\dots(l-m+4)}{8(2l+1)(2l+3)(2l+5)(2l+7)} X_{l+4,m}$$

$$+ \frac{1}{16(2l-5)(2l-3)(2l-1)(2l+1)} X_{l-4,m+4}$$

$$- \frac{1}{4(2l-5)(2l-1)(2l+1)(2l+3)} X_{l-2,m+4}$$

$$+ \frac{3}{8(2l-3)(2l-1)(2l+3)(2l+5)} X_{l,m+4}$$

$$- \frac{1}{4(2l-1)(2l+1)(2l+3)(2l+7)} X_{l+2,m+4}$$

$$+ \frac{1}{16(2l+1)(2l+3)(2l+5)(2l+7)} X_{l+4,m+4} \tag{4.29}$$

Equations (4.24) and (4.29) yield

$$VX_{l,m} = K(\alpha_1^2\alpha_2^2 + \alpha_1^2\alpha_3^2 + \alpha_2^2\alpha_3^2)X_{l,m}$$

$$= R_{l,m}X_{l,m} - \frac{K(l+m)(l+m-1)\dots(l+m-7)}{16(2l-5)(2l-3)(2l-1)(2l+1)} X_{l-4,m-4}$$

$$+ \frac{K(l-m+1)(l-m+2)(l+m)(l+m-1)\dots(l+m-5)}{4(2l-5)(2l-1)(2l+1)(2l+3)} X_{l-2,m-4}$$

$$- \frac{3K(l+m)(l+m-1)\dots(l+m-3)(l-m+1)(l-m+2)\dots(l-m+4)}{8(2l-3)(2l-1)(2l+3)(2l+5)} X_{l,m-4}$$

$$+ \frac{K(l+m)(l+m-1)(l-m+1)(l-m+2)\dots(l-m+6)}{4(2l-1)(2l+1)(2l+3)(2l+7)} X_{l+2,m-4}$$

$$- \frac{K(l-m+1)(l-m+2)\dots(l-m+8)}{16(2l+1)(2l+3)(2l+5)(2l+7)} X_{l+4,m-4}$$

$$- \frac{7K(l+m)(l+m-1)\dots(l+m-3)}{8(2l-5)(2l-3)(2l-1)(2l+1)} X_{l-4,m}$$

$$- \frac{K(l+m)(l+m-1)(l^2-l-2-7m^2)}{2(2l-5)(2l-1)(2l+1)(2l+3)} X_{l-2,m}$$

$$
\begin{aligned}
&- \frac{K(l-m+1)(l-m+2)(l^2+3l-7m^2)}{2(2l-1)(2l+1)(2l+3)(2l+7)} X_{l+2,m} \\
&- \frac{7K(l-m+1)(l-m+2)\dots(l-m+4)}{8(2l+1)(2l+3)(2l+5)(2l+7)} X_{l+4,m} \\
&- \frac{K}{16(2l-5)(2l-3)(2l-1)(2l+1)} X_{l-4,m+4} \\
&+ \frac{K}{4(2l-5)(2l-1)(2l+1)(2l+3)} X_{l-2,m+4} \\
&- \frac{3K}{8(2l-3)(2l-1)(2l+3)(2l+5)} X_{l,m+4} \\
&+ \frac{K}{4(2l-1)(2l+1)(2l+3)(2l+7)} X_{l+2,m+4} \\
&- \frac{K}{16(2l+1)(2l+3)(2l+5)(2l+7)} X_{l+4,m+4}
\end{aligned}
\tag{4.30}
$$

where

$$
\begin{aligned}
R_{l,m} = K \Bigg\{ & \frac{1}{4(2l-1)^2(2l+3)^2} [8(l^2+l+m^2-1)(2l^2+2l-2m^2-1) \\
&+ (l+m)(l+m-1)(l-m+1)(l-m+2) \\
&+ (l+m+1)(l+m+2)(l-m)(l-m-1)] \\
&+ \frac{1}{16(2l-3)(2l-1)^2(2l+1)} [(l+m)(l+m-1)\dots(l+m-3) \\
&+ (l-m)(l-m-1)\dots(l-m-3) \\
&- 16(l+m)(l+m-1)(l-m)(l-m-1)] \\
&+ \frac{1}{16(2l+1)(2l+3)^2(2l+5)} [(l-m+1)(l-m+2)\dots(l-m+4) \\
&+ (l+m+1)(l+m+2)\dots(l+m+4) \\
&- 16(l-m+1)(l-m+2)(l+m+1)(l+m+2)] \Bigg\}
\end{aligned}
\tag{4.31}
$$

Equation (4.31) is simplified on being expressed as a single fraction

$$
R_{l,m} = \frac{K[(XA+YB+ZC)+2(XD+YE+ZF)m^2 - 14(4X+Y+Z)m^4]}{16(2l-3)(2l-1)^2(2l+1)(2l+3)^2(2l+5)}
\tag{4.32}
$$

where

$$
\begin{aligned}
A &= 72l^4 + 144l^3 - 40l^2 - 112l + 32 \\
B &= -14l^4 + 20l^3 + 6l^2 - 12l \\
C &= -14l^4 - 76l^3 - 138l^2 - 92l - 16 \\
D &= -8l^2 - 8l + 36 \\
E &= 22l^2 - 34l + 19 \\
F &= 22l^2 + 78l + 75 \\
X &= (2l-3)(2l+1)(2l+5) = 8l^3 + 12l^2 - 26l - 15 \\
Y &= (2l+3)^2(2l+5) = 8l^3 + 44l^2 + 78l + 45 \\
Z &= (2l-3)(2l-1)^2 = 8l^3 - 20l^2 + 14l - 3
\end{aligned}
\tag{4.33}
$$

Factoring $(2l-1)(2l+1)(2l+3)$ from the numerator using

$$
\begin{aligned}
XA + YB + ZC &= 4(2l-1)(2l+1)(2l+3)(11l^4 + 22l^3 - 43l^2 - 54l + 36) \\
XD + YE + ZF &= 6(2l-1)(2l+1)(2l+3)[6l(l+1) - 5] \\
4X + Y + Z &= 6(2l-1)(2l+1)(2l+3)
\end{aligned}
\tag{4.34}
$$

gives

$$
R_{l,m} = \frac{K\{(11l^4 + 22l^3 - 43l^2 - 54l + 36) + 3m^2[6l(l+1) - 5 - 7m^2]\}}{4(2l-3)(2l-1)(2l+3)(2l+5)}
\tag{4.35}
$$

Equation (4.30) can be expressed compactly as

$$
VX_{l,m} = R_{l,m}X_{l,m} + \sum_{s=-2}^{2} \sum_{t=-1}^{1} f_{l,m,l+2s,m+4t} X_{l+2s,m+4t}
\tag{4.36}
$$

where $f_{l,m,l,m} = 0$. On setting

$$
\begin{aligned}
Q_{l,m} &= b[2K - l(l+1)\beta^{-1} - 10R_{l,m}] \\
&= -l(l+1)b\beta^{-1} + \frac{bK\{9l(l+2)(l^2-1) - 15m^2[6l(l+1) - 5 - 7m^2]\}}{2(2l-3)(2l-1)(2l+3)(2l+5)}
\end{aligned}
\tag{4.37}
$$

Equation (4.9) with the aid of Eq. (4.36) becomes

$$A_{FP}X_{l,m} = Q_{l,m}X_{l,m} + b \sum_{s=-2}^{2} \sum_{t=-1}^{1} \cdot [s(2l+1-2s) - 10] f_{l,m,l+2s,m+4t} X_{l+2s,m+4t} \tag{4.38}$$

or explicitly

$$\begin{aligned}
A_{FP}X_{l,m} = {} & b\left\{-l(l+1)\beta^{-1} + K\,\frac{9l(l+2)(l^2-1) - 15m^2[6l(l+1) - 5 - 7m^2]}{2(2l-3)(2l-1)(2l+3)(2l+5)}\right\} X_{l,m} \\
& - \frac{bK(l-4)(l+m)(l+m-1)\ldots(l+m-7)}{4(2l-5)(2l-3)(2l-1)(2l+1)} X_{l-4,m-4} \\
& + \frac{bK(2l-11)(l-m+1)(l-m+2)(l+m)(l+m-1)\ldots(l+m-5)}{4(2l-5)(2l-1)(2l+1)(2l+3)} X_{l-2,m-4} \\
& + \frac{15bK(l+m)(l+m-1)\ldots(l+m-3)(l-m+1)(l-m+2)\ldots(l-m+4)}{4(2l-3)(2l-1)(2l+3)(2l+5)} X_{l,m-4} \\
& - \frac{bK(2l+13)(l+m)(l+m-1)(l-m+1)(l-m+2)\ldots(l-m+6)}{4(2l-1)(2l+1)(2l+3)(2l+7)} X_{l+2,m-4} \\
& + \frac{bK(l+5)(l-m+1)(l-m+2)\ldots(l-m+8)}{4(2l+1)(2l+3)(2l+5)(2l+7)} X_{l+4,m-4} \\
& - \frac{7bK(l-4)(l+m)(l+m-1)\ldots(l+m-3)}{2(2l-5)(2l-3)(2l-1)(2l+1)} X_{l-4,m} \\
& - \frac{bK(2l-11)(l+m)(l+m-1)(l^2-l-2-7m^2)}{2(2l-5)(2l-1)(2l+1)(2l+3)} X_{l-2,m} \\
& + \frac{bK(2l+13)(l-m+1)(l-m+2)(l^2+3l-7m^2)}{2(2l-1)(2l+1)(2l+3)(2l+7)} X_{l+2,m} \\
& + \frac{7bK(l+5)(l-m+1)(l-m+2)\ldots(l-m+4)}{2(2l+1)(2l+3)(2l+5)(2l+7)} X_{l+4,m} \\
& - \frac{bK(l-4)}{4(2l-5)(2l-3)(2l-1)(2l+1)} X_{l-4,m+4} \\
& + \frac{bK(2l-11)}{4(2l-5)(2l-1)(2l+1)(2l+3)} X_{l-2,m+4} \\
& + \frac{15bK}{4(2l-3)(2l-1)(2l+3)(2l+5)} X_{l,m+4} \\
& - \frac{Kb(2l+13)}{4(2l-1)(2l+1)(2l+3)(2l+7)} X_{l+2,m+4} \\
& + \frac{Kb(l+5)}{4(2l+1)(2l+3)(2l+5)(2l+7)} X_{l+4,m+4}
\end{aligned} \tag{4.39}$$

From Eqs. (4.17) and (4.22), we have

$$
\begin{aligned}
X_{l,m}\partial_0 V &= \frac{\partial V}{\partial \alpha_3} X_{l,m} = 2K\alpha_3(\alpha_1^2 + \alpha_2^2)X_{l,m} \\
&= -\frac{2K(l+m)(l+m-1)}{(2l+1)(2l-1)}\alpha_3 X_{l-2,m} \\
&\quad + \frac{4K[l(l+1)+m^2-1]}{(2l-1)(2l+3)}\alpha_3 X_{l,m} \\
&\quad - \frac{2K(l-m+2)(l-m+1)}{(2l+1)(2l+3)}\alpha_3 X_{l+2,m} \\
&= -\frac{2K(l+m)(l+m-1)(l+m-2)}{(2l+1)(2l-1)(2l-3)} X_{l-3,m} \\
&\quad + \frac{2K(l+m)}{(2l-1)(2l+1)}\left\{\frac{2[l(l+1)+m^2-1]}{(2l+3)} - \frac{(l+m-1)(l-m-1)}{(2l-3)}\right\} X_{l-1,m} \\
&\quad + \frac{2K(l-m+1)}{(2l+1)(2l+3)}\left\{\frac{2[l(l+1)+m^2-1]}{(2l-1)} - \frac{(l+m+2)(l-m+2)}{(2l+5)}\right\} X_{l+1,m} \\
&\quad - \frac{2K(l-m+3)(l-m+2)(l-m+1)}{(2l+1)(2l+3)(2l+5)} X_{l+3,m}
\end{aligned}
\tag{4.40}
$$

Rearranging and simplifying gives

$$
\begin{aligned}
X_{l,m}\partial_0 V = \frac{\partial V}{\partial \alpha_3} X_{l,m} &= -\frac{2K(l+m)(l+m-1)(l+m-2)}{(2l+1)(2l-1)(2l-3)} X_{l-3,m} \\
&\quad + \frac{2K(l+m)(l^2+3m^2-3)}{(2l-3)(2l+1)(2l+3)} X_{l-1,m} \\
&\quad + \frac{2K(l-m+1)[l(l+2)+3m^2-2]}{(2l-1)(2l+1)(2l+5)} X_{l+1,m} \\
&\quad - \frac{2K(l-m+3)(l-m+2)(l-m+1)}{(2l+1)(2l+3)(2l+5)} X_{l+3,m}
\end{aligned}
\tag{4.41}
$$

The recurrence relations for $(\alpha_1 \pm i\alpha_2)X_{l,m}$ are given by Eqs. (A.19) and (A.20), namely,

$$(\alpha_1 + i\alpha_2)X_{l,m} = -\frac{1}{(2l+1)} X_{l-1,m+1} + \frac{1}{(2l+1)} X_{l+1,m+1} \tag{4.42}$$

$$(\alpha_1 - i\alpha_2)X_{l,m} = \frac{(l+m)(l+m-1)}{(2l+1)} X_{l-1,m-1} - \frac{(l-m+2)(l-m+1)}{(2l+1)} X_{l+1,m-1} \tag{4.43}$$

From Eqs. (4.27) and (4.42), we have

$$\begin{aligned}
&-2i\alpha_1\alpha_2(\alpha_1 + i\alpha_2)X_{l,m} \\
&= \frac{(l+m)(l+m-1)(l+m-2)(l+m-3)}{2(2l+1)(2l-1)} (\alpha_1 + i\alpha_2)X_{l-2,m-2} \\
&\quad - \frac{(l+m)(l+m-1)(l-m+2)(l-m+1)}{(2l-1)(2l+3)} (\alpha_1 + i\alpha_2)X_{l,m-2} \\
&\quad + \frac{(l-m+2)(l-m+1)(l-m+4)(l-m+3)}{2(2l+1)(2l+3)} (\alpha_1 + i\alpha_2)X_{l+2,m-2} \\
&\quad - \frac{1}{2(2l+1)(2l-1)} (\alpha_1 + i\alpha_2)X_{l-2,m+2} \\
&\quad + \frac{1}{(2l-1)(2l+3)} (\alpha_1 + i\alpha_2)X_{l,m+2} \\
&\quad - \frac{1}{2(2l+1)(2l+3)} (\alpha_1 + i\alpha_2)X_{l+2,m+2} \\
&= \frac{(l+m)(l+m-1)(l+m-2)(l+m-3)}{2(2l-3)(2l-1)(2l+1)} (-X_{l-3,m-1} + X_{l-1,m-1}) \\
&\quad - \frac{(l+m)(l+m-1)(l-m+2)(l-m+1)}{(2l-1)(2l+1)(2l+3)} (-X_{l-1,m-1} + X_{l+1,m-1}) \\
&\quad + \frac{(l-m+2)(l-m+1)(l-m+4)(l-m+3)}{2(2l+1)(2l+3)(2l+5)} (-X_{l+1,m-1} + X_{l+3,m-1}) \\
&\quad - \frac{1}{2(2l-3)(2l-1)(2l+1)} (-X_{l-3,m+3} + X_{l-1,m+3}) \\
&\quad + \frac{1}{(2l-1)(2l+1)(2l+3)} (-X_{l-1,m+3} + X_{l+1,m+3}) \\
&\quad - \frac{1}{2(2l+1)(2l+3)(2l+5)} (-X_{l+1,m+3} + X_{l+3,m+3})
\end{aligned} \tag{4.44}$$

Rearranging and simplifying gives

$$
\begin{aligned}
-2i\alpha_1\alpha_2(\alpha_1 + i\alpha_2)X_{l,m}
&= -\frac{(l+m)(l+m-1)(l+m-2)(l+m-3)}{2(2l+1)(2l-3)(2l-1)} X_{l-3,m-1} \\
&+ \frac{(l+m)(l+m-1)[3l^2+l-6-m(2l+3)+3m^2]}{2(2l-3)(2l+1)(2l+3)} X_{l-1,m-1} \\
&- \frac{(l-m+2)(l-m+1)[3l^2+5l-4+m(2l-1)+3m^2]}{2(2l-1)(2l+1)(2l+5)} X_{l+1,m-1} \\
&+ \frac{(l-m+2)(l-m+1)(l-m+4)(l-m+3)}{2(2l+1)(2l+3)(2l+5)} X_{l+3,m-1} \\
&+ \frac{1}{2(2l-1)(2l+1)(2l-3)} X_{l-3,m+3} \\
&- \frac{3}{2(2l-3)(2l+1)(2l+3)} X_{l-1,m+3} \\
&+ \frac{3}{2(2l-1)(2l+1)(2l+5)} X_{l+1,m+3} \\
&- \frac{1}{2(2l+3)(2l+1)(2l+5)} X_{l+3,m+3}
\end{aligned} \tag{4.45}
$$

From Eqs. (4.27) and (4.43), we have

$$
\begin{aligned}
2i\alpha_1\alpha_2(\alpha_1 - i\alpha_2)X_{l,m}
&= -\frac{(l+m)(l+m-1)(l+m-2)(l+m-3)}{2(2l-1)(2l+1)} (\alpha_1 - i\alpha_2)X_{l-2,m-2} \\
&+ \frac{(l+m)(l+m-1)(l-m+2)(l-m+1)}{(2l-1)(2l+3)} (\alpha_1 - i\alpha_2)X_{l,m-2} \\
&- \frac{(l-m+2)(l-m+1)(l-m+4)(l-m+3)}{2(2l+1)(2l+3)} (\alpha_1 - i\alpha_2)X_{l+2,m-2} \\
&+ \frac{1}{2(2l-1)(2l+1)} (\alpha_1 - i\alpha_2)X_{l-2,m+2} - \frac{1}{(2l-1)(2l+3)} \\
&\cdot (\alpha_1 - i\alpha_2)X_{l,m+2} + \frac{1}{2(2l+1)(2l+3)} (\alpha_1 - i\alpha_2)X_{l+2,m+2} \\
&= -\frac{(l+m)(l+m-1)(l+m-2)(l+m-3)}{2(2l-3)(2l-1)(2l+1)}
\end{aligned}
$$

$$
\begin{aligned}
&\cdot [(l+m-4)(l+m-5)X_{l-3,m-3} - (l-m+2)(l-m+1)X_{l-1,m-3}] \\
&+ \frac{(l+m)(l+m-1)(l-m+2)(l-m+1)}{(2l-1)(2l+1)(2l+3)} \\
&\cdot [(l+m-2)(l+m-3)X_{l-1,m-3} - (l-m+4)(l-m+3)X_{l+1,m-3}] \\
&- \frac{(l-m+2)(l-m+1)(l-m+4)(l-m+3)}{2(2l+1)(2l+3)(2l+5)} \\
&\cdot [(l+m)(l+m-1)X_{l+1,m-3} - (l-m+6)(l-m+5)X_{l+3,m-3}] \\
&+ \frac{1}{2(2l-3)(2l-1)(2l+1)} [(l+m)(l+m-1)X_{l-3,m+1} \\
&- (l-m-2)(l-m-3)X_{l-1,m+1} - \frac{1}{(2l-1)(2l+1)(2l+3)} \\
&\cdot [(l+m+2)(l+m+1)X_{l-1,m+1} - (l-m)(l-m-1)X_{l+1,m+1}] \\
&+ \frac{1}{2(2l+1)(2l+3)(2l+5)} [(l+m+4)(l+m+3)X_{l+1,m+1} \\
&- (l-m+2)(l-m+1)X_{l+3,m+1}]
\end{aligned} \tag{4.46}
$$

Rearranging and simplifying gives

$$
\begin{aligned}
&2i\alpha_1\alpha_2(\alpha_1 - i\alpha_2)X_{l,m} \\
&\quad = -\frac{(l+m)(l+m-1)\ldots(l+m-5)}{2(2l-1)(2l+1)(2l-3)} \\
&\qquad \cdot X_{l-3,m-3} \\
&\qquad + \frac{3(l+m)(l+m-1)(l+m-2)(l+m-3)(l-m+2)(l-m+1)}{2(2l+1)(2l+3)(2l-3)} \\
&\qquad \cdot X_{l-1,m-3} \\
&\qquad - \frac{3(l-m+4)(l-m+3)(l-m+2)(l-m+1)(l+m)(l+m-1)}{2(2l-1)(2l+1)(2l+5)} X_{l+1,m-3} \\
&\qquad + \frac{(l-m+6)(l-m+5)(l-m+4)(l-m+3)(l-m+2)(l-m+1)}{2(2l+1)(2l+3)(2l+5)} \\
&\qquad \cdot X_{l+3,m-3} + \frac{(l+m)(l+m-1)}{2(2l-1)(2l-3)(2l+1)} X_{l-3,m+1} \\
&\qquad - \frac{3l^2 + l - 6 + m(2l+3) + 3m^2}{2(2l-3)(2l+1)(2l+3)} X_{l-1,m+1} \\
&\qquad + \frac{3l^2 + 5l - 4 - m(2l-1) + 3m^2}{2(2l-1)(2l+1)(2l+5)} X_{l+1,m+1} \\
&\qquad - \frac{(l-m+2)(l-m+1)}{2(2l+1)(2l+3)(2l+5)} X_{l+3,m+1}
\end{aligned} \tag{4.47}
$$

From Eqs. (4.21) and (4.42), we have

$$2\alpha_3^2(\alpha_1 + i\alpha_2)X_{l,m} = -\frac{2}{(2l+1)}\,\alpha_3^2 X_{l-1,m+1} + \frac{2}{(2l+1)}\,\alpha_3^2 X_{l+1,m+1}$$
$$= -\frac{2}{(2l+1)}\left[\frac{(l+m)(l+m-1)}{(2l-1)(2l-3)}X_{l-3,m+1} + \frac{2l(l-1)-3-4m-2m^2}{(2l-3)(2l+1)}X_{l-1,m+1} + \frac{(l-m)(l-m-1)}{(2l-1)(2l+1)}X_{l+1,m+1}\right] + \frac{2}{(2l+1)}$$
$$\cdot\left[\frac{(l+m+2)(l+m+1)}{(2l+3)(2l+1)}X_{l-1,m+1} + \frac{2(l+1)(l+2)-3-4m-2m^2}{(2l+1)(2l+5)}X_{l+1,m+1} + \frac{(l-m+2)(l-m+1)}{(2l+3)(2l+5)}X_{l+3,m+1}\right] \tag{4.48}$$

Rearranging and simplifying gives

$$2\alpha_3^2(\alpha_1 + i\alpha_2)X_{l,m} = -\frac{2(l+m)(l+m-1)}{(2l-3)(2l-1)(2l+1)}X_{l-3,m+1} + \frac{2(l-m+2)(l-m+1)}{(2l+1)(2l+3)(2l+5)}X_{l+3,m+1} - \frac{2[l(l-1)-3-m(2l+3)-3m^2]}{(2l-3)(2l+1)(2l+3)}X_{l-1,m+1} + \frac{2[l(l+3)-1+m(2l-1)-3m^2]}{(2l-1)(2l+1)(2l+5)}X_{l+1,m+1} \tag{4.49}$$

From Eqs. (4.21) and (4.43), we have

$$2\alpha_3^2(\alpha_1 - i\alpha_2)X_{l,m} = \frac{2(l+m)(l+m-1)}{(2l+1)}\,\alpha_3^2 X_{l-1,m-1} - \frac{2(l-m+2)(l-m+1)}{(2l+1)}\,\alpha_3^2 X_{l+1,m-1}$$

$$= \frac{2(l+m)(l+m-1)}{(2l+1)} \left[\frac{(l+m-2)(l+m-3)}{(2l-1)(2l-3)} X_{l-3,m-1} \right.$$
$$+ \frac{2l(l-1) - 2m^2 + 4m - 3}{(2l-3)(2l+1)} X_{l-1,m-1}$$
$$\left. + \frac{(l-m+2)(l-m+1)}{(2l+1)(2l-1)} X_{l+1,m-1} \right]$$
$$- \frac{2(l-m+2)(l-m+1)}{(2l+1)} \left[\frac{(l+m)(l+m-1)}{(2l+3)(2l+1)} \right.$$
$$\cdot X_{l-1,m-1} + \frac{2(l+1)(l+2) - 2m^2 + 4m - 3}{(2l+1)(2l+5)}$$
$$\left. \cdot X_{l+1,m-1} + \frac{(l-m+4)(l-m+3)}{(2l+3)(2l+5)} X_{l+3,m-1} \right] \tag{4.50}$$

Rearranging and simplifying gives

$$2\alpha_3^2(\alpha_1 - i\alpha_2)X_{l,m}$$
$$= \frac{2(l+m)(l+m-1)(l+m-2)(l+m-3)}{(2l-1)(2l+1)(2l-3)} X_{l-3,m-1}$$
$$+ \frac{2(l+m)(l+m-1)[l(l-1) - 3 + m(2l+3) - 3m^2]}{(2l-3)(2l+1)(2l+3)} X_{l-1,m-1}$$
$$- \frac{2(l-m+2)(l-m+1)[l(l+3) - 1 - m(2l-1) - 3m^2]}{(2l-1)(2l+1)(2l+5)} X_{l+1,m-1}$$
$$- \frac{2(l-m+4)(l-m+3)(l-m+2)(l-m+1)}{(2l+1)(2l+3)(2l+5)} X_{l+3,m-1} \tag{4.51}$$

From Eqs. (4.14), (4.45), and (4.51), we have

$$X_{l,m}\partial_{-1}V = \left(\frac{\partial V}{\partial \alpha_1} - i \frac{\partial V}{\partial \alpha_2} \right) X_{l,m}$$
$$= -2iK\alpha_1\alpha_2(\alpha_1 + i\alpha_2)X_{l,m} + 2K\alpha_3^2(\alpha_1 - i\alpha_2)X_{l,m}$$
$$= \frac{3K(l+m-3)(l+m-2)\ldots(l+m)}{2(2l-3)(2l-1)(2l+1)} X_{l-3,m-1}$$

$$+ \frac{K(l+m-1)(l+m)[7l^2 - 3l - 18 + 3m(2l+3) - 9m^2}{2(2l-3)(2l+1)(2l+3)}$$
$$\cdot X_{l-1,m-1}$$
$$- \frac{K(l-m+2)(l-m+1)[7l^2 + 17l - 8 - 3m(2l-1) - 9m^2]}{2(2l-1)(2l+1)(2l+5)}$$
$$\cdot X_{l+1,m-1} - \frac{3K(l-m+4)(l-m+3)(l-m+2)(l-m+1)}{2(2l+1)(2l+3)(2l+5)}$$
$$\cdot X_{l+3,m-1} + \frac{K}{2(2l-1)(2l+1)(2l-3)} X_{l-3,m+3}$$
$$- \frac{3K}{2(2l-3)(2l+1)(2l+3)} X_{l-1,m+3} + \frac{3K}{2(2l-1)(2l+1)(2l+5)}$$
$$\cdot X_{l+1,m+3} - \frac{K}{2(2l+3)(2l+1)(2l+5)} X_{l+3,m+3} \tag{4.52}$$

From Eqs. (4.14), (4.47), and (4.49), we have

$$\left(\frac{\partial V}{\partial \alpha_1} + i \frac{\partial V}{\partial \alpha_2} \right) X_{l,m}$$
$$= 2iK\alpha_1\alpha_2(\alpha_1 - i\alpha_2)X_{l,m} + 2K\alpha_3^2(\alpha_1 + i\alpha_2)X_{l,m}$$
$$= -\frac{K(l+m)(l+m-1)\ldots(l+m-5)}{2(2l-1)(2l+1)(2l-3)} X_{l-3,m-3}$$
$$+ \frac{K(l-m+1)(l-m+2)\ldots(l-m+6)}{2(2l+1)(2l+3)(2l+5)} X_{l+3,m-3}$$
$$+ \frac{3K(l+m)(l+m-1)(l+m-2)(l+m-3)(l-m+2)(l-m+1)}{2(2l+1)(2l+3)(2l-3)}$$
$$\cdot X_{l-1,m-3}$$
$$- \frac{3K(l-m+4)(l-m+3)(l-m+2)(l-m+1)(l+m)(l+m-1)}{2(2l-1)(2l+1)(2l+5)}$$
$$\cdot X_{l+1,m-3} - \frac{3K(l+m)(l+m-1)}{2(2l-3)(2l-1)(2l+1)} X_{l-3,m+1}$$
$$- \frac{K[7l^2 - 3l - 18 - 3m(2l+3) - 9m^2]}{2(2l-3)(2l+1)(2l+3)} X_{l-1,m+1}$$

$$+\frac{K[7l^2+17l-8+3m(2l-1)-9m^2]}{2(2l-1)(2l+1)(2l+5)}X_{l+1,m+1}$$

$$+\frac{3K(l-m+2)(l-m+1)}{2(2l+1)(2l+3)(2l+5)}X_{l+3,m+1} \qquad (4.53)$$

Equation (4.10), with the aid of Eqs. (4.41), (4.52), and (4.53), followed by further simplification, yields the recurrence relation for the gyroscopic operator

$$\begin{aligned}
&G_{FP}X_{l,m}\\
&=-\frac{aK(l-m+1)(l+m-6)(l+m-5)\dots(l+m)}{4(2l-1)(2l+1)(2l-3)}X_{l-3,m-4}\\
&+\frac{3aK(l-m+1)(l-m+2)(l-m+3)(l+m-4)(l+m-3)\dots(l+m)}{4(2l+1)(2l+3)(2l-3)}\\
&\cdot X_{l-1,m-4}\\
&-\frac{3aK(l+m-2)(l+m-1)(l+m)(l-m+1)(l-m+2)\dots(l-m+5)}{4(2l-1)(2l+1)(2l+5)}\\
&\cdot X_{l+1,m-4}\\
&+\frac{aK(l+m)(l-m+1)(l-m+2)\dots(l-m+7)}{4(2l+1)(2l+3)(2l+5)}X_{l+3,m-4}\\
&+\frac{7aKm(l+m-2)(l+m-1)(l+m)}{2(2l-3)(2l-1)(2l+1)}X_{l-3,m}\\
&+\frac{3aKm(l+m)(3l^2-5-7m^2)}{2(2l-3)(2l+1)(2l+3)}X_{l-1,m}\\
&+\frac{3aKm(l-m+1)(3l^2+6l-2-7m^2)}{2(2l-1)(2l+1)(2l+5)}X_{l+1,m}\\
&+\frac{7aKm(l-m+1)(l-m+2)(l-m+3)}{2(2l+1)(2l+3)(2l+5)}X_{l+3,m}\\
&+\frac{aK}{4(2l-1)(2l+1)(2l-3)}X_{l-3,m+4}-\frac{3aK}{4(2l-3)(2l+1)(2l+3)}\\
&\cdot X_{l-1,m+4}+\frac{3aK}{4(2l-1)(2l+1)(2l+5)}X_{l+1,m+4}\\
&-\frac{aK}{4(2l+3)(2l+1)(2l+5)}X_{l+3,m+4} \qquad (4.54)
\end{aligned}$$

The recurrence relations for the alignment operator Eq. (4.39) and the gyroscopic operator Eq. (4.54) yield the recurrence relation for the Fokker–Planck operator [cf. Eq. (2.75)]

$$\begin{aligned} L_{FP}X_{l,m} &= A_{FP}X_{l,m} + iG_{FP}X_{l,m} \\ &= \sum_{s=-4}^{4} \sum_{t=-1}^{1} g_{l+s,m+4t,l,m} X_{l+2s,m+4t} \end{aligned} \tag{4.55}$$

On shifting indices, the coefficients that arise from the alignment operator are

$$g_{l,m,l-4,m-4} = \frac{bK(l+1)}{4(2l-7)(2l-5)(2l-3)(2l-1)}$$

$$g_{l,m,l-2,m-4} = -\frac{bK(2l+9)}{4(2l-5)(2l-3)(2l-1)(2l+3)}$$

$$g_{l,m,l,m-4} = \frac{15bK}{4(2l-3)(2l-1)(2l+3)(2l+5)}$$

$$g_{l,m,l+2,m-4} = \frac{bK(2l-7)}{4(2l-1)(2l+3)(2l+5)(2l+7)}$$

$$g_{l,m,l+4,m-4} = -\frac{bKl}{4(2l+3)(2l+5)(2l+7)(2l+9)}$$

$$g_{l,m,l-4,m} = \frac{7bK(l+1)(l-m-3)(l-m-2)\ldots(l-m)}{2(2l-7)(2l-5)(2l-3)(2l-1)}$$

$$g_{l,m,l-2,m} = \frac{bK(2l+9)(l-m-1)(l-m)(l^2-l-2-7m^2)}{2(2l-5)(2l-3)(2l-1)(2l+3)}$$

$$\begin{aligned} g_{l,m,l,m} &= -l(l+1)b\beta^{-1} + bK \\ &\quad \cdot \frac{9l(l+2)(l^2-1) - 15m^2[6l(l+1) - 5 - 7m^2]}{2(2l-3)(2l-1)(2l+3)(2l+5)} \end{aligned}$$

$$g_{l,m,l+2,m} = -\frac{bK(2l-7)(l+m+1)(l+m+2)(l^2+3l-7m^2)}{2(2l-1)(2l+3)(2l+5)(2l+7)}$$

$$g_{l,m,l+4,m} = -\frac{7bKl(l+m+1)(l+m+2)\ldots(l+m+4)}{2(2l+3)(2l+5)(2l+7)(2l+9)}$$

$$g_{l,m,l-4,m+4} = \frac{bK(l+1)(l-m-7)(l-m-6)\ldots(l-m)}{4(2l-7)(2l-5)(2l-3)(2l-1)}$$

$$g_{l,m,l-2,m+4} = -\frac{bK(2l+9)(l+m+1)(l+m+2)(l-m-5)(l-m-4)\dots(l-m)}{4(2l-5)(2l-3)(2l-1)(2l+3)}$$

$$g_{l,m,l,m+4} = \frac{15bK(l+m+1)(l+m+2)\dots(l+m+4)(l-m-3)(l-m-2)\dots(l-m)}{4(2l-3)(2l-1)(2l+3)(2l+5)}$$

$$g_{l,m,l+2,m+4} = \frac{bK(2l-7)(l-m-1)(l-m)(l+m+1)(l+m+2)\dots(l+m+6)}{4(2l-1)(2l+3)(2l+5)(2l+7)}$$

$$g_{l,m,l+4,m+4} = -\frac{bKl(l+m+1)(l+m+2)\dots(l+m+8)}{4(2l+3)(2l+5)(2l+7)(2l+9)} \tag{4.56}$$

and the coefficients that arise from the gyroscopic operator are

$$g_{l,m,l-3,m-4} = -\frac{iaK}{4(2l-5)(2l-3)(2l-1)}$$

$$g_{l,m,l-1,m-4} = \frac{3iaK}{4(2l-3)(2l-1)(2l+3)}$$

$$g_{l,m,l+1,m-4} = -\frac{3iaK}{4(2l-1)(2l+3)(2l+5)}$$

$$g_{l,m,l+3,m-4} = \frac{iaK}{4(2l+3)(2l+5)(2l+7)}$$

$$g_{l,m,l-3,m} = \frac{7iaKm(l-m-2)(l-m-1)(l-m)}{2(2l-5)(2l-3)(2l-1)}$$

$$g_{l,m,l-1,m} = \frac{3iaKm(l-m)(3l^2-5-7m^2)}{2(2l-3)(2l-1)(2l+3)}$$

$$g_{l,m,l+1,m} = \frac{3iaKm(l+m+1)(3l^2+6l-2-7m^2)}{2(2l-1)(2l+3)(2l+5)}$$

$$g_{l,m,l+3,m} = \frac{7iaKm(l+m+1)(l+m+2)(l+m+3)}{2(2l+3)(2l+5)(2l+7)}$$

$$g_{l,m,l-3,m+4} = \frac{iaK(l+m+1)(l-m-6)(l-m-5)\dots(l-m)}{4(2l-5)(2l-3)(2l-1)}$$

$$g_{l,m,l-1,m+4} = -\frac{3iaK(l+m+1)(l+m+2)(l+m+3)(l-m-4)(l-m-3)\ldots(l-m)}{4(2l-3)(2l-1)(2l+3)}$$

$$g_{l,m,l+1,m+4} = \frac{3iaK(l-m-2)(l-m-1)(l-m)(l+m+1)(l+m+2)\ldots(l+m+5)}{4(2l-1)(2l+3)(2l+5)}$$

$$g_{l,m,l+3,m+4} = -\frac{iaK(l-m)(l+m+1)(l+m+2)\ldots(l+m+7)}{4(2l+3)(2l+5)(2l+7)} \qquad (4.57)$$

Expressions for the matrix elements are calculated from Eq. (2.76), namely

$$d_{l,m,l+s,m+4t} = \sqrt{\frac{(2l+2s+1)(l+m)!(l-m+s-4t)!}{(2l+1)(l+m+s+4t)!(l-m)!}}\, g^{*}_{l,m,l+s,m+4t} \qquad (4.58)$$

Application to Eq. (4.56) gives expressions for the matrix elements arising from the alignment operator

$$d_{l,m,l-4,m-4} = \frac{bK(l+1)}{4(2l-5)(2l-3)(2l-1)}\sqrt{\frac{(l+m-7)(l+m-6)\ldots(l+m)}{(2l-7)(2l+1)}}$$

$$d_{l,m,l-2,m-4} = -\frac{bK(2l+9)}{4(2l-5)(2l-1)(2l+3)} \cdot \sqrt{\frac{(l-m+1)(l-m+2)(l+m-5)(l+m-4)\ldots(l+m)}{(2l+1)(2l-3)}}$$

$$d_{l,m,l,m-4} = \frac{15bK\sqrt{(l+m-3)(l+m-2)\ldots(l+m)(l-m+1)(l-m+2)\ldots(l-m+4)}}{4(2l-3)(2l-1)(2l+3)(2l+5)}$$

$$d_{l,m,l+2,m-4}$$

$$= \frac{bK(2l-7)}{4(2l-1)(2l+3)(2l+7)} \cdot \sqrt{\frac{(l+m-1)(l+m)(l-m+1)(l-m+2)\ldots(l-m+6)}{(2l+1)(2l+5)}}$$

$$d_{l,m,l+4,m-4}$$

$$= -\frac{bKl}{4(2l+3)(2l+5)(2l+7)} \cdot \sqrt{\frac{(l-m+1)(l-m+2)\ldots(l-m+8)}{(2l+1)(2l+9)}}$$

$$d_{l,m,l-4,m}$$

$$= \frac{7bK(l+1)}{2(2l-5)(2l-3)(2l-1)} \times \sqrt{\frac{(l+m-3)(l+m-2)\ldots(l+m)(l-m-3)(l-m-2)\ldots(l-m)}{(2l-7)(2l+1)}}$$

$$d_{l,m,l-2,m}$$

$$= \frac{bK(2l+9)(l^2-l-2-7m^2)}{2(2l-5)(2l-1)(2l+3)} \cdot \sqrt{\frac{(l-m-1)(l-m)(l+m-1)(l+m)}{(2l-3)(2l+1)}}$$

$$d_{l,m,l,m}$$

$$= Q_{l,m} = -l(l+1)b\beta^{-1} + bK \cdot \frac{9l(l+2)(l^2-1) - 15m^2[6l(l+1)-5-7m^2]}{2(2l-3)(2l-1)(2l+3)(2l+5)}$$

$$d_{l,m,l+2,m}$$

$$= -\frac{bK(2l-7)(l^2+3l-7m^2)}{2(2l-1)(2l+3)(2l+7)} \cdot \sqrt{\frac{(l-m+1)(l-m+2)(l+m+1)(l+m+2)}{(2l+1)(2l+5)}}$$

$$d_{l,m,l+4,m} = -\frac{7bKl}{2(2l+3)(2l+5)(2l+7)} \times \sqrt{\frac{(l-m+1)(l-m+2)\dots(l-m+4)(l+m+1)(l+m+2)\dots(l+m+4)}{(2l+1)(2l+9)}}$$

$$d_{l,m,l-4,m+4} = \frac{bK(l+1)}{4(2l-5)(2l-3)(2l+1)} \sqrt{\frac{(l-m-7)(l-m-6)\dots(l-m)}{(2l-7)(2l+1)}}$$

$$d_{l,m,l-2,m+4} = -\frac{bK(2l+9)}{4(2l-5)(2l-1)(2l+3)} \cdot \sqrt{\frac{(l+m+1)(l+m+2)(l-m-5)(l-m-4)\dots(l-m)}{(2l-3)(2l+1)}}$$

$$d_{l,m,l,m+4} = \frac{15bK\sqrt{(l+m+1)(l+m+2)\dots(l+m+4)(l-m-3)(l-m-2)\dots(l-m)}}{4(2l-3)(2l-1)(2l+3)(2l+5)}$$

$$d_{l,m,l+2,m+4} = \frac{bK(2l-7)}{4(2l-1)(2l+3)(2l+7)} \cdot \sqrt{\frac{(l-m-1)(l-m)(l+m+1)(l+m+2)\dots(l+m+6)}{(2l+1)(2l+5)}}$$

$$d_{l,m,l+4,m+4} = -\frac{bKl}{4(2l+3)(2l+5)(2l+7)} \cdot \sqrt{\frac{(l+m+1)(l+m+2)\dots(l+m+8)}{(2l+1)(2l+9)}} \tag{4.59}$$

Application of Eq. (4.58) to Eq. (4.57) gives expressions for the matrix elements arising from the gyroscopic operator

$$d_{l,m,l-3,m-4} = \frac{aiK}{4(2l-3)(2l-1)} \cdot \sqrt{\frac{(l-m+1)(l+m-6)(l+m-5)\ldots(l+m)}{(2l-5)(2l+1)}}$$

$$d_{l,m,l-1,m-4} = -\frac{3iaK}{4(2l-3)(2l+3)} \times \sqrt{\frac{(l-m+1)(l-m+2)(l-m+3)(l+m-4)(l+m-3)\ldots(l+m)}{(2l-1)(2l+1)}}$$

$$d_{l,m,l+1,m-4} = \frac{3iaK}{4(2l-1)(2l+5)} \cdot \sqrt{\frac{(l+m-2)(l+m-1)(l+m)(l-m+1)(l-m+2)\ldots(l-m+5)}{(2l+1)(2l+3)}}$$

$$d_{l,m,l+3,m-4} = -\frac{iaK}{4(2l+3)(2l+5)} \cdot \sqrt{\frac{(l+m)(l-m+1)(l-m+2)\ldots(l-m+7)}{(2l+1)(2l+7)}}$$

$$d_{l,m,l-3,m} = -\frac{7iaKm}{2(2l-3)(2l-1)} \cdot \sqrt{\frac{(l-m-2)(l-m-1)(l-m)(l+m-2)(l+m-1)(l+m)}{(2l-5)(2l+1)}}$$

$$d_{l,m,l-1,m} = -\frac{3iaKm(3l^2-5-7m^2)}{2(2l-3)(2l+3)} \sqrt{\frac{(l-m)(l+m)}{(2l-1)(2l+1)}}$$

$$d_{l,m,l+1,m} = -\frac{3iaKm(3l^2+6l-2-7m^2)}{2(2l-1)(2l+5)} \sqrt{\frac{(l-m+1)(l+m+1)}{(2l+1)(2l+3)}}$$

$$d_{l,m,l+3,m} = -\frac{7iaKm}{2(2l+3)(2l+5)} \times \sqrt{\frac{(l-m+1)(l-m+2)(l-m+3)(l+m+1)(l+m+2)(l+m+3)}{(2l+1)(2l+7)}}$$

$$d_{l,m,l-3,m+4} = -\frac{iaK}{4(2l-3)(2l-1)} \cdot \sqrt{\frac{(l+m+1)(l-m-6)(l-m-5)\ldots(l-m)}{(2l-5)(2l+1)}}$$

$$d_{l,m,l-1,m+4} = \frac{3iaK}{4(2l-3)(2l+3)} \cdot \sqrt{\frac{(l+m+1)(l+m+2)(l+m+3)(l-m-4)(l-m-e)\ldots(l-m)}{(2l-1)(2l+1)}}$$

$$d_{l,m,l+1,m+4} = -\frac{3iaK}{4(2l-1)(2l+5)} \times \sqrt{\frac{(l-m-2)(l-m-1)(l-m)(l+m+1)(l+m+2)\ldots(l+m+5)}{(2l+1)(2l+3)}}$$

$$d_{l,m,l+3,m+4} = \frac{iaK}{4(2l+3)(2l+5)} \cdot \sqrt{\frac{(l-m)(l+m+1)(l+m+2)\ldots(l+m+7)}{(2l+1)(2l+7)}} \tag{4.60}$$

The differential recurrence relations for the Fourier coefficients are then [cf. Eq. (2.12)]

$$\dot{a}_{l,m} = \sum_{s=-4}^{4} \sum_{t=-1}^{1} d^*_{l,m,l+s,m+4t} a_{l+s,m+4t} \tag{4.61}$$

or explicitly

$$\dot{a}_{l,m} = -l(l+1)b\beta^{-1} a_{l,m} + bK \frac{9l(l+2)(l^2-1) - 15m^2[6l(l+1) - 5 - 7m^2]}{2(2l-3)(2l-1)(2l+3)(2l+5)} a_{l,m}$$
$$+ \frac{bK(l+1)}{4(2l-5)(2l-3)(2l-1)} \sqrt{\frac{(l+m-7)(l+m-6)\ldots(l+m)}{(2l-7)(2l+1)}} a_{l-4,m-4}$$
$$- \frac{bK(2l+9)}{4(2l-5)(2l-1)(2l+3)} \cdot \sqrt{\frac{(l-m+1)(l-m+2)(l+m-5)(l+m-4)\ldots(l+m)}{(2l+1)(2l-3)}} a_{l-2,m-4}$$

$$+\frac{15bK\sqrt{(l+m-3)(l+m-2)\dots(l+m)(l-m+1)(l-m+2)\dots(l-m+4)}}{4(2l-3)(2l-1)(2l+3)(2l+5)}a_{l,m-4}$$

$$+\frac{bK(2l-7)}{4(2l-1)(2l+3)(2l+7)}$$

$$\cdot\sqrt{\frac{(l+m-1)(l+m)(l-m+1)(l-m+2)\dots(l-m+6)}{(2l+1)(2l+5)}}\,a_{l+2,m-4}$$

$$-\frac{bKl}{4(2l+3)(2l+5)(2l+7)}$$

$$\cdot\sqrt{\frac{(l-m+1)(l-m+2)\dots(l-m+8)}{(2l+1)(2l+9)}}\,a_{l+4,m-4}$$

$$+\frac{7bK(l+1)}{2(2l-5)(2l-3)(2l-1)}$$

$$\times\sqrt{\frac{(l+m-3)(l+m-2)\dots(l+m)(l-m-3)(l-m-2)\dots(l-m)}{(2l-7)(2l+1)}}\,a_{l-4,m}$$

$$+\frac{bK(2l+9)(l^2-l-2-7m^2)}{2(2l-5)(2l-1)(2l+3)}$$

$$\cdot\sqrt{\frac{(l-m-1)(l-m)(l+m-1)(l+m)}{(2l-3)(2l+1)!}}\,a_{l-2,m}$$

$$-\frac{bK(2l-7)(l^2+3l-7m^2)}{2(2l-1)(2l+3)(2l+7)}$$

$$\cdot\sqrt{\frac{(l-m+1)(l-m+2)(l+m+1)(l+m+2)}{(2l+1)(2l+5)}}\,a_{l+2,m}$$

$$-\frac{7bKl}{2(2l+3)(2l+5)(2l+7)}$$

$$\times\sqrt{\frac{(l-m+1)(l-m+2)\dots(l-m+4)(l+m+1)(l+m+2)\dots(l+m+4)}{(2l+1)(2l+9)}}\,a_{l+4,m}$$

$$+\frac{bK(l+1)}{4(2l-5)(2l-3)(2l-1)}\sqrt{\frac{(l-m-7)(l-m-6)\dots(l-m)}{(2l-7)(2l+1)}}\,a_{l-4,m+4}$$

$$-\frac{bK(2l+9)}{4(2l-5)(2l-1)(2l+3)}$$

$$\cdot\sqrt{\frac{(l+m+1)(l+m+2)(l-m-5)(l-m-4)\dots(l-m)}{(2l-3)(2l+1)}}\,a_{l-2,m+4}$$

$$+ \frac{15bK\sqrt{(l+m+1)(l+m+2)\dots(l+m+4)(l-m-3)(l-m-2)\dots(l-m)}}{4(2l-3)(2l-1)(2l+3)(2l+5)} a_{l,m+4}$$

$$+ \frac{bK(2l-7)}{4(2l-1)(2l+3)(2l+7)}$$

$$\cdot \sqrt{\frac{(l-m-1)(l-m)(l+m+1)(l+m+2)\dots(l+m+6)}{(2l+1)(2l+5)}} a_{l+2,m+4}$$

$$- \frac{bKl}{4(2l+3)(2l+5)(2l+7)}$$

$$\cdot \sqrt{\frac{(l+m+1)(l+m+2)\dots(l+m+8)}{(2l+1)(2l+9)}} a_{l+4,m+4}$$

$$- \frac{aiK}{4(2l-3)(2l-1)}$$

$$\cdot \sqrt{\frac{(l-m+1)(l+m-6)(l+m-5)\dots(l+m)}{(2l-5)(2l+1)}} a_{l-3,m-4}$$

$$+ \frac{3iaK}{4(2l-3)(2l+3)}$$

$$\sqrt{\frac{(l-m+1)(l-m+2)(l-m+3)(l+m-4)(l+m-3)\dots(l+m)}{(2l-1)(2l+1)}} a_{l-1,m-4}$$

$$- \frac{3iaK}{4(2l-1)(2l+5)}$$

$$\cdot \sqrt{\frac{(l+m-2)(l+m-1)(l+m)(l-m+1)(l-m+2)\dots(l-m+5)}{(2l+1)(2l+3)}} a_{l+1,m-4}$$

$$+ \frac{iaK}{4(2l+3)(2l+5)} \sqrt{\frac{(l+m)(l-m+1)(l-m+2)\dots(l-m+7)}{(2l+1)(2l+7)}} a_{l+3,m-4}$$

$$+ \frac{7iaKm}{2(2l-3)(2l-1)}$$

$$\cdot \sqrt{\frac{(l-m-2)(l-m-1)(l-m)(l+m-2)(l+m-1)(l+m)}{(2l-5)(2l+1)}} a_{l-3,m}$$

$$+ \frac{3iaKm(3l^2-5-7m^2)}{2(2l-3)(2l+3)} \sqrt{\frac{(l-m)(l+m)}{(2l-1)(2l+1)}} a_{l-1,m}$$

$$+ \frac{3iaKm(3l^2+6l-2-7m^2)}{2(2l-1)(2l+5)} \sqrt{\frac{(l-m+1)(l+m+1)}{(2l+1)(2l+3)}} a_{l+1,m}$$

$$+\frac{7iaKm}{2(2l+3)(2l+5)}$$

$$\cdot\sqrt{\frac{(l-m+1)(l-m+2)(l-m+3)(l+m+1)(l+m+2)(l+m+3)}{(2l+1)(2l+7)}}\,a_{l+3,m}$$

$$+\frac{iaK}{4(2l-3)(2l-1)}\sqrt{\frac{(l+m+1)(l-m-6)(l-m-5)\ldots(l-m)}{(2l-5)(2l+1)}}\,a_{l-3,m+4}$$

$$-\frac{3iaK}{4(2l-3)(2l+3)}$$

$$\cdot\sqrt{\frac{(l+m+1)(l+m+2)(l+m+3)(l-m-4)(l-m-3)\ldots(l-m)}{(2l-1)(2l+1)}}\,a_{l-1,m+4}$$

$$+\frac{3iaK}{4(2l-1)(2l+5)}$$

$$\cdot\sqrt{\frac{(l-m-2)(l-m-1)(l-m)(l+m+1)(l+m+2)\ldots(l+m+5)}{(2l+1)(2l+3)}}\,a_{l+1,m+4}$$

$$-\frac{iaK}{4(2l+3)(2l+5)}\sqrt{\frac{(l-m)(l+m+1)(l+m+2)\ldots(l+m+7)}{(2l+1)(2l+7)}}\,a_{l+3,m+4}$$

(4.62)

Expressions for the differential recurrence coefficients are calculated from Eq. (2.77), namely,

$$e_{l,m,l+s,m+4t}=\frac{(2l+2s+1)(l+m)!(l-m+s-4t)!}{(2l+1)(l+m+s+4t)!(l-m)!}\,g^{*}_{l,m,l+s,m+4t} \quad (4.63)$$

Application to Eq. (4.56) gives expressions for the differential recurrence coefficients arising from the alignment operator

$$e_{l,m,l-4,m-4}=\frac{Kb(l+1)(l+m-7)(l+m-6)\ldots(l+m)}{4(2l-5)(2l-3)(2l-1)(2l+1)}$$

$$e_{l,m,l-2,m-4}=-\frac{Kb(2l+9)(l-m+1)(l-m+2)(l+m-5)(l+m-4)\ldots(l+m)}{4(2l-5)(2l-1)(2l+1)(2l+3)}$$

$$e_{l,m,l,m-4}=\frac{15bK(l+m-3)(l+m-2)\ldots(l+m)(l-m+1)(l-m+2)\ldots(l-m+4)}{4(2l-3)(2l-1)(2l+3)(2l+5)}$$

$$e_{l,m,l+2,m-4}=\frac{bK(2l-7)(l+m-1)(l+m)(l-m+1)(l-m+2)\ldots(l-m+6)}{4(2l-1)(2l+1)(2l+3)(2l+7)}$$

$$e_{l,m,l+4,m-4} = -\frac{bKl(l-m+1)(l-m+2)\ldots(l-m+8)}{4(2l+1)(2l+3)(2l+5)(2l+7)}$$

$$e_{l,m,l-4,m} = \frac{7bK(l+1)(l+m-3)(l+m-2)\ldots(l+m)}{2(2l-5)(2l-3)(2l-1)(2l+1)}$$

$$e_{l,m,l-2,m} = \frac{bK(2l+9)(l+m-1)(l+m)(l^2-l-2-7m^2)}{2(2l-5)(2l-1)(2l+1)(2l+3)}$$

$$e_{l,m,l,m} = Q_{l,m} = -l(l+1)b\beta^{-1} + bK\frac{9l(l+2)(l^2-1)-15m^2[6l(l+1)-5-7m^2]}{2(2l-3)(2l-1)(2l+3)(2l+5)}$$

$$e_{l,m,l+2,m} = -\frac{bK(2l-7)(l-m+1)(l-m+2)(l^2+3l-7m^2)}{2(2l-1)(2l+1)(2l+3)(2l+7)}$$

$$e_{l,m,l+4,m} = -\frac{7bKl(l-m+1)(l-m+2)\ldots(l-m+4)}{2(2l+1)(2l+3)(2l+5)(2l+7)}$$

$$e_{l,m,l-4,m+4} = \frac{bK(l+1)}{4(2l-5)(2l-3)(2l-1)(2l+1)}$$

$$e_{l,m,l-2,m+4} = -\frac{bK(2l+9)}{4(2l-5)(2l-1)(2l+1)(2l+3)}$$

$$e_{l,m,l,m+4} = \frac{15bK}{4(2l-3)(2l-1)(2l+3)(2l+5)}$$

$$e_{l,m,l+2,m+4} = \frac{bK(2l-7)}{4(2l-1)(2l+1)(2l+3)(2l+7)}$$

$$e_{l,m,l+4,m+4} = -\frac{bKl}{4(2l+1)(2l+3)(2l+5)(2l+7)} \tag{4.64}$$

Application of Eq. (4.63) to Eq. (4.57) gives expressions for the differential recurrence coefficients arising from the gyroscopic operator

$$e_{l,m,l-3,m-4} = \frac{aiK(l-m+1)(l+m-6)(l+m-5)\ldots(l+m)}{4(2l-3)(2l-1)(2l+1)}$$

$$e_{l,m,l-1,m-4} = -\frac{3iaK(l-m+1)(l-m+2)(l-m+3)(l+m-4)(l+m-3)\ldots(l+m)}{4(2l-3)(2l+1)(2l+3)}$$

$$e_{l,m,l+1,m-4} = \frac{3iaK(l+m-2)(l+m-1)(l+m)(l-m+1)(l-m+2)\ldots(l-m+5)}{4(2l-1)(2l+1)(2l+5)}$$

$$e_{l,m,l+3,m-4} = -\frac{iaK(l+m)(l-m+1)(l-m+2)\ldots(l-m+7)}{4(2l+1)(2l+3)(2l+5)}$$

$$e_{l,m,l-3,m} = -\frac{7iaKm(l+m-2)(l+m-1)(l+m)}{2(2l-3)(2l-1)(2l+1)}$$

$$e_{l,m,l-1,m} = -\frac{3iaKm(l+m)(3l^2-5-7m^2)}{2(2l-3)(2l+1)(2l+3)}$$

$$e_{l,m,l+1,m} = -\frac{3iaKm(l-m+1)(3l^2+6l-2-7m^2)}{2(2l-1)(2l+1)(2l+5)}$$

$$e_{l,m,l+3,m} = -\frac{7iaKm(l-m+1)(l-m+2)(l-m+3)}{2(2l+1)(2l+3)(2l+5)}$$

$$e_{l,m,l-3,m+4} = -\frac{iaK}{4(2l-3)(2l-1)(2l+1)}$$

$$e_{l,m,l-1,m+4} = \frac{3iaK}{4(2l-3)(2l+1)(2l+3)}$$

$$e_{l,m,l+1,m+4} = -\frac{3iaK}{4(2l-1)(2l+1)(2l+5)}$$

$$e_{l,m,l+3,m+4} = \frac{iaK}{4(2l+1)(2l+3)(2l+5)} \tag{4.65}$$

The differential recurrence relations for the expectation values $x_{l,m} = \langle X_{l,m} \rangle$ [cf. Eq. (2.14)], hence the decay modes [cf. Eq. (2.17)] are

$$\dot{x}_{l,m} = \sum_{s=-4}^{4} \sum_{t=-1}^{1} e_{l,m,l+s,m+4t} x_{l+s,m+4t} \tag{4.66}$$

or explicitly from Eqs. (4.64) and (4.65)

$$\begin{aligned}
\dot{x}_{l,m} &= b\left\{-l(l+1)\beta^{-1} + K\,\frac{9l(l+2)(l^2-1) - 15m^2[6l(l+1)-5-7m^2]}{2(2l-3)(2l-1)(2l+3)(2l+5)}\right\} x_{l,m} \\
&+ \frac{Kb(l+1)(l+m-7)(l+m-6)\ldots(l+m)}{4(2l-5)(2l-3)(2l-1)(2l+1)} x_{l-4,m-4} \\
&- \frac{Kb(2l+9)(l-m+1)(l-m+2)(l+m-5)(l+m-4)\ldots(l+m)}{4(2l-5)(2l-1)(2l+1)(2l+3)} x_{l-2,m-4} \\
&+ \frac{15bK(l+m-3)(l+m-2)\ldots(l+m)(l-m+1)(l-m+2)\ldots(l-m+4)}{4(2l-3)(2l-1)(2l+3)(2l+5)} x_{l,m-4} \\
&+ \frac{bK(2l-7)(l+m-1)(l+m)(l-m+1)(l-m+2)\ldots(l-m+6)}{4(2l-1)(2l+1)(2l+3)(2l+7)} x_{l+2,m-4}
\end{aligned}$$

$$
\begin{aligned}
&- \frac{bKl(l-m+1)(l-m+2)\dots(l-m+8)}{4(2l+1)(2l+3)(2l+5)(2l+7)} x_{l+4,m-4} \\
&+ \frac{7bK(l+1)(l+m-3)(l+m-2)\dots(l+m)}{2(2l-5)(2l-3)(2l-1)(2l+1)} x_{l-4,m} \\
&+ \frac{bK(2l+9)(l+m-1)(l+m)(l^2-l-2-7m^2)}{2(2l-5)(2l-1)(2l+1)(2l+3)} x_{l-2,m} \\
&- \frac{bK(2l-7)(l-m+1)(l-m+2)(l^2+3l-7m^2)}{2(2l-1)(2l+1)(2l+3)(2l+7)} x_{l+2,m} \\
&- \frac{7bKl(l-m+1)(l-m+2)\dots(l-m+4)}{2(2l+1)(2l+3)(2l+5)(2l+7)} x_{l+4,m} \\
&+ \frac{bK(l+1)}{4(2l-5)(2l-3)(2l-1)(2l+1)} x_{l-4,m+4} \\
&- \frac{bK(2l+9)}{4(2l-5)(2l-1)(2l+1)(2l+3)} x_{l-2,m+4} \\
&+ \frac{15bK}{4(2l-3)(2l-1)(2l+3)(2l+5)} x_{l,m+4} \\
&+ \frac{bK(2l-7)}{4(2l-1)(2l+1)(2l+3)(2l+7)} x_{l+2,m+4} \\
&- \frac{bKl}{4(2l+1)(2l+3)(2l+5)(2l+7)} x_{l+4,m+4} \\
&+ \frac{iaK(l-m+1)(l+m-6)(l+m-5)\dots(l+m)}{4(2l-3)(2l-1)(2l+1)} x_{l-3,m-4} \\
&- \frac{3iaK(l-m+1)(l-m+2)(l-m+3)(l+m-4)(l+m-3)\dots(l+m)}{4(2l-3)(2l+1)(2l+3)} x_{l-1,m-4} \\
&+ \frac{3iaK(l+m-2)(l+m-1)(l+m)(l-m+1)(l-m+2)\dots(l-m+5)}{4(2l-1)(2l+1)(2l+5)} x_{l+1,m-4} \\
&- \frac{iaK(l+m)(l-m+1)(l-m+2)\dots(l-m+7)}{4(2l+1)(2l+3)(2l+5)} x_{l+3,m-4} \\
&- \frac{7iaKm(l+m-2)(l+m-1)(l+m)}{2(2l-3)(2l-1)(2l+1)} x_{l-3,m} \\
&- \frac{3iaKm(l+m)(3l^2-5-7m^2)}{2(2l-3)(2l+1)(2l+3)} x_{l-1,m}
\end{aligned}
$$

$$
\begin{aligned}
&- \frac{3iaKm(l-m+1)(3l^2+6l-2-7m^2)}{2(2l-1)(2l+1)(2l+5)} x_{l+1,m} \\
&- \frac{7iaKm(l-m+1)(l-m+2)(l-m+3)}{2(2l+1)(2l+3)(2l+5)} x_{l+3,m} \\
&- \frac{iaK}{4(2l-3)(2l-1)(2l+1)} x_{l-3,m+4} \\
&+ \frac{3iaK}{4(2l-3)(2l+1)(2l+3)} x_{l-1,m+4} \\
&- \frac{3iaK}{4(2l-1)(2l+1)(2l+5)} x_{l+1,m+4} \\
&+ \frac{iaK}{4(2l+1)(2l+3)(2l+5)} x_{l+3,m+4}
\end{aligned} \tag{4.67}
$$

B. Calculation of the Matrix Elements and Differential Recurrence Coefficients Using the Product Formula for Spherical Harmonics

To obtain compact expressions for the matrix elements, we express the $\partial_i V$ in Eqs. (4.13) and (4.14) in terms of spherical harmonics [Appendix A] in a manner similar to the way in which Eq. (4.6) may be derived from Eq. (4.1). The free energy per unit volume V in Eqs. (4.6) and the $\partial_i V$ in Eq. (4.10) are then replaced by their corresponding spherical harmonic expansions. The ensuing products are then expanded using the *product formula for spherical harmonics* [34] in Eq. (A.22), namely

$$
X_{s,t} X_{l,m} = \sum_{r=-s}^{s} \begin{bmatrix} s & l & l+r \\ t & m & -m-t \end{bmatrix} X_{l+r,m+t} \tag{4.68}
$$

where the product coefficients are expressed in terms of the *Wigner 3j-symbols* [34] by

$$
\begin{bmatrix} s & l & l+r \\ t & m & -m-t \end{bmatrix} = (-1)^m \frac{2l+2r+1}{4\pi} \sqrt{\frac{(l+r-m-t)!(l+m)!(s+t)!}{(l+r+m+t)!(l-m)!(s-t)!}}
\cdot \begin{pmatrix} s & l & l+r \\ 0 & 0 & 0 \end{pmatrix} \begin{pmatrix} s & l & l+r \\ t & m & -m-t \end{pmatrix} \tag{4.69}
$$

The *vector addition rules* imply that unless

$$
r \equiv s (\text{mod } 2) \tag{4.70}
$$

the leading $3j$-symbol in Eq. (4.69) vanishes, so that the summation in Eq. (4.68) is taken over values of r for which Eq. (4.70) is satisfied.

The free energy per unit volume arising from for cubic anisotropy is given by Eq. (4.6), namely,

$$V = \frac{K}{5}\left[1 - X_{4,0} - \frac{1}{336}\,(X_{4,4} + 8!X_{4,-4})\right] \tag{4.71}$$

The $\partial_i V$ are given by Eqs. (4.13) and (4.14), namely,

$$\partial_0 V = \frac{\partial V}{\partial \alpha_3} = 2K(\alpha_3 - \alpha_3^3) \tag{4.72}$$

$$\partial_{\pm 1} V = \frac{\partial V}{\partial \alpha_1} \pm i\,\frac{\partial V}{\partial \alpha_2} = \pm 2iK\alpha_1\alpha_2(\alpha_1 \mp i\alpha_2) + 2K\alpha_3^2(\alpha_1 \pm i\alpha_2) \tag{4.73}$$

We require expressions for Eqs. (4.72) and (4.73) in terms of spherical harmonics as in Eq. (4.71). From the definitions of the direction cosines in Eq. (2.72), we have

$$\alpha_1\alpha_2 = \sin^2\theta\sin\varphi\cos\varphi \qquad \alpha_3 = \cos\theta \qquad \alpha_1 \pm i\alpha_2 = e^{\pm i\varphi}\sin\theta \tag{4.74}$$

Eqs. (4.72) and (4.73) are thus expressed in spherical polar coordinates by

$$\partial_0 V = 2K(\cos\theta - \cos^3\theta) \tag{4.75}$$

$$\partial_{\pm 1} V = \pm 2iK\sin^3\theta\sin\varphi\cos\varphi\, e^{\mp i\varphi} + 2K\cos^2\theta\sin\theta\, e^{\pm i\varphi} \tag{4.76}$$

Using $\sin\varphi\cos\varphi = (1/4i)(e^{2i\varphi} - e^{-2i\varphi})$, we obtain the expansion

$$\pm 2iK\sin^3\theta\sin\varphi\cos\varphi\, e^{\mp i\varphi} = -\frac{K}{2}\,[\sin^3\theta e^{\mp 3i\varphi} + \sin\theta(\cos^2\theta - 1)e^{\pm i\varphi}] \tag{4.77}$$

Equation (4.76) with the aid of Eq. (4.77) becomes

$$\partial_{\pm 1} V = \frac{K}{2}\,(3\cos^2\theta + 1)\sin\theta e^{\pm i\varphi} - \frac{K}{2}\,\sin^3\theta e^{\mp 3i\varphi} \tag{4.78}$$

Using the following expressions for the associated Legendre functions from

Eq. (A.11),

$$P_1^0 = \cos\theta \qquad P_1^1 = \sin\theta \qquad P_3^0 = \tfrac{1}{2}(5\cos^3\theta - 3\cos\theta)$$
$$P_3^1 = \tfrac{1}{2}\sin\theta(15\cos^2\theta - 3) \qquad P_3^3 = 15\sin^3\theta \tag{4.79}$$

Eqs. (4.75) and (4.78) become

$$\partial_0 V = \frac{4K}{5}\,(P_1^0 - P_3^0) \tag{4.80}$$

$$\partial_{\pm 1} V = \frac{K}{5}\,(P_3^1 + 4P_1^1)e^{\pm i\varphi} - \frac{K}{30}\,P_3^3 e^{\mp 3i\varphi} \tag{4.81}$$

The symmetry relation Eq. (A.2) gives

$$P_1^1 = -2P_1^{-1} \qquad P_3^1 = -12P_3^{-1} \qquad P_3^3 = -6!P_3^{-3} \tag{4.82}$$

Equations (4.80) and (4.81), with the aid of Eq. (4.82), yield the spherical harmonic expansions

$$\partial_1 V = \frac{K}{5}\,(X_{3,1} + 4X_{1,1} + 5!X_{3,-3}) \tag{4.83}$$

$$\partial_0 V = \frac{4K}{5}\,(X_{1,0} - X_{3,0}) \tag{4.84}$$

$$\partial_{-1} V = -\frac{4K}{5}\left(3X_{3,-1} + 2X_{1,-1} + \frac{1}{24}\,X_{3,3}\right) \tag{4.85}$$

The calculation of the recurrence relation for

$$L_{FP}X_{l,m} = A_{FP}X_{l,m} + iG_{FP}X_{l,m} \tag{4.86}$$

will be accomplished using the alignment representation formula Eq. (4.9), namely,

$$A_{FP}X_{l,m} = b\left[2K - l(l+1)\beta^{-1} + \frac{l(l+1) - 20}{2}\,V\right]X_{l,m} + \frac{b}{2}\,\Lambda^2 V X_{l,m} \tag{4.87}$$

and the gyroscopic representation formula Eq. (4.10), namely,

$$G_{FP}X_{l,m} = \frac{a(l+m)(l-m+1)}{2} X_{l,m-1}\partial_1 V - amX_{l,m}\partial_0 V + \frac{a}{2} X_{l,m+1}\partial_{-1} V \tag{4.88}$$

Equation (4.87), with the aid of Eq. (4.71), becomes

$$\begin{aligned} A_{FP}X_{l,m} = &-l(l+1)b\beta^{-1}X_{l,m} - \frac{bK}{10}\{[l(l+1)-20]X_{4,0}X_{l,m} + \Lambda^2 X_{4,0}X_{l,m}\} \\ &- \frac{bK}{3360}\{[l(l+1)-20]X_{4,4}X_{l,m} + 8![l(l+1)-20]X_{4,-4}X_{l,m} \\ &+ \Lambda^2 X_{4,4}X_{l,m} + 8!\Lambda^2 X_{4,-4}X_{l,m}\} \end{aligned} \tag{4.89}$$

Equation (4.88), with the aid of Eqs. (4.83–4.85) becomes

$$\begin{aligned} G_{FP}X_{l,m} = &\frac{aK(l+m)(l-m+1)}{10} \\ &\cdot (X_{3,1}X_{l,m-1} + 4X_{l,1}X_{l,m-1} + 5!X_{3,-3}X_{l,m-1}) \\ &- \frac{4aKm}{5}(X_{1,0}X_{l,m} - X_{3,0}X_{l,m}) \\ &- \frac{2aK}{5}\left(3X_{3,-1}X_{l,m+1} + 2X_{1,-1}X_{l,m+1} + \frac{1}{24}X_{3,3}X_{l,m+1}\right) \end{aligned} \tag{4.90}$$

From Eq. (4.68), the products are

$$X_{1,t}X_{l,m} = \sum_{r=-1}^{0} \begin{bmatrix} 1 & l & l+2r+1 \\ t & m & -m-t \end{bmatrix} X_{l+2r+1,m+t} \tag{4.91}$$

$$X_{3,t}X_{l,m} = \sum_{r=-2}^{1} \begin{bmatrix} 3 & l & l+2r+1 \\ t & m & -m-t \end{bmatrix} X_{l+2r+1,m+t} \tag{4.92}$$

$$X_{4,t}X_{l,m} = \sum_{r=-2}^{2} \begin{bmatrix} 4 & l & l+2r \\ t & m & -m-t \end{bmatrix} X_{l+2r,m+t} \tag{4.93}$$

Equation (4.89), with the aid of Eq. (4.93), becomes

$$A_{FP}X_{l,m} = -l(l+1)\beta^{-1}bX_{l,m} + \frac{bK}{10}\sum_{r=-2}^{2}$$

$$\cdot\,[20 - l(l+1) + (l+2r)(l+2r+1)]\left\{\begin{bmatrix} 4 & l & l+2r \\ 0 & m & -m \end{bmatrix} X_{l+2r,m}\right.$$

$$+ \frac{1}{336}\left(\begin{bmatrix} 4 & l & l+2r \\ 4 & m & -m-4 \end{bmatrix} X_{l+2r,m+4}\right.$$

$$\left.\left. + 8!\begin{bmatrix} 4 & l & l+2r \\ -4 & m & -m+4 \end{bmatrix} X_{l+2r,m-4}\right)\right\} \tag{4.94}$$

Equation (4.90), with the aid of Eqs. (4.91) and (4.92), becomes

$$G_{FP}X_{l,m} = \frac{aK(l+m)(l-m+1)}{10}\sum_{r=-2}^{1}$$

$$\cdot\left(\begin{bmatrix} 3 & l & l+2r+1 \\ 1 & m-1 & -m \end{bmatrix} X_{l+2r+1,m}\right.$$

$$+ 4\begin{bmatrix} 1 & l & l+2r+1 \\ 1 & m-1 & -m \end{bmatrix} X_{l+2r+1,m}$$

$$\left. + 5!\begin{bmatrix} 3 & l & l+2r+1 \\ -3 & m-1 & -m+4 \end{bmatrix} X_{l+2r+1,m-4}\right)$$

$$- \frac{4aKm}{5}\sum_{r=-2}^{1}\left(\begin{bmatrix} 1 & l & l+2r+1 \\ 0 & m & -m \end{bmatrix} X_{l+2r+1,m}\right.$$

$$\left. - \begin{bmatrix} 3 & l & l+2r+1 \\ 0 & m & -m \end{bmatrix} X_{l+2r+1,m}\right) - \frac{2aK}{5}\sum_{r=-2}^{1}$$

$$\cdot\left(3\begin{bmatrix} 3 & l & l+2r+1 \\ -1 & m+1 & -m \end{bmatrix} X_{l+2r+1,m}\right.$$

$$+ 2\begin{bmatrix} 1 & l & l+2r+1 \\ -1 & m+1 & -m \end{bmatrix} X_{l+2r+1,m}$$

$$+\frac{1}{24}\begin{bmatrix}3 & l & l+2r+1\\ 3 & m+1 & -m-4\end{bmatrix}X_{l+2r+1,m+4}\Bigg)$$

$$= aK\sum_{r=-2}^{1}\{12(l+m)(l-m+1)\begin{bmatrix}3 & l & l+2r+1\\ -3 & m-1 & -m+4\end{bmatrix}$$

$$\cdot X_{l+2r+1,m-4}+\Bigg[\frac{(l+m)(l-m+1)}{10}$$

$$\cdot\left(\begin{bmatrix}3 & l & l+2r+1\\ 1 & m-1 & -m\end{bmatrix}+4\begin{bmatrix}1 & l & l+2r+1\\ 1 & m-1 & -m\end{bmatrix}\right)$$

$$+\frac{4m}{5}\left(-\begin{bmatrix}1 & l & l+2r+1\\ 0 & m & -m\end{bmatrix}+\begin{bmatrix}3 & l & l+2r+1\\ 0 & m & -m\end{bmatrix}\right)$$

$$-\frac{2}{5}\Bigg(3\begin{bmatrix}3 & l & l+2r+1\\ -1 & m+1 & -m\end{bmatrix}$$

$$+2\begin{bmatrix}1 & l & l+2r+1\\ -1 & m+1 & -m\end{bmatrix}\Bigg)\Bigg]X_{l+2r+1,m}$$

$$-\frac{1}{60}\begin{bmatrix}3 & l & l+2r+1\\ 3 & m+1 & -m-4\end{bmatrix}X_{l+2r+1,m+4}\Bigg\} \tag{4.95}$$

Equation (4.86), with the aid of Eqs. (4.94) and (4.95), becomes [cf. Eq. (2.75)]

$$L_{FP}X_{l,m}=\sum_{s=-4}^{4}\sum_{t=-1}^{1}g_{l+s,m+4t,l,m}X_{l+s,m+4t} \tag{4.96}$$

where the coefficients arising from the alignment operator are

$$g_{l+2r,m-4,l,m}=12bK[20-l(l+1)+(l+2r)(l+2r+1)]$$

$$\cdot\begin{bmatrix}4 & l & l+2r\\ -4 & m & -m+4\end{bmatrix}$$

$$g_{l+2r,m,l,m}=-l(l+1)\beta^{-1}b\delta_{r,0}+\frac{bK}{10}$$

$$\cdot[20-l(l+1)+(l+2r)(l+2r+1)]$$

$$\cdot\begin{bmatrix}4 & l & l+2r\\ 0 & m & -m\end{bmatrix}$$

$$g_{l+2r,m+4,l,m} = \frac{bK}{3360}\,[20 - l(l+1) + (l+2r)(l+2r+1)\,\cdot \begin{bmatrix} 4 & l & l+2r \\ 4 & m & -m-4 \end{bmatrix} \tag{4.97}$$

and the coefficients arising from the gyroscopic operator are

$$\begin{aligned}
g_{l+2r+1,m-4,l,m} &= 12iaK(l+m)(l-m+1) \cdot \begin{bmatrix} 3 & l & l+2r+1 \\ -3 & m-1 & -m+4 \end{bmatrix} \\
g_{l+2r+1,m,l,m} &= iaK\Bigg\{ \frac{(l+m)(l-m+1)}{10} \cdot \left(\begin{bmatrix} 3 & l & l+2r+1 \\ 1 & m-1 & -m \end{bmatrix} + 4 \begin{bmatrix} 1 & l & l+2r+1 \\ 1 & m-1 & -m \end{bmatrix} \right) \\
&\quad + \frac{4m}{5}\left(-\begin{bmatrix} 1 & l & l+2r+1 \\ 0 & m & -m \end{bmatrix} + \begin{bmatrix} 3 & l & l+2r+1 \\ 0 & m & -m \end{bmatrix} \right) \\
&\quad - \frac{2}{5}\left(3\begin{bmatrix} 3 & l & l+2r+1 \\ -1 & m+1 & -m \end{bmatrix} + 2\begin{bmatrix} 1 & l & l+2r+1 \\ -1 & m+1 & -m \end{bmatrix} \right) \Bigg\} \\
g_{l+2r+1,m+4,l,m} &= -\frac{iaK}{60} \begin{bmatrix} 3 & l & l+2r+1 \\ 3 & m+1 & -m-4 \end{bmatrix}
\end{aligned} \tag{4.98}$$

hence on shifting indices, we have

$$\begin{aligned}
g_{l,m,l+2r,m+4} &= 12bK[20 + l(l+1) - (l+2r)(l+2r+1)] \cdot \begin{bmatrix} 4 & l+2r & l \\ -4 & m+4 & -m \end{bmatrix} \\
g_{l,m,l+2r,m} &= -l(l+1)\beta^{-1} b\delta_{r,0} \\
&\quad + \frac{bK}{10}\,[20 + l(l+1) - (l+2r)(l+2r+1)] \cdot \begin{bmatrix} 4 & l+2r & l \\ 0 & m & -m \end{bmatrix}
\end{aligned}$$

$$g_{l,m,l+2r,m-4} = \frac{bK}{3360}[20 + l(l+1) - (l+2r)(l+2r+1)] \cdot \begin{bmatrix} 4 & l+2r & l \\ 4 & m-4 & -m \end{bmatrix}$$

$$g_{l,m,l+2r+1,m+4} = 12iaK(l+m+2r+5)(l-m+2r-2) \cdot \begin{bmatrix} 3 & l+2r+1 & ll \\ -3 & m+3 & -m \end{bmatrix}$$

$$g_{l,m,l+2r+1,m} = iaK\left\{\frac{(l+m+2r+1)(l-m+2r+2)}{10} \cdot \left(\begin{bmatrix} 3 & l+2r+1 & l \\ 1 & m-1 & -m \end{bmatrix} + 4\begin{bmatrix} 1 & l+2r+1 & l \\ 1 & m-1 & -m \end{bmatrix}\right)\right.$$
$$+ \frac{4m}{5}\left(-\begin{bmatrix} 1 & l+2r+1 & l \\ 0 & m & -m \end{bmatrix} + \begin{bmatrix} 3 & l+2r+1 & l \\ 0 & m & -m \end{bmatrix}\right.$$
$$- \frac{2}{5}\left(3\begin{bmatrix} 3 & l+2r+1 & l \\ -1 & m+1 & -m \end{bmatrix}\right.$$
$$\left.\left.\left. + 2\begin{bmatrix} 1 & l+2r+1 & l \\ -1 & m+1 & -m \end{bmatrix}\right)\right)\right\}$$

$$g_{l,m,l+2r+1,m-4} = -\frac{iaK}{60}\begin{bmatrix} 3 & l+2r+1 & l \\ 3 & m-3 & -m \end{bmatrix} \tag{4.99}$$

The product coefficients in Eq. (4.69) on shifting indices becomes

$$\begin{bmatrix} s & l+r & l \\ -t & m+t & -m \end{bmatrix} = (-1)^{m+t}\frac{2l+1}{4\pi}\sqrt{\frac{(l-m)!(l+r+m+t)!(s-t)!}{(l+m)!(l+r-m+t)!(s+t)!}} \cdot \begin{pmatrix} s & l+r & l \\ 0 & 0 & 0 \end{pmatrix}\begin{pmatrix} s & l+r & l \\ -t & m+t & -m \end{pmatrix} \tag{4.100}$$

The product coefficients in Eq. (4.99) are then calculated using Eq. (4.100). Expressions for the nonzero matrix elements $d_{l,m,l+s,m+t}$ and the nonzero differential recurrence coefficients $e_{l,m,l+s,m+t}$ are then calculated from Eqs. (2.76) and (2.77), respectively, to give

$$d_{l,m,l+s,m+t} = (-1)\sqrt{\frac{(2l+2s+1)(l+m)!(l-m+s-t)!}{(2l+1)(l+m+s+t)!(l-m)!}}\, g^{*}_{l,m,l+s,m+t} \tag{4.101}$$

$$e_{l,m,l+s,m+t} = \frac{(2l+2s+1)(l+m)!(l-m+s-t)!}{(2l+1)(l+m+s+t)!(l-m)!}\, g^{*}_{l,m,l+s,m+t} \tag{4.102}$$

The Fourier coefficients and the expectation values then satisfy the differential recurrence relations, respectively, [cf. Eqs. (2.12) and (2.14)]

$$\dot{a}_{l,m} = \sum_{s=-4}^{4} \sum_{t=-1}^{1} d^{*}_{l,m,l+s,m+4t} a_{l+s,m+4t} \tag{4.103}$$

$$\dot{x}_{l,m} = \sum_{s=-4}^{4} \sum_{t=-1}^{1} e_{l,m,l+s,m+4t} x_{l+s,m+4t} \tag{4.104}$$

V. BROWN'S HIGH ENERGY BARRIER APPROXIMATION

A. Introduction

The purpose of the following section, is to provide a detailed derivation of Brown's high energy barrier approximation formula [5]. The Fokker–Planck and Langevin approaches that were utilized in Sections II and III are based on rigorous mathematical considerations [36]. The fundamental test of the validity of these models is the discrete orientation assumption, which, as we recall from Section I.C, states that when the energy barriers separating the stable magnetization orientations are very large in comparison to thermal energy kT (high energy barriers) then the distribution of magnetization orientations $W(\mathbf{r}, t)$ must asymptotically approach the discrete distribution $n_i(t)$. Each $n_i(t)$ is to be interpreted [5] as being the number of representative points that are sufficiently close to the ith stable magnetization orientation (minima of the free energy) at time t. It is therefore natural to assume that the discrete orientation assumption is equivalent to supposing that high energy barriers reduce the non–axially symmetric Fokker–Planck equation given by Eq. (1.83) to the master equation given by Eq. (1.12). This yields an approximation formula for the transition probabilities. On specializing Eq. (1.12) to the case of two orientations, and using Eqs. (1.15) and (1.102), we obtain an approximation to the longest relaxation (Néel) time.

B. Simplification of the Fokker–Planck Equation and the Current Density near a Stationary Point of the Free Energy

Suppose the free energy per unit volume $V = V(\mathbf{r})$ has a bistable structure with minima at $\mathbf{n}_1$ and $\mathbf{n}_2$ separated by a potential barrier that contains a saddle point at $\mathbf{n}_0$. In addition, suppose that the $\mathbf{n}_i$ are coplanar. Denote the plane containing the $\mathbf{n}_i$ by Π. For each $i = 0, 1, 2$ define an orthogonal triad of unit vectors $E_i = (\mathbf{e}_1^{(i)}, \mathbf{e}_2^{(i)}, \mathbf{e}_3^{(i)})$ with $\mathbf{e}_1^{(i)} \perp \Pi$ and $\mathbf{e}_2^{(i)}, \mathbf{e}_3^{(i)} \in \Pi]$ as shown in Fig. 11. Let

$$X_i^T = (\alpha_1^{(i)}, \alpha_2^{(i)}, \alpha_3^{(i)}) \tag{5.1}$$

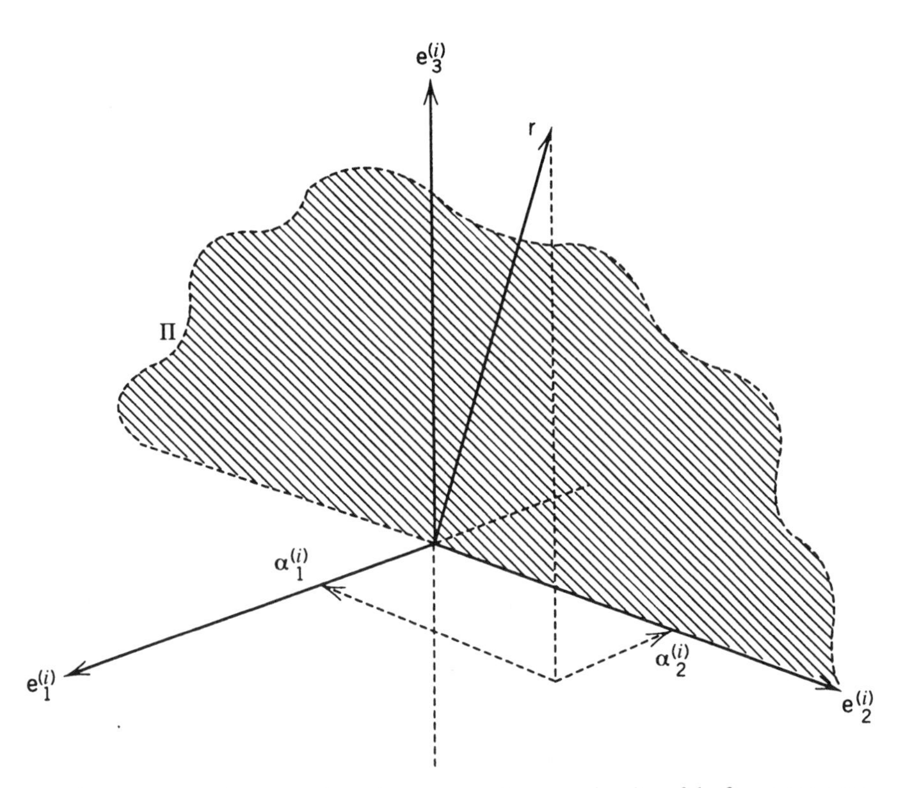

Figure 11. Local orthogonal coordinate system for approximation of the free energy near its *i*th stationary point.

denote the coordinate vectors of $\mathbf{r}$ w.r.t. E_i respectively. The Fokker–Planck equation is given by Eq. (1.74), namely,

$$\dot{W} = a\mathbf{r}\cdot(\Lambda V \times \Lambda W) + b\Lambda\cdot(W\Lambda V) + b\beta^{-1}\Lambda^2 W \tag{5.2}$$

Suppose $\mathbf{r}$ is close to the stationary point $\mathbf{n}_i$ of the potential, then $\mathbf{r} = E_i X_i$ and $V(\mathbf{r})$ can be approximated to the second order of small quantities by the Taylor series

$$V = V_i + \tfrac{1}{2}[c_1^{(i)}(\alpha_1^{(i)})^2 + c_2^{(i)}(\alpha_2^{(i)})^2] \tag{5.3}$$

Hence the tangential gradients are

$$\begin{aligned} \Lambda V &= (\partial V/\partial \alpha_1^{(i)})\mathbf{e}_1^{(i)} + (\partial V/\partial \alpha_2^{(i)})\mathbf{e}_2^{(i)} \\ &= c_1^{(i)}\alpha_1^{(i)}\mathbf{e}_1^{(i)} + c_2^{(i)}\alpha_2^{(i)}\mathbf{e}_2^{(i)} \\ \Lambda W &= (\partial W/\partial \alpha_1^{(i)})\mathbf{e}_1^{(i)} + (\partial W/\partial \alpha_2^{(i)})\mathbf{e}_2^{(i)} \end{aligned} \tag{5.4}$$

The first (gyroscopic) term in Eq. (5.2), with the aid of Eq. (5.4), becomes

$$\begin{aligned} a\mathbf{r}\Lambda V \times \Lambda W &= a\mathbf{r}\begin{vmatrix} \mathbf{e}_1^{(i)} & \mathbf{e}_2^{(i)} & \mathbf{e}_3^{(i)} \\ c_1^{(i)}\alpha_1^{(i)} & c_2^{(i)}\alpha_2^{(i)} & 0 \\ \partial W/\partial \alpha_1^{(i)} & \partial W/\partial \alpha_2^{(i)} & 0 \end{vmatrix} \\ &= a[c_1^{(i)}\alpha_1^{(i)}(\partial W/\partial \alpha_2^{(i)}) - c_2^{(i)}\alpha_2^{(i)}(\partial W/\partial \alpha_1^{(i)})]\mathbf{r}\mathbf{e}_3^{(i)} \end{aligned} \tag{5.5}$$

For high energy barriers, we write $\mathbf{r}\mathbf{e}_3^{(i)} = \alpha_3^{(i)} \cong 1$ so that

$$a\mathbf{r}\cdot(\Lambda V \times \Lambda W) \cong a[c_1^{(i)}\alpha_1^{(i)}(\partial W/\partial \alpha_2^{(i)}) - c_2^{(i)}\alpha_2^{(i)}(\partial W/\partial \alpha_1^{(i)})] \tag{5.6}$$

The second (alignment) term in Eq. (5.2), with the aid of Eq. (5.4), becomes

$$\begin{aligned} b\Lambda\cdot(W\Lambda V) &= b\Lambda\cdot(c_1^{(i)}\alpha_1^{(i)}W\mathbf{e}_1^{(i)} + c_2^{(i)}\alpha_2^{(i)}W\mathbf{e}_2^{(i)}) \\ &= b[c_1^{(i)}\alpha_1^{(i)}(\partial W/\partial \alpha_1^{(i)}) + c_2^{(i)}\alpha_2^{(i)}(\partial W/\partial \alpha_2^{(i)}) \\ &\quad + (c_1^{(i)} + c_2^{(i)})W] \end{aligned} \tag{5.7}$$

The Fokker–Planck equation, Eq. (5.2), with the aid of Eqs. (5.6) and (5.7), becomes

$$\begin{aligned}\dot{W} &= (bc_1^{(i)}\alpha_1^{(i)} - ac_2^{(i)}\alpha_2^{(i)})\partial W/\partial\alpha_1^{(i)} \\ &\quad + (ac_1^{(i)}\alpha_1^{(i)} + bc_2^{(i)}\alpha_2^{(i)})\partial W/\partial\alpha_2^{(i)} + b(c_1^{(i)} + c_2^{(i)})W \\ &\quad + b\beta^{-1}\Lambda^2 W \end{aligned} \tag{5.8}$$

Equation (5.2) can also be expressed as a continuity equation

$$\dot{W} = -\mathbf{\Lambda}\cdot\mathbf{J} \tag{5.9}$$

where the current density is

$$\mathbf{J} = -aW\mathbf{r}\times\Lambda V - bW\Lambda V - b\beta^{-1}\Lambda W \tag{5.10}$$

The components of the current density are

$$\begin{aligned}J_k^{(i)} &= \mathbf{J}\cdot\mathbf{e}_k^{(i)} \\ &= -aW(\mathbf{r}\times\Lambda V)\cdot\mathbf{e}_k^{(i)} - bW\Lambda V\cdot\mathbf{e}_k^{(i)} - b\beta^{-1}\Lambda W\cdot\mathbf{e}_k^{(i)}\end{aligned} \tag{5.11}$$

From Eq. (5.4), the cross products are

$$(\mathbf{r}\times\Lambda V) = \begin{vmatrix} \mathbf{e}_1^{(i)} & \mathbf{e}_2^{(i)} & \mathbf{e}_3^{(i)} \\ \alpha_1^{(i)} & \alpha_2^{(i)} & \alpha_3^{(i)} \\ c_1^{(i)}\alpha_1^{(i)} & c_2^{(i)}\alpha_2^{(i)} & 0 \end{vmatrix} \tag{5.12}$$

hence

$$(\mathbf{r}\times\Lambda V)\cdot\mathbf{e}_1^{(i)} = -c_2^{(i)}\alpha_2^{(i)}\alpha_3^{(i)} \cong -c_2^{(i)}\alpha_2^{(i)} \tag{5.13}$$

$$(\mathbf{r}\times\Lambda V)\cdot\mathbf{e}_2^{(i)} = c_1^{(i)}\alpha_1^{(i)}\alpha_3^{(i)} \cong c_1^{(i)}\alpha_1^{(i)} \tag{5.14}$$

whence

$$J_1 \cong (ac_2^{(i)}\alpha_2^{(i)} - bc_1^{(i)}\alpha_1^{(i)})W - b\beta^{-1}\partial W/\partial\alpha_1^{(i)} \tag{5.15}$$

$$J_2 \cong -(ac_1^{(i)}\alpha_1^{(i)} - bc_2^{(i)}\alpha_2^{(i)})W - b\beta^{-1}\partial W/\partial\alpha_2^{(i)} \tag{5.16}$$

C. Behavior in Quasi-Equilibrium Near the Anisotropic Minima

We assume that equilibrium has been independently established near the ith anisotropic minimum. In this case the distribution can be approximated by the stationary solution of the Fokker–Planck equation (Maxwell–Boltzmann distribution) given by Eq. (1.1), which with the aid of Eq. (5.3), becomes

$$W \cong A_i' e^{-\beta[V - V_i]} \cong A_i' \exp -\tfrac{1}{2}\beta[c_1^{(i)}(\alpha_1^{(i)})^2 + c_2^{(i)}(\alpha_2^{(i)})^2] \tag{5.17}$$

In quasi-equilibrium (cf. Fig. 5) the normalization constant differs for each potential well. The probability of an arbitrary particle being in the ith well is given approximately by

$$\begin{aligned} n_i/n &\cong A_i' \int_{-\infty}^{\infty}\int_{-\infty}^{\infty} \exp -\frac{1}{2}\,\beta[c_1^{(i)}(\alpha_1^{(i)})^2 + c_2^{(i)}(\alpha_2^{(i)})^2]d\alpha_1^{(i)}d\alpha_2^{(i)} \\ &= A_i' \int_{-\infty}^{\infty} \exp -\frac{1}{2}\,\beta c_1^{(i)}(\alpha_1^{(i)})^2 d\alpha_1^{(i)} \int_{-\infty}^{\infty} \exp -\frac{1}{2}\,\beta c_2^{(i)}(\alpha_2^{(i)})^2 d\alpha_2^{(i)} \\ &= \frac{2\pi A_i'}{\beta\sqrt{c_1^{(i)}c_2^{(i)}}} \end{aligned} \tag{5.18}$$

Equation (5.17) becomes

$$W \cong \frac{n_i\beta\sqrt{c_1^{(i)}c_2^{(i)}}}{2\pi n}\, e^{-\beta[V - V_i]} \tag{5.19}$$

D. Behavior Near a Saddle Point

To simplify the notation, let $c_i = c_i^{(0)}$ and $\mu_i = \alpha_i^{(0)}$. Consider the transformation [5] (in [5] Brown writes $c_2 = -c_2'$) with $c_2' > 0$.

$$W = Ue^{-\beta(V - V_0)} = U \exp -\tfrac{1}{2}\beta(c_1\mu_1^2 + c_2\mu_2^2) \tag{5.20}$$

The first derivatives are

$$\frac{\partial W}{\partial \mu_i} = \exp -\frac{1}{2}\,\beta(c_1\mu_1^2 + c_2\mu_2^2)\left(\frac{\partial U}{\partial \mu_i} - \beta c_i\mu_i U\right) \tag{5.21}$$

and the second derivatives

$$\frac{\partial^2 W}{\partial \mu_i^2} = \exp - \frac{1}{2}\,\beta(c_1\mu_1^2 + c_2\mu_2^2)\left(\frac{\partial^2 U}{\partial \mu_i^2} - \beta c_i\mu_i\,\frac{\partial U}{\partial \mu_i} - \beta c_i U\right)$$

$$- \beta c_i\mu_i \exp - \frac{1}{2}\,\beta(c_1\mu_1^2 + c_2\mu_2^2)\left(\frac{\partial U}{\partial \mu_i} - \beta c_i\mu_i U\right)$$

$$= \exp - \frac{1}{2}\,\beta(c_1\mu_1^2 + c_2\mu_2^2)\left[\frac{\partial^2 U}{\partial \mu_i^2} - 2\beta c_i\mu_i\,\frac{\partial U}{\partial \mu_i}\right.$$

$$\left. + \beta c_i(\beta c_i\mu_i^2 - 1)U\right] \tag{5.22}$$

The Laplacian becomes (compare the Kramers method [31])

$$\Lambda^2 W = \frac{\partial^2 W}{\partial \mu_1^2} + \frac{\partial^2 W}{\partial \mu_2^2}$$

$$= \exp - \frac{1}{2}\,\beta(c_1\mu_1^2 + c_2\mu_2^2)\left[\frac{\partial^2 U}{\partial \mu_1^2} - 2\beta c_1\mu_1\,\frac{\partial U}{\partial \mu_i} + \beta c_1(\beta c_1\mu_1^2 - 1)U\right.$$

$$\left. + \frac{\partial^2 U}{\partial \mu_2^2} - 2\beta c_2\mu_2\,\frac{\partial U}{\partial \mu_2} + \beta c_2(\beta c_2\mu_2^2 - 1)U\right]$$

$$= \exp - \frac{1}{2}\,\beta(c_1\mu_1^2 + c_2\mu_2^2)\left\{\Lambda^2 U - 2\beta\left(c_1\mu_1\,\frac{\partial U}{\partial \mu_1} + c_2\mu_2\,\frac{\partial U}{\partial \mu_2}\right)\right.$$

$$\left. + \beta[\beta(c_1^2\mu_1^2 + c_2^2\mu_2^2) - (c_1 + c_2)]U\right\} \tag{5.23}$$

Introducing Eqs. (5.20–5.23) into the Fokker–Planck equation, Eq. (5.8), namely,

$$\dot{W} = (bc_1\mu_1 - ac_2\mu_2)\partial W/\partial \mu_1$$
$$+ (ac_1\mu_1 + bc_2\mu_2)\partial W/\partial \mu_2 + b(c_1 + c_2)W + b\beta^{-1}\Lambda^2 W \tag{5.24}$$

and factoring the exponential term gives

$$\dot{U} = (bc_1\mu_1 - ac_2\mu_2)\left(\frac{\partial U}{\partial \mu_1} - \beta c_1\mu_1 U\right)$$
$$+ (ac_1\mu_1 + bc_2\mu_2)\left(\frac{\partial U}{\partial \mu_2} - \beta c_2\mu_2 U\right) + b(c_1 + c_2)U$$
$$+ b\left\{\beta^{-1}\Lambda^2 U - 2\left(c_1\mu_1\frac{\partial U}{\partial \mu_1} + c_2\mu_2\frac{\partial U}{\partial \mu_2}\right)\right.$$
$$\left. + [\beta(c_1^2\mu_1^2 + c_2^2\mu_2^2) - (c_1 + c_2)]U\right\} \tag{5.25}$$

which becomes (just as [31] for diffusion in phase space)

$$\dot{U} = -(bc_1\mu_1 + ac_2\mu_2)\frac{\partial U}{\partial \mu_1} + (ac_1\mu_1 - bc_2\mu_2)\frac{\partial U}{\partial \mu_2} + b\beta^{-1}\Lambda^2 U \tag{5.26}$$

To obtain the stationary solution ($\partial U/\partial t = 0$), we introduce the nonorthogonal transformation [31]

$$z = q\mu_1 + \mu_2 \tag{5.27}$$

The derivatives transform as follows

$$\frac{\partial U}{\partial \mu_1} = \frac{dU}{dz}\frac{\partial z}{\partial \mu_1} = q\frac{dU}{dz} \qquad \frac{\partial^2 U}{\partial \mu_1^2} = q\frac{d^2U}{dz^2}\frac{\partial z}{\partial \mu_1} = q^2\frac{d^2U}{dz^2}$$
$$\frac{\partial U}{\partial \mu_2} = \frac{dU}{dz}\frac{\partial z}{\partial \mu_2} = \frac{dU}{dz} \qquad \frac{\partial^2 U}{\partial \mu_2^2} = \frac{d^2U}{dz^2}\frac{\partial z}{\partial \mu_2} = \frac{d^2U}{dz^2} \tag{5.28}$$

hence

$$\Lambda^2 U = (1 + q^2)d^2U/dz^2 \tag{5.29}$$

and Eq. (5.26) reduces to the ordinary differential equation as in [31]

$$b\beta^{-1}(1 + q^2)\frac{d^2U}{dz^2} + Lz\frac{dU}{dz} = 0 \tag{5.30}$$

when

$$Lz = (ac_1\mu_1 - bc_2\mu_2) - q(bc_1\mu_1 + ac_2\mu_2)$$
$$= -c_2(qa + b)\left[\frac{c_1(qb - a)}{c_2(qa + b)}\mu_1 + \mu_2\right] \tag{5.31}$$

The multiplicative constant is thus

$$L = -c_2(qa + b) \tag{5.32}$$

and clearly

$$\frac{c_1(qb - a)}{c_2(qa + b)} = q \tag{5.33}$$

This can be rearranged as a quadratic equation as in [31]

$$ac_2q^2 + b(c_2 - c_1)q + ac_1 = 0 \tag{5.34}$$

Hence

$$q = \frac{b(c_1 - c_2) - \sqrt{b^2(c_2 - c_1)^2 - 4a^2c_1c_2}}{2ac_2} \tag{5.35}$$

(The positive radical corresponds to the solution that becomes infinite as $|\mu_2| \to \infty$ and so must be rejected [31].) Furthermore

$$1 + q^2 = 1 + \frac{bq}{ac_2}(c_1 - c_2) - \frac{c_1}{c_2}$$
$$= \left(1 - \frac{bq}{a}\right)\left(1 - \frac{c_1}{c_2}\right) \tag{5.36}$$

Equation (5.30) can be expressed as a pair of first-order equations

$$dU/dz = R \tag{5.37}$$

$$b\beta^{-1}(1 + q^2)\frac{dR}{dz} = c_2(qa + b)zR \tag{5.38}$$

Applying the method of separation of variables, Eq. (5.38) becomes

$$b\beta^{-1}(1+q^2)\int \frac{dR}{R} = c_2(qa+b)\int z\,dz \tag{5.39}$$

Hence

$$R = \exp(-\tfrac{1}{2}\beta c'' z^2) + C \tag{5.40}$$

where C is a constant of integration and

$$\begin{aligned} c'' &= -\frac{c_2(qa+b)}{b(1+q^2)} = \frac{ac_1(1-bq/a)}{bq(1+q^2)} \\ &= \frac{ac_1c_2}{bq(c_2-c_1)} \end{aligned} \tag{5.41}$$

Substituting Eq. (5.40) into Eq. (5.37) and integrating gives

$$U = C_1\int_0^z \exp -\frac{1}{2}\beta c''\xi^2 d\xi + C_2 \tag{5.42}$$

Here C_1 and C_2 are constants of integration. The restriction (to the latitude containing the saddle point) is

$$U|_{\mu_2=0} = C_1\int_0^{q\mu_1} \exp -\frac{1}{2}\beta c''\xi^2 d\xi + C_2 \tag{5.43}$$

which on being differentiated becomes

$$\frac{d}{d\mu_1}\left(U|_{\mu_2=0}\right) = qC_1 \exp -\frac{1}{2}\beta c'' q^2\mu_1^2 \tag{5.44}$$

The restriction of the derivative of Eq. (5.42) is

$$\begin{aligned} \left.\frac{\partial U}{\partial \mu_2}\right|_{\mu_2=0} &= \left.\frac{\partial z}{\partial \mu_2}\frac{dU}{dz}\right|_{\mu_2=0} = C_1\frac{d}{dz}\int_0^z \exp -\frac{1}{2}\beta c''\xi^2 d\xi\Big|_{\mu_2=0} \\ &= C_1\exp -\frac{1}{2}\beta c'' q^2\mu_1^2 \end{aligned} \tag{5.45}$$

The restriction of Eq. (5.42) (to the longitude containing the saddle point) is

$$U|_{\mu_1=0} = C_1 \int_0^{\mu_2} \exp - \frac{1}{2} \beta c'' \xi^2 d\xi + C_2 \tag{5.46}$$

Suppose $U|_{\mu_1=0} = D_1$ for μ_2 large and positive, and $U|_{\mu_1=0} = D_2$ for μ_2 large and negative then

$$\frac{C_1}{2}\sqrt{\frac{2\pi}{\beta c''}} + C_2 = D_1 \qquad -\frac{C_1}{2}\sqrt{\frac{2\pi}{\beta c''}} + C_2 = D_2 \tag{5.47}$$

Solving for C_1 and C_2 gives

$$C_1 = (D_1 - D_2)\sqrt{\frac{\beta c''}{2\pi}} \qquad C_2 = \frac{D_1 + D_2}{2} \tag{5.48}$$

Let primed quantities denote quantities evaluated at two points P_i, sufficiently close to but on opposite sides of the saddle point, and suppose that the approximations in Eq. (5.19) are satisfied at these points, then

$$W_i' = D_i e^{-\beta(V' - V_0)} = \frac{n_i \beta \sqrt{c_1^{(i)} c_2^{(i)}}}{2\pi n} e^{-\beta(V' - V_i)} \tag{5.49}$$

Hence

$$D_i = \frac{n_i \beta \sqrt{c_1^{(i)} c_2^{(i)}}}{2\pi n} e^{-\beta(V_0 - V_i)} \tag{5.50}$$

In particular,

$$C_1 = \frac{\beta}{2\pi n}\sqrt{\frac{\beta c''}{2\pi}} \left[n_1 \sqrt{c_1^{(1)} c_2^{(1)}} e^{-\beta(V_0 - V_1)} - n_2 \sqrt{c_1^{(2)} c_2^{(2)}} e^{-\beta(V_0 - V_2)} \right] \tag{5.51}$$

The restriction of the component of current density in Eq. (5.16) is

$$J_2^{(0)}|_{\mu_2=0} = -(ac_1\mu_1 + bc_2\mu_2)W|_{\mu_2=0} - b\beta^{-1}\partial W/\partial\mu_2|_{\mu_2=0} \tag{5.52}$$

Introducing the transformation in Eq. (5.20), and applying Eqs. (5.21) and (5.45), gives

$$J_2^{(0)}|_{\mu_2=0} = -\exp -\frac{1}{2}\,\beta c_1\mu_1^2\left(b\beta^{-1}\,\frac{\partial U}{\partial \mu_2} + ac_1\mu_1 U\right)\Bigg|_{\mu_2=0}$$
$$= -b\beta^{-1}C_1\exp -\frac{1}{2}\,\beta(c_1+q^2c'')\mu_1^2$$
$$-\,ac_1\mu_1 U|_{\mu_2=0}\exp -\frac{1}{2}\,\beta c_1\mu_1^2 \tag{5.53}$$

The current across the potential barrier is

$$I_2 \cong \int_{-\infty}^{\infty} J_2^{(0)}|_{\mu_2=0}d\mu_1 = -b\beta^{-1}C_1\int_{-\infty}^{\infty}\exp -\frac{1}{2}\,\beta(c_1+q^2c'')\mu_1^2 d\mu_1$$
$$-\frac{ac_1}{2}\int_{-\infty}^{\infty} U|_{\mu_2=0}\exp -\frac{1}{2}\,\beta c_1\mu_1^2 d\mu_1^2$$
$$= -b\beta^{-1}C_1\sqrt{\frac{2\pi}{\beta(c_1+q^2c'')}}$$
$$+\,a\beta^{-1}\int_{-\infty}^{\infty} U|_{\mu_2=0}\,\frac{d}{d\mu_1}\left(\exp -\frac{1}{2}\,\beta c_1\mu_1^2\right)d\mu_1 \tag{5.54}$$

Integrating the second term by parts and substituting for $(d/d\mu_1)(U|_{\mu_2=0})$ from Eq. (5.44) gives

$$I_2 = -b\beta^{-1}C_1\sqrt{\frac{2\pi}{\beta(c_1+q^2c'')}} + a\beta^{-1}\Bigg[\exp -\frac{1}{2}\,\beta c_1\mu_1^2(U|_{\mu_2=0})|_{\infty}^{\infty}$$
$$-\int_{-\infty}^{\infty}\exp -\frac{1}{2}\,\beta c_1\mu_1^2\,\frac{d}{d\mu_1}\,(U|_{\mu_2=0})d\mu_1\Bigg]$$
$$= -b\beta^{-1}C_1\sqrt{\frac{2\pi}{\beta(c_1+q^2c'')}} - a\beta^{-1}qC_1\int_{-\infty}^{\infty}\exp -\frac{1}{2}\,\beta(c_1+q^2c'')\mu_1^2 d\mu_1$$
$$= -b\beta^{-1}C_1\left(1+\frac{aq}{b}\right)\sqrt{\frac{2\pi}{\beta(c_1+q^2c'')}} \tag{5.55}$$

Substituting for C_1 from Eq. (5.51) and rearranging gives

$$I_2 = \frac{b}{2\pi n}\left(1 + \frac{aq}{b}\right)\sqrt{\frac{c''}{c_1 + q^2 c''}}\left[n_2\sqrt{c_1^{(2)}c_2^{(2)}}e^{-\beta(V_0 - V_2)} - n_1\sqrt{c_1^{(1)}c_2^{(1)}}e^{-\beta(V_0 - V_1)}\right]$$

$$= \frac{b}{2\pi n}\left(1 + \frac{aq}{b}\right)\sqrt{\frac{-c_2}{c_1}}\left[n_2\sqrt{c_1^{(2)}c_2^{(2)}}e^{-\beta(V_0 - V_2)} - n_1\sqrt{c_1^{(1)}c_2^{(1)}}e^{-\beta(V_0 - V_1)}\right] \tag{5.56}$$

E. Brown's High Energy Barrier Approximation Formula

For an ensemble of identical noninteracting particles $nI_2 = \dot{n}_1 = -\dot{n}_2$; hence Eq. (5.56) becomes

$$\dot{n}_1 = \frac{b}{2\pi}\left(1 + \frac{aq}{b}\right)\sqrt{\frac{-c_2}{c_1}}\left[n_2\sqrt{c_1^{(2)}c_2^{(2)}}e^{-\beta(V_0 - V_2)} - n_1\sqrt{c_1^{(1)}c_2^{(1)}}e^{-\beta(V_0 - V_1)}\right] \tag{5.57}$$

Equation (5.57) is of the form [cf. Eq. (1.13)]

$$\dot{n}_1 = -\dot{n}_2 = \nu_{21}n_2 - \nu_{12}n_1 \tag{5.58}$$

hence the transition probabilities are

$$\nu_{ij} = \frac{b}{2\pi}\left(1 + \frac{aq}{b}\right)\sqrt{\frac{-c_2}{c_1}}\sqrt{c_1^{(i)}c_2^{(i)}}e^{-\beta(V_0 - V_i)} \tag{5.59}$$

On substituting for q from Eq. (5.35), Brown's high energy barrier approximation to the lowest eigenvalue p_1 of the Fokker–Planck equation [5] becomes (a generalization of [31] to curvilinear coordinates)

$$p_1 = \nu_{12} + \nu_{21} = \frac{b}{2\pi}\left(1 + \frac{aq}{b}\right)\sqrt{\frac{-c_2}{c_1}}\left(\sqrt{c_1^{(1)}c_2^{(1)}}\, e^{-\beta(V_0 - V_1)}\right.$$
$$\left. + \sqrt{c_1^{(2)}c_2^{(2)}}\, e^{-\beta(V_0 - V_2)}\right)$$
$$= \frac{b(-c_1 - c_2 + \sqrt{(c_2 - c_1)^2 - 4\alpha^{-2}c_1c_2})}{4\pi\sqrt{-c_1c_2}}\left(\sqrt{c_1^{(1)}c_2^{(1)}}\, e^{-\beta(V_0 - V_1)}\right.$$
$$\left. + \sqrt{c_1^{(2)}c_2^{(2)}}\, e^{-\beta(V_0 - V_2)}\right) \tag{5.60}$$

where $\alpha = b/a$ (compare Eq. 25 of Kramers [31]).

VI. APPLICATION TO THE CASE WHERE THE FREE ENERGY ARISES FROM UNIAXIAL ANISOTROPY WITH CONSTANT UNIFORM EXTERNAL MAGNETIC FIELD AT AN ANGLE OBLIQUE TO THE EASY AXIS

A. Coordinate System and Stationary Points of the Free Energy

The free energy per unit volume is derived from Eqs. (1.2), (3.1), and (1.4) given by

$$\begin{aligned} V &= K(1 - \mathbf{r}\mathbf{e}_3^2) - HM_s\mathbf{r}\mathbf{h} \\ &= K(1 - \alpha_3^2) - HM_s\gamma_3\alpha_3 - HM_s\gamma_1\alpha_1 \\ &\equiv K\sin^2\theta - HM_s\cos\psi\cos\theta - HM_s\sin\psi\sin\theta\cos\varphi \end{aligned} \tag{6.1}$$

where $\mathbf{r}$, $\mathbf{e}_3$, and $\mathbf{h}$ are the directions of the magnetization orientation, easy axis, and external field direction respectively. Let $E = (\mathbf{e}_1, \mathbf{e}_2, \mathbf{e}_3)$ be the fixed triad of orthogonal unit vectors, orientated as shown in Fig. 1. The coordinate vectors [Appendix C] of $\mathbf{r}$, $\mathbf{h}$, and $\mathbf{e}_3$ w.r.t. E are respectively

$$X = \begin{pmatrix}\alpha_1\\ \alpha_2\\ \alpha_3\end{pmatrix} = \begin{pmatrix}\sin\theta\cos\varphi\\ \sin\theta\sin\varphi\\ \cos\theta\end{pmatrix} \quad Y = \begin{pmatrix}\gamma_1\\ 0\\ \gamma_3\end{pmatrix} = \begin{pmatrix}\sin\psi\\ 0\\ \cos\psi\end{pmatrix} \quad U = \begin{pmatrix}0\\ 0\\ 1\end{pmatrix} \tag{6.2}$$

For the present time we will suppose that $0 < \psi < \pi/2$. The bistable structure of the free energy (see Figs. 12–15) is subject to a restriction that is expressed

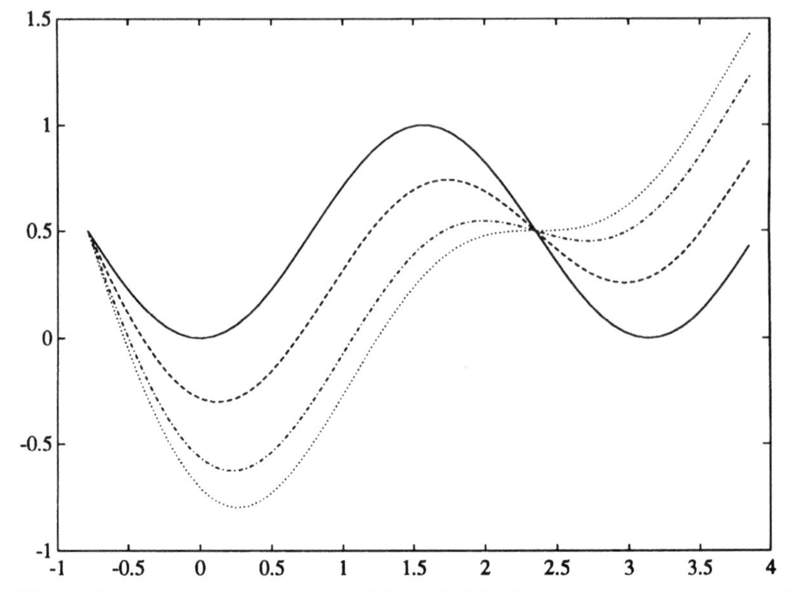

Figure 12. Free energy density V vs θ for uniaxial anisotropy ($\varphi = 0$, $K = 1$, $\psi = \pi/4$); $h = 0.0$ (solid), $h = 0.2$ (dashed), $h = 0.4$ (dashdots), $h = 0.5$ (dots).

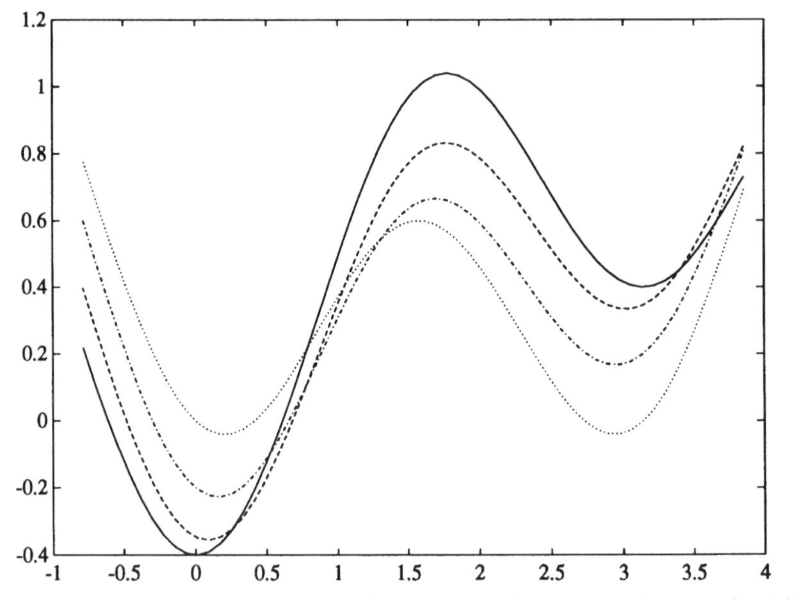

Figure 13. Free energy density V vs θ for uniaxial anisotropy ($\varphi = 0$, $K = 1$, $h = 0.2$); $\psi = 0$ (solid), $\psi = \pi/6$ (dashed), $\psi = \pi/3$ (dashdots), $\psi = \pi/2$ (dots).

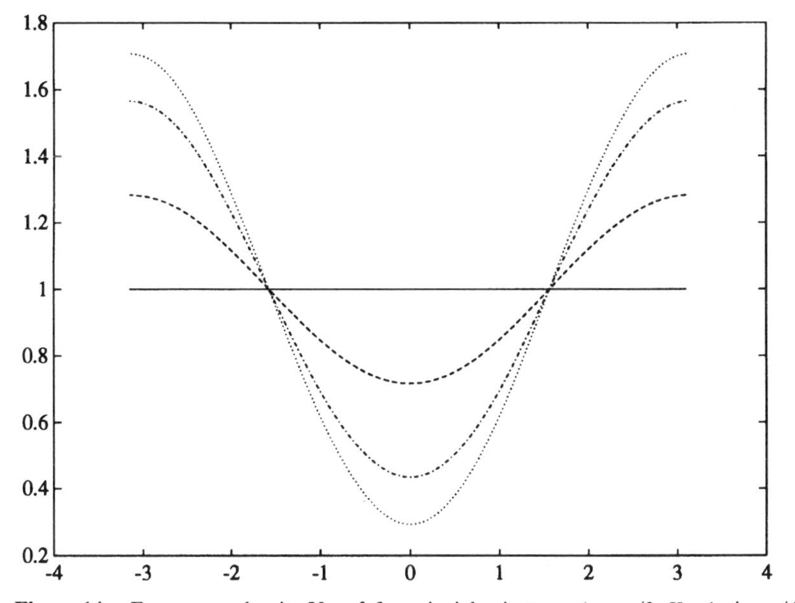

Figure 14. Free energy density V vs θ for uniaxial anisotropy ($\varphi = \pi/2$, $K = 1$, $\psi = \pi/4$); $h = 0.0$ (solid), $h = 0.2$ (dashed), $h = 0.4$ (dashdots), $h = 0.5$ (dots).

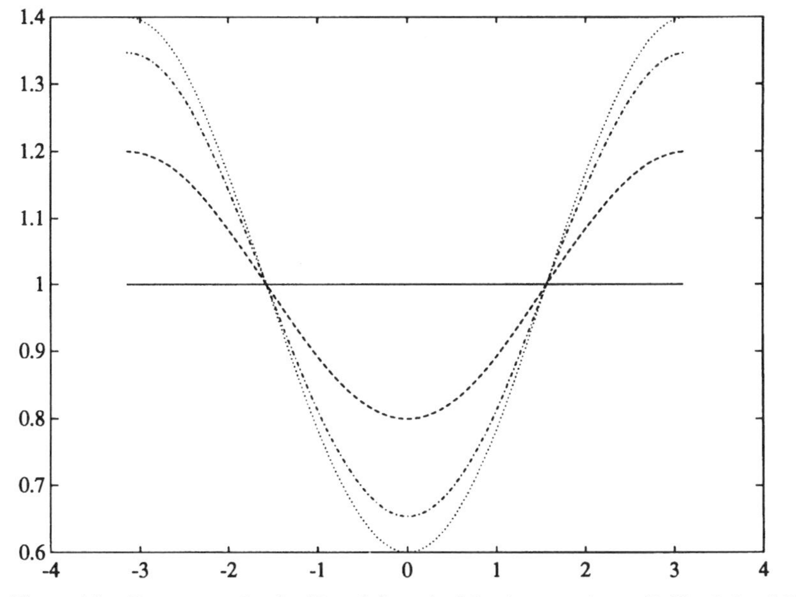

Figure 15. Free energy density V vs θ for uniaxial anisotropy ($\varphi = \pi/2$, $K = 1$, $h = 0.2$); $\psi = 0$ (solid), $\psi = \pi/6$ (dashed), $\psi = \pi/3$ (dashdots), $\psi = \pi/2$ (dots).

as $0 < h < h_c$, where $h = HM_s/2K$ is the ratio of field to barrier height parameter, and the critical value h_c is dependent on ψ. The form of this dependency will be derived in Section VI.C. Equation (6.1) can be expressed in terms of coordinate vectors in Eq. (6.2) as follows

$$V = K[1 - (X^T U)^2] - HM_s X^T Y \tag{6.3}$$

The stationary condition $\partial V/\partial\varphi = 0$ implies that all stationary points (w.r.t. angular variables) occur for $\sin\varphi = 0$, i.e., $\varphi = 0, \pi$, in which case $\mathbf{r}$ is in the plane containing $\mathbf{n}$ and $\mathbf{h}$. Let $\mathbf{n}_i$ denote the direction of the ith stationary point w.r.t. E. The coordinate vector of $\mathbf{n}_i$ w.r.t. E can be expressed as

$$X^{(i)} = \begin{pmatrix} \beta_1^{(i)} \\ 0 \\ \beta_3^{(i)} \end{pmatrix} = \begin{pmatrix} \sin\theta_i \\ 0 \\ \cos\theta_i \end{pmatrix} \tag{6.4}$$

Anisotropic maxima are characterized by

$$\partial^2 V/\partial\varphi^2 = HM_s \sin\psi \sin\theta \cos\varphi < 0 \tag{6.5}$$

and hence occur for $\varphi = \pi$. Anisotropic minima and saddle points are characterized by

$$\partial^2 V/\partial\varphi^2 = HM_s \sin\psi \sin\theta \cos\varphi > 0 \tag{6.6}$$

and hence occur for $\varphi = 0$. The stationary condition $\partial V/\partial\theta = 0$ from Eq. (6.3) is expressed as a trigonometric equation

$$\cos\theta \sin\theta + h(\cos\psi \sin\theta \pm \sin\psi \cos\theta) = 0 \tag{6.7}$$

where the positive sign corresponds to $\varphi = \pi$ and the negative sign to $\varphi = 0$. Writing

$$x = \cos\theta \tag{6.8}$$

so that $\sin\theta = \sqrt{1 - \cos^2\theta} = \sqrt{1 - x^2}$ and defining new parameters

$$u = h\cos\psi \qquad \nu = h\sin\psi \tag{6.9}$$

Eq. (6.7) becomes

$$\sqrt{1 - x^2}\,(x + u) \pm \nu x = 0 \tag{6.10}$$

Squaring gives the quartic equation

$$(x + u)^2(1 - x^2) = \nu^2 x^2 \tag{6.11}$$

The roots are arranged as follows

$$-1 \le x_{-1} \le x_1 \le x'_{-1} \le x'_1 \le 1 \tag{6.12}$$

By means of Eq. (6.8) they correspond to solutions of Eq. (6.7) given by $\theta_2, \theta_0, \theta'_0$, and θ_1 with $0 \le \theta_1 \le \theta_0 \le \theta_2 \le \pi < \theta'_0 < 2\pi$. The stationary point $\mathbf{n}_3$ corresponding to the solution θ'_0 is orientated in the $\varphi = \pi$ direction and is specified by the polar angle $\theta_3 = 2\pi - \theta'_0$. It is thus an anisotropic maximum. Furthermore, from geometrical considerations it is obvious that the stationary point $\mathbf{n}_0$ corresponding to the solution θ_0 is a saddle point, and the remaining stationary points $\mathbf{n}_1$ and $\mathbf{n}_2$ corresponding to the solutions θ_1 and θ_2, respectively, are anisotropic minima.

For the present we concern ourselves only with the stationary points that occur for $\varphi = 0$, in which case the stationary condition Eq. (6.7) becomes

$$\cos\theta \sin\theta + h(\cos\psi \sin\theta - \sin\psi \cos\theta) = 0 \tag{6.13}$$

or more compactly

$$\sin 2\theta + 2h \sin(\theta - \psi) = 0 \tag{6.14}$$

On replacing θ with θ_i, and the trigonometric functions with the direction cosines given in Eqs. (6.2) and (6.4), then Eq. (6.7) becomes

$$\beta_1^{(i)}\beta_3^{(i)} + h(\gamma_3\beta_1^{(i)} - \gamma_1\beta_3^{(i)}) = 0 \qquad i = 0, 1, 2 \tag{6.15}$$

The nature of the stationary point will depend on the sign of the quantity

$$\begin{aligned} \partial^2 V/\partial\theta^2|_{\varphi=0} &= 2K[\cos^2\theta - \sin^2\theta + h(\cos\psi\cos\theta + \sin\psi\sin\theta)] \\ &= 2K[\cos 2\theta + h\cos(\theta - \psi)] \end{aligned} \tag{6.16}$$

hence in terms of direction cosines the anisotropic minima satisfy the condition

$$(\beta_3^{(i)})^2 - (\beta_2^{(i)})^2 + h(\gamma_3\beta_3^{(i)} + \gamma_1\beta_1^{(i)}) > 0 \qquad i = 1, 2 \tag{6.17}$$

and the saddle point satisfies the condition

$$(\beta_3^{(0)})^2 - (\beta_2^{(0)})^2 + h(\gamma_3\beta_3^{(0)} + \gamma_1\beta_1^{(0)}) < 0 \tag{6.18}$$

B. Orthogonal Transformation of the Coordinate System and Approximation of the Free Energy near the Stationary Points

Define orthogonal transformations

$$T_i = P_{\pi/2}Q_{\theta_i} \qquad i = 0, 1, 2 \tag{6.19}$$

where $P_{\pi/2}$ and Q_{θ_i} are defined by the transformation equations in Eqs. (C.24) and (C.25), respectively. The transformations in Eq. (6.19) are precisely the orthogonal transformations Eq. (C.33) specified by the Eulerian angles $(\psi, \theta, \varphi) = (\pi/2, \theta_i, 0)$. The representing matrices of the T_i are given by Eq. (C.34), namely,

$$S_i = S_{\pi/2,\theta_i,0} = \begin{bmatrix} 0 & 1 & 0 \\ -\cos\theta_i & 0 & \sin\theta_i \\ \sin\theta_i & 0 & \cos\theta_i \end{bmatrix} = \begin{bmatrix} 0 & 1 & 0 \\ -\beta_3^{(i)} & 0 & \beta_1^{(i)} \\ \beta_1^{(i)} & 0 & \beta_3^{(i)} \end{bmatrix} \tag{6.20}$$

The T_i give rise to orthogonal coordinate transformations

$$E_i = T_iE = ES_i^T \qquad i = 0, 1, 2 \tag{6.21}$$

where $E_i = (\mathbf{e}_1^{(i)}, \mathbf{e}_2^{(i)}, \mathbf{e}_3^{(i)})$. The coordinate vectors of $\mathbf{r}$, $\mathbf{n}$, and $\mathbf{h}$ with respect to E_i are respectively from Eq. (C.23)

$$X_i = \begin{bmatrix} \alpha_1^{(i)} \\ \alpha_2^{(i)} \\ \alpha_3^{(i)} \end{bmatrix} = S_iX = \begin{bmatrix} 0 & 1 & 0 \\ -\beta_3^{(i)} & 0 & \beta_1^{(i)} \\ \beta_1^{(i)} & 0 & \beta_3^{(i)} \end{bmatrix} \begin{bmatrix} \alpha_1 \\ \alpha_2 \\ \alpha_3 \end{bmatrix}$$

$$= \begin{bmatrix} \alpha_2 \\ -\beta_3^{(i)}\alpha_1 + \beta_1^{(i)}\alpha_3 \\ \beta_1^{(i)}\alpha_1 + \beta_3^{(i)}\alpha_3 \end{bmatrix} \tag{6.22}$$

$$U_i = S_iU = \begin{bmatrix} 0 & 1 & 0 \\ -\beta_3^{(i)} & 0 & \beta_1^{(i)} \\ \beta_1^{(i)} & 0 & \beta_3^{(i)} \end{bmatrix} \begin{bmatrix} 0 \\ 0 \\ 1 \end{bmatrix} = \begin{bmatrix} 0 \\ \beta_1^{(i)} \\ \beta_3^{(i)} \end{bmatrix} \tag{6.23}$$

$$Y_i = \begin{bmatrix} \gamma_1^{(i)} \\ \gamma_2^{(i)} \\ \gamma_3^{(i)} \end{bmatrix} = S_i Y = \begin{bmatrix} 0 & 1 & 0 \\ -\beta_3^{(i)} & 0 & \beta_1^{(i)} \\ \beta_1^{(i)} & 0 & \beta_3^{(i)} \end{bmatrix} \begin{bmatrix} \gamma_1 \\ 0 \\ \gamma_3 \end{bmatrix}$$

$$= \begin{bmatrix} 0 \\ -\beta_3^{(i)}\gamma_1 + \beta_1^{(i)}\gamma_3 \\ \beta_1^{(i)}\gamma_1 + \beta_3^{(i)}\gamma_3 \end{bmatrix} \tag{6.24}$$

The stationary condition in Eq. (6.15) can be expressed compactly in terms of $\gamma_2^{(i)} = \gamma_3\beta_1^{(i)} - \gamma_1\beta_3^{(i)}$ as

$$\beta_1^{(i)}\beta_3^{(i)} + h\gamma_2^{(i)} = 0 \qquad i = 0, 1, 2 \tag{6.25}$$

The conditions in Eqs. (6.17) and (6.18) can be expressed compactly in terms of $\gamma_3^{(i)} = \beta_1^{(i)}\gamma_1 + \beta_3^{(i)}\gamma_3$ as

$$(\beta_3^{(i)})^2 - (\beta_2^{(i)})^2 + h\gamma_3^{(i)} > 0 \qquad i = 1, 2 \tag{6.26}$$

for the anisotropic minima and

$$(\beta_3^{(0)})^2 - (\beta_2^{(0)})^2 + h\gamma_3^{(0)} < 0 \tag{6.27}$$

for the saddle point. The free energy per unit volume in Eqs. (6.1) and (6.3) can be expressed in terms of the new coordinates as follows

$$\begin{aligned} V &= K[1 - (X^T U)^2] - HM_s X^T Y \\ &= K\{1 - [(S_i^T X_i)^T S_i^T U_i]^2\} - HM_s (S_i^T X_i)^T S_i^T Y_i \\ &= K[1 - (X_i^T U_i)^2] - HM_s X_i^T Y_i \\ &= K[1 - (\beta_1^{(i)}\alpha_2^{(i)})^2 - \beta_1^{(i)}\beta_3^{(i)}\alpha_2^{(i)}\alpha_3^{(i)} - (\beta_3^{(i)}\alpha_3^{(i)})^2] \\ &\quad - HM_s(\gamma_2^{(i)}\alpha_2^{(i)} + \gamma_3^{(i)}\alpha_3^{(i)}) \end{aligned} \tag{6.28}$$

Using the relation

$$(\alpha_3^{(i)})^2 = 1 - (\alpha_1^{(i)})^2 - (\alpha_2^{(i)})^2 \tag{6.29}$$

and the binomial approximation to the second order of small quantities (namely $\alpha_1^{(i)}, \alpha_2^{(i)} \ll 1$)

$$(\alpha_3^{(i)}) = \sqrt{1 - (\alpha_1^{(i)})^2 - (\alpha_2^{(i)})^2} \cong 1 - \tfrac{1}{2}[(\alpha_1^{(i)})^2 + (\alpha_2^{(i)})^2] \tag{6.30}$$

Equation (6.28) to the second order of small quantities, becomes

$$\begin{aligned} V = {} & K[1 - (\beta_3^{(i)})^2 - 2h\gamma_3^{(i)}] - 2K[\beta_1^{(i)}\beta_3^{(i)} + h\gamma_2^{(i)}]\alpha_2^{(i)} \\ & + K[(\beta_3^{(i)})^2 + h\gamma_3^{(i)}](\alpha_1^{(i)})^2 \\ & + K[(\beta_3^{(i)})^2 - (\beta_1^{(i)})^2 + h\gamma_3^{(i)}](\alpha_2^{(i)})^2 \end{aligned} \tag{6.31}$$

From the stationary condition Eq. (6.25), the $\alpha_2^{(i)}$ term vanishes; hence

$$V = V_i + \tfrac{1}{2}[c_1^{(i)}(\alpha_1^{(i)})^2 + c_2^{(i)}(\alpha_2^{(i)})^2] \tag{6.32}$$

where

$$\begin{aligned} V_i &= K[1 - (\beta_3^{(i)})^2 - 2h\gamma_3^{(i)}] \\ &= K[1 - \cos^2\theta_i - 2h\cos(\theta_i - \psi)] \end{aligned} \tag{6.33}$$

is the value of the potential at the ith stationary point and

$$\begin{aligned} c_1^{(i)} &= 2K[(\beta_3^{(i)})^2 + h\gamma_3^{(i)}] \\ &= 2K[\cos^2\theta_i + h\cos(\theta_i - \psi)] > 0 \end{aligned} \tag{6.34}$$

$$\begin{aligned} c_2^{(i)} &= 2K[(\beta_3^{(i)})^2 - (\beta_2^{(i)})^2 + h\gamma_3^{(i)}] \\ &= 2K[\cos^2\theta_i - \sin^2\theta_i + h\cos(\theta_i - \psi)] \end{aligned} \tag{6.35}$$

From Eq. (6.26), $c_2^{(i)} > 0$, $i = 1, 2$ for the anisotropic minima, in which case Eq. (6.32) describes an elliptic paraboloid (which has the appearance of a plum pudding bowl). From Eq. (6.27), $c_2^{(0)} < 0$ for the saddle point, in which case Eq. (6.32) describes a hyperbolic paraboloid (which has the appearance of a horse's saddle).

We will now compare the non–axially symmetric approximation formula in Eq. (5.60) for the transverse ($\psi = \pi/2$) case with the axially symmetric approximation formula in Eq. (1.140). We recall that when $\psi = \pi/2$, the stationary points are at $\theta_1 = \sin^{-1} h$, $\theta_0 = \pi/2$, and $\theta_2 = \pi - \sin^{-1} h$. The barrier heights are thus

$$\beta(V_0 - V_i) = \sigma(1 - h)^2 \tag{6.36}$$

The coefficients in Eqs. (6.34) and (6.35) are

$$\begin{aligned} c_1^{(1)} &= c_1^{(2)} = 2K & c_2^{(1)} &= c_2^{(2)} = 2K(1 - h^2) \\ c_1 &= 2Kh & c_2 &= -2K(1 - h) \end{aligned} \tag{6.37}$$

Inserting Eqs. (6.36) and (6.37) into Eq. (5.60) and evaluating $\tau = 1/p_1$ gives

$$\tau = \frac{\pi\beta\sqrt{h}}{b\sigma(1 - 2h + \sqrt{1 + 4\alpha^{-2}h(1 - h)})\sqrt{(1 + h)}}\, e^{\sigma(1 - h)^2} \tag{6.38}$$

If $\alpha \to \infty$, Eq. (6.38) eliminates the error in Fig. 10 because

$$\tau = \frac{\pi\beta\sqrt{h}}{2b\sigma(1 - h)\sqrt{1 + h)}}\, e^{\sigma(1 - h)^2} \tag{6.39}$$

which becomes asymptotic to Eq. (1.140) as $h \to 1$ (see [43]).

C. Bistable Structure of the Free Energy

In this section we calculate the dependency of the critical value h_c of the ratio of field to barrier height parameter $h = HM_s/2K$ on the angle of inclination $\psi = \cos^{-1} \mathbf{h}\mathbf{e}_3$ where $\mathbf{h}$ and $\mathbf{e}_3$ denotes the directions of the external field and easy axis, respectively [39, 40, 41]. Suppose initially $0 < h < h_c$, then the free energy has stationary points in the $\varphi = 0$ direction specified by distinct polar angles θ_i, $i = 0, 1, 2$ as discussed in Section VI.A. As h increases in value and approaches h_c, the saddle point at θ_0 and the anisotropic minimum at θ_2 approach each other. Eventually when $h = h_c$, the free energy loses its bistable character in that the stationary points intersect at a critical angle $\theta_0 = \theta_2 = \theta_c$ and form a point of inflection which is characterized by

$$\partial V/\partial\theta|_{\varphi=0} = \partial^2 V/\partial\theta^2|_{\varphi=0} = 0 \tag{6.40}$$

Equation (6.40), with the aid of Eqs. (6.14) and (6.18), yields the pair of simultaneous trigonometric equations

$$\sin 2\theta = -2h \sin(\theta - \psi) \tag{6.41}$$

$$\cos 2\theta = -h \cos(\theta - \psi) \tag{6.42}$$

Denoting the solution by θ_c and dividing Eq. (6.41) by Eq. (6.42) gives

$$\tan 2\theta_c = 2 \tan(\theta_c - \psi) \tag{6.43}$$

Using the trigonometric formulae

$$\tan 2\theta = \frac{2 \tan \theta}{1 - \tan^2 \theta} \qquad \tan(\theta - \psi) = \frac{\tan \theta - \tan \psi}{1 + \tan \theta \tan \psi} \tag{6.44}$$

and writing $t = \tan \theta_c$, Eq. (6.43) becomes

$$\frac{t}{1 - t^2} = \frac{t - \tan \psi}{1 + t \tan \psi} \tag{6.45}$$

Cross-multiplying gives the cubic equation

$$t^3 + \tan \psi = 0 \tag{6.46}$$

the only real root of which is $t = -(\tan \psi)^{1/3}$. Hence

$$\tan \theta_c = -(\tan \psi)^{1/3} \tag{6.47}$$

The stationary condition Eq. (6.41) in terms of h becomes

$$h = -\frac{\sin 2\theta}{2 \sin(\theta - \psi)} = \frac{\sin \theta \cos \theta}{\cos \theta \sin \psi - \sin \theta \cos \psi}$$

$$= \left(\frac{\sin \psi}{\sin \theta} - \frac{\cos \psi}{\cos \theta} \right)^{-1} \tag{6.48}$$

For $\pi/2 \leq \theta \leq \pi$, we have

$$\sin \theta = -\frac{\tan \theta}{\sqrt{1 + \tan^2 \theta}} \qquad \cos \theta = -\frac{1}{\sqrt{1 + \tan^2 \theta}} \tag{6.49}$$

Eq. (6.48) can be expressed in terms of $\tan\theta$

$$h = -\frac{1}{\sqrt{1+\tan^2\theta}}\left(\frac{\sin\psi}{\tan\theta} - \cos\psi\right)^{-1} \tag{6.50}$$

and hence from Eq. (6.47)

$$\begin{aligned} h_c &= -\frac{1}{\sqrt{1+\tan^2\theta_c}}\left(\frac{\sin\psi}{\tan\theta_c} - \cos\psi\right)^{-1} \\ &= \frac{1}{\sqrt{1+(\tan\psi)^{2/3}}(\sin\psi(\tan\psi)^{-1/3}+\cos\psi)} \qquad (6.51) \\ &= [(\sin\psi)^{2/3} + (\cos\psi)^{2/3}]^{-3/2} \qquad (6.52) \end{aligned}$$

For $0 \le \psi \le \pi/2$, we have

$$\sin\psi = \frac{\tan\psi}{\sqrt{1+\tan^2\psi}} \qquad \cos\psi = \frac{1}{\sqrt{1+\tan^2\psi}} \tag{6.53}$$

Equation (6.51) then gives an expression for h_c in terms of $\tan\psi$ [38]

$$\begin{aligned} h_c &= \frac{1}{\sqrt{1+(\tan\psi)^{2/3}}}\left(\frac{\tan\psi}{\sqrt{1+\tan^2\psi}}(\tan\psi)^{-1/3} + \frac{1}{\sqrt{1+\tan^2\psi}}\right)^{-1} \\ &= \sqrt{1+\tan^2\psi}(1+(\tan\psi)^{2/3})^{-3/2} \qquad (6.54) \end{aligned}$$

D. Geometrical Interpretation of the Effect of Changes in h and ψ on the Positions of the Stationary Points

To construct such a picture, we define the polynomials

$$f(x) = (x+u)^2(1-x^2) \tag{6.55}$$
$$g(x) = v^2x^2 \tag{6.56}$$

Equation (6.55) has zeros at $-u$ (which is of second order) and at ± 1 and has a minimum at $-u$ and maxima at $(-u \pm \sqrt{u^2+8/4})$. Equation (6.56) has a zero at $x = 0$ (which is of second order) and a minimum at $x = 0$. Equation

(6.11) can then be expressed as

$$f(x) = g(x) \tag{6.57}$$

The roots of Eq. (6.57), namely,

$$-1 \le x_{-1} \le x_1 \le x'_{-1} \le x'_1 \le 1 \tag{6.58}$$

are then geometrically speaking the points of intersection of the *graphs* of f and g, as shown in Fig. 18. The effect of changes in h and ψ (hence u and v) on the roots becomes very clear on observing the effect of such changes on the graphs of f and g.

Consider a simulation of the effect of increasing h for a fixed $0 < \psi < \pi/2$ (so that $u > 0$) on the graphs of f and g, as shown in Figs. 16–19. When $h = 0$ (Fig. 16), we have $u, v = 0$. Hence $g = 0$; f is symmetrical about its minima at the origin, with maxima at $\pm 1/\sqrt{2}$; and $x_{-1} = -1$, $x_1 = x'_{-1} = 0$, and $x'_1 = 1$. When h is small [Fig. 17], we have

$$-1 < x_{-1} < -\frac{u + \sqrt{u^2 + 8}}{4} < x_1 < -u < x'_{-1} < 0 < \frac{-u + \sqrt{u^2 + 8}}{4} < x'_1 < 1 \tag{6.59}$$

As h increases [Fig. 18], the most noticeable effects are

1. The well in the graph of f about its minimum at $-u$ moves away from the origin toward the point $x = -1$.
2. The graph of g becomes increasingly more concave upward.
3. The roots at x_{-1} and x_1 approach the maximum of f at $-(u + \sqrt{u^2 + 8})/4$.

The strict inequality is observed until eventually when h reaches the critical value (Fig. 19) at

$$h_c(\psi) = (\cos^{2/3} \psi + \sin^{2/3} \psi)^{-3/2} \tag{6.60}$$

(which has a minimum value of $h_c = 1/2$ at $\psi = \pi/4$ and maxima value of $h_c = 1$ at $\psi = 0, \pi/2$), the points x_{-1} and x_1 coincide, giving

$$-1 < x_{-1} = -\frac{u + \sqrt{u^2 + 8}}{4} = x_1 < -u < x'_{-1} < 0 < \frac{-u + \sqrt{u^2 + 8}}{4} < x'_1 < 1 \tag{6.61}$$

Similar effects occur when $\pi/2 < \psi < \pi$ and when $\psi = 0, \pi/2$.

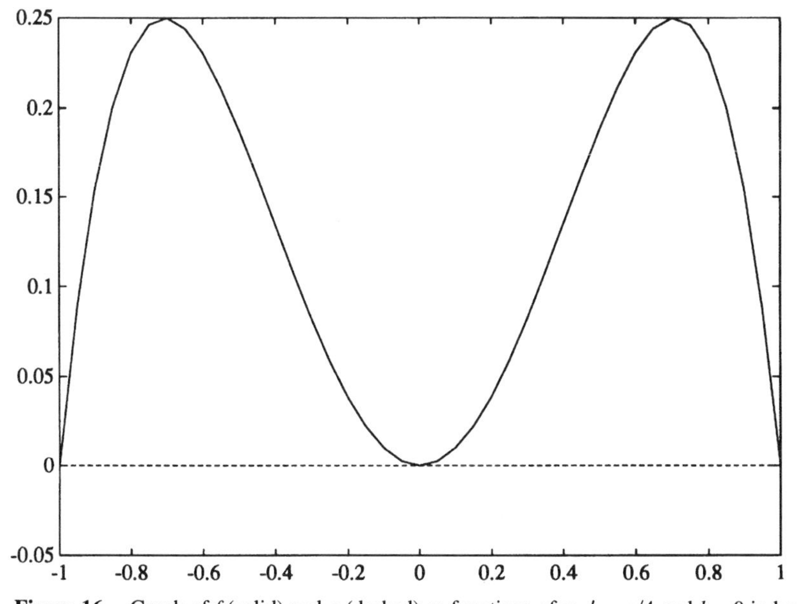

Figure 16. Graph of f (solid) and g (dashed) as functions of x; $\psi = \pi/4$ and $h = 0$ in both cases.

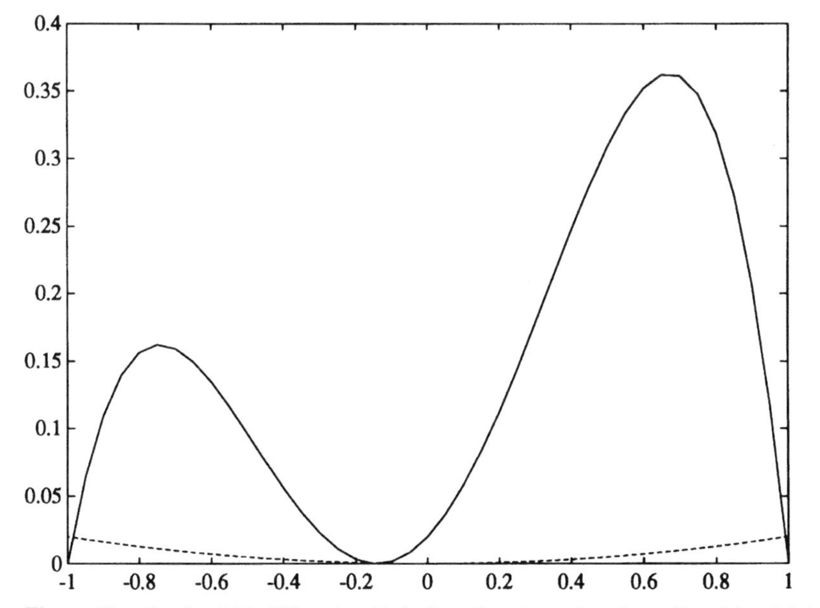

Figure 17. Graph of f (solid) and g (dashed) as functions of x; $\psi = \pi/4$ and $h = 0.2$ in both cases.

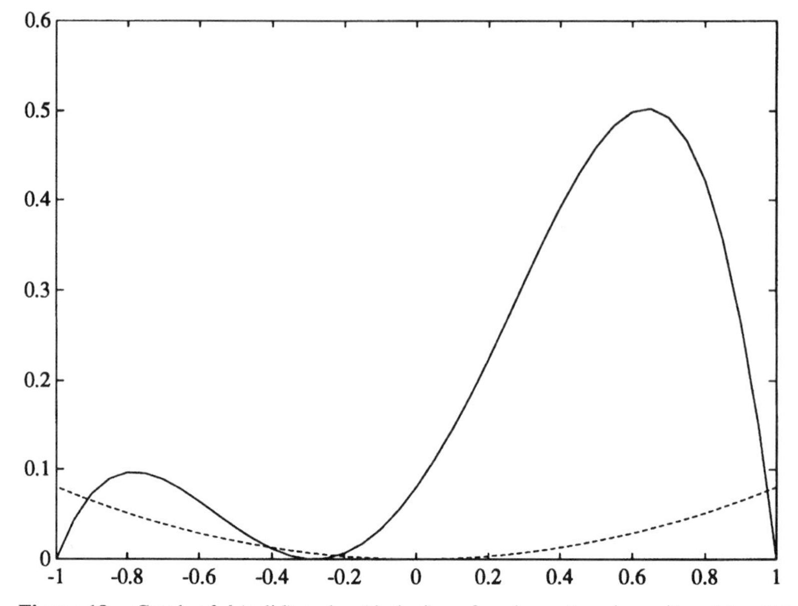

Figure 18. Graph of f (solid) and g (dashed) as functions of x; $\psi = \pi/4$ and $h = 0.4$ in both cases.

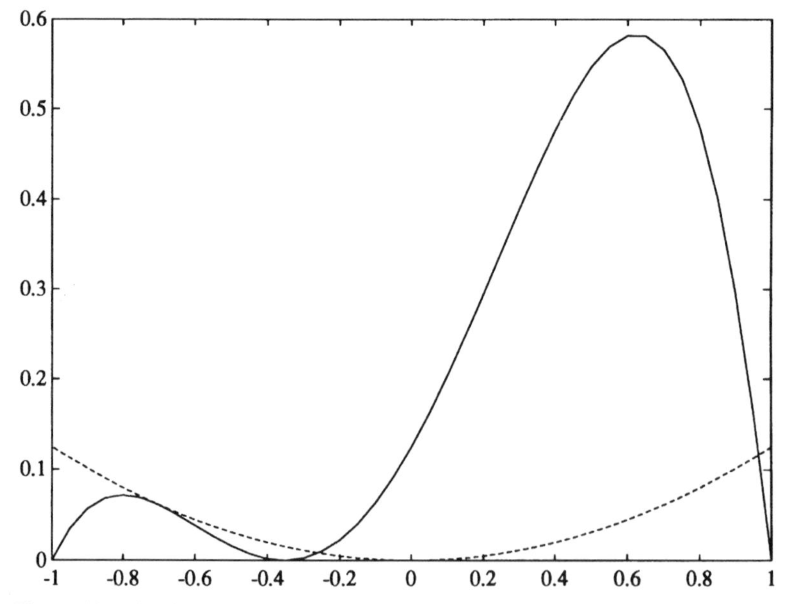

Figure 19. Graph of f (solid) and g (dashed) as functions of x; $\psi = \pi/4$ and $h = 0.5$ in both cases.

Consider a simulation of the effect of increasing ψ for a fixed $0 < h < 1/2$ (Fig. 20). When $\psi = 0$, we have $u = h$, $v = 0$; hence $g = 0$ and the minimum of f occurs at $-h$, with maxima at $(-h \pm \sqrt{h^2 + 8})/4$, and Eq. (6.57) becomes

$$(x + h)^2(1 - x^2) = 0 \tag{6.62}$$

Thus $x_{-1} = -1$, $x_1 = x'_{-1} = -h$ and $x'_1 = 1$. As ψ increases the most noticeable effects are

1. The "well" in the graph of f about its minimum at $-u$ moves away from the point $x = -h$ toward the point $x = 0$.
2. The graph g becomes increasingly more concave upward.
3. The roots at x_1 and x'_{-1} move away from and then reapproach the minimum of f at $-u$.

When $\psi = \pi/2$, we have $u = 0$, $v = h$, and f becomes symmetrical about its minima, which is now at the origin with maxima at $\pm 1/\sqrt{2}$. Equation (6.57) becomes

$$(x^2 + h^2 - 1)x^2 = 0 \tag{6.63}$$

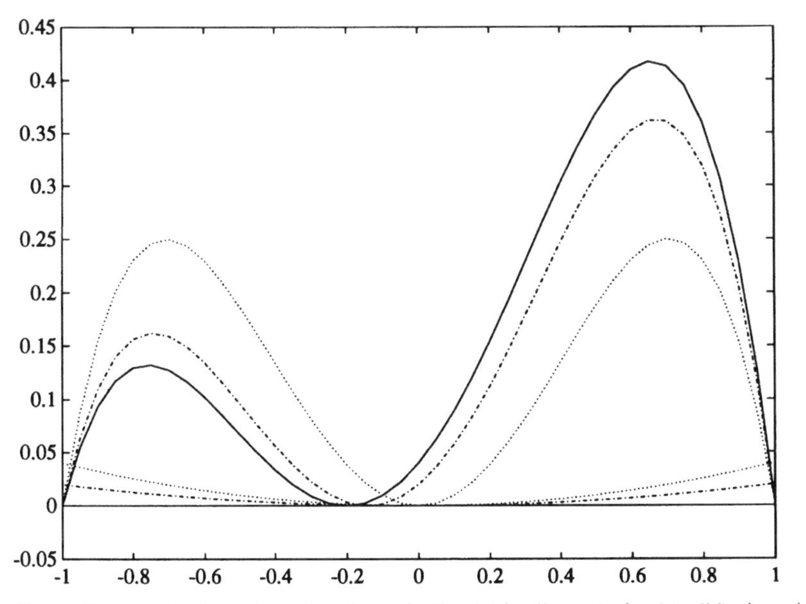

Figure 20. Graph of f and g as functions of x ($h = 0.2$ in all cases); $\psi = 0$ (solid), $\psi = \pi/4$ (dashdots), $\psi = \pi/2$ (dots).

Thus $x_{-1} = -\sqrt{1-h^2}$, $x_1 = x'_{-1} = 0$, $x'_1 = \sqrt{1-h^2}$. A similar effect occurs for $\pi/2 < \psi \leq \pi$.

Suppose $1/2 \leq h < 1$, $0 \leq \psi < \pi/4$, and $h(\psi) < h_c(\psi)$; then the roots satisfy the inequality in Eq. (6.59). In addition to the effects listed above, an increase in ψ leads to a decrease in the value of $h_c(\psi)$, so that the roots at x_{-1} and x_1 approach the maximum at $-(u+\sqrt{u^2+8})/4$ in a manner similar to the case where h is increased to h_c. The strict inequality is observed until eventually, when h_c reaches h, the points x_{-1} and x_1 coincide, and the roots satisfy the inequality in Eq. (6.61). A similar effect occurs for $\pi/4 < \psi \leq \pi/2$.

E. High Energy Barrier Approximation for $\psi = \pi/4$

From the geometrical discussion of the previous section, the critical angle θ_c lies in the second quadrant, i.e., $\pi/2 \leq \theta_c \leq \pi$, and from Eq. (6.47), satisfies the trigonometric equation

$$\tan\,\theta_c = -(\tan\,\pi/4)^{1/3} = -1 \tag{6.64}$$

Hence

$$\theta_c = 3\pi/2 \tag{6.65}$$

From Eq. (6.52) the critical value of the ratio field to barrier height parameter is

$$h_c = [(\sin\,\pi/4)^{2/3} + (\cos\,\pi/4)]^{-3/2} = 1/2 \tag{6.66}$$

Hence $0 \leq h \leq 1/2(= h_c)$, and

$$u = v = h/\sqrt{2} \tag{6.67}$$

The stationary condition for $\psi = \pi/4$ is, from Eq. (6.7), expressed as the trigonometric equation

$$\cos\,\theta\,\sin\,\theta = \frac{h}{\sqrt{2}}\,(\cos\,\theta - \sin\,\theta) \tag{6.68}$$

On writing

$$x = \cos\,\theta \tag{6.69}$$

Eq. (6.68) [*viz.* Eq. (6.10)] becomes

$$x\sqrt{1-x^2} + \frac{h}{\sqrt{2}}\left(\sqrt{1-x^2} - x\right) = 0 \tag{6.70}$$

which can be expressed as a pair of simultaneous equations as follows. Let

$$w = \sqrt{2}\sin(\theta - \pi/4) \tag{6.71}$$

Hence from Eq. (6.69)

$$w = \sqrt{1-x^2} - x \tag{6.72}$$

Equation (6.70) expressed in terms of w is

$$\frac{1}{2}(1-w^2) + \frac{h}{\sqrt{2}}w = 0 \tag{6.73}$$

Equation (6.72) on rearranging and squaring becomes

$$(w+x)^2 = 1 - x^2 \tag{6.74}$$

Equations (6.73) and (6.74) yield, respectively, the pair of simultaneous quadratic equations

$$2x^2 + 2wx + (w^2 - 1) = 0 \tag{6.75}$$

$$w^2 - h\sqrt{2}w - 1 = 0 \tag{6.76}$$

The roots of Eq. (6.76) are

$$w_{\pm 1} = \frac{h \pm \sqrt{h^2+2}}{\sqrt{2}} \tag{6.77}$$

Hence the simultaneous solutions of Eqs. (6.75) and (6.76) are the four roots of the pair of quadratic equations obtained on replacing w in Eq. (6.75) with $w_{\pm 1}$ in Eq. (6.77), i.e.,

$$x_{\pm 1} = -\frac{w_1}{2} \pm \frac{1}{\sqrt{2}}\sqrt{1 - \frac{w_1^2}{2}}$$

$$= -\frac{h + \sqrt{h^2+2}}{2\sqrt{2}} \pm \frac{\sqrt{1 - h^2 - h\sqrt{h^2+2}}}{2} \tag{6.78}$$

$$x'_{\pm 1} = -\frac{w_{-1}}{2} \pm \frac{1}{\sqrt{2}}\sqrt{1 - \frac{w_{-1}^2}{2}}$$

$$= -\frac{h - \sqrt{h^2+2}}{2\sqrt{2}} \pm \frac{\sqrt{1 - h^2 + h\sqrt{h^2+2}}}{2} \tag{6.79}$$

From Eq. (6.71), the roots in Eqs. (6.78) and (6.79) correspond [by means of Eq. (6.69)] to the solutions of the pair of trigonometric equations

$$\sin(\theta - \pi/4) = w_{\pm 1}/\sqrt{2} \tag{6.80}$$

Given that

$$\begin{aligned}\sin(\theta - \pi/4) &= \sin[\pi - (\theta - \pi/4)] \\ &= \sin[(3\pi/2 - \theta) - \pi/4]\end{aligned} \tag{6.81}$$

then if θ is a solution, so too is $3\pi/2 - \theta$. Let

$$\alpha_{\pm 1} = \mp\pi/4 + \sin^{-1}(w_{\pm 1}/\sqrt{2}) \tag{6.82}$$

The solutions of Eq. (6.80) are then

$$\theta_1 = \alpha_{-1} \quad \text{which corresponds to the root } x'_1 \tag{6.83}$$

$$\theta_0 = \pi/2 + \alpha_1 \quad \text{which corresponds to the root } x_1 \tag{6.84}$$

$$\theta_2 = \pi - \alpha_1 \quad \text{which corresponds to the root } x_{-1} \tag{6.85}$$

$$\theta'_0 = 3\pi/2 - \alpha_{-1} \quad \text{which corresponds to the root } x'_{-1} \tag{6.86}$$

The stationary points of the free energy that occur for $\varphi = 0$ are specified by polar angles that are equal to the solutions Eqs. (6.83–6.86). The anisotropic maximum that occurs for $\varphi = \pi$ is specified by the polar angle $\theta_3 = 2\pi - \theta_0' = \pi/2 + \alpha_{-1}$. At $h = 0$, $w_{\pm 1} = \pm 1$; hence

$$x_{-1} = -1 \qquad x_1 = x_{-1}' = 0 \qquad x_1' = 1 \qquad \alpha_{\pm 1} = 0 \tag{6.87}$$

At $h = 1/2$, $w_{\pm 1} = \pm(\sqrt{2})^{\pm 1}$; hence

$$x_{-1} = x_1 = -1/\sqrt{2} \quad x_{\pm 1}' = (\sqrt{2} \pm \sqrt{6})/4 \quad \alpha_{\pm 1} = (2\pi \pm \pi)/12 \tag{6.88}$$

The roots in Eqs. (6.78) and (6.79) thus satisfy the following bounds

$$-1 \le x_{-1} \le -\frac{1}{\sqrt{2}} \le x_1 \le x_{-1}' \le 0 \tag{6.89}$$

$$-\frac{\sqrt{6}-\sqrt{2}}{4} \le x_{-1}' \le 0 \tag{6.90}$$

$$\frac{\sqrt{6}-\sqrt{2}}{4} \le x_1' \le 1 \tag{6.91}$$

The solutions in Eqs. (6.83–6.86) satisfy the bounds

$$0 \le \theta_1 \le \pi/12 \tag{6.92}$$

$$\pi/2 \le \theta_0 \le 3\pi/4 \le \theta_2 \le \pi \tag{6.93}$$

$$17\pi/12 \le \theta_0' \le 3\pi/2 \tag{6.94}$$

Furthermore from Eq. (6.71)

$$\begin{aligned}\sqrt{1 - x_{\pm 1}^2} &= w_1 + x_{\pm 1} \\ &= \frac{w_1}{2} \pm \frac{1}{\sqrt{2}}\sqrt{1 - \frac{w_1^2}{2}} \\ &= \frac{h + \sqrt{h^2 + 2}}{2\sqrt{2}} \pm \frac{\sqrt{1 - h^2 - h\sqrt{h^2 + 2}}}{2}\end{aligned} \tag{6.95}$$

$$\sqrt{1 - x'^2_{\pm 1}} = w_{-1} + x'_{\pm 1}$$

$$= \frac{w_{-1}}{2} \pm \frac{1}{\sqrt{2}} \sqrt{1 - \frac{w^2_{-1}}{2}}$$

$$= \frac{h - \sqrt{h^2 + 2}}{2\sqrt{2}} \pm \frac{\sqrt{1 - h^2 + h\sqrt{h^2 + 2}}}{2} \tag{6.96}$$

Note that the negative value of the square root in Eq. (6.96) arises, since the solution corresponding to x'_{-1} is θ'_0, which from Eq. (6.94) lies in the third quadrant where the values of $\sin\theta$ are negative.

The free energy per unit volume from Eq. (6.1) expressed in terms of $x = \cos\theta$ is

$$V = K[1 - x^2 - h\sqrt{2}(x + \sqrt{1 - x^2})] \tag{6.97}$$

The values of the free energy per unit volume at the stationary points are

$$V_1 = K\left[1 - x'^2_1 - h\sqrt{2}\left(x'_1 + \sqrt{1 - x'^2_1}\right)\right]$$

$$= \frac{K}{2}\left[1 + (\sqrt{2}w_{-1} - 4h)\sqrt{1 - \frac{w^2_{-1}}{2}}\right]$$

$$= \frac{K}{2}\left[1 - (\sqrt{h^2 + 2} + 3h)\sqrt{\frac{1 - h^2 + h\sqrt{h^2 + 2}}{2}}\right] \tag{6.98}$$

$$V_0 = K\left[1 - x^2_1 - h\sqrt{2}\left(x_1 + \sqrt{1 - x^2_1}\right)\right]$$

$$= \frac{K}{2}\left[1 + (\sqrt{2}w_1 - 4h)\sqrt{1 - \frac{w^2_1}{2}}\right]$$

$$= \frac{K}{2}\left[1 + (\sqrt{h^2 + 2} - 3h)\sqrt{\frac{1 - h^2 - h\sqrt{h^2 + 2}}{2}}\right] \tag{6.99}$$

$$V_2 = K\left[1 - x_{-1}^2 - h\sqrt{2}\left(x_{-1} + \sqrt{1 - x_{-1}^2}\right)\right]$$

$$= \frac{K}{2}\left[1 - (\sqrt{2}w_1 - 4h)\sqrt{1 - \frac{w_1^2}{2}}\right]$$

$$= \frac{K}{2}\left[1 - (\sqrt{h^2+2} - 3h)\sqrt{\frac{1 - h^2 - h\sqrt{h^2+2}}{2}}\right] \tag{6.100}$$

The ratios of barrier height to thermal energy are thus

$$\beta(V_0 - V_2) = \sigma(\sqrt{h^2+2} - 3h)\sqrt{\frac{1 - h^2 - h\sqrt{h^2+2}}{2}} \tag{6.101}$$

$$\beta(V_0 - V_1) = \frac{\sigma\sqrt{h^2+2}}{2}\left[\sqrt{\frac{1 - h^2 - h\sqrt{h^2+2}}{2}} + \sqrt{\frac{1 - h^2 + h\sqrt{h^2+2}}{2}}\right] \tag{6.102}$$

Equations (6.101) and (6.102) are in agreement with Pfeiffer [39, 40], who, in addition, obtained the following simple approximation formulae

$$\beta(V_0 - V_1) \cong \sigma(1 - h)^{1.43} \tag{6.103}$$

$$\beta(V_0 - V_2) \cong \sigma[(1 - h)^{1.43} + 2h\sqrt{2}] \tag{6.104}$$

The remaining quantities of interest are from Eqs. (6.34) and (6.35)

$$c_1^{(i)} = 2K\left[\cos^2\theta_i + \frac{h}{\sqrt{2}}(\cos\theta_i + \sin\theta_i)\right] \tag{6.105}$$

$$c_2^{(i)} = 2K\left[-1 + 2\cos^2\theta_i + \frac{h}{\sqrt{2}}(\cos\theta_i + \sin\theta_i)\right] \tag{6.106}$$

Hence

$$c_1^{(1)} = 2K\left[x_1'^2 + \frac{h}{\sqrt{2}}\left(x_1' + \sqrt{1 - x_1'^2}\right)\right]$$

$$= K\left[1 + (2h - \sqrt{2}w_{-1})\sqrt{1 - \frac{w_{-1}^2}{2}}\right]$$

$$= K\left[1 + (h + \sqrt{h^2 + 2})\sqrt{\frac{1 - h^2 + h\sqrt{h^2 + 2}}{2}}\right] \quad (6.107)$$

$$c_1 = c_1^{(0)} = 2K\left[x_1^2 + \frac{h}{\sqrt{2}}\left(x_1 + \sqrt{1 - x_1^2}\right)\right]$$

$$= K\left[1 + (2h - \sqrt{2}w_1)\sqrt{1 - \frac{w_1^2}{2}}\right]$$

$$= K\left[1 + (h - \sqrt{h^2 + 2})\sqrt{\frac{1 - h^2 - h\sqrt{h^2 + 2}}{2}}\right] \quad (6.108)$$

$$c_1^{(2)} = 2K\left[x_{-1}^2 + \frac{h}{\sqrt{2}}\left(x_{-1} + \sqrt{1 - x_{-1}^2}\right)\right]$$

$$= K\left[1 + (-2h + \sqrt{2}w_1)\sqrt{1 - \frac{w_1^2}{2}}\right]$$

$$= K\left[1 + (\sqrt{h^2 + 2} - h)\sqrt{\frac{1 - h^2 - h\sqrt{h^2 + 2}}{2}}\right] \quad (6.109)$$

$$c_2^{(1)} = 2K\left[-1 + 2x_1'^2 + \frac{h}{\sqrt{2}}\left(x_1' + \sqrt{1 - x_1'^2}\right)\right]$$

$$= 2K(h - \sqrt{2}w_{-1})\sqrt{1 - \frac{w_{-1}^2}{2}}$$

$$= 2K\sqrt{h^2 + 2}\sqrt{\frac{1 - h^2 + h\sqrt{h^2 + 2}}{2}} \tag{6.110}$$

$$c_2 = c_2^{(0)} = 2K\left[-1 + 2x_1^2 + \frac{h}{\sqrt{2}}\left(x_1 + \sqrt{1 - x_1^2}\right)\right]$$

$$= 2K(h - \sqrt{2}w_1)\sqrt{1 - \frac{w_1^2}{2}}$$

$$= -2K\sqrt{h^2 + 2}\sqrt{\frac{1 - h^2 - h\sqrt{h^2 + 2}}{2}} \tag{6.111}$$

$$c_2^{(2)} = 2K\left[-1 + 2x_{-1}^2 + \frac{h}{\sqrt{2}}\left(x_{-1} + \sqrt{1 - x_{-1}^2}\right)\right]$$

$$= 2K(\sqrt{2}w_1 - h)\sqrt{1 - \frac{w_1^2}{2}}$$

$$= 2K\sqrt{h^2 + 2}\sqrt{\frac{1 - h^2 - h\sqrt{h^2 + 2}}{2}} \tag{6.112}$$

Brown's high energy barrier approximation to the lowest eigenvalue of the Fokker–Planck equation is then computed using Eq. (5.60).

F. High Energy Barrier Approximation for an Arbitrary Value of ψ

Expanding Eq. (6.11) gives

$$x^4 + 2ux^3 - (1 - h^2)x^2 - 2ux - u^2 = 0 \tag{6.113}$$

whose roots are the four roots of the pair of quadratic equations

$$w^2 + (u \pm E)w + \frac{1}{2}(y_1 \mp F) = 0 \tag{6.114}$$

where

$$E = \sqrt{y_1 + u^2 + 1 - h^2} \tag{6.115}$$

$$F = \sqrt{y_1^2 + 4u^2} \tag{6.116}$$

and y_1 is a *real* root of the cubic equation

$$y^3 + (1 - h^2)y^2 - 4u^2v^2 = 0 \tag{6.117}$$

The *discriminant* of Eq. (6.117) is

$$D = 4u^2v^2[u^2v^2 - (1/27)(1 - h^2)^3] \tag{6.118}$$

For $D \geq 0$, we have

$$y_1 = -\frac{1}{3}(1 - h^2) + \left[2u^2v^2 - \frac{(1 - h^2)^3}{27} + D^{1/2}\right]^{1/3} + \left[2u^2v^2 - \frac{(1 - h^2)^3}{27} - D^{1/2}\right]^{1/3} \tag{6.119}$$

The roots of Eq. (6.114) hence Eq. (6.113) are

$$x_{\pm 1} = -\frac{E + u}{2} \pm \frac{\sqrt{(E + u)^2 + 2(F - y_1)}}{2} \tag{6.120}$$

$$x'_{\pm 1} = \frac{E - u}{2} \pm \frac{\sqrt{(E - u)^2 - 2(F + y_1)}}{2} \tag{6.121}$$

The values of the free energy per unit volume at the stationary points are calculated using the formulae

$$V_0 = K\left(1 - x_1^2 - 2ux_1 - 2v\sqrt{1 - x_1^2}\right) \tag{6.122}$$

$$V_1 = K\left(1 - x_1'^2 - 2ux_1' - 2v\sqrt{1 - x_1'^2}\right) \tag{6.123}$$

$$V_2 = K\left(1 - x_{-1}^2 - 2ux_{-1} - 2v\sqrt{1 - x_{-1}^2}\right) \tag{6.124}$$

The ratios of barrier height to thermal energy, with the aid of Eqs. (6.120) and (6.121), are

$$\begin{aligned}
\beta(V_0 - V_1) &= \sigma\left[x_1'^2 - x_1^2 + 2u(x_1' - x_1) + 2v\left(\sqrt{1 - x_1'^2} - \sqrt{1 - x_1^2}\right)\right] \\
&= \sigma\left[-F + \frac{(E+u)\sqrt{(E-u)^2 - 2(F+y_1)}}{2}\right. \\
&\quad + \frac{(E-u)\sqrt{(E+u)^2 + 2(F-y_1)}}{2} \\
&\quad + 2v\sqrt{1 - \frac{(E-u)^2 - (F+y_1) + (E-u)\sqrt{(E-u)^2 - 2(F+y_1)}}{2}} \\
&\quad \left. - 2v\sqrt{1 - \frac{(E+u)^2 + (F-y_1) - (E+u)\sqrt{(E+u)^2 + 2(F-y_1)}}{2}}\right]
\end{aligned} \tag{6.125}$$

$$\begin{aligned}
\beta(V_0 - V_2) &= \sigma\left[x_1'^2 - x_{-1}^2 + 2u(x_1' - x_{-1}) + 2v\left(\sqrt{1 - x_1'^2} - \sqrt{1 - x_{-1}^2}\right)\right] \\
&= \sigma\left[-F + \frac{(E+u)\sqrt{(E-u)^2 - 2(F+y_1)}}{2}\right. \\
&\quad - \frac{(E-u)\sqrt{(E+u)^2 + 2(F-y_1)}}{2} \\
&\quad + 2v\sqrt{1 - \frac{(E-u)^2 - (F+y_1) + (E-u)\sqrt{(E-u)^2 - 2(F+y_1)}}{2}} \\
&\quad \left. - 2v\sqrt{1 - \frac{(E+u)^2 + (F-y_1) - (E+u)\sqrt{(E+u)^2 + 2(F-y_1)}}{2}}\right]
\end{aligned} \tag{6.126}$$

The coefficients in Eqs. (6.34) and (6.35) are then calculated as follows

$$c_1^{(0)} = 2K\left(x_1^2 + ux_1 + v\sqrt{1 - x_1^2}\right)$$
$$= K\left[(E+u)^2 - u(E+u) + (F - y_1) - E\sqrt{(E+u)^2 + 2(F - y_1)}\right.$$
$$\left. + 2v\sqrt{1 - \frac{(E+u)^2 + (F - y_1) - (E+u)\sqrt{(E+u)^2 + 2(F - y_1)}}{2}}\right] \quad (6.127)$$

$$c_2^{(0)} = 2K\left(2x_1^2 - 1 + ux_1 + v\sqrt{1 - x_1^2}\right)$$
$$= K\left[-2 + 2(E+u)^2 + 2(F - y_1) - u(E+u) - (2E+u)\sqrt{(E+u)^2 + 2(F - y_1)}\right.$$
$$\left. + 2v\sqrt{1 - \frac{(E+u)^2 + (F - y_1) - (E+u)\sqrt{(E+u)^2 + 2(F - y_1)}}{2}}\right] \quad (6.128)$$

$$c_1^{(1)} = 2K\left(x_1'^2 + ux_1' + v\sqrt{1 - x_1'^2}\right)$$
$$+ K\left[(E-u)^2 - (F + y_1) + u(E-u) + E\sqrt{(E-u)^2 - 2(F + y_1)}\right.$$
$$\left. + 2v\sqrt{1 - \frac{(E-u)^2 - (F + y_1) + (E-u)\sqrt{(E-u)^2 - 2(F + y_1)}}{2}}\right] \quad (6.129)$$

$$c_2^{(1)} = 2K\left(x_1'^2 - 1 + ux_1' + v\sqrt{1 - x_1'^2}\right)$$
$$+ K\Bigg[-2 + 2(E-u)^2 - 2(F+y_1) + u(E-u)$$
$$+ (2E-u)\sqrt{(E-u)^2 - 2(F+y_1)}$$
$$+ 2v\sqrt{1 - \frac{(E-u)^2 - (F+y_1) + (E-u)\sqrt{(E-u)^2 - 2(F+y_1)}}{2}}\Bigg] \quad (6.130)$$

$$c_1^{(2)} = 2K\left(x_{-1}^2 + ux_{-1} + v\sqrt{1 - x_{-1}^2}\right)$$
$$= K\Bigg[(E+u)^2 + (F-y_1) - u(E+u) + E\sqrt{(E+u)^2 + 2(F-y_1)}$$
$$+ 2v\sqrt{1 - \frac{(E+u)^2 + (F-y_1) + (E+u)\sqrt{(E+u)^2 + 2(F-y_1)}}{2}}\Bigg] \quad (6.131)$$

$$c_2^{(2)} = 2K\left(2x_{-1}^2 - 1 + ux_{-1} + v\sqrt{1 - x_{-1}^2}\right)$$
$$+ K\Bigg[-2 + 2(E+u)^2 + 2(F-y_1) - u(E+u)$$
$$+ (2E+u)\sqrt{(E+u)^2 + 2(F-y_1)}$$
$$+ 2v\sqrt{1 - \frac{(E+u)^2 + (F-y_1) - (E+u)\sqrt{(E+u)^2 + 2(F-y_1)}}{2}}\Bigg] \quad (6.132)$$

Brown's high energy barrier approximation to the lowest eigenvalue of the Fokker–Planck equation is then again computed using Eq. (5.60).

APPENDIX A. ASSOCIATED LEGENDRE FUNCTIONS AND SPHERICAL HARMONICS

The associated Legendre functions (of the first kind) [34, 41] are defined

$$P_l^m = \frac{1}{2^l l!} (1 - x^2)^{m/2} \frac{d^{l+m}}{dx^{l+m}} (x^2 - 1)^l \tag{A.1}$$

They satisfy the symmetry relation

$$P_l^{-m} = (-1)^m \frac{(l - m)!}{(l + m)!} P_l^m \tag{A.2}$$

the recurrence relations

$$xP_l^m = \frac{(l + m)}{(2l + 1)} P_{l-1}^m + \frac{(l - m + 1)}{(2l + 1)} P_{l+1}^m \tag{A.3}$$

$$\begin{aligned}\sqrt{1 - x^2}P_l^m &= \frac{(l + m)(l + m - 1)}{(2l + 1)} P_{l-1}^{m-1} - \frac{(l - m + 1)(l - m + 2)}{(2l + 1)} P_{l+1}^{m-1} \\ &= -\frac{1}{(2l + 1)} P_{l-1}^{m+1} + \frac{1}{(2l + 1)} P_{l+1}^{m+1}\end{aligned} \tag{A.4}$$

$$(l + m)P_{l-1}^m - (l - m)xP_l^m = \sqrt{1 - x^2}P_l^{m+1} \tag{A.5}$$

$$(l - m + 1)P_{l+1}^m - (l - m + 1)xP_l^m = -\sqrt{1 - x^2}P_l^{m+1} \tag{A.6}$$

$$P_{l-1}^m - xP_l^m = -(l - m + 1)\sqrt{1 - x^2}P_l^{m-1} \tag{A.7}$$

$$P_{l+1}^m - xP_l^m = (l + m)\sqrt{1 - x^2}P_l^{m-1} \tag{A.8}$$

$$\sqrt{1 - x^2}P_l^{m+1} = 2mxP_l^m - (l + m) \cdot (l - m + 1)\sqrt{1 - x^2}P_l^{m-1} \tag{A.9}$$

and the differential recurrence relations

$$
\begin{aligned}
(1-x^2)\frac{d}{dx}P_l^m &= -\sin\theta\frac{d}{d\theta}P_l^m = (l+m)P_{l-1}^m - lxP_l^m \\
&= (l+1)xP_l^m - (l-m+1)P_{l+1}^m \\
&= (2l+1)^{-1}[(l+1)(l+m)P_{l-1}^m \\
&\quad - l(l-m+1)P_{l+1}^m]
\end{aligned}
\tag{A.10}
$$

Expressions for the associated Legendre functions are obtained from the definition Eq. (A.1) or from the recurrence relations in Eqs. (A.3–A.9). In particular,

$$
\begin{aligned}
&P_0^0 = 1 \quad P_1^0 = x \quad P_1^1 = \sqrt{1-x^2} \\
&P_2^0 = \tfrac{1}{2}(3x^2-1) \quad P_2^1 = 3x\sqrt{1-x^2} \quad P_2^2 = 3(1-x^2) \\
&P_3^0 = \tfrac{1}{2}(5x^3-3x) \quad P_3^1 = \tfrac{1}{2}\sqrt{1-x^2}(15x^2-3) \\
&P_3^2 = 15x(1-x^2) \quad P_3^3 = 15(1-x^2)^{3/2} \\
&P_4^0 = \tfrac{1}{8}(35x^4-30x^2+3) \quad P_4^1 = \tfrac{1}{2}\sqrt{1-x^2}\,(35x^3-15x) \\
&P_4^2 = \tfrac{1}{2}(1-x^2)(105x^2-15) \quad P_4^3 = 105x(1-x^2)^{3/2} \quad P_4^4 = 105(1-x^2)^2
\end{aligned}
\tag{A.11}
$$

The eigenfunctions of Λ^2 are the spherical harmonics

$$X_{l,m}(\theta,\varphi) = e^{im\varphi}P_l^m(\cos\theta) \tag{A.12}$$

in that

$$\Lambda^2 X_{l,m} = -l(l+1)X_{l,m} \tag{A.13}$$

From Eq. (A.2) the spherical harmonics satisfy the symmetry relation

$$X_{l,-m} = (-1)^m\frac{(l-m)!}{(l+m)!}X_{l,m}^* \tag{A.14}$$

The spherical harmonics can be defined in terms of the direction cosines by [37]

$$X_{l,m} = \frac{1}{2^l l!} (\alpha_1 + i\alpha_2)^m \frac{d^{l+m}}{d\alpha_3^{l+m}} (\alpha_3^2 - 1)^l \tag{A.15}$$

The following recurrence relations are obtained from Eqs. (A.3) and (A.4)

$$\begin{aligned}\alpha_1 X_{l,m} = {} & \frac{(l+m)(l+m-1)}{2(2l+1)} X_{l-1,m-1} - \frac{(l-m+2)(l-m+1)}{2(2l+1)} X_{l+1,m-1} \\ & - \frac{1}{2(2l+1)} X_{l-1,m+1} + \frac{1}{2(2l+1)} X_{l+1,m+1}\end{aligned} \tag{A.16}$$

$$\begin{aligned}i\alpha_2 X_{l,m} = {} & -\frac{(l+m)(l+m-1)}{2(2l+1)} X_{l-1,m-1} + \frac{(l-m+2)(l-m+1)}{2(2l+1)} X_{l+1,m-1} \\ & - \frac{1}{2(2l+1)} X_{l-1,m+1} + \frac{1}{2(2l+1)} X_{l+1,m+1}\end{aligned} \tag{A.17}$$

$$\alpha_3 X_{l,m} = \frac{(l+m)}{(2l+1)} X_{l-1,m} + \frac{(l-m+1)}{(2l+1)} X_{l+1,m} \tag{A.18}$$

Equations (A.16) and (A.17) may be expressed more compactly as

$$(\alpha_1 + i\alpha_2) X_{l,m} = -\frac{1}{(2l+1)} X_{l-1,m+1} + \frac{1}{(2l+1)} X_{l+1,m+1} \tag{A.19}$$

$$\begin{aligned}(\alpha_1 - i\alpha_2) X_{l,m} = {} & \frac{(l+m)(l+m-1)}{(2l+1)} X_{l-1,m-1} \\ & - \frac{(l-m+2)(l-m+1)}{(2l+1)} X_{l+1,m-1}\end{aligned} \tag{A.20}$$

The following recurrence relation is obtained by successive application of Eq. (A.18)

$$\begin{aligned}\alpha_3^2 X_{l,m} = {} & \frac{(l+m)(l+m-1)}{(2l-1)(2l+1)} X_{l-2,m} \\ & + \frac{[2(l+m)(l-m) + 2l - 1]}{(2l-1)(2l+3)} X_{l,m} \\ & + \frac{(l-m+2)(l-m+1)}{(2l+1)(2l+3)} X_{l+2,m}\end{aligned} \tag{A.21}$$

The *product formula for spherical harmonics* [34] is

$$X_{s,t}X_{l,m} = \sum_{r=-s}^{s} \begin{bmatrix} s & l & l+r \\ t & m & -m-t \end{bmatrix} X_{l+r,m+t} \tag{A.22}$$

where the product coefficients are expressed in terms of the *Wigner 3*j*-symbols* [34] by

$$\begin{bmatrix} s & l & l+r \\ t & m & -m-t \end{bmatrix} = (-1)^m \frac{2l+2r+1}{4\pi} \sqrt{\frac{(l+r-m-t)!(l+m)!(s+t)!}{(l+r+m+t)!(l-m)!(s-t)!}} \cdot \begin{bmatrix} s & l & l+r \\ 0 & 0 & 0 \end{bmatrix} \begin{bmatrix} s & l & l+r \\ t & m & -m-t \end{bmatrix} \tag{A.23}$$

The *vector addition rules* imply that unless

$$r \equiv s(\text{mod}2) \tag{A.24}$$

the leading Wigner 3*j*-symbol in Eq. (A.23) vanishes, so that the summation in Eq. (A.22) is taken over values of r for which Eq. (A.24) is satisfied.

APPENDIX B. DERIVATION OF BROWN'S FOKKER–PLANCK EQUATION FROM THE GENERAL FOKKER–PLANCK EQUATION AND THE GILBERT–LANGEVIN EQUATION

The *stochastic process* [9, 14] $\{X(t_n), n = 1, 2, \ldots\}$ is described by the family of distribution functions

$$f_1(\mathbf{r}_1, t_1) \qquad f_2(\mathbf{r}_1, t_1; \mathbf{r}_2, t_2), \ldots \tag{B.1}$$

For a *stationary process*, the underlying probability distribution during a given interval of time depends only on the length of that interval and not on when that interval began. This means that a shift in the time axis does not influence the f_n, hence Eq. (B.1) can be replaced by

$$f_1(\mathbf{r}_1) \qquad f_2(\mathbf{r}_1, \mathbf{r}_2, t), \ldots \qquad t = |t_2 - t_1| \tag{B.2}$$

For a *Markov process*, the conditional probabilities satisfy

$$P_n(\mathbf{r}_n, t_n | \mathbf{r}_{n-1}, t_{n-1}, \ldots \mathbf{r}_1, t_1) = P(\mathbf{r}_n, t_n | \mathbf{r}_{n-1}, t_{n-1}) \tag{B.3}$$

All desired properties of the system can be calculated from the *transition probability* $P = P_2$, which is defined by

$$f_2(\mathbf{r}_1, \mathbf{r}_2, t) = f_1(\mathbf{r}_1)P(\mathbf{r}_2, t|\mathbf{r}_1) \tag{B.4}$$

and satisfies the following conditions

$$P(\mathbf{r}_2, t|\mathbf{r}_1) \geq 0$$

$$\int P(\mathbf{r}_2, t|\mathbf{r}_1)d\Omega(\mathbf{r}_2) = 1$$

$$f_1(\mathbf{r}_2) = \int f_1(\mathbf{r}_1)P(\mathbf{r}_2, t|\mathbf{r}_1)d\Omega(\mathbf{r}_1)$$

$$\lim_{t\to\infty} P(\mathbf{r}_2, t|\mathbf{r}_1) = f_1(\mathbf{r}_2) \tag{B.5}$$

In general P satisfies an integral equation that has the same form as the Boltzmann integral equation in the kinetic theory of gasses and under certain conditions [17] it can be expressed as a partial differential equation, namely the Fokker–Planck equation [18]

$$\frac{\partial P}{\partial t} + \frac{\partial S_i}{\partial x_i} = 0 \tag{B.6}$$

where

$$S_i = D_i P - \frac{\partial}{\partial x_i} D_{i,j} P \tag{B.7}$$

are the components of "probability current" and $D_i, D_{i,j}$ are the first- and second-order *Kramers–Moyal coefficients*, respectively. When $P = P(x_1, x_2, t)$

$$D_i = \lim_{\tau\to 0} \frac{1}{\tau} \overline{(x_i(t+\tau) - x_i)}_{x_i = x_i(t)}$$

$$D_{i,j} = \lim_{\tau\to 0} \frac{1}{2\tau} \overline{([x_i(t+\tau) - x_i][x_j(t+\tau) - x_i])}_{x_i = x_i(t)} \tag{B.8}$$

The Gilbert–Langevin equation expressed in spherical polar coordinates [7]

$x_i = \theta,\ x_2 = \varphi$ is

$$\dot{x}_i(t) = h_i(x_1, x_2, t) + g_{i,j}(x_1, x_2, t)\lambda_j(t) \tag{B.9}$$

where $\lambda_j, j = 1, 2, 3$ are the rectangular components of the random field term that arises from thermal agitation and

$$\begin{aligned} h_1 &= -b\,\frac{\partial V}{\partial x_1} + \frac{a}{\sin x_1}\,\frac{\partial V}{\partial x_2} & h_2 &= -\frac{b}{\sin x_1}\,\frac{\partial V}{\partial x_1} - \frac{a}{\sin^2 x_1}\,\frac{\partial V}{\partial x_2} \\ g_{1,1} &= M_s(b\cos x_1 \cos x_2 + a\sin x_2) & g_{2,1} &= M_s(-b\sin x_2 + a\cos x_1 \cos x_2) \\ g_{1,2} &= M_s(b\cos x_1 \sin x_2 - a\cos x_2) & g_{2,2} &= M_s(b\cos x_2 + a\cos x_1 \sin x_2) \\ g_{1,3} &= -bM_s \sin x_1 & g_{2,3} &= -aM_s \sin x_1 \end{aligned} \tag{B.10}$$

Stratonovich [36] found that the Kramers–Moyal coefficients in Eq. (B.8) can in general be expressed as

$$D_i = h_i + g_{j,k}\,\frac{\partial}{\partial x_j}\,g_{i,k} \qquad D_{i,j} = g_{i,k}g_{j,k} \tag{B.11}$$

Substituting for the $g_{i,j}$ from Eq. (B.10) and writing $M_s^2(a^2 + b^2) = \eta^{-1}b$, gives the following expressions for the Kramers–Moyal coefficients in the case of spherical polar coordinates:

$$\begin{aligned} &D_1 = h_1 + \eta^{-1}b\cot x_1 \quad D_2 = h_2 \\ &D_{1,1} = \eta^{-1}b \qquad D_{2,2} = \eta^{-1}b\csc^2 x_1 \quad D_{1,2} = D_{2,1} = 0 \end{aligned} \tag{B.12}$$

The components of probability current in Eq. (B.7) become

$$\begin{aligned} S_1 &= D_1P + \frac{\partial}{\partial x_1}\,D_{1,1}P = h_1P + \eta^{-1}b\cot x_1P - \eta^{-1}b\,\frac{\partial P}{\partial x_1} \\ S_2 &= D_2P + \frac{\partial}{\partial x_2}\,D_{2,2}P = h_2P - \eta^{-1}b\,\frac{\partial}{\partial x_2}\,(\csc^2 x_1P) \end{aligned} \tag{B.13}$$

Brown's Fokker–Planck equation for the distribution

$$W(\theta, \varphi, t)\sin\theta d\theta d\varphi \equiv P(x_1, x_2, t)dx_1dx_2 \tag{B.14}$$

is obtained from Eq. (B.6) by making the transformation $P = W \sin\theta$ and imposing the equilibrium condition $\beta = \eta$ giving

$$\frac{\partial W}{\partial t} = \frac{1}{\sin\theta}\frac{\partial}{\partial\theta}\left\{\sin\theta\left[\left(b\frac{\partial V}{\partial\theta} - a\frac{\partial V}{\partial\varphi}\right)W + \beta^{-1}b\frac{\partial W}{\partial\theta}\right]\right\}$$
$$+ \frac{\partial}{\partial\varphi}\left\{\frac{1}{\sin\theta}\left(b\frac{\partial V}{\partial\theta} + a\frac{\partial V}{\partial\varphi}\right)W + \beta^{-1}b\frac{\partial}{\partial\varphi}\left(\frac{1}{\sin^2\theta}W\right)\right\} \tag{B.15}$$

APPENDIX C. ORTHOGONAL TRANSFORMATIONS

Define a *triad of orthogonal unit vectors* in R^3

$$E = (\mathbf{e}_1, \mathbf{e}_2, \mathbf{e}_3) \tag{C.1}$$

where the dot products are expressed as

$$\mathbf{e}_i\mathbf{e}_j = \delta_{i,j} \qquad 1 \le i \qquad j \le 3 \tag{C.2}$$

then E is an *orthonormal basis* for R^3. Any vector $\mathbf{x} \in R^3$ can thus be uniquely represented in terms of E by the equation

$$\mathbf{x} = \sum x_i\mathbf{e}_i \tag{C.3}$$

The scalars x_i are called the *coordinates* of $\mathbf{x}$ w.r.t E and are arranged as a *coordinate vector*

$$X = [x_i]_{1\le i\le 3} = \begin{bmatrix} x_1 \\ x_2 \\ x_3 \end{bmatrix} \tag{C.4}$$

For the purpose of introducing the concept of an orthogonal transformation it is convenient to express Eq. (C.3) in terms of Eqs. (C.1) and (C.4) by

$$\mathbf{x} = EX \tag{C.5}$$

When $\|\mathbf{x}\| = 1$, the x_i are called the *direction cosines* of $\mathbf{x}$ w.r.t E. Choosing another vector $\mathbf{y} \in R^3$ having coordinate vector $Y = [y_i]_{1\le i\le 3}$ w.r.t E, then the dot product can then be expressed as

$$\mathbf{xy} = X^T Y \tag{C.6}$$

In particular, we have

$$x_i = \mathbf{xe}_i \tag{C.7}$$

Define a *linear transformation*

$$T : R^3 \rightarrow R^3 \qquad T\mathbf{e}_i = \sum A_{j,i}\mathbf{e}_j \tag{C.8}$$

where the scalars $A_{i,j}$ are called the *matrix elements* of T. They are arranged as a matrix

$$A = [A_{i,j}]_{1 \leq i,j \leq 3} \tag{C.9}$$

which is called the *representing matrix* or *matrix representation* of T. The transformation equations in Eq. (C.8) can be expressed in terms of Eqs. (C.1) and (C.9) as

$$TE = EA^T \tag{C.10}$$

T is called an *orthogonal transformation* if

$$T\mathbf{x}T\mathbf{y} = \mathbf{xy} \qquad \text{for all } \mathbf{x}, \mathbf{y} \in R^3 \tag{C.11}$$

which, with the aid of Eqs. (C.6) and (C.9), becomes

$$(AX)^T AY = X^T Y \tag{C.12}$$

Hence the representing matrix is orthogonal, i.e.,

$$A^{-1} = A^T \tag{C.13}$$

The triad of vectors

$$TE = (T\mathbf{e}_1, T\mathbf{e}_2, T\mathbf{e}_3) \tag{C.14}$$

is also an orthonormal basis for R^3 since

$$T\mathbf{e}_i T\mathbf{e}_j = \mathbf{e}_i\mathbf{e}_j = \delta_{i,j} \tag{C.15}$$

Suppose T_1 and T_2 are orthogonal transformations, T_i: $R^3 \to R^3$, then by Eq. (C.11) the composition T_2T_1 is also orthogonal, since

$$T_2T_1\mathbf{x}T_2T_1\mathbf{y} = T_1\mathbf{x}T_1\mathbf{y} = \mathbf{xy} \tag{C.16}$$

Let A_1 denote the representing matrix for T_1 with respect to E, i.e.,

$$T_1E = EQ_1^T \tag{C.17}$$

and A_2 the representing matrix for T_2 with respect to T_1E, i.e.,

$$T_2(T_1E) = (T_1E)A_2^T \tag{C.18}$$

then the representing matrix for T_2T_1 with respect to E, is given by the matrix product A_2A_1 since

$$T_2T_1E = (T_1E)A_2^T = (EA_1^T)A_2^T = E(A_1^TA_2^T) = E(A_2A_1)^T \tag{C.19}$$

We observe that

$$(A_2A_1)^{-1} = A_1^{-1}A_2^{-1} = A_1^TA_2^T = (A_2A_1)^T \tag{C.20}$$

Hence the representing matrix A_2A_1 satisfies orthogonally as required. The orthogonal transformation T gives rise to an *orthogonal coordinate transformation*

$$\mathbf{e}_i' = T\mathbf{e}_i = \sum A_{j,i}\mathbf{e}_j \tag{C.21}$$

which can be expressed in matrix form

$$E' = EA^T \tag{C.22}$$

Application of the orthogonal coordinate transformation to Eq. (C.2) gives the following expression for a vector expressed in terms of the new coordinate system

$$\mathbf{x} = EX = EA^{-1}AX = EA^TAX = E'X' \tag{C.23}$$

where

$$X' = AX = [x'_i]_{l \leq i \leq 3} \tag{C.24}$$

is the coordinate vector of $\mathbf{x}$ with respect to E'

Consider the orthogonal transformations P_φ and Q_θ defined on $\{\mathbf{r} \in R^3, \|\mathbf{r}\| = 1\}$, where P_φ is defined by the finite rotation through an azimuthal angle φ about the $\mathbf{e}_3$ axis in the $\mathbf{e}_1\mathbf{e}_2$-plane (Fig. 21), and Q_θ is defined by the finite rotation through a polar angle θ about the $\mathbf{e}_2$ axis in the $\mathbf{e}_1\mathbf{e}_3$-plane, (Fig. 22). The angle of rotation in both cases being positive for anticlockwise rotation. The transformation equations are

$$P_\varphi\mathbf{e}_1 = \cos\varphi\mathbf{e}_1 + \sin\varphi\mathbf{e}_2 \quad P_\varphi\mathbf{e}_2 = -\sin\varphi\mathbf{e}_1 + \cos\varphi\mathbf{e}_2$$
$$P_\varphi\mathbf{e}_3 = \mathbf{e}_3 \tag{C.25}$$
$$Q_\theta\mathbf{e}_1 = \cos\theta\mathbf{e}_1 - \sin\theta\mathbf{e}_3 \quad Q_\theta\mathbf{e}_2 = \mathbf{e}_2$$
$$Q_\theta\mathbf{e}_3 = \sin\theta\mathbf{e}_1 + \cos\theta\mathbf{e}_3 \tag{C.26}$$

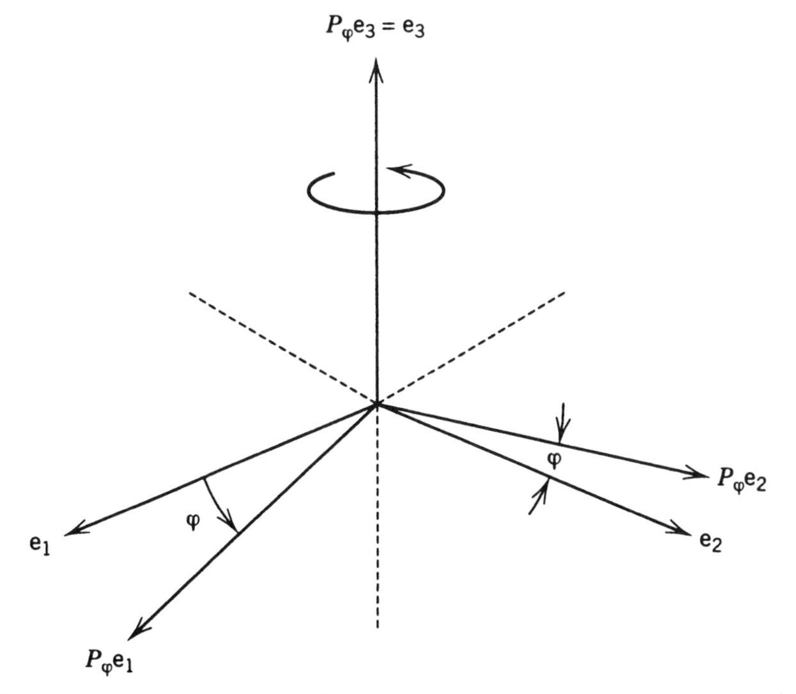

Figure 21. Orthogonal transformation for rotation through an angle φ about the $\mathbf{e}_3$ axis.

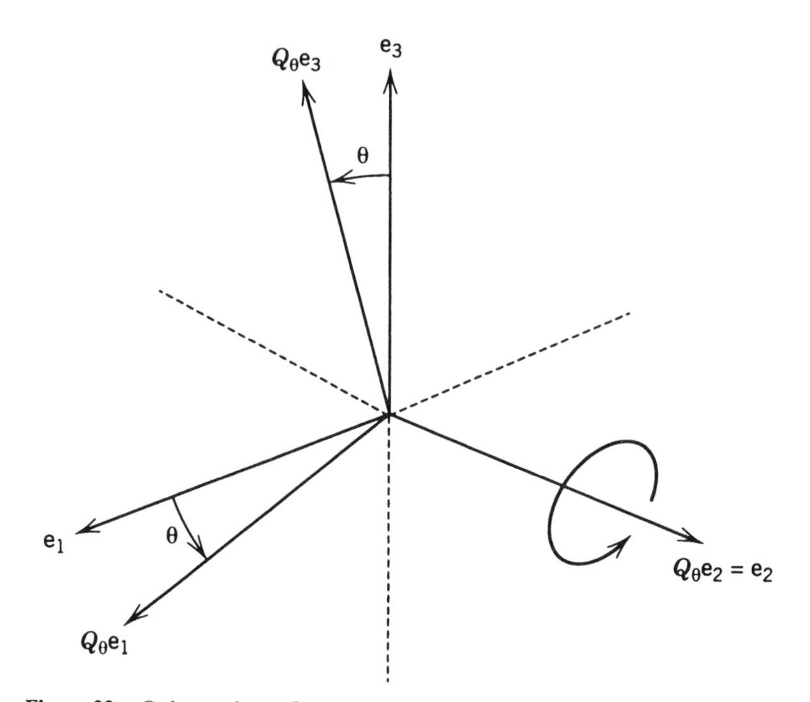

Figure 22. Orthogonal transformation for rotation through an angle θ about the $\mathbf{e}_2$ axis.

Hence the representing matrices of P_φ and Q_θ are, respectively,

$$A_\varphi = \begin{bmatrix} \cos\varphi & \sin\varphi & 0 \\ -\sin\varphi & \cos\varphi & 0 \\ 0 & 0 & 1 \end{bmatrix} \tag{C.27}$$

$$B_\theta = \begin{bmatrix} \cos\theta & 0 & -\sin\theta \\ 0 & 1 & 0 \\ \sin\theta & 0 & \cos\theta \end{bmatrix} \tag{C.28}$$

We observe that

$$C_\varphi^{-1} = C_\varphi^T = C_{-\varphi} \quad \det C_\varphi = 1 \tag{C.29}$$

where C is A or B. The *orthogonal transformation to spherical polar coordinates* is defined by the composition

$$T_{\theta,\varphi} = Q_\theta P_\varphi \tag{C.30}$$

From Eq. (C.19) the representing matrix for $T_{\theta,\varphi}$ is thus

$$S_{\theta,\varphi} = B_\theta A_\varphi = \begin{bmatrix} \cos\theta\cos\varphi & \cos\theta\sin\varphi & -\sin\theta \\ -\sin\varphi & \cos\varphi & 0 \\ \sin\theta\cos\varphi & \sin\theta\sin\varphi & \cos\theta \end{bmatrix} \tag{C.31}$$

The transformation equations are

$$T_{\theta,\varphi}E = E' = ES^T_{\theta,\varphi} \tag{C.32}$$

The new triad of orthogonal unit vectors is denoted

$$E' = (\mathbf{e}'_1, \mathbf{e}'_2, \mathbf{e}'_3) = (\mathbf{e}_\theta, \mathbf{e}_\varphi, \mathbf{e}_r) \tag{C.33}$$

In general, any orthogonal transformation may be specified by three parameters (φ, θ, ψ) and written as a composition of the form

$$T_{\psi,\theta,\varphi} = P_\psi Q_\theta P_\varphi \tag{C.34}$$

The angles (φ, θ, ψ) are called the *Eulerian angles* of S. The presenting matrix for $T_{\psi,\theta,\varphi}$ is

$$S_{\psi,\theta,\varphi} = A_\psi B_\theta A_\varphi$$
$$= \begin{bmatrix} \cos\psi\cos\theta\cos\varphi - \sin\psi\sin\varphi & \cos\psi\cos\theta\sin\varphi + \sin\psi\cos\varphi & -\cos\psi\sin\theta \\ -\sin\psi\cos\theta\cos\varphi - \cos\psi\sin\varphi & -\sin\psi\cos\theta\sin\varphi + \cos\psi\cos\varphi & \sin\psi\sin\theta \\ \sin\theta\cos\varphi & \sin\theta\sin\varphi & \cos\theta \end{bmatrix} \tag{C.35}$$

The transformation equations are

$$T_{\psi,\theta,\varphi}E = \overline{E} = ES^T_{\psi,\theta,\varphi} \tag{C.36}$$

The new triad of orthogonal unit vectors $\overline{E} = (\overline{\mathbf{e}}_1, \overline{\mathbf{e}}_2, \overline{\mathbf{e}}_3)$ results from the three successive rotations

$$\begin{aligned} &\text{rotation } \varphi \text{ about the } \mathbf{e}_3 \text{ axis giving } E' = (\mathbf{e}'_1, \mathbf{e}'_2, \mathbf{e}'_3) \\ &\text{rotation } \theta \text{ about the } \mathbf{e}'_2 \text{ axis giving } E'' = (\mathbf{e}''_1, \mathbf{e}''_2, \mathbf{e}''_3) \\ &\text{rotation } \psi \text{ about the } \mathbf{e}''_3 \text{ axis giving } \overline{E} = (\overline{\mathbf{e}}_1, \overline{\mathbf{e}}_2, \overline{\mathbf{e}}_3) \end{aligned} \tag{C.37}$$

NOTE ADDED IN PROOF

A useful introduction to the theory of ferromagnetism is given in the new book of Aharoni [44].

ACKNOWLEDGMENTS

The authors would like to thank Prof. D. S. F. Crothers, Prof. A. W. Wickstead, Dr. J. L. Dormann, Dr. Yu. P. Kalmykov, Dr. J. T. Waldron, and Miss. E. Kennedy for their assistance.

REFERENCES

1. P. Debye, *Polar Molecules*, Chemical Catalog, New York, 1929 (reprinted by Dover, New York, 1954).
2. A. J. Martin, G. Meier, and A. Saupe, *Symp. Faraday. Soc.*, **5,** 119 (1971).
3. J. L. Déjardin, *J. Chem. Phys.*, **95,** 2787 (1991).
4. A. Abragam, *The Principles of Nuclear Magnetism*, Oxford University Press, London (1961).
5. W. F. Brown Jr., *IEEE Trans. Mag.*, **15,** 1196 (1979).
6. W. T. Coffey, Yu. P. Kalmykov, and J. T. Waldron, *The Langevin Equation*, World Scientific, Singapore (1996).
7. W. F. Brown Jr., *Phys. Rev.*, **130,** 1677 (1963).
8. A. Aharoni, *Phys. Rev.*, **177,** 793 (1969).
9. N. G. van Kampen, *Stochastic Processes in Physics and Chemistry*, North-Holland, Amsterdam, 1981.
10. C. P. Bean and J. D. Livingston, *J. App. Phys.*, **30,** 120S (1959).
11. I. Eisenstein and A. Aharoni, *Phys. Rev. B*, **16,** 1279 (1977).
12. I. Eisenstein and A. Aharoni, *Phys. Rev. B*, 16, 1285.
13. L. Néel, *Ann. Géophys.*, **5,** 99 (1949).
14. W. T. Coffey, P. J. Cregg, and Yu. P. Kalmykov , *Adv. Chem. Phys.*, **83,** 263 (1993).
15. T. L. Gilbert, *Phys. Rev.*, **100,** 1243 (1955).
16. L. D. Landau and E. M. Lifshitz, *Phys. Z. Sowjetunion*, **8,** 153 (1935), (reprinted in collected works of Landau, No. 8, Pergamon Press, London, 1965).
17. M. C. Wang and G. E. Uhlenbeck, *Rev. Mod. Phys.*, **17,** 323 (1945).
18. H. Risken, *The Fokker-Planck Equation*, 2nd ed., Springer-Verlag, Berlin, 1989.
19. A. Einstein, *Investigations on the Theory of the Brownian Movement*, Methuen, London, 1926 (reprinted by Dover, New York, 1956).
20. A. Aharoni, *IEEE Trans. Mag.*, **29,** 2596 (1993).
21. N. A. Lisov and S. E. Pescheny, *J. Mag. Mag. Mat.*, **130,** 275 (1994).
22. L. Weil, *J. Chim. Phys.*, **51,** 715 (1954).
23. A. P. Amuljavichus and I. P. Suzdalev, *Sov. Phys. JETP*, **37,** 859 (1973).
24. I. Eisenstein and A. Aharoni, *J. App. Phys.*, **47,** 321 (1976).
25. A. Aharoni and J. P. Jakubovics, *Phil. Mag. B*, **53,** 133 (1986).

26. A. Aharoni and J. P. Jakubovics, *IEEE Trans. Mag.*, **24,** 1892 (1988).
27. W. T. Coffey, D. S. F. Crothers, Yu. P. Kalmykov, E. S. Massawe, and J. T. Waldron, *Phys. Rev. E*, **49,** 1869 (1994).
28. W. T. Coffey, D. S. F. Crothers, Yu. P. Kalmykov, and J. T. Waldron, *Phys. Rev. B*, **51,** 15947 (1995).
29. W. T. Coffey, D. S. F. Crothers, J. L. Dormann, L. J. Geoghegan, Yu. P. Kalmykov, J. T. Waldron, and A. W. Wickstead, *J. Mag. Mag. Mat.*, **145,** L263 (1995).
30. R. Courant and D. Hilbert, *Methods of Mathematical Physics*, Interscience, New York, 1953, Vol. 1.
31. H. A. Kramers, *Physica*, **7,** 284 (1940).
32. A. Aharoni, *Phys. Rev.*, **135A,** 447 (1964).
33. C. N. Scully, P. J. Cregg, and D. S. F. Crothers, *Phys. Rev. B*, **45,** 474 (1992).
34. A. R. Edmonds, *Elementary Theory of Angular Momentum in Quantum Mechanics*, Princeton University Press, Princeton, NJ, 1957.
35. W. T. Coffey and L. J. Geoghegan, *J. Mol. Liq.*, **69,** 53 (1996).
36. R. L. Stratonovich, *Conditional Markov Processes and Their Application to the Theory of Optimal Control*, Elsevier, New York, 1968.
37. Yu. P. Kalmykov, *J. Mol. Liq.*, **69,** 117, 1996.
38. E. C. Stoner and E. P. Wohlfahrt, *Phil. Trans. Roy. Soc., London*, **A240,** 599 (1948).
39. H. Pfeiffer, *Phys. Stat. Sol.*, **118,** 295 (1990).
40. H. Pfeiffer, *Phys. Stat. Sol.*, **122,** 377 (1990).
41. H. Bateman, *Partial Differential Equations*, Cambridge University Press, London, 1932.
42. W. T. Coffey, D. S. F. Crothers, J. L. Dormann, L. J. Geoghegan, Yu. P. Kalmykov, J. T. Waldron, and A. W. Wickstead, *Phys. Rev. B*, **52,** 15951 (1995).
43. E. C. Kennedy, Private Communication (1997).
44. A. Aharoni, *An Introduction to the Theory of Ferromagnetism*, Oxford University Press, London, 1996.

AUTHOR INDEX

Numbers in parentheses are reference numbers and indicate that the author's work is referred to although his name is not mentioned in the text. Numbers in *italic* show the pages on which the complete references are listed.

Abella, I. D., 70(313), *164*
Abragam, A., 478(4), *640*
Abraham, D. B., 371(230), *392*
Abraham, F., 398(45), *468*
Abramowitz, M., 306(104), *315*
Ackerson, B. J., 355(152), *389*
Adams, J. E., 21(153), *159*
Afinrud, P. A., 62(296), *163*
Agarwal, G. S., 135(419), *167*
Agladze, K., 465(212), *474*
Agrawal, P. M., 23(157), *159*
Aharoni, A., 479(8), 482(11–12), 495(20,24–26), 504(8,26,32), 543(11,20), *640–641*
Ahlrichs, R., 21(151), *159*
Alber, G., 152(483), *169*
Albrecht, A. C., 173–174(16), *226*
Alder, B., 317(1), 321(19–20), 347(111), 356(155), 359(184), 374(240), *384–385*, *388–390*, *392*, 397(36), 438(170), *467*, *472*
Alexander, F. J., 347(111), 374(240), *388*, *392*
Alfano, J. C., 182(40), *226*, 266(63), *314*
Alicka, R., 240(21), 248(21), 302(21), 304(21), *313*
Alimi, R., 120(353), 145(472), *165*, *168*
Allen, J., 255(41), *313*
Allen, L., 133(400), *166*, 255(39), *313*
Allen, M. P., 320(6,10), 321(10), 357(159–160,165), *385*, *389–390*, 435(150), *471*
Alley, W. E., 356(155), 359(184), *389–390*
Allinger, N. L., 23(155), *159*
Almeida, R., 37(221), *161*
Alvarellos, J., 40(241), *162*
Amarouche, M., 141–142(458), *168*
Ambeokar, V., 121(364), *165*
Amellal, A., 460(206), *473*
Amuljavichus, A. P., 495(23), *640*
Anderson, K., 13(129), *159*
Angeloni, L., 115–116(341), *165*
Apkarian, V., 236(17), 301(17), *312*
Apkarian, V. A., 6(66), *157*
Arai, M., 360(188), *390*
Arce, J. C., 141(442), *167*
Argoul, F., 464(208), *474*
Arickx, F., 40(242), *162*
Armstrong, R. C., 353(140), *389*
Arneodo, A., 464(208), *474*
Arnold, L., 422(121), *470*
Arnold, M., 304(79), *314*
Ashkenazi, G., 152(485), *169*, 220(66), *227*, 236(12), 237(18), 253(18), 265(60), 301(12), *312–314*
Ashurst, W. T., 324(32), *386*
Askar, A., 39(232), *161*
Auerbach, S. M., 305(97–98), *315*
Averbuck, I. Sh., 300(66), *314*

Babloyantz, A., 402(60), *468*
Bacic, Z., 36–37(203,205), 38(203), *161*
Bader, J. S., 121(366), *165*, 173(21), 174(21), 192(21), 211(21), 220(78), *226*, *228*
Baer, M., 304(91), *315*
Baer, R., 262(57), *314*, 305(96), *315*
Bagchi, B., 173–174(25), 198(56), *226–227*
Bakker, A. F., 370(223), *391*
Bala, P., 39(239), 141–142(460), *162*, *168*
Balakrishnan, V., 397(21), *467*
Balescu, R., 395(10), *466*
Balint-Kurti, G. G., 270(65), 304(93), *314–315*
Ball, R. C., 353(138), *389*
Ballhausen, C. J., 10(111), 20(111), *158*
Bañares, L., 115(343), 118(343), *165*
Banavar, J. R., 320(15), 324(38), 369(211–212), 370(222,228,232–233), *385–386*, *391–392*

Banin, U., 87(332), *164*, 175(34), 216(34), 220(66), 222(34), *226–227*, 232(1), 233(2,4), 235(11), 237(1–2,4,19), 258(2), 284(2), 294(4), 301(4,11), 305(2), *312–313*
Baras, F., 374(241), *392*, 396(15), 397(19), 404(19), 417(107), 429(133), 433(133), 442(193), 443(194–195), 444(193), 446(195), 450(203), 460(206), 464(207,209), *466–467*, *470–471*, *473–474*
Barbara, P., 266(63–64), 302(64), *314*
Barbara, P. F., 62(279), 68(279), *163*, 172(5), 182(40), *225–226*
Barbara, P. J., 3(7,18), 4(18), 137(18), *155–156*
Bardeen, C. J., 3(15), 70(15), 115(15), 117(15), *155*
Barenblatt, G. I., 442(189), *473*
Barker, J. A., 347(112), 377(256), *388*, *392*
Barnett, R. N., 120(352), 145(470), *165*, *168*
Baronavski, A. P., 62(291), 71(291), *163*
Bartana, A., 87(332), *164*, 175(34), 216(34), 220(66), 222(34), *226–227*, 233(2), 237(2,19), 258(2), 259(45), 261(45), 284(2), 305(2,45), *312–313*
Bateman, H., 609(41), *641*
Baudet, C., 375(252), *392*
Baudrauk, A. D., 40(246), *162*
Bauer, C., 21(143), *159*
Baumert, T., 3(21), 6(21,59,69), 62(21,59,69,286–287), 71(21,59,69,286–287), 87(330), 88(21,330), *156–157*, *163–164*
Baus, M., 320(7), 362(7), 369(218), *385*, *391*
Bavli, R., 31(187), *160*
Bazley, N. W., 442(190), *473*
Bean, C. P., 479(10), 481(10), *640*
Bear, M., 10–11(109–110), 12(109,117), *158*
Bearpark, M. J., 21(152), 142(152), *159*
Beck, M., 173(18), *226*
Becker, P. C., 62(281), 63(281), 66–67(281), 115–116(281), *163*
Beddard, G., 62(300), *163*
Behrman, E. C., 121(374), *166*
Bellemans, A., 332(68), 345(101), *387–388*
Ben Am, B. I., 397(19), 404(19), *467*
Benjamin, I., 235(11), 266(64), 301(64), *312*, *314*
Ben-Nun, M., 152(485), *169*, 236(12), 301(12), *312*
Bentley, J. A., 37(212), 38(225), *161*
Benzi, R., 341(88), *387*, 438(169), *472*
Berendsen, H. J. C., 332(69), 357(166), *387*, *390*
Berg, M., 172(9–10), *225*
Berk, H. L., 375(249), *392*
Berker, A., 358(169), *390*
Berkowitz, M. L., 173(22), *226*
Berman, M., 304(95), *315*
Bernardi, F., 5(29,33), 21(152), 142(152), *156*, *159*
Berne, B. J., 220(78), *228*, 332(70), 352(126), *387*, *389*, 425(129), 438(129), *470*
Berry, R. S., 433(143), *471*
Bersohn, R., 152(481), *169*
Bersuker, I. B., 20(140), 22(140), 27(140,177), 125(140,177), *159–160*
Besseling, R., 304(88), *314*
Bhatanagar, P. J., 341(87), *387*
Bhattacharya, D. K., 368(201–202), *391*, 448(200), *473*
Bianco, R., 302(76–77), *314*
Bigot, J.-Y., 3(15), 62–63(281), 66–67(281), 70(15), 115(15,281,340), 116(281), 117(15), *155*, *163*, *165*
Bilgert, T. L., 489(15), *640*
Billing, G. D., 141–142(449), *168*, 236(13), 301(13), *312*
Binder, K., 320(9,16), *385*
Bird, G. A., 345(102–103), 346(107), 367(197), *388*, *391*, 436(155), 437(159–160), 438(155,162), *471–472*
Bird, R. B., 353(140–141), *389*
Birks, J. B., 3(2), *155*
Bisseling, R. H., 39–40(237), *162*
Bittner, E. R., 236(16), 301(16), *312*
Blais, N. C., 141(443), *167*
Blanché, A., 401(52), *468*
Bloch, F., 31(186), 66(186), 75(186), 132–133(398), *160*, *166*, 301(70), *314*
Blomberg, 412(84), *469*
Blum, K., 6(78), 132(78), *157*
Blümel, R., 139(430), *167*
Boddington, T., 442(191), *473*
Boissonade, J., 439(180–181), 451(180–181), 465(211–212), *472–474*
Bonacic-Koutecky, V., 3(3), 5(3,28), 27(3), *155–156*
Bondybey, V. E., 54(268), *162*
Boon, J. P., 343(96), 360(190), *388*, *390*, 438(167,171), 439(174–175,177), *472*
Boorstein, S. A., 141–142(448), *168*
Borckmans, P., 395(3), 414(101), 418(114a), 429(131–132), *466*, *470–471*

Borgis, D., 142(467), *168*
Borkovec, M., 301(72), *314*
Born, M., 8–9(95), *158*
Bosma, W. B., 7(83), 87(329), 108(329), *157*, *164*
Bossis, G., 352(125), *389*
Bowman, R. M., 6(60), *157*, 234(7), *312*
Boxer, S. G., 172–173(4), *225*
Boyd, J. P., 36–37(199), *161*
Bradforth, S. E., 173(11),175(34), 216(34), 222(34), *225-226*
Brady, J. F., 352(125), *389*
Braitbart, O., 53(263), *162*
Bramley, M. J., 24(171–173), 36(172), 37(172–173), 38(226), *160–161*
Braun, M., 220(79), *228*
Brenig, L., 412(80,82), 437(158), *469*, *472*
Brenner, D. W., 436(154), 464(154), *471*
Brenner, K. H., 173(14), 181(14), *226*
Breton, J., 3(14), *155*
Breuer, H. P., 353(130–131), *389*
Brey, J. J., *389*
Briels, W. J., 357(164), *390*
Briggs, J. S., 6(57), *157*
Brito, R., 439(174), *472*
Broeckhove, J., 40(242), *162*
Brown, D., 36–38(203), *161*, 324(38), 357(159), *386*, *389*
Brown, R., 173(24), *226*
Brown, R. C., 141(444), *167*
Brown, W. F. Jr., 478(5,7), 482(5), 488(5,7), 490(7), 495(7), 499–500(7), 504(7), 543(5,7), 589(5), *640*
Brumer, P., 145(476), *168*
Brun, T. A., 136(428), *167*
Buch, V., 120(350), *165*
Buchanan, S., 3(19), *156*
Buchleitner, A., 139(430), *167*
Buchner, M., 174(33), 212(33), *226*
Buenker, R. J., 13(120–121), *158*
Bühler, B., 3(20), 6(59), 62(59), 71(59), *156–157*
Bulirsch, R., 38–39(229), *161*
Burges, C., 438(169), *472*
Burkert, U., 23(155), *159*
Bussemaker, H., *387*
Buzza, S. A., 62(290), 71(290), *163*

Cacelli, I., 141–142(457), *168*
Cakmak, A. S., 39(232), *161*
Cali, A., 438(169), *472*
Califano, S., 62(283), *163*
Campolieti, G., 145(476), *168*
Campos-Martinez, J., 120(356,359), *165*
Cao, J., 218(72), *227*
Cardoso, O., 375(254), *392*
Carlson, R. J., 6(61,64), 63(61,64), 75(61,64), 88(61,64), *157*, 262(50), *313*
Carmichael, H. J., 136(427), *167*, 173(18), *226*
Caroli, B., 413(89), *469*
Caroli, C., 413(89), *469*
Carr, R., 21(148), *159*
Carrington, T., 5(26), 24(172), 36(172), 37(172–173,210,213), 38(226), *156*, *160–161*
Carter, E. A., 21(149), *159*, 173(21), 174(21), 192(21), 211(21), *226*
Carter, S., 21(144), 38(223), *159*, *161*
Castellucci, E., 53(263), *162*
Castets, V., 465(212), *474*
Castin, Y., 136(421–422), 137(422), *167*
Castleman, A. W. Jr., 3(6), 62(290), 71(290), *155*, *163*
Cazabat, A. M., 370(225–226,229,231), *391–392*
Cederbaum, L. S., 5(35,40), 10(35), 11–12(113), 13(118), 14(35,135–137), 20(35), 26–27(35,175), 33(35,189), 34(189), 36(189,202), 37(202), 38(215), 39(35), 40(189,215), 41(215), 46(35,256–257), 50(35), 53(40,189,256), 120(355,357–358), 122(35), 145–146(189), 148(35), *156*, *158–162*, *165*
Celani, P., 5(33), *156*
Cerjan, C., 39–40(237), *162*, 252(33), 304(88), *313–314*
Certain, P. R., 36(201), *161*
Chakravarty, S., 41(252), 120–121(252), 130(252), 148(252), *162*
Chambers, C. C., 23(162), *159*
Chandler, C., 304(80), *314*
Chandler, D., 173(21), 174(21), 192(21), 211(21), *226*
Chandler, D. W., 62(293), 71(293), 115–116(293), 119(293), 121(366,375–376), *163*, *165–166*
Chandrasekhar, S., 371(234), *392*
Chao, T. H., 28(183), 34–35(183), *160*
Chapman, S., 6(56), 141(446), *157*, *167*
Chaturvedi, S., 402(61–62), *468*
Chebira, A., 62(276), 68(276), 115–117(276), *163*
Chen, E. C. M., 266(61), *314*

Chen, Y., 62(282), 63(303), 100(303), 116(303), *163–164*
Cheng, L. T., 257(43), *313*
Chenoweth, D. R., 438(160), *472*
Chernov, N. I., 328(52), *386*
Chesnoy, J., 62(276,280), 63–64(280), 68(276), 115(276,280), 116(276), 117(276,280), *163*
Cheung, L. M., 13(127), *158*
Child, M. S., 6(56), *157*, 304(92), *315*
China, G. R., 262(50), *313*
Cho, M., 63(302), 70(316), 75–76(316), 80(316), 88(316), 113(316), *164*, 174(30), 192(30), 195(30), 212(30), 213(61), 216(64), *226–227*, 262(54), *313*
Choi, K.-J., 4(23), *156*
Choi, S. E., 6(53), 37(211), *157*, *161*
Chopard, B., 433(145), 466(215), *471*, *474*
Chou, D.-P., 442(193), 444(193), *473*
Chynoweth, S., 358(169), *390*
Ciccotti, G., 320(9,17), 323(31), 326(42), 330(56), 331(58), 332(67–69), 354(142), 356(157), 357(166–167), *385–387*, *389–390*, 398(37–38), *467*
Cichocki, B., 344(99), *388*
Cimiraglia, R., 13(124), *158*
Cina, J. A., 6(61,64), 46(259), 63(61,64), 75(61,64), 88(61,64), *157*, *162*, 173(12), 175(35), 186(49), 188(49), 210(49), *226–227*, 257(42), *313*
Clarke, J. H. R., 324(38), *386*
Clary, D. C., 21(150), *159*
Clause, P. J., 331(63), 368(63), *387*
Clementi, E., 345(104), 349(119–120), 353(127), 364(195), 367(198), 368(203), *388–389*, *391*, 398(42), 438(164,166), 454(204), *468*, *472–473*
Coalson, R. D., 58(273), 120(356,359), 121(369–371), 122(369), 125(369), *163*, *165–166*,185(47), 210(47), *227*
Coffey, W. T., 478(6), 482(14), 488(14), 495(27), 499(14,28–29), 500(35,42), 504(29), 506(28), 507(35), 515(35), 543(14,28), 593(42), 628(42), 631(14), *640–641*
Cohen, C., 359(178), *390*
Cohen, E. G. D., 328(54), 341(89), 384(54), 360(187), *386–387*, *390*
Cohen, J. M., 141–142(459), *168*
Cohen-Tannoudji, C., 133(401), 135(401), *166*, 255(41), *313*
Coker, D. F., 141(440,446), 142(440,465), *167–168*
Cooper, J., 173(18), *226*
Corey, G. C., 24(173), 37(173), 38(222), *160–161*
Cornell, S., 433(145), 465(215), *471*, *474*
Coullet, P. H., 418(109), *470*
Courant, R., 504(30), 632(30), *641*
Cregg, P. J., 482(14), 488(14), 499(14), *640*
Cross, P. C., 22(154), 25–26(154), *159*
Crothers, D. S. F., 495(27), 499(28), 504(29,33), 506(28), 543(28), 593(42), 628(42), *640–641*
Császár, A. G., 24(170), *160*
Cui, S. T., 368(200), *391*
Cukrowski, A. S., 440(187), *473*
Cullum, J. K., 36(196), 95(196), *160*
Curtiss, C. F., 353(141), *389*
Cyr, D. R., 62(295), 71(295), 115(295), 119(295), *163*

Dab, D., 438(171), 439(175,177), 466(171, 217), *472*, *474*
Dahler, J. S., 440(185), *473*
Dalibard, J., 136(421–422), 137(422), *167*
Damen, T. C., 172(3), *225*
Daniel, C., 155(488,490), *169*
Danovich, D., 265(59), *314*
Dantus, M., 3(11), 6(60), 62(11,288), 71(288), 72(11), *155*, *157*, *163*, 234(7), *312*
Dateo, C. E., 37(221), *161*
Davies, E. B., 251(30), 301(74), 302(30), *313-314*
Davis, P. J., 358(169), *390*
De Boeij, W. P., 63(304), 68(304), *164*, 217(80), *228*, 262(55), *314*
Debye, P., 478(1), 481(1), *640*
Decius, J. C., 22(154), 25–26(154), *159*
Deconinck, J., 371(231), *392*
de Gennes, P. G., 358(171), 369(209), 370(226), *390–391*
Déjardin, J. L., 478(3), *640*
De Kepper, P., 465(211–212), *474*
Deleener, M., 322(28), *386*
Delledonne, M., 432(137), *471*
Delos, J. B., 14(134), 141–142(448), *159*, *168*
Denker, H., 410(74), 411(75), *469*
de Pablo, J., 359(176), *390*
de Pasquale, F., 414(97), *469*
de Pazzis, O., 336(80), *387*

De Raedt, B., 122(390), *166*
De Raedt, H., 40(245), 122(390), *162*, *166*
Descalzi, O., 397(32), *467*
deSmedt, Ph., 323(29), *386*
Desouter-Lecomte, M., 15(138), *159*
Deutch, R., 397(24), *467*
Dewel, G., 395(3), 419(114a), 429(131–132), *466*, *470–471*
De Wit, A., 450(202), *473*
Dexheimer, L., 81(327), 87(327), *164*
Dexheimer, S. L., 260(47), *313*
Dhont, J. K. G., 357(164), *390*
d'Humières, D., 338(81), *387*
Dickinson, A. S., 36(201), *161*
Dienes, G. J., 331(59), *386*
Diestler, D. J., 141–142(454), *168*
Dijkstra, M., 357(164), *390*
DiMagno, T. J., 62(278), 68(278), *163*
Dittrich, T., 139(431), *167*
Dixon, M., 331(65), *387*
Dixon, R. N., 155(489), *169*
Dobbyn, A. J., 6(71), *157*
Domcke, W., 5(31,35–36,39–40,44–45), 6(50), 7(39,44,88–94), 10(35), 11(115), 13(130–132), 14(35, 130–132), 15–17(45), 19(45), 20(35), 26(35,175), 27(35,39,45,175,180–182), 28–29(39), 31(50,91), 32(91), 33(35–36,182,189), 34(182,189), 35(39), 36(39,93,189,202), 37(202), 39(35,39), 40(36,189), 41(91), 42(39,91,94), 43(39), 45(94), 46(35–36,256–257), 50(35,45,88–89,91,180–181), 53(36,39–40,88,189,256,264–265), 54(36,39), 56–57(36), 58(39), 64–65(92), 66(50,92), 67(92), 72(91,319–320), 76(50), 77(88–93), 78(90), 83(44,77,92), 85(92), 86(77,89–93), 87(50,93–94), 88(50,91), 90(50), 92(90), 95(36,45,182), 96(337), 97(182), 98(182,337), 102(89–90), 103(89), 105(44,91–92), 106–107(91), 108(50), 115(88–89), 116(88,93), 117(36,88–89), 119(320), 121(378–380, 384–386), 122(35,378–379,384), 123(384), 124–125(386), 126(379–380, 384–385), 127(384–385), 128(45,384–385), 129(384–385), 130(380), 135–136(418), 138(418,429), 142(39), 145(39,88,189), 146(189), 148(35,39), *156–162,164–167*, 212(65), 219(65), *227*, 262(56), *314*
Donaldson, A. B., 442(190), *473*
Doorn, S. K., 58(272), *163*
Dorfman, J. R., 360(187), 384(259–260), *390*, *392*
Dormann, J. L., 504(29), 593(42), 628(42), *641*
Dorsey, A. T., 41(252), 120–121(252), 130(252), 148(252), *162*
Dougherty, T. P., 62(298), *163*
Doxas, I., 375(249), *392*
Dressler, K., 3(19), *156*
Droz, M., 433(145), 466(215), *471*, *474*
Du, M., 6(61,64), 62(278), 63(61,64,302), 68(278), 75(61,64), 88(61,64), *157*, *163–164*, 174(28), *226*, 262(50), *313*
Dubrulle, B., 339(85), *387*
Dufty, J. W., 356(154), *389*
Dujardin, G., 53(263), *162*
Dulos, E., 465(212), *474*
Dum, R., 136(423), *167*
Dunn, T. J., 172–173(2), *225*
Dunnweg, B., 334(71), *387*
Dunweg, B., 358(170), *390*
Dupont, G., 460(206), *473*
Duppen, K., 3(16), 115(345), *156*, *165*
Düren, R., 21(141), 38(141), 41(141), *159*
Durga Prasad, M., 50(260), 120(260), *162*
Durup, J., 141–142(458), *168*
Dussan, E. B., 369(210), *391*
Dyer, R. B., 58(272), 62(297), *163*
Dykman, M., 420(114b), 429(114b), *470*

Ebel, G., 68(310), 87–88(310), *164*
Ebeling, W., 399(47), 416(104), 433(144), 435(153), *468*, *470–471*
Eberly, J. H., 66(308), 69(308), *164*, 173(13), 181(13), *226*, 255(39), *313*
Eberly, Z. H., 133(400), *166*
Ebner, C., 368(204), *391*
Edmonds, A. R., 504(34), 505(34), 581(34), *641*
Edwards, D. M. F., 199(57), *227*
Edwards, S. F., 324–325(34), *386*
Egger, R., 121(367–368,375–377), *165–166*
Ehrenfest, P., 143(469), *168*
Ehrentraut, H., 359(180), *390*
Eiding, J., 53(264–265), *162*
Einstein, A., 491(19), *640*
Eisenstein, I., 482(11–12), *640*
Elbert, S. T., 5(30), 13(127), *156*, *158*
Elphick, C., 418(109), *470*
Elsässer, T., 117(348), 154(348), *165*

Engel, V., 6(52,57,62), 38(221), 40(52), 62(274), 71(52,62), 72(321–322), 87–88(330), 108(322), 119(322), *157*, *161*, *163–165*, 220(79), *228*
Engerholm, G. G., 36(200), *161*
Englman, R., 125(393), *166*
Englund, J. C., 413(93), *469*
Epstein, I. R., 397(18), *467*
Ermolaev, V. L., 4(24), *156*
Ernst, M. H., 338(82), 343(93), *387*, 439(174), *472*
Erpenbeck, J. J., 322(26), 343(95), 355(149), *385*, *388–389*
Escher, C., 433(143), *471*
Evans, D. J., 320(13), 321(13), 326(40–41), 330(55), 331(58,66), 355(147), 358(169), 368(200), *385–387*, *389–391*
Evans, G. T., 192(53), *227*, 357(165), *390*
Eyink, G. L., 328(52), *386*
Ezra, G. S., 27(178), *160*

Faeder, J., 266(62), *314*
Fain, B., 6(55), 7(55,79), 68(55,312), 92(55), 133(406–407), *157*, *164*, *166*
Fan, R., 192(54), *227*
Fang, J.-Y., 152(485), *169*, 220–221(76), *228*
Fano, U., 6(75), *157*
Farantos, S. C., 21(144), *159*
Fee, R. S., 218(75), *227*
Feistel, R., 416(104), *470*
Feit, M. D., 36(206), 38(218), 39(218,237), 40(237), *161–162*, 254(37–38), 304(88), *313–314*
Felderhof, B. U., 344(99), *388*
Feller, W., 409(72), 430(134), *469*, *471*
Felts, A. K., 133–134(414), *167*
Feng, C.-G., 442(191), *473*
Fernández, F. M., 40(243), *162*
Ferrario, M., 356(157), *389*
Ferretti, A., 40(249), 50(249), 155(491), *162*, *169*
Ferwerda, H. A., 75(333), 92(333), *164*
Feyen, B., 40(242), *162*
Feynman, R. P., 120(360–361), 121(372), 125(372), *165–166*
Fife, P., 433(142), *471*
Figuerido, F. E., 134(416), *167*
Finger, K., 155(488), *169*
Fischer, G., 27(176), *160*
Fischer, I., 62(294), 71–72(294), *163*
Fisher, M. P. A., 41(252), 120–121(252), 130(252), 148(252), *162*
Fisher, R. A., 433(142), *471*
Flannery, B. P., 38–39(230), *161*
Fleck, J. A. Jr., 36(206), 38–39(218), *161*, 254(37–38), *313*
Fleming, G. R., 6(61,64), 62(277–278), 63(61,64,302), 68(277–278), 70(316), 75(61,64,316), 76(316), 80(316), 88(61,64,316), 113(316), 115(277), 118(277), 133–134(412,415), 139(415), *157*, *163–164*, *167*, 172(1), 173(11), 174(1,27), 175(34), 182(37–38), 213(61), 216(34), 217(27), 222(34), *225–227*, 262(50–52,54), *313*
Fleming, P. D. III, 359(178), *390*
Foggi, P., 62(283), *163*
Fonseca, T., 174(29), 211(29), *226*
Forster, D., 359(177), *390*
Fourkas, J. T., 172(10), *225*
Fox, R. F., 397(20), *467*
Fragnito, H. L., 62(281), 63(281), 66–67(281), 115(281,340), 116(281),*163*, *165*
Frankowicz, M., 397(17), 404(17,66), *467–468*
Fraysse, N., 370(225,229), *391–392*
Frederick, J. H., 25(174), 142(468), *160*, *168*
Freed, K. F., 5(43), 50(43), *156*, 218(68), *227*
Freeman, J., 414(96), *469*
Freidman, H. L., 173–174(26), *226*
Freisner, R. A., 301(73), 304(88), *314*
Frenkel, D., 320(17), 323(29–30), 342(90–92), 343(97), 357(164–165), *385–388*, *390*
Freudenberg, T., 62(292), 71(292), *163*
Frezzotti, A., 348(114–115), *388*
Fried, L. E., 7(81), 152(480), *157*, *169*
Friedman, H. L., 352(124), *388*
Friesner, R., 236(15), 301(15), *312*
Friesner, R. A., 38(225), 39(237), 133–134(412–414), 138(413), 141–142(439), *161–162*, *167*, 221(77), *228*
Frigerio, A., 249(27), *313*
Frisch, U., 334(75), 338(81,83), 339(85), *387*
Fritzsche, S., 440(187), *473*
Fujii, M., 72(318), *164*
Fujimara, Y., 6(58), 68(311), 87–88(331), *157*, *164*

Gadea, F. X., 13–14(125), 141–142(458), *158*, *168*
Gallavotti, G., 328(54), 384(54), *386*

Gallico, R., 330(56), *386*
Galloy, C., 15(138), *159*
Gammon, R. W., 362(193), *391*, 438(168,173), *472*
Garcia, A. L., 345(104), 347(110–111), 349(119–120), 353(127), 364(195), 374(240–241), *388–389*, *391–392*, 437(156), 438(164,166), 454(204), *471–473*
Garcia-Vela, A., 145(472), *168*, 301(67), *314*
Gardecki, J. A., 172(7), *225*
Gardiner, C. W., 133(402), 136(424), *166–167*, 397(28), 402(61–62), 408(71), 413(92), 420(117), 424(117), 425(126), 458(126), *467–470*
Gardner, M., 335(79), *387*
Garg, A., 41(252), 120(252), 121(252,364), 130(252), 148(252), *162*, *165*
Garraway, B. M., 40(244), *162*, 255(40), *313*
Garret, B. C., 210(58), *227*
Garvia-Vela, A., 120(353), *165*
Gaspard, P., 320(18), 384(258–259), *385*, *392*
Geoghan, L. J., 504(29), 507(35), 515(35), 593(42), *641*
George, T. F., 140–141(435), *167*
Gerber, G., 3(20–21), 6(21,59,69), 62(21,59,69,286), 71(21,59,69,286), 87(330), 88(21,330), 120(350,353), *156–157*, *163–165*
Gerber, R. B., 145(472), *168*, 301(67–68), *314*
Gershgoren, Gordon E., 234(6), *312*
Gertner, B. J., 266(64), 302(64), *314*
Geysermans, 464(209), *474*
Gill, W., 442(190), *473*
Gillan, M. J., 328(48), 331(65), *386–387*
Gillespie, D. T., 353(129), *389*, 401(49–50), 454(49), *468*
Gisin, N., 136(426), *167*
Glansdorff, P., 395(9), *466*
Glauber, R. G., 411(79), *469*
Glownia, J. H., 65(307), 152(307), *164*
Gnanakaran, S., 234(10), 251(10), *312*
Goldstein, S., 328(53), *386*
Gollub, J. P., 375(253), *392*
Gordon, L. K., 38–39(228), 79(326), 95(326), *161*, *164*
Gordon, M. S., 21(145), *159*
Gorecki, J., 444(198), *473*
Gorini, V., 248(25), 249(27), *313*
Gosling, E. M., 324–325(33), *386*
Goto, A., 72(318), *164*
Gottlieb, D., 36–37(197), *160*
Grabert, H., 412(85–86), *469*
Gragg, R. F., 413(93), *469*
Graham, R., 139(430), 141–142(455), *167–168*, 397(23,30,32), 403(23), 418(111–112), *467*, *470*
Grant, E. R., 27(178), *160*
Granucci, G., 155(491), *169*
Gray, P., 442(191), 449(201), *473*
Gray, S. K., 39(238), *162*
Greene, B. I., 62(284), 71(284), *163*
Greer, J. C., 21(151), *159*
Gregory, A. R., 20(139), *159*
Gregory, J. K., 21(150), *159*
Gregory, T. A., 4(22), *156*
Grest, G. S., 334(71), 355(151), *387*, *389*
Grishanin, B. A., 142(468), *168*
Groenenboom, G. C., 28(184), *160*
Grosfils, P., 343(96), *388*, 439(174), *472*
Gross, E. P., 341(87), *387*
Grosser, M., 62(286), 71(286), *163*
Grossman, F., 218(71), *227*
Grossman, S., 402(54), 410(73), 416(106), *468–470*
Gruebele, M., 6(63), *157*, 234(7), *312*
Gryko, J., 444(198), *473*
Gu, Y., 133(405), 135(405), *166*
Guan, Y., 23(156,158), 29(158), *159*
Guayacq, J. P., 14(133), *159*
Gudkov, Y. P., 81(328), *164*
Guenza, M., 218(68), *227*
Guldberg, A., 39–40(237), *162*, 304(88), *314*
Guo, H., 21(142), 38(217), 41(254), 62(288), 71(288), *159*, *161–163*, 240(22), *313*
Gustafson, T. K., 76(324), *164*
Guyon, E., 375(252), *392*
Gwinn, W. D., 36(200), *161*

Haaslacher, B., 334(75), 338(83), *387*
Hadder, J. E., 25(174), *160*
Hagen, M. J. H., 342(92), *387*
Hager, J., 72(317), *164*
Haken, H., 395(2), 420(116), *466*, *470*, 397(23), 403(23), *467*
Hamer, N., 6–7(55), 68(55), 92(55), *157*
Hamilton, I. P., 36(198), *161*
Hamm, P., 3(19), 62(299), *156*, *163*
Hammerich, A. D., 34(191), 39(237), 40(237,250), 68(250), 87–88(250), 120(358), *160*, *162*, *165*, 236(14), 259(44), 301(14), 304(88), *312–314*
Hammes-Schiffer, S., 141(445), *167*

Han, C., 62(296), *163*
Hanamura, E., 133(403), *166*
Handy, N. C., 21(153), 23(165–166), 24(165–166,170–171), 38(223), *159–161*
Hänggi, P., 139(431), *167*, 301(72), *314*, 397(25), 412(81), 413(86), *467*, *469*
Hanley, H. J. M., 320(3), 334(3), *385*
Hannon, L., 368(203), *391*, 398(42), *468*
Hansen, A. E., 10(111), 20(111), *158*
Hansen, J. P., 435(149), *471*
Hanson, D. K., 62(278), 68(278), *163*
Hanusse, P., 401(52), 432(141), *468*, *471*
Hardy, J., 336(80), *387*
Harris, D. O., 36(200), *161*
Harris, S., 341(87), *387*
Hartke, B., 21(149), *159*, 304(81), *314*
Hartmann, S. R., 70(313), *164*
Harvey, J. K., 346(108), *388*
Hase, W. L., 140(433), *167*
Hassager, O., 353(140), *389*
Haug, K., 120(351), *165*
Hauge, E. H., 343(93), *387*
Hautman, J., 370(224), *391*
Hayden, C. C., 62(293,295), 71(293,295), 115(293,295), 116(293), 119(293,295), *163*
Hayoun, M., 369(215), *391*
Heather, R., 38(214), 41–42(214), 68(309), 97(309), *161*, *164*, 186(48), *227*
Heilweil, E. J., 62(298), *163*
Heitz, M. C., 155(490), *169*
Heller, E. J., 27(179), 78(179), 95(179), 96(336), 120(349), 141(444), 145(474), 152(479), *160*, *165*, *167–169*, 185(42–44), 218(69–70), *226–227*, 263(58), *314*
Hendekovic, J., 13(122), *158*
Henderson, D., 347(112), 377(256), *388*, *392*
Henderson, J. R., 38(224), *161*
Henneker, W. H., 20(139), *159*
Hénon, M., 338(81), 339(85), *387*
Herek, J. L., 62(287), 71(287), *163*
Herman, M. F., 141(442), 145(473), *167–168*
Hertel, I. V., 21(151), *159*
Herzberg, G., 5(25), *156*
Heslot, F., 370(225,229), *391–392*
Hess, S., 355(145–146,148,150,153), 359(179–180), *389–390*
Hesselink, W. H., 70(314), *164*
Heumann, B., 21(141), 38(141), 41(141,255), *159*, *162*
Hibbs, A. R., 120(361), *165*
Hilbert, D., 504(30), 632(30), *641*
Hildebrand, F. B., 38–40(227), *161*
Hirata, F., 173–174(26), *226*
Hirsch, G., 13(120), *158*
Hirsch, M. D., 62(275), 68(275), *163*
Hochstrasser, R. M., 58(271), 62(296), 115(342,344), 118(342,344), *163,165*, 234(9–10), 251(10), 302(9), *312*
Hoffman, D. J., 304(79–80,82–83), 305(103), *314–315*
Hoffman, D. K., 39(239), *162*
Hoheisel, C., 324(37), 359(185), *386*, *390*
Höhnerbach, M., 141–142(455), *168*
Holian, B. L., 320(5), 321(25), 327(45), 328(5,48), 331(61–62), 350(123), 356(157), *385–389*, 398(40–41), 438(170,172), 466(172), *468*, *472*
Hoogerbrugge, P. J., 353(135), *389*
Hoorelbeke, S., 371(231), *392*
Hoover, W. G., 320(12), 324(32), 326(44), 328(48–51), 331(62), 332(69), 353(134), *385–387*, *389*, 398(37–38,41), *467–468*
Horng, M. L., 172(7), *225*
Horsthemke, W., 375(246–247), *392*, 412(80,82), 418(110), 424(125), 465(210), *469–470*, *474*
Horton, W., 375(249), *392*
Houard, J., 427(130), 432(139), *470–471*
Hounkonnou, M. N., 357(168), *390*
Hrusak, J., 265(59), *314*
Huang, K., 8–9(95), *158*
Huang, Y., 304(80,82–83), 305(103), *314–315*
Huber, W., 53(265), *162*
Hudson, B., 27(182), 33–34(182), 36(182), 95(182), 97–98(182), *160*
Hulin, J. P., 375(252), *392*
Hunziker, L., 62(282), *163*
Hupp, J. T., 3(18), 4(18), 137(18), *156*
Hutson, J. M., 6(71), *157*
Huxley, P., 21(144), *159*
Hynes, J. T., 173(21), 174(21), 192(21), 211(21), *226*, 266(64), 302(64,76–77), *314*

Iacucci, G., 332(67), *387*
Ianiro, N., 328(53), *386*
Icsevgi, A., 65(306), *164*
Imre, D. G., 6(68), 62(68), *157*
Indekeu, J. O., 370(223), *391*
Innes, K. K., 95(334), *164*
Ippen, E. P., 63(301), *164*
Irving, J. H., 324–325(35), *386*
Israelachvili, J. N., 357(162), *390*

Ito, M., 72(318), *164*
Iung, C., 34(193–194), 38(193), *160*

Jackson, J. D., 261(48), *313*
Jacobsen, E., 192(54), *227*
Jakubovics, J. P., 495(25–26), 504(25), *640*
Jang, S., 135(420), *167*
Jansen, H. K., 408(71), *469*
Jansen, J. P., 320(11), *385*
Janssen, M. H. M., 62(288), 71(288), *163*
Jarezba, W., 62(279), 68(279), *163*, 172(5), *225*
Jean, J. M., 7(85), 62(277), 68(85,277), 115(277), 118(277), 133–134(85,412,415), 138(412), 139(415), *157*, *163*, *167*, 173(19), 220(19), *226*
Jenssen, M., 435(153), *471*
Jia, Y., 217(27), *226*
Jiang, X.-P., 38(214), 41–42(214), *161*
Jimenez, R., 172(1), 173(11), 174(1), 183(38), *225–226*
Joanny, J. F., 370(227), *391*
Johnson, A. E., 62(279), 68(279), *163*, 220(81), *228*, 233(3), 300(3), *312*
Jolicard, G., 39–40(237), *162*, 304(88), *314*
Jonas, D. M., 175(34), 216(34), 222(34), *226*, 262(52), *313*
Joo, T., 217(27), *226*
Jortner, J., 5(42), 121(365), *156*, *165*
Jungwirth, P., 145(472), *168*, 301(68), *314*

Kaburaki, H., 374(241), *392*
Kac, M., 348(117–118), *388*, 437(157), *471*
Kadanoff, L. P., 343(94), 357(158), 359(182), *388–390*
Kaiser, W., 117(348), 154(348), *165*
Kalbfleish, T., 192(54), *227*
Kalmykov, Yu. P., 478(6), 482(14), 488(14), 495(27), 499(14,28), 504(29), 506(28), 537(37), 543(28), 589(37), 593(42), 628(42), 633(37), *640–641*
Kamenetskii, D. A. Frank, 441(188), *473*
Kapral, R., 395(8), 397(22), 438(22,171–172), 439(177), 466(22,171–172,214), *466–467*, *472*, *474*
Karapiperis, T., 433(145), *471*
Karki, L., 3(18), 4(18), 137(18), *156*
Karlin, S., 401(48), 404(67), *468*
Karplus, M., 185(47), 210(47), *227*, 440(184), *473*
Karrlein, W., 39–40(237), *162*, 304(88), *314*
Karttunen, M., 371(230), *392*
Kaski, K., 371(230), *392*
Kawazaki, K., 353(136–137), *389*
Kay, K. G., 145(475), *168*
Keiser, J., 402(55), *468*
Keizer, J. E., 397(20), *467*
Kennard, E. H., 448(199), *473*
Kennedy, E. C., 546(43), *641*
Kestemont, E., 362(194), 372(235–236), *391–392*, 398(43–44), 438(165), *468*, *472*
Keyes, T., 212(60), *227*
Khidekel, V., 7(79), *157*
Khundkar, L. R., 6(51), 62(51), 71(51), *157*, 234(8), *312*, 397(16), *467*
Kilian, A., 359(179), *390*
Kim, C. S., *389*
Kim, S., 359(176), *390*
Kimura, Y., 182(40), *226*
Kinugawa, T., 145(477), *168*
Kipnis, C., 328(53), *386*
Kirkpatrick, T. R., 360(187), *390*
Kirkwood, J. G., 324–325(35), *386*
Kitahara, K., 402(59), 420(115), 423(122), 425(128), 432(138), *468*, *470–471*
Klein, M. L., 328(47), 370(224), *386*, *391*
Klessinger, M., 3(5), 5(34), 27(5), *155–156*
Klomp, U. C., 358(169), *390*
Kluk, E., 145(473), *168*
Knee, J. L., 62(285), 71–72(285), *163*
Knessel, C., 411(77), 414(94), *469*
Know, W. H., 3(7), *155*
Knudson, S. K., 141–142(448), *168*
Kobe, K., 6(70), 62(70,289), 71(70,289), *157*, *163*
Koelman, J. M. V. A., 353(135), *389*
Kohen, D., 251(28), *313*
Kolmogorov, A., 433(142), *471*
Kondepudi, D. K., 404(65), *468*
Kono, H., 6(58), 68(311), *157*, *164*
Koo, H. M., 355(153), *389*
Koplick, J., 320(15), 324(38), 369(211–212), 370(222,228,232–233), *385–386*, *391–392*
Köppel, H., 5(35–38,40), 6(72), 7(88), 10(35), 11–12(113), 13(118), 14(35,135,137), 20(35), 26–27(35), 33(35–36,38,188–189), 34(37–38,189–190), 36(189,202), 37(37,202), 38(215), 39(188), 39(35), 40(36,188–190,215), 41(215), 42(37,215), 46(35–38,72,256–257), 47(38), 48(72), 50(35,38,88), 53(36–38,88,188–190,256,264), 54(36), 56–57(36,188), 83(88), 86(88), 95(36),

Köppel, H., (*continued*) 115(88), 116(88,190), 117(36,38,88), 122(35), 142(37), 145–146(189–190), 148(35), *156–162*
Koroteev, N. I., 142(468), *168*
Kosloff, D., 38–40(219), *161*, 304(90), *314*
Kosloff, R., 31(185), 36(207–209), 37(208), 38(219), 39(207–209,219,235,237), 40(207,219,235,237,247,250), 68(250), 87(250,332), 88(250), 120(354,358), *160–162*, *164–165*, 175(34), 216(34), 220(66), 222(34), *226–227*, 232(1), 233(2,4), 237(1–2,4,18–19), 241(23), 247(23), 249(26), 252(33–34), 253(35), 258(2), 259(44–45), 262(45,53,57), 265(60), 284(2), 294(4), 301(4,14), 303(34,78), 304(35,78,81,84,88,90,95), 305(2,34,45,96,100), *312–315*
Kossokowski, A., 248(25), *313*
Kotler, Z., 120(354), *165*
Kouri, D. J., 39(239), *162*, 304(79,80,82–83), 305(83,103), *314–315*
Kowalczyk, P., 64(305), 68–69(305), 86(305), *164*
Kramers, H. A., 504(31), 5432(31), 631(31), *641*
Kremer, K., 334(71), 358(170), *387*, *390*
Krempl, S., 121(380,384–386), 122–123(384), 124–125(386), 126(380,384–385), 127–129(384–385), 130(380), *166*
Krinsky, V. I., 395(7), *466*
Krook, M., 341(87), *387*
Kubia, L., 435(152), *471*
Kubin, L., 353(139), *389*
Kubo, R., 133(403), *166*, 402(59), 457(205), *468*, *473*
Kubo, R. J., 79(325), 95(325), *164*
Kuharski, A., 121(366), *165*
Kühling, H., 6(70), 62(70,289), 71(70,289), *157*, *163*
Kühn, O., 7(86), 133–134(86,411), 138–139(86), *157*, *167*
Kühne, T., 234(5), *312*
Kulander, K. C., 39(234), 41(253), *161–162*
Kum, O., 353(134), *389*
Kumar, P. V., 172(1), 174(1), *225*
Kuntz, P. J., 141–142(438), *167*
Kuppermann, A., 46(258), *162*
Kuramoto, Y., 396(12), 418(108), 419(113), 424(123–124), *466*, *470*
Kurkijarvi, J., 324(36), *386*
Kurnit, N. A., 70(313), *164*
Kurtz, T. G., 402–403(63), *468*
Laane, J., 28(183), 34–35(183), *160*
Ladanyi, B. M., 174(29,31–33), 211(29), 212(31–33), *226*
Ladd, A. J. C., 326(44), 328(49), 331(64), 343(97), 344(98), *386–388*, 439(176), *472*
Laird, B. B., 251(32), 301(32), *313*
Lakshminarayan, C., 62(285), 71–72(285), *163*
Lallemand, P., 343(96), *388*, 439(174), *472*
Lamb, W. E., 65(306), *164*
Lambry, J.-C., 3(14), *155*
Lami, A., 40(249), 50(249), *162*
Landau, L. D., 360(189), *390*, 399(46), 411(78), *468–469*, 490(16), *640*
Landi, K., 240(21), 248(21), 302(21), 304(21), *313*
Landman, U., 120(352), 145(470), *165*, *168*
Lang, M. J., 217(27), *226*
Lantelme, F., 352(124), *388*
Lathowers, L., 40(242), *162*
Law, B. M., 361–362(191), *390*, 438(168), *472*
Lawniczak, A., 438(171–172), 439(177), 466(171–172), *472*
Leach, S., 53(263), *162*
Lebowitz, J. L., 328(52), *386*
Lee, S.-Y., 6–7(49,54), 62(49), 74(49), 76(49), 79(49), 83(49), 87(49), 92(49), 96(336), 133(49), 152(54), *156*, *165*
Lees, A. W., 324–325(34), *386*
Lefever, R., 411(76), 414(100–101), 418(110), 432(138), *469–471*
Leforestier, C., 34(193–194), 37(212), 38(193,220,225), 39(233,237), 40(237), *160–162*, 304(88–89), 305(97), *314–315*
Leger, L., 370(227), *391*
Leggett, A. J., 41(252), 120–121(252), 130(252), 148(252), *162*
Lekkerker, H. N. W., 357(164), *390*
Lemarchand, A., 397(19,31), 404(19), 412(83), 434(146–148), 438(147), 466(148), *467*, *469*, *471*
Lemarchand, H., 397(31), 412(83), 415(103), 420(118), 424(118), 434(146–147), *467*, *469–471*
Lemoine, D., 37(222), *161*
Lenderick, E., 115(345), 118(345), *165*
Lengsfield, B. H., 9(96), 10(96,100–101), 13(96), *158*
Le Quéré, F., 38(220), *161*
Lesne, A., 434(147–148), 466(148), *471*
Lesyng, B., 39(239), 141–142(460), *162*, *168*
Levesque, D., 321(24), 324(36), *385–386*
Levinson, P., 370(229), *392*

Levy, R. M., 134(416), *167*
Lewis, J. D., 28(183), 34–35(183), *160*
Li, W. B., 362(193), *391*, 438(168,173), *472*
Li, Y., 438(171), 466(171), *472*
Li, Z., 6(66), 152(486), *157*, *169*, 220–221(76), *228*
Lian, T., 62(296), *163*
Librovich, V. B., 442(189), *473*
Lichten, W., 10(105), *158*
Lie, G., 398(42), 438(164,166), 448(200), 454(204), *468*, *472–473*
Lie, G. C., 345(104), 349(119–120), 353(127), 364(195), 368(201–203), *388–389*, *391*
Lie, T. J., 414(96), *469*
Liem, S. Y., 324(38), *386*
Lifshitz, E. M., 360(189), *390*, 399(46), 411(78), *468–469*, 490(16), *640*
Light, J. C., 6(53), 36(198,203), 37(203–205,211), 38(203), 39–40(236), 124(392), *157*, *161–162*, *166*
Lill, J. V., 36(198), *161*
Lin, H.-B., 4(23), *156*
Lin, S. H., 6(55), 7(55,79), 68(55,312), 92(55), 133(406–407), *157*, *164*, *166*
Lindblad, G., 240(20), 248(20), 304(20), *313*
Lindenberg, K., 414(96), *469*
Lindner, P., 358(175), *390*
Link, A., 358(175), *390*
Lipkin, N., 39–40(237), *162*, 304(88), *314*
Lipsky, S., 4(22), *156*
Lisov, N. A., 495(21), *640*
Littlejohn, R. G., 186(50), *227*
Liu, L., 240(22), *313*
Liu, Q., 6(67), *157*
Livingston, J. D., 479(10), 481(10), *640*
Loettgers, A., 21(143), *159*
Logan, J., 348(118), *388*, 437(157), *471*
Longuet-Higgins, H. C., 5(25), 10(104), 14–15(104), 20(104), 22(104),26(104), 46(104), *156*, *158*
Loose, W., 355(143,145–146,150), *389*
Lord, G., 346(108), *388*
Loring, R. F., 173(17), 218(74), *226-277*
Lorquet, J. C., 15(138), *159*
Lotharreichel, 305(101), *315*
Louisell, W. H., 6(77), 80(77), 135(77), *157*
Lovett, R., 369(218), *391*
Lowe, C. P., 342(92), *387*
Lucy, L. B., 353(132), *389*
Ludowise, P., 63(303), 100(303), 116(303), *164*
Ludwise, P., 62(282), *163*
Lukka, T. J., 24(169), *160*
Luo, H., 323(31), *386*
Lynch, G. C., 23(156), *159*
Lyubimtsev, V. A., 4(24), *156*

Ma, J., 172(9), *225*
MacCurdy, C. W., 304(94), *315*
Mackri, N., 301(71), *314*
Madden, P. A., 199(57), *227*
Magnus, W., 124(391), *166*, 254(36), *313*
Mahieu, M., 397(35), 440(35), 446(35), *467*
Mahr, H., 62(275), 68(275), *163*
Mak, C. H., 121(367–368,375–377), *165–166*
Makarov, D. E., 121(381), 130(381), *166*
Makhviladze, G. M., 442(189), *473*
Makri, N., 121(381–383), 125(382,394–395), 130(381–383), *166*, 218(73), *227*
Malek Mansour, M., 345(104), 347(110), 349(119–120), 353(127–128), 362(192), 363(194), 364(195), 373(238), 374(241), *388–389*, *391–392* 397(19,29), 398(44), 402(53,56), 404(19,66), 411(29), 412(82), 416(29), 417(107), 422(120), 423(29), 424(29,125), 425(127), 427(130), 430(135), 431(136,138–140), 438(164–166), 443(194), 450(203), 454(204), 464(207), *467–473*
Malloy, T. B., 28(183), 34–35(183), *160*
Malmquist, P.-A., 13(129), *159*
Malrieu, J.-P., 13(124), *158*
Mandelshtam, V. A., 305(102), *315*
Mangel, M., 411(77), 414(94), *469*
Manneville, P., 374(245), *392*, 396(13), *466*
Manolopoulos, D. E., 145(478), *169*
Mäntele, W., 62(299), *163*
Manthe, U., 5(37–38), 33(38), 34(37–38), 37(37), 38(215), 40–41(215), 42(37), 46(37–38), 47(38), 50(38), 53(37–38), 117(38), 120(355,357–358), 142(37), *156*, *161*, *165*
Mantro, A., 6(61,64), 63(61,64), 75(61,64), 88(61,64), *157*
Manx, T., 155(490), *169*
Manz, J., 3(6,10), *155*
Mareschal, M., 320(4–5), 323(31), 328(5,48), 331(63), 345(104), 346(107), 347(110), 348(116), 349(120–122), 350(123), 360(4), 363(194), 368(63), 369(215,217), 370(219), 372(235–236), 373(238), 378(257), *385–388*, *391–392*, 398(39–40,43–44),434(146–148), 438(163,165,169,170,172), 450(202), 460(206), 466(148), *468*, *471–473*

Maroncelli, M., 172(1,6–7), 173(20), 174(1,20), 175(20), 182(37), 192(20), 195(20), 199(20), 211(20), 218(75), 219(20), *225–227*
Martens, C. C., 6(66), 152(486), *157*, *169*, 220–221(76), *228*, 236(17), 301(17), *312*
Martin, A. J., 478(2), 579(2), *640*
Martin, G., 353(139), *389*
Martin, J.-L., 3(14), *155*
Martin, P. C., 359(182), *390*
Martinez, T. J., 152(485), *169*, 236(12), 301(12), *312*
Martyna, G. J., 328(47), 332(70), *386–387*
Marzocchi, P. M., 115–116(341), *165*
Masiar, P., 438(172), 466(172), *472*
Maslen, P. E., 266(62), *314*
Massawe, E. S., 495(27), *641*
Massobrio, C., 328(48), 331(58), *386*
Masters, A. J., 356(157), *389*
Matheson, I. S., 408(71), 420(117), 424(117), *469–470*
Mathies, R. A., 3(17), 6(49,54), 7(49,80), 62(49), 64(80), 66–67(80), 74(49), 75(80), 76(49,80), 79(49), 81(327), 83(49), 87(49,327), 92(49,80), 115(17,340,346–347), 118(17,346–347), 133(49), 152(54), *156–157*, *164–165*, 260(46–47), 261(46), *313*
Matkowsky, B. J., 411(77), *469*
Matro, A., 173(12), *226*, 262(50), *313*
Matskowsky, B. J., 414(94), *469*
Matsuo, K., 402(59), 413(87), *468–469*
Mauricio, F., 397(17), 404(17), *467*
Maux, J., 155(487), *169*
May, V., 7(86), 133(86,408–411), 134(86,410–411), 138(86,408–410), 139(86), *157*, *166–167*
Mazo, R. M., 432(138), *471*
Mazur, P., 440(183), *473*
Mazurenko, Y. T., 81(328), *164*
McCammon, J. A., 39(239), 141–142(460), *162*, *168*
McCarty, P., 375(247), *392*
McClintock, P. V. E., 404(65), *468*
McCullough, E. A., 39(231), *161*
McCurdy, C. W., 141–142(450), *168*
McDonald, I. R., 199(57), *227*, 320(11,17), 324–325(33), 332(67), *385–387*, 435(149), *471*
McDonald, R. A., 331(60), *387*
McGregor, J., 404(67), *468*
McNamara, G., 343(94), 357(158), *388–389*, 438(170), *472*
McNeil, K. J., 407(69), 420(117), 424(117), 425(126), 458(126), *469–470*
McQuarrie, D., 357(161), *390*, 397(26), 460(26), *467*
McQuarrie, D. A., 191(51), *227*
Mead, C. A., 11(114,116) 12(116), 13(118), 16(114), 46(114), *158*
Medvedev, E. S., 3(4), 5(4), 50(4), *155*
Meiburg, E., 346(106), 366(196), *388*, *391*, 398(42), *468*
Meier, C., 6(57), 72(322), 87–88(330), 108(322), 119(322), *157*, *164–165*, 220(79), *228*
Meier, G., 478(2), 579(2), *640*
Meijer, E. J., 357(164), *390*
Mele, E. J., 40(240), *162*
Melinger, J. S., 173–174(16), *226*
Menou, M., 37(212), 38(225), *161*
Messiah, A., 185(41), *226*
Messina, M., 210(58), *227*
Metiu, H., 6(52,62), 31(187), 38(214,221), 40(52,241), 41–42(214), 62(274), 68(309), 71(52,62), 97(309), 120(351), 141–142(456), 145(456), *157*, *160–165*, *168*, 186(48), *227*, 423(122), *470*
Meyer, H.-D., 14(136), 39–40(237), 51(261), 120(355,357–358), 141(451–452), 142(451–452,462), 144(451), 146(462), 152(451), *159*, *162*, *165*, *168*, 304(88), *314*
Meyer, M., 62(299), *163*, 323(31), 369(215), 370(219), *386*, *391*
Meyer, R. E., 438(612), *472*
Meyer, W., 13(119,128), *158–159*
Micha, D. A., 40(243), 141–142(453,459), *162*, *168*
Michel, H., 3(19), *156*
Michl, J., 3(3,5), 5(3,28), 27(3,5), *155–156*
Michopoulos, Y., 358(169), *390*
Migus, A., 3(15), 70(15), 115(15), 117(15), *155*
Mikhailov, A. S., 433(144), *471*
Miller, W. H., 21(153), 23(164), 140(435–436), 141(435–436,450–452), 142(450–452,463), 144(451), 145(463), 152(451,463), *159–160*, *167–168*, 305(98), *315*
Misewich, J. A., 65(307), 152(307), *164*
Mitsunaga, M., 6–7(47), 83(47), 115(47), *156*
Mo, G., 368(205–206), *391*
Moffat, H. K., 374(244), *392*

Moiseyev, N., 40(247), *162*, 305(100), *315*
Mokhtari, A., 62(276,280), 63–64(280), 68(276), 115(276,280), 116(276), 117(276,280), *163*
Molmer, K., 136(421–422), 137(422), *167*
Monaghan, J. J., 353(133), *389*
Montanero, J. M., 348(113), *388*
Moonmaw, W. R., 95(334), *164*
Moore, P., 192(54), *227*
Moran, B., 326(44), 328(50–51), *386*, 398(41), *468*
Moreau, M., 434(147), *471*
Morgan, M., 62(282), *163*
Morgen, M., 63(303), 100(303), 116(303), *164*
Morriss, G. P., 320(13), 321(13), 326(40–41), 330(55), 355(147), 358(169), *385–386*, *389–390*
Moss, F., 404(65), *468*
Mott, N. F., 141(447), *168*
Mou, C. Y., 432(138), *471*
Mourou, G. A., 3(7), *155*
Moyal, E., 243(24), *313*
Mukamel, S., 5(42), 6(46,48), 7(48,81–84,87), 63(84,302), 68(48), 70(48,84,316), 73–74(84), 75(316), 76(48,316), 79(84), 80(46,84,316), 83(48), 87(329), 88(316), 92(48,84), 108(329), 113(84,316), 133(84,404), 135(404), 152(480,484), *156–157*, *164*, *166*, *169* 173(17), 175(34), 185(45–46), 186(46), 193(46), 216(34,64), 217(67), *226–227*, 261(49), 262(54), 302(75), *313–314*
Mulder, B. M., 357(165), *390*
Müller-Dethlefs, K., 72(323), *164*
Muntz, E. P., 438(161), *472*
Murao, T., 95–96(335), 134(417), *164*, *167*
Murrell, J. N., 21(144), *159*
Myers, A. B., 220(81), *228*, 233(3), 300(3), *312*

Nagai, T., 353(136), *389*
Nagypál, I., 397(18), *467*
Nakajima, S., 132(396), *166*
Nakashima, K., 353(136), *389*
Nanbu, K., 347(109), *388*
Napier, D. G, 440(186), *473*
Naraschewski, M., 136(425), *167*
Néel, L., 481(13), 487(13), *640*
Nelson, E., 122(388), *166*
Nelson, K. A., 257(43), *313*
Neria, E., 120(352), 145(470–471), *165*, *168*, 173(23), *226*
Neuhauser, D., 304(85–87,91), *314–315*
Neuheuser, T., 10(102), 13(102), *158*
Nicolaides, C. A., 13(126), *158*
Nicolis, G., 353(128), 362(192), 384(258), *389*, *391–392*, 395(1), 396(14), 397(1,17,19,21,29), 400(1), 402(1,53,56,60), 404(17,19,66), 407(69), 408–409(70), 411(29,76), 414(100–101), 416(29), 417–418(14), 420(1,118), 422(119–120), 423(29), 424(29,118), 432(136,138,140), 442(193), 443(195), 444(193), 446(195), 452(14,19), 460(206), 466(216–217), *466–471*, *473-474*
Nieminen, J. A., 371(230), *392*
Nienhuis, G., 173(15), *226*
Nikitin, E. E., 141(437), *167*
Nitzan, A., 34(191), 120(352,354), 145(470–471), *160*, *165*, *168*, 173(23), *226*, 395(5), 397(24), *466–467*
Noak, F., 62(292), 71(292), *163*
Norris, J. R., 62(278), 68(278), *163*
Nosé, S., 326(43), 327(46), *386*

Oberthur, R. C., 358(175), *390*
Oelschlägel, B., 139(431), *167*
Ohmine, I., 213(61), *227*
Ohta, T., 416(105), 419–420(105), *470*
Okuzono, T., 353(137), *389*
Olivucci, M., 5(29,33), 21(152), 142(152), *156*, *159*
O'Malley, T. F., 10(106–107), 11(107), *158*
O'Neil, S. V., 266(62), *314*
Onuchic, J. N., 121(364), *165*
Onuki, A., 348(118), *388*, 437(158), *472*
Oppenheim, I., 360(186), *390*, 412(85), 414(95), *469*
Orban, J., 345(101), *388*
Orel, A. E., 41(253), *162*
Orszag, S., 340(86), *387*
Orszag, S. A., 36–37(197), *160*
Ortoleva, P., 395(5), 397(24), 432(137), 439(179), *466–467*, *471–472*
Osherov, V. I., 3(4), 5(4), 50(4), *155*
Ouyang, Q., 395(7), *466*
Owrutsky, J. C., 62(291), 71(291), *163*

Pacault, A., 395(6), *466*
Pacher, T., 11–12(113), 13(118), 14(135,137), *158–159*
Packard, N. H., 335(76), *387*

Palit, D. K., 115(344), 118(344), *165*, 234(9), *312*
Paolini, G. V., 332(68), *387*
Paolucci, S., 438(160), *472*
Papanikolas, J. M., 266(62), *314*
Papazyan, A., 172(7), *225*
Parinello, M., 21(148), *159*
Park, T. J., 39–40(236), *162*
Parkins, A. S., 136(424), *167*
Parlant, G., 10(103), *158*
Parson, R., 266(62), *314*
Paskin, A., 331(59), *386*
Passino, S. A., 175(34), 216(34), 222(34), *226*
Patterson, C. W., 350(123), *388*
Pearson, J. E., 450(203), 465(210), *473–474*
Pechukas, P., 124(392), 140–141(434), *166–167*
Pecora, R., 352(126), *389*, 425(129), 438(129), *470*
Pedersen, S., 115(343), 118(343), *165*
Peeters, P., 325(39), *386*, 397(21), *467*
Pelissier, M., 13–14(125), *158*
Percival, I. C., 136(426,428), *167*
Perelman, N. F., 300(66), *314*
Perera, A., 434(147), *471*
Perera, L., 173(22), *226*
Peric, M., 13(121), *158*
Perng, B. C., 173–174(26), *226*
Perraud, J. J., 465(212), *474*
Persico, M., 13(123–124), 155(491), *158*, *169*
Pescheny, S. E., 495(21), *640*
Peskin, U., 40(247), *162*, 305(100), *315*
Peteanu, L. A., 3(17), 81(327), 87(327), 115(17,346–347), 118(17,346–347), *156*, *164–165*, 260(47), *313*
Petrongolo, C., 13(120), *158*
Petrovsky, I., 433(142), *471*
Petruccione, F., 353(130–131), *389*
Petsalakis, I. D., 13(126), *158*
Pettini, L., 62(283), *163*
Peyerimhoff, S. D., 10(102), 13(102,121), *158*
Pfeiffer, H., 609(39–40), *641*
Pfeiffer, L., 172(3), *225*
Piasecki, J., 322(27), *385*
Pierleoni, C., 330(57), 357(168), 358(172–174), *386*, *390*
Pigurnov, P. N., 81(328), *164*
Piskunov, N., 433(142), *471*
Plöhn, H., 121(380,384), 122–123(384), 126(380,384), 127–129(384), 130(130), *166*
Podlipchuk, V. Yu., 435(153), *471*
Podolsky, B., 23(167), *160*
Polinger, V. Z., 20(140), 22(140), 27(140), 125(140), *159*
Pollard, W. T., 6(49,54), 7(49,80), 62(49), 64(80), 66–67(80), 74(49), 75(80), 76(49,80), 77(80), 79(49), 81(327), 83(49), 87(49,327), 92(49,80), 115(340), 133(49,413–414), 134(413–414), 139(413), 152(54), *156*, *164–165*, *167*, 221(77), *228*, 260(46–47), 261(46), 302(73), *313–314*
Pomeau, Y., 318(2), 322(27), 334(75), 336(80), 339(83), 375(250–252), *384–385*, *387*, *392*
Popielawski, J., 440(187), *473*
Portella, M. T., 3(15), 70(15), 115(15), 117(15), *155*
Portnow, J., 439(178), *472*
Posch, H. A., 328(48), 353(134), *386*, *389*
Preiskorn, A., 23(159), *159*
Present, R. D., 440(182), 443(196), *473*
Press, W. H., 38–39(230), 40(251), *161–162*
Preston, R. K., 141(441), *167*
Price, W., 63(303), 100(303), 116(303), *164*
Prigogine, I., 395(1,9–10), 397(1,34–36), 400(1), 414(100), 402(1), 420(1), 422(119), 432(138), 440(35), 446(35), *466–467*, *470–471*
Procaccia, I., 360(186), *390*, 397(33), *467*
Promel, M., 466(217), *474*
Proppe, B., 155(487), *169*
Provata, A., 466(216–217), *474*
Pshenichnikov, M. S., 3(16), 63(304), 68(304), *156*, *164*, 217(80), *228*, 262(55), *314*
Pugliano, N., 115(344), 118(344), *165*, 234(9–10), 251(10), 302(9), *312*
Puhl, A., 373(238), *392*, 398(44), *468*
Pumir, A., 375(250), *392*
Purnell, J., 62(290), 71(290), *163*
Pusey, P. N., 355(152), 357(163), *389–390*

Qin, Y., 23(161), *159*
Quack, M., 34(192), 54(192), *160*
Quian, J., 38(216), 41(216), *161*
Quyang, Q., 465(211,213), *474*

Racz, Z., 433(145), *471*
Radloff, W., 62(292), 71(292), *163*
Radzewicz, C., 172–173(2), *225*

Radzewski, C., 64(305), 68–69(305), 86(305), *164*
Raff, L. M., 23(157), 140(432), 144(432), *159*, *167*
Rahman, A., 321(21), 359(185), *385*, *390*
Raineri, F. O., 173–174(26), *226*
Rama Krishna, M. V., 58(273), *163*
Rapaport, D. C., 320(14), 334(72–74), 367(198–199), 373(237), 374(239), *385*, *387*, *391–392*, 398(42–43), *468*
Ratner, M. A., 34(191), 40(250), 68(250), 87–88(250), 120(350), *160*, *162*, *165*
Ravelo, R., 327(45), *386*
Raymer, M.G., 173(18), *226*
Redfield, A. G, 251(29), *313*
Redfield, A. G., 6(76), 132–133(76), *157*
Reid, P. J., 3(18), 4(18), 137(18), *156*
Reischl, B., 6(73), *157*
Repinec, S. T., 58(271), 115(342), 118(342), *163*, *165*
Resat, H., 173–174(26), *226*
Résibois, P., 318(2), 322(27–28), 359(183), *384–386*, *390*
Ribbing, C., 155(490), *169*
Rice, S. A., 6(61), 31(185), 63(61), 75(61), 88(61), *157*, *160*, 262(50,53), *313*
Richetti, P., 464(208), *474*
Richter, P. H., 397(33), 416(106), *467*, *470*
Righini, R., 62(283), *163*
Rips, I., 121(365), *165*
Risken, H., 413(88), *469*, 491(18), 493(18), 496(18), 505–506(18), 535(18), 632(18), *640*
Ritsch, H., 136(423–424), *167*
Ritze, H.-H., 62(292), 71(292), *163*
Rivet, J. P., 338(81), 339(85), *387*
Robb, M. A., 5(29,33), 21(152), 142(152), *156*, *159*
Robbins, M. O., 355(144), 369(207,213–214), *389*, *391*
Robertson, D. H., 436(154), 464(154), *471*
Robinson, G. W., 5(41), *156*
Robles, S. J., 3(14), *155*
Roebber, J., 192(54), *227*
Romero-Rochin, J. A., 262(50), *313*
Romero-Rochin, V., 6(61), 46(259), 63(61), 75(61), 88(61), *157*, *162*
Roncero, O., 39–40(237), *162*, 304(88), *314*
Ronis, D., 360(186), *390*
Roos, B. O., 10(98), 13(98,129), *158–159*
Root, L., 251(32), 301(32), *313*
Rosenberger, R., 368(205–206), *391*
Rosenthal, S. J., 62(277–278), 68(277–278), 115(277), 118(277), *163*, 174(28), 182(38), *226*
Rosker, M. J., 3(11–13), 6(12–13), 62(11–13), 63(13), 71(11–12), 115(13), 117(13), *155*
Rosmus, P., 21(143), *159*
Ross, I. G., 95(334), *164*
Ross, J., 395(5), 397(24,33), 419(114b), 423(122), 440(183), *466–467*, *470*, *473*
Rossky, P., 236(15–16), 301(15–16), *312*
Rossky, P. J., 141(439), 142(439,466), 152(482), *167–169*, 172(8), 182(39), *225–226*
Rothenfusser, C., 3(21), 6(21), 62(21), 71(21), 88(21), *156*
Rothman, D. H., 344(100), *388*
Röttgerman, C., 3(21), 6(21), 62(21), 71(21), 88(21), *156*
Roulet, B., 413(89), *469*
Roux, J.-C., 464(208), 465(211), *474*
Rowlinson, J. S., 369(216), *391*
Roy, S., 198(56), *227*
Ruedenberg, K., 5(30), 13(127), *156*, *158*
Ruggiero, A. J., 6(61), 62(277), 63(61), 68(277), 75(61), 88(61), 115(277), 118(277), *157*, *163*, 262(50), *313*
Ruhman, S., 87(332), *164*, 175(34), 216(34), 220(66), 222(34), *226–227*, 232(1), 233(2,4), 234(6), 235(11), 237(1–2,4,18–19), 258(2), 265(60), 284(2), 294(4), 301(4,11), 304(81), 305(2), *312*, *314*
Rull, L. F., 320(7), 362(7), *385*
Rupert, L. A. M., 370(220), *391*
Rusenbluth, M. N., 375(249), *392*
Rutz, S., 6(70), 62(70,289), 71(70,289), *157*, *163*
Ryckaert, J. P., 320(7), 326(42), 332(68), 362(7), 330(57), 356(157), 357(166–168), 358(172–174), *385–387*, *389–390*

Saalfrank, P., 155(488), *169*, 305(96), *315*
Sadlej, A. J., 13(129), *159*
Sagues, F., 375(246), *392*
Saito, S., 213(61), *227*
Salem, L., 5(27), *156*
Salomons, E., 348(116), 349(121–122), 350(123), 369(217), *388*, *391*, 438(163), *472*
Sánta, I., 62(283), *163*
Santos, A., 348(113), 369(208), *388–389*, *391*

Saupe, A., 478(2), 579(2), *640*
Saven, J. G., 192(52), *227*
Sawada, S., 120(351), 141–142(456), 145(456), *165*, *168*
Saxe, P., 10(100), *158*
Schack, R., 136(428), *167*
Schatz, G. C., 21(147), 23(147), 41(254), *159*, *162*
Scheer, H., 62(299), *163*
Schenter, G. K., 210(58), *227*
Schenzle, A., 136(425), *167*, 418(112), *470*
Scherer, N. F., 6(61,64), 63(61,64,302), 70(316), 75(61,64,316), 76(316), 80(316), 88(61,64,316), 113(316), *157*, *164*, 262(50–52,54), *313*
Schieve, W. C., 413(93), *469*
Schiffer, M., 62(278), 68(278), *163*
Schimansky-Geier, L., 399(47), 433(144), *468*, *471*
Schinke, R., 2(1), 21(141,143), 27(1), 38(121), 41(121,255), 42(1), 68(310), 87–88(310), *155*, *159*, *162*, *164*
Schirmer, J., 14(136), *159*
Schizgal, B., 440(183,186), *473*
Schlag, E. W., 72(323), *164*
Schlijper, A. G., 370(220), *391*
Schlögl, F., 405(68), 433(143), *468*, *471*
Schmidt, B., 155(487–488), *169*
Schmidt, M., 62(278), 68(278), *163*
Schmidt, S., 3(19), *156*
Schneider, R., 5(36), 7(89), 27(180), 33(36), 40(36), 46(36), 50(89,180), 53(36,264), 54(36), 56–57(36), 77(89), 86(89), 95(36), 102–103(89), 115(89), 117(36,89), *156*, *158*, *160*, *162*
Schoen, M., 359(185), *390*
Schoenlein, R. W., 3(15,17), 70(15), 115(15,17,346–347), 117(15), 118(17,346–347), *155–156*, *165*
Schofield, S. A., 34(195), *160*
Schön, G., 6(72), 46(72), 48(72), *157*
Schraner, R., 410(73), 416(106), *469–470*
Schreiber, E., 6(70), 62(70,289), 71(70,289), *157*, *163*
Schreiber, M., 7(86), 133(86,408–411), 134(86,410–411), 138(86,408–410), 139(86), *157*, *166–167*
Schuler, K. E., 414(95), *469*
Schulman, L. S., 120(362), *165*
Schuss, Z., 411(77), 414(99), *469*
Schwartz, B. J., 142(466), 152(482), *168–169*, 182(39), *226*
Scott, S. K., 449(201), *473*
Scully, C. N., 504(33), *641*
Seel, M., 7(91), 31–32(91), 41–42(91), 50(91), 72(91,320), 77(91), 86(91), 88(91), 105–107(91), 119(320), *158*, *164*
Seeley, G., 212(60), *227*
Segrè, P. N., 362(193), *391*, 438(168,173), *472*
Seidner, L., 5(39), 6(50), 7(39,94), 27(39,181), 29(39), 31(50), 35–36(39), 39(39), 42(39,94), 43(39), 45(94), 50(181), 53–54(39), 58(39), 66(50), 76(50), 87(50,94), 88(50), 90(50), 108(50), 142(39), 145(39), 148(39), *156*, *158*, *160*, 262(56), *314*
Seki, K., 425(128), *470*
Sengers, J. V., 361–362(191), 362(193), *390–391*, 438(168,173), *472*
Senison, R. J., 58(271),115(342), 118(342), *163*, *165*
Sepúlveda, M. A., 145(474), 152(484), *168–169*, 218(70–71), *227*
Seshadri, V., 414(96), *469*
Sewell, T. D., 23(160), *159*
Sgarra, C., 348(114), *388*
Shah, J., 172(3), *225*
Shaik, S., 265(59), *314*
Shaik, S. S., 46(257), *162*
Shampine, L. F., 38–39(228), *161*
Shank, C. V., 3(15,17), 62(281), 63(281,301), 66–67(281), 70(15), 81(327), 87(327), 115(15,17,281,340,346–347), 116(281), 117(15), 118(17,346–347), *155–156*, *163–165*, 260(47), *313*
Shemetulskis, N. E., 218(74), *227*
Shen, H., 40(246), *162*
Shen, Y. R., 6(74), 62(74), 64–65(74) 73–74(74), 76–77(74), 81(74), 133(74), *157*
Shepard, R., 10(97), 13(97), *158*
Shibata, F., 134(417), *167*
Shouman, A. R., 442(190), *473*
Showalter, K., 395(8), *466*
Shraiman, B. I., 375(248), *392*
Sidis, V., 11–12(112), *158*
Siebrsnd, W., 20(139), *159*
Siegert, A., 31(186), 66(186), 75(186), *160*
Siggia, E. D., 360(188), *390*
Sikkenk, J. H., 370(223), *391*
Silva, C., 3(18), 4(18), 137(18), *156*, 182(40), *226*
Simon, J. D., 172(8), *225*
Sinai, Ya. G., 328(52), *386*

Singer, K., 324–325(33), *386*
Sirko, L., 139(430), *167*
Skinner, J. L., 192(52), *227*, 251(31–32), 301(32), *313*
Slichter, C. P., 132–133(399), *166*
Smale, S., 396(11), *466*
Smilansky, U., 139(430), *167*
Smirnov, V. A., 81(328), *164*
Smit, B., 323(29–30), 370(220), *386*, *391*
Smith, B. R., 21(152), 142(152), *159*
Smith, F. T., 10–11(108), *158*
Smith, J. M., 62(285), 71–72(285), *163*
Smith, M. A., 72(317), *164*
Smith, T. J., 175(35), *226*, 257(42), *313*
Smith, T. J. Jr., 46(259), *162*
Smithey, D. T., 173(18), *226*
Smits, C., 357(164), *390*
Smulevitch, G., 115–116(341), *165*
Sobolewski, A. L., 5(31,45), 13–14(130), 15–17(45), 19(45), 27(45,181), 50(45,181), 72(319), 95(45), 128(45), *156*, *159–160*, *164*
Socolar, J., 40(240), *162*
Solomon, T. H., 375(253), *392*
Sonnino, G., 362(194), *391*, 438(165), *472*
Sorokin, P. P., 65(307), 152(307), *164*
Spicae, B., 142(465), *168*
Spiegelmann, F., 13(124), *158*
Springer, J., 358(175), *390*
Staib, A., 72(319), 142(467), *164*, *168*
Stanley, R. J., 172–173(4), *225*
Steckler, R., 21(145), *159*
Stegun, I. A., 306(104), *315*
Steiger, A., 38–39(218), *161*, 254(37), *313*
Steinrück, H.-P., 53(264), *162*
Stengle, M., 13–14(132), *159*
Stenholm, S., 255(40), *313*
Stert, V., 62(292), 71(292), *163*
Stevens, M. J., 355(144), *389*
Stiller, W., 444(197), *473*
Stock, G., 5(44), 6(50), 7(89–90,92–94), 27(181–182), 31(50), 33–34(182), 36(182), 39(93), 42(94), 45(94), 50(89,181), 53(262,266–267), 56(262,266), 64–65(92), 66(50,92), 67(92), 76(50), 77(89–90,92–93), 78(90), 83(92), 85(92), 86(89–90,92–93), 87(50,87–88,93), 88(50,93), 90(50), 92(90), 95(182), 96(337), 97(182), 98(182,337), 102(89–90,338), 103(89,338), 105(92), 108(50), 114(338), 115(89), 116(93),117(89),141(461), 142(266,461,463–464), 144(461,464), 145(262,266,462–463),146(266,461,464), 148(262,464),149(262),152(463–464),153 (262), *156*, *158*, *160*, *162*, *165*, *168*, 212(65), 219(65), *227*, 262(56), *314*
Stock, H., 301(69), *314*
Stoer, J., 38–39(229), *161*
Stolow, A., 62(294), 71–72(294), *163*
Stoner, E. C., 536(38), 630(38), *641*
Stotutt, U., 355(146), *389*
Stoutland, P. O., 62(297), *163*
Stratonovich, R. L., 533(36), *641*
Stratt, R. M., 174(30–33), 192(30), 195(30), 212(30–33), 213(61), *226–227*
Straub, G. K., 331(61–62), *387*, 398(41), *468*
Stround, C. K., 304(94), *315*
Stroutland, P. O., 58(272), *163*
Stucherbrukhov, H. A., 54(269), *162*
Stump, M., 21(143), *159*
Succi, S., 341(88), *387*, 438(169), 439(175), *472*
Sudarshan, E. C. G., 248(25), *313*
Sugawara, M., 87(331), *164*
Sukumar, N., 10(102), 13(102), *158*
Sulpice, E., 397(31), 434(146), *467*, *471*
Sun, M., 368(204), *391*
Sundberg, R. L., 185(44), *227*
Suominen, K.-A., 40(244), *162*, 255(40), *313*
Sutcliffe, B. T., 24(168), *160*
Suzdalev, I. P., 495(23), *640*
Suzuki, M., 122(389), *166*, 414(98–99), *469–470*
Sweetser, J. N., 172–173(2), *225*
Swinney, H. L., 375(255), *392*, 395(7), 464(208), 465(213), *466*, *474*
Swinney, T., 27(182), 33–34(182), 36(182), 95(182), 97–98(182), *160*
Szarka, A. Z., 58(271), 115(342,344), 118(342,344), *163*, *165*, 234(9–10), 251(10), 302(9), *312*

Tabeling, P., 375(254), *392*
Takagahara, T., 133(403), *166*
Takeno, T., 442(190), *473*
Tal-Ezer, H., 39–40(235), *162*, 237(18), 252(33), 303(78), 304(84,95), *313–315*
Talkner, P., 301(72), *314*, 413(86), *469*
Tam, W. Y., 375(255), *392*
Tang, C. L., 3(13), 6–7(47), 62–63(13), 83(47), 115(13,47), 117(13), *155–156*
Tang, E., 236(15), 301(15), *312*

Tang, H., 262(53), *313*
Tanimura, Y., 7(87), *157*
Tannor, D. J., 31(185), 38(216), 39(239), 41(216), *160–162*, 185(44), *227*, 251(28), 259(44–45), 262(45), 263(58), 305(45), *313–314*
Tartaglia, P., 414(97), *469*
Taylor, G. I., 374(242), *392*
Taylor, H., 401(48), *468*
Taylor, H. S., 305(102), *315*
Tel, T., 397(30), *467*
Tenenbaum, A., 330(56), *386*
Tennyson, J., 38(224), *161*
Terpstra, J., 75(333), 92(333), *164*
Teukolsky, S. A., 38–39(230), *161*
Thakur, K. A. M., 266(63), *314*
Thalweiser, R., 3(20–21), 6(21,59,69), 62(21,59,69,286), 71(21,59,69,286), 88(21), *156–157,163*
Theodorakopoulos, G., 13(126), *158*
Theodosopulu, M., 422(121), *470*
Thomas, H., 397(25), 413(86), *467*, *469*
Thompson, D. L., 23(156–162), 29(156,158,162), 140(432), 144(432), *159*, *167*
Thompson, P. A., 369(207,213–214), *391*
Thorson, W. R., 14(134), 141–142(448), *159*, *168*
Tielsen, J., 397(20), *467*
Tier, C., 411(77), 414(99), *469*
Tij, M., 369(208), *391*
Tildesley, D. J., 320(6,10), 321(10), *385*, 435(150), *471*
Tirapegui, E., 418(109), *470*
Tobias, D. J., 328(47), *386*
Todd, D. C., 62(277), 68(277), 115(277), 118(277), *163*
Tombesi, P., 414(97), *469*
Tomita, H., 416(105), 419–420(105), 457(205), *470*, *473*
Tomita, K., 416(105), 419–420(105), *470*
Tomsovic, S., 145(474), *168*, 218(70), *227*
Topaler, M., 125(394–395), *166*, 218(73), *227*
Toplar, M., 301(71), *314*
Topp, M. R., 4(23), *156*
Toxvaerd, S., 370(221), *391*
Trefethen, L. M., 305(99), *315*
Tremblay, A. M. S., 360(188), *390*
Tretyakov, A., 466(217), *474*
Tribel, O., 466(217), *474*
Triechel, M., 234(10), 251(10), *312*
Tromp, J. W., 24(173), 37(173), *160*
Trotter, M. F., 122(387), *166*
Trozzi, C., 354(142), *389*
Truhlar, D. G., 11–12(116), 14(116), 21(145), 141(443), *158–159*, *167*
Truong, T. N., 39(239), *162*
Tsai, D. H., 331(60), *387*
Tsuzuki, T., 418(108), 419(113), *470*
Tuckerman, M., 332(70), *387*
Tuekolsky, S. A., 40(251), *162*
Tully, J. C., 141(441,445), *167*
Turner, J. S., 401(51), 432(138), *468*, *471*
Turner, J. W., 397(19), 401(51), 404(19), 408–409(70), 411(75), 415(102), 418(102), 432(138), 466(216), *467–471*, *474*
Turq, P., 352(124), 370(219), *388*, *391*

Ueyama, H., 397(25), *467*
Uhlenbeck, G. E., 491(17), *640*
Umlauf, E., 3(19), *156*
Ungar, L. W., 175(35), 186(49), 188(49), 195(55), 210(49), 212(59), 221(59), *226–227*, 257(42), *313*
Untch, A., 21(143), *159*
Uzer, T., 54(270), *162*

Vachev, V. D., 142(468), *168*
Valignat, M. P., 370(229,231), *392*
Valuev, A. A., 435(153), *471*
Vanden Bout, D., 172(9), *225*
Van den Broeck, C., 374(243), *392*, 397(29), 411(29), 416(29), 417(107), 423(29), 424(29,125), 425(127), 427(130), 430(135), 437(158), *467*, *470–472*
Van der Hoef, M. A., 342(90–91), *387*
Van Der Ziel, A., 402(64), *468*
van Grondelle, R., 173(11), *225*
van Gunsteren, W. F., 332(69), *387*
Van Hove, L., 359(181), *390*
Van Kampen, N. G., 397(27), 402(57–58), 404(58), 411(75), 413(90–91), *467–469*, 479(9), 482(9), 631(9), *640*
van Leeuwen, J. M. J., 370(223), *391*
van Leeuwen, M. J., 343(93), *387*
Van Leuven, P., 40(242), *162*
van Megen, W., 357(163), *390*
Vannitsem, S., 378(257), *392*
Van Nypelseer, A., 432(138), *471*
van Os, N. M., 370(220), *391*
Van Vliet, K. M., 402(64), *468*

Varandas, A. J. C., 21(144,146), *159*
Vaz Pires, M., 15(138), *159*
Velasco, S., 397(17), 404(17), *467*
Vergassola, M., 341(88), *387*, 438(169), *472*
Verlet, L., 321(22–24), 324(36), *385–386*
Vernon, F. L., 121(372), 125(372), *166*
Verosky, J. M., 39(238), *162*
Vetterling, W. T., 38–39(230), *161*
Vibok, A., 270(65), 304(93), *314–315*
Vidal, C., 395(3,6), 412(83), *466*, *469*
Vigil, R. D., 465(213), *474*
Villani, G., 40(249), 50(249), 155(491), *162*, *169*
Villeneuve, D. M., 62(294), 71–72(294), *163*
Vöhringer, P., 234(5), *312*
von Mourik, F., 173(11), *225*
von Neissen, W., 53(264), *162*
Von Neumann, J., 335(78), *387*
Vos, M. H., 3(14), *155*
Vossnack, E. O., 370(223), *391*
Voter, A. F., 327(45), *386*
Voth, G. A., 218(72), *227*
Vrakking, M. J. J., 62(294), 71–72(294), *163*

Wagner, J., 397(20), *467*
Wagner, W., 345(105), *388*
Wainwright, T. E., 317(1), 321(19–20), *384–385*, 397(36), *467*
Wake, G. C., 442(190,192), *473*
Waldeck, J. R., 120(359), *165*
Waldron, J. T., 478(6), 495(27), 499(28), 504(29), 506(28), 543(28), 593(42), 628(42), *640–641*
Walgraef, D., 396(15), *466*, 419(114a), 429(131–132), *470–471*
Walhout, P. K., 182(40), *226*, 266(63), *314*
Walker, G. C., 62(279), 68(279), *163*
Walkup, R. E., 65(307), 152(307), *164*
Wall, M. R., 304(87), *314*
Wallace, S. C., 72(317), *164*
Walls, D. F., 407(69), 408(71), 420(117), 424(117), 425(126), 458(126), *469–470*
Walmsley, I. A., 64(305), 68–69(305), 86(305), *164*, 172–173(2), *225*
Walter, H., 139(430), *167*
Walton, A. R., 145(478), *169*
Wang, H., 172(3), *225*
Wang, J.-K., 6(67), *157*
Wang, M. C., 491(17), *640*
Wang, Q., 81(327), 87(327), 115(347), 118(347), *164–165*, 260(47), *313*
Wang, X.-J., 320(18), *385*
Wangsness, R. K., 132–133(398), *166*
Warshel, A., 23(163), *160*
Watanabe, T., 374(241), *392*
Webster, F., 236(15), 301(15), *312*
Webster, F. J., 141–142(439), *167*
Wei, H., 37(213), *161*
Weide, K., 21(141), 38(141), 41(141), *159*
Weider, T., 355(145–146), *389*
Weil, L., 495(22), *640*
Weimar, J. R., 439(175), *472*
Weiss, G. H., 414(95), *469*
Weiss, R. M., 23(163), *160*
Weiss, U., 120(363), 121(363,368), *165–166*
Weiss, V., 3(21), 6(21), 62(21), 71(21), 88(21), *156*
Wentworth, W. E., 266(61), *314*
Werner, H.-J., 5(45), 10(99), 13(99,119,128), 15–17(45), 19(45), 21(143), 27(45), 50(45), 95(45), 128(45), *156*, *158–159*
West, B. J., 414(96), *469*
Weyers, K., 62(292), 71(292), *163*
Whetten, R. L., 27(178), *160*
White, C. T., 436(154), 464(154), *471*
Whitnell, R. M., 6(65), 36(203–204), 37–38(203), *157*, *161*
Wickstead, A. W., 504(29), 593(42), 628(42), *641*
Widom, B., 369(216), *391*
Wiersma, D. A., 3(9,16), 63(304), 68(304), 70(314–315), 75(333), 92, (333), 113(315), 115(345), 118(345), *156*, *164–165*, 217(80), *228*, 262(55), *314*
Wiesenfeld, J. J., 62(284), 71(284), *163*
Willemsen, J. F., 324(38), 369(211–212), *386*, *391*
Williams, C. J., 38(216), 41(216), *161*
Williams, S. O., 6(68), 62(68), *157*
Willoughby, R. A., 36(196), 95(196), *160*
Wilson, E. B., 22(154), 25–26(154), *159*
Wilson, K. R., 6(65), *157*
Winterstetter, M., 121(378–380,384–386), 122(378–379,384), 123(384), 124–125(386), 126(379–380,384–385), 127–129(384–385), 130(380), *166*
Wise, F. W., 3(13), 62–63(13), 115(13), 117(13), *155*
Wódkiewicz, K., 66(308), 69(308), *164*, 173(13–14), 181(13–14), *226*
Wogelsang, R., 324(37), *386*
Wohlfahrt, E. P., 536(38), 630(38), *641*
Wolfram, S., 335(76–77), *387*

Wolfseder, B., 135–136(418), 137(418,429), *167*
Wolinski, K. J., 13(129), *159*
Wolynes, P. G., 34(195), 121(374), *160*, *165*
Wood, W. W., 322(26), 343(95), *385*, *388*
Woodruff, W. H., 58(272), 62(297), *163*
Wöste, L., 3(10), 6(70), 62(70,289), 71(70,289), *155*, *157*, *163*
Woywod, C., 5(31,45), 7(93), 13–14(130–132), 15–17(45), 19(45), 27(45,182), 33–34(182), 36(182), 39(93), 50(45), 77(93), 86–87(93), 95(45,182), 97–98(182), 116(93), 128(45), *156*, *158*, *160*
Wu, G. X.-G., 397(22), 438(22), 466(22,214), *467*, *474*
Wu, W. X., 133(406), *166*
Wu, Y. M., 46(258), *162*
Wyatt, R. E., 34(193–195), 37(212), 38(193), 39(231), 40(248), *160–162*, 304(89), *314*

Xantheas, S., 5(30), *156*
Xhrouet, E., 397(34), *467*
Xiao, L., 141–142(440), *167*
Xiaolin, Chu, 420(114b), 429(114b), *470*
Xie, X., 62(278), 68(278), *163*, 174(28), *226*
Xu, Z., 359(176), *390*
Xue, W., 355(151), *389*
Xystris, N., 440(185), *473*

Yakhot, V., 340(86), *387*
Yamanaka, T., 95–96(335), *164*
Yamashita, K., 265(60), *314*
Yamazaki, I., 95–96(335), *164*
Yan, Y. J., 6(65), 7(81–83), 87(329), 108(329), 133(404), 135(404), *157*, *164*, *166*, 173(17), 175(34), *226*, 261(49), *313*
Yan, Y. X., 257(43), *313*
Yang, D., 62(277), 68(277), 115(277), 118(277), *163*
Yang, J. X., 370(228), *391*
Yao, G., 40(248), *162*
Yarkony, D. R., 5(32), 9(96), 10(96,100–101,103), 13(96,103), *156*, *158*
Yee, T. K., 76(324), *164*
Yeh, C. Y., 68(312), *164*
Yip, S., 356(155), 360(190), *389–390*, 438(167), 439(179), 442(193), 444(193), *472–473*
Yokokawa, M., 374(241), *392*
Yoshihara, K., 95–96(335), *164*
Young, W., 375(250–251), *392*
Youvan, D. C., 3(14), *155*
Yu, Y., 217(27), *226*
Yuan, X.-F., 353(138), *389*

Zadkov, V. N., 142(468), *168*
Zadoyan, R., 6(66), *157*
Zadoyan, Z. L., 236(17), 301(17), *312*
Zaleski, S., 344(100), *388*, 438(169), *472*
Zanetti, G., 343(94), 357(158), *388–389*
Zappoli, B., 323(31), *386*
Zeldovich, Y. B., 442(189), *473*
Zewail, A. H., 3(7–8), 6(51,60,63,67), 62(8,51,287–288), 71(8,51,287–288), 115(343), 118(343), 152(481), *155*, *157*, *163*, *165*, *169*, 234(7–8), *312*, 397(16), *467*
Zgierski, M. Z., 20(139), *159*
Zhabotinski, A. M., 395(4), *466*
Zhu, W., 304(80,82–83), 305(83), *314*
Ziegler, L. D., 6(64), 63(64), 75(64), 88(64), *157*, 192(54), *227*, 262(51), *313*
Zinth, W., 3(19), *156*, 62(299), *163*
Zobey, O., 152(483), *169*
Zoller, P., 136(423), *167*
Zülicke, Ch., 433(144), *471*
Zurek, M., 62(299), *163*
Zurek, W., 413(93), *469*
Zwanzig, R., 132(397), *166*
Zwerger, W., 41(252), 120–121(252), 130(252), 148(252), *162*

SUBJECT INDEX

Ab initio computation:
derivative couplings, diabatic electronic states, 13–20
Hamiltonian models, potential energy (PE) surfaces, 20–23
photoinduced dynamics, 7–8
real-time detection, internal-conversion process, 95–108
"Ab initio" molecular dynamics, microscopic simulations, 435
Absorption spectrum:
iodine photodissociation dynamics:
interpretation, 262–263
overview, 231–235
"pump" pulse photodissociation, 274
short-time fluorescence Stokes shift dynamics, 200–201
Action angle variables, iodine photodissociation grid and eigenstate representations, degrees of freedom (DOF), 241
Ad hoc velocity rescaling technique, complex flow simulations, non-equilibrium molecular dynamics (NEMD), 326
Adiabatic electronic representations:
femtosecond experiments, 117–119
nonadiabatic photoisomerization, 142–145
non-Born-Oppenheimer dynamics, 8–11
ultrafast intramolecular non-Born-Oppenheimer dynamics:
electronic population probabilities, 49–54
wave packets, 42–49
Aftereffect solution, Fokker-Planck equation, correlation time calculations, 496–499
Alignment representation formula, Fokker-Planck equation:
cubic anisotropic free energy:
spherical harmonic product formula, 583–589
spherical harmonic recurrence relations, 568–581
high energy barrier approximation, Brown's model, 591–592
Langevin spherical harmonic representation, 537–539
spherical harmonic representation, 517–524
cubic anisotropic free energy, 568–589
Anisotropic minimum, Fokker-Planck equation:
discrete orientation model, 483–487
quasi-equilibrium behavior, 593
Approximation minimization, Fokker-Planck equation, axial symmetry, 504–505
Arrhenius law, chemical instability simulations:
continuously stirred tank reactor (CSTR), 449
exothermic chemical reaction, 442–443
overview, 397
Asymptotic values:
reactive fluid stochastic theory, pitchfork bifurcation fluctuation, 408–414
ultrafast intramolecular non-Born-Oppenheimer dynamics, electronic population probabilities, 51–54
Autocorrelation function:
Fokker-Planck equation, correlation time calculations, 497–499
real-time detection, internal-conversion process, 102–108
Axial symmetry. *See also* Non-axial symmetry
Fokker-Planck equation, 499–505

Basis-set dilemma:
iodine photodissociation dynamics, conjugated basis-sets, 242
phase space picture, 243–245
single basis picture, 243
multidimensional problems, 119–120
time-dependent Schrödinger computation, state vector representation, 38
Bénard instability, complex flow simulations, 353
Bénard transitions, complex flow simulations, 372–373
"Berry's phase," electronic wave functions, 16

BGK (Bhatnagar, Gross, Krook) model, complex flow simulations, lattice gases, 341
Bifurcation analysis:
chemical instability simulations, overview, 396
Hopf bifurcation:
continuously stirred tank reactor (CSTR), 454–457
one-dimensional system fluctuations, 433–434
reactive fluid stochastic theory, 414–420
pitchfork bifurcations:
continuously stirred tank reactor (CSTR), 451–454
reactive fluid stochastic theory, 405–414
spatial correlations, 427–430
reactive fluid stochastic theory, 399–434
Hopf bifurcation, 414–420
pitchfork bifurcation fluctuation, 405–414
pitchfork bifurcation spatial correlations, 427–430
spatially extended systems, 462–463
Biorthonormality, Fokker-Planck equation, matrix elements and differential recurrence coefficients, 505–513
Bird algorithm. *See also* Direct simulation Monte Carlo (DSMC) technique
Boltzmann dynamics, 437–439
continuously stirred tank reactor (CSTR), 449–457
microscopic simulation, 464
Bistable structure, free energy, uniaxial anisotropy, 609–611
Bloch equations:
molecular pump-probe femtosecond spectroscopy, 6–8
nonlinear polarization, 80–81
reduced density-matrix formalism, 133–139
Boltzmann distribution. *See also* Lattice Boltzmann technique
Bird algorithm, 437–439
chemical instability simulations, overview, 397–398
complex flow simulations:
direct simulation Monte Carlo (DSMC) technique, 345–352
non-equilibrium molecular dynamics (NEMD), 331
hard-sphere chemistry, microscopic simulation, 440–441
iodine photodissociation dynamics:
background, 244–245
"probe" pulse states, 281–282
"pump" pulse photodissociation, 266–272
reactive fluids, molecular dynamic microscopic simulations, 436
short-time fluorescence Stokes shift dynamics, 190–191
Born-Oppenheimer approximation, iodine photodissociation dynamics, degrees of freedom (DOF), 240–241
Born-Oppenheimer (BO) approximations. *See also* Non-Born-Oppenheimer dynamics
adiabatic approximation, 9–10
iodine photodissociation dynamics, degrees of freedom (DOF), 240–241
photoinduced dynamics, 3–6
time-dependent Schrödinger computation, state vector representation, 34–38
Boundary conditions:
chemical instability simulations:
exothermic chemical reaction, 442–448
microscopic simulations, 435
complex flow simulations:
channel flows, 368–369
non-equilibrium molecular dynamics (NEMD), 331
Rayleigh-Bénard instability, 376–377
two-phase flows, 369–370
exothermic chemical reaction, 441
microscopic simulation, 464
reactive fluids, molecular dynamic microscopic simulations, 436
spatially extended systems, microscopic simulation, 459–463
Boussinesq equations, complex flow simulations, 373–374, 380
Brillouin equation, complex flow simulations, non-equilibrium steady state, 362
Brownian dynamics:
colloidal suspensions, 343–344
complex flow simulations, 352–353
Brown's model, Fokker-Planck equation:
derivation from general Fokker-Planck and Gilbert-Langevin equations, 631–634
high energy barrier approximation, 589–601
overview, 488–495
Brusselator model:
microscopic simulations, continuously stirred tank reactor (CSTR), 450
stochastic theory of reactive fluids, Hopf bifurcation, 414–420

Bulrisch-Stoer technique, time-dependent Schrödinger equation, state vector propagation, 39, 41
Burger's equation, complex flow simulations, 353

Carnahan-Stirling equation, complex flow simulations, Rayleigh-Bénard instability, 377
CBA (consistent Boltzmann algorithm) method, complex flow simulations, direct simulation Monte Carlo (DSMC) technique, 347
CCP5 program library, complex flow simulations, 321
Cellular automata, complex flow simulations, lattice gas models, 335
Center frequency, Stokes shift dynamics, simulation conditions, 201–206
Chain rule for partial derivatives, Fokker-Planck equation, spherical harmonic representation, 518–524
Channel flow, complex flow simulations, 368–369
Chapman-Kolmogorov equation, stochastic theory of reactive fluids, one-dimensional system fluctuations, 430–434
Chapmann-Enskog method:
- complex flow simulations, non-equilibrium molecular dynamics (NEMD), 331
- hard-sphere chemistry, microscopic simulation, 440–441

Chebychev polynomial expansion:
- iodine photodissociation dynamics:
  - approximating functions, 303–306
  - contour mapping, 308–310
  - evolution operator, 253
- time-dependent Schrödinger equation, state vector propagation, 39–40

Chemical clocks, reactive fluid stochastic theory, 415–420
Chemical instability simulations:
- microscopic simulations, 434–463
  - Bird algorithm for Boltzmann dynamics, 437–439
  - continuously stirred tank reactor (CSTR), 449–457
  - exothermic chemical reaction, 441–448
  - hard-sphere chemistry, 439–441
  - reactive fluids, 435–436
  - spatially extended systems, 457–463
- overview, 393–399
- reaction-diffusion master equation, 420–434
  - Langevin limit, 422–424
  - one-dimensional systems, 430–434
  - pitchfork bifurcation spatial correlations, 427–430
  - spatial correlations, 424–427
- stochastic history of reactive fluids:
  - global view, 399–420
  - Hopf bifurcation, 414–420
  - pitchfork bifurcation fluctuation, 405–414
  - reaction-diffusion master equation, 420–434

Chemically reactive dynamics, ultrafast intramolecular non-Born-Oppenheimer dynamics, nuclear observables, 58–61
Chromophore-solvent systems, photoisomerization dynamics, 148–152
Cis-trans photoisomerization dynamics:
- femtosecond timescales, 3
- real-time detection, nonadiabatic processes, 108–114
- ultrafast intramolecular non-Born-Oppenheimer dynamics:
  - femtosecond experiments, 118–119
  - nuclear observables, 60–61

Classical law of divergence, reactive fluid stochastic theory:
- pitchfork bifurcation fluctuation, 408–414
- reaction-diffusion master equation, 427

Classical mechanics:
- femtosecond experiments, 152–154
- internal conversion:
  - benzene cation conversion, 145–148
  - photoisomerization, 145–154
- photoisomerization, chromophore in solution, 148–152
- short-time fluorescence Stokes shift dynamics:
  - linear response analysis, 191–192
  - quantum operator approximation, 190–191
  - simulation techniques, 209–215

Classical-path techniques:
- nonadiabatic photoisomerization, 141–145
- photoisomerization, chromophore in solution, 148–152

Coalescence mechanism, complex flow simulations, 371
Coherent mode of magnetization, Fokker-Planck equation, Brown's model, 495

Collision frequency:
  complex flow simulations, direct simulation Monte Carlo (DSMC) technique, 345–348
  microscopic simulations:
    continuously stirred tank reactor (CSTR), 449–457
    exothermic chemical reaction, 445–448
Collisions per particle (CCP) parameter:
  continuously stirred tank reactor (CSTR), 454
  exothermic chemical reaction, 444–448
  spatially extended systems, microscopic simulation, 460–463
Colloidal suspensions, complex flow simulation:
  fluid rheology, 357–359
  lattice gas models, 343–344
Complete-active-space self-consistent-field (CASSCF) wave function:
  derivative couplings, diabatic electronic states, 13–14
  real-time detection, internal-conversion process, 98–108
Complex flow simulations:
  direct simulation Monte Carlo method, 344–352
  fluid rheology:
    complex fluids, 357–359
    simple fluids, 353–357
  instabilities, 366–383
  lattice gases, 334–344
  molecular dynamics (MD) technique, 320–324
  non-equilibrium molecular dynamics (NEMD), 324–334
  non-equilibrium steady states, 359–366
  overview, 317–320
  particle simulations, 352–357
  Rayleigh-Bénard instability, 371–383
    background, 318
Condon principle, real-time detection, internal-conversion process, 101–108
Configuration-interaction (CI), derivative couplings, diabatic electronic states, 13–14
Conical intersections:
  adiabatic-to-diabatic transformation, 14–16
  ultrafast intramolecular non-Born-Oppenheimer dynamics, 41–42
    electronic population probabilities, 50–54
    wave packets, 45–49
Conjugated basis-sets, iodine photodissociation dynamics, 242
Continuously stirred tank reactor (CSTR):
  microscopic simulation, 449–457, 464–465
    Hopf bifurcation, 454–457
    multiple steady-state transitions, 451–454
  reactive fluid stochastic theory:
    Hopf bifurcation, 415–420
    pitchfork bifurcation, 405–406
Continuous-wave (CW) spectrscopy:
  nonlinear polarization, 78–81
  real-time detection, *vs.* femtosecond spectroscopy, 95–108
Coordinate representation:
  Fokker-Planck equation:
    orthogonal transformations, 634–640
    uniaxial anisotropic free energy, 601–606
      orthogonal transformation, 606–609
  iodine photodissociation dynamics:
    Hamiltonian calculations, 247–248
    impulsive excitation, two-level approximation, 254–256
Correlation functions, complex flow simulations:
  non-equilibrium steady state, 362–363
  real-time path-integrals, 127–131
Correlation time, Fokker-Planck equation:
  eigenvalues and amplitudes, 495–499
  matrix elements and differential recurrence coefficients, 514–515
Couette shear flow, complex flow simulations:
  fluid rheology, 353–354
  non-equilibrium molecular dynamics (NEMD), 325–326, 368
Coulombic potentials, short-time fluorescence Stokes shift dynamics, 199–201
Coupled-surface wave-packet calculations, time-dependent Schrödinger computation, state vector representation, 38
Cubic anisotropy, Fokker-Planck equation:
  discrete orientation model:
    $K < 0$, 486–487
    $K > 0$, 483–485
  ferromagnetic relaxation, 480
  free energy, 543–589
    spherical harmonic product formula, 581–589
    spherical harmonic recurrence relations, 543–581
Curling mode of magnetization, Brown's Fokker-Planck equation, 495
Current density, high energy barrier approximation, Brown's Fokker-Planck equation, 590–592

Cusp bifurcation, reactive fluid stochastic theory, pitchfork bifurcation fluctuation, 406–407

Deborah number, complex flow simulations, fluid rheology, 354–355, 357–359
Debye relaxation, Fokker-Planck equation, ferromagnetic relaxation, 481
Decay modes, Fokker-Planck equation:
  matrix elements and differential recurrence coefficients, 507–508
  uniaxial anisotropy, free energy parameters, 532–533
Deconvolution, Stokes shift dynamics:
  optical Kerr effect, 216
  pulse overlap, 181–182
  testing procedures, 206–209
  vs. excited-state spectrum, 217–221
Degrees of freedom (DOF), iodine photodissociation dynamics:
  "hole" dynamics, 276–277
  "probe" pulse interpretation, 281
  "pump" pulse interpretation, 266–267
  quantum mechanics, 239–241
Density-density space-time correlation function. *See* Van Hove function
Density functional theory:
  foam simulations, 353
  potential energy (PE) surfaces, Hamiltonian models, 21–23
Density-matrix formalism:
  iodine photodissociation dynamics:
    degrees of freedom (DOF), 242
    Hamiltonian eigenstate representation, 248
    "probe" pulse states, 281–282
    quantum mechanics, 245–248
  nonlinear polarization:
    perturbation theory, 74–76
    spectroscopic characterization, 80–81
  reduced approach, 131–139
  short-time fluorescence Stokes shift dynamics:
    classical Franck approximation, 183–185
    excited-state *vs.* deconvolved spectrum, 217–221
    semiclassical treatment, 186–188
Density profiles, complex flow simulations, Rayleigh-Bénard instability, 379–380
Derivative couplings:
  diabatic electronic states, 11–20
  Fokker-Planck equation, saddle point behavior, 593–600
  non-Born-Oppenheimer dynamics, nonadiabatic coupling, 10–11
  potential energy (PE) surfaces, 21–23
Diabatic orbitals, derivative couplings, 13–15
Diabatic representation:
  explicit construction techniques, 11–20
  femtosecond experiments, 117–119
  internal-conversion processes, 145–148
  nonadiabatic photoisomerization, semiclassical techniques, 142–145
  non-Born-Oppenheimer dynamics, 8–11
  potential energy (PE) surfaces, Hamiltonian models, 22–23
  ultrafast intramolecular non-Born-Oppenheimer dynamics, electronic population probabilities, 50–54
Diatomic molecules, photoinduced dynamics, 6–8
Difference-potential-only approximation:
  short-time dynamics, 176
  short-time fluorescence Stokes shift dynamics, 196–197
Differential recurrence coefficients, Fokker-Planck equation:
  correlation time calculations, 514–515
  cubic anisotropic free energy, 543–581
  non-axially symmetric problems, 505–513
  spherical harmonic representation:
    associated Legendre functions, 628–631
    cubic anisotropic free energy, 543–581
    overview, 515–524
    uniaxial anisotropic free energy, 524–533
      Gilbert-Langevin equation, 533–543
  uniaxial anisotropic free energy:
    Gilbert-Langevin equation, 533–543
    spherical harmonic representation, 524–533
Diffusion, chemical instability simulations:
  overview, 394–395
  complex flow simulations, lattice gas models, 341–342
  reactive fluid stochastic theory, pitchfork bifurcation spatial correlations, 427–430
Dimensionality, complex flow simulations, 323
Directional dependence, nonlinear polarization, nonperturbative approach, 89–90
Direction cosines, Fokker-Planck equation, orthogonal transformations, 634–640
Direct-product-basis expansion technique, time-dependent Schrödinger computation, state vector representation, 34–38

Direct simulation Monte Carlo (DSMC) technique.
  See also Bird algorithm
  complex flow models:
    channel flow studies, 369
    comparative analysis, 344–352
    flow studies, 366–371
  molecular dynamic microscopic simulations:
    Boltzmann dynamics, 437–439
    reactive fluids, 436
Discrete basis expansion, iodine photodissociation grid and eigenstate representations, degrees of freedom (DOF), 241
Discrete orientation model, Fokker-Planck equation, 482–487
  Néel model, 488
  negative anisotropy K < 0, 485–487
  positive anisotropy K > 0, 483–485
Discrete variational representation (DVR), time-dependent Schrödinger computation, state vector representation, 36–38
Discretization techniques:
  real-time path-integrals, vibronic-coupling models, 122–125
  time-dependent Schrödinger computation, 32
Dispersed pump-probe (PP) techniques, femtosecond spectroscopy, transient transmittance, 66–68
Dissipation, reduced density-matrix formalism, 134–139
Dissipative super-operators:
  iodine photodissociation dynamics, 248–251
  model parameters, 301–302
  "probe" pulse photodissociation, 289–290
Distribution of magnetization orientations, Fokker-Planck equation, ferromagnetic relaxation, 479
Distribution of orientations function, Fokker-Planck equation, 478
Doorway operators, short-time fluorescence Stokes shift dynamics:
  classical Franck approximation, 185
  gate-pulse duration, 216–217
  observed fluorescence spectrum, 222–223
  semiclassical aspects, 188
  temporally separated pulses, 182
Dynamical structure factor (DSF), complex flow simulations:
  direct simulation Monte Carlo (DSMC) technique, 348–349
  non-equilibrium steady state, 360–366
Dynamic case, Fokker-Planck equation, correlation time calculations, 497–499

Ehrenfest force, nonadiabatic photoisomerization, semiclassical techniques, 143–145
Eigenstate representation:
  iodine photodissociation dynamics:
    degrees of freedom (DOF), 241
    Hamiltonian calculations, 248
    "probe" pulse states, 281–282
  Redfield theory, 134–138
Eigenvalues, Fokker-Planck equation, correlation time calculations, 495–499, 514–515
Einstein theory of equilibrium fluctuations, complex flow simulations, non-equilibrium steady state, 360–366
Electromagnetic field, iodine photodissociation dynamics:
  atomic unit parameters, 311–312
  "pump" pulse photodissociation, 267–268
  "push" pulse states, 292–293
Electronic population:
  real-time detection, internal-conversion process, 104–108
  ultrafast intramolecular non-Born-Oppenheimer dynamics, 49–54
Emission directions, nonlinear polarization, perturbation theory, 76–81
Energy barriers, Fokker-Planck equation, ferromagnetic relaxation, 481
Energy of collision, chemical instability simulations, exothermic chemical reaction, 443–448
Energy pooling, iodine photodissociation dynamics, dissipative super-operators, 249–251
Energy relaxation, iodine photodissociation dynamics, 300–301
Enskog kinetic equation, complex flow simulations:
  direct simulation Monte Carlo (DSMC) technique, 347–348
  Rayleigh-Bénard instability, 377
Equilibrium correlation functions, short-time fluorescence Stokes shift dynamics:
  classical linear response, 191–192
  excited-state dynamics, 221
Eulerian angles, Fokker-Planck equation, orthogonal transformations, 639–640

Evans-Gillan technique, complex flow simulations, non-equilibrium molecular dynamics (NEMD), 331
Evolution operator, iodine photodissociation dynamics, 251–253
Excitation-pulse duration, Stokes shift dynamics, simulation conditions, 202–206
Excited-state dynamics, short-time fluorescence Stokes shift:
  curve crossing, 220–221
  instantaneous normal mode (INM) analysis, 212–215
  semiclassical analysis, 187–188
Excited-state linear response:
  short-time fluorescence Stokes shift dynamics, 189–190
  classical linear response, 191–192
  simulation approximations, 210–212
Exothermic chemical reaction, microscopic simulation, 441–448
Exothermic gas-phase reaction, hard-sphere chemistry, microscopic simulation, 440–441
Extensitivity parameter, reactive fluid stochastic theory, pitchfork bifurcation fluctuation, 408–414
Extensitivity parameter, reactive fluid stochastic theory, pitchfork bifurcation fluctuation, 408–414

Faber polynomial, iodine photodissociation dynamics, 305
Factorizability, time-dependent Schrödinger computation, state vector representation, 35–36
Fast Fourier transform (FFT):
  iodine photodissociation dynamics:
    conjugated basis-sets, 242
    "probe" pulse states, 281–282
  time-dependent Schrödinger computation, state vector representation, 38
FCHC models, complex flow simulations, lattice gases, 337–338
Femtochemistry, photoinduced processes, 3–8
Femtosecond spectroscopy:
  basic concepts, 61–73
  classical modeling, 152–154
  molecular pump-probe techniques, 6–8
  nonstationary state preparation and detection, 61–63
  photoinduced processes, 3–8
  spectroscopic signals, 64–73
    photon echo, 70
    time-resolved fluorescence, 68–70
    time-resolved ionization and fragment detection, 70–73
    transient transmittance, 64–68
  ultrafast intramolecular non-Born-Oppenheimer dynamics, 41–61
    nuclear observables, 54–61
  ultrafast nonadiabatic excited-state processes, 94–119
    molecular experiments, 114–119
    real-time detection, internal-conversion, 94–108
    real-time detection, nonadiabatic photoisomerization, 108–114
Ferromagnetic particle relaxation, Fokker-Planck equation, 478–482
FHP models, complex flow simulations:
  lattice gases, 337–344
  two-phase flows and spinodal decomposition kinetics, 344
Fick's diffusion coefficient, stochastic theory of reactive fluids, reaction-diffusion master equation, 421, 426-427
Finite-difference equations, complex flow simulations, lattice gas models, 335
Finite subspace, diabatic electronic states, 12
First-order differential equations, time-dependent Schrödinger computation, state vector representation, 33–38
Flow studies, complex flow simulations, 366–371
Fluctuation-dissipation theorem, stochastic theory of reactive fluids, pitchfork bifurcation fluctuation, 411–414
Fluid rheology, complex flow simulation, 353–359
  complex fluids, 357–353
  future research, 383
  simple fluids, 353–357
Fluid shearing, complex flow simulations, nonequilibrium molecular dynamics (NEMD), 324–325
Fluorescence upconversion:
  Stokes shift dynamics, 172–176
  system excitation theory, 176–180
Fokker-Planck equation:
  axial symmetry, 499–505
  background, 478

Fokker-Planck equation (*continued*)
Brown's model, 488–495
background, 589–590
current density at free energy stationary point, 590–592
general Fokker-Planck and Gilbert-Langevin derivation, 631–634
formula, 600–601
high energy barrier approximation, 589–601
quasi-equilibrium behavior near anisotropic minima, 593
saddle point behavior, 593–600
correlation time calculations:
eigenvalues and amplitudes, 495–499
matrix elements and difference recurrence coefficients, 514–515
cubic anisotropy free energy, 543–589
spherical harmonic recurrence relations, 543–581
spherical harmonics product formula, 581–589
discrete orientation model, 482–487
negative anisotropy $K < 0$, 485–487
positive anisotropy $K > 0$, 483–485
ferromagnetic relaxation, 478–482
free energy applications:
bistable structure, 609–611
coordinate system and stationary points, 601–606
high energy barrier approximation, $\psi = \pi/4$, 616–623
high energy barrier approximation, ψ arbitrary value, 623–628
orthogonal transformation of coordinate system, 606–609
stationary point geometric interpretations, 611–616
Néel's model, 487–488
non-axially symmetric problems:
correlation time calculations, 514–515
differential recurrence coefficients and matrix elements, 505–513
Gilbert-Langevin equation, spherical harmonics, 533–543
numerical computation results, 543
spherical harmonic representation, 515–524
orthogonal transformations, 606–609, 634–640
reactive fluid stochastic theory, reaction-diffusion master equation, 423–424
spherical harmonic representation:
cubic anisotropy free energy:
product formula, 581–589
recurrence relations, 543–581
Gilbert-Langevin equation, 524–533
Legendre functions, 628–631
non-axially symmetric problems, 515–524
uniaxial anisotropy applications, 524–533
stochastic theory of reactive fluids, 403–404
Hopf bifurcation, 418–420
pitchfork bifurcation, spatial correlation, 428–430
pitchfork bifurcation fluctuation, 410–414
uniaxial anisotropy applications:
free energy cases, 601–628
spherical harmonic representation, 524–533
Fourier coefficients, Fokker-Planck equation:
cubic anisotropic free energy, 574–581
matrix elements and differential recurrence coefficients, 506–507
uniaxial anisotropy, free energy parameters, 530–533
Fourier-Laplace series, Fokker-Planck equation, matrix elements and differential recurrence coefficients, 506
Fourier transforms:
complex flow simulations, non-equilibrium steady state, 360–366
femtosecond spectroscopy, transient transmittance, 66–68
Hamiltonian models, nonadiabatic photoisomerization, 28–29
iodine photodissociation dynamics:
absorption spectrum, 263
grid and eigenstate degrees of freedom (DOF), 241
laser pulse absorption, 260–262
spatially extended systems, microscopic simulation, 459–460
time-dependent Schrödinger computation, state vector representation, 38
Franck approximation, Stokes shift dynamics, 175
classical aspects, 183–185
gate-pulse duration, 217
semiclassical aspects, 185–188, 217–218
Franck-Condon progression:
Hamiltonian models, 26–27
iodine photodissociation dynamics:
absorption spectrum, 263
ground surface "hole" and excited surface density, 258–259

short-time fluorescence Stokes shift dynamics, 200–201
instantaneous normal mode (INM) analysis, 215
Frank-Kamenetskii parameter, exothermic chemical reaction, 441–448
Free energy parameters, Fokker-Planck equation:
cubic anisotropic parameters, 581–589
spherical harmonic recurrence relations, 543–581
high energy barrier approximation, Brown's model, 590–592
spherical harmonic recurrence relations, 543–581
uniaxial anisotropy, 524–543
bistable structure, 609–611
coordinate system and stationary points, 601–606
geometrical changes of polynomials, 611–616
Gilbert-Langevin equation, 533–543
high energy barrier approximation, arbitrary value of ψ, 623–628
high energy barrier approximation for $\psi = \pi/4$, 616–623
orthogonal transformation near stationary points, 606–609
spherical harmonic representation, 524–533
Frequency-resolved fluorescence, femtosecond spectroscopy, 68–70
Full-width half-maximum (FWHM) linewidt, real-time detection, internal-conversion process, 98

"Game of life," complex flow simulations, lattice gas models, 335
Gamma function, reactive fluid stochastic theory, pitchfork bifurcation fluctuation, 409–414
Gate-pulse duration:
Stokes shift dynamics, doorway conditions, 216–217
simulation conditions, 203–206
Gaussian distribution:
pitchfork bifurcation fluctuation, 409–414
stochastic theory of reactive fluids, 402–405
Gaussian least constraint principle, complex flow simulations, non-equilibrium molecular dynamics (NEMD), 326
Gaussian pulses:
femtosecond spectroscopy, classical modeling, 152–154
fluorescence upconversion, 176–180
iodine photodissociation dynamics:
dissipative super-operators, 249–251
photo-induced "hole" dynamics, 277
nonlinear polarization, 84–87
short-time fluorescence Stokes shift dynamics:
semiclassical treatment, 186–188
separated pulses, 193–194
Stokes shift dynamics:
deconvolution testing, 206–209
pump-probe analysis and, 217–218
Gaussian white noise:
complex flow simulations:
non-equilibrium steady state, 361–366
particle simulations, 352–353
reactive fluid stochastic theory,master equation, 423–424
Geometric-phase effects, ultrafast intramolecular non-Born-Oppenheimer dynamics, 46–49
Gibbs canonical ensembles:
Fokker-Planck equation, ferromagnetic relaxation, 479
stochastic theory of reactive fluids, pitchfork bifurcation fluctuation, 411–414
Gilbert's equation, Fokker-Planck equation:
Brown's model, 489–491
derivation from general Fokker-Planck and Gilbert-Langevin equations, 631–634
spherical harmonic representation, 533–543
Glauber equation, stochastic theory of reactive fluids, pitchfork bifurcation fluctuation, 411–414
G matrix of vibrational spectroscopy, kinetic-energy operator, 23–25
Golden-Rule expression, iodine photodissociation dynamics, absorption spectrum, 263
Green-Kubo integrands, complex flow simulations:
background, 323
fluid rheology, 355–356, 358–359
lattice gas models, 341–342
non-equilibrium molecular dynamics (NEMD), 324
Rayleigh-Bénard instability, 377–378
viscosity studies, 383

Grid representations, iodine photodissociation dynamics, degrees of freedom (DOF), 241
Ground-state linear response:
  short-time fluorescence Stokes shift dynamics, 189
  simulation approximations, 210–212
Ground surface dynamics, iodine photodissociation:
  atomic unit parameters, 310–311
  "hole" creation, 256–259
  photo-induced "hole" dynamics, 277–279
Gyroscopic representation formula, Fokker-Planck equation:
  cubic anisotropic free energy, 548–581
    spherical harmonic product formula, 584–589
  high energy barrier approximation, Brown's model, 591–592
  Langevin spherical harmonic representation, 537–539
  spherical harmonic representation, 520–524
    cubic anisotropic free energy, 584–589

Hamiltonian models:
  complex flow simulations, non-equilibrium molecular dynamics (NEMD), 328–329
  iodine photodissociation dynamics:
    coordinate-dependent, two-level approximation, 254–256
    dissipative super-operators, 248–251
    evolution operator, 251–253
    quantum mechanics, 245–248
  non-Born-Oppenheimer dynamics:
    adiabatic electronic representations, 8–11
    kinetic-energy operator, 23–25
    nonadiabatic photoisomerization, 27–29
    normal-mode Taylor expansion, 25–27
    PE surfaces, 20–23
  nonlinear polarization, perturbation theory, 73–76
  photochemical funnels, 7–8
  time-dependent Schrödinger computation, 29–41
    state vector representation, 32–38
Hamilton-Jacobi formulation, stochastic theory of reactive fluids, 402
Hard-sphere fluid model:
  chemical instability microscopic simulations, 439–441
    exothermic chemical reaction, 443–448
  complex flow simulations:
    fluid rheology, 357–359
    non-equilibrium steady state, 363–366
    phase-space diagrams, 321–323
    Rayleigh-Bénard instability, 377
Harmonic oscillator:
  iodine photodissociation dynamics, dissipative super-operators, 249–251
  nonlinear polarization:
    multitime correlation functions, 81–83
    nonperturbative approach, 90–94
Hartree-Fock theory, potential energy (PE) surfaces, Hamiltonian models, 21–23
Heisenberg's uncertainty principle, iodine photodissociation dynamics, phase-space picture, 243–245
Heterodyne detection:
  femtosecond spectroscopy, 63
  nonlinear polarization, spectroscopic characterization, 77–81
High energy barrier approximation:
  Fokker-Planck equation, axial symmetry, 503–505
  Brown's model, 589–601
  uniaxial anisotropic free energy:
    arbitrary value of ψ, 623–628
    $\psi = \pi/4$, 616–623
Hilbert adjoint, Fokker-Planck equation, matrix elements and differential recurrence coefficients, 506
Hilbert space, iodine photodissociation dynamics, quantum statics, 239–243
"Hole" dynamics, iodine photodissociation:
  degrees of freedom, 276–277
  ground surface dynamics, 256–259
  initial state, 277
  interpretation, 279–280
  phase space diagrams, 277–279
  "push" pulse photodissociation, 290–297
Homodyne detection:
  femtosecond spectroscopy, 63
    photon echo, 70
  nonlinear polarization, spectroscopic characterization, 77–81
Hopf bifurcation:
  continuously stirred tank reactor (CSTR), 454–457
  reactive fluid stochastic theory, 414-420
    one-dimensional system fluctuations, 433–434
Hopping criterion, nonadiabatic photoisomerization, 141

HPP (Hardy, Pomeau, de Pazzis) model:
lattice gases, 336–344
sound propagation, 337–338
Hydrodynamical effects, exothermic chemical reaction, 445–448
Hysteresis, reactive fluid stochastic theory, pitchfork bifurcation fluctuation, 407

Impulsive excitation, iodine photodissociation dynamics, 254–259
coordinate-dependent two-level approximation, 254–256
ground surface dynamical "hole," 256–259
laser pulse power absorption, 259–260
Impulsive limit, nonlinear polarization, 83–86
Impulsive limits.
See also Photodissociation dynamics; Resonant impulsive stimulated Raman scattering
nonlinear polarization, 83–86
Impulsive stimulated Raman scattering (ISRS), iodine photodissociation dynamics, ground surface "hole" and excited surface density, 257–259
Infinite-dimensional systems, real-time path-integrals, 130–131
Inhomogeneous fluctuations, stochastic theory of reactive fluids, pitchfork bifurcation, 414
Instantaneous normal mode (INM) analysis, short-time fluorescence Stokes shift dynamics, excited-state dynamics, 212–215
Integral feedback, complex flow simulations, non-equilibrium molecular dynamics (NEMD), 326–327
Interaction representation, Hamiltonian models, time-dependent Schrödinger computation, 31–38
Intermediate scattering function (ISF), complex flow simulations, non-equilibrium steady state, 360–366
Internal-conversion processes:
classical modeling, 145–154
benzene cation C to B to X process, 145–148
real-time detection, femtosecond spectroscopy, 94–108
Iodine photodissociation dynamics:
absorption spectrum, 262–263
applications, 264–298
electronic potential energy surfaces, 265–266
"hole" dynamics, 276–280
"probe" pulse dynamics, 281–290
"pump" pulse photodissociation, 266–272
"push" pulse vibrational excitation, 290–297
critical evaluation, 298–303
dissipative super-operators, 248–251
evolution operator, 251–253
experimental background, 231–235
Hamiltonian operator, 246–248
coordinate representation, 247–248
eigenstate representation, 248
impulsive excitation, 254–259
coordinate-dependent two-level approximation, 254–256
ground surface "hole" dynamics, 256–259
laser pulse power absorption, 259–262
transient spectroscopies, 260–262
weak pulse transient absorption, 259–260
numerical methods, 303–310
approximating functions of operators, 303–305
Leja's interpolation points, 306–307
Newton's interpolation method, 306
operator applications, 307–310
Raman spectrum, 263–264
statics, 239–245
conjugated basis sets, 242–243
degrees of freedom, 239–241
density matrix, 242
grid and eigenstate representations, 241
initial states, 244–245
phase space picture, 243–244
single basis visualization, 243
state vector, 241–242
system parameters, 310–312
electromagnetic parameters, in atomic units, 311–312
potential surfaces, in atomic units, 310–311
theoretical background, 235–237
time scales, 298–301
Ionization experiments:
femtosecond spectroscopy, 70–73
real-time detection, internal-conversion process, 105–108
Isserlis's theorem, Fokker-Planck equation, Brown's model, 490–491

Jacoby coordinates, iodine photodissociation dynamics:
  "probe" pulse dynamics, 281–282
  "pump" pulse photodissociation, 266–276
Jahn-Teller effect:
  femtosecond spectroscopy, 6–8
  Hamiltonian models, normal-mode Taylor expansion, 26–27
  internal-conversion processes, classical techniques, 146–148
  nonadiabatic coupling, 10
  potential energy (PE) surfaces, Hamiltonian models, 22–23
  ultrafast intramolecular non-Born-Oppenheimer dynamics:
    electronic population probabilities, 53–54
    wave packets, 46–49
Jump Markov processes, Boltzmann dynamics:
  Bird algorithm, 437–439
  stochastic theory of reactive fluids, 400, 402–404

Kinetic-energy operators:
  Hamiltonian models, 23–25
    nonadiabatic photoisomerization, 27–29
    normal-mode Taylor expansion, 25–27
  time-dependent Schrödinger computation, state vector representation, 34–38
Kinetic potential:
  exothermic chemical reaction, 447–448
  reactive fluid stochastic theory, pitchfork bifurcation, 406
Kolmogorov flow, complex flow simulations, non-equilibrium molecular dynamics (NEMD), 325, 329–334
Kramers-Moyal coefficients, Fokker-Planck equation, 632–634
Kramers theory:
  stochastic theory of reactive fluids, pitchfork bifurcation, 413–414
  transition state, Fokker-Planck axial symmetry, 500–505

Lagrange multipliers:
  complex flow simulations, non-equilibrium molecular dynamics (NEMD), 326
  kinetic-energy operator, 23–25
Lanczos tridiagonalization, time-dependent Schrödinger equation, state vector propagation, 39–40
Landau-Ginzburg potential:
  chemical instability simulations, 397
  stochastic theory of reactive fluids:
    one-dimensional system fluctuations, 430–434
    pitchfork bifurcation, spatial correlation, 429–430
Landau-Lifshitz equation:
  Boltzmann dynamics, Bird algorithm, 438–439
  complex flow simulations, 352–353
  Fokker-Planck equation, Brown's model, 490
  reactive fluid stochastic theory, 403–404
  reaction-diffusion master equation, 423–424
Landau-Placzek formula, complex flow simulations:
  direct simulation Monte Carlo (DSMC) technique, 348–349
  lattice gas models, 343–344
  non-equilibrium steady states, 359–366
Langevin equations:
  complex flow simulations, 353
  continuously stirred tank reactor (CSTR), 453–454
  Fokker-Planck equation:
    derivation from general Fokker-Planck and Gilbert-Langevin equations, 631–634
    spherical harmonic representation, 533–543
  microscopic simulation, 464
  stochastic theory of reactive fluids, 403–404
    Hopf bifurcation, 416–420
    pitchfork bifurcation, spatial correlation, 428–430
    pitchfork bifurcation fluctuation, 410–414
    reaction-diffusion master equation, 422–424
Laplacian transforms:
  Fokker-Planck equation:
    Brown's model, 493–495
    correlation time calculation, 515
    cubic anisotropic free energy, 547–581
    discrete orientation model, 487
    matrix elements and differential recurrence coefficients, 411–413
    saddle point behavior, 594–600
    spherical harmonic representation, 515–524
    uniaxial anisotropy, free energy parameters, 525–533
  kinetic-energy operator, 24–25

Laser-induced fluorescence, femtosecond spectroscopy, ionization and fragment detection, 71–73
Laser pulse, iodine photodissociation dynamics:
  power absorption, 259
  "pump" pulse photodissociation, 267–268
Lattice Boltzmann technique:
  Bird algorithm, 438–439
  complex flow simulations:
    background, 319
    lattice gas models, 340–344
Lattice gas cellular automata (LGCA):
  Bird algorithm, 438–439
  complex flow simulations, colloidal suspensions, 343–344
  microscopic simulation, 465–466
Lattice gas models:
  background, 319–320
  techniques, 334–344
Lees-Edwards boundary conditions, complex flow simulations, non-equilibrium molecular dynamics (NEMD), 331
Legendre functions, Fokker-Planck equation:
  spherical harmonic representation, 519–520
  associated functions, 628–631
Leja interpolation points, iodine photodissociation dynamics:
  evolution operator, 253
  techniques, 306–307
Lennard-Jones interaction potentials:
  complex flow simulations:
    fluid rheology, 357–359
    liquid-liquid interface, 370–371
    phase-space diagrams, 321–323
    two-phase flows, 369–370
  short-time fluorescence Stokes shift dynamics, 199–201
LEPS functional form, iodine photodissociation dynamics, potential energy (PE) surfaces, 265–266
Light-matter interaction, iodine photodissociation dynamics, 253–264
Linear response analysis, short-time fluorescence Stokes shift dynamics, 175–176
  classical linear response, 191–192
  equilibrium and non-equilibrium functions, 221
  excited-state response, 189–190
  ground-state response, 189
  instantaneous normal mode (INM) analysis, 212–215
  separated pulses, 194–195
  simulation techniques, 209–212
Linear transformations, Fokker-Planck equation, orthogonal transformations, 635–640
Liouville-von Neumann equation, iodine photodissociation dynamics:
  degrees of freedom (DOF), 240–241
  laser pulse absorption, 261–262
  photo-induced "hole" dynamics, 277
  quantum mechanics, 245–248
Liquid-liquid interface, complex flow simulations, 370–371
Long-time dynamics, Stokes shift, very short and very long time behavior, 223–225
Low energy barrier approximation, Fokker-Planck equation, axial symmetry, 500–505
Lyapunov instability, complex flow simulations, non-equilibrium molecular dynamics (NEMD), 332–334

Macroscopic coexistence, reactive fluid stochastic theory, pitchfork bifurcation fluctuation, 407
Magnetic aftereffect behavior, Fokker-Planck equation, 482
Magnetization orientation, Fokker-Planck equation:
  Brown's model, 489–495
  ferromagnetic relaxation, 478–482
  Langevin spherical harmonic representation, 535–537
Magnus expansion, iodine photodissociation dynamics, evolution operator, 251–253
Markovian approximation, reduced density-matrix formalism, 132–138
Markov processes:
  complex flow simulations, direct simulation Monte Carlo (DSMC) technique, 345
  Fokker-Planck equation:
    general Fokker-Planck and Gilbert-Langevin derivations, 631–634
    ferromagnetic relaxation, 479
  stochastic theory of reactive fluids, 400–404
    reaction-diffusion master equation, 420–421
Mass action law, chemical instability simulations, 394
Master equation:
  chemical instability simulations, 397–398

Master equation (*continued*)
complex flow simulations, 353
Fokker-Planck equation, discrete orientation model, 482–487
reactive fluid stochastic theory, 399–405
"extrema statistics," 414
Hopf bifurcation, 415–420
one-dimensional system fluctuations, 432–434
pitchfork bifurcation fluctuation, 408–414
reaction-diffusion master equation, 420–434
spatially extended systems, microscopic simulation, 458–463
Matrix elements, Fokker-Planck equation:
orthogonal transformations, 635–640
correlation time calculations, 514–515
cubic anisotropy free energy:
spherical harmonic product formulas, 581–589
spherical harmonic recurrence relations, 543–581
non-axially symmetric problems, 505–513
uniaxial anisotropy free energy, spherical harmonic representation, 524–533
Maxwell-Boltzmann distribution:
Fokker-Planck equation, quasi-equilibrium behavior, 593
microscopic simulations:
continuously stirred tank reactor (CSTR), 449–457
hard-sphere chemistry, 439–441
Maxwell distribution:
complex flow simulations, non-equilibrium steady state, 363–364
exothermic chemical reaction, 446–448
hard-sphere chemistry, microscopic simulation, 440–441
iodine photodissociation dynamics, laser pulse absorption, 260–262
microscopic simulation, 464
Mean field results, reactive fluid stochastic theory, 399
pitchfork bifurcation spatial correlations, 427–430
Mean jump frequency, stochastic theory of reactive fluids, reaction-diffusion master equation, 421
Mean square displacement (MSD), complex flow simulations, Rayleigh-Bénard instability, 380–383
Microreversibility, reactive fluid stochastic theory, reaction-diffusion master equation, 425–427
Mode coupling effects:
Hamiltonian models, nonadiabatic photoisomerization, 28–29
real-time path-integrals, 127–131
ultrafast intramolecular non-Born-Oppenheimer dynamics, 45–49
Model systems, short-time fluorescence Stokes shift dynamics, 198–201
Molecular dynamics (MD):
complex flow simulations:
flow studies, 366–367, 370–371
non-equilibrium molecular dynamics (NEMD), 324–334
non-equilibrium steady state, 366
Rayleigh-Bénard instability, 372–383, 381–383
vs. direct simulation Monte Carlo (DSMC) technique, 346–352
wetting phenomenon, 370–371
iodine photodissociation dynamics, 235–237
microscopic simulations, chemical instabilities, 434–466
continuously stirred tank reactor (CSTR), 450–457, 465–466
exothermic chemical reaction, 442–448
overview, 397–398
reactive fluids, 435–436
short-time fluorescence Stokes shift dynamics:
classical linear response, 191–192
ground- and excited-state linear response approximations, 210–212
quantum classical approximations, 190–191
Molecular Hamiltonians, time-dependent Schrödinger computation, 31–32
Molecular mechanics, potential energy (PE) surfaces, 23
Monte Carlo-Gibbs ensemble technique, complex flow simulations, 323
Monte Carlo simulations, complex flow simulations, wetting phenomenon, 371
Monte Carlo techniques.
See also Direct simulation Monte Carlo (DSMC) technique
complex flow simulations:
background, 319–320
wetting phenomenon, 371
real-time path-integral technique, 121

Monte Carlo wave-function propagation, reduced density-matrix formalism, 136–139
Mössbauer effect, Fokker-Planck equation, Brown's model, 495
Multiconfiguration self-consistent-field (MCSCF) wave function, derivative couplings, diabatic electronic states, 13
Multiconfiguration time-dependent Hartree (MCTDH) method, multidimensional problems, 120
Multi-Poissonian distribution, stochastic theory of reactive fluids, 401–402
Multireference configuration interaction (MRCI):
  derivative couplings, diabatic electronic states, 13
  real-time detection, internal-conversion process, 98–108
Multitime correlation functions, nonlinear polarization, 81–83
Multivariate master equation, stochastic theory of reactive fluids, 421

Nakajima-Zwanzig equation, reduced density-matrix formalism, 132–138
Navier-Stokes equations:
  Boltzmann dynamics, Bird algorithm, 438–439
  complex flow simulations:
    background, 319
    direct simulation Monte Carlo (DSMC) technique, 350–352
    lattice gas models, 338–344
    non-equilibrium molecular dynamics (NEMD), 330–334
    non-equilibrium steady state, 361–366
    Rayleigh-Bénard instability, 373–374, 379–380
    wetting phenomenon, 371
Néel relaxation, Fokker-Planck equation:
  ferromagnetic relaxation, 481
  granulometric applications, 495
  model parameters, 487–488
  numerical computation, 543
Newton interpolation polynomial, iodine photodissociation dynamics:
  absorption spectrum, 263
  approximating functions, 304–305
  contour mapping, 308–309
  evolution operator, 253
  Raman spectrum, 264
  techniques, 306
Newton's law, complex flow simulations, fluid rheology, 354
Nonadiabatic photoisomerization. *See also* Ultrafast nonadiabatic excited-state processes
  Hamiltonian model, 27–29
  real-time detection, 108–114
  semiclassical description, 140–145
Non-axial symmetry, Fokker-Planck equation:
  correlation time calculations, 514–515
  differential recurrence relations, 505–524
  matrix elements and differential recurrence coefficients, 505–513
  spherical harmonic representation, 515–524
Non-Born-Oppenheimer dynamics:
  femtosecond chemistry and, 6–8
  intramolecular concepts, 8–29
    adiabatic and diabatic electronic representations, 8–11
    explicit diabatic state construction, 11–20
    model Hamiltonians, 20–29
  ultrafast intramolecular aspects, 41–61
    electron population dynamics, 49–54
    nuclear observables, 54–61
    wave packets, 42–49
Non-equilibrium classical and semiclassical approximations, short-time Stokes shift dynamics, 210, 221
Non-equilibrium Ising model, complex flow simulations, background, 319
Non-equilibrium molecular dynamics (NEMD), complex flow simulations:
  flow studies, 368–371
  techniques, 324–334
Non-equilibrium phase transitions, reactive fluid stochastic theory, pitchfork bifurcation fluctuation, 407–408
Non-equilibrium steady states:
  complex flow simulations, 359–366, 383–384
  stochastic theory of reactive fluids, pitchfork bifurcation, 411–414
Nonlinear behavior, complex flow simulations, fluid rheology, 354–355, 358–359
Nonlinear master equation, reactive fluid stochastic theory, one-dimensional system fluctuations, 432–434
Nonlinear polarization:
  nonperturbative approach, 87–94
    directional dependence, 89–90
    one-dimensional harmonic oscillator, 90–94

Nonlinear polarization (*continued*)
perturbative approach, 73–87
conditions, 75
multitime correlation functions, 81–83
pump-probe (PP) signals, 83–87
spectroscopic characterization, 76–81
Non-overlapping laser pulses, nonlinear polarization:
perturbation theory, 75
spectroscopic characterization, 76–81
Nonperturbative theory, nonlinear polarization, 87–94
directional dependence, 89–90
one-dimensional harmonic oscillator, 90–4
Nonstationary states, femtosecond spectroscopy, 61–63
Normal-mode Taylor expansion, Hamiltonian models, 25–27
Nosé-Hoover equations, complex flow simulations:
non-equilibrium molecular dynamics (NEMD), 326–334
non-equilibrium steady state, 364–366
Rayleigh-Bénard instability, 375–376
wetting phenomenon, 371
Nuclear coordinates, potential energy (PE) surfaces, Hamiltonian models, 22–23
Nuclear dynamics, ultrafast observables, 54–61
Numerical computation, Fokker-Planck equation, uniaxial anisotropy, 543
Nyquist frequency, iodine photodissociation grid and eigenstate representations, degrees of freedom (DOF), 241

One-dimensional chemical systems, stochastic theory of reactive fluids, critical fluctuations, 430–434
Ornstein-Zernike equilibrium fluctuations:
complex flow simulations, 360
reaction-diffusion master equation, 425–427
Orthogonal transformations, Fokker-Planck equation:
formulae, 634–640
high energy barrier approximation, Brown's model, 590–592
uniaxial anisotropic free energy, coordinate system near stationary points, 606–609
Out-of-equilibrium fluid states, complex flow simulations, non-equilibrium steady state technique, 359–366
Passive mass transport, complex flow simulations, Rayleigh-Bénard instability, 374–375
Path-class techniques, real-time path-integrals, 128–131
Path-integral techniques:
nonadiabatic photoisomerization, 140–145
real-time detection, 120–131
vibronic coupling discretization, 122–125
Peak emission frequency, Stokes shift dynamics, 175–176
deconvolution testing, 206–209
semiclassical expression, 218–221
separated pulses, 194–195
short-time analysis, 192–193
simulation conditions, 202–206
very short and very long time behavior, 223–225
Péclet number, complex flow simulations, Rayleigh-Bénard instability, 374–375, 378
Periodic boundary deformation, complex flow simulations, non-equilibrium molecular dynamics (NEMD), 325–326
Periodic forcing, complex flow simulations, non-equilibrium molecular dynamics (NEMD), 324–325
Perturbation theory:
Fokker-Planck equation, axial symmetry, 500–505
nonlinear polarization, 73–87
conditions, 74
multitime correlation function, 81–83
pump-probe (PP) signals, 83–87
spectroscopic characterization, 76–81
Phase-averaged detection, nonlinear polarization, perturbation theory, 75
Phase diffusion, stochastic theory of reactive fluids, Hopf bifurcation, 419–420
Phase-space diagrams:
complex flow simulations, 321–323
iodine photodissociation dynamics, 243–245, 301
reactive fluid stochastic theory, reaction-diffusion master equation, 421–422
Photochemical funnels, photoinduced dynamics, 5–8
photoisomerization, 7–8
nonadiabatic Hamiltonian model, 27–29
Photodissociation dynamics:
iodine in solution:
absorption spectrum, 262–263

applications, 264–298
approximating functions of operators, 303–305
conjugated basis sets, 242
critical evaluation, 298–303
degrees of freedom, 239–241
density matrix, 242
dissipative super-operators, 248–251
dynamics, 245–264
electric potential energy surfaces, 265–266
electromagnetic field parameters, atomic units, 311–312
evolution operator, 251–253
experimental background, 231–235
grid and eigenstate representation, 241
Hamiltonian operator, 246–248
"hole" dynamics, 276–281
impulsive excitation, 254–259
initial states, 244–245
laser pulse, 259–262
Leja's interpolation points, 306–307
Newton's interpolation method, 306
numerical methods, 303–310
operator applications, 307–310
potential surfaces parameters, atomic units, 310–311
"probe" pulse, nascent iodine, 281–290
"pump" pulse techniques, 266–276
"push" pulse, vibrational excitation, 290–297
Raman spectrum, 263–264
state vector, 241–242
statics, 239–245
system parameters, 310–312
theoretical background, 235–237
visualization techniques, 243–244
ultrafast intramolecular non-Born-Oppenheimer dynamics, 41
Photoinduced processes:
photoisomerization, chromophore in solution, 150–152
theory, 2–6
Photoisomerization:
chromophore in solution, 148–152
nonadiabatic:
Hamiltonian model, 27–29
real-time detection, 108–114
photochemical funnels, 7–8
time-dependent Schrödinger computation, 34–38
ultrafast intramolecular non-Born-Oppenheimer dynamics:
nuclear observables, 57–61
wave packets, 43–49
Photon echo signals:
femtosecond spectroscopy, 70
real-time detection, nonadiabatic photoisomerization, 112–115
Picosecond relaxation:
femtosecond experiments, 117–119
photoisomerization, chromophore in solution, 149–152
Pitchfork bifurcation:
continuously stirred tank reactor (CSTR), 451–454
reactive fluid stochastic theory, 405–420
fluctuation, 408–414
spatial correlation, 427–430
Podolsky formalism, kinetic-energy operator, 23–25
Poiseuille flow, complex flow simulations, 368–369
Poisson processes:
iodine photodissociation dynamics, dissipative super-operators, 249–251
reactive fluid stochastic theory:
one-dimensional system fluctuations, 432–434
reaction-diffusion master equation, 426–427
Polyatomic molecules, femtosecond experiments, 118–119
Polyatomic systems, Hamiltonian models, potential energy (PE) surfaces, 21–23
Polymer systems, complex flow simulations, fluid rheology, 357–359
Population probabilities:
internal-conversion processes, classical techniques, 146–148
photoisomerization, chromophore in solution, 148–152
reduced density-matrix formalism, 138–139
Potential energy (PE) functions:
ab initio computation, 7–8
derivative couplings, 11
Hamiltonian models, 20–23
nonadiabatic photoisomerization, 28–29
normal-mode Taylor expansion, 25–27
iodine photodissociation dynamics, 265–266
applications, 265–266
atomic unit parameters, 303
critical evaluation, 302–303

Potential energy (PE) functions (*continued*)
nonadiabatic photoisomerization:
real-time detection, 108–114
semiclassical techniques, 140–145
photoinduced dynamics, 3–8
time-dependent Schrödinger computation, state vector representation, 33–38
ultrafast intramolecular non-Born-Oppenheimer dynamics, 41–61
Power absorption, iodine photodissociation dynamics:
laser pulse, 259–262
transient weak pulse absorption, 259–260
Predictor-corrector integrator, time-dependent Schrödinger equation, state vector propagation, 39–40
Probability distributions, ultrafast intramolecular non-Born-Oppenheimer dynamics, wave packets, 46–49
"Probe" pulse dynamics, iodine photodissociation, 281–290
degrees of freedom, 281
initial state, 281–282
propagation results, 284–287
solvent potential and interaction, 282–284
transient spectrum:
dissipation conditions, 289
individual states, 287–289
probe wavelength differences, 289–290
Process-dependent noise, stochastic theory of reactive fluids, pitchfork bifurcation fluctuation, 412–414
Product formula for spherical harmonics, Fokker-Planck equation, 581–589
Projection-operator techniques, reduced density-matrix formalism, 132–139
Propagation results, iodine photodissociation dynamics:
"probe" pulse photodissociation, 284–287
"pump" pulse photodissociation, 268–272
"push" pulse states, 294
Pseudo-spectral techniques. *See* Discrete variational representation (DVR)
Pulse duration, Stokes shift dynamics, simulation conditions, 201–206, 219–221
Pulse overlap, Stokes shift dynamics, 173–176
deconvolution, 181–182
short-time analysis, 192–193
simulation conditions, 201–206
temporally separated pulses, 182–183
Pulse separation, Stokes shift dynamics, 193–197
difference-potential-only approximation, 196–197
rigid-cage approximation, 195–196
short-time expansion, 194–195
wave packet picture, 193–194
Pump-probe (PP) techniques:
femtosecond spectroscopy, 61–63, 115–119
transient transmittance, 64–68
iodine photodissociation dynamics, 253–264
absorption spectrum, 262–263
impulsive excitation, 254–259
photo-induced "hole" dynamics, 278–280
power absorption of laser pulse, 259–262
Raman spectrum, 263–264
nonlinear polarization, 84–87
nonperturbative approach, 87–94
perturbative approach, 84–87
real-time detection:
internal-conversion process, 99–108
nonadiabatic photoisomerization, 108–114
short-time fluorescence Stokes shift dynamics:
stimulated-emission (SE) amplitude, 222
vs. time-dependent correlation, 216–221
"Pump" pulse iodine photodissociation, 266–276
absorption spectrum, 274
degrees of freedom, 266–267
electromagnetic field, 267–268
electromagnetic field parameters, 311–312
initial state, 267
propagation results, 268–272
Raman spectrum, 274–276
"Push" pulse, iodine photodissociation:
background, 233–235
vibrational excitation, 290–297
propagation results, 294–297
solvent-electromagnetic field interaction, 292–293
statics, 291–292

Quantum-jump method, reduced density-matrix formalism, 136–139
Quantum mechanics:
benzene cation C to B to X conversion, 145–148
iodine photodissociation dynamics, 236–237
conjugated basis sets, 242
critical evaluation, 298
degrees of freedom, 239–241
density matrix, 242
dissipative super-operators, 248–251

dynamics, 245–253
evolution operator, 251–253
grid and eigenstate representations, 241
Hamiltonian operator, 246–248
initial states, 244–245
phase space picture, 243–244
single basis visualization, 243
state vector, 241–242
statics, 239–245
multidimensional problems, 119–120
real-time path-integral approach, 120–131
reduced density-matrix approach, 131–139
nonadiabatic photoisomerization, 141
short-time fluorescence Stokes shift dynamics, 190–191
Quantum state diffusion formalism, reduced density-matrix, 136–139
Quasi-diatomic systen (ICN), photoinduced dynamics, 6–8
Quasi-equilibrium, Fokker-Planck equation:
anisotropic minima behavior, 593
axial symmetry, 500–505

Rabi frequency, iodine photodissociation dynamics, ground surface "hole" and excited surface density, 257–258
Radiation-matter coupling, photoinduced dynamics, 4–5
Raman spectrum:
femtosecond spectroscopy, 62
iodine photodissociation dynamics:
experimental relevance, 298–302
interpretation, 263–264
"pump" pulse photodissociation, 274–276
nonlinear polarization, 79–81
multitime correlation functions, 81–83
nonperturbative approach, 92–94
pump-probe (PP) techniques, 84–87
real-time detection, nonadiabatic photoisomerization, 110–114
Random walk modeling, microscopic simulation, spatially extended systems, 465
Rate constants, chemical instability simulations, 394
Rayleigh-Bénard instability, complex flow simulations, 371–383
Rayleigh number, complex flow simulations, 372, 376, 378
Rayleigh peak predictions, complex flow simulations, non-equilibrium steady state, 362–363
Reaction-diffusion master equation:
chemical instability simulations, 395
reactive fluid stochastic theory, 420–434
Langevin limit, 422–424
one-dimensional critical fluctuations, 430–434
one-dimensional system fluctuations, 432–434
pitchfork bifurcation spatial correlations, 427–430
spatial correlation onset, 424–427
Reaction-path Hamiltonian model, potential energy (PE) surfaces, 21–23
Reactive fluids:
molecular dynamic microscopic simulations, 435–436
stochastic theory, 399–434, 405–434
Hopf bifurcation, 414–420
pitchfork bifurcation fluctuation, 405–414
reaction-diffusion master equation, 420–434
Reactive hard-sphere collisions, microscopic simulations, 439–441
Real-time detection:
femtosecond spectroscopy, 3
internal-conversion processes, 94–108
nonadiabatic photoisomerization, 108–114
path-integral approach, 120–131
evaluation of techniques, 125–131
vibronic coupling discretization, 122–125
Recurrence relations, Fokker-Planck equation, spherical harmonic representation:
associated Legendre functions, 628–631
cubic anisotropic free energy, 543–581
Redfield relaxation tensor, reduced density-matrix formalism, 132–138
Reduced density-matrix (RDM) theory, multidimensional problems, 131–139
Reduced-dimensionality technique, photoinduced dynamics, 5–6
Relaxation mechanisms, Fokker-Planck equation:
correlation time calculations, 498–499
ferromagnetic relaxation, 481–482
Resonance Raman spectra, real-time detection, internal-conversion process, 95–108
Resonant impulsive stimulated Raman scattering (RISRS):
ground surface "hole" and excited surface density, 256–259
iodine photodissociation dynamics, 232–235
"pump" pulse photodissociation, 274–276

Reynolds numbers:
Boltzmann dynamics, Bird algorithm, 438–439
complex flow simulations:
direct simulation Monte Carlo (DSMC) technique, 347
lattice gas models, 339–344
Rigid cage approximation, short-time Stokes shift dynamics, 175–176, 195-196, 219–221
Rigid coupling assumption, Fokker-Planck equation, Brown's model, 494–495
Néel model, 488
Rotating-wave approximation (RWA):
fluorescence upconversion, 179–180
iodine photodissociation dynamics:
degrees of freedom (DOF), 240–241
power absorption, 259–262
nonlinear polarization:
nonperturbative approach, 89–90
perturbation theory, 75
spectroscopic characterization, 76–81
Runge-Kutta techniques, time-dependent Schrödinger equation, state vector propagation, 39–41

S_2 absorption spectrum, real-time detection, internal-conversion process, 95–108
Saddle point behavior, Fokker-Planck equation:
high energy barrier approximation, 593–600
uniaxial anisotropic free energy:
bistable structure, 609–611
coordinate system and stationary points, 601–606
orthogonal transformation, 606–609
Scaling laws, complex flow simulations, fluid rheology, 358–359
Schlögl models:
continuously stirred tank reactor (CSTR), 453–454
reactive fluid stochastic theory:
one-dimensional system fluctuations, 430–434
pitchfork bifurcation, spatial correlation, 427–430
reaction-diffusion master equation, 426–427
Schrödinger equation.
See also Time-dependent Schrödinger equation
iodine photodissociation dynamics:
laser pulse absorption, 260–262
quantum mechanics, 245–248
photochemical funnels, 7–8
stochastic theory of reactive fluids, pitchfork bifurcation
fluctuation, 413–414
time-dependent computation, 29–41
state vector propagation, 38–41
state vector representation, 32–38
time-independent computation, non-Born-Oppenheimer dynamics, 9–11
Schwartz-Christoffel conformal mapping algorithm, iodine photodissociation dynamics, 305
Self-diffusion phenomenon, complex flow simulations, Rayleigh-Bénard instability, 374–375, 377–378
Semenov model, exothermic chemical reaction, 441–443
Semiclassical techniques:
nonadiabatic transitions, 140–145
short-time fluorescence Stokes shift dynamics, 185–188
background, 175–176
non-equilibrium approximations, 209
pump-probe analysis and, 217–218
Shear rate, complex flow simulations, fluid rheology, 354–359
Short-time dynamics:
kinetic-energy operator, 24–25
peak emission frequencies, 175–176
Stokes shift, background, 172–176
classical approximation, 190–191
classical Franck approximation, 183–185
classical linear response, 191–192
deconvolution procedure tests, 206–209
difference-potential-only approximation, 196–197
doorway-window observations, 222–223
excited-state linear response, 189–190
ground and excited-state linear response approximations, 210–212
ground-state linear response, 189
instantaneous normal mode treatment, 212–215
model system, 198–201
non-equilibrium classical and semiclassical approximations, 210
overlapping pulses, 192–193
peak emission frequency, 194–195
peak emission frequency, very short and long time behaviors, 223–225

pulse duration, overlap, and center frequency, 201–206
pulse intensities, deconvolution, 181–182
rigid-cage approximation, 195–196
semiclassical wave packet treatment, 185–188
separated pulses, 193–197
simulations, 197–215
stimulated emission amplitude, pump-probe signal, 222
system excitation and fluorescence detection, 176–180
temporally separated pulses, 182–183
time-resolved fluorescence spectrum, 180–181
transition energy approximations, 209–212
very short and very long time behavior, 223–225
wave packet picture, 193–194
Simulations:
chemical instability simulations:
microscopic simulations, 434–463
Bird algorithm for Boltzmann dynamics, 437–439
continuously stirred tank reactor (CSTR), 449–457
exothermic chemical reaction, 441–448
hard-sphere chemistry, 439–441
reactive fluids, 435–436
spatially extended systems, 457–463
overview, 393–399
reaction-diffusion master equation, 420–434
Langevin limit, 422–424
one-dimensional systems, 430–434
pitchfork bifurcation spatial correlations, 427–430
spatial correlations, 424–427
stochastic history of reactive fluids:
global view, 399–420
Hopf bifurcation, 414–420
pitchfork bifurcation fluctuation, 405–414
reaction-diffusion master equation, 420–434
complex flows:
direct simulation Monte Carlo method, 345–352
increased complexity, 357–359
instabilities, 366–383
lattice gases, 334–344
molecular dynamics (MD) technique, 320–324
non-equilibrium molecular dynamics, 324–334
non-equilibrium steady states, 359–366
overview, 317–320
particle simulations, 352–357
Rayleigh-Bénard instability, 371–383
short-time fluorescence Stokes shift dynamics, 197–215
classical, wave packet and linear response approximations, 209–212
deconvolution procedures, 206–209
excited-state dynamics, 212–215
model systems, 198–201
pulse duration, overlap, and center frequency, 201–206
Slodd equations, complex flow simulations, non-equilibrium molecular dynamics, 331–334
Smoothing techniques, derivative couplings, diabatic electronic states, 13–14
Soft-sphere models, complex flow simulations:
non-equilibrium steady state, 365–366
phase-space diagrams, 322–323
Solvation dynamics:
complex flow simulations, fluid rheology, 358–359
iodine photodissociation:
background, 236–237
"probe" pulse states, 282–284
"push" pulse states, 292–293
Stokes shift analysis, 172–176
Solvent particles, spatially extended systems, 458–463
Sound propagation, complex flow simulations, HPP lattice gas model, 337–338
Sparsity properties, time-dependent Schrödinger computation, state vector representation, 36
Spatial correlation functions, spatially extended systems, 461–463
Spatial correlations:
reactive fluid stochastic theory:
one-dimensional system fluctuations, 431–434
pitchfork bifurcation, 427–434
reaction-diffusion master equation, 424–427
spatially extended systems, 461–463
Spatially extended systems, microscopic simulation, 457–463, 465

Spatial Markovianity, stochastic theory of reactive fluids, one-dimensional system fluctuations, 430–434
Spectroscopic processes, nonlinear polarization, perturbation theory, 76–81
Spherical harmonic representation, Fokker-Planck equation:
  associated Legendre functions, 628–631
  cubic anisotropic free energy, 543–589
    product formulas, 581–589
    recurrence relations, 543–581
  matrix elements and differential recurrence coefficients, 505–513
  representation formula, 515–524
  uniaxial anisotropic free energy:
    Gilbert-Langevin equation, 533–543
    matrix elements and differential recurrence coefficients, 524–533
    numerical computation results, 543
Spin-boson Hamiltonian, real-time path-integral technique, 120–121
Spinodal decomposition kinetics, complex flow simulations, lattice gas models, 344
Splie-operator technique, time-dependent Schrödinger equation, state vector propagation, 39–40
Spontaneous magnetization, Fokker-Planck equation, ferromagnetic relaxation, 478–482
Stable magnetization orientations, Fokker-Planck equation, ferromagnetic relaxation, 479
State vectors:
  iodine photodissociation dynamics:
    degrees of freedom (DOF), 241–242
    Hamiltonian eigenstate representation, 248
    "pump" pulse photodissociation, 267
  time-dependent Schrödinger computation:
    propagation techniques, 38–41
    representation techniques, 32–38
Static case, Fokker-Planck equation, correlation time calculations, 497–499
Stationary points, Fokker-Planck equation:
  general Fokker-Planck and Gilbert-Langevin derivation, 631–634
  uniaxial anisotropic free energy, 601–606
    bistable structure, 609–611
    geometrical interpretations, 611–616
    high energy barrier approximation, Fokker-Planck equation, 616–628
    orthogonal transformation, 606–609
Statistical correlations, real-time path-integrals, 127–131
Steady-state transitions, microscopic simulations, continuously stirred tank reactor (CSTR), 451–454
Stimulated-emission (SE) polarization:
  femtosecond spectroscopy, classical modeling, 152–154
  nonlinear polarization, 78–81
    multitime correlation functions, 81–83
    pump-probe signals, 85–87
  real-time detection:
    internal-conversion process, 99–108
    nonadiabatic photoisomerization, 112–114
  short-time fluorescence Stokes shift dynamics, pump-probe signal, 222
Stochastic analysis:
  chemical instability simulations:
    Hopf bifurcation, 414–420
    overview, 397–398
    pitchfork bifurcation fluctuation, 405–414
    reaction-diffusion master equation, 420–434
    reactive fluids, 399–434
  general Fokker-Planck and Gilbert-Langevin derivations, 631–634
  stochastic theory of reactive fluids, 402–404
Stoichiometric coefficient, chemical instability simulations, 394
Stokesian dynamics, complex flow simulations, 352–353
Stokes shift dynamics:
  complex flow simulations, 352–353
  short-time fluorescence:
    background, 172–176
    classical approximation, 190–191
    classical Franck approximation, 183–185
    classical linear response, 191–192
    deconvolution procedure tests, 206–209
    difference-potential-only approximation, 196–197
    doorway-window observations, 222–223
    excited-state linear response, 189–190
    ground and excited-state linear response approximations, 210–212
    ground-state linear response, 189
    instantaneous normal mode treatment, 212–215
    model system, 198–201
    non-equilibrium classical and semiclassical approximations, 210
    overlapping pulses, 192–193

peak emission frequency, 194–195
peak emission frequency, very short and long time behaviors, 223–225
pulse duration, overlap, and center frequency, 201–206
pulse intensities, deconvolution, 181–182
rigid-cage approximation, 195–196
semiclassical wave packet treatment, 185–188
separated pulses, 193–197
simulations, 197–215
stimulated emission amplitude, pump-probe signal, 222
system excitation and fluorescence detection, 176–180
temporally separated pulses, 182–183
time-resolved fluorescence spectrum, 180–181
transition energy approximations, 209–212
wave packet picture, 193–194
Stratonovich sense, Fokker-Planck equation, Langevin spherical harmonic representation, 535–536
Stress-stress time autocorrelation function, complex flow simulation:
fluid rheology, 356
lattice gas models, 342–343
Strong intermolecular couplings, photoinduced dynamics, 4–8
Sturm-Liouville problem, Fokker-Planck equation, axial symmetry, 499–500
Subtraction technique, complex flow simulations, non-equilibrium molecular dynamics, 331–332
Superparamagnetic behavior, Fokker-Planck equation, 481–482
Surface-hopping techniques, nonadiabatic photoisomerization, semiclassical techniques, 140–145
Symmetry relations, Fokker-Planck equation:
associated Legendre functions, 628–631
matrix elements and differential recurrence coefficients, 506–507
System-bath interactions, iodine photodissociation dynamics, dissipative super-operators, 251, 302
System-environment coupling, photoinduced dynamics, 4
System excitation, short-time fluorescence Stokes shift dynamics, 176–180
Taylor expansion:
Hamiltonian models:
nonadiabatic photoisomerization, 28–29
normal-mode expansion, 25–27
high energy barrier approximation, Brown's model of Fokker-Planck equation, 591–592
time-dependent Schrödinger computation, 33–38
transition-dipole-moment functions, 16–20
Temperature:
complex flow simulations:
channel flow studies, 369
lattice gas models, 343–344
non-equilibrium molecular dynamics, 328–334
Rayleigh-Bénard instability, 379–380
exothermic chemical reaction, 442
Temporally separated pulses, Stokes shift dynamics, 182–183
Tensor propagation method, real-time path-integrals, 130–131
Thermal equilibrium, Fokker-Planck equation, ferromagnetic relaxation, 479
Thermal instability, exothermic chemical reaction, 441
Thermostatting technique, complex flow simulations, non-equilibrium molecular dynamics, 325–327
Three-mode model system, short-time fluorescence Stokes shift dynamics, 175–176
Time-dependent correlation:
real-time detection, internal-conversion process, 94–108
short-time fluorescence Stokes shift dynamics, 172–176
instantaneous normal mode (INM) analysis, 214–215
vs. pump-probe measurements, 216–221
Time-dependent Schrödinger equation:
internal-conversion processes, classical techniques, 146–148
iodine photodissociation dynamics, 303–304
molecular dynamics, 29–41
state vector propagation, 38–41
state vector representation, 33–38
nonadiabatic photoisomerization, semiclassical techniques, 143–145
nonlinear polarization, perturbation theory, 73–76
reduced density-matrix formalism, 136–139

Time-dependent Schrödinger equation (*continued*)
ultrafast intramolecular non-Born-Oppenheimer dynamics, electronic population probabilities, 53–54
Time-dependent self-consistent field (TDSCF) approximation:
iodine photodissociation dynamics, 301
multidimensional problems, 120
nonadiabatic photoisomerization, semiclassical techniques, 144–145
Time-resolved fluorescence:
femtosecond spectroscopy, 68–70
classical modeling, 152–154
nonlinear polarization, stimulated-emission (SE) polarization, 86–87
real-time detection, internal-conversion process, 101–108
Stokes shift dynamics, 180–181
model systems, 198–201
Time-resolved ionization and fragment detection, femtosecond spectroscopy, 70–73
Time-resolved photoelectron spectrum, real-time detection, internal-conversion process, 105–108
Torsional probability densities, ultrafast intramolecular non-Born-Oppenheimer dynamics, wave packets, 43–49
Transient RISRS (TRISRS):
iodine photodissociation dynamics, 233–235
experimental relevance, 300–302
Transient spectrum:
femtosecond spectroscopy, 64–68, 118–119
iodine photodissociation dynamics:
"push" pulse states, 295–297
spectroscopic techniques, 260–262
weak pulses, 259–260
nonlinear polarization, 86–87
"probe" pulse photodissociation, 287–290
dissipation conditions, 289
individual states, 287–288
probe wavelength differences, 289–290
real-time detection, internal-conversion process, 98–108
Transition-dipole-moment functions, electronic wave functions, 16–20
Transition probabilities, Fokker-Planck equation:
general Fokker-Planck and Gilbert-Langevin derivation, 632–634
stochastic theory of reactive fluids, 400–404
Trans photoisomerization dynamics, ultrafast intramolecular non-Born-Oppenheimer dynamics, nuclear observables, 58–61
Triatomic systems, Hamiltonian models:
potential energy (PE) surfaces, 20–21
ultrafast intramolecular non-Born-Oppenheimer dynamics, 41–42
Trotter formula, iodine photodissociation dynamics, coordinate-dependent, two-level approximation, 254–256
Tuning modes:
Hamiltonian models, normal-mode Taylor expansion, 25–27
real-time path-integrals, vibronic-coupling models, 123–125
ultrafast nuclear dynamics, 54–61
Turing structures, microscopic simulation, 465
Two-phase flows:
complex flow simulations, lattice gas models, 344
molecular computation, 369–370
Two-pulse photon-echo emission, nonlinear polarization, nonperturbative approach, 92–94

Ultrafast nonadiabatic excited-state processes, femtosecond spectroscopy, 94–119
molecular experiments, 114–119
real-time detection:
internal-conversion, 94–108
nonadiabatic photoisomerization, 108–114
Ultrashort laser pulses, femtosecond spectroscopy, 61–63
Ultraviolet (UV) pump pulse, iodine photodissociation dynamics, 231–235
Uniaxial anisotropy, Fokker-Planck equation:
axial symmetry, 503–505
ferromagnetic relaxation, 479–481
free energy with external magnetic field, 524–543
bistable structure, 609–611
coordinate system and stationary points, 601–606
geometrical changes of polynomials, 611–616
high energy barrier approximation, arbitrary value of ψ, 623–628
high energy barrier approximation, $\psi = \pi/4$, 616–623
orthogonal transformation of coordinate system, 606–609

Van der Pol oscillator, stochastic theory of reactive fluids, Hopf bifurcation, 417–420
Van Hove function, non-equilibrium steady states, complex flow simulations, 359–366
Vector addition rules, Fokker-Planck equation, product formula for spherical harmonics, 581–589
Velocity-time autocorrelation function, complex flow simulation:
  background, 317–318
  lattice gas models, 342–344
  Rayleigh-Bénard instability, 378–379
Velocity Verlet algorithm, complex flow simulations, non-equilibrium molecular dynamics, 332–334
Verlet algorithm, complex flow simulations, non-equilibrium steady state, 365–366
Vibrational dissipation:
  femtosecond experiments, 117–119
  photo-induced "hole" dynamics, 277–279
  photoisomerization, chromophore in solution, 149–152
  ultrafast intramolecular non-Born-Oppenheimer dynamics, nuclear observables, 54–61
Vibrational excitation, iodine photodissociation dynamics:
  dissipative model, 302
  "push" pulse excitation, 290–297
Vibronically induced anharmonicity, Hamiltonian models, nonadiabatic photoisomerization, 28–29
Vibronic coupling, femtosecond experiments, 116–119
Vibronic coupling theory:
  diabatic representation, 20
  femtosecond experiments, 116–119
  normal-mode Taylor expansion, Hamiltonian models, 26–27
  real-time detection, internal-conversion process, 94–108
  real-time path-integral technique, 121–131
    discretization construction, 122–125
  reduced density-matrix formalism, 135–139
  time-dependent Schrödinger computation, state vector representation, 33–38
  ultrafast intramolecular non-Born-Oppenheimer dynamics, nuclear observables, 57–61
Viscoelasticity, complex flow simulations, fluid rheology, 356
Viscosity, complex flow simulations:
  fluid rheology, 354–359
  future research, 383
  non-equilibrium molecular dynamics (NEMD), 368–371
von Karman streets, complex flow models, flow studies, 368
Vorticity, complex flow simulations, direct simulation Monte Carlo (DSMC) technique, 345–347, 366–367

Wave-function formalism, nonlinear polarization, perturbation theory, 74–76
Wave-packet calculations:
  iodine photodissociation dynamics, "pump" pulse photodissociation, 269–276
  real-time path-integrals, 128–131
  short-time Stokes shift dynamics:
    semiclassical approximations, 186-188, 210
    separated pulses, 193–194
    simulation techniques, 209–215
  ultrafast intramolecular non-Born-Oppenheimer dynamics, 42–49
WCA (Weeks, Chandler, Anderson) potential, complex flow simulations, non-equilibrium steady state, 364–366
Weak-field limit, nonlinear polarization, perturbation theory, 75
Weak pulses, iodine photodissociation dynamics, transient absorption, 259–260
Wetting phenomenon, complex flow simulations, 370–371
Wigner distribution function:
  iodine photodissociation dynamics:
    "hole" in phase space, 257–259
    initial states, 244–245
    phase-space picture, 243–245
    photo-induced "hole" dynamics, 277–278
    "probe" pulse states, 281–287
    "push" pulse photodissociation, 291–297
  short-time fluorescence Stokes shift dynamics, 186–188
Wigner 3*j*-symbols, Fokker-Planck equation:
  spherical harmonics, associated Legendre functions, 631
    product formula, 581–589
Window functions:
  iodine photodissociation dynamics:
    "probe" pulse photodissociation, 287–290
  transient weak pulse absorption, 260

Window functions (*continued*)
short-time fluorescence Stokes shift dynamics:
background, 174–176
classical Franck approximation, 183–185
deconvolution testing, 208–209
gate-pulse duration, 216–217
observed fluorescence spectrum, 222–223
temporally separated pulses, 182

WKB approximation, stochastic theory of reactive fluids, pitchfork bifurcation fluctuation, 413–414

Zero-kinetic-energy (ZEKE) photoelectron signal, femtosecond spectroscopy, 72–73
Zeroth-order computation, real-time path-integrals, 128–131